CAMBRIDGE LIBRARY COLLECTION

Books of enduring scholarly value

Astronomy

From ancient times, humans have tried to understand the workings of the world around them. The roots of modern physical science go back to the very earliest mechanical devices such as levers and rollers, the mixing of paints and dyes, and the importance of the heavenly bodies in early religious observance and navigation. The physical sciences as we know them today began to emerge as independent academic subjects during the early modern period, in the work of Newton and other 'natural philosophers', and numerous sub-disciplines developed during the centuries that followed. This part of the Cambridge Library Collection is devoted to landmark publications in this area which will be of interest to historians of science concerned with individual scientists, particular discoveries, and advances in scientific method, or with the establishment and development of scientific institutions around the world.

Popular Astronomy

French astronomer Camille Flammarion (1842–1925) called the study of the heavens 'the science which concerns us most'. He believed that learning 'what place we occupy in the infinite' could delight and instruct, and might even promote an end to war and strife. Flammarion dedicated the present work to François Arago (1786–1853), author of earlier work on popular astronomy. Since Arago's time, the capabilities of telescopes and other instruments had vastly improved, advancing understanding in areas such as the composition of stars. Flammarion sought to bring this new knowledge to the public in a charming yet 'scrupulously exact' style. His highly illustrated introduction to astronomy succeeded in reaching a wide readership, selling over 100,000 French copies before this English translation appeared in 1894. The 1881 French version and Flammarion's work on the origins of the Earth, *Le Monde avant la création de l'homme* (1886), are also reissued in this series.

Cambridge University Press has long been a pioneer in the reissuing of out-of-print titles from its own backlist, producing digital reprints of books that are still sought after by scholars and students but could not be reprinted economically using traditional technology. The Cambridge Library Collection extends this activity to a wider range of books which are still of importance to researchers and professionals, either for the source material they contain, or as landmarks in the history of their academic discipline.

Drawing from the world-renowned collections in the Cambridge University Library and other partner libraries, and guided by the advice of experts in each subject area, Cambridge University Press is using state-of-the-art scanning machines in its own Printing House to capture the content of each book selected for inclusion. The files are processed to give a consistently clear, crisp image, and the books finished to the high quality standard for which the Press is recognised around the world. The latest print-on-demand technology ensures that the books will remain available indefinitely, and that orders for single or multiple copies can quickly be supplied.

The Cambridge Library Collection brings back to life books of enduring scholarly value (including out-of-copyright works originally issued by other publishers) across a wide range of disciplines in the humanities and social sciences and in science and technology.

Popular Astronomy

A General Description of the Heavens

CAMILLE FLAMMARION
TRANSLATED BY JOHN ELLARD GORE

CAMBRIDGE
UNIVERSITY PRESS

University Printing House, Cambridge, CB2 8BS, United Kingdom

Published in the United States of America by Cambridge University Press, New York

Cambridge University Press is part of the University of Cambridge.
It furthers the University's mission by disseminating knowledge in the pursuit of education, learning and research at the highest international levels of excellence.

www.cambridge.org
Information on this title: www.cambridge.org/9781108067843

This edition first published 1894
This digitally printed version 2014

ISBN 978-1-108-06784-3 Paperback

POPULAR ASTRONOMY

POPULAR ASTRONOMY

A GENERAL DESCRIPTION OF THE HEAVENS

BY

CAMILLE FLAMMARION

TRANSLATED FROM THE FRENCH WITH THE AUTHOR'S SANCTION
BY

J. ELLARD GORE
F.R.A.S. M.R.I.A. &c.
AUTHOR OF 'THE SCENERY OF THE HEAVENS' 'THE VISIBLE UNIVERSE'
'THE WORLDS OF SPACE' ETC.

WITH 3 PLATES AND 288 ILLUSTRATIONS

London
CHATTO & WINDUS, PICCADILLY
1894

PRINTED BY
SPOTTISWOODE AND CO., NEW-STREET SQUARE
LONDON

TO THE IMMORTAL GENIUS OF

COPERNICUS, GALILEO, KEPLER, AND NEWTON

WHO OPENED TO MANKIND THE PATHS TO INFINITUDE

AND TO

FRANÇOIS ARAGO

FOUNDER OF POPULAR ASTRONOMY

THIS WORK IS RESPECTFULLY DEDICATED

BY

CAMILLE FLAMMARION

TRANSLATOR'S PREFACE

M. CAMILLE FLAMMARION is the most popular scientific writer in France. The circulation of his works has been very large, and several societies have been named after him. Of the present work no fewer than one hundred thousand copies were sold in a few years—a sale probably unequalled among scientific books. It was considered of such merit that the Montyon Prize of the French Academy was awarded to it; besides which, the author has received various other honours. The work has also been selected by the Minister of Education for use in the public libraries—a distinction which proves that it is well suited to the general reader. The subject is treated in a very popular style, and the work is at the same time interesting and reliable. It should be found very useful to those who wish to acquire a good general knowledge of Astronomy without going too deeply into the science. In translating this work I have endeavoured to make as close a version as possible, with, of course, due regard to the English idiom. I have reduced the figures given by the author to English measures. Many new illustrations have been added, and for some of these my best thanks are due to Mr. W. F. DENNING, F.R.A.S., Mr. T. G. ELGER, F.R.A.S., and Mr. A. COWPER RANYARD, F.R.A.S. (Editor of 'Knowledge'). I have also given some notes with reference to recent researches and discoveries, so as to bring the work up to date. These notes are given in square brackets.

J. E. G.

CONTENTS

BOOK I

THE EARTH

BOOK II

THE MOON

BOOK III

THE SUN

BOOK IV

THE PLANETARY WORLDS

BOOK V

COMETS AND SHOOTING STARS

BOOK VI

THE STARS AND THE SIDEREAL UNIVERSE

LIST OF ILLUSTRATIONS

SEPARATE PLATES

POPULAR ASTRONOMY

BOOK I

THE EARTH

CHAPTER I

THE EARTH IN THE SKY

THIS work is written for those who wish to hear an account of the things which surround them, and who would like to acquire, without hard work, an elementary and exact idea of the present condition of the universe. Is it not pleasant to exercise our minds in the contemplation of the great spectacles of nature? Is it not useful to know, at least, upon what we tread, what place we occupy in the infinite, the nature of the sun whose rays maintain terrestrial life, of the sky which surrounds us, of the numerous stars which in the darkness of night scatter through space their silent light? This elementary knowledge of the universe, without which we live, like plants, in ignorance of and indifference to the causes of which we perpetually witness the effects, we can acquire not only without difficulty, but with an ever-increasing pleasure. Far from being a difficult and inaccessible science, Astronomy is the science which concerns us most, the one most necessary for our general instruction, and at the same time the one which offers for our study the greatest charms and keeps in reserve the highest enjoyments. We cannot be indifferent to it, for it alone teaches us where we are and what we are; and, moreover, it need not bristle with figures, as some severe *savants* would wish us to believe. The algebraical formulæ are merely scaffoldings analogous to those which are used to construct an admirably designed palace. The figures drop off, and the palace of Urania shines in the azure, displaying to our wondering eyes all its grandeur and all its magnificence.

We do not mean by this that the reading of a work on popular astronomy appeals only to an inattentive mind. Such a book, on the contrary, although of more real interest and more attractive than a novel, should be read with attention, and only on this condition can the ideas it contains impart lasting scientific instruction. But whereas when we reach the last page of a novel we know just as much as when we began the first, we must be either blind or oblivious to all intellectual apprehension if the reading of a scientific work does not greatly extend the sphere of our knowledge, and does not more and more elevate the level of our judgment. We might even say that in our age it should be impossible for anyone's mind to be so little cultivated as to remain in ignorance of the absolute truths revealed by the grand conquests of modern astronomy.

In the first edition of this work, published in 1879, we wrote the following lines:—

'What immense progress the science of the heavens has achieved in recent years! One of the best works written on the subject is undoubtedly the "Astronomie populaire" of François Arago. Our venerated master, the true founder of popular astronomy, left the world in 1853. It is already more than a quarter of a century since we placed our wreaths of immortelles upon his tomb. How quickly the earth turns! and how soon our years pass away! This quarter of a century has nevertheless produced greater progress than the preceding half-century. Astronomy has been transformed in all its branches: the stars have revealed their chemical constitution to the bold and indefatigable investigations of the spectroscope; a comparison of the observations made upon double stars has enabled us to determine the true character of these systems, and the importance of the part which they play in the universe; the suns which shine in the depths of infinity are seen to be animated with rapid motions which carry them through all the directions of immensity; the nebulæ we admire to-day in the telescopic field of the powerful instruments recently constructed are believed to be immense and indescribable collections of suns; the wandering comets have disclosed the secrets of their chemical formation, and their relation to shooting stars; the planets have descended almost to our door, and now, having been brought to an astonishing proximity, we have been able to discover their meteorology, their climatology, and even to draw geographical maps representing their continents and oceans; the sun has revealed his physical constitution, and shows to our eyes his storms and fantastic eruptions, the powerful palpitations of the heart of the planetary organism; the moon shows the photographer her landscapes, and almost allows us to touch her with the finger! Such admirable progress entirely alters our astronomical knowledge, already so imposing. On the one hand the science is enriched and transformed; on the other it has become less dry and less selfish, more philosophical and more popular.

'However, in spite of this splendid progress, it would have seemed to me rash to publish a new "Astronomie populaire," after the important work of Arago, if twenty years of astronomical labours and of free discussion had not directly prepared the way, and if more than two hundred thousand copies of my different works had not already been scattered among all classes of society, and thus given the opportunity for a publication destined to diffuse under a more general form a taste for this magnificent science; when too, so many thousands of readers have by their very great sympathy encouraged the realisation of the project—a realisation which appears desirable and useful, although already excellent works, notably those of MM. Guillemin, Delaunay, Faye, Dubois, Lias (to mention only French authors), have, in recent years, encouraged, under different forms, the popularisation of astronomy. I venture, however, to present this work as one absolutely new in its method of treatment and in its character. Its dearest aim is to be entirely popular without ceasing to be scrupulously exact, and worthy of the incomparable science to which it is devoted.'

What we then wrote we can only repeat to-day. We did not deceive ourselves. An unprecedented success has attended our enterprise, and this rapid success has been confirmed by the high distinction which the French Academy has accorded to the work by awarding it the Montyon prize. We have, then, at last arrived at a scientific era long wished for by the friends of progress. We begin to think that it was unworthy of us to live in the midst of a universe without knowing something about it. We begin to understand that these are the first ideas to be acquired by all education which has the ambition of being serious. The shadows of night gradually vanish; the light dawns in the soul; it is a manifest, eloquent, and unquestionable sign of the actual state of men's minds, and their aspirations towards true science—positive science, towards true philosophy—scientific philosophy. The Author is pleased to report this success, not because it is the first time it has been obtained by a scientific book, not from wretched vanity or puerile pride, but because it is a sign of the times, because it marks the fundamental character of our epoch, and because it is pleasant to see these noble tendencies asserting themselves more and more in the great human family, so slowly progressive.

Moreover, astronomy presents us now with one of those radical transformations which form an epoch in the history of the science.

It ceases to be a figure and becomes alive. The spectacle of the universe is transfigured before our astonished minds. It is no longer inert bodies rolling in silence in eternal night that the finger of Urania shows us in the depths of the heavens; it is life—life immense, universal, eternal, unfolding itself in waves of harmony out to the inaccessible horizon of an eternal infinite.

What marvellous results! What splendours to contemplate! What magnificent fields to traverse! What a series of pictures to admire in these noble and peaceful conquests of the human mind—sublime conquests which cost neither blood nor tears, and where we live in the knowledge of the truth, in the contemplation of the beautiful.

The science of the stars ceases to be the confidential secret of a small number of the initiated ; it penetrates all understandings, it illuminates nature. It shows that without it man would always be ignorant of the place he occupies in the general scheme of things, and that its study, at least to an elementary extent, is indispensable to all serious education. It becomes, in short, truly universal, and everyone feels now the necessity of comprehending the reality.

Of all the truths which astronomy has revealed to us, the first, the most important for us, and the one which should immediately interest us, is its disclosures with reference to the planet we inhabit—its form, its magnitude, its weight, and its position and motions. It is with the study of the earth that it is convenient now to commence the study of the sky, because in reality it was on the situation of our globe in space and on its motions that ancient astronomy was founded, and it is to an exact knowledge of our planet that modern astronomy conducts us. Observation shows that, far from being fixed in the centre of the universe, the earth, borne along in time, presses towards an end which always existed, and rolls with rapidity in space, carrying through the fields of immensity the generations which arise upon its surface.

Mankind was thoroughly deceived during thousands of years with reference to the nature of the earth, its true place in infinity, and the general construction of the universe. Without astronomy it would still be deceived to-day, and indeed it must be admitted that ninety-nine persons out of a hundred have an erroneous idea of our world and of creation, simply because they are ignorant of the elements of astronomy.

The earth appears to us as an immense plain, chequered with a thousand varieties of aspects and reliefs—verdant hills, flowery valleys, mountains more or less elevated, winding rivers in the plains, lakes of fresh water, vast oceans, countries of infinite variety. This earth seems to us fixed, seated for ever upon eternal foundations, crowned with a sky, sometimes clear, sometimes cloudy, spread out to form the firm base of the universe. The sun, the moon, the stars seem to turn above it. According to all these appearances man might easily believe himself to be the centre and the object of creation. Vain presumption, which was believed in so much the more because for a long time there was no one to contradict it.

During the long ages of primitive ignorance, when the whole life of man was spent in dull occupations, the principal exercise of his dawning imagination consisted in securing his existence from the injuries of external nature, in defending it against his enemies, and in adding to his physical welfare. But soon superior minds began to make progress in moral as well as in material civilisation. Intelligence slowly developed, and the day arrived when in the glowing plains of the East—then fertile, now sterile ; at that time peopled, now a desert—some men of merit commenced to observe the course of the stars, and to found the astronomy of observation. These observations were at first simple remarks made by the shepherds of the Himalayas after sunset and before sunrise : the phases of the moon, and diurnal retrograde motion of that body with reference to the sun and stars, the apparent motion of the starry sky accomplished silently above our heads, the movements of the beautiful planets through the constellations, the shooting star which seems to fall from the heavens, eclipses of the sun and moon, mysterious subjects of terror, curious comets which appear with dishevelled hair in the heights of heaven—such were the first subjects of these old observations made during thousands of years. Astronomy is the most ancient of the sciences. Even before the invention of writing and the beginning of history men examined the sky and laid the foundations of a primeval almanack. The primitive observations have been lost in the revolutions of nations ; we possess, however, some fairly good records considering their antiquity, among others those of the Egyptians and Chinese made in the thirtieth century before our era, stating that at the vernal equinox the sun was situated in the constellation Taurus, then the first sign of the Zodiac ; that of an eclipse of the sun made in Egypt in the year 2720 B.C. ; that of a conjunction of the planets in Capricorn, made by the Chinese astronomers in the year 2449 B.C. ; that of a star in the constellation Hydra made in the year 2306 B.C. The Egyptian calendar was instituted about the year 2782 B.C. ; and the Chinese calendar about the year 2637 B.C. At least four thousand years have elapsed since our week of seven days was formed in the plains of Babylon, and for several thousand years also each day has taken the name of one of the moving stars known to the ancients : the Sun, the Moon, Mars, Mercury, Jupiter, Venus, and Saturn.

At the epoch of Homer (about 900 years before our era) it was believed that the earth, surrounded by the river *Okeanos*, filled the lower half of the celestial sphere, while the upper half extended above, and that *Helios* (the sun) extinguished his fires each evening and re-lit them in the morning after bathing in the deep waters of the ocean.

According to the more ancient ideas, founded upon illusions, which the uncultivated mind shares with the infant, there was no continuity between the night sky in which the stars shine and the sky in which the light of day is diffused. He who first ventured to maintain that during the day the sky is studded with stars as it is in the night, and that we do not see them because they are effaced by the light of the sun, was certainly a logician full of genius and boldness.

Many Greek astronomers still believed, 2,000 years ago, that the stars were fires fed by exhalations from the earth.

They were soon forced to remark that the sun, the moon, the planets, and the stars rise and set, and that during the hours which elapse between their setting and their rising it was absolutely necessary that the stars should pass under the earth. *Under the earth!* What a revolution is in these three words! Up to that time they had supposed that the world extended to infinity below our feet, solidly founded for ever, and, without comprehending this infinite extension of matter, they remained in ignorance and believed in the firm solidity of the earth. But when the curves described by the stars above our heads were continued, after they set below the horizon, to start again when they rose, it was necessary to imagine the earth pierced right through with tunnels large enough to permit the passage of the celestial torches. Some represented our abode under the form of a circular table borne upon twelve columns, others under the form of a dome placed on the backs of four bronze elephants; but the idea of supporting the earth on mountains or otherwise only evades the difficulty, for these mountains, elephants, or columns would, of course, require to rest on some lower foundation. As, moreover, the sky seems to turn round us in one piece, the subterfuges invented in order to preserve for the earth something of its original stability at last disappeared by the force of circumstances, and they were obliged to admit that *the earth is isolated in all its parts.*

Hesiod, the contemporary of Homer, believed that the earth was supported like a disc midway between the vault of the sky and the infernal regions, a distance measured once, he claimed, by Vulcan's anvil, which took nine days and nine nights to fall from the sky to the earth, and the same time to fall from the earth to Tartarus. These ideas ruled for a very long time men's conceptions of the construction of the universe.

But the torch of progress was lit and could not be extinguished. The developments of geography proved that our world has the form of a sphere. The earth was then represented as an enormous ball placed in the centre of the universe, and it was supposed that the sun, moon, planets, and stars turned round us, in circles drawn one beyond the other, as appearances seemed to indicate.

For about two thousand years astronomers observed attentively the apparent revolutions of the heavenly bodies, and this attentive study gradually showed them a large number of irregularities and inexplicable complications, until at last they recognised that they were deceived as to the earth's position, in the same way that they had been deceived as to its stability. The immortal Copernicus, in particular, discussed with perseverance the earth's motion, already previously suspected for two thousand years, but always rejected by man's self-love, and when this learned Polish canon bid adieu to our world in the year 1543 he bequeathed to science his great work, which demonstrated clearly the long-standing error of mankind.

The terrestrial globe turns upon itself in twenty-four hours, and this motion produces the appearance of the whole sky turning round us. This is the first truth demonstrated by Copernicus, and the first which we shall examine. It is necessary, indeed, to begin precisely our astronomical

FIG. 1.—THE EARTH, AS SEEN FROM THE MOON

study by a general examination of the earth's position in space and the harmony of its motions.

In fact, this motion of diurnal rotation is not the only one with which the earth is animated. Carried by the force of gravitation, it sails round the sun at a distance of 148 millions of kilometres [or 91,964,240 English miles. The correct mean distance is, however, 92,796,950 miles.—J. E. G.], and a long revolution of 584 millions of miles, which it takes a year to traverse.

In order to accomplish, as it does in 365¼ days, this immense distance round the sun our sphere is obliged to travel a distance of 2,544,000 kilometres a day, or 106,000 kilometres an hour, or 29 kilometres (18 miles) a second! This is an absolutely demonstrated mathematical fact. Six different and independent methods agree in establishing the sun's distance at 93 millions of miles; now the earth sails at that distance in a revolution entirely described in a year. The calculation is easy.

We sail, then, in immensity with a velocity eleven hundred times quicker than that of an express train. As such a train travels eleven hundred times more rapidly than a tortoise, if we could send a locomotive in pursuit of the earth in space it would be exactly the same as if we set a tortoise to run after an express train! This velocity of our globe in its celestial orbit is seventy-five times swifter than that of a cannon-ball.

A being placed in space not far from the ideal orbit which the earth describes in its rapid course would shudder with terror as it approached under the form of a magnificent star, as it came nearer becoming a frightful moon, covering the whole sky with its enormous mass, traversing without a stop the field of his affrighted vision, rolling on itself, passing away like lightning, and diminishing in the yawning depths of space. . . .

Upon this moving globe we live, almost in the same situation as grains of dust adhering to the surface of an enormous cannon-ball shot into immensity. . . . Sharing absolutely in all the motions of the globe, with all that surrounds us, we cannot perceive these motions, and we can only detect them from observations of the stars, which do not participate in the motion. Marvellous sidereal mechanism—the force which transports our planet is exercised without an effort, without friction, and without shocks in the midst of absolute silence in the eternal heavens. Smoother than the barge upon the limpid river, smoother than the gondola moving on the mirror of the Venetian canals, the earth glides majestically in its ideal orbit, showing no perceptible trace of the powerful force which guides it. Thus, but not with such perfection, glides the solitary balloon in the midst of the transparent air. How many times, entrusted to the car of the aerial ship, either during the bright hours of the day above the verdant fields, or in the darkness of night with the melancholy light of the moon and stars —how many times have I compared the glorious course of the balloon in the atmosphere to that of the earth in space!

In spite of appearance, *the earth is, then, a star in the sky* like the moon and the other planets, which are no more luminous than she is in reality, and only shine in the sky because they are illuminated by the sun. Seen from a distance in space, the earth shines like the moon; seen from a greater distance, like a star. Viewed from Venus and Mercury, it is the brightest star in the sky.

The motion of translation of our globe round the sun produces our succession of seasons and years; its motion of rotation upon itself produces the succession of days and nights. Our divisions of time are formed by these two motions. If the earth did not turn, if the universe were motionless, there would be neither hours, nor days, nor weeks, nor months, nor seasons, nor years! . . . But the world moves on.

The two motions which we have just described are the most important so far as we are concerned, but these are not the only motions with which our globe is animated. The earth, in fact, is carried along in the sky, and moves in various ways by *more than ten* different motions, of which the following are the principal:—

In the first place, our globe does not move as a bullet would do in its flight, that is to say, preserving an ideal horizontal axis round which the motion of rotation is performed; it is not transported through space with its axis vertical, like a top spinning vertically and sliding along a smooth floor; its axis of rotation is neither upright nor horizontal, but inclined at a certain angle, and this inclination remains the same during the whole year, in such a way that the earth is carried round the sun, and always preserves the same inclination with reference to him. Its axis of rotation remains parallel to itself during the whole course of its annual revolution, and its northern extremity remains constantly directed to a fixed point in the sky, in the neighbourhood of the pole star. But slowly, in the course of ages, this axis itself turns, as a finger directed to a star might slowly trace a circle in the sky, in such a way that the pole is displaced among the stars, and in the course of 258 centuries describes a complete circle. The present pole star will soon recede from the pole; in twelve thousand years the brilliant star of the Lyre [Vega or *α* Lyræ] will be close to the pole, as it was fourteen thousand years ago. This secular motion is that of the *precession of the equinoxes.* We have here, then, a third motion, very much slower than the other two.

(The reader is requested not to be uneasy if he does not for the moment thoroughly understand all the terms used: the object is here merely to give a general view, and everything will be explained a little further on.)

A fourth motion, due to the action of the moon, causes the earth to move round the sun, neither in a perfectly regular line, nor with a uniform velocity. In reality, it is the centre of gravity of the earth and moon which revolves annually round the sun, and this centre of gravity itself turns in a month, being always on the side turned towards the moon, although in the interior of our globe at about 1,056 miles below the surface. There is therefore a monthly displacement of the earth in space which produces what is called the parallactic inequality of the sun.

A fifth motion, also due to the moon's action, but slower than the preceding, and named *nutation*, causes the axis of the earth to describe rapid small ellipses traced upon the celestial sphere in eighteen years.

A sixth motion produces a slow oscillation in the inclination of the axis, which is at present about 23 degrees, or about a quarter of a right angle; it is now diminishing, to increase again in future ages. [The limits of the variation are about 21° 58′ 36″ and 24° 35′ 58″, and the rate of variation about 46½″ per century—J. E. G.]. This secular oscillation is called the variation in the *obliquity of the ecliptic.*

A seventh motion produces a variation in the curve which our planet describes round the sun, a curve not circular, but elliptical; in the course of ages, the ellipse differs more or less from the circular form. This motion is termed *the variation of the eccentricity.*

In this ellipse, of which the sun occupies one of the foci, the point of nearest approach to that luminary is called the perihelion. The earth passes this point on January 1. An eighth motion displaces this point also. In the year 4000 before our era the earth was at this point on September 21, and in the year 1250 before our era on December 21. The perihelion will be reached on March 21 in the year 6590; on June 22 in the year 11910; and finally, in the year 17000, it will have returned to the point where it was six thousand years ago—period, 210 centuries. This is the secular variation of the *perihelion.*

And yet this is not all.

A ninth order of motion, caused by the variable attraction of the planets, and especially of our neighbour Venus and the powerful Jupiter, disturbs all the preceding motions by producing *perturbations* of different orders.

A tenth displaces the sun by causing it to revolve round the centre of gravity of the whole system, which centre is often at one side of the solar globe, and this displaces at the same time the centre of the earth's annual revolution.

Finally, an eleventh motion, still greater than any of the preceding, carries the sun through space, and with it the earth and all the other planets. Owing to the existence of this motion, *our globe has never passed twice through the same place,* and it can never return to the spot where it is at present. We fall into the infinite, describing a series of spirals which are continually changing.

These motions will be explained in detail in the following chapter. It was important to *state* them at once, so that we may free ourselves from all prejudice as to the supposed importance of our world, and that we may feel that our abode is simply a moving globe carried through space, a veritable sport of cosmical forces, speeding through the eternal void towards an end of which we are ignorant, subject in its unsteady course to the most varied oscillations, balancing itself in the infinite with the lightness of an atom of dust in the sunlight, flying with a dizzy velocity above the unfathomable abyss, and carrying us for thousands of years past, and perhaps for thousands of years to come, in a mysterious destiny, which the most far-seeing mind cannot discern, beyond an horizon always fading into the future.

It is impossible coolly to consider this reality without being struck with the astonishing and inexplicable illusion in which the majority of mankind slumbers. Behold a little globe whirling in the infinite void. Round this globule vegetate 1,450 millions of so-called reasonable beings—or rather talkers—who know not whence they come nor whither they go, each of them, moreover, born to die very soon; and this poor humanity has resolved the problem, not of living happily in the light of nature, but of suffering constantly both in body and mind. It does not emerge from its native ignorance, it does not rise to the intellectual pleasures of art and science, and torments itself perpetually with chimerical ambitions. Strange social organisation! This race is divided into tribes subject to chiefs, and from time to time we see these tribes, afflicted with furious folly, arrayed against each other, obeying the signal of a handful of sanguinary evil-doers who live at their expense, and the infamous hydra of war mows down its victims, who fall like ripe ears of corn on the blood-stained fields. Forty millions of men are killed regularly every century in order to maintain the microscopical divisions of a little globule into several anthills. . . .

When men know something of the earth, and understand the modest position of our planet in infinity; when they appreciate better the grandeur and the beauty of nature, they will be fools no longer, as coarse on the one hand as credulous on the other; but they will live in peace, in the fertile study of Truth, in the contemplation of the Beautiful, in the practice of Good, in the progressive development of the reason, and in the noble exercise of the higher faculties of intelligence.

CHAPTER II

HOW THE EARTH TURNS UPON ITSELF AND ROUND THE SUN

Day and Night—Dates—Hours—Meridians. The Year and the Calendar.

WE now come to study in detail the earth's motions. We will not follow the ordinary method of astronomical treatises, and begin by describing appearances of which we should be obliged afterwards to demonstrate the falsity. Let us begin at once with the reality.

There is nothing more curious than these motions and their consequences on our ordinary life, as well as upon the judgments of our mind; they really constitute the measure of time, and our life is wholly regulated by this measure. Even the duration of our life, the periods into which it is divided, the duties which occupy it, our annual calendar and the epochs of history, are so many effects intimately connected with the motions of the earth. To study these motions is to study the principles of human biology.

What inexhaustible variety distinguishes the planets from each other! On the moon, for example, there are but twelve days and twelve nights in a year, and yet their year is of the same length as ours. Here we count 365 days in a year. On Jupiter the year is nearly twelve times longer than ours, and the day less than half the terrestrial day; hence it follows that there are no less than 10,455 days in the year of that world! On Saturn the disproportion is still more extraordinary; for its year, thirty times longer than ours, contains 25,217 days. And what shall we say of Neptune, whose year lasts for a century and a half—165 of our rapid years! If biology is there regulated in the same proportions, a young girl of seventeen years on Neptune would really have lived 2,800 of our years; she would have lived nearly a thousand years before Christ was born in Judea; she would have been contemporary with Romulus, Julius Cæsar, Constantine, Clovis, Charlemagne, François I., Louis XIV. Robespierre, . . . and she would be still only seventeen! Lethargic *fiancée*, she will marry in three or four hundred years a young man of her dreams, aged himself more than three thousand terrestrial years. . . .

The succession of day and night naturally formed the first measure of time. It is the natural fact which strikes us most, and it was only later on that the succession of the seasons was noticed, their duration estimated, and the length of the year recognised. The phases of the moon are more rapid and striking than the seasons, and time was divided into days and months long before it was divided into years. The ancient poems of India have even preserved for us the last echoes of the fears of primeval man at the approach of night. The sun, the good sun has completely disappeared in the west; is it very certain that he will return to-morrow morning in the east? If he should return no more! no more light, no more heat; the frozen night, gloomy night covers the world! How shall we recover the lost fire? How replace the beneficent sun and his celestial light? The stars from the height of the heavens shed their melancholy light; the moon pours out in the vacuities of the atmosphere that rosy, silvery light which diffuses such a charm upon the sleep of nature; but this is not the sun, this is not the day. . . . Ah, see the dawn which brightens slowly! behold the light, behold the day! Sun! King of the heavens, be blessed! Oh! never forget to return.

Fig. 2.—Diagram of Day and Night

What is the day? What is the night? Two contrary effects produced by the earth's motion of rotation combined with the shining of the sun. If our globe did not turn round, the day-star would be fixed, there would be eternal day on one half of the globe and eternal night on the other.

Our globe is isolated in space, and there is neither top nor bottom in the universe. Let us consider it, then, for a moment at any hour; for example, that from which we count noon. We find ourselves then on the central line of the hemisphere illuminated by the sun. The terrestrial globe (see fig. 2) produces of itself a shadow opposite to the solar light: the countries situated on the opposite hemisphere are then plunged in darkness, that is night. The earth turns; twelve hours later we shall be in our turn in the middle of the shadow, or at midnight. Turn up the figure, and you will then see the sun under your feet and the night above your heads. But this shadow produced by the earth does not extend throughout the universe, as the first impression of its meaning might lead us to think; it is but the size of the earth (7,912 miles), and all which is outside it remains illuminated in space, where there is as much light at midnight as at noonday. The moon and planets constantly receive the light of the sun. Besides, as the sun is larger than the earth, and even very much larger, this shadow which the earth throws behind it has the form of a horn, or a cone, and it terminates in a point at a distance of about 300,000 leagues (744,000 miles). Sometimes the moon, of which the distance is only 96,000 leagues, passes through the earth's shadow, and then we ascertain from the eclipse of that globe that our shadow is circular. This is, indeed, one of the first proofs we have had of the globular form of our floating island.

Let us take as the model of the earth a small ball pierced with a needle (fig. 3), and suppose we make it turn between two fingers. The needle represents the *axis*. The two diametrically opposite points at which the needle comes out are the two poles. Here are two important ideas, and as we see, very easy to retain. We know now what is the axis of the globe: that it is the ideal line which pierces it, and round which its motion of rotation is performed; we also know what is meant by the poles. Very well. Bring back the ball to our side in a way to see the head of the needle right in front, and let us suppose that it turns like the earth; we shall see that the globe turns in a direction contrary to the motion of the hands of a watch.

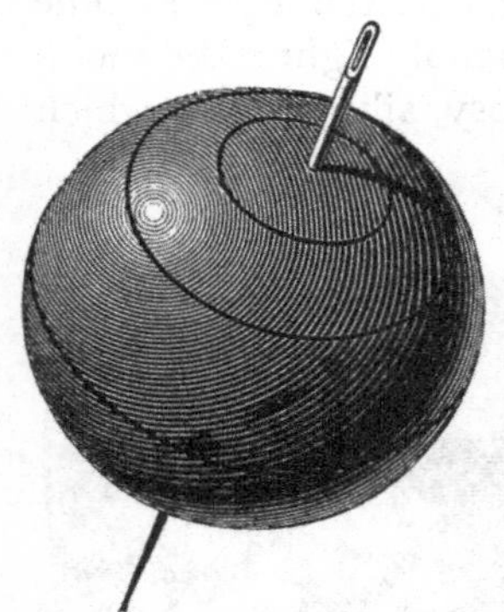

FIG. 3.—THE TERRESTRIAL GLOBE TURNING UPON ITS AXIS

Our fig. 4 shows how the different countries of the globe pass in turn through day and night. In the position represented in this figure Paris is found just below the sun, and we reckon noon. The countries situated to the left of France are to the east of her; they leave the shadow before her and have passed before her under the sun; so that when it is noon at Paris, it is 1 P.M. at Vienna, 2 P.M. at Suez, 3 P.M. at Teheran, 4 P.M. at Bokhara, 5 P.M. at Delhi in India, &c. All the countries on the same hour-line have the same hour at the same time. These hour lines are the *longitudes*. They are great circles which diverge from the pole. If we cut the sphere in two, at an equal distance from both poles, by a plane perpendicular to the axis, we trace in this manner the *equator*: it is the great circle which bounds our figure. In order to measure distances between the pole and the equator, we draw round the pole as a centre connective circles which bear the name of *latitudes*.

When it is noon at Paris it is noon at the same time all along the line drawn from the North to the South Pole passing through Paris, Bourges, Carcassonne, Barcelona, Algiers, Gambia (South Africa), &c. It is the same for each longitude. The differences of the hours are regulated by the differences of longitude. We have written on the figure the numbers corresponding to different cities scattered round the world. When it is noon at Paris, these different points have the hour written opposite to them.

		H. M.
1.	Paris	Noon
2.	Vienna	0 56 P.M.
3.	St. Petersburg	1 52 „
4.	Suez	2 0 „
5.	Teheran	3 16 „
6.	Bokhara	4 3 „
7.	Delhi	5 0 „
8.	Ava	6 14 „
9.	Pekin	7 37 „
10.	Jeddo	9 10 „
11.	Okhotsk	9 23 „
12.	Aleutian Isles	0 45 A.M.
13.	Petropolowski	1 35 „
14.	San Francisco	3 41 „
15.	San Diego	4 2 „
16.	Mexico	5 14 „
17.	New Orleans	5 50 „
18.	Cuba	6 21 „
19.	New York	6 55 „
20.	Quebec	7 6 „
21.	Cape Farewell	8 55 „
22.	Rickiavik	10 23 „
23.	Mogador	11 12 „
24.	Lisbon	11 14 „
25.	Madrid	11 36 „
26.	London	11 51 „

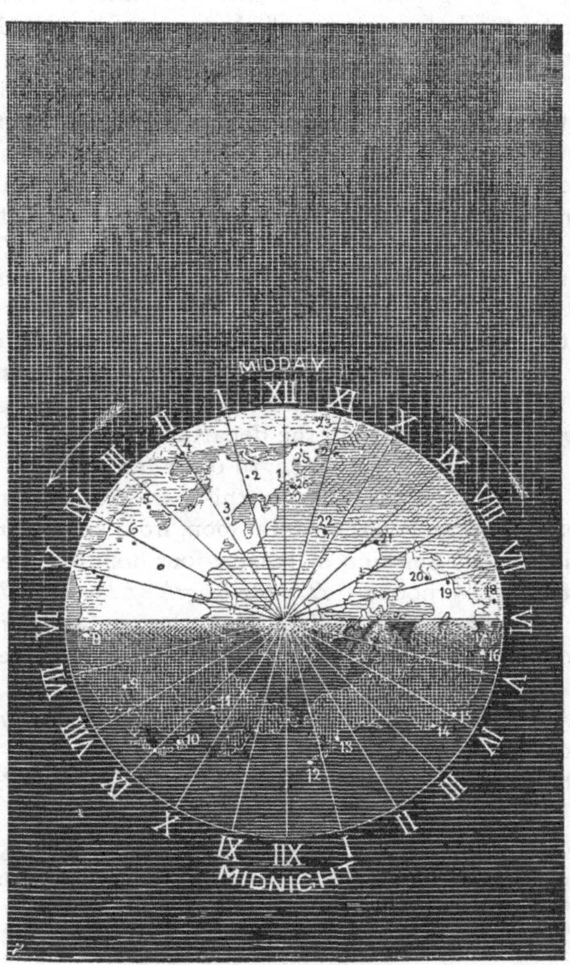

FIG. 4.—HOURS OF THE DAY AND NIGHT

Geographical France from the ocean to the Rhine has but a width which is passed over by the sun in 49 minutes.[1]

[1] One does not generally imagine the small space necessary to sensibly change the hours. Rouen and Paris differ by five minutes, so that a watch regulated at Paris is five minutes fast when it is carried to Rouen; and in Paris even two points very much closer—for example, the Luxembourg and the Polytechnic School—differ by three seconds of time, so that a clock well regulated at the Luxembourg is slower than a clock equally well regulated at the Polytechnic School. For at the latitude of Paris the circumference of the globe is 26,350,000 metres, and 305 metres give a difference of one second of time. The noonday sun takes 37 seconds to pass over

Charles V. boasted of the extent of his dominions—'on which the sun never set.' What influence have the States of Charles V. had in the progress of mankind? It is neither the size nor the weight of man which constitutes his greatness. If France has played for more than a thousand years a preponderating part in the enfranchisement of the human mind, she owes it to the independence of character of her children, and their constant rise towards progress.

Let us remark in passing a very curious consequence of these differences of the hours. The city of New York, for example, is 5 hours 5 minutes behind Paris, and San Francisco is 8 hours 19 minutes behind. If, then, we send from Paris to these two cities a telegraphic despatch, which can be transmitted directly, as the velocity of electricity is, we may say, instantaneous, the message would be received in New York 5 hours and 5 minutes, and at San Francisco 8 hours and 19 minutes *before* the hour at which it had been sent. Starting, for example, from Paris on January 1, 1880, at 4 A.M., it would arrive at New York on December 31, 1879, at $10^h\ 55^m$ P.M., and at San Francisco at $7^h\ 41^m$ P.M., thus arriving at its destination on the eve of its departure and in the preceding year. The arrival stamp would be earlier than the departure stamp.

What is the exact length of the day?

From a high antiquity this period has been divided into twenty-four parts, counting either from noon, from sunset, from midnight, or from sunrise. This duration of twenty-four hours is the time which separates two consecutive noons. *It is the length of the civil day.*

Everyone has noticed that the sun rises in the morning in the east, mounts slowly in the sky, attains its greatest elevation at noon, descends slowly, continuing the same oblique circle, and sets in the evening in the west. If one has the east to the left, and the west to the right, he has the south in front and the north behind; while if we look towards the south, then the north pole is behind us. The *meridian* is a great circle of the

Paris. Versailles is at 51 seconds from the meridian of the Observatory, Mantes at 2 minutes 28 seconds, &c. I need not say that the differences count in the direction of east to west, the direction north to south having nothing to do with the diurnal motion. The velocity of the globe is 463 metres at the equator.

A traveller who made the tour of the globe from east to west in twenty-four hours, leaving at noon, for example, would have the sun always above his head, always the same hour and always the same day. Every time that he repassed his point of departure the residents would count one day more. Every traveller making the circuit of the world from east to west is behindhand a day in returning. Every traveller in making this circuit from west to east is a day in advance.

The day changes its name at midnight in each country, but at the same moment the same name extends all over the surface of the globe. Where does Sunday end? By a sort of tacit convention between mariners and geographers this is on the meridian which traverses Behring's Straits and Polynesia, but following a rather irregular curve (see our *Revue mensuelle d'Astronomie populaire* for the year 1883, fig. 85, p. 226).

celestial sphere which is ideally drawn, leaving the north pole, passing exactly over our heads and continued to the south pole. The sun crosses this circle at noon. Between two transits of the sun across the meridian there are twenty-four hours.

Constant observation of the sky has shown that this number does not represent the exact duration of the earth's motion of rotation. The sun does not return to the meridian every day at exactly the same instant; sometimes it is behindhand, sometimes in advance. If, on the contrary, we observe a star, we find that it rises like the sun, that it sets in the west, and that it crosses the meridian as he does, but with absolute punctuality—to even a second. Between two consecutive transits of a star across the

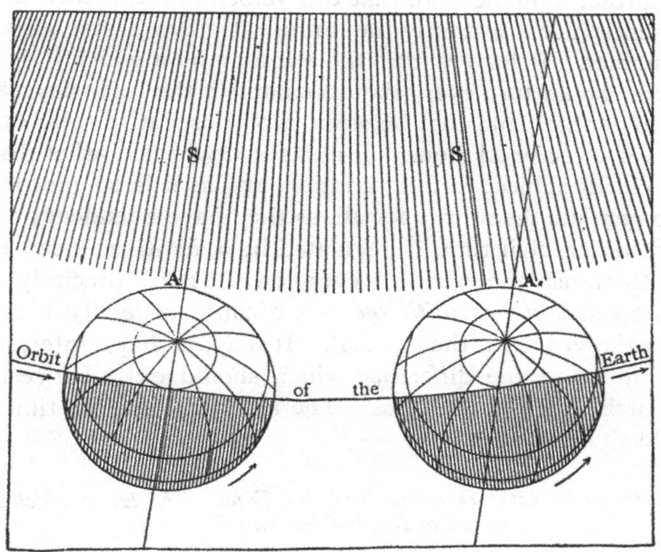

FIG. 5.—DIFFERENCE BETWEEN THE LENGTH OF THE DAY AND THE TIME OF THE ROTATION OF THE EARTH

meridian we can always count 86,164 seconds, never a second more and never a second less; these 86,164 seconds are not exactly 24 hours, but 23 hours 56 minutes 4 seconds. Such is the *precise* and constant *duration of the earth's motion of rotation.*

The difference between this duration and that of a solar day is very easily explained, if we reflect on the way in which the earth turns on itself and around the sun. Let us consider the terrestrial globe at any moment. It revolves round the sun from left to right (fig. 5) in a length of orbit which it takes a year to pass over, and at the same time turns upon itself every day in the direction indicated by the arrow. At noon the point A (left-hand position) is exactly in front of the sun: when the earth has accomplished a complete revolution, the next day, it will have been carried to the right-hand position, and the meridian A will be found exactly as it was the day before. But the transfer of the earth towards the right will cause the sun, by perspective, to move back towards the left, and in order that the point A may again return in front of the sun, and there may be noon again, it is necessary that the earth should still turn for 3 minutes 56 seconds; and this on every day of the year. It is this which makes the solar or civil day longer than the diurnal rotation of the globe (also termed the sidereal day).

There are in a year 365¼ solar days, but there are in reality 366¼ rotations, exactly one more.

Let us further remark here that the velocity of the earth in its orbit round the sun is not constantly the same: it moves faster in winter and slower in summer. It follows that the extra amount which it is necessary that the earth should turn every day to complete the solar day varies from one season to another, and that between two consecutive solar noons there are not always exactly 24 hours. But, as it would be rather disagreeable to have our clocks subject to this variation, and as it is so much the more necessary to regulate them once for all, as they have a tendency themselves to get out of order, civil time is regulated by a fictitious mean sun which is supposed to transit the meridian every day at noon precisely. *A well-regulated watch does not go with the sun*, because in reality it agrees only four times in the year with the sun dial. It may, perhaps, interest many of our readers to know the difference which should exist between a well-regulated watch and the sun dial. The following are the times which a watch should show at solar noon:—

Difference between the Civil Time and the Solar Time. The Hour which a Watch shows at Noon of the Sun Dial

Date	H. M.	Date	H. M.
Jan. 1	0 4 P.M.	July 15	0 5 P.M.
„ 15	0 10 „	„ 26	0 6 „
Feb. 1	0 14 „	Aug. 15	0 4 „
„ 11	0 14½ „	„ 31	Noon
Mar. 1	0 12 „	Sept. 15	11 55 A.M.
„ 15	0 9 „	Oct. 1	11 49 „
April 1	0 4 „	„ 15	11 46 „
„ 15	Noon	Nov. 3	11 43 „
May 1	11 57 A.M.	„ 16	11 44 „
„ 15	11 55 „	Dec. 1	11 49 „
June 1	11 57 „	„ 15	11 55 „
„ 15	Noon	„ 25	Noon
July 1	0 3 P.M.		

We see that at the dates April 15, June 15, August 31, and December 25, the civil time is the same as that of the sun dial; while on February 11 the latter is more than fourteen minutes behind the former, on May 15 it advances nearly five minutes, on July 26 is six minutes slow, and on November 3 it is seventeen minutes fast. The regulation of watches and of public clocks is not very ancient. It was made after the First Empire, in 1816. However, as early as the time of Louis XIV. the corporation of Paris watchmakers had taken for their arms a pendulum with this proud motto: '*Solis mendaces arguit horas*,' which shows that solar hours are false.

Everybody knows what sun dials are, on which the shadow of a style exposed to the sun indicates approximately the solar time. The most usual form is that of a vertical surface on a wall exposed in such a way that the sun shines upon it, and which receives the shadow of the style, thus tracing the hour lines, with which the shadow successively coincides. But we can construct a sun dial on any plane surface, vertical, horizontal, or inclined, and even on a curved surface of any form or position we may desire. The only condition to be fulfilled in order to construct a sun dial is that it should receive the sun's rays during a portion of the day.

Sun dials from their nature necessarily show solar time. If we wish to use them for the purpose of setting a clock which shows mean time it is necessary to have recourse to a table of the equation of time, which we have given above.

However, we can give to sun dials an arrangement which will furnish directly the relative indications of mean time. The most usual arrangement consists in tracing on a fixed sun dial, with a pierced plate, a curved line designed to indicate every day the instant of mean noon. This curved line, which is called the *meridian of mean time*, has the form of an elongated 8, as in fig. 6, or better still in fig. 7.

Every day, at the instant of mean noon, the little illuminated spot *a* should be found on the curve; so that in observing the moment when this bright spot crosses it we have the mean noon, just the same as if we had the true noon in observing the moment when it crosses the horary line of noon.

During many centuries time was only measured by sun dials and water clocks, or clepsydras. Water, running out regularly from a reservoir, is received in a vase which shows every hour. A float placed upon the liquid carries a little figure of a boy, which rises regularly and points to the hours. The ancient astronomers of China, Asia, Chaldea, and Greece measured in this way the hours of the night, the transits of stars across the meridian, and the duration of eclipses.

A rather curious thing is that the diurnal rotation of the earth upon itself and its annual revolution round the sun are two facts absolutely independent of each other, and which have not between them any common measure. There is not an exact number of days in the year. A complete revolution of our globe round the central luminary is performed not in 365 days exactly, nor in 366, but in 365¼ days. It follows, therefore, that we are obliged to make every fourth year a year of 366 days, the three others being 365. Still, this quarter is not itself accurate. There is not quite a quarter of a day to be added to 365 in order to form an exact year, so that if during several centuries we maintained regularly a bissextile year in every four we should go too quick, and we should soon be perceptibly in advance of nature. It was this which happened

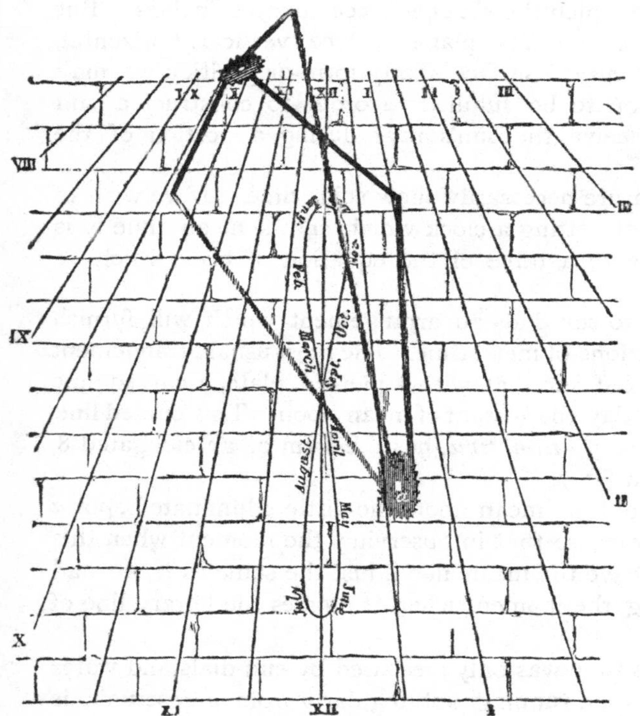

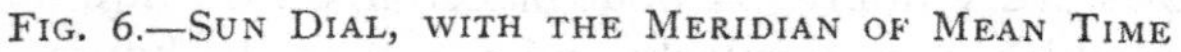
FIG. 6.—SUN DIAL, WITH THE MERIDIAN OF MEAN TIME

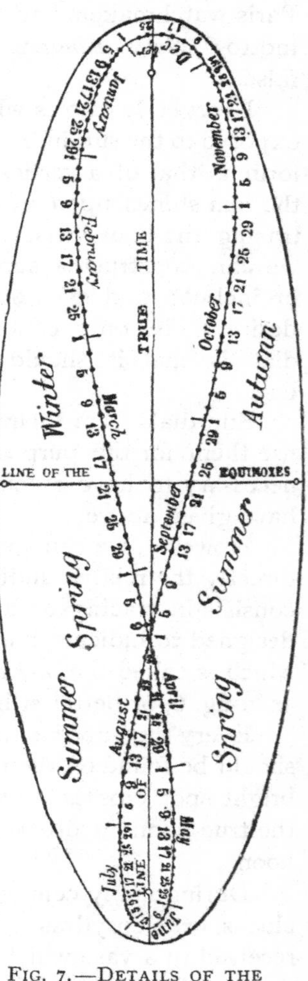

FIG. 7.—DETAILS OF THE MERIDIAN

at last, and which occasioned in 1582 the reform of the calendar decided upon by Pope Gregory XIII. In that year they had to cut off ten days which had accumulated since the time of Julius Cæsar, who in the century preceding the Christian era had added a quarter of a day to the year, considered up to that time to be 365 days exactly, and made one leap year in four. The astronomers of the sixteenth century corrected their predecessors; October 5, 1582, was called the 15th in most Catholic countries, and it was decided, in order to avoid the recurrence of a similar difference, that the bissextile years should be suppressed in four centuries. Thus the years 1700, 1800, and 1900 are leap years, according to the old chronology, though not in the new; but the year 2000 will be a leap year.

There are countries behindhand, like Russia, which for religious and political motives have not yet adopted this reform, and which prefer to be in disagreement with Nature rather than in accord with the Pope; they have left the years 1700 and 1800 leap years, and are at present in arrears by twelve days. In 1900 they will be thirteen days behind, if they continue to follow the calendar of Julius Cæsar. The exact duration of the year is 365 days 5 hours 48 minutes 46 seconds, or 365·2422 days.

Such is the length of 'the tropical year'—that is to say, the revolution of the seasons, which constitutes for us the principal apparent motion of the sun, and expresses by its effects the phenomena of nature. It is for us the true year, the meteorological year, the civil year.. But it is not exactly the precise duration of the earth's revolution round the sun. On account of the precession of the equinoxes, of which we have already spoken in the preceding chapter, and which we will soon explain in detail, while the earth returns at the end of the year to the point of the vernal equinox, it is still at a distance of more than twenty minutes from the point in space to which it should return in order to perform a complete revolution round the sun. The astronomical revolution of the earth, or its 'sidereal year,' is 365 days 6 hours 9 minutes 11 seconds [9·3 seconds, Harkness, 1891.—J. E. G.]—365·2564 days.

The earth revolves in a circle round the sun (it is really an ellipse, which very nearly approaches a circle). Such a figure has neither beginning nor end, so that nature herself is not bound to mark where the year commences or where it ends. Besides, the year, like the day, begins and ends at no particular point.

In the time of Charlemagne they began the year at Christmas in France, and in all the countries subject to the jurisdiction of the great emperor. This day was doubly celebrated as the festival of the birth of Christ and as the day of the renewing of the year. This old custom has left imperishable traces in Saxon habits, for even now among the Germans and English Christmas is celebrated with much more show than the 1st of January. It might have been more logical, and more pleasing at the same time, to close the year with the winter and commence it with the return of the sun—that is to say, to fix the beginning of the year at the vernal equinox, on March 21, or to leave it at March 1, as it was for two thousand years. Far from that, they have chosen exactly the most disagreeable season which can be imagined, and it is in the midst of cold, rain, snow, and frost that they have placed the wishes for a happy new year! It is now more than three hundred years since they adopted this custom in France, for it was the date of an edict of the melancholy little king Charles IX. (1563). It was only adopted in England in the year 1752, and it gave rise to a veritable riot. The ladies found themselves older, not only by eleven days, but still more by three months, since the date of the year was changed to January 1 instead of March 25, and they did not forgive the promoter of the reform for this deceit; the workmen, on the other hand, losing apparently a quarter's pay in their year, rebelled before they could understand that it was only in appearance, and the people followed Lord Chesterfield in the streets of London with repeated cries of 'Give us back our three months!' But the English almanacs of the day assured them that all nature was in agreement, and that 'the cats themselves, which were in the habit of falling upon their noses at the moment when the new year commenced, had been seen going through the same performance on the new date.' The Neapolitans also affirmed, on the other hand, that in 1583 the blood of St. Januarius was liquefied ten days sooner—on September 9 instead of the 19th. These superstitious or puerile arguments are similar to those of the Romans, who pretended to deceive destiny by calling 'twice six' *bissextus*, instead of the seventh, the intercalated day in February every four years. By this subterfuge February had not always only twenty-eight days, and they thus avoided a sacrilege and great public misfortune. This supplementary day being thus hidden between two others, the gods did not see it!

Not only is this fixing of the beginning of the year on January 1 illogical and disagreeable, but it further adds to the irregularities of the calendar by changing the meaning of the denominations of the months of the year. The Roman year commenced on March 1, and the twelve months were regulated thus:—

Month	Origin	Month	Origin
1. Mars	God Mars	7. September	Seventh
2. Aprilis	Aphrodite (Venus) or aperire (to open)	8. October	Eighth
3. Maia	Goddess Maia	9. November	Ninth
4. Junius	Goddess Juno	10. December	Tenth
5. Quintilis	Fifth	11. Januarius	God Janus
6. Sextilis	Sixth	12. Februarius	God of the dead

The first month was consecrated to the god of war, supreme patron of the Romans, and the last to the memory of the dead; Quintilis and Sextilis became Julius and Augustus, to honour the memory of Julius Cæsar and Augustus. Tiberius, Nero, and Commodus attempted to consecrate to themselves the following months, but, fortunately for the honour of nations, this attempt did not succeed. At present the month to which we have given the name of the seventh month of the year, *September*, is found to be the ninth month; October (the eighth) is the tenth; November (the ninth) is the eleventh, and December (the tenth) has become the twelfth and last. Can any more absurd designations be conceived? And all this because we have changed the beginning of the year from March, when spring commences, to January, which is generally the most gloomy and most melancholy month in the year.

Thus the names of the months have nothing in common with the Christian calendar (since they are heathen), nor with their proper origin (because they are changed), and they have not even the climatological character of those of the republican calendar of the great Revolution of '89, so euphemistically and happily imagined! How well these names correspond to the pictures of nature! They have the same termination as the months of each season, and are connected with meteorological facts or agricultural matters. Vendémiaire (the first month of the year) corresponds with the vintage, Pluviôse (the fifth month) with the time of rains; Frimaire (the third month) is the epoch of cold; in Germinal (the seventh month), Floréal, and Prairial, the nymphs appear dancing in the joyful sun of spring; Fructidor announces the fruits, Messidor the harvest. The following shows the correspondence between these and the Gregorian calendar:—

Vendémiaire,	from	Sept. 21	to	Oct. 20	Germinal,	from	Mar. 20	to	April 18
Brumaire	„	Oct. 21	„	Nov. 19	Floréal	„	April 19	„	May 18
Frimaire	„	Nov. 20	„	Dec. 19	Prairial	„	May 19	„	June 18
Nivôse	„	Dec. 20	„	Jan. 18	Messidor	„	June 19	„	July 17
Pluviôse	„	Jan. 19	„	Feb. 17	Thermidor	„	July 18	„	Aug. 16
Ventôse	„	Feb. 18	„	Mar. 19	Fructidor	„	Aug. 17	„	Sept. 20

These dates change with those of the equinox. Each month has thirty days, and they added five or six supplementary days, depending on whether the year was leap year or not. This was a complication so much the more whimsical that they gave these days the name of *sans-culottides* (the inhabitants of our planet cannot remain reasonable for two hours at a time).

We may also add that these denominations, suggested by our climate, do not correspond with the southern hemisphere, nor even with our own hemisphere, and that on this account the republican calendar cannot be universal.[1]

[1] Our present calendar is at once illogical and incomplete. If we retain the names of the months, the year should commence on March 1. If we begin the year on January 1, we should change the names of the months. Why is a month of twenty-eight days placed between two of thirty-one? No date is complete on this system; even a date of the month passes successively through all the days of the week, and an historical or intimate fact which happens on Sunday has for its anniversary a totally different day. We could correct this irregularity and obtain in future uniform and always similar years by deciding that January 1 shall count no longer in the year and be merely a festival, and that in leap years there shall be two. Such is the principle of the reform

There are, moreover, persons who prefer that the years should not be counted at all. Such was at least the advice of the two ladies of the Court of Louis XV. who were in the habit of deciding together in the last week of each year 'the age which they should be the following year.'

of the calendar to which I had the honour to award a prize in 1887 as president of the Astronomical Society of France. (See the review *L'Astronomie*, 1884, pp. 325 and 413; 1885, p. 294; 1887, pp. 212, 297, and 384.)

CHAPTER III

HOW THE EARTH GOES ROUND THE SUN

The Inclination of the Axis—Seasons—Climate

WE have just studied the diurnal rotation of the globe and its effects, and the discussion of the number of days in the year has already led us to the study of the annual motion of the earth round the sun. We will continue the analysis of these motions. They form, indeed, the foundation of a general knowledge of nature.

The moving planet upon which the game of our destiny is played sails in space, tracing its course round the light-giving sun. Day follows night, spring winter; the infant is born into the light, the old man sleeps in the darkness of the grave; fruits drop from the trees; flowers spring up again; human generations follow each other with rapidity; centuries pass, and the world goes round for ever.

From the motion of our planet round the centre of its heat and light climates and seasons result. In the polar regions the low-lying sun affords but a feeble heat and a pale light—desolate zones, where the traveller has often, in place of sunshine, a long twilight feebly illuminated by the rays of the aurora borealis—whilst in tropical regions a fiery sun darts its rays vertically upon the head, and the ground bathed in this glowing temperature produces an exuberant vegetation.

The orbit described by our globe in its annual voyage of circumnavigation round the sun is not circular, but elliptical, as we have already remarked. Everyone knows how to draw an ellipse. The simplest method is that which gardeners make use of. They fix two stakes, to which are attached the ends of a cord longer than the distance which separates the stakes; then they stretch the cord by means of a point, and trace the ellipse on the ground by simply following the curve produced by the motion. The nearer the stakes are to each other the nearer the ellipse approaches a circle; the farther they are separated the more elongated is the curve [the length of the cord being the same in both curves. —J. E. G.]. Now, it is found that all the heavenly bodies follow in their motions, not circles, but ellipses. The points represented by the stakes are called the foci of the ellipse (F, F′ in fig. 8). The centre is at O; the diameter AA′ is called the *major axis*, and the diameter BB′ the *minor axis*.

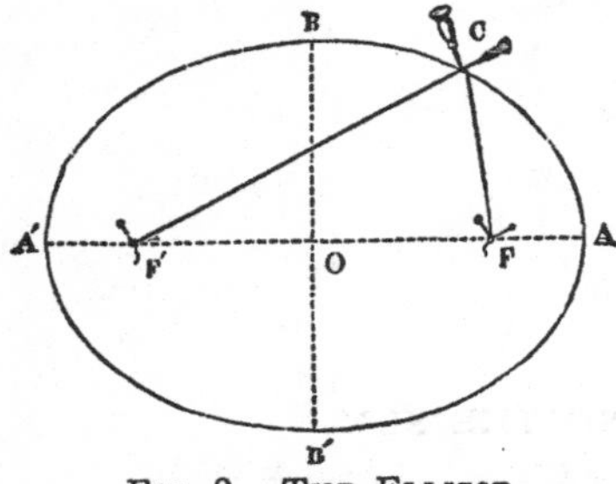

FIG. 8.—THE ELLIPSE

If we consider the earth's orbit round the sun we shall find that the sun occupies one of the foci of the ellipse described by our globe in its orbit, and that the other focus remains empty. It follows that the distance of our globe from the sun varies during the course of the year. It is on January 1 that we are at the greatest proximity, and on July 1 at the greatest distance. The first point is named the *perihelion* and the second *aphelion*. The differences of distances are as follow:—

Perihelion distance	145,700,000 kilometres
Mean distance	148,250,000 "
Aphelion distance	151,800,000 "

[The most recent results are as follows—based on a solar parallax of 8″·80905, as deduced by Professor Harkness in 1891:

Perihelion distance	94,353,155 miles
Mean distance	92,796,950 "
Aphelion distance	91,240,745 "

—J. E. G.]

We see that the earth is 6,100,000 kilometres nearer the sun on January 1 than it is on July 1. [The difference in English measure is about three millions of miles. See measures given above.—J. E. G.] The difference of temperature between summer and winter is caused, as we shall see presently, by the inclination of the earth's axis. In winter the solar rays glance upon our hemisphere, scarcely warming it; the days are short and the nights are long: in summer, on the contrary, the solar rays reach us more perpendicularly; the days are long and the nights short. But while our northern hemisphere is in winter the southern hemisphere is in summer, and *vice versâ*. As the difference of distance of the earth from the sun in January and July is also sufficiently perceptible, the summers of the southern hemisphere are warmer than ours and their winters colder. The denominations of winter, summer, spring, and autumn, applying inversely to the two hemispheres, do not agree for the whole earth. In place of speaking of the winter solstice, the summer solstice, the vernal equinox, the autumnal equinox, it would be better to speak of the December solstice, the June solstice, the equinox of March, and the equinox of September. These denominations apply to the whole earth—to Australia, South America, and South Africa as well as to Europe.

Our readers will very easily understand the way in which the earth goes round the sun by examining the following diagram (fig. 9). They will see

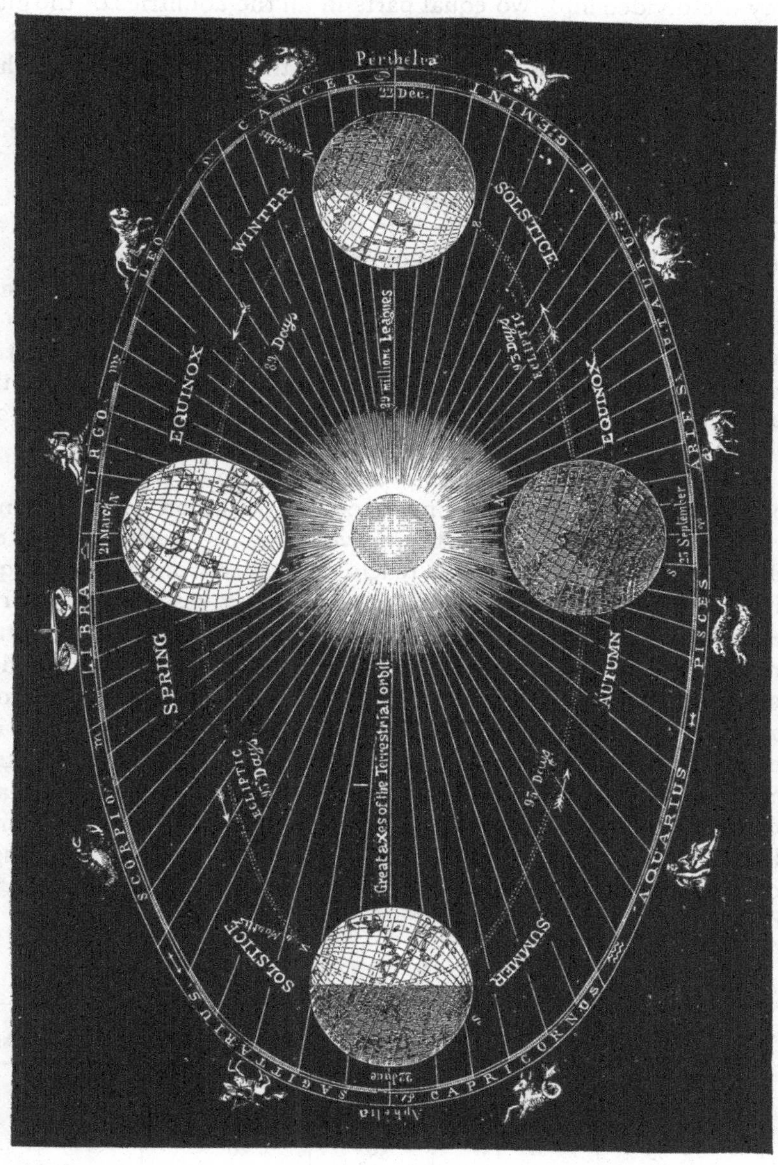

Fig. 9.—Diagram of the Rotation of the Earth round the Sun

at a glance that it always keeps its axis of rotation constantly in the same direction, always parallel to itself, and that, as it is not upright, but inclined, the pole is for six months illuminated by the sun, and for six months it is in darkness. At the two equinoxes the illuminated hemisphere passes exactly through the two poles ; so that, as we see, the twenty-four hours of the day are divided into two equal parts in all the countries of the globe. But as we advance towards summer, the inclination of the axis causes the solar light to encroach more and more beyond the pole, so that northern countries have longer and longer days and shorter and shorter nights. It is the contrary if we examine the positions of the earth during the winter. We see, for example, that Paris (about the fifth circle of latitude) has in December but eight hours of day, and remains sixteen hours in the night. The nearer we approach the pole the greater is the difference, and at the pole itself there are six months of day and six months of night.

This figure has been drawn to show the annual motion of our planet round the sun. It was necessary, then, to give a certain importance to the terrestrial globe, and merely to indicate, so to say, the sun by its position, for were we to represent that body in the proportion of its volume to that of the earth it would be necessary to give it a diameter of 7·54 feet and to place it at a distance of 820 feet! . . .

The inclination of the earth's axis produces, then, a difference in the duration of day and night, depending upon the situation of the country which we inhabit. At the equator they have constantly twelve hours day and twelve hours night. When we reach a distance from the pole equal to the inclination of the axis—that is to say, 23 degrees 27 minutes from the pole, or, which is the same thing, 66 degrees 33 minutes of latitude (there are 90 degrees of latitude from the equator to the pole)—the sun does not set on the day of the summer solstice, but only glides along at midnight above the northern horizon, and as a compensation it does not rise above the horizon on the day of the winter solstice. For those countries up to the pole, the sun does not set or does not rise for a number of days, a number which increases up to the pole itself, where we find six months day and six months night. The following is a little table of the length of the days, according to the latitude, (I.) from the equator up to the polar circle, (II.) from the polar circle up to the pole :—

I.

Latitude	Length of the Longest Day	Length of the Shortest Day	Latitude	Length of the Longest Day	Length of the Shortest Day
	H. M.	H. M.		H. M.	H. M.
0°	12 0	12 0	40°	14 51	9 9
5°	12 17	11 43	45°	15 26	8 34
10°	12 35	11 25	50°	16 9	7 51
15°	12 53	11 7	55°	17 7	6 53
20°	13 13	10 47	60°	18 30	5 30
25°	13 34	10 26	65°	21 9	2 51
30°	13 56	10 4	66° 33′	24 0	0 0
35°	14 22	9 38			

II.

Northern Latitudes	The Sun does not Set for about	The Sun does not Rise for about
66° 33′	1 day	1 day
70°	65 days	60 days
75°	103 „	97 „
80°	134 „	127 „
85°	161 „	153 „
90°	186 „	179 „

France is included between the 42nd and 51st degree of latitude, and Paris is placed at 48° 50′. The length of the longest day is 15 hours 58 minutes, and that of the shortest 8 hours 2 minutes. It is necessary to add to this calculation the geometrical influence of atmospheric refraction, of which we shall speak further on (Chap. VI.), and which raises the stars above their real position. We see the sun rising before it is really above the horizon, and we still see it when it has really set. It follows that the longest day at Paris is 16 hours 7 minutes, and the shortest 8 hours 11 minutes. The illumination of the atmosphere by the dawn and by twilight still further increases the duration of the day. The atmosphere remains illuminated so long as the sun has not sunk to 18 degrees below the horizon. A rather curious fact follows from this: on June 21 at Paris the sun descends obliquely in the north-west, after he sets, to reappear next morning in the north-east, and at midnight, when he is due north, he has only sunk 17° 42′, so that night is not perfect at Paris at the summer solstice.

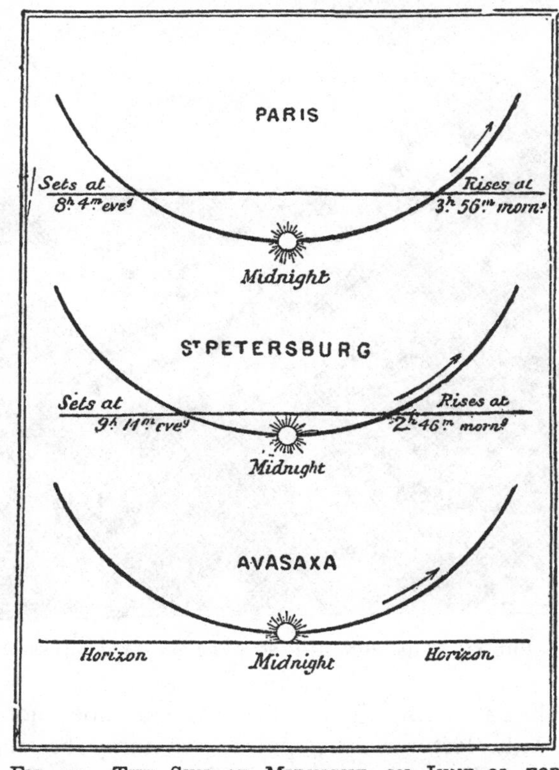

FIG. 10.—THE SUN AT MIDNIGHT, ON JUNE 21, FOR PARIS, ST. PETERSBURG, AND MOUNT AVASAXA

This effect is the more perceptible the farther we advance to the north. At St. Petersburg on June 21 there is sufficient light at midnight to read.

It follows from the same effect of atmospheric refraction that it is not necessary to go to the polar circle in order to see the sun not setting and grazing the horizon at midnight. At the sixty-sixth degree of latitude in Sweden and Finland we can enjoy this spectacle, so strange to us—*the midnight sun.* It has even become the fashion in recent years to make a voyage to Tornea, a little town on the frontiers of Russia and Sweden, on the Gulf of Bothnia, and to be present on June 21 on Mount Avasaxa,

FIG. 11.—THE SUN SHINING OVER AVASAXA (RUSSIA) AT MIDNIGHT, ON JUNE 21

only 227 metres high, when the sun does not set at the summer solstice. A lady lecturer on popular astronomy has made this excursion, and she has kindly sent us a sketch (fig. 11), drawn from nature.[1]

The duration of twilight increases as we approach the extremities of the earth's axis, so that the long polar night is much shorter than it would be without the refraction of the atmosphere, and much less melancholy than it would be without the aurora borealis.

North of the 67th degree of latitude the sun does not rise at the winter

[1] See our *Revue Mensuelle d'Astronomie Populaire*, 1st year, 1882, p. 377.

solstice. Two days, three days, a whole week passes by without his disc reappearing above the southern horizon at noon. Only a pale gleam indicates that he glides along below the horizon. Farther north he continues two months, three months, without appearing, and the world remains shrouded in a dark and icy night, illuminated only by the moon, or by the fitful gleams of the northern lights. One of the last expeditions made for the discovery of the North Pole—that of the English navigators Nares and Stephenson (May 29, 1875, to Nov. 2, 1877)—which advanced farther than any of its predecessors—up to 82° 24′—had 142 days of solar privation—nearly five months of night! From November 6 to February 6 the night was complete and dark. Even on November 8 the darkness was so complete at *noon* that it was impossible to read. But soon the moon brought a reflection of the vanished sun, and turned round the pole without ever setting for ten times twenty-four hours. The thermometer went down to 58° Centigrade below zero! (It has been seen still lower at Werchojansk, in Siberia—68°). These low temperatures are never accompanied with wind; otherwise no human being could stand such cold. O icy solitudes of the pole, you have already received heroes who have lain down for ever in your gloomy shroud! The road to the pole is already marked with martyrs; but it is not the odious war of man against man: it is the triumph of mind over matter, the conquest of nature by genius!

The effect produced by the inclination of the earth on the sun's apparent motion has caused a division of the terrestrial surface into five zones—(1) the tropical zone, situated on both sides of the equator up to the tropics at 23° 27′ latitude, which includes all places on the earth where the sun is seen to pass through the zenith at certain times of the year; (2) the temperate zones, in which the sun does not reach the zenith, but sets every day; (3) the frigid zones, or polar caps traced round each pole at the latitude of 66° 33′, in which the sun remains either constantly above or below the horizon during several consecutive days at the time of the solstices. As their designation indicates, moreover, the first is hot, because it receives the rays of the sun almost perpendicularly; the second are temperate, because the rays of the sun are received more obliquely; the last are truly frigid, because the solar rays merely glide, so to speak, along their surface.

The areas of these zones are very unequal: the torrid zone embraces forty hundredths of the total surface of the terrestrial spheroid, the two temperate zones fifty-two hundredths, and the two frigid zones eight hundredths. Thus the two temperate zones, the most favourable to human habitability and to the development of civilised life, form more than half the surface of the earth; the frigid zones, so to say uninhabitable, form a very small fraction.

Let us return now to the motion of the earth round the sun.

The attraction of the sun diminishes in intensity with distance, and the motion of translation of the earth being governed by this attraction, the enormous ball which carries us sails more slowly at aphelion than at perihelion, in July than in January! The total length of the immense curve described each year by the globe is 584 millions of miles, a circuit performed in 365 days 6 hours, which gives 66,666 miles an hour, 1,111 a minute, or about 18·5 miles per second, as the mean velocity. This velocity diminishes to 28,900 metres on July 1 (about 18 miles a second), and increases to 30,000 metres (about 18·64 miles a second) on January 1. Thus in one day, while it performs one rotation upon itself, the earth moves in the sky about 200 times its own diameter. In an hour it is displaced eight and one-third times its diameter of 12,742 kilometres [7,912 miles] (fig. 12). Seventy-five times faster than that of a cannon ball, this motion is so tremendous that were the earth suddenly stopped in its course the shock would be transmitted by recoil, so to say, to all the constituent molecules of the terrestrial globe, as if each received a stunning blow; the whole earth would become instantaneously luminous and burning, and an immense conflag-

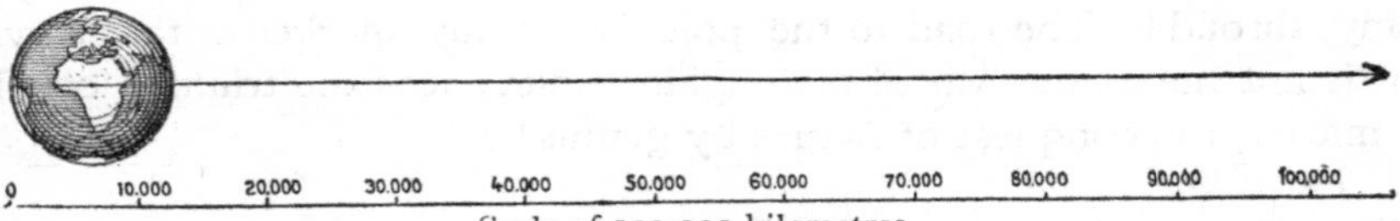

0 10.000 20,000 30.000 40.000 50.000 60.000 70.000 80.000 90.000 100,000

Scale of 100,000 kilometres

FIG. 12.—DISTANCE TRAVELLED BY THE EARTH IN ONE HOUR RELATIVELY TO ITS DIAMETER

ration would devour the world. The earth could no more be stopped than the sun in his course. Such an event would certainly be the greatest in history; though, indeed, it could not be spoken of as historical, for no one would survive to record it.

We see in our fig. 9 that the curve described by the earth from spring to autumn is a little longer than the opposite part passed over between autumn and spring. The spring and summer last a little longer than the autumn and winter, so much the more that the earth itself goes a little less quickly in its orbit in summer than in winter. We will give, however, the respective duration of the seasons to a tenth of a day nearly:—

	Days
Spring	92·9
Summer	93·6
Autumn	89·7
Winter	89·0
	365¼

The astronomical seasons begin at the equinoxes and solstices—that is to say, on March 20, June 21, September 22, and December 21—to a day nearly, according to the years. Geometrically, these dates should rather mark the middle of the seasons, because after June 21 the days begin to shorten and after December 21 to lengthen. The temperature, on the contrary, continues to increase after the June solstice, on account of the accumulation of heat day by day, and it decreases after the December solstice for the contrary reason. The annual maximum of temperature happens about July 15, and the minimum about January 12. Even the maximum diurnal temperature occurs after noon (about 2 P.M.), and the minimum about four o'clock in the morning.

The earth's axis of rotation prolonged in imagination to the apparent vault of the sky marks the pole, the point round which the star sphere appears to turn, in a direction opposite to that of the earth's rotation. The nearest star to this point has received the name of Pole Star. *When we look at the North Pole*, this diurnal motion is performed in a direction opposite to that of the hands of a watch; all the stars the distance of which from the pole is less than the altitude of the pole above the horizon never set; they skim the northern horizon, and rise again to the right, or east, of the spectator. We show in fig. 13 the most conspicuous of these stars. This little celestial chart will be of great use to us—in the first place to show the motion of the starry heavens round the pole, and again to fix at once in our mind the forms of the constellations which are perpetually visible in our latitudes. We have shown only the principal stars, so as to avoid confusion. It is easy to identify rapidly these Northern constellations: the Little Bear, nearest the pole; the Great Bear, chiefly composed of seven remarkable stars, otherwise known as the Waggon or Plough, always easily recognised; the Dragon (Draco), formed by a winding line of stars which begins between the two bears; Cepheus, Cassiopeia, Perseus, Cameleopardalis. We shall learn further on to know these constellations and all the others; but it would be very useful for our readers to attempt at once to recognise these stars in the sky, by looking towards the north, and accustoming themselves to identify them by practice on the first fine evening.[1]

We notice on this chart the place of the pole. All the stars shown turn

[1] It is customary to designate the stars by the letters of the Greek alphabet. Those of our readers who do not know this alphabet doubtless imagine that there is here an insuperable difficulty. It is no such thing, very fortunately. It can be learned very easily. The following are the letters and their names. With a little attention anybody will read in ten minutes the names of the stars on the accompanying chart:

α	Alpha	η	Eta	ν	Nu	τ	Tau
β	Beta	θ	Theta	ξ	Xi	υ	Upsilon
γ	Gamma	ι	Iota	ο	Omicron	φ	Phi
δ	Delta	κ	Kappa	π	Pi	χ	Chi
ε	Epsilon	λ	Lambda	ρ	Rho	ψ	Psi
ζ	Zeta	μ	Mu	σ	Sigma	ω	Omega

The brightest star of each constellation has received the first letter (α), and has often a proper name, as Sirius, Vega, Arcturus, Capella, &c.

D

in twenty-four hours in the direction indicated by the arrows. The position represented is that of the sky on December 21 at 10 P.M., and is the same on March 20 at 4 P.M., June 21 at noon, and September 22 at 4 A.M. If

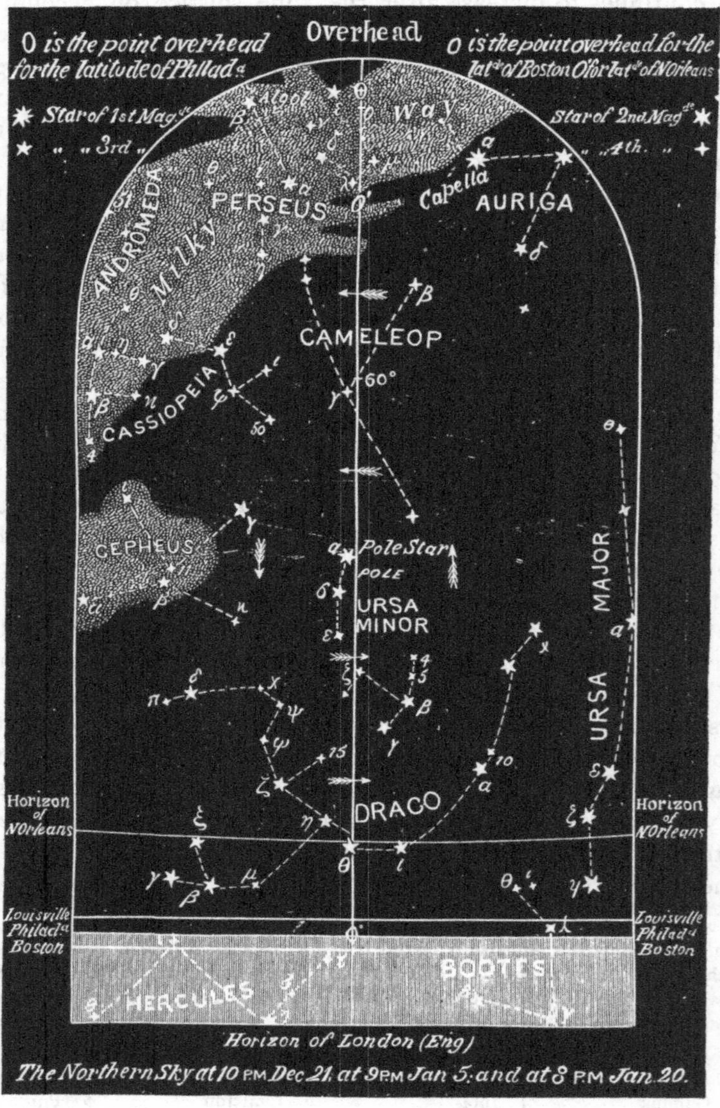

FIG. 13.—STARS SURROUNDING THE POLE, AND WHICH NEVER SET IN ENGLAND

we turn the paper upside down we shall have the aspect of the sky on June 21 at 10 P.M., on September 22 at 4 P.M., on December 21 at 10 A.M., and on March 20 at 4 A.M. If we place the left side of the page at the bottom we shall have the aspect of the sky on March 20 at 10 P.M., June 21 at 4 P.M., September 22 at 10 A.M., and December 21 at 4 A.M. It would be the opposite if we looked at the map by placing at the bottom the right side of the page.

Every day, every hour, the aspect of the sky changes. Thus, an hour after that of the position drawn in the map the Great Bear is a little higher, two hours after it is higher still, and six hours later it soars to the top of the sky; then it descends, and if the night is sufficiently long we see it twelve hours afterwards occupying the part of the sky diametrically opposite to that which it occupied at the beginning of the observation. It thus easily indicates the hour during the night. As we see, it never descends below the horizon; the ancients remarked this fact, and it was sung in particular by Homer among the Greeks and by Ovid among the Romans.

All the stars turn in $23^h\ 56^m$ round the pole in a direction contrary to that of the diurnal motion of the globe passing once a day across the meridian—that is to say, across the imaginary line drawn from north to south [through the zenith] dividing the sky into two equal parts. All coming from the east, the stars mount slowly in the sky, reach the highest point of their course, and descend towards the west, as the sun himself does every day. The most important instrument of every observatory is the *meridian telescope* or *meridian circle*, an instrument so named because it is fixed in the plane of the meridian, and cannot be turned aside; moving in the same plane, it can be pointed to all possible altitudes, and is intended to observe the transits of stars across the meridian. The exact instant at which this transit takes place is determined by the aid of vertical wires which cross the field of view of the telescope and behind which the star passes.

To this telescope is attached a perfectly vertical circle, which serves to measure the altitude of the stars or their distance from the pole of the equator, while the telescope serves to determine the exact instant of their passage across the meridian. We may say that the meridian telescope indicates the vertical line on which the star lies, and that the circle shows the horizontal line, so that the exact position of the star at the intersection of the two lines indicates its real position on the celestial sphere, as the position of a town on the earth is determined by its longitude and latitude.

These instruments can only catch the stars at the moment when they cross the meridian, and they cannot be directed to other points of the sky. A natural addition to such appliances is, in all observatories, an instrument mounted in such a way that it can be pointed to all the regions of space. Such an instrument is represented in fig. 14. It is called an equatorial, because the clockwork with which it is fitted causes it to turn like the earth in a plane parallel to the equator: the instrument may be pointed to any star, and it follows this star from east to west in its diurnal motion. It is as if the earth ceased to turn for the astronomer engaged on the study of the star. There are at the Paris Observatory several instruments of this

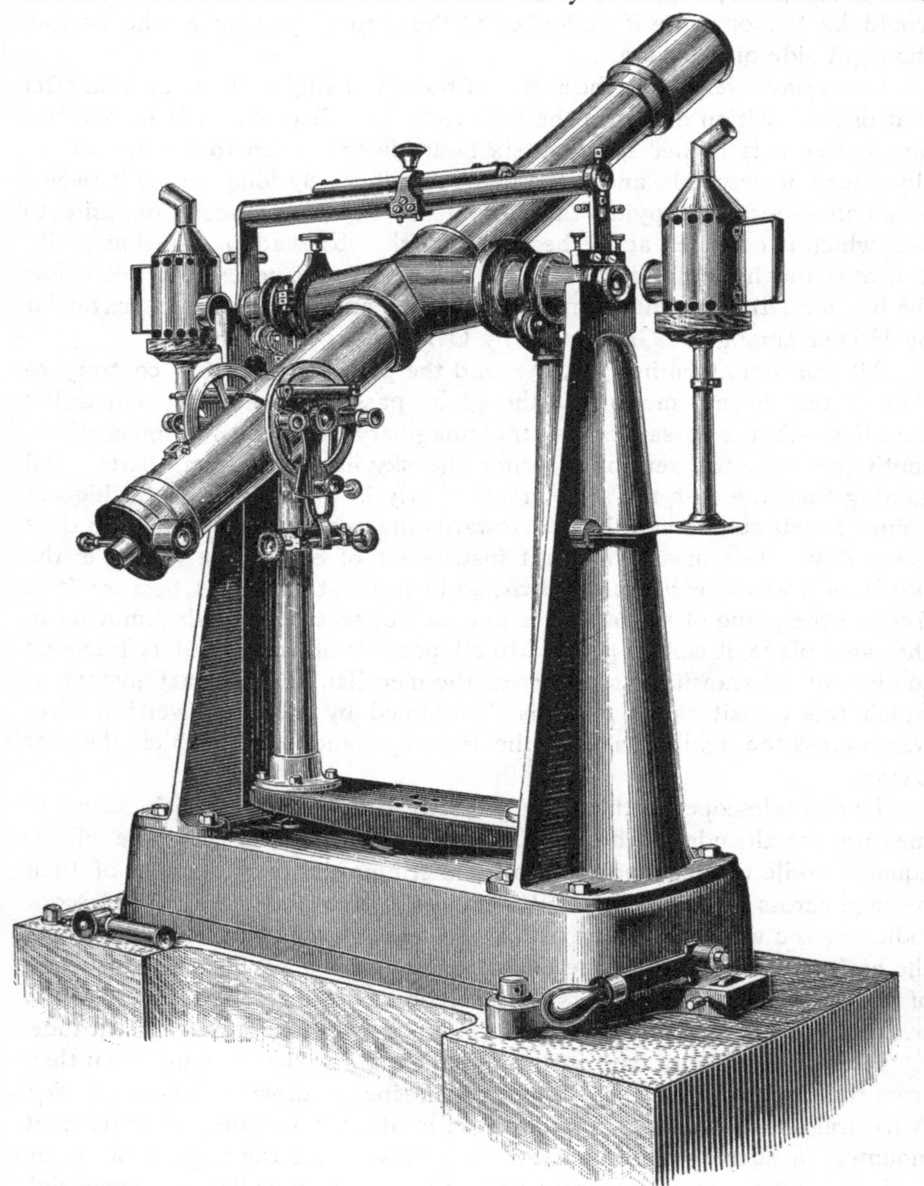

FIG. 14.—TRANSIT INSTRUMENT BY TROUGHTON & SIMMS

sort. The largest measures 38 centimetres (about 15 inches) in diameter and 9 metres (about 29½ feet) long (with it I have made numerous measures of double stars, which will be considered further on); another of $12\frac{1}{2}$ inches in diameter and $16\frac{1}{2}$ feet long; two others measure $9\frac{1}{2}$ inches and 10 feet.

FIG. 15.—REFLECTING EQUATORIAL TELESCOPE BY JOHN BROWNING, F.R.A.S.

We will not further enlarge here on optical instruments, to which we will devote a special description at the end of this volume.

CHAPTER IV

THE ELEVEN PRINCIPAL MOTIONS OF THE EARTH

The Precession of the Equinoxes

As changeable as the iridescent bubble which the breath of a child blows by the aid of a simple drop of soapy water, and allows to fly away through the air in the rays of the cheerful sun, the terrestrial globe floats in space, the veritable sport of cosmical forces which carry it whirling through the vast heavens. We have just estimated the velocity of its motion round the sun, and the manner of its diurnal rotation on itself. These two motions are not the only movements with which our turning ball is animated. We have already mentioned briefly nine others which are superadded in its perpetual balancing; it is now necessary to analyse these more in detail and to understand them well.

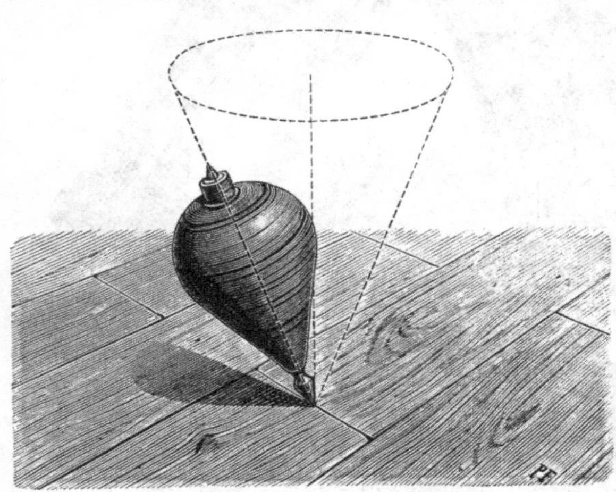

Fig. 16.—Diagram of the Displacement of the Axis of the Earth by the Precession of the Equinoxes

And in the first place, the axis round which the diurnal rotation is performed, and which remains, as we have seen, directed during the whole year towards the same point in the sky, towards the pole, is not absolutely fixed. It is slowly displaced, describing a cone of 47 degrees opening, a motion analogous to that of a top, which, while spinning rapidly upon itself, moves with its axis inclined, and traces in space a cone in the shape of a funnel, which we can represent geometrically (fig. 16). The celestial pole being the point where the supposed prolongation of the terrestrial axis ends, there follows a secular displacement of this point among the stars. It is not always the same star which bears the name of the *Pole star.* At present it is the star at the extremity of the tail of the Little Bear which is nearest to the pole and which has received this characteristic name. It will continue to approach the pole up to the year 2105; but then the pole will recede from it, not to return for 25,000 years. The direction of the motion of precession is 25,765 years [25,695 years.—Stockwell]. The circle is not a closed curve, as it has hitherto been represented in all astronomical works. During this lapse of time the sun, moving towards the constellation Hercules, changes the starry perspectives.

We can easily explain this secular motion by the celestial chart which represents it. This chart (fig. 17), which contains a few more stars than the first, should be studied carefully. Its special object is to show the progress of the pole during the course of the revolution of which we are speaking. We have indicated the dates of the successive positions of the pole from 6,000 years before the Christian era to the year 28,000. We see that six thousand years before our era the pole passed in the neighbourhood of two small stars of the fifth magnitude.[1] The nearest bright star then was θ (Theta) Draconis, a star of the fourth magnitude. About the year 4500 the pole passed not far from a rather bright star of the third magnitude, ι (Iota) of the same constellation. About the year 2700 another star of about the same brightness, α (Alpha) Draconis, became the pole star, by which name it was known in China and in Egypt. The ancient Chinese astronomers have recorded it in their annals of the time of the Emperor Hoang-Ti, who reigned about the year 2700 B.C. The Egyptians, who built the great pyramids more than forty centuries ago, constructed the passages which permit us to penetrate into the interior exactly in the direction of the north, and at an inclination of 27 degrees, which is precisely the altitude which the pole star of that day, α Draconis, attained at its lower transit across the meridian. The pole afterwards passed near the fifth magnitude

[1] From a remote antiquity the light of the stars has been divided into six magnitudes. These magnitudes only represent the apparent brightness, and not the real dimensions of the stars, which depend upon their light and their distances. The brightest stars form the first magnitude, then follow the second, the third magnitude, &c. The smallest stars visible to the naked eye form the sixth order. We can count in the whole sky—

19 stars of the first magnitude		530 stars of the fourth magnitude	
59 ,, ,, second ,,		1,620 ,, ,, fifth ,,	
182 ,, ,, third ,,		4,900 ,, ,, sixth ,,	

[These numbers are variously given by different observers. See *The Visible Universe*, p. 330.—J. E. G.]

We shall advance further in our study of the stars and constellations. The reader is advised to identify at first the positions and magnitudes of the northern stars as represented in the two charts in the text.

star, ι Draconis, then between β (Beta) Ursæ Minoris and κ (Kappa) Draconis. This was at the time of Chiron's sphere, the most ancient sphere known, constructed about the epoch of the Trojan War, 1300 B.C. Afterwards we see the pole gradually approaching the tail of the Little Bear.

At the beginning of our era no bright star indicated the place of the

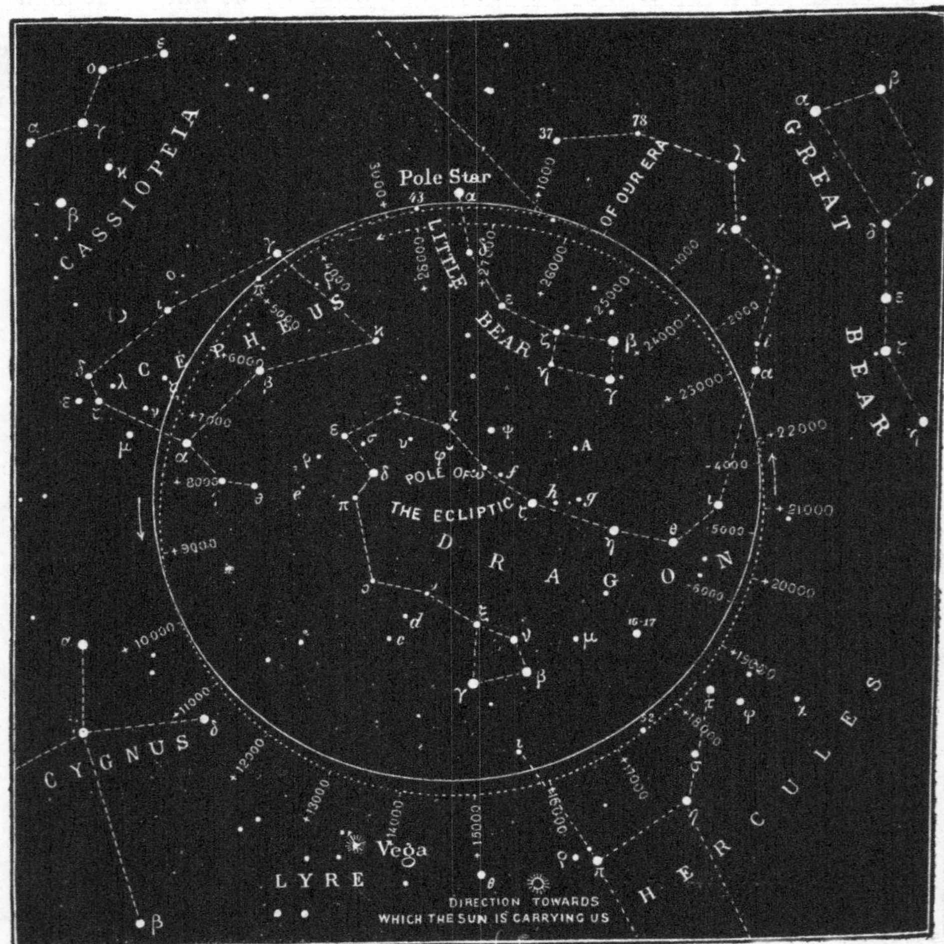

FIG. 17.—SECULAR DISPLACEMENT OF THE POLE, FROM 6,000 YEARS BEFORE OUR ERA TO THE YEAR 28,000—A CYCLE OF 25,765 YEARS

pole. About the year 800 it passed near a small star of the Giraffe [Cameleopardalis] (a double star which bears the numbers 4339 and 4342 in the catalogue). But the present pole star, of the second magnitude, is in reality one of the brightest of those which lie near the path of the pole, and it has enjoyed its title for more than a thousand years ; it will still retain it till about the year 3500, at which epoch we see the path of the pole's motion approaching a star of the third magnitude; this is γ (Gamma) Cephei. In the year 6000 it will pass between the two stars β (Beta) and ι (Iota) of the same constellation; in the year 7400 it will approach α (Alpha) of the same brightness. The year 10000 will give the title of pole star to the fine star α (Alpha) Cygni, of the second magnitude (almost of the first) [measured 1·47 at Harvard]; and in the year 13600 (11,700 years hence) it will be near the brightest star of our northern sky, Vega of the Lyre, which for 3,000 years or so will be the pole star of future generations, as it was 14,000 years ago for our ancestors.

During this period the aspect of the celestial sphere alters with the motion of the pole. The sky of different countries changes. For example, some thousands of years ago the Southern Cross was visible in Europe. On the other hand, some thousands of years hence the sparkling Sirius will have disappeared from the European sky. The constellations of the southern sky show themselves to us for some centuries, then withdraw from our view, while our northern stars become visible to the inhabitants of the south. The revolution of 257 centuries exhausts all these aspects.

Immense and slow revolution of the skies! What events occur on our globe during the course of one of these periods! The last time that the pole occupied the place which it does to-day, 25,765 years ago, none of the present countries existed. None of the nations who dispute to-day for supremacy on the planet had then left the cradle of nature. Already, doubtless, there were men upon the earth, but the social unions which they formed have left no trace of the degree of civilisation to which they had attained, and it is very probable that these uncultured and savage beings were then in the midst of the primitive Stone Age, of which so many proofs have recently been collected. Where shall we be in our turn when, after another period of equal duration, the pole will have again returned to its present position? French, English, Germans, Italians, Spaniards, may then join hands in a common obscurity. None of our contemporary nations will have resisted the transforming work of Time. Other nations, other languages, other religions will have long since replaced the present state of things. One day a traveller wandering on the banks of the Seine will be attracted by a heap of ruins, seeking the place where Paris had, during so many ages, shed its light. Perhaps he will find the same difficulty in recovering places formerly famous that the antiquary now finds in identifying the site of Thebes or of Babylon. Our nineteenth century will be then, *in antiquity*, very much further back than are for us the ages of the Pharaohs and the ancient Egyptian dynasties. A new human race intellectually superior to ours will have won its way to the sunlight; and we shall perhaps be very surprised, you and I, O studious and thoughtful readers! to meet each other, side by side—blanched and carefully labelled skeletons—

installed in a glass case of a museum, by a naturalist of the two hundred and seventy-sixth century, as curious specimens of an ancient race, rather wild, but already endowed with a certain aptitude for the study of the sciences. Vanity of vanities! O noisy ambitions of a day, who pass our life disputing about tinsel, about empty titles and many-coloured decorations, ask yourselves what philosophy must think of your ephemeral vainglory, when it compares your puerile rivalries with the majestic work of nature, which bears us all to the same destiny!

Thus the whole starry sky proceeds in a combined motion, which causes it to slowly turn round an axis ending in the pole of the ecliptic. The ecliptic is the path which the sun seems to follow in the sky in its motion round the earth. We have seen that in reality it is our globe which moves round the radiant star. By an effect of perspective, which is easily explained, the sun appears to move in the contrary direction, and to make the circuit of the sky in a year; it is the track of this apparent motion of the sun which is called the *ecliptic*, a name derived from the fact that eclipses take place when the moon is found, like the sun, in the plane of this great circle of the celestial sphere. The pole of the ecliptic is the central point of this great circle on the sphere, the point on which we should place the point of a compass opened to a right angle in order to trace at ninety degrees of distance the circle of the ecliptic.

It follows from this general motion that the stars do not remain for two consecutive years at the same points of the sky, and that they move all together in order to accomplish, during this long period, a complete revolution. We are obliged constantly to re-draw our celestial charts, and to make the paper, so to say, to slide upon the stars. The maps made in 1860 would not have suited for 1880, and those which we now draw would no longer agree with the sky in the year 1900. There are very precise mathematical formulæ for calculating the effects of this motion, and determining the exact positions of the stars at any date past or future.

This motion does not belong to the sky, any more than the diurnal motion and the annual motion. It is the earth itself which is animated, and which performs, during this long period, an oblique rotation upon itself, in a direction contrary to that of its motion of diurnal rotation; this motion is caused by the combined attraction of the sun and moon on the equatorial protuberance of the earth. If the earth were a perfect sphere, this retrograde motion would not exist. But it is flattened at its poles and protuberant at its equator. The molecules of this equatorial pad retard a little the motion of rotation; the action of the sun and moon makes them retrograde, and they draw away in this motion the globe to which they are adherent.

We have, then, a *third* motion of the earth, the secular motion of the *precession of the equinoxes*, so named because it causes each year an advance of the vernal equinox on the real revolution of the earth round the sun.

The positions of the stars on the celestial sphere are fixed by starting from a line drawn from the pole to a point on the equator cut by the ecliptic at the moment of the vernal equinox. This point advances each year from east to west; the equinox occurs successively in all points of the equator. The mean velocity is fifty seconds of arc per annum. We shall explain farther on what is meant by *degrees*, *minutes*, and *seconds of arc.*

The stars situated in the region of the sky which the sun seems to pass over in his apparent annual motion were divided, at an unknown epoch, but which we know to be very ancient, into twelve groups called the *Zodiacal Constellations.* The first, in which the sun was found at the moment of the equinox, two thousand years ago, took the name of *The Ram*; the second, going from west to east, was called *The Bull*; the third group is that of *The Twins*; the three following are *The Crab*, *The Lion*, and *The Virgin*; the six others are *The Balance*, *The Scorpion*, *The Archer*, *The Goat*, *The Water Bearer*, and *The Fishes.* The following classical lines from the poet Ausonius present them under a mnemonic form :—

Sunt Aries, Taurus, Gemini, Cancer, Leo, Virgo,
Libraque, Scorpius, Arcitenens, Caper, Amphora, Pisces.

[The last four are usually called Sagittarius, Capricornus, Aquarius, and Pisces.—J. E. G.]

The vernal equinox falls at present in the constellation Pisces, near the end, and will soon pass into Aquarius. We have sketched (fig. 18) the stars of the twelve zodiacal constellations. The line of the ecliptic is the middle line of the zodiac. The equator is inclined to this line, as we have already remarked in speaking of the earth's motion of rotation. The two bands of six constellations each placed one above the other in this map should be supposed to continue by placing the extremities of the figures in juxtaposition, and rolled like a cylinder round the eye of the observer: this is the zodiacal zone of the immense celestial sphere. We have given above each figure the months of the year in which the sun passes successively into each of the constellations.

We may picture to ourselves the march of the equinoctial sun along the constellations of the zodiac as we have pictured the secular march of the pole amongst the northern stars. At the beginning of our era the equinox occurred in the first degrees of the Ram; 2,150 years previously it coincided with the first stars of the Bull, which had been the equinoctial sign since the year 4300 B.C. It was probably during this epoch that the first observers of the sky established the zodiacal constellations, because in all the ancient religious myths the Bull is associated with the productive work of the sun upon the seasons and the produce of the earth, while we find no trace of an analogous association with the Twins. This was already a legend eighteen centuries ago, when Virgil greeted the celestial Bull opening with his golden horns the cycle of the year :—

Candidus auratis aperit cum cornibus annum
Taurus, et averso cedens Canis occidit astro.

The stars of Taurus, especially the Pleiades, were for the Egyptians, the Chinese, and even for the early Greeks, the stars of the equinox. The annals of astronomy have preserved a Chinese observation of the Pleiades as marking the vernal equinox in the year 2357 before our era.

This secular advance of the equinox is not exactly uniform, and it follows that the tropical year is not absolutely invariable. Thus it is now shorter by 16 seconds than in the time of Hipparchus, and 30 seconds

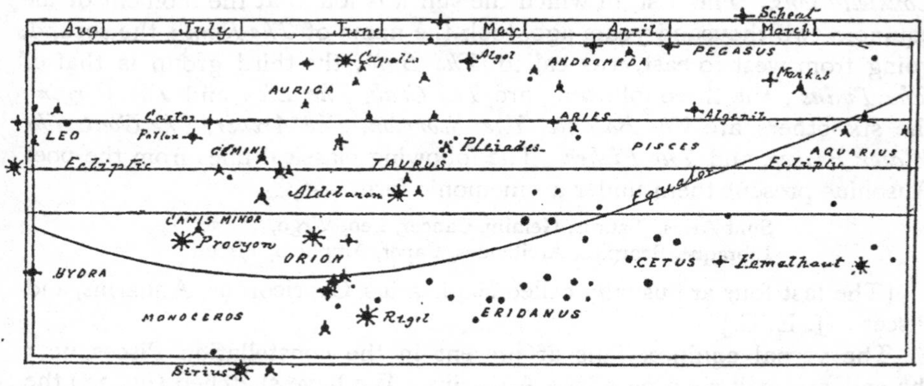

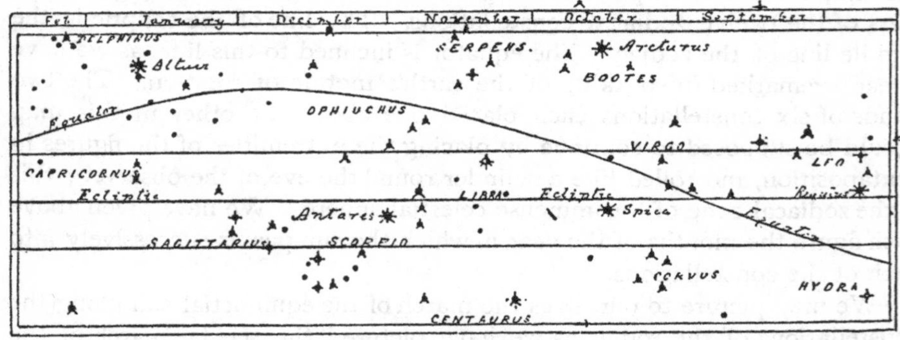

FIG. 18.—PRINCIPAL STARS AND CONSTELLATIONS OF THE ZODIAC

shorter than at the time when the city of Thebes in Egypt was the capital of the world. At the beginning of this century it was 365 days 5 hours 48 minutes 51 seconds. It is diminishing. At present it is 365 days 5 hours 48 minutes 46 seconds. Its maximum duration took place in the year 3040 B.C.; its minimum length will be in the year 7600, when it will be 76 seconds less than in the year 3040 B.C. A centenarian of our day has really lived twenty minutes less than a centenarian of the age of Augustus, and an hour less than a centenarian of the year 2500 before our era.

The ancients imagined that the political state of the globe was also periodical, and that they could name the great year which should restore to the earth the same nations, the same facts, the same history, as in the sky the course of ages restores the same aspects of the stars. They took in general thirty thousand years for this great year: doubtless the period of the equinoxes, which they believed had after that period given birth to this subsequent position. As they believed that human destinies depend upon planetary influences, it was natural also to believe that the same configurations of stars ought to reproduce the same events. But to restore the planets to the same relative position, thirty thousand years would not nearly suffice. In order to restore the Moon, Saturn, Jupiter, Mars, Venus, and Mercury, to the same degree of the zodiac, *two hundred and fifty thousand centuries* would be necessary! How long should it be if we add to this calculation the planets Uranus and Neptune, and also the small planets invisible to the naked eye? The astrologers believed that at the creation of the world all the planets were on the same line. There are even some savants who have attempted to calculate the day and the hour on which the first man was created. According to a work which I have before me, this event, so interesting to us all, took place on September 21 of the year 0, at nine o'clock in the morning!

These durations of celestial periods exceed the ordinary idea of time which man has when he wonders at the age of a centenarian. These sidereal events, which are only reproduced after thousands of centuries, and which appear to us very rare occurrences, are, on the contrary, frequent phenomena of eternity. These periods of millions of ages are but the seconds of the eternal clock.

CHAPTER V

CONTINUATION AND CONCLUSION OF THE ELEVEN PRINCIPAL MOTIONS OF THE EARTH

WE now come to a *fourth* motion of the earth.

We have already referred to this above. By its monthly motion round our globe the moon entirely displaces this globe in space, for, in fact, the earth and the moon turn as a couple round their common centre of gravity. The moon weighs eighty times less than the earth, and this centre of gravity is found eighty times nearer the centre of the earth than it is to the centre of the moon. It is therefore at a distance of 4,680 kilometres (about 2,900 miles) from the centre of our world, and we turn monthly round this point. At new moon, our satellite lying between the sun and us, we are a little nearer the sun. At full moon we are, on the contrary, a little farther away; at first quarter we are behindhand, restrained by the moon as by a bridle; at last quarter we are in advance, attracted, on the contrary, towards the moon, which precedes us. There is, then, a fourth kind of terrestrial motion, which becomes evident to us by a periodical variation in the size and position of the sun.

That luminary appears a little larger at new moon than it is at full moon, and between the first and last quarter it seems to be displaced by the 290th part of its diameter.

We will now consider a *fifth* sort of motion, due likewise to the influence of our satellite.

While the imaginary axis round which the diurnal rotation is performed turns slowly in space in such a way as to pass over in 25,765 years the cycle of the precession of the equinoxes, the influence of the moon causes this axis to describe a small gyratory motion, in virtue of which the pole traces on the celestial sphere a small ellipse, 18 seconds long by 14 seconds in width, directed towards the pole of the ecliptic, and described in eighteen years and a half. This is a motion, so to say, microscopic, but it is none the less real, and affects the apparent positions of all the stars. The result of these two motions, one on a circle traced at 23½ degrees from the pole of the ecliptic (as we have seen, p. 38), the other on a small ellipse, sliding as it were on the circumference of the preceding circle, causes the regular curve which we have drawn in fig. 17 to undulate slightly. This fifth inequality in the motion of our planet has received the name of *nutation.* It is due, like precession, to the equatorial bulging of the globe, upon which the attraction of the moon acts.

Thus there is grafted on the secular movement of the pole a winding motion, of which the fluctuation lasts eighteen and a half years, and goes on indefinitely. This is another reason why the pole never returns exactly to the point of departure.

Here is a *sixth* species of motion.

We have seen that the axis of our planet is inclined 23 degrees 27 minutes to the perpendicular to the plane in which it moves round the sun, and which is called the plane of the ecliptic. We turn obliquely; but this obliquity *varies* also from century to century. Eleven hundred years before our era it was measured by the Chinese astronomers and found to be 23 degrees 54 minutes. We will explain soon the value of these measures. In the year 350 B.C. it was also measured at Marseilles by Pytheas, and found to be 23 degrees 49 minutes. All modern measures confirm this diminution, which has been, it is shown, 27 minutes in three thousand years. It decreases at present at the rate of 48 seconds per century, or 1 minute in 125 years. If this diminution were constant it would be 60 minutes, or 1 degree, in 7,500 years, and in 177,000 years we should have the pleasure of having the globe perpendicular, of seeing the seasons gradually disappearing, and of enjoying a *perpetual* spring. This is, however, but the dream of sanguine Utopians.

Ancient traditions have preserved for us the ideal remembrance of an *age of gold*, which mankind enjoyed from its cradle. Then, said they, the prolific earth yielded its treasures without cultivation, then all animals were the humble servants of man, then the trees were covered with savoury fruits, the flowers always full-blown, the air fragrant with perfumes, the sun always shining, and neither storms nor cold ever disturbed the delightful harmony of creation. We may even read in Milton's 'Paradise Lost,' canto x., the history of the consequences of the fall of Adam or of Eve; among others, the arrival of strong angels sent by the Almighty and 'pushing by force the axis of the globe in order to incline it' [Milton's words are: They with labour pushed oblique the centric globe.'—J. E. G.], so as to give to us unfortunate descendants of that happy couple the most disagreeable and roughest seasons possible.

Celestial mechanics show that these are but dreams: there is merely a slight swinging of the equator on the ecliptic, of which the variation cannot exceed 2 degrees 37 minutes. The diminution continues for some time, then it is arrested and a contrary motion comes into operation. This sixth motion of the earth is named the variation in *the obliquity* of the ecliptic.

The diminution is at present less than half a second per annum. Here is the exact state of the obliquity of the ecliptic for the present epoch :—

1800	23° 27′ 55′	1880	23° 27′ 18″
1850	23° 27′ 31″	1890	23° 27′ 13″
1870	23° 27′ 22″	1900	23° 27′ 9″

In consequence of this variation the curve of the secular path of the pole drawn in fig. 17 diminishes and increases alternately in amplitude and forms a spiral, which at the present epoch is diminishing, but which later on will expand again. This spiral, which opens and closes in turn, recalls the motion of the balance-spring of a watch. Here is a new irregularity in the motions of the earth.

We have seen (p. 26, fig. 9) that the orbit followed by the earth round the sun is not circular but elliptical. This figure of the terrestrial orbit is not, however, constant; the ellipse is sometimes more and sometimes less elongated. At present the eccentricity is 168 ten thousandths; a hundred thousand years ago it was nearly three times greater, 473 ten thousandths; in twenty-four thousand years more it will, on the contrary, have sunk to its minimum (33 ten thousandths), and the terrestrial orbit will be almost a perfect circle; then it will increase again. This *variation of the eccentricity* may be considered as a *seventh* motion affecting the earth in its secular destiny. In twenty-four thousand years there will thus be neither perihelion nor aphelion; then the planet will be at nearly the same distance from the sun at the first point as at the second. This variation of the orbit passed over by our planet round the fire which warms and lights it exercises upon the seasons and climates a perceptible secular influence.

An *eighth* motion, caused by the general influence of the planets, produces a rotation of the major axis of the terrestrial orbit, *the line of apsides*, which moves from perihelion to aphelion the length of this orbit itself, in such a way that this major axis does not maintain for two consecutive years the same absolute direction. Four thousand years before our era the earth arrived at the perihelion on September 21, the day of the autumnal equinox; in the year 1250 of our era it passed this point on the day of the winter solstice, December 21; then our winters, happening in the part of the ellipse nearest the sun, were less cold than they are now, and our summers, falling in the portion of the orbit the most distant, were less hot than they are now. As the difference of distance between perihelion and aphelion is more than one million of leagues (over three million miles), and that of the heat received a fifteenth, this variation should have a real influence on the intensity of the seasons. The perihelion is reached at present on January 1. Our winters tend to become colder, and our summers warmer. In the year 11900 our summers will be the warmest and our winters the coldest possible. But we know that there are every year local causes of variations. At last, in the year 17000 the perihelion will have returned to the point where it was 4000 years B.C., that is to say, to the Autumnal Equinox. This cycle is one of 21,000 years. Several geologists have thought that this period might correspond to a renewal of the continents, and a renovation of the globe ; but this is merely an hypothesis.

To all these complications there must now be added that which is produced by the attraction of the different planets according to their positions relatively to the earth. All bodies attract each other in the direct ratio of their mass, and in the inverse ratio of the square of their distance—that is to say, their distance multiplied by itself. Jupiter at 155 millions of leagues from us influences our globe and displaces it : the earth is really disturbed in its course very little, say some few yards only, but on the whole it is affected and subject to the perpetual and variable influence (according to the variations of distance) of Jupiter, Venus, Saturn, Mars, and even to that of the most distant or faintest stars. This *ninth* irregularity connected with the motions of the earth is known and studied under the name of *perturbations*.

When all the planets are together on the same side of the sun they attract that body towards themselves and displace it from the geometrical focus, so that its centre of gravity no longer coincides with the centre of figure of the solar globe. Now, as the earth revolves annually round the centre of gravity and not round the centre of figure, there is here again a new complication (a *tenth*) affecting the elliptical motion of our planet round the sun. This centre of gravity round which the earth annually turns is often outside the solar globe.

Here we have a series of arguments doubtless a little technical, and, I fear, as devoid of ornament as 'the discourse of an Academician,' as Alfred de Musset said. I am rather afraid to find myself, even in these first pages of my book, in the situation of the Academician Berthoud, whose scientific dissertations on clock-making were learned, but, to tell the truth, wearisome. I hope my readers will not imitate the hearers of Berthoud at the Institute ! One day, while that famous watchmaker explained his theory of the escapement, a witty scientist wrote the following quatrain :

Berthoud, when of the escapement
You trace for us the theory,
Happy he who can adroitly
Escape from the Academy.

Then he passed the note to his neighbour and went out. His neighbour, wearied like him, read the paper and profited by the advice, so that step by step the desertion was complete. At last there remained only the reader of the paper, with the president and secretaries, whose dignity kept them in their chairs.

As for us who wish to learn the real state of the universe, it was important to begin by an examination of the situation of the earth and its motions in space. The terms which we have not accurately understood will be explained in the following chapters, and then no doubt should remain in the mind.

But we have not yet finished with the motions of our world, and we must still explain here an *eleventh*, more important and more considerable than all the preceding motions put together, because it represents the true astral motion of the sun, the earth, and all the planets in space.

The sun is not motionless in space. He moves on and draws with him the earth and the whole planetary system. We have detected his motion by that of the stars. When we travel on the railway, with the velocity of the new Pegasus of modern science, through countries diversified with fields, meadows, woods, hills, and villages, we see all the objects flying past us in a direction opposite to that of our motion. Well, by carefully watching the stars, we observe an analogous fact in celestial objects. The stars appear animated with motions which draw them apparently towards a certain region of the sky—that which is behind us. On each side of us they seem to fly past, and the constellations which are in front of us appear to enlarge so as to open for us a passage. Calculation has shown that these perspective appearances are caused by the translation of the sun, the earth, and all the planets towards a region of the sky marked by the constellation Hercules. We travel towards that region with a velocity which it is difficult to determine exactly, but which appears to be from 400 to 500 millions of miles per annum. We leave the starry latitudes where Sirius sparkles, and we sail towards those where shine the stars of Lyra and of Hercules. The earth has never passed twice over the same course.

On a clear summer night, when the beauties of the sky multiply their shining eyes in the dark and silent vault, look among the constellations for the brilliant Vega of the Lyre, a star of the first magnitude, which sparkles on the border of the Milky Way. Not far from that, in this whitish region, Cygnus, the Swan, extends like an immense cross; opposite to Cygnus, with reference to Vega, stands at a certain distance the Northern Crown, easily recognised by its form, composed of six principal stars curved like a coronet.

Between Vega and the Crown (see fig. 19) there are a certain number of stars of the third and fourth magnitude.[1] These belong to the constellation Hercules: it is this point of the sky towards which we are borne in the universal destiny of the worlds. If this motion continues in a straight line we shall approach in some millions of centuries the regions illuminated by these distant suns. [Recent researches tend to show that the point towards which the sun is moving, or the 'apex of the sun's motion,'

[1] These are not shown in the figure.

as it is termed, lies nearer to Vega than the point found by Struve.—J. E. G.]

I have had the curiosity to picture to myself this fall into the infinite. As there is neither top nor bottom in the universe, we may, in order better to imagine this motion in the midst of the stars, and to fix

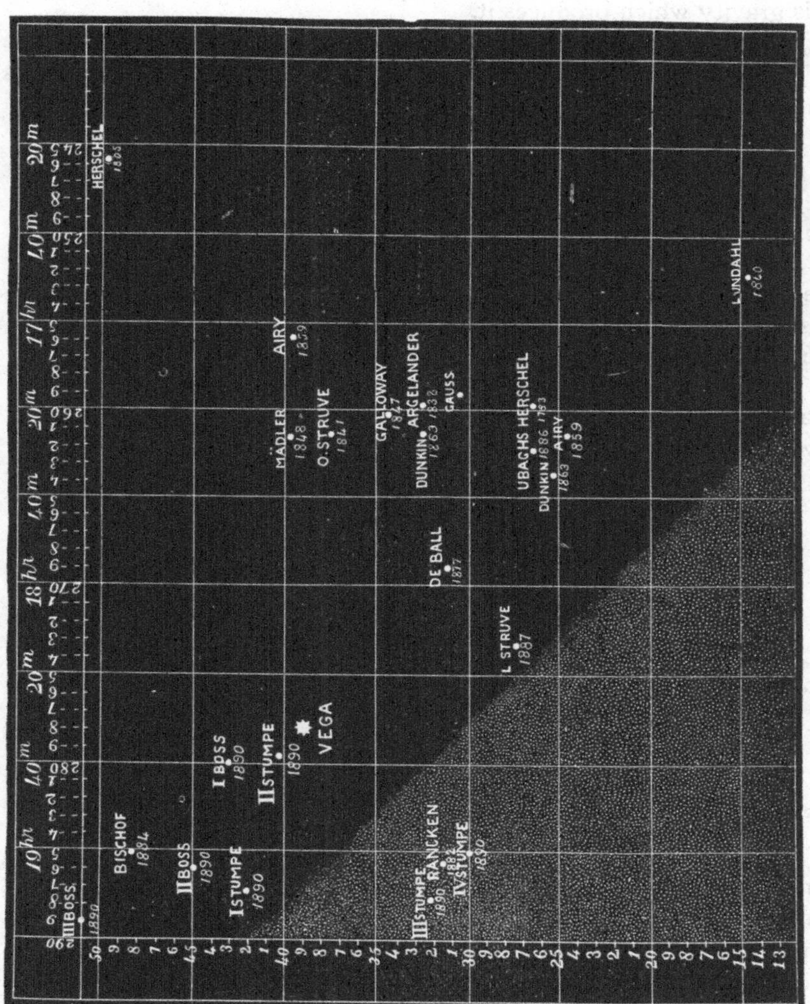

FIG. 19.—DIAGRAM SHOWING THE POSITION ON THE HEAVENS OF THE APEX OF THE SUN'S WAY, ACCORDING TO DIFFERENT OBSERVERS. THE SHADED AREA INDICATES THE MILKY WAY

(From Proctor's 'Old and New Astronomy.')

it with reference to the general plan of the planetary system, take the ecliptic as a point of comparison. All the planets and their satellites turning round the sun in the zodiac with a small inclination to the ecliptic, we may ask ourselves whether the solar system, comparable with a disc tossed into space, travels in the direction of its extent—in its horizon, we might say—or whether it falls flat, or glides obliquely. One might doubtless reply that, since it falls, it matters little whether it falls flat or sideways. However, the subject is not the less interesting. If, then, we take as the horizontal the plane of the ecliptic, and for the vertical the pole of the ecliptic, we can trace the figure of our fall in space—a real fall, since it is gravity which produces it.

Now, this point makes an angle of 38 degrees with the pole of the ecliptic. The direction of the motion of the solar system in space is represented by the large straight arrow in fig. 20; we fall neither flat, nor in the sense of a planetary disc, but obliquely through the open void, like the vulture which describes in the air his immense spirals, and we advance with great velocity towards the inaccessible abyss.

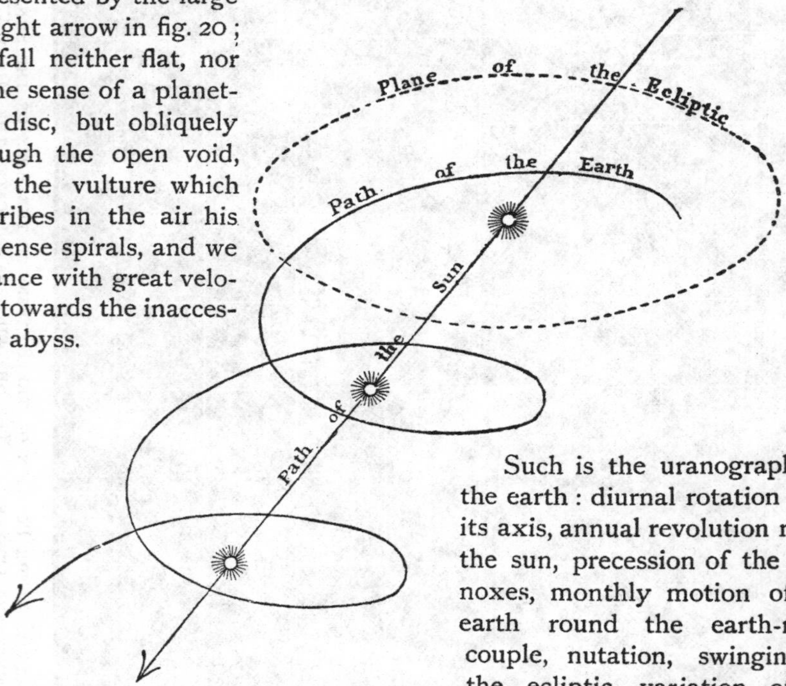

FIG. 20.—SPIRAL COURSE OF THE EARTH IN SPACE

Such is the uranography of the earth: diurnal rotation upon its axis, annual revolution round the sun, precession of the equinoxes, monthly motion of the earth round the earth-moon couple, nutation, swinging of the ecliptic, variation of the eccentricity, displacement of the perihelion, planetary perturbations, derangement of the centre of gravity of the sun, translation of the solar system, unknown sidereal actions, cause our little world to whirl round and round, and roll rapidly through space, lost in the myriads of worlds, of suns, and of systems with which the immensity of the heavens is peopled. The study of the earth teaches us to understand the sky, and in the microscopical atom we inhabit the vibrations of the infinite are revealed.

These ideas constitute the basis of modern astronomy, and we have now made the first and most difficult step in the accurate knowledge of the universe.

CHAPTER VI

THE EARTH, PLANET AND WORLD

Theoretical and Practical Demonstration of the Motions of our Globe. Life on the Earth

THE wise man affirms nothing which he cannot prove, says an old proverb. Astronomy is the most accurate of the sciences. All the truths which it teaches are absolutely demonstrated, and cannot be disputed by any mind which gives itself the trouble, or rather the pleasure, to gain information in the study of this admirable science.

There are, doubtless, mathematical demonstrations of a transcendent order which cannot be rendered popular. But, very fortunately for the general understanding, the fundamental proofs of the earth's situation in space, and the nature of its motions, can be explained in a form accessible to all, and as easy to understand as the popular reasonings of the most simple logic. It is this which we purpose to do in the following pages. It is especially important for us to exactly comprehend the reality, and to recognise our terrestrial home as a planet and as a world.

'Astronomers will have enough to do,' wrote in 1815 a member of the Institute, Mercier, who was not, however, wanting in intellect; 'they will never make me believe that I turn like a fowl on a spit.' The personal opinion of the intelligent author of the *Tableau de Paris* does not, indeed, prevent the earth from turning, for, willing or unwilling, we go round.

I know even now many persons, apparently educated, who doubt the motion of the earth, and who for one reason or another imagine that astronomers may perhaps be deceived, that the system of Copernicus is no better demonstrated than that of Ptolemy, and that, in the future, science may make such progress that our present theories may be upset, in the same way that modern science has overthrown the ideas of the ancients. It seems certain that these persons have not given themselves the pleasure of seriously studying the question. It is, then, interesting from all points of view to collect in one body of arguments *the positive proofs which we have of the motions of the earth.*

I will not insult my readers by insisting on the proofs of the earth's rotundity. For three hundred years the circuit of the world has been made in almost every direction; we have measured the size and determined the figure of our globe by well-known methods; even the elements of geography are universally taught, and no one can doubt that the earth is round like a sphere.

The principal difficulty which still prevents certain minds from admitting that our globe can be suspended like a balloon in space, and completely isolated in every way from a bearing-surface, proceeds from an erroneous idea of gravity. The history of ancient astronomy shows us a deep anxiety on the part of the first observers, who began to conceive the reality of this isolation, but did not know how to prevent this heavy globe on which we walk from *falling*. The early Chaldeans made the earth hollow like a boat; they could then float it on the abyss of air. The ancient Greeks placed it on pillars, and the Egyptians upon the backs of four elephants, as we have already remarked; the elephants were installed upon a tortoise, and the tortoise swam on the sea. Some of the ancients also thought the earth rested on pivots placed at the two poles. Others considered that it must extend indefinitely under our feet. All these systems were conceived under a false idea of gravity. In order to free oneself from this ancient illusion it is necessary to know that '*gravity is but an effect produced by attraction to a centre.*' The objects situated all round the terrestrial globe tend towards a centre, and all round the globe all the verticals are directed towards this centre. The terrestrial globe attracts everything to itself, like a magnet. The fear of the earth falling is therefore nonsense; where could it fall to? It would be necessary that another more powerful body should attract it. All the verticals are directed towards the centre of the globe. If we imagine a series of men standing upright all round the globe, each with a plumb-line in his hand, all these plumb-lines will be directed towards the centre, which is thus the *bottom*, the *below*; while all the heads represent the *top*, the *above* (fig. 21). When we consider our globe isolated in space we can do nothing which can give force to the objection which fears to see it falling we know not where. *There is neither top nor bottom in the universe.* If the earth existed alone, it would remain for ever at the point where it had been placed, without being able to displace itself in any way.

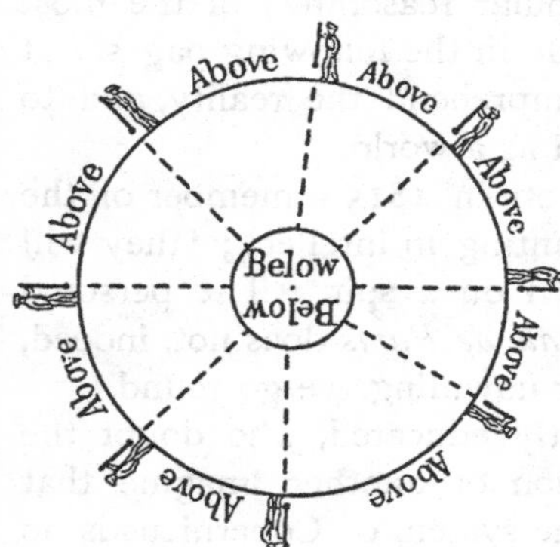

FIG. 21.—ALL OVER THE GLOBE, WEIGHT IS DIRECTED TOWARDS THE CENTRE

Let us now examine the question of motion. We see all the stars turning round the earth in twenty-four hours. There are but *two* suppositions

tions to be made in order to explain this fact—either the stars turn round from east to west, or the terrestrial globe turns upon itself from west to east. In both cases the appearances would be the same to us, and absolutely the same, if we remember that the displacement of the celestial bodies which do not participate in the earth's motion is the only index of this motion, our ethereal ship having no obstacle to meet with in its course. If, for example, a man in a boat which glides in the middle of a river's course had never left it, had been born in this boat and received an education which convinced him that the appearances are real, and that, as he sees it, the banks, the trees, the hills move slowly on each side of him, that man would evidently have the greatest difficulty in disabusing his mind of this opinion, and all the reasonings in the world would not at once convince him of his error. A certain amount of reflection would be necessary to make him understand that the villages did not move past him.

How, then, can we voyagers on the terrestrial ship arrive at certainty on this same point, and know whether it is really the sky which turns round the earth, or the earth which turns on itself?

In the first case, see what it is necessary to admit. The nearest body to us, the moon, is at a distance of 238,000 miles from the earth. It would, then, have to pass over in twenty-four hours a circle of 476,000 miles in diameter—that is to say, of 1,495,000 miles in circumference. It would be necessary for it to move with a velocity of 25,125 leagues an hour—that is to say, more than 1,000 miles a minute, or 17 miles a second. The distance of the moon is not open to dispute: it has been more exactly measured by triangulation than the distance from Paris to Rome. But still this is nothing.

The sun, at 93 millions of miles from us, would have in the same interval of twenty-four hours to move over a circumference of 575 millions of miles round the earth. It would be necessary for him to fly with a velocity of 24 millions of miles an hour—that is to say, 400,000 miles per minute, or over 6,000 miles per second. In fact, it would have to move over in a day the distance which our globe travels in a year. And this body is 1,300,000 times larger than the earth! The logical improbability of such an hypothesis, as well as its mechanical impossibility, will make itself felt by the aspect alone of fig. 22, which we give here in anticipation of our studies on the sun; in this figure the eloquent proportions of the day star are represented on an exact scale. The diameter of this body is 108 times greater than that of our planet. As to its distance, it has been accurately determined by six different methods, independent of each other. From the aspect alone of this proportion it is impossible for the simplest common-sense to wish to make the sun turn round the earth. As Cyrano de Bergerac said, it is as if, in order to roast a lark, we should place it on a spit, and, in place of turning the spit, we should turn, round the fixed lark, the fireplace, the kitchen, the house, and the whole town.

The planets, the distances of which are likewise determined with mathematical precision, also participate in the diurnal motion. They would then be carried through space with a velocity more inconceivable still. The farthest planet known to the ancients, Saturn, nine and a half times farther from us than the sun, would be obliged, in order to revolve in twenty-four hours round the earth, to describe a circumference of 4,960 millions of miles in length, and to rush through space with a velocity of more than 49,000 miles in each second!

The farthest known planet of our system, Neptune, would have to pass over 17,000 millions of miles in twenty-four hours, that is, 292 millions of leagues an hour!

And the stars? The nearest of them lies at a distance of 275,000 times the sun's distance from the earth—that is to say, at 25 billions of miles from us. In order to turn round the earth in twenty-four hours this star would have to pass over in the same interval of time a circumference measuring 160 billions of miles in length; for this its velocity would be 2,666,000 millions of leagues per hour, 44,400 millions per minute, or, in short, 1,835 *millions of miles* per second! And this for the nearest star to us!

Sirius, situated very much farther, would need to accomplish an indescribable circumference round us with a rapidity incomparably more fantastic still. And most of the other stars are immeasurably farther off, situated at all imaginable distances. And they are all immensely larger and heavier than the earth. And they extend to infinity!

Thus we see the two hypotheses: either to oblige the whole universe to turn round us every day, or to suppose our globe animated with a motion of rotation, and thus spare the whole universe this incomprehensible labour.

When we see the extent of the heavens, peopled with millions and millions of far-off stars at distances the most tremendous, when we reflect on the smallness of the earth in comparison with all these enormous distances, it becomes impossible to imagine that all this could turn at the same time with a motion uniform, regular, and constant in twenty-four hours round an atom such as the earth. Not only is the diurnal motion of all the stars in twenty-four hours round us a very improbable hypothesis, but we may add that it would be absurd, and that it is necessary to be blind in order to lend oneself to such an idea. Now, all the planets, which are at such different distances, and of which the proper motions are so different from each other; all the comets, which seem to bear scarcely any resemblance to the other celestial bodies, add to the difficulty. All these bodies, which are independent one of the other, and at distances which the imagination has a difficulty in conceiving, unite to turn all together, as if in one piece, round an axis or axle-tree, which even changes its plane! This equality in the motions of so many bodies, so unequal in all respects, ought alone to indicate to thinkers that there is nothing real in these motions ; and

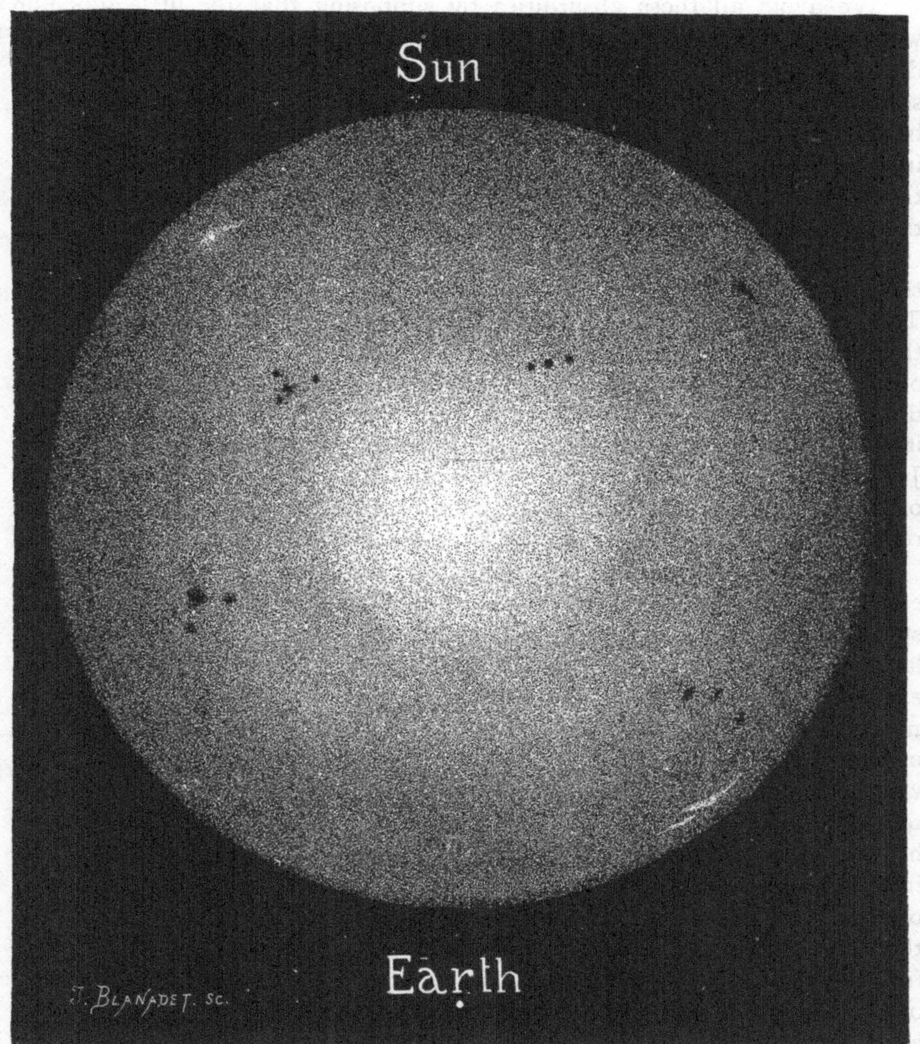

FIG. 22.—COMPARATIVE MAGNITUDES OF THE SUN AND OF THE EARTH

when they reflect on it, it proves the rotation of the earth in a way which leaves not the smallest doubt, and to which there is no room for reply.

We may add to this that the stars are millions and millions of times larger than the earth ; that they are not united among themselves by any solid bond which could attach them to a motion of supposed celestial vaults; that they are all situated at the most diverse distances; and that this frightful complication of the system of the heavens should of itself prove its non-existence and its mechanical impossibilty.

We avoid all these absurdities by supposing that the little globe on which we are simply makes one turn in twenty-four hours. In default of direct demonstrations—which are not wanting, however, as we shall see—simple common-sense settles the question. In turning on itself, the earth has simply to pass over its own equatorial circumference of 24,800 miles in twenty-four hours, which is 1,525 feet per second for a town situated on the equator, 1,000 for Paris, and less and less in proportion as we approach the poles, where the circle it has to pass over is very small.

On the other hand, again, analogy confirms directly the hypothesis of the motion of the earth, and changes into certainty its high probability. The telescope has shown in the planets earths analogous to ours, moved themselves by a motion of rotation round their axes, a motion of rotation of [about] twenty-four hours for the neighbouring planets, and of a less duration still for the distant worlds of our system. The sun turns upon itself in twenty-five days, Venus and Mars in [about] twenty-four hours, Jupiter in ten hours, &c. Thus both simplicity and analogy are in favour of the earth's motion. We may add now that this motion is rigorously required and determined by all the laws of celestical mechanics.

One of the greatest difficulties opposed to this motion was this: if the earth turns under our feet, if we could rise in space and find the means of sustaining ourselves for some seconds or more, we should fall, after this lapse of time, at a point more to the west than the point of departure. For example, he who could at the equator find means of sustaining himself motionless in the atmosphere for half a minute, ought to fall again at over eight miles to the west of the place whence he started. This would be an excellent way of travelling, and Cyrano de Bergerac pretended to have employed it when, being raised in the air by a balloon of his own manufacture, he fell some hours after his departure in Canada instead of descending in France. Some sentimentalists, Buchanan among others, have given to the objection a somewhat sentimental form by saying that if the earth turned round, the turtle-dove would not dare to go away from its nest, because she would soon unavoidably lose sight of her young ones. The reader has already replied to this objection by reflecting that all that belongs to the earth participates (as we have said) in the motion of rotation, and up to the highest limits of the atmosphere our globe carries along everything in its course.

When we play at balls or at billiards in a ship borne in rapid motion on a smooth sea, the shock of the balls takes place with the same force in one direction as in another, and when we drop a stone from the top of the mast of a ship in motion it falls exactly at the foot of the mast, just as if the ship were at rest. The motion of the vessel is communicated to the mast, to the stone, and to everything on the floating abode; there is nothing but the resistance of the liquid plain cleft by the ship which permits the passengers to perceive the motion. It is the same on the railway and in a balloon. But as the earth does not encounter any strange obstacle, there is absolutely nothing in nature which can by its resistance, by its motion, or by its shock, enable us to perceive the motion. This motion is common to all terrestrial bodies: if they are raised in the air, they have received beforehand the motion of our globe, its direction and its velocity; and even when they are at the highest point of the atmosphere they continue to move as the earth does.[1]

We verify the same law in a balloon. I remember myself one day passing over the town of Orleans. I had taken care to write a despatch addressed to the leading journal of that town, and I had expected when we arrived above a promenade to let it fall, by affixing a stone for a counterpoise. What was my surprise to see this stone, while descending, suspended beneath the balloon as if it had slipped the length of a cord. The balloon sails rather fast. Instead of falling on the spot I had chosen, or even in the town, the despatch, following a diagonal, fell into the Loire. I had not reflected on one of the oldest questions of my bachelor's degree, the independence of simultaneous motions. Very fortunately, the balloon, having crossed the Loire, had towards evening descended sufficiently near the earth to allow us to hail an inhabitant of the town, who was following the Orleans road, on his way home, seated in a cabriolet, which advanced at a slow pace. It was nightfall, and the Angelus was wafted from the village bells. Much surprised was this traveller on hearing himself hailed from the height of heaven. He seemed at first to believe neither his ears nor his eyes. But the horse was promptly stopped, and we had sufficient time to announce our passing, which next morning was published in the newspapers.

A cannon-ball shot up perpendicularly towards the zenith would fall again into the gun, although during the time of its ascent and descent it may have advanced with the earth for several miles. The reason is evident: this ball in rising into the air has lost nothing of the velocity which the motion of the globe has imparted to it. These two impressions are not opposed to each other; it can move one mile in height while it moves six towards the east. Its motion in space is the diagonal of a parallelogram, of which one side is one mile and the other six; it will fall again by its natural gravity, following another diagonal (curved on

[1] The horsewoman in the circus carried on a fast horse has the same experience. When she jumps up from her steed she continues to move on, and falls directly on the saddle, as if the horse had remained standing.

account of acceleration), and it will return to the cannon, which has not ceased to remain vertically below it.

. . . .

The direct observation of various phenomena has absolutely demonstrated the motion of the earth by unexceptionable proofs.

If the globe turns, it develops a certain centrifugal force. This force will be nothing at the poles ; it will have its maximum at the equator, and it will be so much the more as the object to which it applies is itself at a greater distance from the axis of rotation. Well, the earth bulges at the equator and is flattened at the poles, and it is *ascertained* that objects at the equator lose one 289th of their weight on account of the centrifugal force.

The oscillations of the pendulum further confirm the fact. A pendulum of 39 inches in length, which at Paris makes in a vacuum 86,137 oscillations in twenty-four hours, would, if carried to the poles, make 86,242, and at the equator not more than 86,017 in the same time.

The length of a seconds pendulum at Paris is 994 millimetres (about 39·134 inches). At the equator it is only 991 millimetres (about 39·016 inches).

A stone which falls from a fifth storey in Paris passes over 16·08 feet in the first second of its fall. At the pole, where there is no centrifugal force, the fall is a little more rapid, 16·13 feet. At the equator it falls at the rate of 16·04 feet, a velocity one inch less than it has at the poles. The figure of the earth, which is flattened at the poles, accounts for one part of this difference, the centrifugal force for the other part.

There is a curious remark to be made here. At the equator this force is $\frac{1}{289}$ of gravity. Now, as gravity increases proportionately to the square of the velocity of rotation, and as 289 is the square of 17 (17 × 17=289), if the earth turned round 17 times faster, bodies placed at the equator *would have no weight*. [The centrifugal force, not gravity, varies as 'the square of the velocity of rotation.'—J. E. G.]

As the velocity of rotation is so much the more the farther we are from the centre of the earth, a stone placed on the surface of the ground is animated with a velocity towards the east a little greater than a stone at the bottom of a pit. Now, as this excess of velocity cannot be destroyed, if we allow a small leaden ball to fall into a pit, it will not exactly follow the vertical line, but will deviate a little towards the east. The deviation depends upon the depth of the pit : it is at the equator 1 inch for 252 feet of depth. In the pits of the Freiberg mine (Saxony) they have found an eastern deviation of 1·1 inch for 518 feet. It is evident that this is an experimental proof of the earth's motion. We have at the Paris Observatory a pit which descends to the Catacombs, a depth of 92 feet, and passes through the building up to the upper terrace, of which the height is likewise 92 feet. We have, then, a pit of 184 feet in depth. From the time of Cassini the above experiment has been made here in order to give an experimental proof of the motion of the earth. A ball of lead let fall from the top of the tower of Notre-Dame does not follow the vertical exactly, but drops at 10·6 of an inch towards the east, the difference between the velocity at the foot and at the summit. (It is difficult to succeed with the experiment on account of the motions of the air.)

The natural philosophy of the globe has also furnished its quota of proofs to the theory of the earth's motion, and we may say that all the branches of science which are connected closely or remotely with cosmography are united in the unanimous confirmation of this theory. Even the figure of the terrestrial spheroid shows that this planet has been a fluid mass animated with a certain velocity of rotation, a conclusion at which geologists have arrived as the result of their special researches.

Other facts, as the currents of the atmosphere and the ocean, the polar currents and the trade winds—the directions of which towards the east imperceptibly incline towards the north-east—find also their cause in the rotation of the globe.

This is the place to mention the brilliant experiment made by Foucault in the Pantheon. Unless we deny the evidence, this experiment indisputably demonstrates the motion of the earth. It consists, as we know, in fitting a steel wire by its upper extremity in a metallic plate solidly fixed in a vault. To this wire is suspended a rather heavy ball of metal. A point is attached below the ball, and fine sand is arranged to receive the trace of this point when the pendulum is in motion. Now, it happens that this trace is not always in the same line: several lines crossing at the centre succeed each other, and show a deviation of the plane of the oscillations from the east towards the west. In reality, the plane of the oscillations remains fixed; the earth turns beneath from west to east. The explanation is based on the fact that *the torsion of the wire does not prevent the plane of the oscillations from remaining invariable.* Anyone can verify this by a very simple experiment. Take a ball suspended by a wire of a yard or two in length, attach the wire to the ceiling with a screw, make the pendulum oscillate, and while it moves cause the screw to turn. The wire will be twisted more or less, but the direction of its oscillations will not vary on that account.

Such is the principle of the celebrated experiment designed by Foucault, and carried out by him under the dome of the Pantheon in 1851.

If we imagine a pendulum of great length suspended over one of the poles of the earth and set in motion, the plane of its oscillations remaining invariable notwithstanding the torsion of the wire, the earth would turn round under it, and the plane of oscillation of the pendulum would appear to turn in twenty-four hours round the vertical, in a direction contrary to the true motion of rotation of the earth.

If the pendulum be suspended at a point on the equator, there would be no variation ; but for all places situated between the equator and the poles the invariability of the plane of oscillation is shown by a deviation in a direction contrary to the earth's motion.

Such are the position and absolute proofs of the motion of rotation of the earth on its axis. The proofs of the motion of translation round the sun are not the less convincing.

And, in the first place, all the other planets revolve round the sun, and the earth is but a planet. In order to explain the apparent motions of the five planets known to the ancients (Mercury, Venus, Mars, Jupiter, and Saturn) on the hypothesis of the earth's immobility, astronomers were obliged to strangely complicate the system of the world, and to imagine seventy-two crystal spheres fitted one within the other! All the planets revolved round the sun in the same time as the earth. Changes of perspective, easily understood, follow from the long circuit described annually by the earth round the sun. When we advance, a certain planet appears to move back. When we go to the left, the other seems to go to the right. In certain cases the combination of the two motions apparently arrests the planet in its course, and renders it motionless on the celestial sphere.

In the theory of the translation of the earth round the sun, these variations explain themselves and are calculated in advance. On the contrary hypothesis, they created an intolerable complication, a confusion such that, in the thirteenth century, the King-astronomer Alphonso X. of Castile dared to say that 'if God had consulted him when He created the world, he could have given Him some advice as to its construction in a simpler and less complicated manner,' an imprudent speech, which very nearly cost the King his crown. Since the thirteenth century the study we have made of the numerous comets which plough through space in all directions has shown that, however erratic they may be themselves, these wandering stars protest against the ancient system, for, as Fontenelle said, they would long since have split all the crystals of the heavens. The calculation of the orbits of comets, of which the accuracy is proved by the return of these bodies to the points of the sky indicated, would be impossible on the hypothesis of the earth's immobility.

The planet Uranus, discovered at the end of last century beyond the orbit of Saturn ; the planet Neptune, discovered in the middle of the present century farther off still—have proved also that they revolve round the sun, and not round the earth ; and the discovery of the last, made by pure induction from the mathematical theory, has been truly the finishing stroke to the last partisans of the ancient system, since it is by relying on the laws of universal gravitation that the mathematician has announced the existence of a body distant more than two thousand millions of miles from us, and revolving round the sun in 165 years. We may add, further, that more than two hundred small planets [over 380, April 1894] have been discovered since the beginning of this century between Mars and Jupiter, and that they all, without exception, likewise revolve round the sun. Thus the solar system forms a single family, of which the gigantic and powerful sun is the centre and regulator.

This is not all. We *see* the earth's annual motion of translation reflected in the sky. The stars are not separated from us by infinite distances. Some are relatively rather close, and lie only at some billions of miles from here. Now, the earth, in revolving round the sun, describes in space an ellipse of 575 millions of miles. Well, then, if we examine attentively one of the nearest stars during the course of the year, taking for a point of comparison a very distant star, we see that the nearer one undergoes a perspective change in its position, caused by the motion of the earth, and instead of remaining fixed during the whole year at the same point, it appears to move, following an ellipse described in a direction contrary to that of our annual motion. It is, in fact, solely by measuring these little ellipses described by the stars in the depths of the heavens that astronomers have been able to calculate their distances. In the days of Copernicus, Tycho Brahé, and Galileo, the apparent immobility of the stars was one of the strongest arguments invoked against the annual motion of the earth. This argument has been upset, like all the others, by the progress realised in the ever-growing precision of astronomical observations.

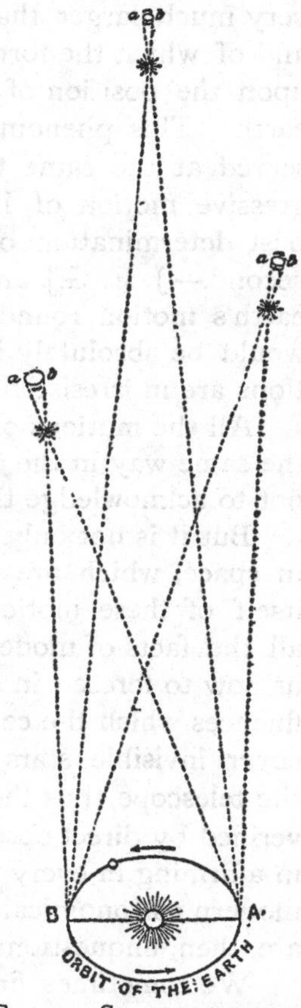

FIG. 23.—SMALL APPARENT ELLIPSES DESCRIBED BY THE STARS AS A RESULT OF THE ANNUAL MOVEMENT OF THE EARTH

But still this is not all. The annual motion of the earth round the sun is likewise reflected on the celestial vault by another phenomenon, which is called 'the aberration of light.' Let us see on what it depends. The rays of light reach us from the stars in a straight line with a velocity about ten thousand times faster than that of the earth in its orbit. If the earth were fixed, we should receive these rays directly and without alteration ; but we run under the luminous rays as, for example, we run under a vertical shower ; the faster we run the more we must incline our umbrella, if we do not wish to get wet. If we are on the railway, the combination of the horizontal velocity of the train with the vertical velocity of the drops of rain causes the rain to trace oblique lines on the door of the carriage. Well, we can compare our telescopes pointing to the stars to our umbrellas pointing in the direction of the drops of rain. The motion of the earth is such that we are obliged to incline our telescopes in order to receive the luminous rays of the stars. Each star traces annually on the celestial sphere an ellipse very much larger than that which is due to the perspective of distance, and of which the form and size depend, no longer upon the distance, but upon the position of the star with reference to the annual motion of the earth. This phenomenon is of great importance in astronomy. It has served at the same time to show the precision of the theory of the progressive motion of light at the rate of 186,416 miles per second [the best determination of the velocity of light is 186,337±49·722 miles per second.—J. E. G.], and it has furnished a direct proof of the reality of the earth's motion round the sun. If the earth were at rest, these motions would be absolutely inexplicable. We see, then, that all these demonstrations are in irresistible agreement.

All the motions of the earth which we have described above are read in the same way in the observation of the sky, and one must be wilfully blind not to acknowledge that they are so.

But it is not only the motions of our planet, but also those of our sisters in space, which are now absolutely demonstrated; the theoretical cause itself of these motions, *attraction* or *universal gravitation*, is proved by all the facts of modern astronomy. The knowledge of this cause enables us now to foresee in advance the smallest perturbations, the minutest influences which the celestial bodies exercise on each other, and even to discover invisible stars; thus Neptune was discovered without the aid of the telescope, thus the satellite of Sirius was discovered: bodies afterwards verified by direct observation. *All* the facts of the science agree in proving, in affirming in every way, in demonstrating more and more the truth of modern astronomical theories. *No one* attempts to contradict them. They are, then, unquestionable and absolute certainties.

We sometimes find a real difficulty in getting our convictions accepted by certain persons, rebels to all demonstration. Thus, for example, an old proverb asserts 'that it would be much easier to give an intellect to a fool than to persuade him that he is wanting in one.' Very fortunately, the problem with which we are occupied is not one of such laborious solution. We believe that we are not optimists in hoping that, after the account of all the preceding arguments, there no longer remains any room for the least doubt in the mind of a single one of our readers.

Let us now stop for a moment to contemplate the earth in its living unity.

This globe which carries us has a diameter of 7,918 miles [equatorial diameter=7926·25 miles, polar diameter=7899·84 miles.—J. E. G.].

But it is not absolutely spherical, being slightly flattened at the poles ($\frac{1}{292}$) [$\frac{1}{300\cdot205}$.—Harkness]. The diameter from one pole to the other is a little less than that which is taken from a point on the equator to a point diametrically opposite, and the difference is 43 kilometres (26·7 miles). On a globe of one metre in diameter (39·37 inches) the difference between the two diameters would be only 3⅓ millimetres (0·13 inch). On the same globe the most elevated mountain on our world, Gaorisankar (Mount Everest), in the Himalayas, of which the height is 8,840 metres (29,000 feet), would be only seven-tenths of a millimetre. Thus our globe is proportionately much more uniform than an orange, as smooth in reality as a billiard-ball. As to the material size of a man with reference to the world which he inhabits, he would be so small on a globe of 40 feet in diameter that ten thousand might lie side by side in a space of the size of an *o* as here printed! And yet, who knows? There are, perhaps, in Infinitude, worlds and men also liliputians! (No one is ignorant of the fact that ants *think*; I had the curiosity one day to weigh the brain of an ant, and I found as the result 15 hundredths of a milligramme, or a tenth and a half of a milligramme; and yet this brain thinks!) [A milligramme is 0·0154 grain.—J. E. G.]

When we rise above the surface of the globe the horizon widens in proportion to the relation which exists between our elevation and the size of the sphere. At 1,000 yards high we soar above a circle (or rather a spherical cap) of which the radius measures 69½ miles—that is to say, we take in an extent 139 miles in diameter. The horizon of Paris prolonged to Marseilles would pass at a height of more than 30 kilometres above that town (18½ miles).

We may further add, that our globe is surrounded by an atmosphere, at the bottom of which we breathe and live, composed of gas (oxygen, nitrogen, carbonic acid), and the vapour of water which rises from the seas, lakes, and rivers, and the lands watered by the rain. It is this atmosphere which, not being absolutely transparent, reflects the light of day and gives the colour to that celestial azure which seems to spread above us in the atmospheric sky. It is this illumination of the molecules of the air by the light of day which prevents us seeing the stars by day as well as at night. The most brilliant—Venus, Jupiter, Sirius—sometimes succeed in piercing this azure veil. They can thus be detected, if specially searched for, by the aid of a telescope, or merely with the assistance of a simple blackened tube, or even with the naked eye (Venus and Jupiter) if the sight is very keen. Our atmosphere is not unlimited: at 30 miles high it becomes almost nothing, and long before that height it would be unfit for respiration. No one has ever ascended to a greater height than 6 miles in a balloon. It is probable that above this aerial atmosphere there is another lighter still, hydrogenous; for the study of twilights, shooting stars, and the aurora borealis seems to extend the extreme limit up to 300 kilometres (about 186 miles). It may be extended mathematically much farther still. I have calculated that it is only at a height of 42,300 kilometres (about 26,284 miles) round our globe that the centrifugal force developed by the whirling of the earth would cast off into space the molecules of air which may exist in that region. It is there that equilibrium is established, and here might circulate a satellite revolving round us in the same time as the rotation of the earth, or twenty-three hours fifty-six minutes.[1]

The atmosphere plays a rather important part in astronomical observations, for it diverts the luminous rays which reach us from the stars, and makes us see them above their real position. This is what is called *refraction* (fig. 24). At a point situated diametrically above our heads, named the zenith, the deviation is nothing, because the luminous rays arrive per-

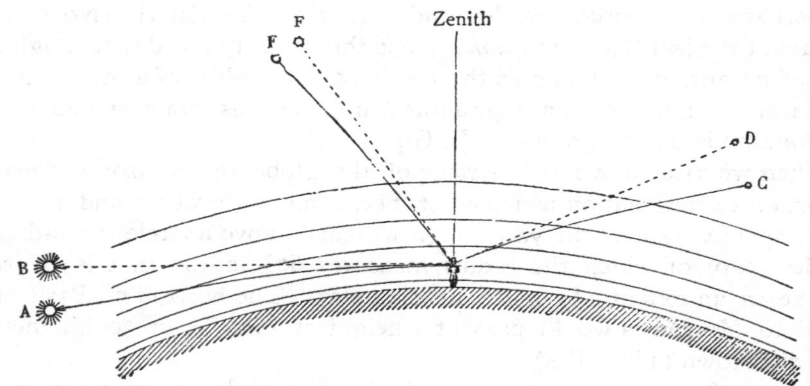

FIG. 24.—REFRACTION: APPARENT ELEVATION IN THE POSITIONS OF STARS CAUSED BY THE ATMOSPHERE

pendicularly to the strata of air. It increases as they recede from the zenith and approach the horizon. The sun, for example, situated in the direction A, is seen in the direction B; the star C is seen at D; the star E is seen at F, &c. At the horizon itself the refraction is enormous, for it raises the stars by a quantity equal to the apparent diameter of the sun and moon, so that when we see these bodies rising, they are in reality still set and below the prolonged plane of the observer's horizon. It is for this reason also that the setting sun appears oval when, on beautiful summer evenings, we witness the magnificent and luminous sunsets on the seashore. It is necessary, therefore, to apply a correction to all astronomical observations in order to restore the stars to their real positions.

The terrestrial globe, measuring 7,918 miles in diameter, represents

[1] See our work, *L'Atmosphère, météorologie populaire*, Book I. chap. iii.

a volume of a million millions of cubic kilometres. As it is an insignificant morsel of matter, we have been able to weigh it (by the Cavendish experiment). It weighs five and a half times more than if it were entirely formed of water, which corresponds to a weight of *five sextillions* of tons (a kilogramme is about 2·2 lbs., or 5 kilogrammes = 11 lbs.). The atmosphere weighs about eleven hundred thousand times less, or pretty nearly *five quadrillions* of tons. The atmospheric pressure is 10,330 kilogrammes per square metre [about 14·69 lbs. per square inch.—J. E. G.].

The surface of the earth is 511 millions of square kilometres [about 187 millions of square miles.—J. E. G.], of which 384 are covered by the waters of the ocean, so that there remain but 127, or one quarter only, for the habitable land.

Our planet is alive with a certain stellar life which we cannot yet sufficiently understand. Magnetic currents circulate in it, and incessantly, under their mysterious influence, the magnetic needle seeks the north with its restless and agitated finger. The intensity and direction of these currents vary day by day, year by year, century by century. About two centuries ago, in 1666, the compass as observed at Paris pointed exactly to the north. After that it turned towards the west—that is to say, towards the left if we look towards the north. The deviation was 8 degrees in 1700, 17 degrees in 1750, 22 degrees in 1800. It further increased a half degree up to 1814, when it commenced to return towards the north. The deviation was 22 degrees in 1835, 20 in 1854, 19 in 1863, 18 in 1870, 17 in 1878, and 16 in 1888 (and 15½ degrees in 1893). It still continues to decrease, and it is probable that it will point again to the north about the year 1962. Here is an important secular variation which has caused many maritime disasters to pilots who are ignorant of it. We may add that every day this curious needle deviates from its magnetic meridian towards the east at eight o'clock in the morning, and towards the west at one o'clock in the afternoon. The extent of this variation varies year by year, and, what is truly surprising, this variation appears to correspond with the number of spots visible on the sun; it is in the years when there are most spots that this fluctuation is most marked. The number of auroræ boreales seems likewise connected with the state of the day star. Indeed, the magnetic needle enclosed in a cellar of the Paris Observatory *follows* the aurora borealis, which lights its aërial fires in Sweden and Norway. It is restless, agitated—I might say feverish; more than that, infatuated—and its disturbance only ceases when the distant meteor has disappeared. What book like the book of Nature! And how strange it is that it has so few readers!

The life of the planet shows itself externally by the plants which adorn its surface, by the animals which people it, by mankind which inhabits it. We are acquainted with one hundred and twenty thousand vegetable species, and three hundred thousand species of animals. There is but one human species, for mankind is the incarnation of mind.

The human population of our planet is composed, according to the latest statistics, of 1,450 millions of inhabitants. An infant is born every second; every second a human being dies. The number of births is, however, a little greater than the number of deaths, and the population increases according to a variable proportion.

We may estimate at 400,000 millions the number of men who have lived upon the earth since the advent of mankind. If we could raise from the dead all these men and women, old men and children, and lay them side by side, they would already cover the entire surface of France. But all these different bodies have been composed successively of the same elements; the molecules which we breathe, drink, eat, and incorporate in our organism have already formed part of our ancestors.

A universal exchange works incessantly between all beings: death keeps nothing. The molecule of oxygen which escapes the ruin of the old oak laid low by the weight of centuries is incorporated in the blonde head of the new-born infant, and the molecule of carbonic acid which escapes from the heaving breast of the dying man on his bed of pain flourishes again in the brilliant corolla of the rose. Thus the most absolute brotherhood governs the laws of life; thus eternal life is organised by death eternal. The mind alone lives and contemplates. Dust returns to dust. The worlds sail through space illuminated with radiance, and smile in a life incessantly renewed.

From age to age living beings are replaced by others, and, on the continents as in the seas, if life always flourishes, it is not the same hearts which beat, it is not the same eyes which see, it is not the same lips which smile. Death lays successively in the tomb men and their affairs; but from our ashes, as from the ruins of empires, the flame of life is incessantly renewed. The earth gives to man his fruits, his flocks, his treasures; life circulates, and the spring-time always returns. We might almost believe that our own existence, so weak and so transient, is but a constituent part of the long existence of the planet, like the annual leaves of the perennial tree, and that, fellow-creatures with the mosses and mildew, we vegetate for an instant on the surface of this globe only to subserve the processes of an immense planetary life which we do not comprehend.

The human species is affected, in a less degree than the plants and animals, by the circumstances of the soil and by the meteorological conditions of the atmosphere. By the activity of the mind, by the progress of intelligence, which raises it little by little, as well as by that marvellous flexibility of organisation which adapts it to all climates, it escapes more easily from the forces of nature; but it does not the less participate in an essential manner with the life which animates the whole of our globe. It is by these secret relations that the problem, so obscure and so controverted, of the possibility of a common origin for the different human races enters into the sphere of ideas which embrace the physical description of the world.

There are families of nations more susceptible of culture, more civilised, more enlightened ; but we can say with Humboldt that not one is more noble than the other. All are alike made for liberty, for that liberty which in a state of society little advanced belongs only to the individual, but among nations endowed with the power of proper political institutions is the right of the entire community. An idea which is revealed by history, extending each day its salutary empire—an idea which better than any other proves the fact, so often disputed, but oftener still badly understood, of the general perfection of the species—this is the idea of humanity. It is this which tends to overthrow the barriers which prejudices and interested views of every sort have raised among men, and makes us look upon mankind as a whole, without distinction of race, of religion, of nations, or of colour, as a great family of brothers, as a single body, marching towards one and the same end—the free development of the moral forces. This end is the final end, the crowning end of social life, and at the same time the direction imposed on man by his peculiar nature for the indefinite enlargement of his existence. He regards the earth as far as it extends, the heavens as far as he can penetrate them, illuminated with stars ; his intelligence is elevated above all other terrestrial beings.

> Os homini sublime dedit, coelumque tueri
> Jussit, et erectos ad sidera tollere vultus.

Progress and liberty! Already the child aspires to overleap the mountains and seas which circumscribe his limited abode ; and then, yearning for home, he longs for the return journey. It is this, in fact, which is, in man, touching and beautiful, this double aspiration towards that which he desires and that which he has lost ; it is this which preserves him from danger and attaches him in an exclusive manner to the present moment. And thus rooted in the depths of human nature, governed at the same time by his sublimer instincts, this benevolent and fraternal union of the entire species becomes one of the grand ideas which preside over history. Our humanity has not yet reached the age of reason, since it knows not yet how to find its way; it has not yet issued from the environment of the brute's coarse instincts, and the most advanced nations are still essentially soldiers —that is to say, slaves ; but it is destined to become instructed, enlightened, intellectual, *free*, and great in the light of heaven.—At our side, on the floating islands which accompany us through space, and in the bosom of the inaccessible depths of Infinitude, other worlds, our sisters, also bear living creatures, who rise at the same time in an indefinite progress, and towards a perfection which shines above all destinies like a star in the depths of the heavens.

CHAPTER VII

HOW WAS THE EARTH FORMED?

Age of our Planet; its Past; its Future. The Origin and the End of Worlds

THE preceding pages have taught us the place we occupy in the universe, and have led us to recognise the earth as a star in the sky. Such was, in fact, the first point of view under which it was important to consider our globe, in order to free ourselves for ever from the vainglorious sentiment which has hitherto made us consider the earth as the basis and the centre of creation, and from that local patriotism in pursuance of which we prefer our country to the rest of the world. We shall soon occupy ourselves with other worlds in following the logical order of situations and distances.

Our celestial programme spreads itself before us. The moon will be the first halting-place on our great voyage. We shall stop at its surface, in order to contemplate its strange nature and to study its history; this is the nearest body to us, and it forms, so to say, part of ourselves, since it faithfully accompanies the earth in its course, and gravitates round us at a mean distance of 96,000 leagues. [The mean distance in English miles is 238,854.—J. E. G.]

We will then transport ourselves to the sun, the centre of the planetary family, and we shall attempt to be present at the titanic combats to which the dissociated elements devote themselves in this fiery furnace, the beneficent rays of which shed life upon all the worlds.

Each of the planets will then be the object of a special excursion, from Mercury, the nearest to the centre, to Neptune, the present frontier of the solar republic. The satellites, eclipses, and comets will also engage our attention in order to complete the full knowledge we wish to acquire.

But this will still be only a small part of our study, for in a bound we shall pass from the frontiers of solar Neptune to the stars, each of which is a brilliant sun with its own light, and the centre probably of a system of inhabited planets. Here we shall really penetrate into the domain of Infinitude. Suns will succeed suns, systems will succeed systems. It is no longer by thousands that they are counted, but by millions; and it is no longer by millions of miles that sidereal distances are measured, nor even by thousands of millions, but by millions of millions, or *billions*. Thus, for example, the star of the first magnitude, Alpha of the Centauri, lies at twenty-five billions of miles from us, and Sirius at ninety-seven billions. [Recent measures seem to indicate that the distance of Sirius is not more than double that of Alpha Centauri.—J. E. G.] Now, these suns count among the nearest. Beyond these lie other universes, which the piercing vision of the telescope begins to grasp in the inaccessible depths of immensity. The description of the great instruments of the observatories, by the aid of which these splendid discoveries have been made, will then be given, and we shall consider the choice of some more modest instruments, which may serve the amateur for the practical study of popular astronomy.

Before undertaking this wonderful voyage, which promises to be fertile in surprises of every kind, before leaving for ever and dropping into the night of space this earth where we live, and which serves us as an obser-tory from which to study the universe, it will not be without interest to consider for a little, from the point of view of the life which embellishes it, the conditions under which this life appeared, the origin of beings and of the planet itself, as well as the destinies which await us and all the inhabitants of this world.

This wonderful life, vegetable, animal, and human, which swarms all round this globe from the poles to the equator, and which animates the ocean depths as well as the surface of the continents; this life, varied and incessantly renewed, has not always been such as we see it to-day. From age to age it is modified, transformed. The conditions of habitation are changed, and with them the species. There was a time when none of the species now living existed on the surface of the globe. There was a time when life itself did not exist in any form whatever. Even the figure of the terrestrial globe, its flattening at the poles, the arrangement of the lands, the mineral nature of the lower primitive strata, the volcanoes which still smoke and throw out their fiery lavas, earthquakes, the regular increase of temperature as we descend into the interior of the globe—all these facts agree in proving that in primitive times the earth was uninhabitable and uninhabited, and that it was at first in the condition of the sun—hot, luminous, and incandescent. On the other hand, if we examine the annual motion of our planet round the sun, as well as the orbits of the other planets, we remark that they all revolve near the plane of the solar equator, all in the same direction, which is precisely that in which the sun turns on itself (certain small planets deviate, however, from this general plane; but their number in the same zone and their singularly small size show that they have been subjected to peculiar perturbations). It is difficult to resist the impression that the origin of the planets is connected in some way or other with the sun round which they gravitate, like children indissolubly attached to their father. This idea had already, in the last century, struck Buffon, Kant, and Laplace. It still strikes us to-day with the same force, notwithstanding certain difficulties of detail which have not yet been explained. As we were not personally present at the creation of the world, direct observation cannot apply here, and we can only form an idea by having recourse to the method of induction. Well, the most probable hypothesis, the most scientific theory, is that which represents the sun as a condensed nebula. This carries us back to an unknown epoch, when this nebula occupied the present place of the solar system, and, even more, an immense lens-shaped mass of gas turning slowly on itself and having its exterior circumference in the zone which marks the orbit of Neptune; farther still, for Neptune does not form the true limit of the system. But perhaps the planets themselves have gradually increased their distance (from the sun).

Let us imagine, then, an immense gaseous mass placed in space. Attraction is a force inherent in every atom of matter. The denser portion of this mass will insensibly attract towards it the other parts, and in the slow fall of the more distant molecules towards this more attractive region a general motion is produced, incompletely directed towards this centre, and soon involving the whole mass in the same motion of rotation. The simplest form of all, even in virtue of this law of attraction, is the spherical form ; it is that which a drop of water takes, and a drop of mercury if left to itself.

The laws of mechanics show that, as this gaseous mass condenses and shrinks, the motion of rotation of the nebula is accelerated. In turning it becomes flattened at the poles, and gradually takes the form of an immense lens-shaped mass of gas. It has begun to turn so quickly as to develop at its exterior circumference a centrifugal force superior to the general attraction of the mass, as when we whirl a sling ; the inevitable consequence of this excess is a rupture of the equilibrium, which detaches an external ring. This gaseous ring will continue to rotate in the same time and with the same velocity ; but the nebulous mother will be henceforth detached, and will continue to undergo progressive condensation and acceleration of motion. The same feat will be reproduced as often as the velocity of rotation surpasses that by which the centrifugal force remains inferior to the attraction. It may have happened also that secondary centres of condensation would be formed even in the interior of the nebula.

The telescope shows us in the depths of space nebulæ of which the forms correspond to these transformations. Such are, among others, the three which we reproduce here. The first (fig. 25) is found in the constellation Canes Venatici, and shows an example of a central condensation commencing a solar focus at the centre of a spherical or lenticular nebula; the second (fig. 26) is found in Aquarius, and presents a sphere surrounded by a ring seen edgeways, recalling singularly the formation of a world, such as Saturn; the third (fig. 27) belongs to the constellation Pegasus, and is remarkable for the zones

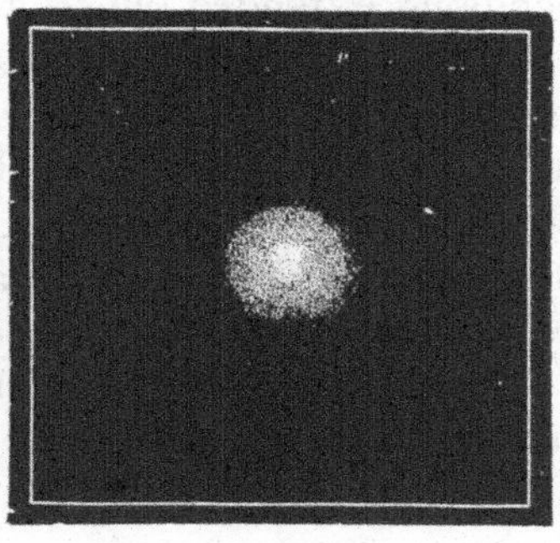

FIG. 25.—NEBULA: PRIMORDIAL CONDENSATION

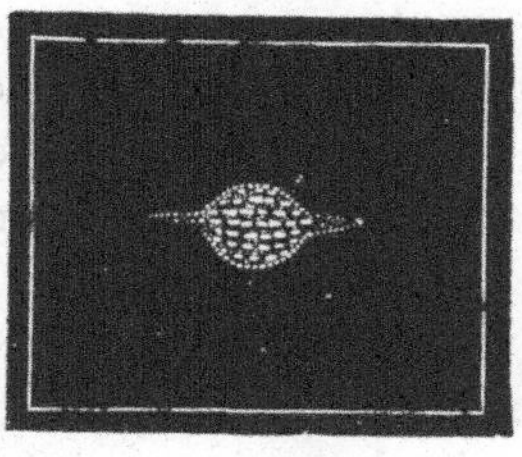

FIG. 26.—NEBULA: TYPE OF A WORLD IN CREATION

already detached from the central nucleus, a veritable sun surrounded by gaseous spirals. [Recent photographs of the great nebula in Andromeda, and others, confirm the above conclusions.—J. E. G.]

We shall see others not less remarkable, later on. Spectrum analysis shows that these nebulæ are not formed of stars close together, but truly of gas, in which nitrogen and hydrogen prevail. [Hydrogen, but *not* nitrogen.—J. E. G.]

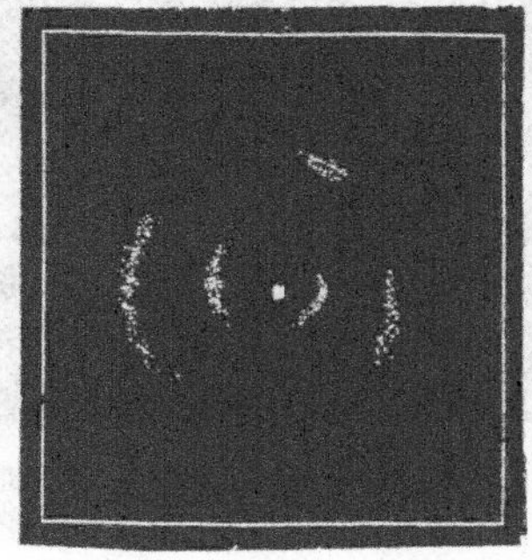

FIG. 27.—NEBULA: REMAINS OF DETACHED RINGS

In our system the rings of Saturn still subsist.

The successive formation of the planets, their situation near the plane of the solar equator, and their motions of translation round the same centre, are explained by the theory which we are discussing. The most distant known planet, Neptune, would be detached from the nebula at the epoch when this nebula extended as far as the planet, out to nearly three thousand millions of miles, and would turn in a slow revolution requiring a period of 165 years for its accomplishment. The original ring could not remain in the state of a ring unless it was perfectly homogeneous and regular; but such a condition is, so to say, unrealisable, and it did not delay in condensing itself into a sphere. Successively, Uranus, Saturn, Jupiter, the army of small planets, Mars, would thus be detached or formed in the interior of this same nebula. Afterwards came the earth, of which the birth goes back to the epoch when the sun had arrived at the earth's present position; Venus and Mercury would be born later. Will the sun give birth to a new world? This is not probable.

For this purpose it would be necessary that its rotation should be enormously accelerated ; it should be 219 times more rapid.

The moon would be formed in a similar way, at the expense of the terrestrial equator, when the earth, still nebulous, extended out to its orbit, then, perhaps, nearer to us.

The relative densities of the planets strengthen this theory. The moon, formed from matter, floating, so to say, on the terrestrial nebula, is very much lighter than the earth. The superior planets, Neptune, Uranus, Saturn, and Jupiter, are much less dense than the inferior planets, Mars, the Earth, Venus, and Mercury. Besides, we find in the chemical composition of the different worlds, and even in that of comets, of shooting stars, and of meteoric stones, the same materials which compose the earth, and which exist also in the gaseous state in the sun.

FIG. 28.—THEORY OF THE FORMATION OF WORLDS : BIRTH OF THE EARTH

Thus the earth was formed by the slow condensation of a gaseous ring detached from the sun (fig. 28), which, continuing afterwards to contract and to condense, gave birth later on to Venus and to Mercury. The terrestrial nebula had from that time an independent existence. It proceeded slowly to form an immense gaseous globe turning upon itself ; thus condensed, heated by the molecular and constant clashing together of all the materials which compose it, the new-born earth shone with a feeble glimmer in the gloomy night of space.

From a gaseous condition it became liquid, then solid, and doubtless it continues to cool and contract even now. But its mass increases from age to age by the meteoric stones and shooting stars which continually fall upon it (more than a hundred thousand millions per annum.)

It is no longer by years or by centuries that we must reckon in order to describe the immeasurable time which nature has employed in the genesis of the world's system. Millions added to millions scarcely mark the seconds of the eternal clock. But our mind, which embraces time as well as space, henceforth sees worlds being born ; it sees them at first shining with a feeble nebulous gleam, afterwards resplendent like the suns, cooling, covered with spots, then with a solid crust, subject to upheavals and tremendous disasters, by the frequent slippings of the crust into the furnace, marked with numerous scars, slowly gaining strength in cooling, to receive henceforth heat and light externally from the sun, to be peopled with living beings, to become the seat of industrious humanities (who in their turn have transformed the surface), and, after having served as the abodes of superior life and thought, slowly lose their fertility, imperceptibly wear away, like the living being itself, arrive at old age, at decrepitude, and at death, to roll henceforth, like travelling tombs, in the silent deserts of eternal night. This is the perpetual evolution of things.

Secular metamorphosis of worlds and beings! How many times has the face of the earth been renewed since the far-off epoch of its fiery genesis from the equatorial frontiers of the solar nebula? For how many centuries has the earth revolved round the sun? For how many ages has the sun himself shone? On the hypothesis that the nebulous matter was originally of extreme tenuity, has been calculated the quantity of heat which must have been produced by the fall of all those molecules towards the centre, to the condensation of which was due the birth of the solar system. Supposing the specific heat of the condensing mass was that of water, the heat of the condensation would be sufficient to produce an elevation of temperature of 28 *millions of degrees Centigrade* (Helmholtz and Tyndall). It has been known for some time past that heat is but a mode of motion: it is an infinitesimal vibratory motion of the atoms. We can now convert at will all motion into heat, and all heat into motion. The motion of condensation has sufficed, and more than sufficed, to produce the present temperature of the sun and the original temperature of all the planets. If that brilliant star continues to condense, as is probable, a condensation which would shorten its diameter by $\frac{1}{2000}$ of its present length would produce a quantity of heat sufficient to cover the loss by emission during two thousand years. At the present rate of emission the solar heat produced by the earlier condensation of its mass would still last for 20 *millions of years.* The length of time required by the condensation to which the primitive nebula was subjected in order to constitute our planetary system entirely defies our imagination. To count it by thousands of millions of centuries would not be an exaggeration. The experiments of Bischof on basalt seem to prove that in order to pass from the liquid state to the solid state, to cool from 2,000 degrees to 200, our globe has required 350 millions of years. The sun has existed for many more millions of centuries. What is the whole history of mankind compared with such periods ?—a wave upon the ocean.

During thousands of centuries the terrestrial globe rolled through space in the condition of a great chemical laboratory. A perpetual deluge of boiling water fell from the clouds upon the burning soil, and rose in vapour in the atmosphere, again to fall. When the temperature became lower than that of boiling water, the vapour liquefied and was precipitated. In the midst of these frightful tempests the terrestrial crust, broken open a thousand times by the convulsions of the central fire, vomited out flames, and closed again. The first lands which emerged from the universal ocean were islets of arid and sterile granite. Later on, from the bosom of the waves, the first semi-fluid combinations of carbon formed the earliest rudimentary attempts at life, protoplasm, a substance which scarcely merits the name of organism, which is no longer a simple mineral, but is still neither vegetable nor animal. The primitive plants, the seaweeds, which float inert in the ocean, were already in progress. The primitive animals, the zoophytes, elementary molluscs, corals, medusas, were also progressing. Imperceptibly, age by age the planet loses its roughness, the conditions of life are improved, beings multiply and become different from the primitive stock, acquiring organs, at first rough and rudimentary, afterwards developed and perfected.

The primeval age, in which the new-born life was represented by the seaweeds, crustacea, and vertebrates still destitute of a head, seems to have occupied alone 53 hundredths of the time which has elapsed since the earth became habitable.

The primary period which succeeded it had for its type the establishment of the coal vegetation and the reign of fishes, and appears to have occupied the following 31 hundredths.

The secondary period, during which the splendid coniferous vegetation ruled the vegetable world, while enormous saurian reptiles dominated the animal world, lasted for the following 12 hundredths. The earth was then peopled with fantastic beings, devoting themselves to perpetual combats in the midst of ungovernable elements.[1]

We have here, then, according to the comparative thickness of the geological formations which have been deposited in these successive epochs, 96 hundredths of past time occupied by a living nature absolutely different from that which now embellishes our globe, a nature relatively formidable and coarse, and as distinct from what we know as that of another world. Who would have then dared to raise the mysterious veil of futurity and divine the future unknown epoch when, by a new transformation, Man should appear upon the planet ?

[1] For a complete description of these vanished ages, see our work, *Le Monde avant la Création de l'Homme.*

The tertiary period, during which appeared only mammals and animal species, which show more or less physical affinity with the human species, came, then, to gather up the inheritance of the primitive ages, and to substitute itself for the preceding period. Its duration did not even reach three-hundredths of the total time.

Finally, the quaternary age saw the birth of the human species, and of cultivated plants. It represents but a hundredth of the scale of time, more probably half a hundredth.

How these grand contemplations enlarge the ideas which we habitually form of nature! We imagine that we go very far back in the past in contemplating the old pyramids still standing on the plains of Egypt, the obelisks engraved with mysterious hieroglyphics, the silent temples of Assyria, the ancient pagodas of India, the idols of Mexico and Peru, the time-honoured traditions of Asia and of the Aryans, our ancestors, the instruments of the stone age, the flint weapons, the arrows, the lances, the knives, the sling-stones of our primitive barbarism—we scarcely dare to speak of ten thousand, of twenty thousand years. But even if we admit a hundred thousand years for the age of our species, so slowly progressive, what is even this compared with the fabulous succession of ages which have preceded us in the history of the planet!

In allotting one hundred thousand years to the quaternary age, the age of our present nature, we see that the tertiary period would have reigned during five hundred thousand years previously, the secondary period during two millions three hundred thousand, the primary period during six millions four hundred thousand, and the primeval period during ten millions seven hundred thousand. Total: twenty millions of years! And what is even this history of life compared with the total history of the globe, since it has taken more than three hundred millions of years to render the earth solid and to reduce its external temperature to 200°? And how many millions must we still add to represent the time which elapsed between the temperature of 200° and that of 70°, the probable maximum possible for organic life!

The study of the worlds opens to us in the order of *time* horizons as immense as those which it opens to us in the order of space. It makes us think of eternity as we think of infinity.

We all admire to-day the beauties of terrestrial nature—the verdant hills, the perfumed meadows, the murmuring brooks, the woods with mysterious shadows, the groves animated with singing-birds, the mountains crowned with glaciers, the immensity of the seas, the warm sunsets in clouds edged with gold and scarlet, and the sublime sunrises on the summit of coloured mountains, when the first rays of morning quiver in the grey vapours of the plain. We admire also the human works which to-day crown those of nature—the bold viaducts thrown from one mountain to another on which the railways run; the ships, marvellous fabrics which traverse the ocean; cities brilliant and animated, palaces and churches, libraries, museums of the mind; the arts of sculpture and of painting, which idealise the reality; the musical inspirations which make us forget the vulgarity of things; the labours of intellectual genius, which searches the mysteries of worlds and carries us into Infinitude; and *we live* with pleasure in the midst of this life so radiant, of which we ourselves form an integral part. But all this beauty, all these flowers, and all these fruits shall pass away.

The earth is born. It shall die.

It shall die, either from old age, when its vital elements shall be worn out, or by the extinction of the sun, on the rays of which its life depends.

It might also die by accident, by the shock of a celestial body it may encounter in its course; but this end of the world is the most improbable of all.

It might, we say, die a natural death by the slow absorption of its vital elements. In fact, it is probable that the water and air are diminishing. The ocean, like the atmosphere, appears to have been formerly much more considerable than in our day. The terrestrial crust is penetrated by the rains which combine chemically with the rocks. The oxygen, nitrogen, and carbonic acid which compose our atmosphere appear also to be subject to a slow absorption. The thinker can foresee, through the mist of ages to come, an epoch still very distant, when the earth, deprived of the atmospheric vapour of water which protects it against the glacial cold of space, by concentrating round it the solar rays, as in a hothouse, will become cold in the sleep of death. From the tops of the mountains the shroud of snows will descend on the high table-lands and valleys, driving before it life and civilisation, and hiding for ever the towns and nations it will meet in its passage. Life and human activity will imperceptibly retire towards the intertropical zone. St. Petersburg, Berlin, London, Paris, Vienna, Constantinople, Rome, will successively fall asleep under their eternal winding-sheet. During many centuries equatorial mankind will vainly undertake Arctic expeditions to find again under the ice the site of Paris, Lyons, Bordeaux, Marseilles. The shores of the seas will have changed, and the geographical map of the world will be transformed. They will live no more, they will breathe no more, except in the equatorial zone, till the day when the last tribe shall come to stand, already dead with cold and hunger, on the shore of the last sea, in the rays of a pale sun, which will henceforth shine here below only on a travelling tomb turning round a useless light and barren heat. Surprised by the cold, the last human family has been touched with the finger of Death, and soon their bones will be buried under the winding-sheet of eternal ice-fields.

The historian of nature may write in the future: Here lies the entire humanity of a world which has lived! Here lie all the dreams of ambition, all the conquests of glorious war, all the resounding affairs of finance, all the systems of imperfect science, and all the asseverations of mortal passions! Here lie all the beauties of the earth! But no gravestone will mark the spot where the poor planet shall have breathed its last sigh!

But perhaps the earth will live long enough to die only by the extinction of the sun. Our fate would be still the same (it would be still death by cold); only, it would be delayed for a much longer time. Some millions of years in the first case; twenty, thirty, or perhaps more in the second. But it is only a question of time. Humanity will be transformed physically and morally long before reaching its apogee, long before it decreases.

The sun will die out. It constantly loses a part of its heat, for the energy which it expends in its radiation is, so to say, inconceivable. The heat emitted by this body would boil per hour 12 millions of cubic miles of water at the temperature of ice! Almost all this heat is lost in space. The quantity which the planets arrest in its passage and utilise for their life is insignificant compared with the quantity lost.

If the sun still condenses with a velocity sufficient to compensate for this loss, or if the rain of meteoric stones, which must incessantly fall on its surface, is sufficient to make up the difference, the star will not yet grow cold; but in the contrary case, its period of cooling down has already commenced. This is more probable, because the spots which periodically appear on its surface can hardly be considered as anything but a manifestation of cooling. The day will come when these spots will be much more numerous than they are at present, and when they will commence to hide a considerable portion of the solar globe. From age to age the darkness will gradually increase, but not regularly, for the first fragments of crust which cover the incandescent liquid surface will be quickly broken up, to be replaced by new formations. Future ages will see the sun dying out and rekindling, until the distant day when the cooling down shall have finally overrun its entire surface, when the last dim and intermittent rays shall vanish for ever, and the enormous red ball shall darken, never again to return and enliven nature with the sweet beneficence of its light.

We have already seen twenty-five stars blazing out in the sky with a spasmodic gleam, and relapsing to an extinction bordering on death [the number of *well-authenticated* cases of 'temporary stars' is much less than twenty-five.—J. E. G.]; already bright stars observed by our fathers have disappeared from the maps of the sky [that any *bright* stars have really disappeared is very doubtful.—J. E. G.]; a great number of red stars have entered on their period of extinction [that red stars are really cooling down is now a disputed question.—J. E. G.]. The sun is but a star; he will meet with the fate of his sisters ; suns, like worlds, are born to die, and in eternity their long career will have endured but 'the space of a morning.'

The sun, then, a dark star, but still warm and electrical, and doubtless feebly illuminated by the quivering lights of the magnetic aurora, will be an immense world inhabited by strange creatures. Round him will continue to turn the planetary tombs until the day when the solar republic shall be entirely effaced from the Book of Life, to give place to other systems of worlds, other suns, other earths, other humanities, and other minds, our successors in the universal and eternal history.

Such are the destinies of the earth and of all the worlds. May we conclude, then, that in these successive endings the universe will one day become an immense and dark tomb? No: otherwise it would already have become so during a past eternity. There is in nature something else besides blind matter ; an intellectual law of progress governs the whole creation ; the forces which rule the universe cannot remain inactive. The stars will rise from their ashes. The collision of ancient wrecks causes new flames to burst forth, and the transformation of motion into heat creates nebulæ and worlds. Universal death shall never reign.

BOOK II

THE MOON

CHAPTER I

THE MOON, SATELLITE OF THE EARTH

Its Apparent Size. Its Distance. How we measure Celestial Distances. How the Moon turns round the Earth

THE light of the moon was the first astronomical illumination. Science commenced with this dawn, and age by age it has conquered the stars and the immense universe. This sweet and calm light releases our thoughts from terrestrial bonds and compels us to think of the sky; then the study of other worlds develops, observations increase, and astronomy is founded. It is not yet the heavens, and it is already more than the earth. The silent star of night is the first halting-place on a voyage towards the infinite.

In ancient times the Arcadians, desirous of being considered as the most ancient of all nations, imagined nothing better to add to their nobility than to carry back their origin to an epoch when the earth had not yet the moon for a companion, and they took as an aristocratic title the name of *Proselenes*—that is to say, *preceding the moon.* Accepting this fable as history, Aristotle relates that the barbarians who originally peopled Arcadia had been driven out and replaced before the appearance of the moon.

Theodorus, still bolder, determines the epoch of the creation of our satellite. 'It was,' says he, 'a little time before the combat of Hercules.' Horace also speaks of the Arcadians in the same way. The rhetorician Menander, ridiculing the pretensions of the Greeks to make themselves as old as the world, wrote in the third century: 'The Athenians pretend to have been born at the same time as the sun, as the Arcadians believed themselves to go back further than the moon, and the inhabitants of Delphos to the Deluge.' After all, the Arcadians are not the only nations who have pretended to have been witnesses of the installation of the moon in the firmament.

We have seen above that the moon is the daughter of the earth, that it was born—millions of years ago—at the limits of the terrestrial nebula, long before the ages when our planet took its spherical form, solidified, and became inhabitable, and that consequently it had shone for a very long time in the sky at the epoch when the first human glances were raised towards its gentle light and considered its course.

The moon is the nearest celestial body to us. It belongs to us, so to say, and accompanies us in our destiny. We may almost touch it with the finger. It is a terrestrial province. Its distance is but thirty times the diameter of our globe, so that thirty earths joined together side by side on the same line would form a suspension bridge sufficient to unite the two worlds. This insignificant distance is scarcely worthy of an astronomical title. Many sailors in ships, travellers on railways, and even many postmen on foot, have passed over a distance greater than that which separates us from the moon. A telegraphic message would be delivered in a few seconds, and a luminous signal would traverse this distance even faster, if we could correspond with the inhabitants of this province, annexed even by nature to our country. It is but the four-hundredth part of the distance which separates us from the sun, and only the one-hundred-millionth part of the distance of the nearest star. It would be necessary to repeat the distance of the moon nearly one hundred million times in order to reach the stellar regions. Our satellite is, then, from all points of view, the first halting-place on a celestial voyage.

At the time of the invention of air-balloons, in 1783, when for the first time men had the pleasure of travelling in the air, the discovery of Montgolfier carried away some minds to such an extent that they already imagined voyages from the earth to the moon, and the possibility of a direct communication between the two worlds. In one of the numerous and curious prints of the time we see a balloon reaching the lunar region, and on the disc of the moon is drawn a sketch of the Paris Observatory and a multitude of amateur astronomers.

Without absolutely denying that the progress of human inventions may one day permit us to make the voyage, it would not be in a balloon that the journey could be performed, since the terrestrial atmosphere is far from filling the space which extends from the earth to the moon. Besides, although near, this province does not exactly touch us; its real distance is 384,000 kilometres, or 96,000 leagues (238,000 miles).

How can it be proved, we may ask, that these figures are accurate? How can we be certain that astronomers are not deceived in their calculations? Who can affirm even that they do not sometimes impose on a kind public? Here is an excellent objection, and one which starts with a sceptical spirit, anxious not to be led into error. Doubt is one of the chief characteristics of the human mind. United with curiosity, it represents the most fertile cause of progress. And exact science, far from forbidding doubt, approves of it, and desires to reply to it. We will accordingly proceed at once, by the same method which has guided us in treating of the motion of the earth, to reply to the objections, to clear up the doubts, and to prove that the assertions of astronomy are demonstrated and unquestionable truths. Perhaps some rather indolent minds will still prefer to retain their doubts rather than satisfy themselves of the reality. That is their business, and the persistency of their obsolete ideas will not prevent the world from turning.

In order to measure the stars we make use of angles, and not a particular measure, like the yard, for example. In fact, the apparent size of an object depends upon its real dimensions and its distance. To say, for example, that the moon appears 'as large as a plate' (which I have often heard said among the hearers of my popular lectures) does not give a sufficient idea of what they mean by it. We often see persons, struck with the brightness of a shooting-star or fireball, describe their observation by stating that the meteor was a yard long by 4 inches in diameter at the head. Such expressions do not satisfy all the conditions of the problem.

When we do not know the distance of an object—and this is generally the case with the stars—there is but one way of expressing its apparent size; that is, to measure the angle which it occupies. If, later on, we can measure the distance, we can, by combining this distance with the apparent size, find the real dimensions.

The measurement of all distance and size is intimately connected with that of the angle. For a given distance, the real size corresponds exactly

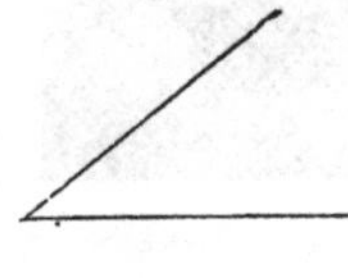

FIG. 29.—ANGLE

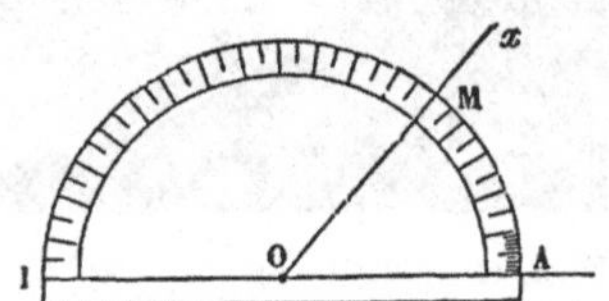

FIG. 30.—MEASUREMENT OF ANGLES

with the measured angle. We can easily understand, then, that the measurement of angles is the first step in celestial geometry. Here the old proverb applies: 'It is but the first step which costs us dear.' In fact, the consideration of an angle is neither poetical nor attractive. But it is not on that account absolutely disagreeable and tedious. Indeed, everyone knows that fig. 29, for example, is an angle, and everyone knows also that the measurement of an angle is expressed in parts of the circumference. A line O*x* (fig. 30) movable round the centre O, can measure any angle, from A to M and to B, and even beyond the semicircle by continuing to turn. The complete circumference has been divided into 360 equal parts, which are called *degrees*. Thus, a half circumference represents 180 degrees; the quarter, or a right angle, represents 90 degrees; half a right angle is an angle of 45 degrees, &c. On the semicircle A M B we have drawn divisions of 10 degrees, and for the first ten degrees, at the point A, we have been able to draw divisions of one degree.

A degree, then, is simply the 360th part of the circumference (fig. 31). We have, then, an independent measure of the distance. On a table of 360 inches round a degree is an inch, seen from the centre of the table; on a piece of water 36 yards round a degree would be marked by 3·6 inches, &c.

The angle does not change with the distance, and a degree, measured on the sky or on this book, is always a degree.

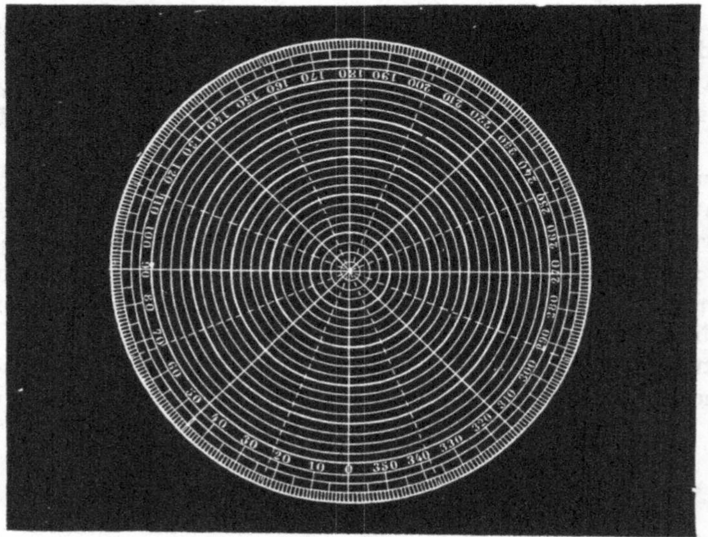

FIG. 31.—DIVISION OF THE CIRCUMFERENCE INTO 360 DEGREES

As we have often to measure angles much smaller than a degree, it is convenient to divide this angle into 60 parts, which have been named *minutes*. Each of these parts has likewise been divided into sixty others, named *seconds*. These denominations have no connection with the minutes and seconds of the measure of time, and they are troublesome on account of this equivocation.

The degree is written briefly by a small zero placed at the head of the figure (°), the minute by a mark thus (′), and the second by two marks thus (″). Thus the present angle of the obliquity of the ecliptic, which we have studied above, and which is 23 degrees 27 minutes 13 seconds, is written 23° 27′ 13″. *This notation should be well understood once for all.*

I ask pardon of my readers (and especially of my lady readers) for these rather dry details, but they are not only necessary—they are *indispensable.* In order to speak a language, it is necessary at least to understand it. As astronomy is composed principally of measures, it is necessary that we should understand these measures. The thing is not difficult; it only demands a moment of serious attention.

One day the Tyrant of Syracuse ordered the illustrious Archimedes to omit the principal mathematics in a lesson on astronomy which promised well but commenced a little severely. 'Let us proceed,' replied Archimedes, without modifying his professorial tone—'let us proceed: there is here no privileged road for kings.'

In astronomy there is no privileged road for anyone, and if we wish to gain information, it is indispensable that we should first understand the principles of geometrical measurements, which besides, we may add, are very interesting in themselves. We have now learned very simply what an angle is. Well, then, the disc of the moon measures 31′ 8″ (31 minutes 8 seconds) in diameter—that is to say, a little more than half a degree. A chaplet of 344 full moons, placed side by side, would be necessary to make the circuit of the sky, from one point of the horizon to a point diametrically opposite.[1]

If, now, we wish to give an account of the relations which connect the real dimensions of objects with their apparent dimensions, it will suffice to notice that every object appears so much smaller the more distant it is, and that when it is distant 57 times its diameter, whatever may be its real

[1] We have said just now that a degree measured on the edge of a table of 360 inches in circumference would be 1 inch. The apparent size of the moon, then, exceeds a little that of a small circle of a ½ inch in diameter, seen at 57 inches from the eye [or ⅕th of an inch seen at a distance of 22½ inches.—J. E. G.] (since a table measuring 360 centimetres in circumference would be 1·14 metre in diameter). Now, it is generally believed that it appears very much larger than this small circle. In reality, however, it is equal, to take a familiar example, to a little wafer of a ½ centimetre in diameter, held at 55 centimetres (about the length of the arm), or a wafer of 1 centimetre seen at 1 metre 10 centimetres, or a globe of 1 foot seen at 110 feet.

We may remark here that when the moon rises or sets it appears enormous, and much larger than when it soars in the heights of the sky. This is a very curious illusion—an illusion of the sight, in fact; for if we measure the lunar disc at the horizon by the aid of a telescope provided with wires, which are made to touch the borders of the moon, we find that *in reality it does not appear larger.* On the contrary, it appears a little larger in the zenith, and this is accounted for, since in the zenith it is a little nearer to us. To what cause is this illusion due? The vapours of the horizon do not play the part which is attributed to them, since the measurements show the contrary. Two causes of enlargement appear to act here. The first is the aspect of the apparent vault of the sky, which appears flattened like the roof of an oven, so that the horizon seems to us more distant than the zenith, and the same angle seems larger in the lower region than in the higher.[1] Try to divide the arc from the horizon to the zenith into two equal parts: you will always place your point too low, and you will suppose 45° at 30°. The Great Bear and Orion appear enormous at the horizon. Another effect is added to this: it is that various objects—trees, houses—being placed before the moon, appear to us more distant still, leading us to suppose it larger than these objects, and so much the more that it is luminous and they are not.

[1] See my work, *L'Atmosphère*, Popular Meteorology, p. 172, fig. 86.

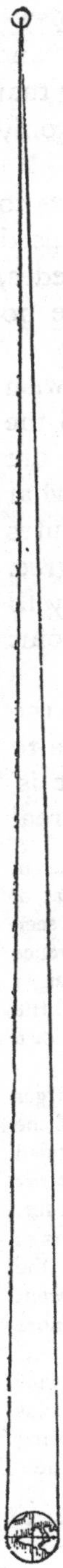

Fig. 32 —Exact Distance of the Moon: Thirty Times the Diameter of the Earth

dimensions, it measures exactly an angle of one degree. For example, a circle of 1 foot in diameter measures exactly one degree if we see it at a distance of 57 feet.

The moon measures a little more than half a degree; we know, then, from this fact alone that it is distant from us a little less than twice 57 times its diameter, or 110 times.

But this idea does not teach us anything yet of the *real distance*, nor of *the real dimensions* of the star of night, if we cannot measure this distance directly.

It is an interesting fact that this distance has been estimated for *two thousand years* with a remarkable approximation to the truth, but it was in the middle of the last century, in 1752, that it was positively settled by two astronomers, observing at two very distant points, one at Berlin, the other at the Cape of Good Hope. These two astronomers were two Frenchmen, Lalande and Lacaille.

Let us consider for a moment fig. 32. The moon is at the top, the earth at the bottom. The angle formed at the moon will be so much the smaller as it is more distant, and the knowledge of this angle will show *what apparent diameter the earth shows to the moon.*

The name of Lunar *parallax* is given to the angle under which the *semi-diameter* of the earth is seen *from the moon.* Let us form a little table of the relations which connect the angles with the distances :—

An angle of 1 degree corresponds to a distance of	57
,, ½ ,, (or 30 minutes) ,, ...	114
,, 1/10 ,, (or 6 minutes) ,, ...	570
,, 1 minute ,, ...	3,438
,, ½ ,, (or 30 seconds) ,, ...	6,875
,, 20 seconds ,, ...	10,313
,, 10 ,, ,, ...	20,626
,, 1 second ,, ...	206,265

We may picture to ourselves, then, the magnitude of an angle of 1 degree by knowing that it is equal to that of a man of 1·70 metre (about 5 ft. 7 in.) distant 57 times his height, that is to say, at 97 metres (about 318 feet). A sheet of squared paper of 4 inches in the side seen at 19 feet likewise represents a magnitude of 1 degree. A little cardboard square of 1 inch, seen at 286 feet, represents 1 minute. A line of 1 inch in length, drawn on a board distant 3¼ miles, represents the magnitude of one second. Taking a hair of the tenth of a millimetre in thickness, and placing it at 20 metres, the size of this hair seen at that distance also represents a second. Such an angle is therefore of extreme smallness and invisible to the naked eye.

This valuation of angular magnitudes will assist us afterwards in estimating *all celestial distances.* The parallax of the moon being 57 minutes (nearly one degree), *proves* that the distance of that body is 60¼ semi-diameters, or radii, of the earth (60·27). In round numbers it is *thirty* times the diameter of the earth.

As the radius of the earth is 6,371 kilometres (3,956 miles), this distance is, then, 384,000 kilometres, or 96,000 leagues of 4 kilometres. [The mean distance in English *miles* is 238,854.—J. E. G.] This is a fact as certain as that of our own existence.

We have represented this distance of the moon on an exact proportional scale (fig. 32). In this drawing the earth has been sketched with a diameter of 6 millimetres, having on its face the meridian which passes from Berlin to the Cape of Good Hope; the moon, with a diameter equal to three-elevenths of that of our globe—that is to say, 1·6 millimetre—has been placed at 180 millimetres from the earth—that is to say, 30 times its diameter. Such is the *exact proportion* which exists between the earth and the moon as to volume and distance. This distance, thus calculated by geometry, is, we can affirm, determined with a precision greater than that with which we are satisfied in the ordinary measurement of terrestrial distances, such as the length of a road or of a railway. Although the assertion may appear rash in the opinion of many, it is not open to dispute that the distance which separates the earth from the moon at any moment is more exactly known, for example, than the precise length of the road from Paris to Marseilles. We may even add that astronomers put incomparably more precision into their measures than the most scrupulous traders.

Let us attempt, now, to imagine this distance in thought.

A cannon-ball animated with a constant velocity of 1,640 feet a second would take 8 days 5 hours to reach the moon. Sound travels at the rate of 332 metres a second (in air at the temperature of 0°) [about 1,089 feet per second]. If the space which separates the earth from the moon were entirely filled with air, the sound of a lunar volcanic explosion, sufficiently powerful to be heard here, would only arrive 13 days 20 hours after the event; so that if it happened at the epoch of full moon we should see it at the moment it occurred, but we should not hear it till near the time of the following new moon. A railway train which could make the circuit of the world, if travelling continuously, in 27 days, would reach the lunar station after 38 weeks.

But light, which has the most rapid of known motions, darts from the moon to the earth in a second and a quarter.

The knowledge of the moon's distance enables us to calculate its real volume by measurement of its apparent volume. Since the semi-diameter of the earth, as seen from the moon, measures 57 minutes, and the semi-diameter of the moon seen from the earth measures 15′ 34″, the diameters of these two globes are to each other in the same proportion. Making the exact calculation, we find that the diameter of our satellite is to that of the earth in the ratio of 273 to 1,000 : it is a little more than a quarter of the diameter of our world, which measures 12,732 kilometres (7,912 miles). The diameter of the moon is therefore 3,484 kilometres [2,165 miles. The most recent results make it 2,163 miles.—J. E. G.], which gives for its circumference 10,940 kilometres (6,798 miles), for the surface of the lunar globe 145 millions of square miles, and for the volume 5,300 millions of cubic miles. The surface of this neighbouring world is equivalent to about four times that of the European continent, or to the total area of the two Americas. It might have been enough to satisfy the ambition of a Charlemagne or of a Napoleon, and we believe that Alexander regretted that he could not extend his empire to it! But for astronomy it is merely a toy. The moon's volume is the 49th part of the volume of the earth. It would therefore require 49 moons united together to form a globe as large as ours. It would take 62 millions to form one of the size of the sun!

FIG. 33.—COMPARATIVE MAGNITUDE OF THE EARTH AND OF THE MOON

We see, then, that there is nothing so simple, nothing so sure, as these apparently marvellous facts: *the measurement of the distance and volume of a world*. I hope that my readers have clearly understood this method of celestial geometry, so logical and so exact.

As we have said, the mean distance of the moon is 238,000 miles. At this distance the moon revolves round the earth in a period of 27 days

7 hours 43 minutes 11 seconds, with a mean velocity of 1,017 metres per second [3,337 feet a second, or about double the velocity of a cannon-ball.—J. E. G.].

The examination of the moon's motion teaches us, even in the history of its discovery, the fundamental principle of the motion of the heavenly bodies, and the equilibrium of creation. It was, in fact, the discussion of our satellite which led Newton to the discovery of the laws of universal gravitation.

One evening, about two hundred years ago, seated in the orchard of his paternal residence, a young man of twenty-three was meditating. In the midst of the evening silence an apple, they say, happened to fall near him. This simple fact, which would have passed unnoticed by anyone else, seized and captivated his attention. The moon was visible in the sky. He began to reflect on the singular power which urged bodies towards the earth; he asked himself frankly *why the moon does not fall*, and by the power of thought he arrived at one of the most beautiful discoveries—one of which the human mind may be proud. This young man was Newton. The discovery to which he had been led by the fall of an apple is the great law of universal gravitation, the principal basis of all our astronomical theories, now become so precise.

Let us see by what series of reasonings we can realise the identity of terrestrial gravity with the force which moves the stars.

Gravity, which causes bodies to fall to the earth, is not manifested merely near the surface of the ground; it still exists at the top of buildings, and even on the highest mountains, its energy apparently experiencing no appreciable enfeeblement. It is natural, therefore, to think that this gravity would be also felt at very great distances, and if we withdraw from the earth to a distance from its centre equal to sixty times its radius—that is to say, to the moon—it may well happen that the gravity of bodies towards the earth may not have entirely disappeared. Might not this gravity be even the cause which retains the moon in her orbit round the earth? Such is the question which Newton set himself.

Galileo analysed the motions of bodies in their fall to the earth. He recognised that gravity always produces the same effect on them in the same time, whether they be in a state of rest or of motion. In the case of bodies falling vertically without initial velocity, it always increases the velocity by the same quantity in the space of one second, whatever time may have elapsed since the commencement of the fall. In the motion of bodies projected in any direction, it draws down the body below the position which it would have occupied at each instant, in virtue only of its velocity of projection, by a quantity precisely that which it would fall vertically in the same time if the body had been let fall without initial velocity.

A ball projected horizontally would move indefinitely in a straight line and with the same velocity if the earth did not attract it. Owing to gravity, it falls little by little below the straight line along which it has been projected, and the distance which it successively falls below this line is exactly the same as it would have fallen in the same time on a vertical line, if it had been dropped at its point of departure without giving it any impulse. Produce the direction of motion at first imparted to the ball to meet a vertical wall against which it strikes; then measure the distance which separates this point from the point below where the ball strikes the wall: you will have precisely the distance which the ball would have fallen vertically without initial velocity during the time which has elapsed from its departure to its arrival at the wall.

These simple ideas are directly applicable to the moon. At every instant in its motion round the earth we can compare it with a ball projected horizontally. Instead of continuing to move indefinitely on the straight line on which it has been, so to say, projected, it imperceptibly falls below it and approaches us, describing an arc of its almost circular orbit. It falls, then, at each instant towards us, and the distance which it thus falls in a certain time may be easily obtained, as in the case of the ball, by comparing the arc of the curve which it describes in this time with the path which it would have pursued during the same time on a tangent at the first point of this arc, if its motion had not been subject to alteration.

Let us see how the calculation of the distance which the moon falls towards the earth in a second of time is effected.

Our planet being spherical, and the length of the circumference of one of its great circles (meridian or equator) being 24,900 miles, the orbit of the moon, drawn with an opening of a compass equal to 60 times the radius of the earth, would be a length of 60 times 24,900 miles, or 1,494,000 miles.

The moon describes its complete orbit in 27 days 7 hours 43 minutes 11 seconds, which gives a number of seconds equal to 2,360,591. Dividing 1,494,000 by this number, we find that the moon moves each second 0·633 mile, or a little more than 1 kilometre (about $\frac{2}{3}$ of a mile).

In order to find the distance which the moon falls towards the earth in a second, let us suppose that it is at the point marked L (fig. 34) at a certain moment, the earth being at the point marked T. Projected horizontally on the line towards the left, the moon would describe the straight line L A if the earth did not act upon it; but instead of following this tangent it follows the arc L B. The path described in one second is, we have said, 1,017 metres; now, if we measure the distance which separates the point A from the point B, we find the distance which the moon falls towards the earth in one second, since without the attraction of the earth it would be farther off in a straight line. This quantity is 1·353 millimetre—that is to say, about $1\frac{1}{3}$ millimetre [a little over $\frac{1}{20}$ of an inch.—J. E. G.].

Well, then, if we could raise a stone to the height of the moon, and there let it fall, it would fall exactly towards the earth with the same velocity of $1\frac{1}{3}$ millimetre in the first second of its fall. Gravity diminishes in proportion as we recede from the centre of the earth in the ratio of the square of the distance—that is to say, the distance multiplied by itself. Thus, at the earth's surface a falling stone passes over 4 metres 90 centimetres (about 16 feet) in the first second of its fall. The moon is at 60 times the distance of the surface of the earth from its centre. Gravity is then diminished at this point by 60×60, or 3,600. In order to know, then, what distance a stone would fall in a second if raised to this height, we have only to divide 4·90 metres by 3,600. Now, $\frac{4·90 \text{ metres}}{3,600} = 1·353$ millimetre—that is to say, exactiy the amount which the moon deviates per second from a straight line. A stone raised to the height of the moon would take, instead of one second, a minute to fall 4·90 metres (about 16 feet).

Why does the moon not fall altogether? Because it has been projected into space like a ball. Any other bodies projected with the same velocity, at this distance from the earth, would act exactly like the moon. The velocity of its motion (more than a kilometre a second) produces, like a stone in a sling, a centrifugal force, of which the tendency is to increase its distance from us *precisely by the same amount* by which it tends to approach us on account of gravitation, and this makes it always remain at the same distance.

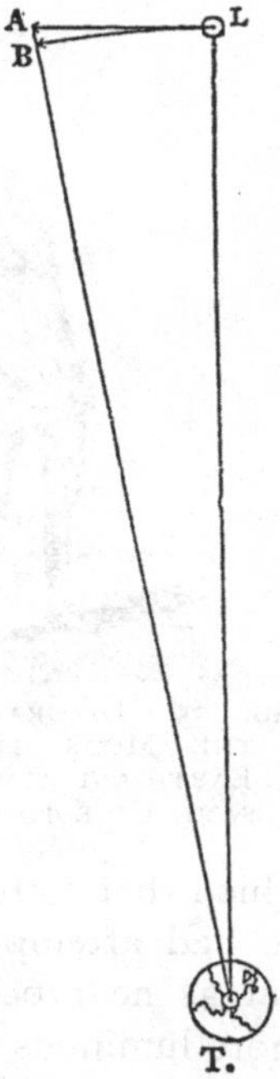

FIG. 34.—EXPLANATION OF THE MOVEMENT OF THE MOON

Even the velocity of the moon's motion round the earth arises from the power of our planet. The earth is the hand which turns the moon in the sling. If our planet had more power, more energy, than it has, it would turn its satellite more rapidly; if, on the contrary, it were weaker, it would turn this sling more slowly. The velocity of the moon's motion gives exactly the measure of the earth's power.

The elementary sketch (fig. 35) shows what the force is which holds the moon in its motion round us; it is the attraction of the earth, comparable with the tension of the cord. This same figure also shows how the moon always presents the same face to the earth—always the half to which we may suppose the cord attached. While the earth turns freely on itself during its annual voyage round the sun, the moon remains attached to us, as with a chain.

At the time when Newton attempted to make this comparison between gravity at the surface of the earth and the force which keeps the moon in her orbit, the diameter of the terrestrial globe was not known with sufficient exactness. The result did not completely answer his expectations: he found for the distance which the moon falls towards the earth in one second a little less than the twentieth of an inch (it should be a little *more*, about 0·053 inch); but, although the difference was not large, it appeared sufficient to prevent him from inferring the identity which he hoped to find. The cause of his failure was not explained till sixteen years later. In the year 1682, being present at a meeting of the Royal Society of London, he heard mentioned a new measure of the earth made by the French astronomer Picard, and having obtained the result which that astronomer had found, he again took up the calculation which he had attempted sixteen years previously, employing the new data; but as he proceeded the desired precision came with evidence more and more luminous; the thinker became as if mentally dazed, and felt seized with such emotion that he could not continue, and begged one of his friends to finish the calculation.

FIG. 35.—DIAGRAM SHOWING HOW THE MOON TURNS ROUND THE EARTH AND ALWAYS PRESENTS THE SAME FACE TO IT

In fact, the success of the comparison which Newton sought to establish was complete, and left no room for doubt that the force which retains the moon in her orbit is really the same as that which causes bodies to fall to the earth's surface, diminished in intensity in the ratio indicated by the square of the distances.

Newton also found, by methods of calculation of which he was the inventor, that, under the action of a similar force directed towards the sun, each planet should describe an ellipse, having one of its foci at the centre of the sun; and this result agreed with one of the laws of planetary motion established by Kepler by the aid of a long series of observations. He was then warranted in saying that the planets tend or gravitate towards the sun, even as the satellites gravitate towards the planets on which they depend; and that the weight of bodies on the earth is but a particular case of the gravitation manifested in the celestial spaces by the motion of revolution of the planets round the sun and of the satellites round the planets.

What more natural, then, than to generalise this idea by saying that the stars scattered through space tend or gravitate towards each other according to the beautiful law which has taken its place in science under the name of *attraction*, or *universal gravitation !*

The progress of astronomy has absolutely demonstrated the universality of this force,[1] of the cause and intimate essence of which we are still ignorant. It is expressed by this formula, which it is important to remember:

Matter attracts matter in the direct ratio of the mass and the inverse ratio of the square of the distances.

We will develop these laws later on in the chapter on the motion of the planets round the sun (Book III. Chap. I.).

Thus was discovered the enigma of celestial motions. Always preoccupied with his profound researches, the great Newton showed in the ordinary affairs of life an absence of mind which has become proverbial. It is related that one day, wishing to find the number of seconds necessary for the boiling of an egg, he perceived, after waiting a minute, that he held the egg in his hand, and had placed his seconds watch (an instrument of great value on account of its mathematical precision) to boil!

This absence of mind reminds one of the mathematician Ampère, who one day, as he was going to his course of lectures, noticed a little pebble on the road; he picked it up, and examined with admiration the mottled veins. All at once the lecture which he ought to be attending to returned to his mind; he drew out his watch; perceiving that the hour approached, he hastily doubled his pace, carefully placed the pebble in his pocket, and threw his watch over the parapet of the Pont des Arts.

But we must not allow absence of mind to cause us to forget the subject of our chapter. The moon, we have said, turns round the earth in a revolution of which the duration is 27 days 7 hours 43 minutes 11 seconds, with a velocity of about ⅔rds of a mile a second, or 37 miles a minute, and which produces a centrifugal force tending every instant to increase the moon's distance by a quantity exactly equal to that which the attraction of our globe tends to make it approach, so that it finally remains suspended in space, always at the same mean distance. The orbit which it describes round us measures about 600,000 leagues in length (about 1,494,000 miles).

If the moon could be stopped in its course, the centrifugal force would be suppressed; it would then obey solely the attraction of the earth, and it would fall upon us, according to a calculation which I have made, in 4 days 19 hours 54 minutes 57 seconds, or 417,297 seconds. We leave to our

[1] [The law of gravitation, although in all probability universally true throughout all space, does not yet admit of *absolutely rigid* mathematical proof in the case of the binary or revolving double stars.—J. E. G.]

readers the task of imagining the effects which so formidable a fall would produce on the earth's inhabitants.

While the moon turns round the earth, the earth itself revolves round the sun. In an interval of 27 days it accomplishes about a thirteenth of its annual revolution. This translation of the earth, which carries the moon with it in its course, is the reason why the period of the lunar phases, or a lunation, is longer than that of the real revolution of our satellite.

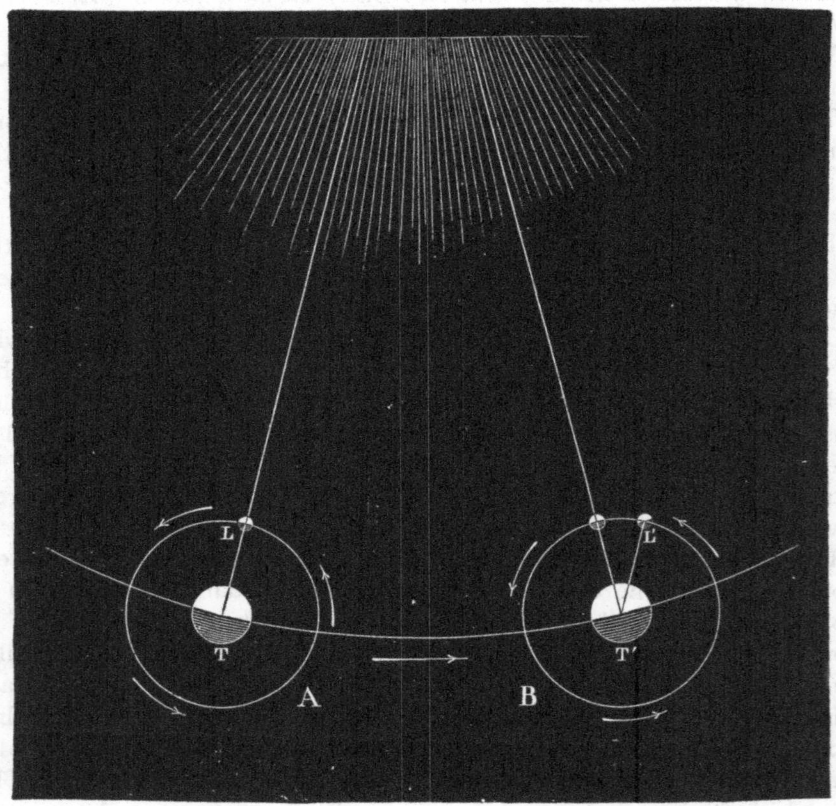

FIG. 36.—DIFFERENCE BETWEEN THE TIME OF THE MOVEMENT OF THE MOON ROUND THE EARTH AND THE TIME OF A LUNATION. (L = MOON, T = EARTH)

The moon is a dark globe, like the earth, which has no light of its own, and is only visible in space because it is illuminated by the sun. Naturally, therefore, one half of it is always illuminated, neither more nor less. The phases vary according to the position of the moon with reference to the sun and to ourselves. When the moon is between us and the sun its illuminated hemisphere is naturally turned towards the luminous body and we do not see it: this is the epoch of new moon. When it forms a right angle with the sun, we see half the illuminated hemisphere; this is the epoch of the quarters. When it passes behind us, relatively to the sun, it shows the whole of its illuminated hemisphere; this is the full moon. In order to explain the difference of duration between the period of the phases and that of the moon's revolution (and this is a difference which beginners have sometimes a certain difficulty in clearly understanding), let us consider our satellite at the moment of new moon. In this position we can picture to ourselves the earth, the moon, and the sun ranged on the same straight line. Let this be, for example, the position which we have represented in the drawing A of fig. 36. The moon is placed exactly between the earth and the sun at the moment of new moon [not *exactly*, except during a total eclipse of the moon, but nearly so.—J. E. G.].

While it turns round us in the direction indicated by the arrow, the combined system of the earth and moon is transported, as it were, in one piece from left to right, and when our satellite has accomplished a complete revolution, at the end of 27 days, the earth and moon are found at the positions T′ L′ respectively (B, fig. 36). The two lines, T L and T′ L′, are parallel. If a star, for example, were found exactly in the direction of the first line, it would be found again in the direction of the second. But, in order that the moon should return again in front of the sun, it must still move for about 2 days 5 hours (2 days 5 hours 0 minutes 52 seconds). The sun has apparently moved back towards the left in consequence of the perspective of our motion. It follows that the duration of a lunation, or of the return to new moon, is 29 days 12 hours 44 minutes 3 seconds. This is called the *Synodical* revolution of the moon. The real revolution is named the *Sidereal* revolution.

There is here, as we see, a difference analogous to that which we have noticed (p. 17) between the duration of the earth's rotation and the duration of the solar day.

The proper motion of the moon from west to east, and the succession of phases, may be considered as the most ancient facts of observation of the sky, and as the first basis of measurement of time and of the calendar.

CHAPTER II

THE PHASES OF THE MOON

The Week. The Measurement of Time

OUR forefathers lived in more intimate communication with nature than we do. They had neither the artificial life, nor the hypocrisy, nor the anxieties created by the factitious necessities of modern existence. It was they who established the first bases of the sciences by the direct observation of natural phenomena. If astronomy is the most ancient of the sciences, the study of the moon was the most ancient of astronomical observations, because it was the simplest, the easiest, and the most useful. The solitary globe of night pours out its calm and clear light in the midst of the silence and contemplation of nature. The succession of its phases provided shepherds as well as travellers with the first measure of time; after that, of day and night, due to the diurnal rotation of our planet. The lunar crescent, with its melancholy light, gave to nature a pastoral calendar.

In the course of about a month our companion makes a complete circuit of the sky in a direction opposite to the diurnal motion; and, while it appears to rise and set like all the other heavenly bodies, moving from east to west, it is retarded every evening by three-quarters of an hour, and seems to keep behindhand among the stars, or to move back towards the east. This motion is very perceptible, and it is sufficient to observe the position of the moon on three consecutive nights in order to recognise it. If it is, for example, near a bright star, it becomes detached from it and recedes, making the circuit of the sky from the right towards the left; at the end of the first day it has receded 13°; the second day it is at 26°; the third at 39°, &c.; finally, after twenty-seven days, it is distant 360°, and consequently has returned to the same point; thus it is found again at the same place which it occupied in the sky just a month previously, having made the complete circuit of the sky from west to east.

The *phases* of the moon are more readily remarked than its motion. When it begins to disengage itself from the rays of the sun in the evening, two days after conjunction, or new moon, it presents the form of a very slender crescent, the convexity of which is always *turned towards the setting sun.* (This fact many painters seem still to forget, for a year does

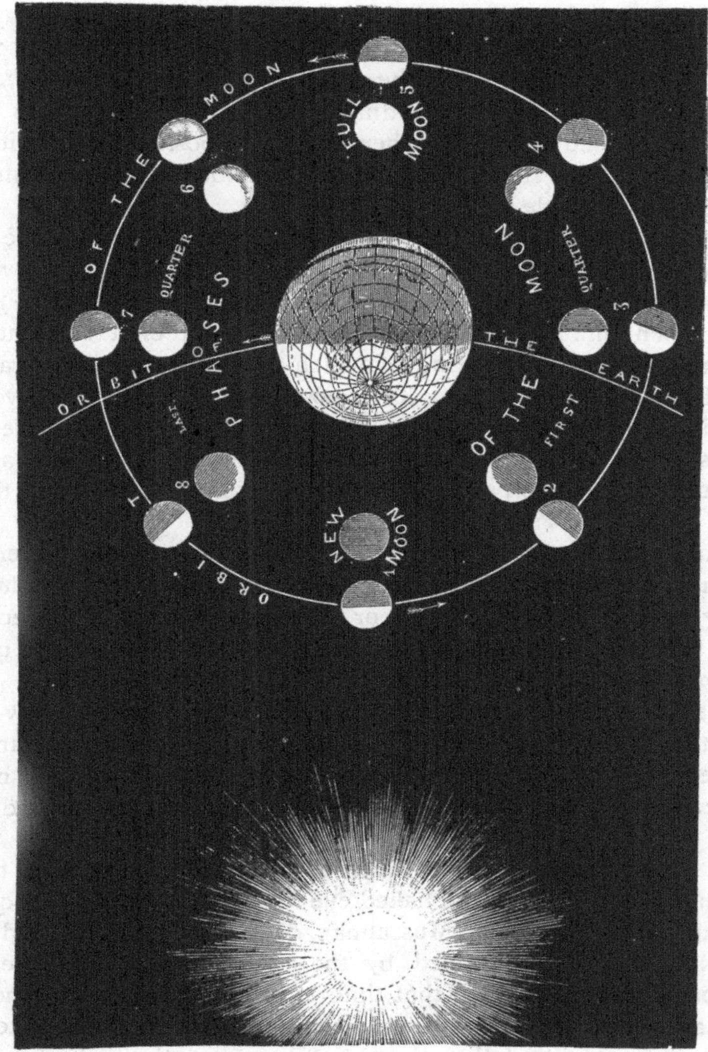

FIG. 37.—PHASES OF THE MOON

not pass without our seeing in picture-galleries moons with the crescent on the wrong side.)

The size of the crescent gradually increases; in the course of five to six days the star of night attains the form of a semicircle; the luminous parts is then limited by a straight line, and we say that the moon is 'dichotomised,' or in quadrature; this is the *first quarter*. It may then be easily seen during the day. Continuing to increase its distance from the sun, it becomes of an oval form, increasing in light for seven to eight days, after which it becomes exactly circular. Its round and luminous disc shines through the whole night; this is the epoch of *full moon*, or *opposition*; we see it crossing the meridian at midnight and setting as soon as the sun rises; everything shows that it is then directly opposite the sun with reference to us, and that it shines because the luminous star illuminates it in front and not at the side.

After the full moon comes the decrease, which gives the same phases and the same forms shown during the increase; it is at first oval, then arrives imperceptibly at the shape of a semicircle (*last quarter*). This semicircle then diminishes and assumes the aspect of a crescent, which becomes narrower every day, and of which the horns are always raised on the side farthest from the sun. The moon, then, is found to have made the circuit of the sky. We see it rising in the morning a little before the day star; it approaches the sun, and is finally lost in its rays; we see it return to *new moon*, also named the 'conjunction,' formerly the *neomenia*.

We have already seen that the series of different aspects under which the moon is presented to us has for its period the time of revolution of that body with reference to the sun, or 29 days 12 hours. The epochs of new and full moon are also called the *syzygies*, and those of the quarters the *quadratures*.

It is evident that the moment when the moon becomes new—or, in other words, the moment when the lunar month commences—cannot be determined by direct observation, except when, at the precise moment of *conjunction*, the moon passes exactly in front of the sun and produces an eclipse.

What is the shortest interval after or before the conjunction that we can perceive the moon with the naked eye? The answer to this question should particularly interest a Mussulman, considering that the end of the fast of Ramadan is determined by the first appearance of the moon. As millions of persons are at that time expecting this phenomenon, it is in the East especially that we ought to obtain the most accurate reply. But it must be admitted that they hardly concern themselves now with astronomical observations in those countries.

Hevelius assures us that in the torrid zone Americus Vespucius had seen on the same day the moon to the east and to the west of the sun; but in Germany, he observed, he had never been able to perceive it sooner than forty hours after its conjunction, or later than twenty-seven hours before, although Kepler has asserted that it may be discerned even at the conjunction, when its latitude is 5°.[1]

When the moon is a crescent, during the first days of the lunation, we notice that the rest of the lunar globe is visible, illuminated by a pale light. This is the *lumière cendrée.* It is caused by the earth itself.

In fact, the earth is illuminated by the sun, and reflects the light into space. When the moon is in conjunction with the sun the earth is in 'opposition,' as seen from the moon; it is the epoch of *full earth* for an observer on our satellite. The light which our globe then sends to the moon exceeds about fourteen times that which the full moon sends to us.

FIG. 38.—THE ASHY LIGHT OF THE MOON

The ancients found great difficulty in explaining the cause of this secondary light: some attributed it to the moon itself, which they thought was either transparent or phosphoric, others to the fixed stars. Kepler asserts that Tycho attributed it to the light of Venus, and that Mœstlin, of whom Kepler declared himself a disciple, was the first who explained, in 1596, the true cause of this ashy light. But it had been already explained by the celebrated painter, Leonardo da Vinci, who died in 1518.

This secondary light almost entirely disappears when the moon is in quadrature—(1) because the earth then sends four times less light to the moon; (2) because the phase of the moon, becoming four to five times larger, prevents us from distinguishing it. For the same reason, this ashy light appears a little brighter after the last quarter—that is to say, in the morning—because, on the one hand, the eastern part of the earth reflects

[1] On December 11, 1885, one of our correspondents, M. L. Decroupet, at Soumagne (Belgium), observed the crescent moon 26 hours 47 minutes after the new moon, and on November 24, 1886, 36 hours before. See our *Revue mensuelle d'Astronomie*, 1886, p. 110, and 1887, p. 70.

more solar light than the western portion, where the waters of the sea absorb the rays; and, on the other hand, the eastern region of the moon is itself a little darker, on account of the dark spots which exist there (we may also remark that in the morning our sight is a little more sensitive, and that the pupil of the eye is a little more dilated after the darkness of the night than it is after the brightness of daylight). This ashy light, reflection of a reflection, resembles a mirror in which we may see the luminous state of the earth. In winter, when a great part of the terrestrial hemisphere is covered with snow, it is perceptibly brighter. Before the geographical discovery of Australia, astronomers suspected the existence of that continent from the ashy light, which was very much brighter than could be produced by the dark reflection from the ocean. This lunar light generally presents a greenish-blue tint; indicating that our planet, seen from a distance, would show this shade.

The luminous crescent seems to be of a larger diameter than the ash-coloured disc of the moon. The English call this appearance 'the old moon in the new moon's arms.' This effect proceeds from irradiation, from the contrast between a bright light and a fainter one beside it. One effaces the other and kills it, as the painters say; the crescent appears enlarged by the flooding of light which expands the disc of the moon; the illuminated atmosphere further increases this illusion.

It was these phases and aspects of the moon which formerly gave birth to the custom of measuring time by months, and by weeks of seven days, on account of the return of the moon's phases in a month, and because the moon appears about every seven days, so to say, under a new form. Such was the first measure of time; there was not in the sky any signal of which the differences, the alternations, and the epochs were more remarkable. Families met together at a time fixed by some lunar phase.

The *neomenia* served to regulate assemblies, sacrifices, and public functions. The ancients counted the moon from the day they first perceived it. In order to discover it easily, they assembled at evening upon the heights. The first appearance of the lunar crescent was watched with care, reported by the high priest, and announced to the people by the sound of trumpets. The new moons which correspond with the renewal of the four seasons were the most solemn; we find here the origin of the 'ember weeks' of the Church, as we find that of most of our festivals in the ceremonies of the ancients. The Orientals, the Chaldeans, Egyptians, and Jews religiously observed this custom.

The festival of the new moon was also celebrated among the Ethiopians, the Sabæans of Arabia Felix, the Persians, and the Greeks. The Olympian games, established by Iphitus, commenced at the new moon. The Romans had also this festival (it is mentioned by Horace). We find it at present among the Turks. The ceremony of the mistletoe among the Gauls was performed at the same epoch, and the Druid carried a crescent, as we see in the ancient pictures. We find the same custom practised by the Chinese, as well as among the Peruvians and in the island of Tahiti. The Tasmanians, a savage people of whom the last representative died in 1876, and whose customs are known for a century past, followed the same practices. Thus the days of new moons were naturally set apart among primitive nations for certain ceremonies.

In the first calendars the public administration had to predict a long time in advance the day of the year on which the neomenias would be celebrated. An oracle had prescribed to the Greeks the sacred observance of the ancient custom. From this we may imagine how important it was for the ancients to discover a period which would bring back the phases of the moon to the same days of the year. This discovery has been preserved for us under the name of *Meton*, who, in the year 433 before our era, announced it to the Greeks assembled to celebrate the Olympian games. Here is what it consists in. Any phase of the moon returns after an interval of 29½ days. Now, it is found that nineteen solar years, or 6,940 days, contain almost exactly 235 lunations. Hence, after nineteen years, the same phases of the moon return on the same days of the year, on the same dates, so that it is sufficient to have registered these dates during nineteen years in order to know them in advance during all the following periods of the same length. This cycle is in fault only one day in 312 years.

This discovery seemed so beautiful to the Greeks that they exhibited the calculation in letters of gold in the public places for the use of the citizens, and they called by the name of the *Golden Number* the current year of this period of nineteen years, which brought back the moon into conjunction with the sun at the same point of the sky on the same day of the solar year. This number remains in the ecclesiastical calendar, which is regulated more by the motion of the moon than by that of the sun.

The *lunar cycle* is, then, a period of nineteen years—of which five are bissextiles—or of 6,940 days, in which there are 235 lunations; so that at the end of nineteen years the new moons return to the same degree of the zodiac, and consequently to the same day of the year as nineteen years previously.[1] We call the first year of the lunar cycle that in which the new

[1] This rule serves to determine in advance the dates of the Church festivals after the date of Easter. The festival of Easter, in fact, is fixed on the Sunday which follows the full moon of the equinox. The computers assume that the vernal equinox always happens on March 21, and give for the date of Easter each year the first Sunday after the full moon which follows March 21. It follows from this that Easter cannot occur earlier than March 22, nor later than April 25, and consequently can occupy thirty-five different places. The movable feasts of the ecclesiastical

moon occurs on January 1, and we call the *Golden Number* the year of the lunar cycle in which we are.

As we have seen above, the week has also the moon for its origin; it is the natural measure created by the four phases of the moon. It is also of very ancient origin: the Egyptians, the Chaldeans, the Jews, the Arabians, and the Chinese have used it from the most remote times. The seven principal stars of ancient mythology, being equal in number to the days of the week, were considered as divine protectors, and the names which those days still bear are derived from the sun, the moon, and the five planets, as it is easy to show.

Dimanche (Sunday) is the day of the Sun.
Lundi (Monday) „ „ „ Moon.
Mardi (Tuesday) „ „ Mars.
Mercredi (Wednesday) „ „ Mercury.
Jeudi (Thursday) „ „ Jupiter.
Vendredi (Friday) „ „ Venus.
Samedi (Saturday) „ „ Saturn.

It is the same in almost all modern languages. In its canonical language, however, the Church has not accepted these heathen names, and it names the seven days thus: Dominica, day second, day third, day fourth, day fifth, day sixth, Sabbath—a Jewish legacy.

The order of the names, which is not that of the brightness of the stars nor of their motions or distances, had an astrological origin, which we find by examining fig. 39. On this diagram let us place the seven wandering bodies known to the ancients, in the order of their distances as recognised at that ancient epoch—that is to say,

The Moon.	Mars.
Mercury.	Jupiter.
Venus.	Saturn.
The Sun.	

Let us place these, we say, at equal distances along the circumference of a circle, and join one to the other by a line. We shall thus produce a cabalistic figure greatly valued by the ancient astrologers, the heptachord, a star of seven rays enclosed in a circle. Well, then let us start from the

calendar advance or recede each year, being regulated by that of Easter taken as a point of departure.

We may add that the moon which the computers use in their calculations is not the true moon, but a mean moon called the ecclesiastical moon. This fictitious moon may arrive at its full a day or two before the true moon. Hence we have differences sometimes inexplicable to the public. Thus, for example, in 1876 the full moon which followed March 21 happened on April 8; that day was Saturday: Easter should, then, have been fixed on the next day, April 9. Now, it was fixed on the 16th, according to the ecclesiastical moon, which, theoretically, was delayed some hours after the true moon.

In a sequel to *Popular Astronomy* (*Les Etoiles*: Paris, E. Flammarion) may be seen the date of Easter from the year 1890 to the year 2150.

moon and follow the line which leads us to Mars; from Mars we take the other line, which brings us to Mercury; from here we follow the line which leads to Jupiter; then from there to Venus, from Venus to Saturn, and from Saturn to the sun, and we return to the moon, having named *the seven days of the week in their true order.*

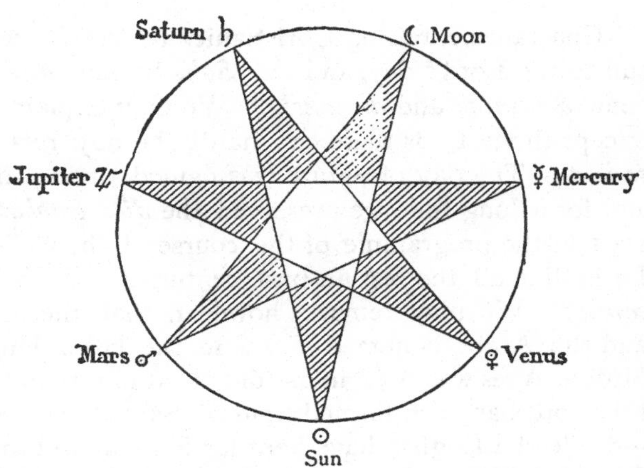

Fig. 39.—Astronomical Origin of the Names of the Days of the Week

Is it thus that the names of the days of the week have been really formed? It is difficult to find their authentic source. Dion Cassius, a Greek historian of the second century, asserts that this custom comes from the Egyptians, and rests on two systems. The first consists in counting the hours of the day and night, allotting the first to Saturn, the second to Jupiter, the third to Mars, &c. (the ancient order commencing with the most distant planet). If we perform this operation for the twenty-four hours, we find that the first hour of the second day returns to the sun, the first hour of the third day to the moon, and so on. Thus each day would be designated by the name of the divinity of the first hour.

Anyone can verify this process, and it is possible that it may be, in fact, the original cause of the designation.

The second system, of which the same author speaks, is a relation founded on music, and having for its basis the interval of a quarter. If, in fact, each planet represents a tone, commencing with Saturn, and omitting Jupiter and Mars, the fourth is given by the sun; then, omitting Venus and Mercury, by the Moon; then, omitting Saturn and Jupiter, by Mars; and so on. But this is a little strained.

Whichever of the three processes may have been used, the interesting point for us to know is that the division of time by periods of seven days is of the highest antiquity and due to the phases of the moon, but that it has not been in use among all nations, for the Greeks and Romans did not make use of it, the first having weeks of ten days (decades), and the second counting by kalends, ides, and nones. But it came into almost general use about the first century of our era, and the Latin etymology has remained:

Dies solis.
Lunæ dies.
Martis dies.
Mercurii dies.
Jovis dies.
Veneris dies.
Saturni dies.

Constantine, raising Christianity to the throne, changed the day of the sun to the Lord's day, and *dies Solis* became *dies dominica*, from which has come *dominche* and *dimanche*. We may explain all the other etymologies except the last, for there can hardly be any relation between *Saturni* and *samedi*. The day of Saturn was named among the Jews the Sabbath Day, and for a long time we preserved the *dies Sabbati*, which is even found in 1791 in the programme of the course of the College of France, written in Latin, like all the rest in past centuries. *Sabbaidi* may well have led to *samedi*. We may remark, however, that the Sun-god of the Assyrians and the Arabs is named *Sams* in the latter language, which during the Middle Ages was used in astronomical nomenclature. Can it be that from this word has been formed Samsdi, Samedi, the Samstag of the Germans? —while the English have kept for Samedi and for Dimanche the names of Saturday (day of Saturn) and Sunday (day of the Sun)? This is a possible explanation.

CHAPTER III

THE MOTION OF THE MOON ROUND THE EARTH

Weight and Density of the Moon. Gravity on the other Planets. How the Moon has been Weighed

THE moon revolving round the earth describes, not a perfect circle, but an ellipse (see p. 26). The eccentricity is small—not more than $\frac{1}{18}$. We may form an exact idea of this ellipse by considering that if we represent the lunar orbit by an ellipse having its major axis eighteen centimetres in length, the distance between the two foci would be one centimetre—that is to say, the distance from the centre to each of the foci would be only half a centimetre.[1]

This eccentricity is expressed geometrically by the number 0·0549. It is greater than that of the terrestrial orbit, which is 0·0167—that is to say, the lunar ellipse differs more from a circle than ours. The moon's distance, then, varies during the whole of its revolution, and of this we can assure ourselves by measuring the apparent diameter of its disc, the variations of which correspond with those of its distances from the earth. When the moon is at the extremity of the longer axis at the point nearest to the focus [occupied by the earth], its distance is a minimum ; it is then in perigee, and its diameter has its greatest value. At the other extremity of the same axis, or at apogee, the distance is, on the contrary, a maximum, and the diameter is smallest ; finally, at each of the extremities of the shorter axis the distance is a mean between the extremes, and it is the same with the magnitude of the disc. The following show the variation of the diameter and of the distance which results from this slightly elongated orbit :—

	Diameter of the Moon	Geometrical Distance	Distance in Kilometres	Distance in Leagues
Maximum distance, or apogee . .	29′ 31″·0	1·0549	405,000	101,250
Mean distance	31′ 8″·2	1·0000	384,000	96,000
Minimum distance, or perigee . .	32′ 56″·7	0·9451	363,000	90,750

[The latest results for these distances in English miles are as follow :—

Maximum distance, or apogee	251,968 miles
Mean distance	238,855 „
Minimum distance, or perigee	225,742 „]

[1] [For centimetres read inches, and the proportion will still hold good.—J. E. G.]

Thus, in fifteen days the distance of the moon varies from 225,742 to 251,968 miles—that is to say, about one-ninth. This difference is, as we see, perceptible in the moon's apparent size; it is noticed in eclipses of the sun, which are sometimes total and sometimes annular, and this difference of distance also acts on the tides.

If we deduct the radii of the earth and moon from the perigee distance, we shall find the smallest distance at which we can be from the *surface* of our satellite. This distance is 355,200 kilometres, or 220,000 miles. With these conditions a telescope magnifying 2,000 times would bring our satellite to forty-four leagues (110 miles).

The motion of the moon in space is still more complicated than that of the earth. Without entering into all the details, we will here notice the most curious particulars.

And (1) the ellipse described round us by this little globe does not remain motionless in its plane; it turns in this plane round the earth, in the direct sense—that is to say, in the same direction in which the moon moves. The major or longer axis of the ellipse thus makes a complete revolution in 3,232 days, or a little less than nine years. We see that this motion is analogous to that which the line of apsides of the terrestrial orbit performs in 21,000 years (which we have explained on p. 48), but more rapid.

(2) The orbit of the moon is not situated in the plane in which the earth moves round the sun—the ecliptic—for in that case, if our satellite revolved round us exactly in the plane in which we ourselves revolve round the sun, there would be an eclipse of the sun at every new moon and an eclipse of the moon at every full moon. The plane in which the moon moves is inclined 5 degrees to ours.[1] The line of intersection along which the two planes cut each other we call the 'line of nodes.' Well, then, this line of intersection does not remain fixed, but makes the circuit of the ecliptic in 6,793 days, or $18\frac{2}{3}$ years.

(3) The inclination itself of the plane of the lunar orbit varies. At the mean it is $5° 8' 48''$; but it is subject to an oscillation, which sometimes reduces it to $5° 0' 1''$, and sometimes raises it to $5° 17' 35''$, all the changes being passed through in 173 days.

It is not indispensable for our astronomical instruction to understand the precise mechanism of all these irregularities, but it is useful to know that they exist. We may add that the motion of our little satellite round us is disturbed by many other inequalities, such as (4) the *equation of the centre*, which causes the moon to oscillate every month, on account of the eccentricity of its orbit; (5) the *evection*, of which the period is 32 days;

[1] Hence in winter the full moon is at an altitude in the sky near the limit attained by the sun in summer, and even at certain times 5° higher. It is the contrary in summer, a season when the moon remains very low. At the December solstice the moon has, in our latitudes, an altitude of 69° to 70°.

(6) the *variation*, of which the period is 15 days; (7) the *annual equation*, of which the period is a year; (8) the *parallactic equation* of 29 days, which enables us to calculate the distance of the sun; without counting inequalities of 206 days, 35 days, 26 days, &c.

The analysis of the moon's motion has revealed the fact that it is accelerated by 12 seconds of arc per century. Half of this acceleration is due to the slow and progressive diminution of the eccentricity of the terrestrial orbit, and half to an imperceptible slackening of the earth's motion of rotation, which appears to increase the length of the day by one second in a hundred thousand years (!), and to apparently shorten the period of revolution of our satellite. If this acceleration should continue, the moon would end by falling on us! but it is only a periodical oscillation. We see how these motions have been studied, and to what precision modern science has attained. We see also how complicated are the fluctuations of this body, apparently so benign; on account of which motions it has become the veritable despair of geometers. Analysis has already discovered in this wandering body *more than sixty different irregularities!*

We sometimes meet in the examinations of the Paris University with professors who, taking a malicious pleasure in embarrassing the students, gain an easy victory by crushing with bad marks the candidates to whom they have set the most arbitrary questions. The complexity of the moon's motions has often served as a trap. But the examiners do not always get the upper hand. Arago relates that at the Polytechnic School Professor Hassenfratz had lost all claim to consideration on account of his character and his incompetency, and one day, fully prepared to embarrass a student, he called him to the table with a look which promised nothing good. But the student (it was M. Leboullenger) was on his guard, and knew that it was important to make a neat reply in order not to be vanquished.

'M. Leboullenger,' said the professor, 'you have seen the moon?' 'No, sir!' 'What! you say that you have never seen the moon?' 'I can only repeat my answer: No, sir.'

Beside himself, and seeing his prey escape him on account of this unexpected reply, M. Hassenfratz addressed the superintendent of the day: 'Sir, here is M. Leboullenger, who asserts that he has never seen the moon.' 'What do you wish me to do?' stoically replied the latter. Repulsed again, the professor returned once more to M. Leboullenger, who remained calm and serious in the midst of the suppressed mirth of the whole lecture-room, and exclaimed with undisguised anger, 'You persist in maintaining that you have never seen the moon?' 'Sir,' replied the student, 'I should mislead you if I said that I had not heard you speak, but I have never seen it!' 'Sir, go back to your place.'

After this farce Hassenfratz was a professor only in name; his teaching could no longer be of any use.

This little scene has diverted us for a moment from the complicated analysis of the moon's motions. In order to complete the account of these motions, and especially that we may form a correct idea of the course of our satellite, let us see what is the effect produced by the combination of the monthly motion of the moon round the earth with the annual motion of the earth round the sun.

If the earth were motionless, the moon would return at the end of its revolution to the point where it was at the commencement, and its orbit would be a closed curve, as in fig. 40. But it does not remain motionless. While the moon, for example, is at A, and moving towards B, going from new moon to first quarter, the earth is displaced towards the right, and in seven days it is carried with the moon seven times 1,600,000 miles in space. The first quarter happens at B (fig. 41). Seven days afterwards the earth is still farther on, and the full moon occurs at C; A week later the last quarter happens at D; and when, after having accomplished its complete revolution, our satellite returns to A, it has in reality described in space, not a closed curve, as in fig. 40, but

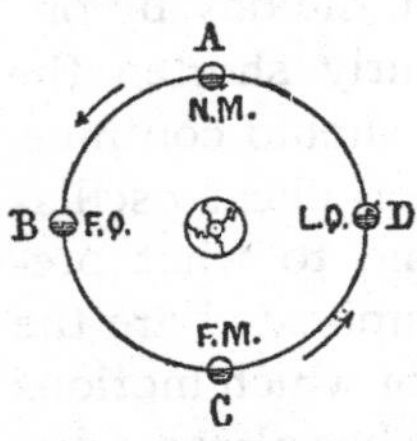

FIG. 40.—MOVEMENT OF THE MOON

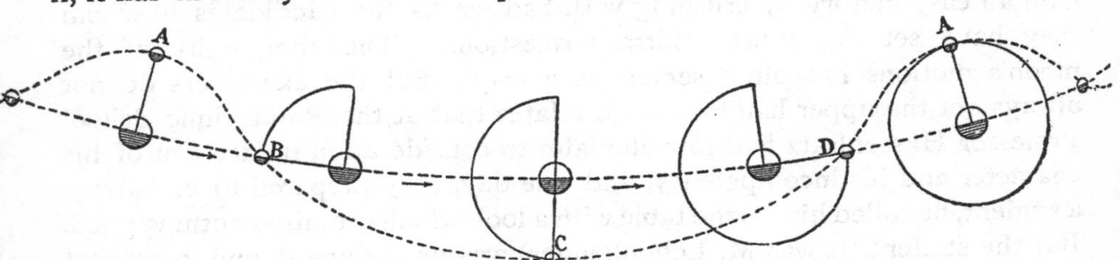

FIG. 41.—COMBINED MOVEMENT OF THE EARTH AND MOON

a line even more elongated than that of fig. 41, traced by joining by a series of points the positions A, B, C, D, A.

A rather curious fact, and one generally forgotten, is that this sinuous curve is so elongated that it scarcely differs from that which the earth annually describes round the sun; and instead of being (as it is always drawn in astronomical treatises) convex towards the sun, at the epoch of every new moon, it is *always concave* to the sun! I have represented it exactly (fig. 42) on the scale of 1 millimetre to 100,000 leagues. In this figure the arc of the terrestrial orbit is drawn with an opening of the compass of 37 centimetres for 37 millions.[1]

The attentive reader will himself add to this motion of the moon round the earth the motion of the sun in space, of which we have already spoken (p. 50), and in virtue of which the moon accompanies the earth in its

[1] This true form of the lunar orbit was drawn for the first time in 1876, in the first edition of our *Terres du Ciel.*

oblique fall towards Hercules, further complicating, by the motions we have just described, the curve shown in fig. 20.

Thus perpetual motion bears the world along! The sun moves through space; the earth moves round him, letting herself be carried along in his flight; the moon moves, circulating round us, while we gravitate round the radiant hearth which precipitates itself into the eternal void. Like a shower of stars the worlds whirl, borne along by the winds of heaven, and are carried down through immensity; suns, earths, satellites, comets, shooting stars, humanities, cradles, graves, atoms of the infinite, seconds of eternity, perpetually transform beings and things; all move on, all wing their flight under the breath divine—while trade goes on, or the investor counts his gold and piles it up, believing that he holds the entire universe in his casket.

O folly of terrestrial mannikins! folly of busy merchants, folly of the miser, folly of the suitor, folly of the pilgrim to Mecca or to Lourdes, folly of the blind! When shall the inhabitant of the earth open his eyes to see where he is, to live the life of the mind, and to base his happiness on intellectual contemplations? When shall he throw off the old man, the animal cover, to free himself from the fetters of the flesh and soar in the heights of knowledge? When shall astronomy shed its light upon all minds?—But the star of night recalls us. We already know its distance, its magnitude, its motions. We shall soon pay a visit to its rugged surface. Before undertaking this voyage there still remains, however, an interesting point to elucidate; it is the weight of this globe, and, in connection with that, the density of the materials which compose it, and the force of gravity at its surface.

How has the moon been weighed?

We can make the processes employed intelligible without entering into too technical details.

The weight of the moon is determined by an analysis of the attractive effects which it produces on the earth. The first and the most evident of these effects is presented by the tides. The water of the seas is raised twice a day by the silent call of our satellite. In studying with precision the height of the tides thus raised, we find

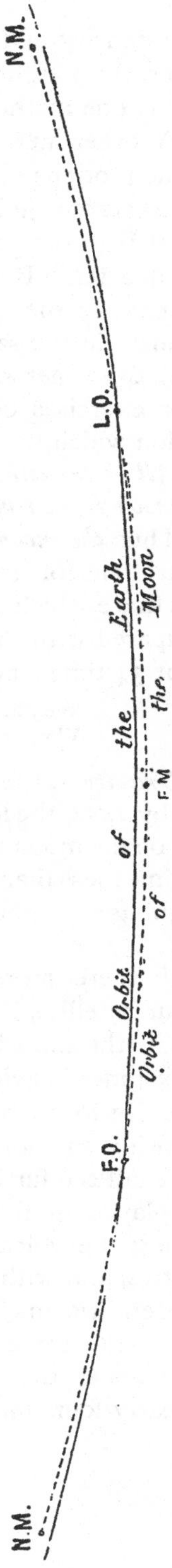

FIG. 42.—TRUE FORM OF THE ORBIT OF THE MOON

the intensity of the force necessary to raise them, and consequently the power, the weight (it is the same thing) of the cause which produces them. This is one method.

Another method is based on the influence which the moon exercises on the motions of the terrestrial globe; when it is in advance of the earth it attracts our globe and makes it move quicker; when it is behind, it retards it. It is in the position of the sun that this effect is shown, at the first and last quarter. It appears displaced in the sky by three-quarters of its parallax, or the 290th part of its diameter. From this displacement we calculate in the same way the mass of the moon.

A third method is based on the calculation of the attraction which the moon exercises on the equator, and which produces the nutation and precession which we have spoken of above (pp. 38 and 47).

All these methods verify each other, and agree in proving that the mass of the moon is 81 *times smaller than that of the earth.*

Thus *the moon weighs* 81 *times less than our globe.* Its weight is about 74 trillions of tons. The materials which compose it are less dense than those which constitute the earth—about 6-tenths of the density of ours. Compared with the density of water, the moon weighs 3·27—that is to say, about $3\frac{1}{4}$ times more than a globe of water of the same size.

Weight of the earth . .	6,000,000,000,000,000,000,000 tons [1]
Weight of the moon . .	74,000,000,000,000,000,000 „

Gravity at the surface of the moon is weaker than with us; if we represent by 1,000 the force which causes objects to adhere to the terrestrial globe, that on the moon would be represented by 164. Hence, objects weigh there six times less than here; they are attracted six times less strongly. A stone weighing one pound, if transported to the moon, would not weigh more than 3 ounces. A man weighing 11 stone on our planet would not weigh there more than 26 pounds. If we imagine a man transported to our satellite, if we suppose, moreover, that his muscular powers would remain the same in this new abode, he would be able to raise weights five to six times heavier than on the earth, and his own body itself would seem to be five to six times lighter. The least muscular effort would enable him to spring to enormous heights or to run with the speed of a locomotive. We shall see further on what a considerable part this weakness of gravity has played in the topographical organisation of the lunar world, by permitting the volcanoes to pile up giant mountains on Cyclopean amphitheatres, and with a powerful hand to toss Alps upon Pyrenees.

Here we may remark a rather curious fact; it is that if the moon, having the same mass, were as large as the earth, since the attraction decreases as the square of the distance, and the radius of the lunar sphere is nearly four times less than that of the terrestrial sphere, the attraction

[1] [This may be expressed as 6×10^{21} tons].

would be diminished nearly 16 times, and instead of being reduced only to the sixth of terrestrial gravity, it would not be more than the 90th. A pound would not then weigh more than 77 grains; a man of 140 terrestrial pounds in weight would not weigh more than about a pound and a half! The muscular effort which we make to jump on a stool would enable us to reach in a bound the height of a mountain, and the least force of volcanic projection would toss the materials so high in the lunar sky that they would never return. . .

There may exist worlds of which the mass is so small and the motion of rotation so rapid that gravity does not exist on their surface, and bodies would there weigh *nothing*. On the other hand, there may exist worlds of which the density is so enormous that objects would have a tremendous and truly inconceivable weight. Let us suppose, for example, that without changing its volume the earth became as heavy as the sun. Then a pound would henceforth weigh 324,000 lbs., and a young, slender and graceful girl, whose present weight is 110 lbs., would be found to weigh *sixteen thousand* tons! Or, in other words, were she made of bronze, she would be crushed into an indefinite number of molecules scattered on the ground. Would nature, notwithstanding her infinite power, be capable of organising beings sufficiently energetic to resist such a gravity?

What marvellous diversity must exist—from this fact alone—among the various worlds which people Infinitude!

Before proceeding further, let us form an exact idea of these curious differences in the intensity of gravity on the planets of the solar system. We will calculate later on the weights and the volumes.

Comparative Intensity of Gravity at the Surface of the Planets

The Sun	27·47	Uranus	0·88	
Jupiter	2·58	Venus	0·86	
Saturn	1·10	Mercury	0·52	[0·272.—J.E.G.]
The Earth	1·00	Mars	0·37	
Neptune	0·95	The Moon	0·16	

Hence it is on the moon that the intensity of gravity is weakest, and on the sun that it is strongest.[1] If transported to the former of these bodies, a terrestrial kilo. would not weigh more than 2½ ounces, while on the sun it would weigh 27 lbs., on Jupiter 2½ lbs., &c. But we shall appreciate better these differences of intensity if we express them by the distance which a body will pass over—a stone, for example—if allowed to fall from the top of a tower. The following is the distance which would be passed over in the first second of the fall on each of the planets we are considering:—

[1] We have said above that the volume of the moon is the forty-ninth part of the earth. If we could distribute the matter of the moon all round the earth as we spread a heap of gravel in a thick layer on the walks of a park or garden, the added burden which would result for our earth would be a thickness of 43 kilometres (about 26¾ miles).

Space passed over by a Falling Body during the First Second of the Fall

	M		M
On the Moon	0·80 (2·62 feet)	On Neptune . .	4·80 (15·75 feet)
On Mars	1·86 (6·10 „)	On the Earth . .	4·90 (16·07 „)
On Mercury	2·55 (8·36 „)	On Saturn . .	5·34 (17·52 „)
On Venus	4·21 (13·81 „)	On Jupiter . .	12·49 (40·98 „)
On Uranus	4·30 (14·10 „)	On the Sun . .	134·62 (441·67 „)

Let us imagine, then, that we let a stone fall from the top of a tower, and suppose its height to be 42·65 feet. At the end of the first second of the fall the stone would have nearly arrived at the foot of the tower on Jupiter, where bodies are attracted with a great intensity. In the same time it would not have reached the middle of the tower on Saturn. It would have passed over 16 feet on the earth, 4 inches less on Neptune, 14·1 feet on Uranus, 13·81 feet on Venus, 8·36 feet on Mercury, 6·10 feet on Mars, and only 2·62 feet on the moon, so weak is the attraction there. As to the sun, in order to represent the same force at its surface, it would be necessary to suppose the tower built on the top of a steep mountain, and overlooking the plain at a height of 440 feet. In one second our block of stone, attracted by an enormous force, would be, at one rapid bound, precipitated through the whole distance.

These calculations are made without taking into account the resistance of the atmosphere, which reduces more or less, according to its density, the velocity of the fall. But gravitation, or weight, is ruled by the same laws throughout the whole universe. Perhaps, however, there may exist in nature forces which we do not know, and which in certain worlds play a part analogous to gravity, but different in their effects from that force. For example, if we could forget the existence of the magnet, we could never imagine that a magnet could attract to itself, in opposition to gravity, objects of iron. There is nothing to forbid us imagining that iron, which enters in a small quantity into our blood and our flesh, may exist in much larger proportions in organisms constituted otherwise than we are, and that, under influences analogous to those of the magnet, these beings might be attracted by a special force independent of gravity. We are not prevented from imagining the existence of natural forces other than those of the magnet, which, in certain worlds, may modify the effects of gravity, and may even push living beings towards the upper regions of the atmosphere. But experimental science can at present only calculate the masses, the volumes, the densities, and the gravity, as we have done. When shall we be able to discover the living beings which exist in these diverse worlds under so many different forms? When shall we see and know them? Nature, O immense, fascinating, infinite Nature! Who can divine, who can hear, the sounds of thy celestial harmony! What can we include in these childish formulæ of our young science? We lisp an alphabet while the eternal Bible is still closed to us. But it is thus when all reading begins, and these first words are surer than all the antique affirmations of ignorance and human vanity.

CHAPTER IV

PHYSICAL DESCRIPTION OF THE MOON

The Mountains, the Volcanoes, the Plains called Seas. Selenography. Map of the Moon. The Ancient Revolutions of the Moon.

THE moon has not ceased to be a problem for the earth. The human mind is insatiable for knowledge. It is of its essence to desire to penetrate into the nature of things, and to make conjectures on all points which it cannot thoroughly comprehend. How pleasant it would be to know what is going on in a world so near to us as the moon! For what is the distance of 238,000 miles which separates us in comparison with the distance of the stars, which is estimated at billions and thousands of billions of miles in the celestial spaces? Our pride, already flattered by knowing that our globe is the ruler of this province, would be infinitely more so if it could be proved that this satellite is peopled with intelligent beings capable of understanding and appreciating our planet, of which the benefits to them are only comparable with those they receive from the sun.

Most of the philosophers of antiquity have given their opinion about the moon; not having sufficient means of observation, they reasoned according to simple common-sense. Some suspected that it had no proper light of its own, and shone with a brightness borrowed from the sun's rays. Such was the opinion of Thales, Anaximander, Anaxagoras, and Empedocles. The latter philosopher, from what Plutarch says, concluded that it was on account of its reflection that the light of the moon reaches us less keenly, and without producing perceptible heat. Proclus, in his 'Commentary on Timæus,' quotes three verses attributed to Orpheus, in which it is said that 'God builds another immense world, which the immortals call *Selene*, and which men call the moon, in which rise a great number of mountains, a great number of towns and habitations.' The doctrine of Xenophanes was exactly similar to that of Orpheus. Anaxagoras speaks of the fields, the mountains, and the valleys of the moon, but without mentioning towns, or habitations.

Pythagoras and his disciples were much more explicit on this question, for they assure us that 'the moon is an earth similar to that which we inhabit, with this difference, that it is peopled with larger animals and more beautiful trees, the lunar beings surpassing in their height and strength fifteen times those of the earth.' Diogenes of Tarsus attributed to Heraclides of Pontus a very singular assertion : according to that historian, Heraclides affirmed that he had knowledge of a lunar inhabitant having descended on the earth, but he abstains from giving the description. A tradition may be added that the Nemean lion had fallen from the moon. Even in the sixteenth century, did not the astrologer Cardan assert that one evening he received a visit from two inhabitants of the moon? They were, he says, two old men, nearly dumb. This singular man was, moreover, so honestly convinced of the astrological dogmas that, his horoscope having predicted the day and hour of his death, he divided his property in consequence, was reduced to poverty, and died of starvation.

Other ancient philosophers took the moon for a mirror reflecting the earth from the height of the sky. Nevertheless, the great question of an atmosphere and waters on the surface of the moon, which is still discussed to-day, was already started in the time of Plutarch. This writer quotes in the following terms the opinion of those who maintained the opposite: 'Is it possible that those who are in the moon can support for long years the sun, during fifteen days every month, darting his rays full upon their heads? It is not supposable that with such great heat, in the midst of such rarefied air, they have winds, clouds, and rains, without which plants could not spring up, nor last when they have grown, when we see that the most terrible hurricanes do not rise from the depths of our atmosphere even to reach the summits of our highest mountains. The air of the moon is itself so rarefied and changeable, owing to its great lightness, that each of its molecules escapes from the aggregation, and nothing can condense into clouds.' This argument differs but little from that used by modern writers who declare that the moon is uninhabitable.

Dissertations with reference to the moon and its inhabitants were then so much the fashion that this philosopher has drawn up a special treatise ('De facie in orbi Lunæ'), in which he states most of the opinions expressed in his time; and Lucian of Samosata has written, as a critic, a lunar voyage as amusing as his witty dialogues of the dead.

During all the Middle Ages, and up to the invention of the telescope, there was almost an end to serious dissertations on our satellite. Galileo, in 1609, used the first telescope, which he constructed for astronomical observations, to study the nature of the moon; he recognised in it a globe, covered with many sinuosities, whose extraordinary and deep valleys are overshadowed by very elevated mountains.

The first drawing made of the moon was certainly a rough representation of the human figure, seeing that the positions of the spots correspond sufficiently well with those of the eyes, the nose, and the mouth, to justify the resemblance. Thus we see everywhere, and in all ages, this human face reproduced. This resemblance is merely due to the accidental geographical configuration of our satellite ; it is, moreover, very vague, and is effaced at once when we examine the moon with a telescope. Other imaginations have seen, instead of a head, an entire body, which some represent as Judas Iscariot, others as Cain carrying a bundle of thorns, others as a hare, &c.

The principal spots on the moon are perceptible to the naked eye, but the number of those which we can distinguish with telescopes is incomparably greater.

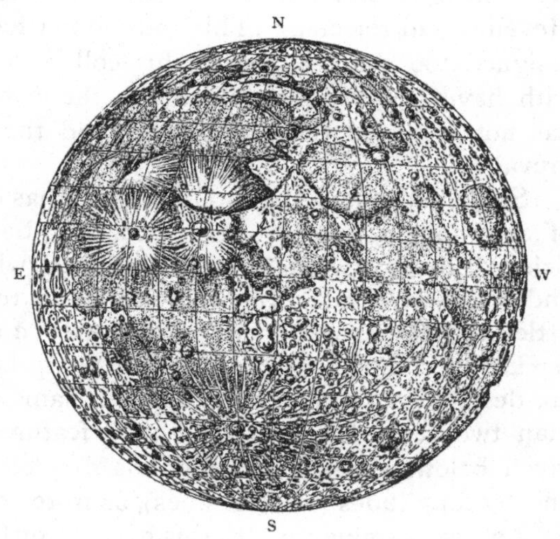

FIG. 43.—APPEARANCE OF THE MOON AS SEEN BY THE NAKED EYE OR THROUGH AN OPERA-GLASS

In order to take in with the naked eye the whole of the lunar disc, we should choose the epoch of the full moon in preference. It is important, at first, to take one's bearings well. Suppose that, for this purpose, we look at the moon at that epoch about midnight—that is to say, at the moment when it crosses the meridian and shines full in the south. The two extreme points of the vertical diameter of the disc give the north and south points of the disc, the north being at the top and the south at the bottom. To the left lies the east point, and to the right the west point. If we view it with the aid of an astronomical telescope, the image is *reversed*; the south lies at the top, and the north at the bottom, the west to the left, and the east to the right. *This latter orientation is that of all maps of the moon.*

Astronomers have succeeded in drawing maps of the moon just as geographers have constructed terrestrial maps, and we may even say that the former have always been more exact than the latter. This is easily explained : we see the moon, we do not see the whole of the earth.

The first map of the moon was drawn in 1647, by the astronomer Hevelius. He was so anxious for scrupulous accuracy that he undertook the task of engraving it himself. When it became necessary to give names to the different spots which the map contained, he hesitated between the names of celebrated persons and those of the different countries of the then known world. He frankly admits that he gave up taking the names of men for fear of making enemies of those who might have been totally forgotten, or who might find that they were given too small a part.' He decided, then, to transport to the moon our seas, towns, and mountains. Riccioli, who made a second map of the moon some time after Hevelius, showed more boldness, and, in a map which was the fruit of the observations of his colleague and friend, Grimaldi, he adopted the nomenclature which Hevelius had rejected. This astronomer has been reproached with having assigned too great a part to his colleagues in the Society of Jesus, and with having placed himself among the favoured scientists. But posterity has not regarded these criticisms, and the nomenclature of Riccioli has prevailed.

Since that epoch the moon's surface has been studied by a great number of astronomers, notably, in our century, by Beer and Mädler, Lohrmann, Schmidt, Neison, Gaudibert [Birt, Elger], who have constructed maps more and more in detail. In order easily to recognise the details we give a little map (fig. 47, page 122), which will be useful for reference.

Let us commence by placing this map before us. The great grey plains are designated by the name of seas, a name which they have borne for more than two centuries. The principal features are marked, and the names given below. Lunar geography is divided by latitudes (horizontal lines) and by longitudes (vertical lines), as in terrestrial geography.

Let us examine rapidly this general surface. We remark at first that the great grey and sombre spots occupy chiefly the northern or lower portion of the disc, while the southern or upper regions are white and mountainous. However, on one side, this luminous tint is found again on the north-west border, as well as towards the centre; and, on the other hand, the spots encroach on the southern regions on the eastern side at the same time that they descend, but less deeply, on the west. We will first follow on the map the distribution of the grey plains or seas, and sketch the lunar geography.

Let us begin our description with the western part of the lunar disc—that which is first illuminated after the new moon, when, as a thin crescent, it is visible in the evening sky, and enlarges day by day, becoming the first quarter on the seventh day of the lunation (this is the right-hand side to the naked eye and the left on the map). There, not far from the border, we distinguish a little spot, of an oval form, isolated from everything else in the midst of the luminous ground. To this has been given the name of the *Mare Crisium*.

It is not necessary to attach to this name of *Sea* [Mare] any special meaning; it is the common denomination under which the earlier observers

Fig. 44.—Photographs of the Moon taken with the Great Refractor of the Lick Observatory, California

have designated all the greyish spots on the moon ; they took these spaces for large tracts of water. But we now know that there is no more water there than in the other lunar regions. They are vast plains. Everything leads us to believe that they are ancient seas, now dried up.

The situation of the Mare Crisium near the western limb of the moon permits us to recognise it with the naked eye in the first phases of the lunation and up to full moon ; for the same reason, it is the first to disappear at the beginning of the decrease.

To the right of the Mare Crisium, and some distance to the north, is seen another spot, larger, and of an irregularly oval form, which is also easily recognised with the naked eye ; this is the *Mare Serenitatis*, or *Sea of Serenity*.

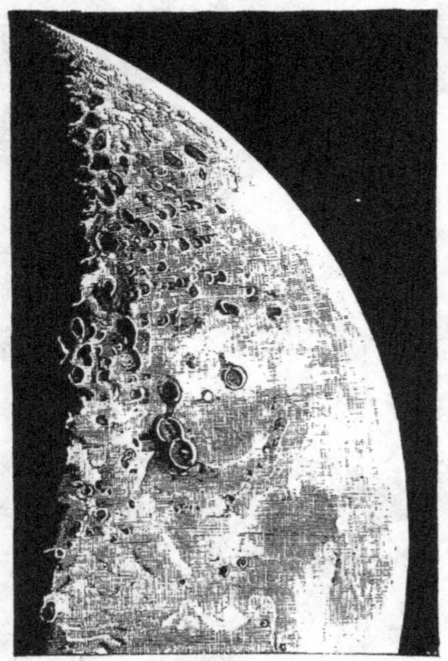

FIG. 45.—LUNAR CRESCENT, SIX DAYS AFTER NEW MOON

Between these two grey plains above we remark another of which the boundaries are less regular, which is named the *Mare Tranquillitatis*, or *Sea of Tranquillity*. It runs out towards the centre of the disc in a gulf which has received the name of the *Mare Vaporum*, or *Sea of Vapours.*

The Sea of Tranquillity separates into two branches, which represent the legs of a human body, as is sometimes imagined. The nearest branch to the border forms the *Mare Fœcunditatis* ; the nearest to the centre is the *Mare Nectaris*.

We distinguish further, below the Sea of Serenity, and in the vicinity of the North Pole, a straight spot, elongated from east to west, and known under the name of the *Mare Frigoris*, or *Sea of Cold*.

Between the Seas of Serenity and of Cold extend the *Lake of Dreams* (*Lacus Somniorum*) and the *Lake of Death* (*Lacus Mortis*), doleful echo of astrology. The marshes of *Putrefaction* and *Fogs* (*Palus Putredinis* and *Palus Nebularum*) occupy the western part of the Sea of Rains (*Mare Imbrium*), of which the northern border forms a round gulf, designated by the name of the *Gulf of Iris* (*Sinus Iridium*).

All the part of the lunar disc situated to the east is uniformly dark. The boundaries of this immense spot disappear and are merged into the luminous parts of the body. The northern part of this spot is formed by the *Sea of Rains* (*Mare Imbrium*), which gives birth to a great gulf opening out into the *Ocean of Tempests* (*Oceanus Procellarum*), where shine two great craters, *Kepler* and *Aristarchus*. The more southern regions of this ill-defined ocean are designated towards the centre by the name of the *Sea of Clouds* (*Mare Nubium*), and towards the limb by that of the *Sea of Humours* (*Mare Humorum*).

It is important from a selenographical point of view that **most of these plains have rounded borders**; examples: Mare Crisium, the Sea of Serenity, and even the vast Sea of Rains, bordered on the south by the Carpathians, on the south-west by the Apennines, on the west by the Caucasus, and on the north-west by the Alps.

Outside these spots, which cover about a third of the lunar disc, the observer can only distinguish with the naked eye confused luminous points. However, in the upper region we can recognise with the naked eye the principal mountain of the moon—the crater *Tycho*, which shines with a vivid white light, and sends out rays to a great distance round it.

Let us not forget the remark made above: the maps of the moon are drawn reversed, as we see it in a telescope; to compare the moon as seen with the naked eye with our map, it is necessary to turn it upside down, to place the north at the top, and the west to the right.

All these lunar lands have been accurately measured. The surface of the hemisphere which we see at the moment of full moon is 1,182,500 square leagues. The mountainous portion, which is the more general, covers 830,000 square leagues, and the region occupied by the grey spots includes 352,500 square leagues.

The angular diameter of the moon being 31′ 8″ (see p. 85), and its real diameter being 2,163 miles (p. 88), a second of arc represents 6,119 feet, and a minute represents 70 miles. The proportion diminishes from the centre to the circumference, since the moon is not flat, but spherical, and the perspective of the projection increases as we approach the borders. On the moon a degree is equivalent to 18·8 miles, the circumference of the lunar globe measuring 6,795 miles.

Such is the first general aspect of lunar geography, or *selenography*.

Let us now sketch the character of the numerous mountains which bristle over this surface.

It is sufficient to observe the moon with a telescope of small power to recognise at once that its surface presents very pronounced inequalities. The moon seen with a small telescope two days before the first quarter gives us an idea of this rudimental aspect. The irregularity of the interior border shows well the rugosity of the surface. We see, moreover, up to a certain distance from this border, circular cavities obliquely illuminated and with very characteristic shadows. These shadows, observed on several consecutive days, increase or diminish in extent and intensity according as the obliquity of the solar rays on the corresponding part of the moon's surface varies in one direction or the other. It has thus been known since the

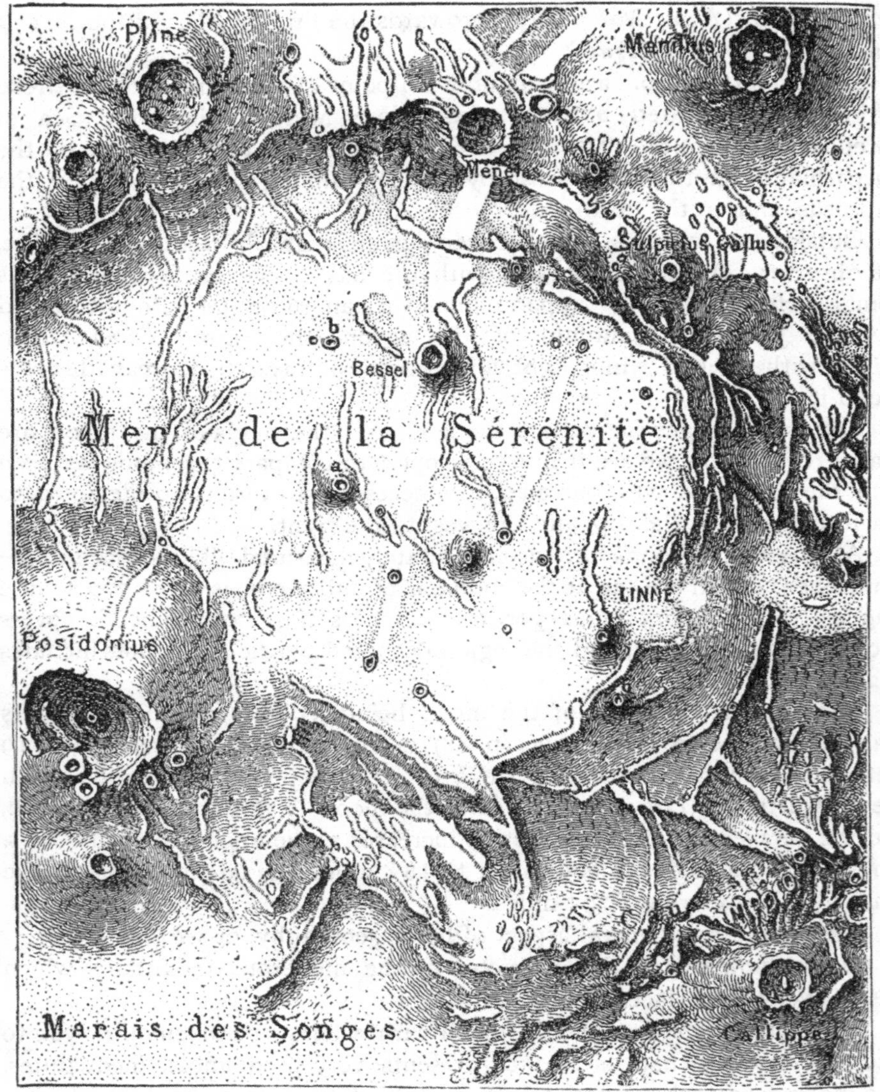

FIG. 46.—LUNAR TOPOGRAPHY. SEA OF SERENITY AND LAKE OF DREAMS

beginning of observations that the moon is a solid globe studded with craters.

I drew in 1865 a very curious lunar region (the Sea of Serenity and its environs), which gives an exact idea of the diversity which exists between the country of plains and the mountainous country in this little neighbouring world. As we shall see further on, the attention of astronomers has been especially directed to this region by a change which has probably happened to the little crater Linné (on the right bank of the sea); but the aspect of this drawing (fig. 46) shows clearly, on the one hand, the sandy, wrinkled, undulating nature of the soil of the lunar 'seas, on the other hand the crateriform nature of all the mountains.

If we wish to appreciate from a geological point of view the whole of the mountainous formations, we must consider the southern region of our satellite.

We distinguish with the naked eye in the lower part of the moon a very brilliant white point, from which rays start. A simple opera-glass will show it admirably. This is the famous mountain *Tycho.* It occupies, with the lines which radiate from it in all directions, the centre of the southern region of the lunar disc, and it is with this that it is natural to commence the description of the mountains of the moon. It is one of the most colossal and majestic of all these formations. It presents a yawning crater in the form of an amphitheatre which measures 85 kilometres (about 53 miles) in diameter, and it may be distinguished by the aid of an astronomical telescope of moderate power.

This annular mountain appears to be the great centre where volcanic action has had the greatest intensity. There the boiling lavas, instead of uniting to form layers, are maintained as they were at the epoch when they felt the volcanic force.

At the time of full moon Tycho is surrounded by a luminous halo so radiant that it dazzles the eye and prevents us from observing the geological curiosities of the crater.

If we wish to form an idea of the aspect of the lunar mountains, we may examine in detail a typical annular mountain, such, for example, as that of Copernicus, which is one of the most beautiful and interesting in the whole moon. This vast amphitheatre measures 90 kilometres (56 miles) in diameter. At the full moon, rays diverge from it as in the case of Tycho. When the sun does not illuminate it fully, we can distinguish the central mountains which rise from the bottom of its crater, and the slopes of the annular amphitheatre which form the enclosure. The interior of the crater, rather steep, moreover, shows itself as a triple enclosure of broken rocks, and a great number of large heaped-up fragments at the foot of the escarpment, as if they were masses detached from the top of the mountain and rolled down. The bottom of the amphitheatre is almost flat; but at the centre are still seen the ruins of the central peak and a quantity of fallen rubbish. (See fig. 48.)

This is indeed the type of all lunar mountains. They are all hollow. The sides of the mountain, which surround each amphitheatre, are cut out almost to the peak to a depth which varies from ten to thirteen thousand feet. There is in the lunar Alps—mountains which are exceeded in height by the Caucasus and Apennines of the Moon—a remarkably large transverse valley, which cuts across the chain in a direction from south-

Crater of Albategnus Crater of Eratosthenes

FIG. 47.—THE FULL MOON, AS SEEN THROUGH A TELESCOPE

I. Glacial Sea; II. Gulf of Dew; III. Gulf of Flowers; IV. Marsh of Fogs; V. Sea of Rains; VI. Carpathian Mountains; VII. Ocean of Tempests; VIII. Midland Sea; IX. Sea of Clouds; IXb. Sea of Vapours; X. Sea of Darkness; XI. Altai Mountains; XII. Sea of Fertility; XIII. Sea of Tranquillity; XIV. Sea of Sleep; XV. Sea of Serenity; XVI. Lake of Dreams; XVII. Lake of Death; XVIII. Humboldt Sea; A. Black Lake; B. Valley of Endymion.

east to north-west. It is bordered by summits more elevated above the bed of the valley than the Peak of Teneriffe is above the level of the sea. We may remark that the height of this peak is 3,700 metres (about 12,000 feet).

The heights of all the mountains of the moon have been measured to within a few yards (we cannot say as much for those of the earth). The following are the highest :—

Mount Leibnitz	. . . . . .	7,610 metres	(about 24,970 feet)
Mount Dörfel	. . . .	7,603 „	(„ 24,941 „)
Crater of Newton	. . . . . .	7,264 „	(„ 23,830 „)
Crater of Clavius	. . . . .	7,091 „	(„ 23,260 „)
Crater of Casatus	. . . . .	6,956 „	(„ 22,820 „)
Crater of Curtius	. . . . .	6,769 „	(„ 22,208 „)
Calippus (Caucasus)	. . . .	6,216 „	(„ 20,390 „)
Mount Huygens (Apennines)	. . .	5,560 „	(„ 18,240 „)
Short (near Newton)	. . . .	5,500 „	(„ 18,045 „)
Crater of Tycho	. . . . . .	5,300 „	(„ 17,390 „)

The mountains Leibnitz and Dörfel lie near the south pole of our satellite. These two chains are sometimes seen in profile during eclipses of the

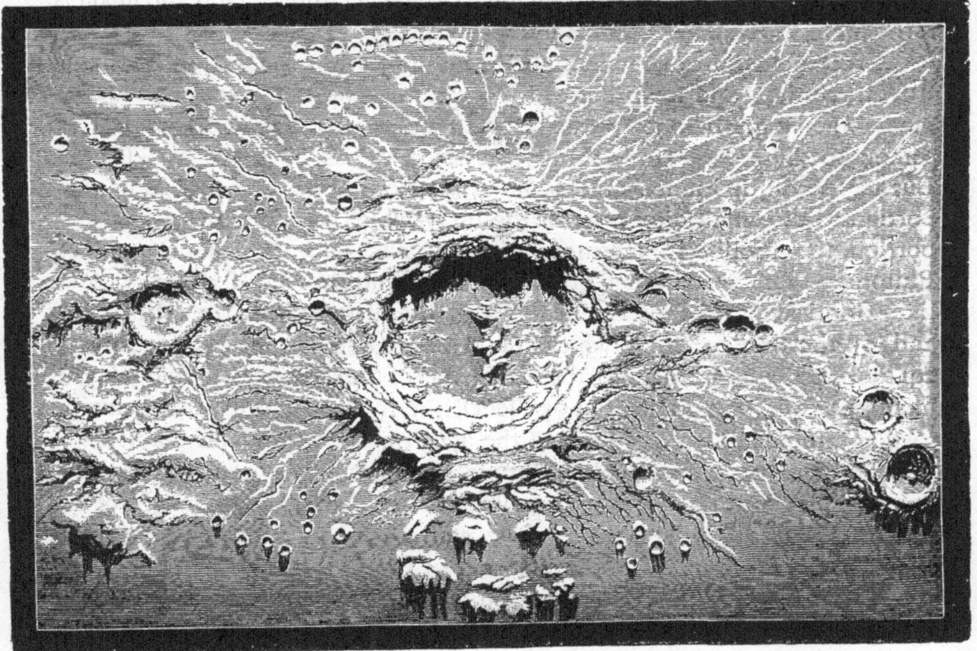

FIG. 48.—THE LUNAR CRATER COPERNICUS. (From a Drawing by Nasmyth)

sun; I observed and drew them during the eclipse of October 10, 1874. At the lunar poles (where, moreover, we see neither snow nor ice) there are mountains so strangely situated that their summits know no night; the sun *never* sets on them. They may be called *the mountains of eternal light.*

How immense are these lunar craters! The largest terrestrial volcanoes in activity do not attain a diameter of a thousand yards. If we consider the ancient amphitheatres due to previous eruptions, we see that in Vesuvius the exterior amphitheatre of Somma measures 11,800 feet; and on Etna, that of the Val del Bove measures 18,000 feet Some amphitheatres formed by extinct volcanoes show larger dimensions; such are, for example, the Pyrenean amphitheatre of Heas, of which the diameter is 29,500 feet, of which the ramparts have a height of 2,790 feet, and of which the steps would afford room for six millions of men; the amphitheatre of Cantal, of which the diameter is 32,800 feet, that of Oisans in Dauphiny, which is not less than 65,000 feet, and finally that of the island of Ceylon, the most spacious on the globe, of which the diameter is estimated at 43½ miles.

But what is even this in comparison with the extent of many craters on the moon? Thus, the circle of Clavius shows a diameter of 130 miles; that of Schickard more than 124 miles; that of Sacrobosco 100 miles; that of Petavius exceeds 93 miles, &c. We may count on our satellite a score of amphitheatres of which the diameter exceeds 60 miles. And the moon is forty-nine times smaller than the earth!

As to the height of the mountains, the highest on the satellite are, it is true, three thousand feet inferior to those of the planet. [Neison, however, makes the height of one of the Leibnitz mountains nearly 36,000 feet, or some 7,000 higher than our Mount Everest, the highest in the world.—J. E. G.] But this small difference renders the lunar mountains stupendous in proportion to the small dimensions of the body which supports them. In proportion, the satellite is much more mountainous than the planet, and the Plutonian giants are in much greater number than here. If we have peaks like Gaorisanker (Mount Everest), the most elevated of the Himalayas and of the whole earth, of which the height, 8,840 metres (29,000 feet), is equal to the 1,440th part of the diameter of our globe, we find on the moon peaks of 7,600 metres (24,935 feet), like those of Dörfel and Leibnitz, of which the height is equivalent to the 470th part of the lunar diameter.

In order that the comparison should be exact, however, it is necessary to suppose the waters of the sea to disappear, and to take the height of the lands reckoning from the bottom of the seas; the height of the Alps above the bottom of the Mediterranean, or that of the Pyrenees above the bed of the Atlantic, is thus notably increased. According to marine soundings, we may estimate that the highest summits of the globe are doubled. The height of the Himalayas, then, above the bed of the seas represents not the 1,440th, but the 720th part of the diameter of the globe.

This correction does not, however, prevent the lunar mountains from being relatively much higher than terrestrial mountains. In order that our mountains should have the same proportionate altitude, it would be necessary that the summits of the Himalayas should be raised to a height of 13 kilometres (about 8 miles). It is, then, as astonishing to see on the moon summits of over four miles high as it would be to see on the earth mountains of double that height.

The mountains of the moon are of volcanic origin. This is an important fact, which is shown directly by the rounded annular form of the great valleys, the amphitheatres, and all the smaller cavities to which has been given, as we have seen, the name of craters.

The existence of these craters, the rugged form of these amphitheatres, their enormous size, and their prodigious number, prove that the moon was formerly, like the earth, and even more so than our world, the seat of tremendous revolutions. It also began in the fluid state, then cooled down and became covered with a solid crust.

This crust has been modelled by geological phenomena, of which traces remain to-day under the form of asperities of very various dimensions. The causes of this series of formations are undoubtedly the expansive forces of the internal gas.

At first, the solid crust of the moon, less thick, was for that reason less resisting, and as it was not yet disturbed, it probably had at all points almost the same homogeneousness and the same thickness. The expansive force of gas and vapours acting, then, perpendicularly to the superficial layers, and following the lines of least resistance, would burst open the crust and produce upheavals of a circular form. It is doubtless to this primitive period that we must attribute the formation of the immense circumvallations, the interior of which is now occupied by the plains called seas. We have already remarked the circular form of the Mare Crisium, and those of the Seas of Serenity, of Rains, and of Humours. Their walls, half ruined by subsequent revolutions, form still the longest series of asperities on the lunar soil, chains of mountains like the Carpathians, the Apennines, the Caucasus, and the Alps, and Mounts Hæmus and Taurus.

Then came new upheavals, which, occurring at an epoch when the crust of the lunar globe had acquired a much greater thickness, or else proceeding from less powerful elastic forces, gave rise to large amphitheatres, but very inferior in dimensions to the primitive formations. Such appear to be the amphitheatres of Shickhard, Grimaldi, and Clavius.

There appeared afterwards the innumerable [1] craters of medium dimensions which swarm all over the moon, and of which many are formed even in the depths of the primitive circumvallations. We can easily understand the reason of the successive diminution of these geological rings. Each of them is due to the upheaval of a bubble; now, the dimensions of these blisters would be in proportion to the intensity of the internal force which produces them, and to the resistance of the solid, or rather pasty, crust of the lunar globe. It is probable that these two causes have co-operated to produce the effects mentioned above, so that, in general, it was the largest circumvallations which were formed first.

We may also remark that the lunar soil shows two very distinct aspects.

[1] The astronomer Schmidt, of Athens, to whom we owe an immense map of the moon, has counted 33,000 craters on the lunar hemisphere turned towards us. We may conclude that there are about 60,000 on the whole lunar globe.

The first, of a white colour, represents that which has been named from the beginning the continental soil; it is that of the mountainous districts which cover nearly all the southern region. Its porous structure, its great reflecting power, and especially its elevation above the plains, clearly distinguish it from the level land, of which the dark colour and the smooth surface give all the appearances of alluvial plains. True seas must have covered these regions. The banks remind us of the action of water. What has happened to these seas? They must have been from the beginning of less importance and shallower than the terrestrial oceans, and it is probable that they have been slowly absorbed by the porous soil on which they rested. Perhaps there may still remain some liquid and some humidity in the low lands.

One of the most remarkable lunar regions is the chain of the Apennines which borders the vast Sea of Rains—of which this part bears the inelegant and very unmerited name of the 'Sea of Putrefaction.' This vast chain of mountains measures no less than 720 kilometres (about 447 miles) in length, and its highest summits exceed 16,000 feet. These heights, illuminated by the sun and casting their great black shadows, are truly wonderful to behold on the day of the first quarter and the day before and after. The great yawning crater which appears below is Archimedes, of which the diameter is 83 kilometres (51½ miles), and the height 1,900 metres (about 6,235 feet). Besides this, we notice two other craters: the first, to the west (the upper one), is Autolycus; the second, below, is Aristillus. Aristarchus is another fine crater (fig. 49).

This same region shows the curious clefts ('rills') which open across certain lunar plains. One commences at the south rampart of Archimedes and extends nearly 150 kilometres (about 93 miles), at first about a kilometre and a half wide, then diminishing; the other begins at the opposite side of the same crater, and runs in a winding form towards the north. These fissures are several miles in depth, and in certain places landslips block up the bottom; their walls are nearly perpendicular. Two other considerable clefts run along the Apennines in the sunlight or in the shadow of the giant mountains, bordered by precipices of frightful depth, peaks projecting their black profiles to a distance of more than 130 kilometres[1] (about 80 miles).

We see that there is an essential difference of form between the lunar and the terrestrial mountains. All the lunar mountains are hollow, and their bottom almost always descends below the mean exterior level, the height of the ramparts, measured externally, not being more than a half or a third of the true depth of the crater. Some terrestrial districts, however,

[1] In our work, *Les Terres du Ciel*, will be found a photograph of this region, a direct photograph of the moon made two days after the first quarter. In this work will also be found *details* of the study of the lunar world and of the different planets of our system.

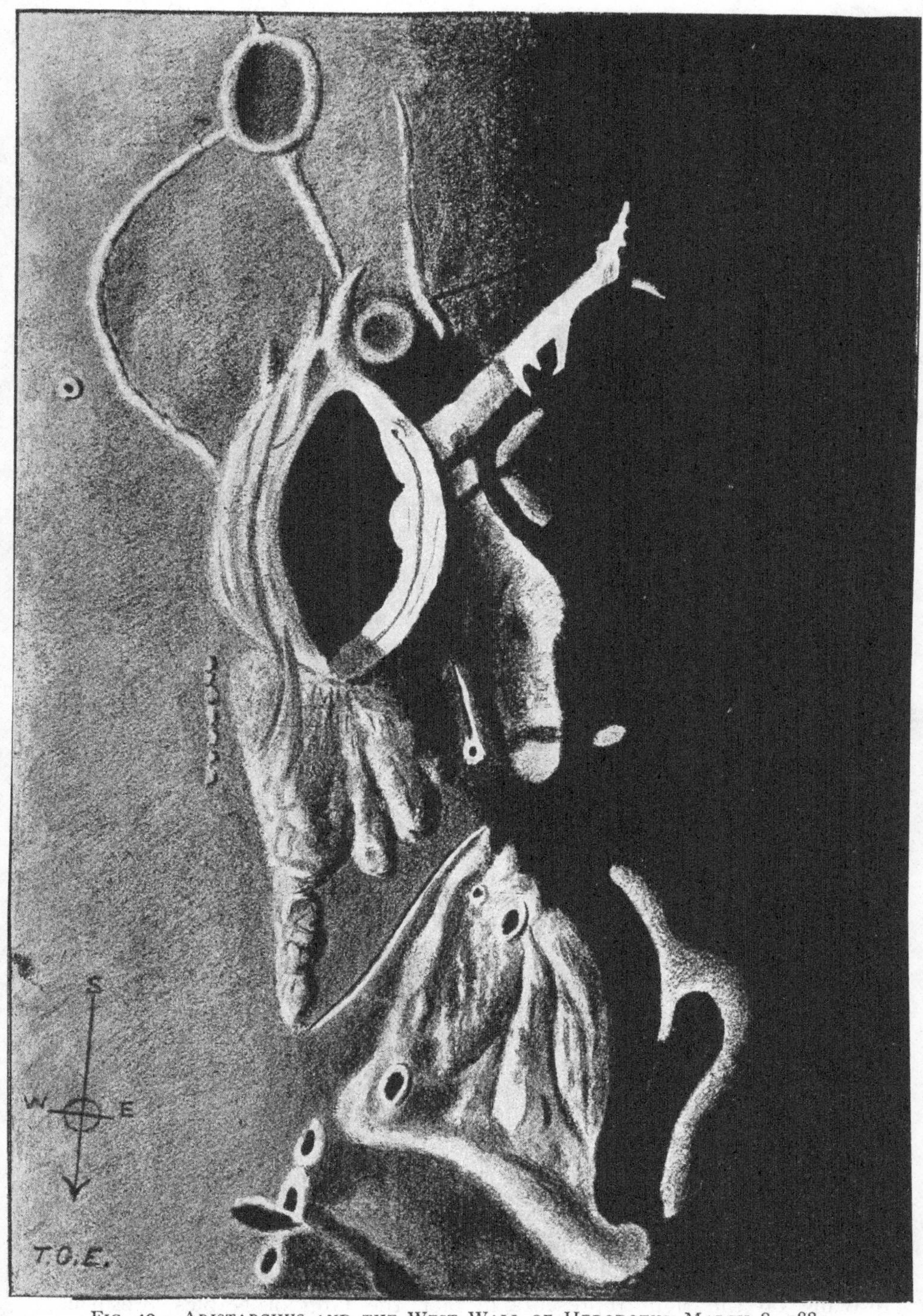

FIG. 49.—ARISTARCHUS AND THE WEST WALL OF HERODOTUS, MARCH 8, 1884
(Drawn by T. G. Elger. Powers 350–400, on $8\frac{1}{2}$-inch Calver Reflector)

show an apparent resemblance to certain parts of the lunar surface, a resemblance which would appear still more marked if these regions could be observed with a telescope. The example which is generally mentioned is Vesuvius with the neighbouring country, known by the name of the Phlegræan fields. This analogy is even so striking that we might call the

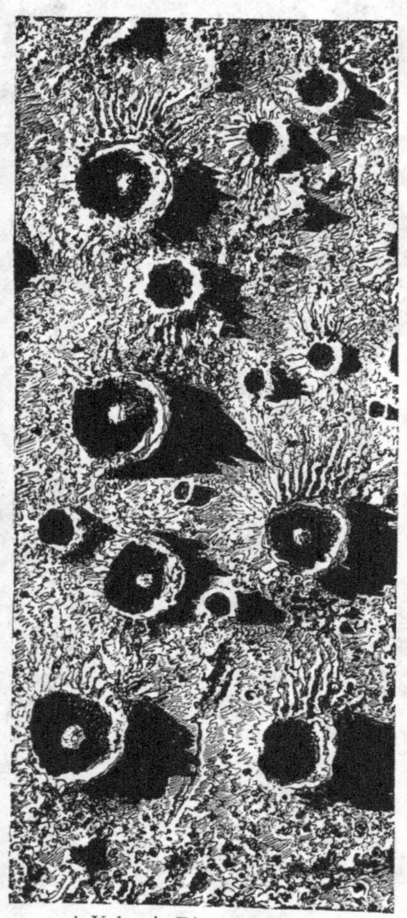

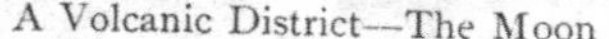
A Volcanic District—The Moon

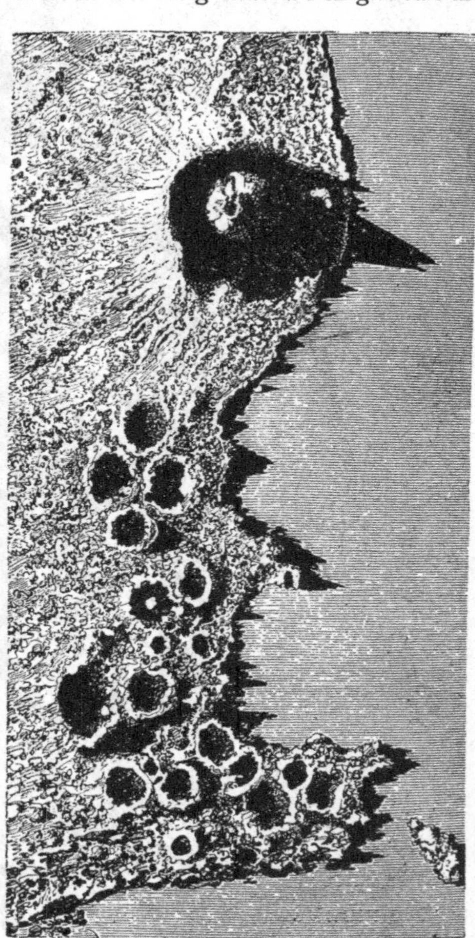

Relief-Map of a Volcanic District near Naples

FIG. 50.—TOPOGRAPHY OF LUNAR AND TERRESTRIAL VOLCANIC DISTRICTS COMPARED

moon a vast Phlegræan field. Our readers may understand this by an examination of fig. 50, drawn from two photographs of models in relief, representing side by side a lunar volcanic district and a terrestrial volcanic district—a comparison due to Messrs. Nasmyth and Carpenter. The drawing on the right represents, in fact, the Gulf of Naples, Vesuvius, Solfatara, Pozzuoli, Cumæ, Baiæ, up to the island of Ischia. It is a model in relief, a skeleton of the animated and luxurious landscape of Naples, placed on an anatomical table, and obliquely illuminated by the solar light like the lunar relief with which it is compared. Vesuvius, which is one of the largest European volcanoes, would be on the moon merely one of the small craters, hardly visible round Copernicus and the other lunar giants. This disproportion might make us even doubt the volcanic character of the lunar craters, if we did not observe, as on the earth, the central cone, which is unquestionably produced by the final efforts of the volcanic mouth shooting forth with its last breath the enfeebled emissions of a dying-out fire.

The type of lunar mountains is represented in our figure as it might be seen on the moon itself. We recognise in certain terrestrial volcanic formations aspects quite lunar. Sometimes, in the Alps, the Jungfrau, seen from Interlaken, is illuminated in such a manner at sunset that it

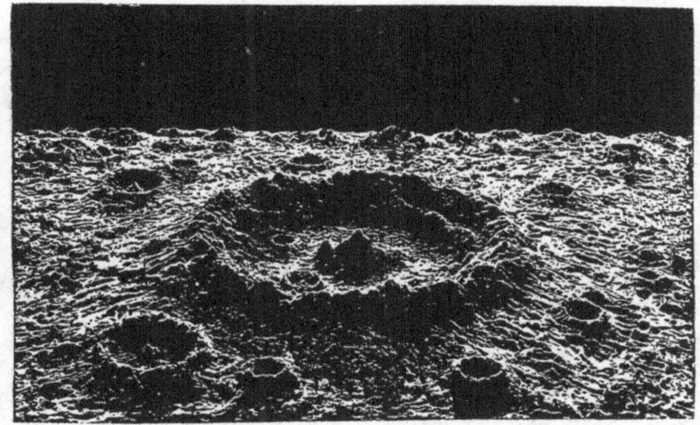

FIG. 51.—A TYPICAL LUNAR MOUNTAIN

singularly reminds one of certain lunar alps. The illusion is almost complete in the extinct crater near Mount Hecla (Iceland), drawn in fig. 52. We here seem transported to the moon at the epoch which preceded the disappearance of the waters.

Without going farther, however, we have even in the centre of France, on the ancient table-lands of Auvergne, cones of extinct volcanoes which represent on a small scale what the lunar world shows us on a large scale almost over its whole surface.

We see, then, that between the earth and the moon there is only a difference of degree, due to the special nature of our satellite, and especially to the weakness of gravity on its surface.

The lunar landscapes in the mountainous districts must present a truly grand and entirely special character. Summits succeed summits, illuminated by the sun in an aërial perspective scarcely perceptible, and in a strange day, which is lit up without extinguishing the stars in a sky constantly crepuscular.[1]

The topographical descriptions we have given, and the considerations which result from them, apply only to the lunar hemisphere which we see. Everybody knows, in fact, that we always see the same face of the moon, and that there is one side of the lunar globe which no inhabitant of the earth has ever seen or ever will see. In revolving round us our satellite *constantly shows us the same side*, as if it were attached to the earth by a

FIG. 52.—AN EXTINCT CRATER IN ICELAND: IMAGE OF A LUNAR LANDSCAPE BEFORE THE DISAPPEARANCE OF WATER

rigid bar. It is not completely freed from our attraction, and it simply turns round the terrestrial globe, as we do ourselves if we make a journey round the world. In the same way as we have our feet always against the earth, so its feet, or its lower hemisphere, is always turned towards the earth. A balloon making the circuit of the world gives an accurate illustration of the motion of the moon round the earth: it performs slowly one turn on itself during its voyage, since when it passes the Antipodes its

[1] Illuminated by twilight.

position is diametrically opposite to that which it had at the point of departure, in the same way that our Antipodes have a position diametrically opposed to ours. Thus the moon performs one rotation on itself in the same time that it completes its revolution round the earth. Otherwise, if it did not turn at all upon itself, if it circulated round us, keeping its absolute orientation fixed, we should see successively all sides of it during its revolution.

From the fact that the moon always presents the same face it has been

FIG. 53.—EXTINCT VOLCANOES IN AUVERGNE: IMAGE OF A LUNAR LANDSCAPE

concluded that it is elongated like an egg in the direction of the earth. One of the astronomers who have been most occupied with the mathematical theory of the moon, Hansen, has even come to the conclusion that the centre of gravity must be situated at a distance of 59 kilometres (about 36½ miles) beyond the centre of figure; that the hemisphere which we look at is in the condition of a high mountain, and that 'the other hemisphere may possess an atmosphere as well as all the elements of vegetable and animal life,' considering that it is situated below the mean level.

We have said that the moon always shows us the same face, but this is only generally speaking, for, as it sometimes moves a little more quickly, sometimes a little more slowly, while sometimes it is a little lower, sometimes a little higher, it allows at times a little of its right side to be seen, and at times a little of its left side, on one day a little beyond its upper pole, and on another day a little beyond its lower pole. This is what they call its rockings, or *librations*, which attain 6° 51′ in latitude and 7° 54′ in longitude. It follows that we thus see a little more than a half: the part always hidden is to the part visible in the ratio of 42 to 58.

The lunar topography is the same on these eight-hundredths of the other hemisphere as on the whole surface of the visible one. It is, then, probable that the other hemisphere does not differ essentially from ours as to its geology. Undoubtedly it would be much more agreeable to know how this hemisphere is really constituted ; but we can hardly hope to go there while we live.

CHAPTER V

THE ATMOSPHERE OF THE MOON

Conditions of Habitability of the Lunar World

WE have seen that the lunar world offers, from a geological point of view, remarkable analogies with ours, with, however, essential differences due to the exaggeration of its volcanic character. We will now enter a little further into the examination of its physical constitution. And, firstly, the aërial atmosphere which envelops our globe and bathes its whole surface in its azure fluid is intimately connected with life; it is that which adorns the arid soil with a sumptuous vegetable carpet, with dark and animated forests, with verdant meadows, with numerous plants enriched with flowers and fruits. Through this descend the fertilising rays of the sun, which form the cloud with flaky outlines; through this the rain pours out its streams, the storm bursts, and the rainbow throws its brilliant bow over the transparent and perfumed landscape. It is this which glides through our life, gives us the aërial fluid which we breathe in our lungs, starts the frail existence of the new-born infant, and receives the last sigh of the dying man stretched upon his bed of pain. The atmosphere is certainly the most important of all the elements which compose a world, and which we call its physical constitution. Without an atmosphere, without this gaseous envelope, whence organised beings incessantly draw what nourishes their own existence, it is impossible for us to conceive anything but immobility and the silence of death. Neither animals nor vegetables, nor even the lowest organisms, seem capable of living and developing elsewhere than in the midst of a fluid, elastic and mobile, of which the molecules are in continual exchange with their own bodies. Undoubtedly, we are very far from knowing all the conditions under which life is manifested, but, without leaving the domain of observed facts to enter that of pure imagination, we are obliged to acknowledge that the existence of an atmosphere seems to us one of the most essential conditions for the existence of organised beings.

I say *seems to us*, for it has not been proved that nature is incapable of producing beings organised to live without air. It has been absolutely denied. We do not contradict it. But the reason of our reservation is not less easy to understand. If, before we observed any of the innumerable living beings which people the waters of our planet, and before hearing their existence spoken of, we learned, all at once, that it is possible to be born, to breathe, and to move in the midst of the waters; if we had confidence in our own experience, which taught us that prolonged immersion in a liquid is deadly, this news would cause us the most profound surprise. Such would be our astonishment should we ever come to demonstrate by unexceptionable proofs the existence of living beings on the surface of the moon. Nature is so varied in its modes of action, so multiple in the manifestations of its power, that we have no right to set any limits to its capabilities.

No question has been more keenly and variously controverted than that of the existence of an atmosphere round the moon. The solution should unequivocally inform us whether our satellite can be inhabited by beings endowed with an organisation *analogous* to ours.

The attentive observation of this neighbouring globe was not long in showing that, if an atmosphere exists round the moon, that atmosphere can never give birth to any cloud like those in the midst of which we live; for these clouds would conceal from us certain portions of the moon's surface, and there would result variations of aspect, white spots more or less extended, and endowed with various motions. But this disc is always presented to us with the same aspect, and nothing ever stands in the way of our constantly perceiving the same details.

We thus know that the atmosphere of the moon, if it exists, always remains perfectly transparent. But we may go further than that. Every atmosphere produces twilights. One half the moon receiving directly the light of the sun, the solar rays which would illuminate the heights of this atmosphere, above the regions still in the night, would diffuse along the dark border a certain light, increasing gradually up to the illuminated hemisphere. The moon, seen from the earth, should thus present an imperceptible shading of light along the terminator circle. Now, there is nothing of the kind; the illuminated part and the dark part are separated from each other by a clearly cut line. This line is more or less sinuous and irregular, on account of the mountains, but it shows no trace of this degradation of light. We see, then, that if the moon has an atmosphere it must be very feeble, since the twilight which it would produce is quite imperceptible.

We will describe another and more precise method of estimating this atmosphere. When the moon, in virtue of its proper motion on the celestial sphere, passes in front of a star, we can fix the precise instant of the disappearance of the star and also the exact instant of its reappearance, and thus find the duration of the occultation of the star. On the other hand, we can accurately determine by calculation what line the star follows behind the lunar disc during its occultation, and deduce the time which the moon takes to advance in the sky by a quantity equal to this line. Now, if the rays of the star were ever so little disturbed in their course by the refraction of an atmosphere, instead of disappearing at the exact instant when the moon touches it, the star would still remain visible for some time after, because the rays would be deflected by the lunar atmosphere; for the same reason, the star would begin to reappear at the opposite side some time before the interposition had completely ceased. The duration of the occultation would thus be necessarily diminished from this cause. But we generally find a complete agreement between calculation and observation. Besides, the light of the star does not suffer any diminution. We know from this that the atmosphere of the moon, if it exists, is less dense at the edge of the lunar hemisphere than the air which remains in the receiver of an air-pump when a vacuum is made.

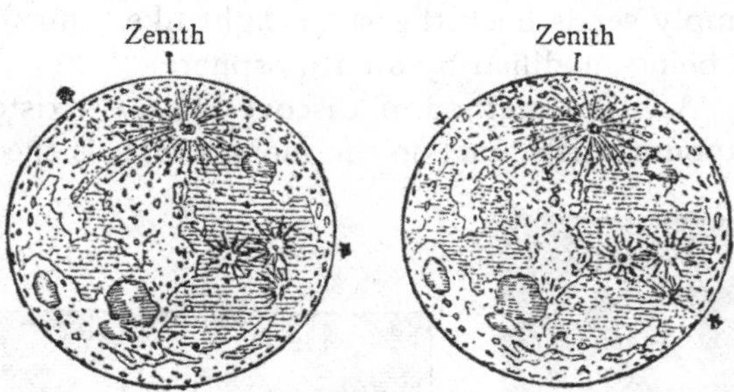

FIG. 54.—OCCULTATION OF STARS BY THE MOON

On the other hand, again, when the moon passes before the sun and eclipses it, its outline is always presented absolutely sharp and without penumbra.

I have observed with care, for this purpose, several eclipses and occultations, notably the occultation of the planet Venus by the moon on October 14, 1874, at three o'clock in the afternoon, in a very clear sky and in full sunshine. The beautiful planet appeared in the telescope as a thin crescent of the same form as that of the moon, then at its fourth day—a little larger relatively, very visible, and clearly defined. The moon took 1 hour 14 minutes to pass over it. The three principal moments of the immersion of Venus behind the lunar disc and those of its emersion are represented in the two drawings of fig. 54. There was not the faintest penumbra nor the slightest deformation indicating the presence of the least lunar atmosphere.

Jupiter, Saturn, and Mars are also from time to time occulted by the moon which passes in front of them. We may mention, among others, the occultation of Saturn on April 9, 1883 (fig. 55). We observed hardly anything then but phenomena of diffraction, which do not prove a lunar atmosphere.

Spectrum analysis, of which we will soon explain the principle and the processes, has been applied with particular care in the search for traces of a lunar atmosphere. If this atmosphere exists, it is evident that the solar rays traverse it once before reaching the lunar soil, and a second time in being reflected towards the earth. The spectrum formed by the light of the moon should then show rays of absorption added by this atmosphere to the solar spectrum. Now, all the observations made prove that the moon simply sends back the solar light like a mirror, without the slightest trace of being modified by an atmosphere.

Another method of discovering the existence of any atmosphere with vapours, mists, &c., on the moon's limb is the examination of the spectrum

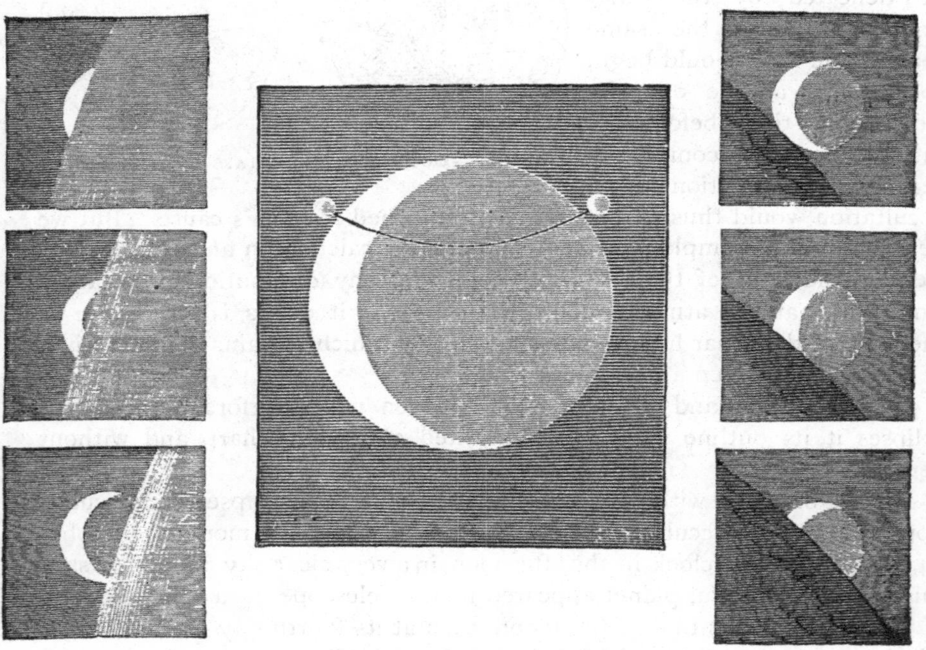

FIG. 55.—OCCULTATION OF VENUS BY THE MOON, OCTOBER 14, 1874

of a star at the moment of occultation. The least gas would modify the colour of this spectrum, as well as certain lines, and the spectrum would not disappear instantaneously without experiencing some modification. We have here, then, a new proof that if a lunar atmosphere exists, it is not perceptible at the moon's limb.

Such are the facts that militate against the existence of a lunar atmosphere. Having explained them, it is now important to declare that they are not sufficient to *prove the total absence of air* at the surface of our satellite; and we may notice certain observations which tend, on the contrary, to show that there may exist there some atmosphere, feeble and small, but real. We generally believe, on account of being taught so, that there can be there merely the shadow of an atmosphere, and that it cannot produce any vital manifestation analogous to ours. This proposition is much too general.

In fact, it is at the border of the lunar disc that occultations of stars occur, and this border is formed by the summits of mountains projected one upon the other ; a low-lying plain rarely arrives at the moon's limb without being hidden. Now, it is precisely in the low grounds, and not on the heights, that we should search for this atmosphere.

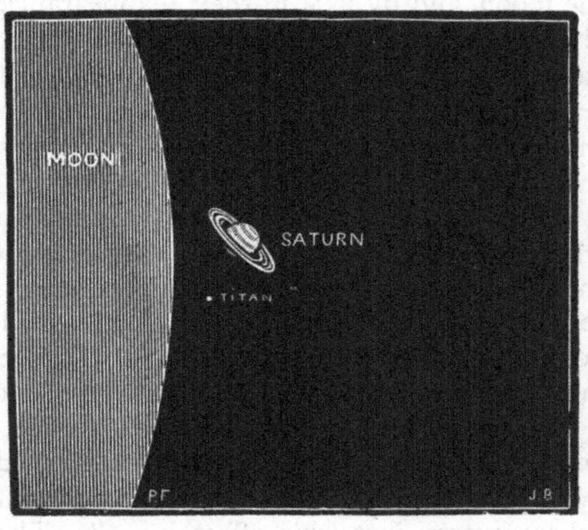

FIG. 56.—OCCULTATION OF SATURN BY THE MOON, APRIL 9, 1883

As early as the end of the last century, Schröter observed that the summits of the lunar mountains, which show themselves on the unilluminated border like detached points, are much less luminous than those which are found at a greater distance from the line of separation of shadow and light, or, which comes to the same thing, according as the illuminating rays have grazed the lunar soil to a greater extent.

One evening, while he was observing the thin crescent of the moon, two and a half days after new moon, it occurred to him to see whether the dark outline of this body, which can only be seen as an ashy gleam, would show itself all at once before the weakening of our twilight, or only by parts. Now, it happened that the dark limb showed itself at once in the prolongation of the two horns of the crescent over a length of 1′ 20″ and a width of about 2″, with a very feeble greyish tint, which gradually lost its intensity and its width going towards the east. At the same moment the other parts of the dark limb were totally invisible, notwithstanding that, being more distant from the dazzling portion of the crescent, they should have been seen first. A gleam reflected by the atmosphere of the moon on the portion of that body which the solar rays had not yet reached directly —a veritable twilight gleam—seems alone capable of explaining this phenomenon.

Schröter found by calculation that the twilight arc of the moon, measured in the direction of the tangent solar rays, would be 2° 34′, and that the atmospheric layers which illuminate the extremity of this arc should be

452 metres (about 1,486 feet) in height. This observation has been repeated several times since.

On the other hand, in discussing 295 carefully observed occultations, the astronomer Airy has concluded that the semi-diameter of the moon is diminished by 2″·0 in the disappearance of stars behind the dark side of the moon, and by 2″·4 in their reappearance, also at the dark limb. The observations relating to occultations at the bright limb give larger values for the semi-diameter than had been expected *à priori*, as much on account of the extreme delicacy of these determinations as on the irradiation of the lunar limb, which extinguishes the light of a star before the contact.

This excess of the telescopic diameter is usually attributed to irradiation, which enlarges it to the eye. 'However, there is nothing to prove that the lunar atmosphere does not add something to the difference,' says, with reason, Mr. Neison [now Nevill], my learned colleague of the Royal Astronomical Society of England; 'and if we compare the diameter so accurately determined by Hansen with that which is inferred from the occultations observed from 1861 to 1870, we find a correction of 1″·70, which it does not appear should be justly attributed to irradiation. It would be more satisfactory to admit that the horizontal refraction of a lunar atmosphere enters into this effect by 1″. The semi-diameter of the moon calculated in total eclipses of the sun, when the irradiation of the moon is nil, and, on the contrary, when the solar light diminishes the diameter of the black moon, agree with this hypothesis.' Such is also the opinion of the Director of the Royal Observatory, England.

On the other hand, the absence of refraction, which we have explained just now, is not absolute. That in occultations stars have been seen projected on the disc of the moon is an unquestionable fact, and one rather frequent; but this fact is produced by diffraction (see the 'Revue d'Astronomie,' 1886, p. 286, article by M. Trépied); nevertheless, certain delays appear rather to be attributed to the refraction of an atmosphere. At the time of an occultation of Jupiter, May 24, 1860, a dark line, which might very well have been produced by an atmosphere, ran along the lunar limb, and was projected on the disc of Jupiter.

The lunar border is not always presented under the same conditions, on account of the librations of the moon, of which we have spoken. We do not always see the same point, and there are, moreover, tremendous variations of temperature, which would have a great influence on the state of the atmosphere.

During the total eclipse of the sun on May 17, 1882, M. Thollon, a skilful spectroscopist, thought he observed a strengthening of the solar spectral rays close to the moon.

Now, what would be the extent of a lunar atmosphere which would produce a horizontal refraction of 1″? Our satellite is in a peculiar condition of density, of gravity, and of temperature. Its surface passes in turn from torrid heat to glacial cold, as we have seen. The maximum temperature of the western limb occurs about the eighth day of the lunation, and its minimum temperature about two days before new moon, while the maximum temperature of the eastern limb happens the day after the last quarter, and its minimum temperature two days before full moon.

The height of the lunar atmosphere might be about 32 kilometres (about 20 miles), according to the calculations of Mr. Neison; its density at the surface, at 0° of temperature and at the ordinary pressure, would be $\frac{23}{10000}$ compared with the density of the terrestrial atmosphere at the level of the sea and at zero. This atmosphere would give a refraction of 1″·27 at the unilluminated lunar limb, supposing a temperature of 30° of frost, 1″·03 at zero, and 0″·86 at the illuminated limb, at a temperature of 30° Centigrade.

Such a state of things would be in accordance with the different observations made in occultations, and no fact contradicts this hypothesis. The extent of this atmosphere will be better understood if we remark that its weight on a surface of a thousand square miles (1,609 metres in the side) would be 882 millions of pounds. It would be, in proportion to the mass of the moon, an eighth of that which the terrestrial atmosphere is in proportion to the earth.

Such an atmosphere is not insignificant, and it may exist.

The density of the air on any planet depends on the attraction of the planet. All weights on the earth would be doubled if the terrestrial attraction were doubled, and diminished by one half if this attraction were reduced by one-half, and so on; now, this fact applies as well to the atmosphere as to all other matter. If the terrestrial gravity were reduced to that of the moon, the atmospheric pressure and the density of the air would be reduced to one-sixth of their present state; a given quantity of air, at the level of the sea, would occupy more space, and the whole atmosphere would expand in a corresponding proportion; it would rise six times higher. If, then, the moon had an atmosphere constituted like ours, this atmosphere would be six times higher than ours; at the mean level of the lunar plains the pressure would be equal to a sixth of that of our air at the level of the sea. Thus, although the Selenites would have as much air per square foot as we have, they would have, nevertheless, an atmosphere irrespirable by us. If, now, we suppose it to be differently constituted, and of a density six times greater than ours, it would only have, on account of the feebleness of the lunar gravity, the density of that which we breathe, and it would rise as high.

I have many times observed, especially in the very disturbed region which extends to the north of the Hyginus Cleft, a variable grey tint which, if it is not a simple optical effect, may be produced either by mist or by vegetation. On the other hand, I have very often had the impression of a twilight effect when observing the vast eastern [? western] plain of the Sea of Serenity on the sixth day of the lunation. To the north the irregular oval amphitheatre of the Caucasus, and to the south the chain of Menelaus, stand out like two luminous points in an ordinary opera-glass. The illuminated border of the plain does not end suddenly by an abrupt line, clearly separating the light from the shadow, but degrades softly, *as if the level had sunk*. This is a veritable penumbra. Calculation shows that the solar disc should produce by its width a penumbra of 32′ of arc of a great circle on the moon, which would make a width of about ten miles. But I have often remarked there a penumbra very much wider. Fig. 56, drawn at the Observatory of Harvard College, United States, gives an idea of this degradation of tint at the illuminated limb.

We may add other facts still. Thus, on February 1, 1887, an English observer, Mr. T. Gwyn Elger, noticed that the shadows of the elongated peaks on the floor of Plato were stumped ('Revue d'Astronomie, 1887, p. 209).

Upon the whole, then, there may (and there should) exist on the moon *an atmosphere of feeble density*, and probably of a composition very different from ours. Perhaps there may also exist certain liquids, such as water, but in a minimum quantity. If it had no air at all there could not exist a single drop of water, seeing that it is the atmospheric pressure which maintains water in the liquid state, and that without it all water would immediately evaporate. It is possible, after all, that the lunar hemisphere which we never see may be richer in fluids than the visible one. But we see that in any case it would be contrary to the real interpretation of facts to assert, as is too often done, that there is absolutely no atmosphere nor any liquid or fluid on the surface of the moon.

We may add now that this world, very near as it is, has conditions of habitability very strange to us. We have already seen that at its surface living bodies, or othres, have scarcely any weight, and that everything must have there the easiest mobility. The atmosphere being, on the other hand, extremely light itself, there would not be a celestial vault as here, no sky, azure or otherwise, never any clouds, but a void, unfathomable and without form, in which an infinite number of stars shine during the day as well as the night. The light and the heat received from the sun are there of the same intensity as here, since the moon and the earth revolve in space at the same distance from the sun (what is 238,000 miles to 93 millions?—almost nothing); but their effects are very different, because the atmosphere is not sufficient to moderate them. In full sunshine the light is intense, harsh, and fatiguing; in the shade it is almost nil, a sinister reflection from illuminated rocks. In the first situation the heat is intolerable; in the second, they experience a glacial cold. Here the atmosphere serves as a greenhouse-protector, preserving the heat received during the day, and the winds moderate the different extremes of temperature. In the moon, on the contrary, all the heat received during the day escapes without hindrance as soon as the sun is absent, and the night brings on a glacial cold. The lunar organisms could only live by being constituted to bear without pain these tremendous contrasts, which would be so perilous to us.

In this singular little world the days and nights are nearly *thirty times longer* than in ours. The revolution and the rotation of the moon on itself relatively to the sun being, as we have seen, 29 days 12 hours 44 minutes—that is to say, about 709 hours, this is also the total duration of the day and night in this strange world: the day properly so called, from the rising to the setting of the sun, lasts 354 hours, and the night as long; the sun takes no less than 177 hours to rise from the eastern horizon to the highest point at noon, and as long to descend to the west. What a long day! And never a single cloud to moderate the heat of this everlasting sun.

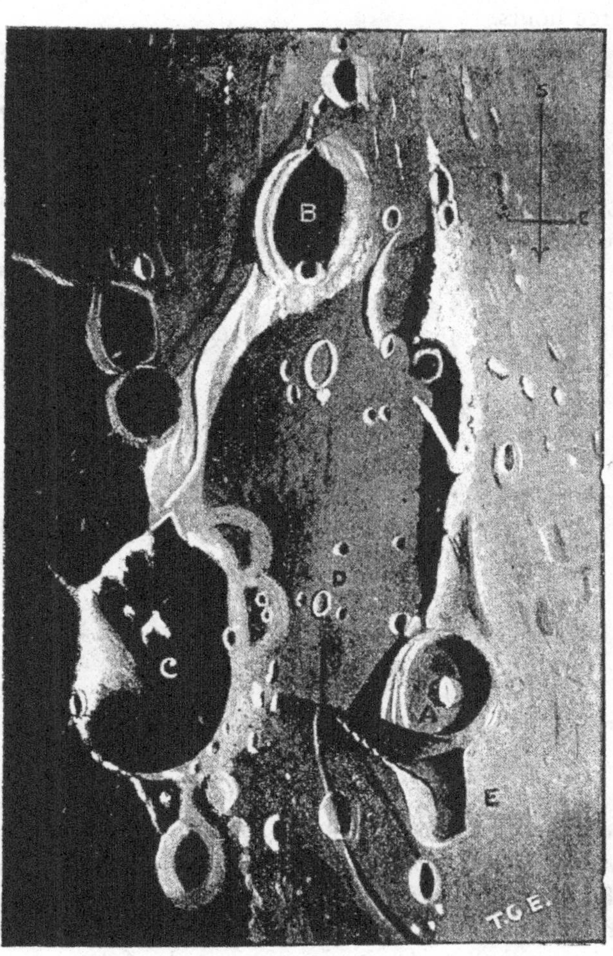

FIG. 57.—VENDELINUS, DECEMBER 17, 1891; 9 HOURS TO 9 HOURS 30 MINUTES. (Drawn by T. G. Elger)

In the whole universe we do not know of days and nights so long [except, possibly, in Japetus, the outer satellite of Saturn.—J. E. G.].

The rarity of the lunar atmosphere permits the stars to shine during the day as in the night. They are seen to turn slowly round the lunar pole, which is near the pole of our ecliptic, and situated in the head of the Dragon, only it turns a little quicker than the sun, in 27 days 7 hours 43 minutes, instead of 29 days 12 hours 44 minutes. Here the solar day exceeds the sidereal day by four minutes; there the difference is fifty-three hours.

But while the lunar day is very much longer than ours, the lunar year is shorter: it consists of 346 terrestrial days, or less than twelve lunar days (11·74). Thus on this neighbouring globe there are scarcely *twelve days in the year!*

A being moving on the moon would feel himself extremely light; he could run with the velocity of the swallow's wing, climb without effort the steepest mountains, jump from precipices, and throw stones or projectiles to wonderful distances. While on the sun's surface the most powerful of our pieces of artillery could scarcely throw a ball a few yards, the solar attraction stopping it almost immediately on its leaving the cannon's mouth, a good lunar slinger would throw a stone over the mountains.

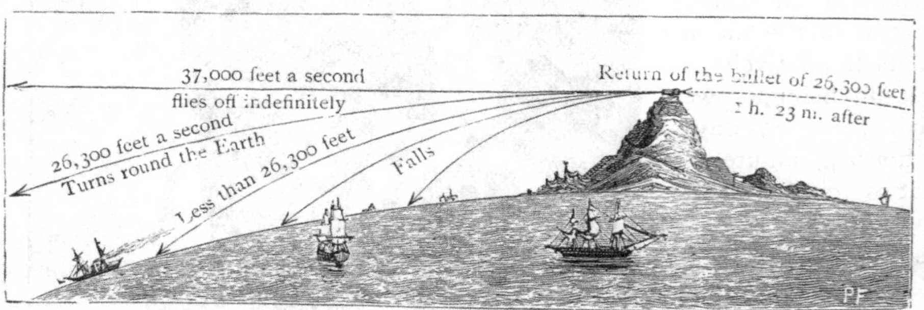

FIG. 58.—SPEED THAT IT WOULD BE NECESSARY FOR A PROJECTILE TO ATTAIN IN ORDER THAT IT SHOULD NEVER RETURN TO THE EARTH

Neglecting the resistance of the air, we find that a ball fired horizontally from a cannon placed on the summit of a high mountain on the earth *would never fall*, if it moved with such a velocity that it would make the circuit of the world in 5,000 seconds—that is to say, in 1 hour 23 minutes 20 seconds, or with a rapidity seventeen times greater than the motion of rotation of a point on the equator, or, in other words, if it were projected with a velocity of 26,300 feet (about 5 miles) a second. The tangential force which would be developed in this furious course would be precisely equal to the intensity of the earth's attraction, and it would remain in dynamical equilibrium. The artilleryman who fired it would thus create a new satellite of the earth.

The drawing (fig. 58) illustrates this idea. A ball projected horizontally from the summit of the mountain, with a velocity of 26,300 feet per second, will fall in this distance 16·1 feet, which is exactly the curvature of the earth, and will consequently follow a line parallel to this curvature, and return along the arrow after 1 hour 23 minutes 20 seconds.

Could we, theoretically, project a ball vertically with such a velocity that it would never return to the earth? The question is certainly original and curious. Where would the attraction of the earth be stopped? Nowhere. The attraction. diminishes in proportion to the square of the distance, but it never becomes equal to zero. To get out of the sphere of attraction of the earth is, then, not possible, at least to penetrate into that of another celestial body. But can we suppose a projectile animated with such a velocity that it would leave the earth for ever? Yes. For this it would be necessary to project it with a velocity of 11,200 to 11,300 metres per second (about 7 miles). A projectile thus fired *would never return to the earth*, and it would not revolve round us, but would escape into the interplanetary spaces.[1]

But we forget the moon. We wish, however, to give an exact idea of the weakness of gravity at its surface by remarking that a cannon-ball which would require on the earth a velocity of 26,300 feet a second in order to revolve round our planet without ever falling again, would require a velocity of only 10,500 feet to play the same part round the moon. Such would be the fate of a projectile fired horizontally with this velocity from the summit of the lunar mountain Leibnitz.

The same considerations show us that a stone shot from a lunar volcano with a velocity of 4,500 metres (about 2·8 miles) in the first second would escape from the lunar attraction, and would never return to that globe. I need not say that if it were directed towards the earth it would come straight to us. In this particular case it would not be necessary to project it with the same force in order to reach us. The sphere of the earth's attraction touches that of the moon's at a distance of 36,780 kilometres (22,854 miles) from the moon, and 347,220 kilometres (215,756 miles) from the earth (for the mean distance of 238,000 miles). A body shot from the moon in the direction of the earth would enter our sphere of attraction if it were projected with the comparatively moderate velocity of 2,500 metres (about 8,200 feet) a second. This velocity is not higher than the velocities of projection observed in certain terrestrial volcanoes—for example,

[1] The formula for this calculation may interest some mathematical readers. The velocity which it would be necessary to give a projectile in order to project it into the infinite is the same as that which would be acquired by a body drawn in from infinity by the attraction of the earth alone.

Let us represent by the letter r the radius of the earth, equal to 6,371 kilometres, and by g the force of gravity, equal to 9·81 metres. We shall have for the velocity of a body falling to the earth from infinity—

$$V = \sqrt{2\,gr.}$$

$$2\,gr. = 125{,}000 \text{ km.}$$

$$\sqrt{2\,gr.} = 11{,}200 \text{ m.}$$

in Cotopaxi—and it is not, perhaps, above those which human power is capable of producing. In the beginning of this century Laplace, Olbers, Poisson, and Biot had even concluded that the uranoliths—stones fallen from the sky—might very well be sent to us by lunar volcanoes.

In order to reach the sphere of the lunar attraction, a terrestrial ball should be projected vertically towards the moon in the zenith with a velocity of 10,900 metres (6·77 miles per second).

When the republican federation of the United States of Europe, Asia, Africa, and America shall be founded (in some thousands of years), and when the last battle shall have been fought between terrestrial brothers, the conquerors will still have the moon on which to exercise their ambition for the science of projectiles, and by sufficiently exciting terrestrial patriotism they will doubtless come to declare war on the moon. Our enemy would then be in a position very superior to ours. All their projectiles would be sure to reach us, while only a portion of ours would fall on them. It would be none the less the most curious of battles. [A much more useful expenditure of power would be to project a large hollow steel globe horizontally from the top of a lofty mountain, thus forming a new satellite, which would act as a useful timepiece for the earth.—J. E. G.]

However this may be, the fact which should strike us most in the physical conditions of the lunar globe is the feebleness of gravity at its surface, and the proportional lightness of any organisms which nature has been able to produce upon this globe.[1]

[1] There must be a very curious state of lightness on the moon, and it is strange that novelists, who have made so many imaginary voyages to our satellite, have not taken more advantage of this special fact. Many people have seen in Paris an interesting fairy scene played under the title of *A Voyage to the Moon*. The libretto did not lack spirit, the get-up was elegant, and the ladies of the *corps de ballet* left little to be desired. How easy it would have been to bring in the lunar lightness! But no one thought of it, any more than they did of the other peculiar astronomical conditions of the moon.

CHAPTER VI

IS THE MOON INHABITED?

ORB of dream and mystery, pale sun of the night, solitary globe wandering in the silent firmament, the moon has in all times and among all nations peculiarly attracted attention and thought. Nearly two thousand years ago Plutarch wrote a treatise with the title, 'On the Face which We See in the Moon'; and Lucian of Samosate made an imaginary voyage into the realms of Endymion.

For two thousand years, and especially in the years which followed the first astronomical discoveries of the telescope, a hundred voyages [1] to this neighbouring world were written by travellers whose brilliant imagination was not always enlightened by sufficient knowledge. The most curious of these scientific romances is still that of Cyrano de Bergerac, who found there men like those of the earth, but with singular manners and customs, which showed, as he thought, nothing in common with ours. From the time of Plutarch they had imagined on the moon beings similar to ourselves, but—I do not know why—fifteen times larger. In the first half of our century—in 1835—there was sold over the whole of Europe a pretended pamphlet by Sir John Herschel, representing the inhabitants provided with the wings of a bat and flying 'like ducks' above the lunar lakes. Edgar Poe has made a voyage to the moon in a balloon with an interesting citizen of Rotterdam, and has made an inhabitant of the moon come down to Rotterdam to give news of the voyage. More recently still Jules Verne projected a carriage-ball [2] to the moon; but it is to be regretted that his celestial travellers did not even catch a glimpse of the Selenites, and learned nothing of things relating to them.

This charming moon has undergone in human opinion the vicissitudes of this opinion itself, as if it had been a political personage—sometimes an admirable abode, at times a terrestrial and celestial paradise, blest region of the sky, enriched with a luxuriant life, and inhabited by superior beings; sometimes a frightful habitation, disinherited from all the gifts of nature, deserted and silent, a veritable tomb travelling forgotten in space. Before

[1] See our work, *Les Mondes imaginaires et les Mondes réels.*

[2] A hollow projectile fitted up as a carriage.

the invention of the telescope, philosophers were naturally disposed to see in it a world analogous to that which we inhabit. When Galileo directed the first telescope towards this globe, and recognised those mountains and valleys analogous to the inequalities of the land which diversify our planet, and the vast grey plains which might be easily taken for seas, the resemblance between this world and ours seemed evident, and they peopled it at once, not with real humanity, but with varied animals. They drew the first maps, and baptised the great spots with the names of seas which they still bear to-day.

In the times of Huygens, Hevelius, Cassini, and Bianchini they constructed telescopes of more than a hundred feet long, of which the last-

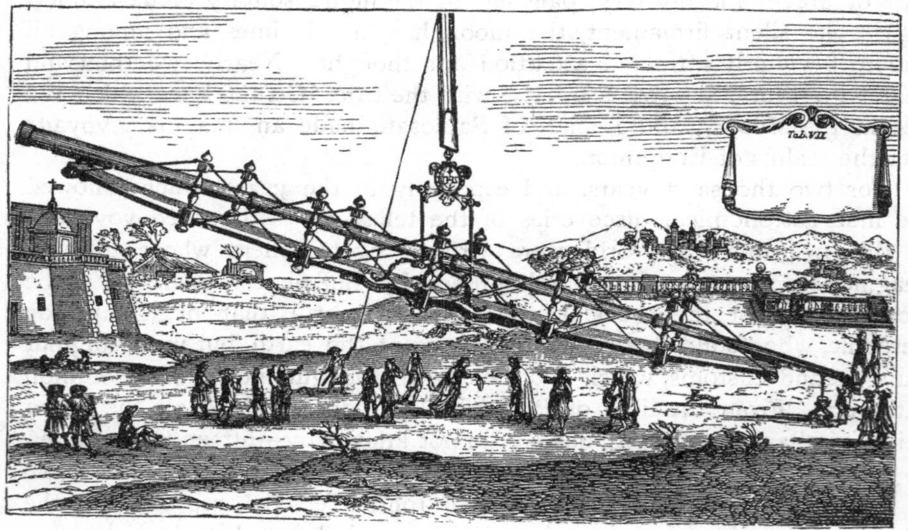

Fig. 59.—Great Telescope of the Seventeenth Century
(From an engraving by Bianchini)

named author has given, in his work on 'Venus,' the curious specimen here reproduced: but these telescopes not being achromatic, were not as good as our present telescopes of five yards in length.

Astronomers, thinkers, even the intelligent public, expected to see a rapid progress in the enlargement of telescopes, and it was even proposed under Louis XIV. to construct 'a telescope of ten thousand feet, to show the animals in the moon.' But the opticians had enough to do; optics did not progress according to the will of the imagination. On the contrary, the more the instruments were improved, the more the analogies at first remarked between the moon and the earth were effaced. The surface of the seas being clearly distinguished, they ascertained that this surface was neither uniform nor smooth, but sandy and rough, and chequered with a thousand inequalities—hills, valleys, amphitheatres, &c. Attentive observation did not succeed in discovering on this body a single true sea, nor a single lake, nor any certain proof of the presence of water in any known form—cloud, snow, or ice. Observation, no less careful, of the stars and planets, at the moment when the moon passes over them and occults them, showed at the same time that these bodies are neither veiled nor refracted when they touch the limb of the lunar disc, and that consequently this globe is not surrounded by any perceptible atmosphere. The analogy which they supposed to have found between the two worlds vanished; the lunar life disappeared in smoke, and they gradually became accustomed to write in all books on astronomy this phrase, already become traditional: *The moon is a dead world.*

This was a rather hasty conclusion. It was, indeed, to be singularly deceived as to the value of telescopic testimony.

My old master and friend, Babinet, asserted that if there were on the moon herds of animals similar to the herds of buffaloes in America, or regiments of soldiers marching in order of battle, or rivers, canals, and railways, or monuments like Notre Dame, the Louvre, and the Paris Observatory, the great telescope of Lord Rosse would enable us to recognise them. It has been said, in fact, that this colossal telescope (the largest which has yet been constructed: fig. 60), of which the length exceeds 52 feet and of which the mirror has a diameter of one metre eighty-three centimetres [six feet], can bear a magnifying power of six thousand times. Now, since to enlarge a distant object or bring it nearer is geometrically the same thing, if, in fact, we could bring the moon six thousand times nearer it would come within forty miles. But Lord Rosse's telescope is not perfect, and, far from being able to bear such a magnifying power as six thousand, we cannot, if we wish to see clearly, exceed two thousand.

The largest telescope after that of Lord Rosse is the great telescope of Lassell, of 1·22 metre in diameter [4 feet] and 36 feet in length [not now in use]. The most powerful telescopes are those of the Observatory of Mount Hamilton (California) [the Lick Observatory] and of the Nice Observatory. The first has an object glass of 97 centimetres (91 clear aperture) [36 inches], and measures 15 metres in length; the second possesses an object glass of 76 centimetres (74 clear aperture) [30 inches], and measures 18 metres in length. These two great instruments were erected in 1887. Now, the most powerful eye-pieces which can be applied to these masterpieces of optical art do not much exceed two thousand, and that in the most favourable atmospheric conditions. Of what use is it to magnify excessively an image which ceases to be clear and cannot be usefully observed? As we have remarked above, the greatest proximity to which we can bring the moon under the best conditions is 90 kilometres [about 56 miles].

Now, I ask, what can be distinguished and recognised at such a distance? The appearance or disappearance of the pyramids of Egypt would probably pass unnoticed. 'We see nothing moving,' is often objected. I can easily believe it. A notable earthquake (or moonquake) would be necessary in order that we should perceive it here, and, further, it would be necessary that exactly at the moment a terrestrial astronomer, favoured with a clear sky and a powerful instrument, should be occupied in examining precisely the region of the disaster. We should not be warned by any sound; the most frightful catastrophe might occur, the entire moon might be convulsed with a thousand thunders, and the faintest echo would not cross the space which separates us.

Hence, when they declare that the moon is uninhabited because they

FIG. 60.—LORD ROSSE'S TELESCOPE

see nothing moving, they are singularly deceived in the value of telescopic testimony. At some miles high in a balloon, with a clear sky and beautiful sunshine, we distinguish with the naked eye towns, woods, fields, meadows, rivers, roads; but we see nothing moving, and the impression felt directly (I have often experienced it in my aërial voyages) is of silence, solitude, and the absence of life. Living beings are no longer visible, and if we did not know that there are harvest men in the fields, flocks in the meadows, birds in the woods, fish in the waters, there is nothing to make us realise their existence. If, then, the earth seems like a dead world when seen from only a few miles' distance, what is it but illusion to assert that the moon is truly a dead world, because we view it at 120 miles or more? for it is only exceptionally that we can use the highest magnifying powers, and in general we do not apply to the observation of the moon powers exceeding a thousand. What, then, can we see of life at such a distance? Assuredly nothing, for forests, plants, cities, all would disappear.

The only means we have of forming a correct opinion on the condition of the lunar world is to observe carefully and draw separately certain districts, then to compare year by year these drawings with the reality, taking into account the difference between the instruments used. It is also necessary to allow a certain amount of variety to the difference in the eyes of observers as well as to the transparency of the atmosphere. We must also take into account the difference of illumination according to the height of the sun, considering that the more oblique the sun is the more the reliefs of the land are visible. The differences observed are very extraordinary. We could not believe them if we did not see them.

Now, this critical method, applied for some years, does not confirm the hypothesis of the death of the lunar world. It teaches us, on the contrary, that geological and even meteorological changes seem still to be at work on the surface of our satellite.

And, in the first place, the lunar surface can hardly be otherwise than changing, as well as the terrestrial surface. On our planet, it is true, we have violent volcanic eruptions and disastrous earthquakes; we have the waves of the ocean, which, eating away the shores under the cliffs and penetrating the mouths of rivers, incessantly modify the outlines of the continents (which I have verified with my own eyes along the French coasts); we have the movements of the soil which raise and lower it below the level of the sea, as anyone may see at Pozzuoli, in Italy, as well as in Sweden and in Holland; we have the sun, the frost, the winds, the rains, the rivers, plants, animals, and men, which incessantly modify the surface of the earth. Nevertheless, on the moon there are two agents which are sufficient to effect still more rapid modifications: these are heat and cold. In every lunation the surface of our satellite is subjected to contrasts of temperature which would suffice to disintegrate vast countries, and, in time, to shake down the highest mountains. During the long lunar night, under the influence of a cold more than glacial, all the substances which compose the soil must contract more or less, according to their nature. After that, the soil becomes heated under the direct radiation of a sun without clouds, and attains a high degree of temperature, notwithstanding the absence or the rarefaction of the atmosphere, as we find it in a balloon or on mountains, and all the minerals, which fifteen days previously were reduced to their smallest dimensions, would be expanded in various proportions. If we consider the effects which winter and summer produce on the earth, we may imagine those which must be produced a hundred-fold on the moon by this succession of contractions and expansions in materials which are less coherent and less massive than those of the earth. And if we add that these contrasts are repeated, not year by year, but month by month, and that all the circumstances which accompany them must be further exaggerated, it will certainly not appear surprising that *topographical variations are now produced* on the surface of the moon, and that, far from despairing of recognising them, we may, on the contrary, expect to observe them.

Besides, we cannot affirm that, independently of variations due to the mineral kingdom, there are not some which may be due to the vegetable or even the animal kingdom, or—who knows?—to some living formations which are neither vegetable nor animal.

But volcanic operations seem still to show themselves. A volcano larger than Vesuvius was formed, or at least increased in a way to become visible, in the course of the year 1875, in the midst of a landscape well known to selenographers. When the moon arrives at its first quarter the sun begins to illuminate the surface of the 'Sea of Vapours,' a region very favourably situated in the centre of the lunar disc. Among many fine craters we notice there one which has received the name of Agrippa. Round this amphitheatre the ground descends in a slope, and comes out on a plain. Across this plain we distinguish a sort of river, intersected at nearly the middle of its course by a little crater named Hyginus. I have very often observed this curious region of the lunar world, and I have made a large number of drawings, of which the most complete are those of July 31, 1873; August 1, October 29, November 27 of the same year; and April 24, 1874. Now, to the north-west of the crater of Hyginus none of the astronomers who have observed and drawn this region have ever seen or described an amphitheatre of 4,900 yards in diameter, which is now visible, and which one of the most industrious of our contemporary selenographers —Herr J. Klein, of Cologne—saw for the first time on May 19, 1876. Not to have seen a thing, even when looking at the place where it may be, is no proof that it does not exist; but when observers have been numerous and attentive, and when the object is very apparent, it is hardly possible to doubt. This is the case with the new amphitheatre, and the doubt that remains proceeds from the numerous irregularities of the land, which are very difficult to draw accurately.

There is in England[1] a society the members of which have sworn allegiance to the moon, and undertake not to forget it for a single month; this is the *Selenographical Society*. It lost no time in publishing in its selenographical journal the details given by Professor Klein and the observations

[1] [Or rather *was*, for the Society has done nothing for several years. However, its place is well filled by the Lunar Section of the British Astronomical Association, which is under the able direction of Mr. T. G. Elger, F.R.A.S., one of the greatest living selenographers.—J. E. G.]

which have confirmed his discovery. For my part, as I have just said, although I have not made our satellite the exclusive object of my observations, I have very often passed long evenings in studying with the telescope

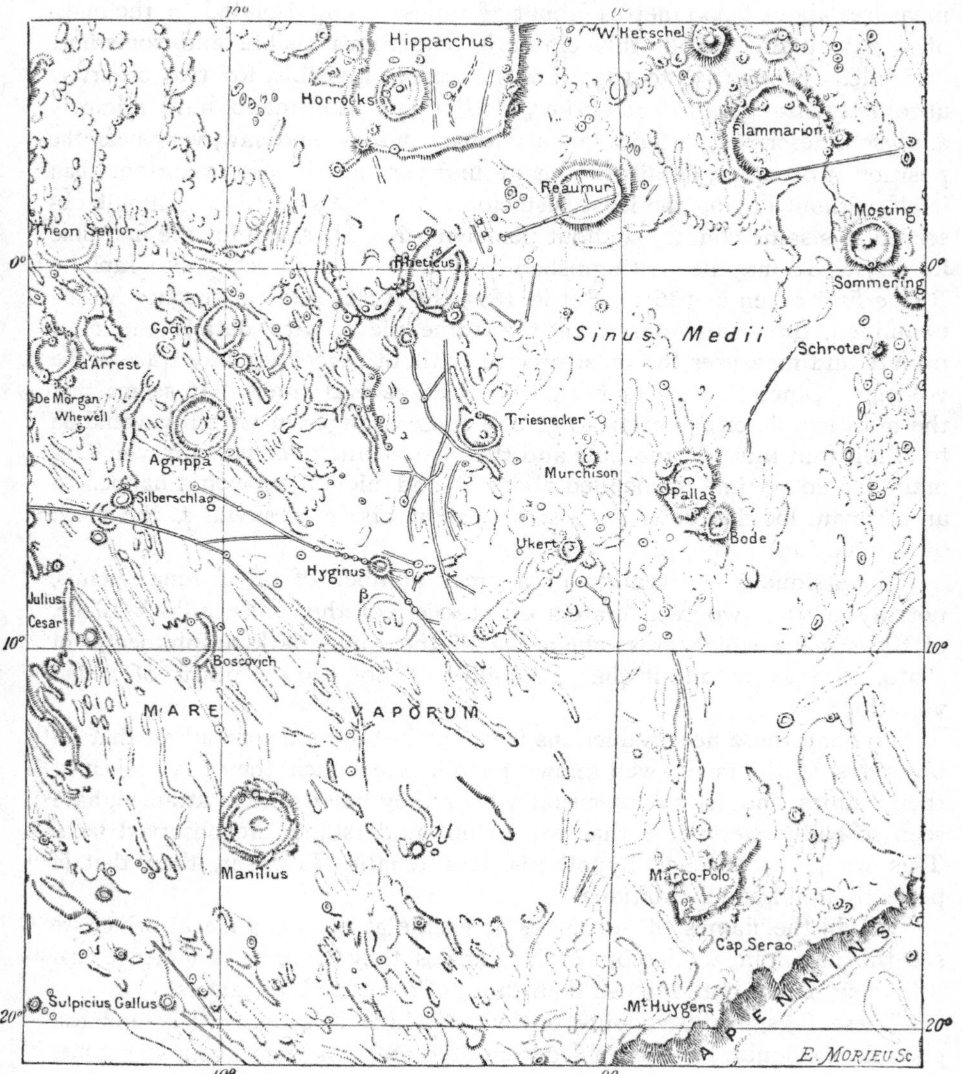

Fig. 61.—Lunar Topography: Neighbourhood of the Cleft of Hyginus

its curious topography, and I have made, among others, in 1873 alone, about thirty drawings of the Valley of Hyginus, which has always particularly attracted me. Now, I cannot recognise on any of my drawings the new crater which I have seen several times since. Fig. 61 represents this region.

The observed change has occurred to the left and below the point marked β on this little map.

In the Sea of Nectar we see a small crater, of which the diameter measures about 6,000 metres [about $3\frac{3}{4}$ miles], rising isolated in the midst of a vast plain. Well, this crater is sometimes visible and sometimes invisible. From 1830 to 1837 it was certainly invisible, for two observers absolutely strangers to each other, Madler and Lohrmann, have minutely analysed, described, and drawn this lunar country, and saw, very near the position it occupies, details of the ground very much less important than itself, without having the least suspicion. In 1842 and 1843 Schmidt observed this same country without perceiving it. He saw it for the first time in 1851. It may be distinguished very well in a direct photograph by Rutherfurd taken in 1865. But in 1875 the English selenographer Neison examined, drew, and described this same place, with details the most minute and measures the most precise, without perceiving any trace of the volcano. Since then it has been seen again several times. It seems that the most simple explanation to give of these changes of visibility would be to admit that this volcano now and then emits smoke or vapours which remain for some time suspended above it and hide it, as would happen to an aëronaut looking down from some height above Vesuvius at the epoch of its eruptions.

The assiduous observation of a great number of other lunar points, notably of the two twin craters of Messier, of the white hillock Linné, of the streaks which cross the arena of the great dark amphitheatre of Plato, lead us to admit the probability, if not the certainty, of actual variations.

To deny these new deductions it would be necessary to admit that all observers of the moon, well known for the care which they have taken in their studies, and for the precision which they have always attained, have seen so badly every time that we do not understand the observed facts. This would be another hypothesis, less tenable, however, than that of perfectly admissible variations.

Would the flames of volcanoes be visible at the distance at which we see the moon in a telescope? No, unless they were of a violence and brilliancy much more intense than those of terrestrial volcanoes.

These fogs, mists, vapours, or smoke, which it becomes less and less possible to doubt, have even led Schröter to think that their sometimes singular situations may seem to indicate some *industrial origin*—furnaces or factories of the inhabitants of the moon! The atmosphere of industrial towns, he remarks, varies according to the hour of the day and the number of fires lighted. We often meet, in the work of this observer, with conjectures 'on the activity of the selenites.' He also believed he observed changes possibly due to modifications in the vegetation or in cultivation.

The attentive and persevering observation of the lunar world is not so destitute of interest as many astronomers imagine. Doubtless, near as it is, this world differs more from ours than does the planet Mars, of which the analogy with the earth is so manifest, and which should be inhabited by beings differing very little from those which constitute terrestrial natural history and even our humanity; but although very different from the earth, it has none the less its value and its interest.

And, moreover, why should we suppose that there is not, on this little globe, a vegetation more or less comparable with that which decorates ours? Thick forests, like those of Africa and South America, may cover vast extents of land without our being able to recognise them. On the moon they have neither spring nor autumn, and we cannot trust to the variations of tint of our northern plants, to the verdure of May and the fall of the yellow leaves in October, to strictly typify that the lunar vegetation should show the same aspects, or should not exist. There winter succeeds summer every fifteen days; the night is winter, the day is summer. The sun remains above the horizon during fifteen times twenty-four hours; such is the duration of the lunar day and of spring. During fifteen days also the sun remains below the horizon; such is the duration of the lunar night and of winter. There, climatic conditions are absolutely different from those which govern the terrestrial vegetation. In the inter-tropical climates, where there is neither winter nor summer, the trees do not change colour. We have also in our climate plants of a persistent foliage [evergreens], shrubs which do not vary with the seasons; and as for the type even of vegetable verdure, the grass of the meadows, it remains as green in winter as in summer. Now we are here presented with a series of questions which remain unanswered. Do there exist on the moon passive beings analogous to our vegetation? If they exist, are they green? If they are green, do they change in colour with the temperature? And if they vary in aspect, can these variations be perceived from here?

What light can telescopic observations afford us on these obscure points? Certainly there is not in all the lunar topography any country so green as a meadow or a terrestrial forest, but there are on certain lands distinct and even changeable tints. The plain named the Sea of Serenity presents a greenish shade crossed by an invariable white zone. The observer Klein has concluded from his observations that the general tint, which is sometimes clearer, is due to a vegetable carpet, which, moreover, may be formed of plants of all sizes, from the mosses and mushrooms up to firs and other cedars, whilst the invariable white trail would represent a desert and sterile zone. The astronomers who are more occupied with lunar photographs are also of opinion that the deep tints of the spots called seas—a tint so little actinic that it makes an impression with difficulty on the sensitive plate (so that a much longer exposure is necessary to photograph the dark than the bright regions)—might be caused by a *vegetable* absorption. This greenish shade of the Sea of Serenity varies slightly, and at times it is very marked. The Sea of Humours shows the same tint, surrounded by a narrow greyish border. The Seas of Fecundity, of Nectar, and of Clouds do not present this aspect, and remain nearly colourless, while certain points are yellowish, as, for example, the crater Lichtenberg and the marsh of Sleep [Palus Somnii]. Have we here the colour of the soil itself, or are these tints produced by vegetation?

It is a rather singular fact that it is the valleys and the plains which change their tint with the elevation of the sun above them. Thus the arena of the great and admirable amphitheatre of Plato *darkens as the sun illuminates it more*, which seems opposed to all imaginable optical effects. After the full moon, the epoch which represents the middle of summer for this lunar longitude, the surface appears much darker than any other point on the lunar disc. The odds are 99 to 1 that it is not the light which produces this effect, and that it is the solar heat, which we do not sufficiently take into account when we are considering the modification of tints observed on the moon, although it may be quite as intimately connected as the light with the action of the sun. It is highly probable that this periodical change of tint on the circular plain of Plato, visible every month to any attentive observer, is due to a modification of a vegetable nature, caused by the temperature. The country to the north-west of Hyginus, of which we have already spoken, presents analogous variations. We see also in the vast fortified plain baptised by the name of Alphonsus three spots which come out pale in the morning from the lunar night, darken as the sun rises, and again become pale in the evening at sunset.

Far, then, from having a right to assert that the lunar globe is destitute of any vegetable life, we have facts of observation which are difficult, not to say impossible, to explain, if we assume a soil purely mineral, and which, on the contrary, are easily explained by admitting a vegetable coating, of whatever form it may be. It is to be regretted that we cannot analyse here the chemical composition of lunar lands as we analyse that of the vapours which surround the sun and the stars; but we should not despair of succeeding, for before the invention of spectrum analysis the possibility of arriving at such wonderful results had not been imagined. However this may be, we have now reason to suppose that the lunar globe was formerly the seat of tremendous geological movements, of which the traces remain visible upon its much disturbed soil, and that these geological movements are not extinct; that these seas have been covered with water, and that this water has not yet, perhaps, absolutely disappeared; that its atmosphere seems reduced to its last stage, but is not annihilated, and that life, which for ages upon ages may have flourished on its surface, is probably not yet extinct.

Lunar beings and things differ unavoidably from terrestrial beings and things. The lunar globe is forty-nine times smaller than the terrestrial globe, and eighty-one times less heavy. A cubic foot of the moon weighs but six-tenths of a cubic foot of the earth. We have also seen that gravity at the surface of this world is six times weaker than at the surface of ours, and that a pound carried there and weighed by a dynamometer would not weigh more than three ounces. The climates and seasons differ essentially from ours. The year consists of twelve days and twelve lunar nights, each lasting 354 hours, the day being the maximum of temperature and summer, the night being the minimum of tem-

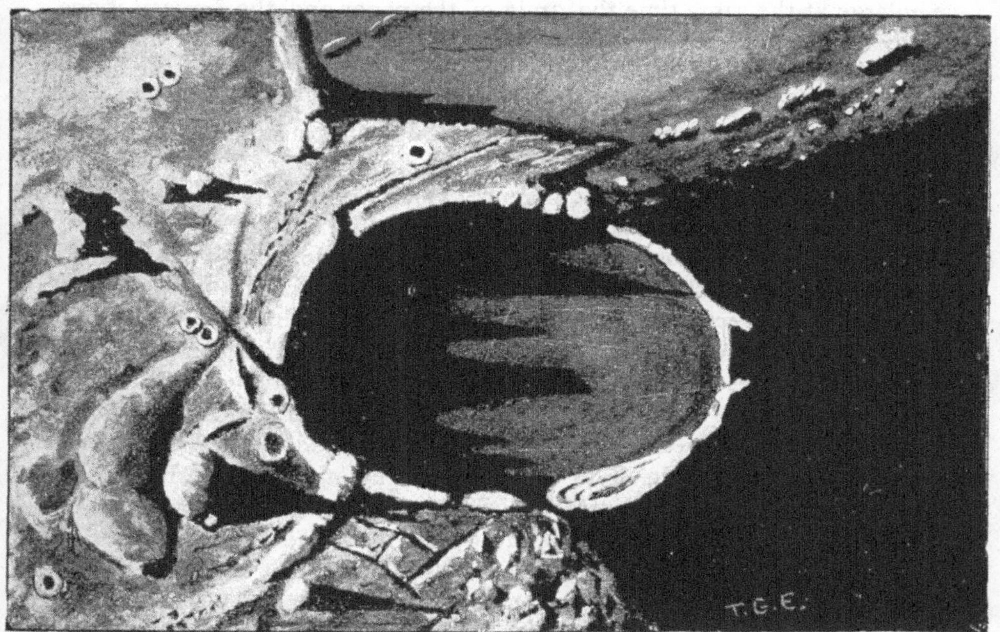

FIG. 62.—THE LUNAR WALLED PLAIN, PLATO, FEBRUARY 1, 1887, 5 HOURS 30 MINUTES TO 6 HOURS 30 MINUTES. (Drawn by T. G. Elger. Power, 340 on 8½-inch Calver Reflector)

perature and winter, with a thermometric difference of probably more than 100° [probably *much* more] if the atmosphere is everywhere extremely rare. We have here a difference more than necessary to constitute on this globe an order of life absolutely distinct from ours.

. . . .

Let us repeat, our best telescopes do not bring the moon nearer to us than 120 miles. Now, at such a distance, not only is it impossible for us to distinguish the inhabitants of the world, but even the material works of those inhabitants remain invisible ; roads, canals, villages, populous cities even, remain concealed by distance. We have, it is true, admirable photographs, and these photographs contain, in a latent state, all that exists on the surface of the moon. *If there are inhabitants, they are there*—they, their habitations, their works, their cultivations, their buildings, their cities. Yes, they are there; and it is difficult to resist a certain emotion when we hold one of these photographs in our hands, and when we are told that the inhabitants of the moon are there (if they exist), and that a sufficiently high magnifying power might permit us to perceive them, as we see with a microscope the strange population of a drop of water. Unfortunately, these photographs, altogether admirable as they are, are not perfect. They may be enlarged a little, five or six times, but we enlarge at the same time the grain of the paper and the imperfections of the image, and all becomes vague and diffuse, less useful, and less pleasing to analyse than the original negative. We should not, therefore, desist from studying with care the smallest details, drawing them exactly, re-observing them year by year, and determining the variations and movements which may be produced.

Those who, relying on the difference which exists between the moon and the earth, deny the possibility of all species of lunar life, do not employ the reasoning of a philosopher, but (we may be pardoned the expression) the reasoning of a fish! . . . Every fish reasoner is naturally convinced that water is the exclusive element for life, and that nothing could live out of water. On the other hand, an inhabitant of the moon would surely be drowned by descending into our atmosphere, so heavy and so dense (each of us supports 33,000 pounds). *To assert* that the moon is a dead world because it does not resemble the earth would be the idea of a narrow mind, imagining itself to know everything, and daring to suppose that science has said its last word.

This lunar life not having been formed on the same plan as terrestrial life, all we can assure ourselves on this question, so ancient and so debated, is that the inhabitants of the moon, if they exist, should be absolutely different from us as to organisation and sense, and certainly much more different from us in their origin than are the inhabitants of Venus and of Mars.

Moreover, we do not lose sight of the fact that the lunar hemisphere which we do not know is lighter than this side, and that, although its topography seems to resemble that of the visible hemisphere, we can say nothing of the fluids and liquids which may exist there. Doubtless the solar heat would bring atmospheric currents from that side to this, but might not this be the secret of the variableness of the effects observed in occultations?

The lunar life should have been anterior to terrestrial life, for the moon, although daughter of the earth, is relatively older than it. The geological, physical, and chemical movements which have so roughly disturbed it, have been, doubtless, as in our world, contemporaries of the primordial genesis of its living organisms; no observation proves that this life has really disappeared.[1]

Let us not leave this lunar world without trying to give an account of the effect produced by *the earth seen from* the moon, and forming an idea of astronomy for an observer situated on our satellite.

Whatever may be the beings inhabiting or having inhabited the moon (whether existing still in their period of decline, as seems probable, or whether the exhausted lunar humanity has lived through thousands of centuries and is already lulled in its last sleep), it will not be the less interesting for us to transport ourselves to this external province, and to give an account of the spectacle of the universe as it is presented when seen from this special station.

Let us suppose that we arrive in the midst of these savage steppes about the beginning of the day. If this is before sunrise the dawn is not announced, for with no atmosphere, or a rare one, there is no sort of twilight; there 'the timorous Aurora opens not to the sun his enchanted palace,' but the zodiacal light, which we can see so rarely here, is constantly visible there, and it is this which is the harbinger of the royal star. In a moment from the black horizon shoot rapid arrows of solar light, which strike the summits of the mountains, while the plains and valleys remain in darkness. The light increases slowly. While with us, in central latitudes, the sun takes but two minutes and a quarter to rise, on the moon it takes nearly an hour, and consequently the light which it sends is very feeble for several minutes, and only increases with extreme slowness. This is a sort of dawn, but of short duration, for when, at the end of half an hour, the solar disc is half risen the light seems nearly as intense to the eye as when it is entirely above the horizon. These lunar sunrises are far from equalling ours in splendour. The illumination of the higher regions of the atmosphere, so soft and delicate, the colouring of the gold and scarlet clouds, the fans of light which cast their rays across the landscapes, and,

[1] This interesting question of the inhabitants of the moon might be settled in our day, as well as a great number of others, by a powerful telescope, the cost of construction of which ought certainly not to exceed 40,000*l.* Studies made with this object in view show that we might now, in the present state of optics, construct an instrument capable of bringing the moon within a few miles of us, and even attempt to establish communication with our celestial neighbours, which would not be bolder or more extraordinary than the telegraph or phonograph.

In fact, what is the size of the smallest object which it would be possible to distinguish on the moon? The diameter of this globe is 2,163 miles, and measures geometrically 31 minutes 24 seconds. One kilometre on the moon measures, then, 0″·54, and one second represents 1,850 metres [about 6,070 feet, or 2,023 yards]. Now, at present, according to the calculations of Mr. Hall, to whom science is a debtor for the curious discovery of the satellites of Mars, we distinguish an angle of three-hundredths of a second—that is to say, a length of 55 metres [180 feet]. We may go further, and distinguish an object of 30 metres [98·4 feet]. At sunrise and sunset the elongated shadow throws into relief heights of 10 metres [33 feet].

above all, that rosy glow which bathes the valleys with such mellow light, are phenomena unknown on our satellite. But, on the other hand, the radiant star shows itself with its prominences and its fiery atmosphere. It rises slowly, like a luminous god in the depths of a sky always black, a sky profound and formless, in which *the stars continue to shine during the day*, as in the night, for they are not hidden by any veil. There the sky is not reflected in the mirror of any sea or any lake.

Aërial perspective does not exist in these lunar landscapes. The most distant objects are as clearly visible as the nearest, and we might almost say that in such a landscape there is but one plane. No more of those vaporous tints which, on the earth, increase the distances and shade off the decreasing light; no more of those vague and charming lights which float over the valleys bathed in the sunlight; no more of that celestial azure which pales from the zenith to the horizon, and casts a transparent blue veil over the distant mountains: a light dry, homogeneous and dazzling, illuminates harshly the rocks of the craters; the sky is not illuminated; everything which is not directly exposed to the rays of the sun remains in night.

In the same way that we never see but one side of the moon, there is only one side of this globe which sees us. The inhabitants of the lunar hemisphere turned towards us admire in their sky a brilliant body having a diameter about four times greater than that of the moon seen from our globe, and with a surface fourteen times larger. This body, which is the earth, is 'the moon of the moon.' It soars almost motionless in the sky. The inhabitants of the centre of the visible hemisphere see it constantly in their zenith. Its height diminishes with the distance of the country from this central point up to the limit of this hemisphere, where they see our world placed like an enormous disc on the mountains. Beyond that they see us no more.

An immense orb in the lunar sky, the earth shows to the selenites the same phases as those which the moon presents to us, but in the inverse order. At the moment of new moon the sun illuminates in full the terrestrial hemisphere turned towards our satellite, and they have the *full earth*. At the epoch of full moon, on the contrary, it is the hemisphere not illuminated which is turned towards our satellite, and they have *new earth*; when the moon shows us the first quarter the earth has its last quarter, and so on.

Independently of its phases, our globe is presented to the moon as turning on itself in 24 hours, or, more accurately, in 24 hours 48 minutes, since the moon does not return to each terrestrial meridian until after that interval. There are variations in this apparent rotation of the earth from 24 hours 42 minutes to 25 hours 2 minutes. But if the lunar astronomers know how to calculate their motion, as we have done for ours, they know that the moon revolves round the earth, and that our planet turns on itself in 23 hours 56 minutes. We are not certain, however, as Kepler was ('Astronomia Lunaris'), that the inhabitants of the moon have given to the earth the name of *Volva* (from *volvere,* to turn). It was this which originated the name of *Subvolves* (under the turning one) for the inhabitants of the hemisphere which faces us, and that of *Privolves* (deprived of the turning one) for those who inhabit the opposite hemisphere. This name of *Volva* was, however, well imagined, for it pictures capitally the terrestrial phenomena which would first strike the mind of the inhabitants of our satellite.

On the visible lunar hemisphere they can observe curious eclipses of the sun, and among them eclipses which may last two hours. The enormous black disc of the earth, surrounded by a luminous nimbus produced by the refraction of light in our atmosphere, passes before the dazzling disc of the sun. They also see sometimes very small *eclipses of the earth*—that is to say, the passage of the circular shadow of the moon along a terrestrial zone.

We say on our planet, 'Deprived of all liquid and aërial envelope, the moon is not subject to any of the meteorological phenomena which we experience here: it has neither rain nor hail, wind nor storm. It is a solid mass, arid, desert, silent, without the smallest vestige of vegetation, and where *it is evident* that no animal can be found to exist. If, however, they will insist that it may have inhabitants, we will consent with pleasure, provided that they compare them with beings deprived of all sensitiveness, of all feeling, of all motion—which reduces them to the condition of gross bodies, of inert substances, of rocks, stones, metals—which, evidently, are the only Selenites possible.'

The academicians of the moon doubtless say, in their turn, with an assurance no less convinced, 'The earth is composed of elements dissimilar and very extraordinary. One, which forms the nucleus of the body and which gives birth to fixed spots, appears to have some consistence; but it is covered with another element, of a strange constitution, which appears to have neither body, nor stability, nor continuance. It has neither colour nor density. It takes all forms, moves in all directions, obeys all shocks, submits to all impulses, is extended, contracted, condensed, appears and disappears, without our being able to imagine such strange metamorphoses. This is the world of instability, the planet of revolutions. It experiences in turn all imaginable disasters. It seems to be matter in fermentation, which tends to dissolve. We only see storms, cyclones, whirlwinds, and acts of violence of all sorts. They assert that there are inhabitants on this planet; but on what point can they live? Is it on the solid element of the body? They would be crushed, suffocated, asphyxiated, drowned by that element which weighs on them from all sides. Is it through the openings in this mobile curtain that they can enjoy, as we do, the pure ether of the heavens? But how can we suppose that they might not at any moment be torn from the soil by the violence of the disorders which torment the surface? Do they wish to place them on the light and mobile stratum which hides from us so often the aspect of the terrestrial nucleus? How can they be maintained upright on this element without solidity? . . There is no necessity for long consideration to *prove conclusively* that this planet is very vast, but that it is no place for animated beings. The whole earth is not worth the soul of a single Selenite. If, however, they will insist that it may have inhabitants, we will consent with pleasure, provided that they compare them with fantastic beings floating at the pleasure of all the forces which contend with each other on this aëriform planet There can only exist there rather coarse animals. Such are, in our opinion, the only inhabitants which can people the earth.'

The scientists of the moon have, as we see, the ability to prove, in the most categorical manner, to the ignorant who surround them, that the earth, not being habitable, should not be inhabited, and that *it is made solely to serve as a clock to the moon and to shine during the night.*

The different parts of the terrestrial surface are far from having a uniform brightness to the eyes of the lunar observer. At both poles of the body they notice two vast white spots, which vary periodically in size. As one increases the other diminishes. They would believe that one always conquers a portion of land equal to that lost by the other, in such a way that as one advances so the other recedes, and conversely, that of the south pole always showing a much larger extent than that of the north pole. On the moon they make a thousand suppositions about these white spots, but they have not divined the cause of them.

The earth is always to a great extent enveloped in clouds. However, attentive observations have enabled them to determine that it has a motion of rotation.

Let us now consider our planet at the hour when America begins to disappear on the eastern border of the terrestrial disc. We see then from the moon, drawn on the dark part, the relief of the high summits of the Cordilleras, represented by a long line of shadows and lights, of which some points are of dazzling whiteness. Then unfolds during some hours on the opposite border an enormous dark spot, which descends, enlarging towards the southern part of the disc, until it occupies nearly the whole hemisphere. This is the great ocean [the Pacific], studded with a multitude of small islands.

We perceive to the north, not far from the northern ice, a greyish spot which has commenced to form towards the south a point (the almost island of Kamschatka) on the dark background of the vast ocean. It afterwards unfolds itself towards the west, descending almost to the equator. Its curved sides show a very varied aspect. This is Asia, the part of the Old World most drawn towards the extreme east. Its tint is far from being uniform: it presents to the north the Siberian district—snows, ice, and hoar frost.

The whole of the centre of the continental spot is occupied by a great band of dazzling whiteness, which seems encircled to the north and the south by very high mountains (the chains of the Altai and the Himalayas). This zone commences at the great desert of Gobi, occupies nearly all the central table-land of High Asia, and is prolonged across Afghanistan and Persia up to the sandy plains of Arabia. The deserts of Nubia and Sahara, which traverse Africa, are even a continuation. Thus this great desert zone cuts the whole of the Old World into two nearly equal parts by a band of sand, causing the solar light to shine far away into the celestial spaces: it is the Milky Way of the earth.

Below the region of sand is a conspicuous portion of the land of Asia, shut in, so to say, between the mountains and the ocean, which reflects on the moon a very clear light. It includes the magnificent countries of China and India, situated to the south of the mountains of Mongolia and Thibet.

Above the Sahara desert we see a small spot, broken up in all directions into numerous branches. It is of a dark tint, like the great spot of the disc which surrounds all the continents. This is the Mediterranean, which serves as a southern boundary to a region of undecided colour, between grey and green. This region, cut up into islands and peninsulas, which seems to the inhabitants of the moon so little worthy of attention, is our Europe, whose civilisation, envied by all nations, is sufficiently powerful to dictate laws to the rest of the world. As to France, it is necessary to have good eyes to distinguish it. Telescopes as powerful as ours might just recognise the form of our shores, the Pyrenees, the Alps, the British Channel, the Rhine, the mouth of the Gironde, that of the Seine, and even the existence of Paris and of our principal towns.

Europe marks the extreme western limit of the ancient continent. When the planetary globe has turned a few more degrees upon its axis all the land will have disappeared; the Selenites will only perceive the dark spot of the Atlantic Ocean; and the first land to reappear will be America, with which we commenced.

The scientists of the moon will thus have merely to watch the earth turning to learn the whole of our geography. This is what we have done ourselves with the planet Mars.

The lunar astronomers have even a great advantage over our geographers: they can study with ease all points of our globe, and carefully examine the mysteries of our most inaccessible countries, such as the polar regions, which are, perhaps, for ever closed to us, and those of Central Africa, which are only now beginning to be revealed. Unconcerned spectators perhaps contemplate in the evening the clear earth with a look of careless reverie, and view these inhospitable regions without suspecting the fatigues and dangers which the earth-dwellers voluntarily undergo in order to acquire the same knowledge. Perhaps, also, when seeing each meridian pass into the shadow at the end of the day, they may think that these moments mark successively the hour of repose and sleep for all the nations of our world.

Thus our globe forms for the moon a permanent celestial clock. The earth's motion of rotation replaces the hand which moves round the dial, and each fixed spot, situated at a different longitude, is the number which marks the hour.

The defenders of final causes have a much better right to declare that the earth was made for the use of the moon than to support the opposite opinion. The moon performs very badly its function with regard to us, and, assisted by clouds, leaves us for three-fourths of the time in darkness. The earth, on the contrary, shines quite clearly every night in the lunar sky, and the full moon constantly occurs to the minute. Could we dare, then, prove to a lunarian that we are not created and placed in the universe expressly for him?

The length of the day and night, the absence of seasons and years, the measure of time by periods of twenty-nine days, divided into a day and a night of fourteen days and a half each, and the constant presence of the earth's orb in the sky, constitute for the lunar inhabitants essential differences which distinguish their world from ours from a cosmological point of view. The constellations, the stars, and the planets are seen by them as we observe them here, but with a more vivid light, a greater richness of tone, and considerably increased in number, owing to the constant clearness of the lunar sky. The invisible hemisphere, which never receives the light of the earth, would form an excellent observatory for astronomical studies.

Such is the lunar world, so near us, and yet so different. The knowledge we have obtained does not, however, satisfy our scientific ambition. When, then, shall science find friends sufficiently devoted to dare to attempt a more complete conquest, to sacrifice to optical experiments—of which the results would be assuredly amazing and unexpected—sums analogous to those which are expended to no purpose in cannon foundries and elsewhere? . . . Marvellous discoveries await the heroes of future astronomy.

Perhaps the present races of lunar humanity are themselves provided with instruments sufficiently powerful to show our cities, our villages, our cultivation, our industrial works, our railways, our assemblies, and ourselves. Perhaps they have viewed our latest battles, and have followed with perplexity the strategical movements of our imperturbable folly. Perhaps the astronomers of this neighbouring province have made signs to us, and have attempted a thousand times to attract our attention and to open communications with us. It cannot be doubted that there were living beings there even before any existed on our planet: the forces of nature nowhere remain sterile, and the epochs which have marked the great lunar geological revolutions, of which we distinctly see the results, have been, as on the earth, the times of organic production. Do these beings still exist?

If we were willing we might know definitely what we assert . . . yes, if we were willing! And what dazzling wonder, what unexpected pleasure, what fantastic ecstasy, the day when we shall distinguish with certainty the proofs of life on this neighbouring continent, when we shall trace here, by the electric light, geometrical figures, which *they* will see and *they* will reproduce. First and sublime communication from heaven to earth! Search through the whole history of our humanity for an event equally amazing. What do I say? Search for facts which reach but to the ankle of this colossus in scientific interest and philosophical consequences, and you will find but pigmies crawling at the foot of a giant!

We dare not attempt it, because we are not sure. And this is the opinion of serious men! Civilised Europe, which dares not spend a million francs to attempt to reach the celestial life, spends with a light heart six thousand million francs a year for a peace army! for imminent war, for the certain destruction of her children! But to lay a hundred thousand dead bodies on the earth is interesting. . . . Oh, folly of follies!

However this may be, the general conclusion of the study we have made of the lunar world is that our conception of nature should embrace *time* as well as *space*. In space, we see across millions and millions of miles; in time we may travel back through centuries and millions of centuries. Our position and our time are important to us, but they are absolutely nothing in nature; for here there is nothing absolute but infinity and eternity. Universal life is the object of creation and the definite result of the existence of matter and of force. But that a world is either inhabited to-day, or that it has been yesterday, or may be to-morrow, is the same thing in eternity. The moon is the world of yesterday; the earth is the world of to-day; Jupiter is the world of to-morrow; the idea of time is thus fixed upon our minds like that of space. But the law of the plurality of worlds rules for ever. What does it matter when humanity appears on such or such a world? The dial of the heavens is eternal, and the inexorable hand which slowly marks their destinies shall turn for ever. It is we who say *yesterday* or *to-morrow*. For nature it is always *to-day*.

Before the epoch when the first terrestrial human glance was raised towards the sun and wondered at nature the universe existed as its exists to-day. Already there were other inhabited planets, other suns shining in space, other systems gravitating under the impulse of the primordial forces of nature; and, in fact, there are stars which are so distant from us that their light does not reach us till after millions of years: the luminous ray which we receive to-day left the bosom of space not only before the existence of man here below, but even before the existence of our planet itself. Our human personality, of which we think so much, and in the image of which we have imagined God and the whole universe to be formed, is of no importance in the harmony of creation. When the last human eyelid closes here below, and our globe—after having been for so long the abode of life with its passions, its labours, its pleasures and its pains, its loves and its hatred, its religious and political expectations and all its vain finalities—is enshrouded in the winding-sheet of a profound night, when the extinct sun wakes no more; well, then—then, as to-day, the universe will be as complete, the stars will continue to shine in the sky, other suns will illuminate other worlds, other springs will bring round the bloom of flowers and the illusions of youth, other mornings and other evenings will follow in succession, and the universe will move on as at present; for creation is developed in infinity and eternity.

CHAPTER VII

THE TIDES

THE waters of the ocean rise and fall every day by a regular motion of *flow* and *ebb.* This movement so hopelessly puzzled the ancients that they called it the grave of human curiosity. Nevertheless, it shows to attentive examination such an evident connection with the motion of the moon that many astronomers of antiquity recognised and affirmed this relation. Thus Cleomedes, a Greek writer of the age of Augustus, says positively in his 'Cosmography' that 'the moon produces the tides.' It is the same with Pliny and Plutarch. But the fact was not proved. Some denied it. In modern times, even Galileo and Kepler did not believe it. It was Newton who commenced the mathematical demonstration, and it was Laplace who completed it, by proving that the tides are caused by the attraction of the moon and the sun.

The surface of the earth is partly covered by the waters of the sea, which, on account of their fluidity, can easily move on this surface in virtue of the moon's attraction. Now, the different portions of these waters spread all round the globe, and consequently, placed at unequal distances from the moon, are not equally attracted by it. Directly below it the waters of the sea are more strongly attracted than the solid part of the earth, considered as a whole; in the opposite region the waters of the sea are, on the contrary, less strongly attracted, because they are more distant. It follows that the waters situated on the side towards the moon are raised, on account of this excess of attraction, and that on the opposite side of the earth the waters tend to remain behind, relatively to the mass of the globe, which is more strongly attracted than they are. Consequently the former accumulate on the side towards the moon, and form a protuberance which would not exist without the presence of that body; in the same way, the latter accumulate on the side opposite to the moon, and form there a similar protuberance (fig. 63). Add to this that the earth, turning on itself in twenty-four hours, brings successively the various parts of its circumference in the direction of the moon (so that the two liquid protuberances of which we have just spoken, in order to occupy always the same position with reference to the moon, continually change their place on the surface of the terrestrial globe), and you will see that at the same point of this surface, at the same port, we should observe successively two high waters, and consequently also two low waters, during the time in which the earth makes a complete rotation relatively to the moon—that is to say, in 24 hours 48 minutes.

Fig. 63.—Explanation of the Tides

The sun produces a similar effect on the waters of the sea; but the enormous mass of that body is more than compensated by the great distance at which it lies from the earth, so that, after all, the tide due to the action of the sun is much smaller than that of which we have spoken, and which is due to the action of the moon. The phenomenon, then, in its general aspects, depends upon the position of the moon with reference to the earth; the action of the sun only modifies it, sometimes accelerating, sometimes retarding the time of high water, sometimes increasing, sometimes diminishing the intensity of the phenomenon, according as the star of day occupies such or such a position in the sky with reference to the star of night. In taking into account these two circumstances relative to the mass and distance, we find that the effect produced by the sun should be to that produced by the moon in the ratio of 1 to 2·05—that is to say, in the general phenomenon of the tides the moon is responsible for two-thirds, and the sun for only one-third. The moon raises the surface of the sea at the equator by fifty centimetres [about 19·7 inches], and the action of the sun being added, the elevation reaches 74 centimetres [29·1 inches]. The height decreases up to the poles, where the amplitude of the oscillations is reduced to zero, and there is no tide, even when the sea is not frozen.

The highest tides are consequently those which happen at the new moons and full moons, for then the actions of the sun and moon co-operate, while at the quadratures they are exerted at right angles to each other. The interval of time included between two consecutive high waters is, on an average, equal to 12 hours 24 minutes; but the high water, instead of occurring at the moment the moon crosses the meridian, does not happen until a certain time after this passage. The oscillation of the surface of the sea is always very much regulated, on the whole, by the diurnal motion of the moon round the earth; but each of the phases of this oscillation is behindhand at the instant it should be produced, owing to theoretical considerations which we are about to explain, and this delay is, moreover, very different in different places.

In our ports the highest tides follow the new and full moon by a day and a half.

The amount by which the surface of the sea is raised and lowered successively is, in general, very much greater than what we have stated, assuming that this surface takes at each instant the figure of equilibrium which agrees with the magnitude and direction of the attractions of the sun and moon. We have seen that the greatest difference of level which can exist, on this hypothesis, between high water and the following low water is only 2·43 feet at the equator, if the sun and moon are at their mean distances. Now, there exist certain localities where the same difference exceeds thirty-two feet in the vertical direction. On imperceptibly sloping shores the horizontal difference between high and low water is several miles: you lie down at one time where the sea reaches your feet and you are lulled to sleep by the noise of the waves; the next morning on awaking, the sea has disappeared and you walk on a dry beach.

In reality, however, the intensity of this force on a mass so considerable as that of the waters of the ocean seems almost infinitely small. A weight of two thousand pounds is diminished by 1¾ grains when the moon is in the zenith or nadir, is increased by 0·8 of a grain when it is at the horizon, and is not altered when it is at 35° above or below the horizon. The attraction of our satellite, then, causes a weight of a thousand kilogrammes to vary by 2½ grains. A man of eleven stone in weight weighs 0·18 of a grain less when the star of night passes over his head than when it is on the horizon. This difference is nearly equivalent to the weight of a grain of wheat. And yet the structure of the continents and the configuration of the coasts have been slowly but irresistibly modified by this many-headed ram which beats twice a day with merciless shock on the sandhills and cliffs.

The waters of the sea, contained in a space limited on both sides by the continents, oscillate in this space, which forms a sort of vessel of small depth relatively to its surface; these oscillations are kept up by the disturbing actions of the moon and sun, of which the intensity and the direction change every instant. When, in consequence of these actions the surface of the sea is forced to rise at a certain side of the basin which contains it, the water is carried to that side, and the velocity with which the change of place is effected is the reason that it does not stop when the surface has attained equilibrium, but continues to move in the same direction until the

FIG. 64.—PATH OF THE TIDE FOR DIFFERENT PORTS

velocity is completely destroyed by the action of gravity, and by the friction against the bottom ; so that the oscillatory movement in the vertical direction thus becomes, on the borders of the sea, of much greater proportions than if the sea were placed at each instant in equilibrium under the action of the forces which are applied to it. We understand from this, not only why the sea is raised and lowered much more than seems to be caused by the actions of the moon and sun, but, further, why, at the time of the syzygies, the high water does not happen at the instant of the passage of the moon over the meridian; at this moment the actions of the sun and moon are in the conditions suitable to sustain the waters of the sea at the greatest height; but the waters which have risen by these actions up to the passage of the moon across the meridian continue still to rise for some time after this passage, in virtue of their acquired velocity.

The crosswise form of certain shores swallows up the water which arrives and causes it to rise to a considerable height. The tides of the Atlantic, for example, produce very intense derived tides in the English Channel, with which the ocean communicates freely. When the sea becomes high on the west of France, in the neighbourhood of Brest, the flood of the high water advances gradually in the English Channel. This little sea becoming suddenly narrowed, the flood rises against the barrier which is thus opposed to its progress, and thence result very high tides on the coasts of the Bay of Cancale, and notably at Granville. From thence the flood continues to advance, and high water takes place successively at Cherbourg, Havre, Dieppe, Calais, &c. This progress of the flood tide is made clear in the following table, which gives, for the different ports of the coasts of France, the delay of high water behind the time of the passage of the moon over the meridian at the epoch of new and full moon, a delay known by the name of *the establishment of the port.* The same table also contains the mean heights of the tides at the same epochs. This is the difference between high water and low water. They call the *unit of height* the half of this difference—that is to say, the elevation above the mean level. This height may be increased, by the influence of the wind, in force and direction.

Name of Port	Establishment of Port	Mean Height of the Tide at the Syzygies
	H. M.	METRES
Bayonne (mouth of the Adour)	4 5	2·80
Royan (mouth of the Gironde)	4 1	4·70
Bordeaux	7 45	4·50
St. Nazaire (mouth of the Loire)	3 45	5·36
Lorient	3 32	4·48
Brest	3 46	6·42
St. Malo	6 10	11·36
Granville	6 40	12·30
Cherbourg	7 58	5·64
Havre (mouth of the Seine)	9 8	7·14
Dieppe	11 8	8·80
Boulogne	11 26	7·92
Calais	11 49	6·24
Dunkirk	12 13	5·36

In consequence of the delay which the tidal wave experiences, *the establishment*—that is to say, the time which elapses between the passage of the moon across the meridian and the moment of high water—varies remarkably at different ports. Thus, while at Gibraltar high water occurs almost exactly at the moment of the passage of the moon across the meridian, the delay is already an hour and five minutes at Cadiz, and three hours on the coast of Spain. It proceeds afterwards as we see it on the map, fig. 64. The general form of the curves shows in a striking manner

Arrival of the Mascaret→ ←Course of the Seine

FIG. 65.—THE 'MASCARET' AT CAUDEBEC

that the velocity of propulsion of the tides diminishes with the depth of the sea.

In the mouths of great rivers, and especially in the Seine, the tide produces a very curious and picturesque effect, justly admired by tourists. It surmounts by force the current of the river, dashes in a cascade, rolls with fury a sheet of water which is sometimes several yards in height, damages all the buildings on the banks, and at the same time upsets all the ships which are not afloat. This singular accumulation is produced in those parts of the river where the bottom gradually rises. There the first waves, propagating themselves in shallower water, are overtaken by those which follow them and which end by falling over the first; for it is a mechanical law that, the deeper the water, the more rapidly does a wave travel. This is what is called *la barre* or *mascaret*[1] [in English, the bar or the bore].

In spreading from east to west, in a direction contrary to the rotation

[1] It is especially at Caudebec that this spectacle should be seen, choosing for the excursion the days of highest tides in March, September, and October. At Caudebec it is a 'bore,' at Aizier an angry sea.

On the day and the hour indicated the wharf, shaded with perennial trees and splendid walks, is crowded with spectators. These are the inhabitants, who are never tired of the grand spectacle of the river transformed, and strangers, come from far to enjoy and to study it.

For a long time before the arrival of the flood impatient eyes search the horizon, and the less experienced think every moment that they see it beginning at the extremity of the bay which forms this bend of the Seine. A low murmur announces its approach when it is still at a distance of several miles; directly all the ships and boats hasten to push off into the river, and abandoning themselves to the current, which continues to descend, are borne along before the flood. The little flotilla seeks the deep spots, which the daily experience of the bargemen of the country indicate as the safest. These spots often vary, on account of the transformation which the motion of the sands produces in the channel. Woe to the imprudent bark which, from indolence or contempt of the danger, remains behind! The inclined planes of the wave falling over into cascades would soon envelop it in their furious whirlpools, and then knowledge and courage would be impotent. Very often sad shipwrecks are the painful result.

The vast sheet of water advances rapidly, raising one after another the ships and boats, which in turn rise on the crest of the waves or are hidden in their troughs. Under a radiant sun, in the midst of a verdure which a zephyr scarcely stirs, there are all the motions, all the agitations, all the fury of a tempest-tossed sea.

Very soon the spectacle changes, to become grander and more singular still. The enormous wave which marches at the head of the tide swells, rises, stands up; it bursts of a sudden, and its summit falls with a crash; an immense roll is formed and unfolds itself, sometimes from one end to the other; it is a cascade which moves, which runs and remounts the river with the speed of a galloping horse. The flood runs along the banks like a wall of foam, overthrowing all obstacles, and rearing itself up each instant like a gigantic plume, to fall again quivering on the bank, which it deluges. The ground sometimes trembles under the feet of the spectators, who see, in less time than it takes to describe it, the boiling mass passing on and pursuing its ungovernable course.

Immediately after the passage of the flood this great tumult calms down, and the river resumes its peaceful aspect. Only the current has changed its direction, and moves back rapidly from the mouth towards the source in the direction of Rouen.

The introduction of the tidal waters into the Seine, on account of the small slope which the bed of the river has, is the primary and necessary cause of this motion of the waters. The difference of level between Rouen and Havre, points distant from each other more than 74 miles, is, following the course of the river, not more than 18·83 feet. Every time the tides in the English Channel attain a greater height, the accumulated waters seek to balance themselves, lean over in the bay, then overflow into the river. The difference of level is further increased in this case by the difference of density, the waters of the ocean being denser than those of the river.

Such is the scientific explanation of this beautiful phenomenon. It is, perhaps, less pleasing than the poetical explanation given by Bernardin de Saint-Pierre. 'The Seine, nymph of Ceres and daughter of Bacchus, walking one day on the seashore, was seen by the old monarch of the waters, who, delighted with her charms, set out to follow her. He had already reached her when Bacchus and Ceres, invoked by the nymph, and being unable otherwise to save her, metamorphosed her into an azure river, which ever since has borne her name, and has everywhere on its banks joy and fertility. Neptune, however, has not ceased to love her, although she has preserved her aversion of the earth, the tides act, by their friction, as a break on this motion, and slacken it, so that the length of the day gradually increases. The reaction of the same cause on the moon consequently delays it, and increases the duration of the month. These effects are of extreme slowness; but in the evolution of the universe ages pass like days. An industrious mathematician, Mr. G. H. Darwin, son of the eminent naturalist, has concluded from his calculations that there was an epoch when the rotation of the earth was effected in 3 hours only, and that at this same epoch the revolution of the moon was likewise performed in 3 hours, and that this epoch is that of the birth of the moon, detached from the terrestrial equator—the earth being then entirely liquid—on account of a tremendous solar tide. This occurred about 54 millions of years ago. The same calculations indicate that the terrestrial day should increase until it becomes 70 times as long as at present, which would give $5\frac{1}{4}$ days per annum, and that the lunar month will likewise arrive at this same duration, but that no less than 150 millions of years will be necessary to reach this state, in which the moon and earth would constantly show each other the same face.

If the moon, which is 81 times less powerful than the earth, produces such tides now, what influence must have been exercised by the earth on the moon when it was liquid and much nearer to us? It is due to this influence that this globe can no longer turn freely on itself; in persistently retarding its primitive motion the earth has ended by so far annulling it that the moon always keeps the same hemisphere towards us. This is to be regretted by everybody.[1]

It is natural to ask here whether the sun and especially the moon, in

to him. Twice a day he pursues her with great bellowings, and each time the Seine escapes into the meadows, and goes back towards its source, contrary to the natural course of rivers.'

One day, after being present at Caudebec at this always curious spectacle of the 'bore' of the Seine, I returned on foot through a charming wood, by the road which leads to Yvetot, where I was overtaken by a countryman, with whom I entered into conversation. On my asking him what he thought, and what the old people of his family thought, of the phenomenon which they had observed for so many years, 'I do not know,' he replied, 'how the wise men explain it; but for us it seems to be nothing else but the well-known antipathy between salt water and fresh. They are not of the same character, you see, and there is on that account a natural hostility which we forget. But this is certain, that the fresh water in falling into the sea worries the salt water, with which it finds a difficulty in mixing. It is easy to follow the difference of colour up to Trouville. Well, then, the salt water ends by becoming angry. It accumulates its anger, and every evening, especially at the equinoxes, when it is already naturally furious, it resolves on hunting the fresh water and sending it back with great velocity. I assure you, sir, that this reason is very much simpler than the attraction of the moon.'

We may add that the tide produces similar effects in other rivers suitable for their production.

In the beautiful bay of Mont Saint-Michel the arrival of the high tides affords to the contemplator of nature one of the grandest spectacles which it is possible to see.

[1] Man makes use of the tides for ships entering and leaving his ports. But it is not necessary to conclude from that, like the Abbot Pluche, author of *Le Spectacle de la Nature* that the tides have been created expressly that ships might enter Havre—and castor oil to clear the obstructed mucous membranes. These are final causes—not divine, but very human.

acting on the atmosphere of the earth, produce an effect analogous to that which these bodies produce on the sea, and which we have just analysed. There cannot be the least doubt on this subject. The sun and moon exercise their actions on the atmospheric air just the same as on the sea, and there result in the atmosphere true tides. But how can we ascertain their existence?

We are not placed so as to see the external surface of the terrestrial atmosphere, as we see the surface of the sea. It is not, then, by observation of the motion of this external surface, sometimes rising, sometimes falling, that the atmospheric tides can be rendered perceptible. Situated at the bottom of the atmosphere, we cannot perceive the existence of the atmospheric tides, as we could not perceive the tides of the ocean if we were placed at the bottom of the sea. Now, it is clear that the only effect we could experience in this case would be a periodical change in the pressure of the water, on account of the alternate increase and decrease in the thickness of the liquid situated above us. The atmospheric tides, then, can only be rendered perceptible by the periodical variations of the pressure of the atmosphere at the place where we are—that is to say, by the alternate increase and decrease in the height of the barometric column, which serves to measure this pressure. Calculation shows that there would only be some hundredth of an inch of difference in the barometer.

Reduced to these terms the question is very clear. The daily observations show that the height of the barometric column experiences at one and the same place accidental variations, which may amount to $1\frac{1}{4}$, 2, and even $2\frac{1}{3}$ inches, and that it frequently varies several hundredths of an inch without any great atmospheric disturbances. If the tides produced in the atmosphere by the action of the moon take part in these variations, it must be acknowledged that this part is very feeble, and that we are not warranted in seeing therein one of the principal causes of those changes of weather which we should have so much interest in being able to predict, and which baffle so much, whatever we may do, all the attempts made with a view to arrive even at a rough outline of this prediction.

Perhaps the moon produces not only oceanic and atmospheric tides, but even subterranean tides. The interior heat of the globe leads us to think, not that the centre of the earth is entirely liquid, but that there is a fluid layer at some distance under our feet. This layer may be acted on by the lunar attraction. Statistical researches have been made with the object of testing the correctness of this argument, and their author, M. Perrey, my lamented colleague of the Academy of Dijon, has found, by classing all the earthquakes in order of date, that they happen oftener at new and full moon, as well as on the days when the moon is in perigee, or at its least distance from the earth.

We now arrive at the much controverted question of the influences of the moon.

CHAPTER VIII

THE INFLUENCES OF THE MOON

IF the saying, 'Vox populi, vox Dei' were still true, we might be assured that the moon exercises on the earth and its inhabitants the most extraordinary influences. According to popular opinion, it should have an action on the changes of the weather, on the state of the atmosphere, on planets, animals, men, women, eggs, seeds, on everything in the world. The moon has entered into all forms of language, from the 'honeymoon' to the 'April moon.' What is there of truth in these traditions? All are certainly not accurate, but all are, perhaps, not false either.

'I am delighted to see you collected round me,' said Louis XVIII. one day to the members composing a deputation from the Bureau des Longitudes, who had gone to present to him the 'Connaissance des Temps' and the 'Annuaire,' 'for you will explain to me what the *April moon* is, and its mode of action on the crops.' Laplace, to whom he more especially addressed these words, was astounded. He, who had written so much on the moon, had never, in fact, thought of the April moon. He consulted all his neighbours by a look, but seeing nobody disposed to speak, he determined to reply himself. 'Sire, the April moon does not hold any place in astronomical theories; we are not, then, able to satisfy the curiosity of your Majesty.' In the evening, during his game, the King was very merry over the embarrassment in which he had placed the members of *his* Bureau des Longitudes. Laplace heard of it, and went to ask Arago if he could enlighten him about this famous April moon, which had been the subject of such a disagreeable mishap. Arago went for information to the gardeners of the Jardin des Plantes, and the following is the result of his inquiry:—

Gardeners give the name of 'April moon' to the moon which commences in April and becomes full either at the end of that month or, more usually, in the course of May. In popular opinion the light of the moon in April and May exercises an injurious action on the young shoots of plants. They are confident of having observed that on the nights when the sky is clear the leaves and buds exposed to this light are blighted—that is to say, are frozen, although the thermometer in the atmosphere stands at several degrees above zero. They add, however, that if a clouded sky arrests the lunar rays, and prevents them reaching the plants, the same effects no longer take place, in circumstances of temperature, moreover, perfectly similar. These phenomena seem to indicate that the light of our satellite may be endowed with a certain freezing effect. Nevertheless, in directing the largest lenses and reflectors towards the moon, and then placing in their focus very delicate thermometers, nothing has ever been perceived which could justify such a singular conclusion. While, on the one hand, scientists have put aside the April moon among popular prejudices, on the other hand agriculturists are convinced of the correctness of their observations. The following is the explanation :—

The physicist Wells first ascertained that at night objects may acquire a temperature different from that of the atmosphere which surrounds them. This important fact is now proved. If we place in the open air small pieces of cotton, eiderdown, &c., we often find that their temperature is six or seven or even eight degrees Centigrade below the temperature of the surrounding atmosphere. Vegetables are in the same case. We cannot, then, judge of the cold which a plant has experienced in the night by the sole indications of a thermometer suspended in the atmosphere. Place a thermometer flat on the ground : its temperature will descend below that of the air, if the sky is very clear. A plant may be much frozen although the air may be constantly maintained at several degrees above zero.

These differences of temperature are only produced in perfectly clear weather. If the sky is cloudy the difference disappears entirely, or becomes imperceptible.

In the nights of April and May the temperature is often only a few degrees above zero.[1] At that time plants exposed to the light of the moon—that is to say, to a clear sky—may be frozen, notwithstanding the thermometer. If the moon, on the contrary, does not shine, if the sky is cloudy, the temperature of the plants not descending below that of the atmosphere, they would not freeze—at least not till the thermometer has marked zero. It is, then, true, as the gardeners assert, that with quite similar thermometric circumstances a plant may be frozen or not, according as the moon is visible or hidden behind clouds. If they are mistaken, it is only in the conclusions—that is, in attributing the effect to the light of the moon. The lunar light is here but an index of a clear atmosphere : it is in consequence of the clearness of the sky that the nocturnal freezing of the plants is effected. The moon contributes in no way to the result. Whether it is set or on the horizon the phenomenon would be the same.

It is thus that *dew* is produced. By the effect of nocturnal radiation bodies exposed in the open air are cooled down, and this cooling condenses on them the vapour of water diffused in the atmosphere. Dew does not

[1] [That is, zero of the Centigrade scale, or 32° Fahrenheit.]

descend from the sky, nor does it rise from the earth. A light covering, a sheet of paper, a cloud, is sufficient to check the radiation and prevent dew, as it would prevent frost.

People also attribute to the moon the power of destroying old buildings. Moonlight seems to prefer ruins and solitudes, and the mind associates with it the devastations caused by the rain and sun. Examine the tower of Nötre Dame in Paris, compare carefully the south side with the north, and you will find that the former is incomparably more worn, more worm-eaten, than the latter. The caretakers would tell you 'it is the moon.' Now, as this body follows the same course in the sky as the sun, it would certainly be very difficult to assign the part due to each; but if we reflect that the rain and the wind come precisely from this same south side, we cannot doubt for a single instant that these, in conjunction with the solar heat, are the destructive agents, and that the moon may be quite innocent.

Another point now: *The moon eats up the clouds.* Such is the widely-spread saying among country people, and especially among sailors.

The clouds, they think, tend to disappear when the rays of the moon strike them. Is it permissible to consider this opinion as a prejudice unworthy of examination when we see a scientist like Sir John Herschel vouching for its correctness?

We have said that the lunar light is not absolutely in the same state at the surface of the earth, where experiments with lenses and reflecting mirrors are generally made, as it is in the aërial heights where the clouds soar. When the moon is full it has experienced for several days, without interruption, the calorific action of the sun. Its temperature is very much raised. The vapour of water which constitutes the clouds may be in that state of unstable equilibrium in which the slightest influence might transform the visible into invisible globules. That there is not less water on that account in the atmosphere I have ascertained many times in a balloon; but the clouds may disappear, because the vapour passes from the visible state to the invisible. It is, then, not impossible that the observations of sailors and many scientists may not be due to simple coincidence, but may be based on a real fact. But we can easily observe in full sunshine that the light clouds diminish and disappear in a few minutes in consequence of a change in their height. In such a case the moon would have nothing to do with it, and would only enable us to see the fact.

We may add that the lunar light emits *chemical* rays. Since the discovery of photography we know that the moon acts on the sensitive plates and imprints itself with the greatest fidelity.

As for *the influence of the moon on the weather*, the luminous or calorific action of our satellite is so feeble that it by no means explains the popular prejudices. At the epoch of new moon the lunar globe sends us neither rays of light nor calorific rays. To the full moon, on the contrary, corresponds the maximum of the effects of this kind. Between these two points it is by imperceptible degrees that the action increases or diminishes. We do not see, then, what can be the cause of the supposed abrupt changes. We have seen above that the atmospherical tides are imperceptible. Moreover, before looking for the reason of these changes, it is necessary that they should be proved by observation, and this has not yet been clearly established by anyone.[1]

Arago found that at Paris the maximum of rainy days happens between the first quarter and the full moon, and the minimum between the last quarter and the new moon; Schübler has found the same result for Stuttgart. But A. de Gasparin found the contrary for Orange and Poitiers, and something else for Montpellier. It is, then, probable that these results depend entirely on the variation of the weather, whatever it may be, and prove nothing with reference to the moon.

In the present state of our knowledge we cannot yet base anything on the phases of the moon. What makes a great many farmers and sailors give the first place to the four phases of the moon for the regulation of the weather is, that they only consider them to within a day or two, before or after; noticing one fact in agreement, and neglecting ten which are not so.

The prediction of the weather for a long time ahead cannot, then, inspire any confidence in so far as it is based on the motions of the moon.

This prediction of the weather cannot, besides, be better based on other grounds.[2] At present it is absolutely futile to venture conjectures on fine or bad weather for a year, a month, or even a week in advance.

The human mind, especially the popular mind, is so constituted that it must believe, even when the object of its belief is proved to be neither real nor rational, and teachers of science should always be prepared to reply to questions. We know the story of the lady who, in an elegant drawing-room, asked an Academician: 'What is there at the back of the moon?' 'Madam, I do not know.' 'But to what is due the persistency of rain this year?' 'Madam, I am ignorant of it.' 'And do you think the inhabitants of

[1] This question, whether the light of the moon produces appreciable calorific and chemical effects, is not without interest from a theoretical point of view, and also when we consider the part which the moon is made to play in the explanation of meteorological phenomena. It has been submitted to the proof of experience.

Photometric measures seem to show that the light of the full moon is 300,000 times less than that of the sun. It would be necessary to suppose the whole sky covered with full moons to equal the intensity of daylight. [According to Professor Young, a sky covered with full moons would only give one-eighth of the light of the sun.—J. E. G.]

According to the most delicate experiments of Melloni, Piazzi Smyth, Lord Rosse, and Marié Davy, the heat of the lunar rays which reach the bottom of the atmosphere, where we breathe, is scarcely *twelve-millionths of a degree!* On the Peak of Teneriffe, under a much less thickness of atmosphere, it has been found equal to one-third of that of a wax candle placed at a distance of 4·75 metres (15·6 feet). This is always extremely feeble.

[2] See our work, *L'Atmosphère, Météorologie populaire.*

Jupiter are formed like us?' 'Madam, I know nothing about it.' 'Why, sir, you are joking! What is the use, then, of being so learned?' 'Madam, to reply sometimes that we are ignorant.'

He had certainly nothing to be ashamed of in admitting his ignorance on questions of which no one could say, *I know it*. To what, then, is due the great success of the almanacks of Mathieu Lænsberg and others? Evidently to the vulgar predictions which are inserted. When men speculate on human credulity they are always sure to succeed. Although the predictions may be clearly falsified, the public do not the less consult the famous almanack. Moreover, in the matter of proverbs, predictions, and superstitions, the memory remains impressed with the one case in a hundred in which the predictions and proverbs are realised, and passes unnoticed the ninety-nine other cases.[1] The position of the personages to whom these predictions refer also plays an important part. Thus, in the almanack for 1774, Mathieu Lænsberg announced that one of the most favoured ladies would play her last part in the month of April. Precisely in this month Louis XV. was attacked with small-pox and Dubarry was expelled from Versailles. Nothing more was needed to give the Liège almanack an increase of popularity.

The principal income of the Academy of Berlin was formerly derived from the proceeds of the sale of its almanack. Ashamed to see figuring in this publication predictions of all sorts, made by chance, or which, at least, were not founded on any accepted principle, a distinguished scientist proposed to suppress them, and to replace them by ideas clear, precise, and certain on subjects which, it seemed to him, should be of more interest to the public; they attempted this reform, but the sale of the almanack was so much diminished, and consequently the funds of the Academy so much reduced, that they were obliged to return to their former plan, and restore the predictions, which the authors themselves did not believe in.

Moreover, the astronomical almanack of France, which has given every year, for more than two centuries, the positions of the sun, the moon, the planets, and the principal stars in the sky—has it not had, like all almanacks, a meteorological rather than an astronomical origin, and does it not lead into error the ignorant public, who judge it by its title, the 'Connaissance des Temps?' Now, this almanack of calculations is in no way concerned with the weather, in the general meaning attached to this word. But its title is deceptive.

[1] Thus, for example, a little book which I have before me asserts that in a battle a ball fired at a Pontifical Zouave was flattened against a medal, 'thus testifying to Divine protection.' We will admit the *fact*, observed among a thousand wounded men. Well, then, some years later, the son of Napoleon III., godson of Pius IX., and bearer of a cross, a medal, and a chaplet of beads, fell under seventeen blows of assegais given by the Zulus. We do not say that this fact is absolutely opposed to the first, which was, however, itself an arbitrary interpretation. Thus credulity is sustained.

There is a good story of a preacher who spoke against lotteries. 'Because they may have dreamt of,' said he, 'three numbers' (and he named them), 'they deprive their families of necessaries and the poor of their portion, to put into the lottery.' At the end of the sermon a good woman approached him: 'My father,' said she, 'I heard the first two numbers; what is the third?'

The public still attribute to the moon influences on the nervous system, on the trees, the seed-time of certain vegetables, the laying of eggs, &c. Of all the questions which I have proposed to the believers in this influence, the result is that *no one* has ever asserted to me that he has himself made *a single conclusive experiment.*

Without being able to deny in an absolute manner the reality of some of the influences which have not been proved, observation and discussion do not warrant us in sharing the popular beliefs. Scientific men are sometimes accused of not yielding to evidence; but here the evidence is far from being real. Without denying anything, science can only admit what is *ascertained.*

CHAPTER IX

ECLIPSES

WE now come to one of the most striking and popular of celestial phenomena. When, in the middle of a fine day, in a sky clear and cloudless, the dazzling disc of the sun, devoured by an invisible dragon, gradually diminishes in extent, becomes a thin thread of pale light, and entirely disappears, why are we not impressed with this mysterious extinction? If we were ignorant that this fact is due to the momentary interposition of the moon in front of the luminous star, and that it is an inevitable result of the regular motion of our satellite, should we not believe in the continuance of this extraordinary night? Should we not imagine it to be the work of some evil spirit, or dread it as the manifestation of Divine anger? This is, in fact, the general impression which is remarked among all ignorant nations and in all ages: for most of them, an invisible dragon devours the sun. The impression caused by an eclipse of the moon is of the same order, in so far as it leads us to fear some derangement in the harmonious regularity of the celestial motions.

Eclipses, like comets, have always been interpreted as the indication of inevitable calamities. Human vanity sees the finger of God making signs to us on the least pretext, as if we were the end and aim of universal creation.

Let us mention, for example, what passed even in France with reference to the announcement of an eclipse of the sun on August 21, 1560. For one, it presaged a great overthrow of States and the ruin of Rome; for another, it implied another universal deluge; for a third, nothing less would result than a conflagration of the globe; finally, for the less excited, it would infect the air. The belief in these terrible effects was so general that, by the express order of the doctors, a multitude of frightened people shut themselves up in very close cellars, well heated and perfumed, in order to shelter themselves from these evil influences. Petit relates that the decisive moment approached, that the consternation was at its height, and that a parish priest of the country, being no longer able to confess his parishioners, who believed their last hour had come, was obliged to tell them in a sermon 'not to be so much hurried, seeing that, on account of the wealth of the penitents, the eclipse had been postponed for a fortnight.' These good parishioners found no more difficulty in believing in the postponement of the eclipse than they had had in believing in its unlucky influence.

History relates a crowd of memorable acts on which eclipses have had the greatest influence. Alexander, before the battle of Arbela, expected to see his army routed by the appearance of a phenomenon of this kind. The death of the Athenian general Nicias and the ruin of his army in Sicily, with which the decline of the Athenians commenced, had for their cause an eclipse of the moon. We know how Christopher Columbus, with his little army, threatened with death by famine at Jamaica, found means of procuring provisions from the natives by depriving them in the evening of the light of the moon. The eclipse had scarcely commenced when they supplied him with food. This was the eclipse of March 1, 1504, observed in Europe at Ulm by Stoffer, and at Nuremburg by Bernard Walter, and which happened at Jamaica at 6 o'clock in the evening. We need not relate other facts of this nature, in which history abounds, and which are known to everyone.

Eclipses no longer cause terror to anyone, since we know that they are a natural and inevitable consequence of the combined motions of the three great celestial bodies—the sun, the earth, and the moon; especially since we know that these motions are regular and permanent, and that we can predict, by means of calculation, the eclipses which will be produced in the future as well as recognise those which have occurred in the past. Thus, an astronomer of the end of last century, Pingré, author of the 'Cometographie,' calculated the precise dates of all the eclipses which have happened during the last three thousand years.

Everyone knows now that it is the moon which, in revolving round the earth, produces sometimes an eclipse of the sun, when it is interposed between the sun and the earth, and sometimes an eclipse of the moon, when it is placed behind the earth with reference to the sun. These two phenomena are of a different kind. In an eclipse of the sun the moon hides the sun wholly or in part for certain points of the earth's surface; the eclipse is presented with such or such a character according as we are placed at such or such a place to observe it. Here it is total or annular; there it is but partial, and the part of the sun concealed is greater or less; elsewhere they perceive no trace of the eclipse. In an eclipse of the moon, on the contrary, our satellite ceases wholly or in part to be illuminated by the sun, because it then passes through the shadow of the earth, and this aspect of the moon is the same for all the inhabitants of the terrestrial hemisphere where our satellite is above their horizon.

We understand at once from this that the calculation of an eclipse of the moon presents fewer complications than that of an eclipse of the sun, since for the former we have only to indicate the general circumstances of the phenomena, which are the same for all observers; while for the latter the indication of the general circumstances is far from sufficing, on account of the differences of aspect according to the region, and the narrowness of the zone in which the eclipse is central. The ancients, who were far from knowing the motion of the moon with as much precision as we do, had not the means of accurately predicting eclipses of the sun. They only predicted eclipses of the moon, relying on the fact that they are reproduced very nearly periodically, presenting the same characters, and with the same interval between them, in the course of 18 years 11 days; so that it was sufficient to have observed and registered all those which were produced in a similar period of time to foretell with certainty those which would be produced in the following period.

Now, on the contrary, with the much more precise knowledge we have of the moon's motion, we are in a position to calculate and foretell for a great number of years, and even centuries in advance, not only the general circumstances of eclipses of the moon, but even the detailed course of eclipses of the sun. We can even, by a retrospective examination, give an account of all the circumstances which an ancient eclipse should have presented in such or such a locality, and find the precise date of certain historical events of which the epoch is a subject of discussion. An eclipse of the sun is a veritable rarity for any given place. (Thus, for example, there has not been one at Paris since May 22, 1724; the nineteenth century has not a single one; in the twentieth century, on April 17, 1912, Paris will be just on the limit of totality; but a true total eclipse, of several minutes' duration, will not be seen in the capital of France till August 11, 1999.) Herodotus relates that at the moment of a battle between the Lydians and the Medes a total eclipse of the sun at once stopped the stupefied combatants and put an end to the war. Till recently, historians gave various dates for this event, from the year 626 before our era down to the year 583; astronomical calculation, however, proves that this battle took place on May 28 of the year 585 B.C.

We will explain these phenomena in a few words.

Eclipses of the sun always happen at the moment of new moon, and eclipses of the moon at the moment of full moon. This circumstance has long since made known the cause to which they are due. At the moment of new moon, the moon, passing between the earth and the sun, can hide from our view a larger or smaller portion of that body. At the moment of full moon, on the contrary, the earth is found between the sun and the moon; it can then prevent the solar rays from reaching the latter body. All is thus easily explained.

If the moon revolved round the earth in the same plane as the earth round the sun, it would be eclipsed in our shadow at every full moon, and it would eclipse the sun at every new moon, as we see by fig. 66. But she passes sometimes above and sometimes below the cone of the shadow, and she can only be eclipsed when she penetrates into this shadow.

We can very easily understand the production of eclipses by an examination of this figure. The sun is represented at the top of the drawing. We see in the lower part the earth, accompanied by the moon. The latter revolves, as we have seen, round the earth. When it passes, at the moment of full moon (the lower part of its orbit), through the shadow of our globe, she no longer receives the light of the sun : this is an *eclipse of the moon*, total or partial according as our satellite is totally or partially immersed in our shadow. On each side of the complete shadow there is a penumbra (which will be explained by following the dotted lines), due to the fact that a part only of the solar light penetrates into this region. A second penumbra, very thin, is produced by the atmosphere which surrounds our globe.

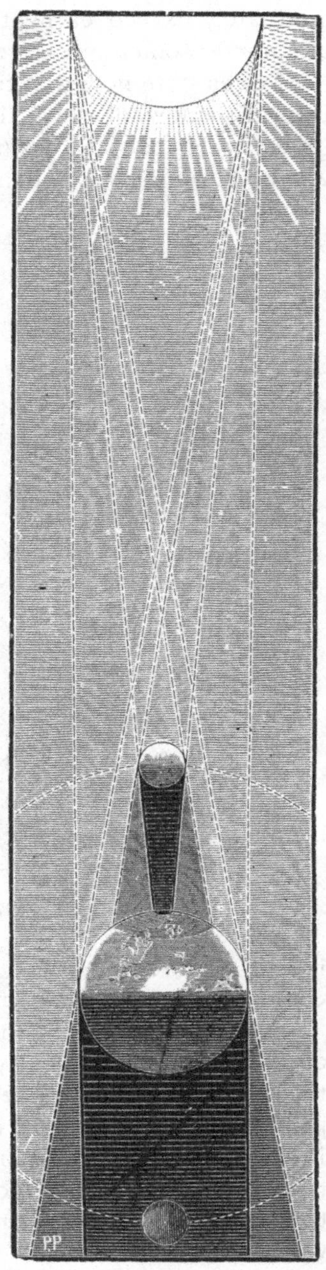

FIG. 66.—THEORY OF ECLIPSES

On the other hand, when at the moment of new moon our satellite passes exactly before the sun, its shadow falls on us, and produces on the surface of our globe a little circle which travels over different countries, following the motion of rotation of the earth. All the countries over which this shadow passes have the sun hidden during a certain time ; this is an *eclipse of the sun—total* if the moon is sufficiently near us for its apparent diameter to exceed that of the sun, *annular* if it is then in the most distant region of its orbit and is smaller than the solar disc, *partial* if the centres of the moon and the sun do not coincide, and if the moon only hides the sun on one side.

Such is the general theory of eclipses. We will now examine the details of the phenomenon, and we will commence with ECLIPSES OF THE MOON.

ECLIPSES OF THE MOON

Although the moon is very small compared with the sun, it subtends pretty nearly the same angle, because it is very much nearer ; it happens, then, on account of the changes of distance of the two bodies from the earth, that they alternately exceed each other in apparent size, and that the moon shows a diameter sometimes larger and sometimes smaller than that of the sun.

Let us state now that the earth throws behind it, opposite to the sun, a conical shadow, of which the length is 108 times the semi-diameter of our globe, or 345,000 leagues (about 857,000 miles). There it ends in a point. At the mean distance of the moon, 238,000 miles, the shadow of the earth

FIG. 67.—AN ECLIPSE OF THE MOON

is a little more than twice (2·2) as wide as the moon. When our nocturnal companion passes through this shadow it is eclipsed.

At the beginning of a total eclipse of the moon we notice a weakening of its light, at first scarcely perceptible, then more and more marked ; at this moment the moon enters, or has entered for some time, into the penumbra. Then a slight indentation is formed on the limb, and by degrees it encroaches on the luminous part of the disc. The form is circular, and this is one of the first proofs observers had of the sphericity of the earth, the shadow having evidently the same form as the profile of the object which produces it.

The colour of the shadow is at first greyish black, which permits us to see nothing of the eclipsed part; but as the shadow overruns the lunar disc it is coloured with a reddish tint, and the details of the principal spots become visible. We often remark round the shadow a thin edging of greyish blue. As soon as the eclipse is total the red colour becomes more intense, and spreads directly over the whole of the disc. The moon may remain eclipsed for nearly two hours. After traversing the whole width of the shadow, it reappears, showing at first a thin luminous crescent, which imperceptibly enlarges. Its proper motion round us taking place from west to east—that is to say, from right to left—it is on its left or eastern side that it penetrates into our shadow and commences to be eclipsed, and it is likewise on that side that it first returns to the sunlight.

The moon scarcely ever disappears in total eclipses. The cause of this fact is in the refraction of the solar rays, which, passing through the lower and densest layers of the earth's atmosphere, project on the moon the purple tints of our sunsets. It becomes, however, sometimes completely invisible; we may quote as examples of this fact the eclipses of 1642, 1761, and 1816, in which it was impossible to find the place of the moon in the sky. At other times the visibility, without being nil, is very imperfect; for example, the eclipse of October 4, 1884. Sometimes, on the contrary, as in 1703 and 1848, the moon remained so illuminated that some people doubted that it was eclipsed at all. The explanation of these circumstances is to be found in the particular state of the atmosphere over the whole of the terrestrial periphery, including the places where the sun is rising and setting at the moment of the eclipse.[1]

We may very easily form an idea of the course followed by astronomers in order to calculate in advance eclipses of the moon.

We have already seen that they call the 'line of nodes' the line of intersection where the plane of the lunar orbit cuts the plane of the ecliptic, and that these two planes have between them an angle of 5 degrees. This line revolves, and returns into the same direction relatively to the sun at

[1] Another phenomenon appears inconsistent with the geometrical theory of eclipses. I refer to the simultaneous presence of the sun and moon above the horizon during the totality of a lunar eclipse. The first of these bodies being setting at the moment the other is rising, it seems that the moon, the earth, and the sun are not in a straight line. This is an appearance merely due to refraction. The sun, already below the horizon, is raised by refraction, and remains visible to us. It is the same with the moon, which has not yet really risen when it seems to have already done so. The eclipses of 1666, 1668, and 1750 may be mentioned as having presented this singular appearance. But it is not necessary to go back so far. On February 27, 1877, the moon rose at Paris at 5 hours 29 minutes, and the sun set at 5 hours 39 minutes, a total eclipse having begun. If the fact is not very often observed, it is the fault of observers. On December 16, 1880, there was a total eclipse of the moon visible at Paris. On that day the moon rose at 4 hours 0 minutes, and the sun set at 4 hours 2 minutes; this was almost in the middle of a total eclipse, which took place from 3 hours 3 minutes to 4 hours 33 minutes. The rarest coincidence is to see at the same time the sun and the moon just at the middle of a total eclipse. In order to see the moon totally eclipsed before the setting of the sun or after he rises, it is sufficient to have the moon at the horizon near the middle of the eclipse.

the end of 223 lunations, or 6,585 days, or 18 years and 11 days. As eclipses are only produced when the full moon and the new moon happen on this line, it is necessary and it is sufficient to register all the eclipses which occur during this period in order to know all those which can be produced indefinitely. This method of prediction of eclipses was already known by the Chaldeans more than two thousand years ago, and designated by the name of *Saros.*

This period is not absolutely mathematical. It may serve to predict that an eclipse will happen at a certain epoch, but it will not determine with precision the importance or the duration of the eclipse, which really differs a little from the preceding eclipse, with which it should be identical if the period were exact. It may even happen that a very small partial eclipse is not reproduced at all at the end of 18 years 11 days, and also that a partial eclipse may appear 18 years 11 days after an epoch when there had

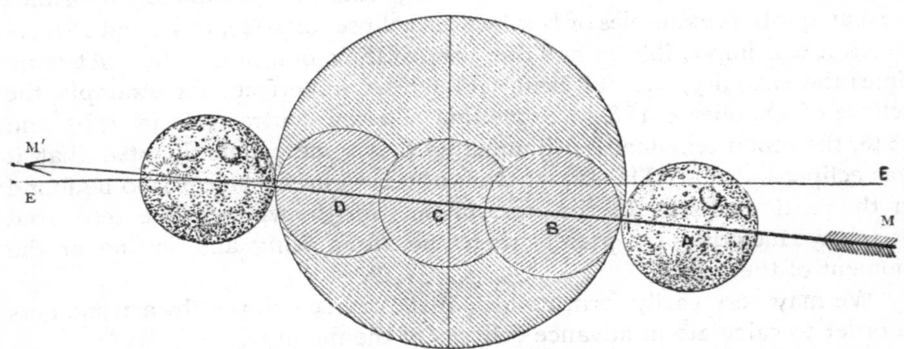

Fig. 68.—Passage of the Moon through the Shadow of the Earth during an Eclipse of the Moon

not been one. Then, too, the employment of this period, which constituted the only means used by the ancients for their predictions, no longer suffices, now that astronomical theories permit us to attain an incomparably greater precision in the determination of the series of eclipses which must occur.

. . . .

The quantity by which the moon is eclipsed is nearly the same (after the lapse of the above period); however, a partial eclipse may become total. Thus, that of October 13, 1856, which was 99 hundredths of the lunar disc, became total on October 25, 1874, being then 105 hundredths—that is to say, a little larger than the lunar disc. The differences of the hours form the great apparent difference for the public, seeing that they may render the eclipse visible or invisible for a given place according as the moon has risen or set.

I have observed since 1858 (the year of my entering the Paris Observatory) all the lunar eclipses which have been visible at Paris. Several have presented certain interesting features.

That of June 1, 1863, was observed in company with my intellectual master, Babinet, and my lamented friend, Goldschmidt. The lunar disc remained constantly visible, coloured with a deep red, although the total eclipse lasted more than an hour. Before and after the totality the illuminated lunar crescent showed a bluish tint, evidently arising from the contrast of its white light with the adjacent red colouring. We distinguished through the whole duration of the eclipse the different tints of the lunar disc. Our satellite passed on that evening through a celestial region crowded with stars, and the motion of the moon in front of them was shown by the march of these small stars along the limb; several appeared successively hidden and uncovered by the hollows of the lunar mountains. In the middle of the eclipse the moon showed almost the same amount of light as the star Alpha of the Eagle, a little more than Spica of the Virgin, and much more than Antares. When it emerged from the earth's shadow, the crescent thus formed appeared very much lit up in its eastern half and very dull in its western half, and the difference persisted nearly to the end of the eclipse. This difference proved without doubt that the solar rays grazing the terrestrial globe were stopped in Greenland by its glacier of 500 metres thick, while in the other section they grazed the North Sea.

In the eclipse of October 4, 1865, the only interesting fact which I remarked was that the rays of Tycho remained perfectly visible in the middle of the eclipse, as well as the eclipsed amphitheatres and craters.

In the eclipse of July 12, 1870, the quantity of light received from the moon was less than that of Saturn, and superior to that of Alpha of the Eagle. During the ten minutes which followed the moment of central eclipse this brightness was considerably increased. The state of the terrestrial atmosphere and refraction played here an important part.

October 25, 1874, 6 A.M.—There were in this month three eclipses in fifteen days, because the moon eclipsed the sun on October 10, occulted Venus on the 14th, and was itself eclipsed on the 25th. If astronomical observations differ very much among themselves, they differ still more, perhaps, in the variety of the meteorological conditions under which we are obliged to make them. Thus it was that, to observe the eclipse of the sun on the 10th, it was necessary to expose one's face to the burning heat of a veritable summer sun, for the occultation of Venus it was necessary to search for the planet in the dazzling southern sky with half-blinded eyes, and the eclipse of the moon on the 25th had to be observed in the cold of an early morning atmosphere worthy of the nights of winter. But all these little corporeal inconveniences are nothing when a cloud does not happen to hide the expected phenomenon, and we can, after all, make a satisfactory observation.

The full moon began to enter the penumbra at 4 hours 55 minutes A.M., but it had already sunk towards the western horizon, and thick vapours, mists, and cloudy trails surrounded it with a sort of whitish veil. The image was far from being clear, although we distinguished very well the whole of the lunar geography. The white and radiant mountain of Aristarchus shone just in the lower portion of the vertical diameter of the disc, and remained perceptible even when this region had entered into the shadow. I did not succeed in distinguishing the penumbra for nearly an hour after the moon entered it. At 5 hours 20 minutes nothing could be seen. It was the same at 5 hours 30 minutes, and at 5 hours 45 minutes the moon appeared perceptibly cut into on the north-east—that is to say, at the top and to the left (direct image).

At 6 o'clock our satellite was eclipsed by about a quarter of its diameter; the earth's shadow was bounded imperceptibly by a shaded tint, and not by a clear and sharply-cut limit. We saw black corpuscles passing in all directions in front of the star of night; these were birds flying at a great height. At 6 hours 25 minutes the cone of the shadow arrived at the centre of the lunar disc, but reaching the low regions of the atmosphere, the star of Diana seemed to die out and sink in a nest of dark clouds forming the horizon. At 6 hours 30 minutes it disappeared; the shadow had then reached the Sea of Serenity and Mount *Manilius.* This was the largest phase of the eclipse which was visible at Paris. Some minutes afterwards, at 6 hours 37 minutes, the sun was radiant on the eastern horizon.

Neither the 'Connaissance des Temps' nor the 'Annuaire du Bureau des Longitudes' accurately announced the conditions of this eclipse. One announced it for the evening; the other supposed that the full moon would rise at 6 o'clock in the morning! In 1887 the 'Annuaire' also announced that the total eclipse of the sun of August 19 would pass over Egypt instead of Russia. These errors are to be regretted, especially when they occur in official publications.

The eclipse I have just spoken of was total, but only half of it was seen at Paris, owing to the setting of the moon.

The eclipse of September 3, 1876, which was but partial, and one-third only, was favoured at Paris with a very clear sky during the first half of its duration; then the sky became clouded. At Havre it appeared surrounded by a halo, which formed an admirable frame.

There was on August 23, 1877, from 10 hours 28 minutes P.M. to 13 minutes after midnight, a very fine total eclipse of the moon, which everybody observed in France and in Europe, for the sky was on that evening of exceptional clearness. During the whole duration of totality (1 hour 45 minutes) the moon remained perfectly visible and of a reddish tint; a double fact, produced, as we have said just now, by the refraction of the rays of the sun through our atmosphere, which was very clear on that day. These are the same rays which after sunset illuminate in the east with a beautiful rose-colour the clouds, and even the buildings. The borders of the moon were brighter than the centre.

I observed the total eclipse of the moon of October 4, 1884, at the Observatory of Juvisy, with a sky cloudy, but otherwise rather favourable. The character of this remarkable eclipse (nearly central, duration 1 hour 32 minutes) was the most complete darkening of the moon during the whole of the totality; in this respect it should be placed some degrees only above those in which our satellite has entirely disappeared. The shadow of the earth appeared bordered by a transparent shadow of about 2′ in width, apparently produced by the atmosphere, and indicating for that atmosphere a height of 360 kilometres (about 224 miles).

The partial eclipse of August 3, 1887, did not show any remarkable peculiarity, except that the part eclipsed remained constantly visible.

The total eclipse of January 28, 1888, was nearly central (duration 1 hour 38 minutes), as may be understood by fig. 68, which represents the course of the moon through the shadow of the earth during this eclipse: A, B, C, D, positions of the moon at entry, exit, and through the shadow; MM′, course of the moon; EE′, ecliptic. It occurred in excellent atmospherical conditions, notwithstanding the season. I observed it at the Nice Observatory. The moon remained perfectly visible, very clear, and coloured with a tone of clear *copper red* during the whole duration of the eclipse. The moon maintained nearly the brightness of the star Procyon. The borders remained clearer than the interior of the disc.

This colouring of the moon during eclipses arises from the refraction of the solar rays which traverse the atmosphere surrounding the terrestrial globe, and illuminate the moon like the rising or setting sun. The degree of colouring varies according to the state of the atmosphere and its transparency.

We now come to ECLIPSES OF THE SUN.

Eclipses of the Sun

The method of which we have spoken will also serve to indicate in advance that at such or such an epoch there will be an eclipse of the sun, but it cannot tell us whether the eclipse will be visible or not at a given place; and in case the eclipse is visible, it cannot tell us the degree of importance which it will have.

This difference arises from the fact that eclipses of the sun and eclipses of the moon are not of the same nature. The latter are due to the star of night really losing its light, and they are visible in all countries which have the moon, at the time, above their horizon. In an eclipse of the sun, on

Fig. 69.—Image of the Solar Disc projected through the Leaves of a Tree

the contrary, the day star loses nothing of its light. The moon passing in front of him hides a portion of its disc from observers, and this portion is larger or smaller according as the observer occupies such or such a position on the earth, which, moreover, rotates on itself, and thus varies the course of the shadow on its surface.

In certain very rare circumstances an eclipse may be even total at one place and annular at another, when the apparent diameters of the sun and moon are nearly equal, because the moon is not at the same distance from all points of the terrestrial surface. It is total for the country which has the eclipse about noon.

We sometimes see isolated clouds casting their shadows in the middle of a plain, while the sun directly illuminates all the other parts. These clouds being usually in motion, their shadow moves over the fields, often with great rapidity. It is in exactly the same way that the shadow of the moon in total eclipses of the sun moves on the surface of the terrestrial globe in going from one border to the other of the illuminated hemisphere. The shadow of a balloon gives a still more correct illustration. The shadow of the moon is sometimes very small; thus, in the eclipse of May 17, 1882, observed in Egypt, it only measured 22 kilometres (13·67 miles) in width. But this width may rise to 50, 100, 200, 300 kilometres. In the eclipse of August 19, 1887, for example, it rose in Russia to 220 kilometres (136·7 miles). This width depends on the difference of the solar

FIG. 70.—IMAGE OF AN ECLIPSE PROJECTED THROUGH THE LEAVES OF A TREE

and lunar discs on the day of the eclipse. The shadow moves with a velocity which depends on the rotation of the earth and the motion of the moon, and it may be estimated if we observe it from the top of a mountain.

Astronomers always determine in advance the general circumstances which each eclipse of the sun should present over the whole surface of the earth, and in order to easily take it into account they construct a map designed to show the course of the eclipse over the globe. Fig. 71 shows the arrangement of these maps; it refers to the annular eclipse of April 1, 1764, which passed precisely over Paris. The line A B C indicates the points where the eclipse begins at the moment the sun rises, and the line A D C those where the eclipse ends at sunrise. For all the points situated on the line A E C, intermediate between the two preceding, the sun rises in the middle of the eclipse. Similarly, the lines A F G, A H G, A I G include respectively the points where sunset occurs at the end, at the commencement, and in the middle of the eclipse. The narrow band L L′, represented by *three parallel curved lines*, marks the course which the cone of the moon's shadow follows in passing over the earth's surface, as we have said just now. We see that this shadow passed to the north of the Cape Verd Islands, over the Canary Islands, and to the south of Madeira ; that it then

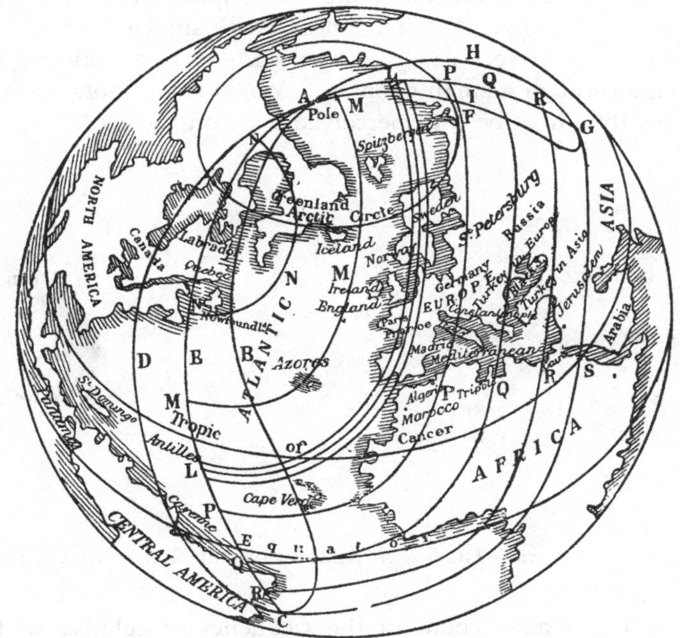

Fig. 71.—Sketch of the Course of an Eclipse of the Sun, and of its Extent in Different Countries

traversed Portugal, Spain, France, the Netherlands, Denmark, and Sweden. The eclipse was central at Lisbon, Madrid, Paris, and in Sweden. On both sides of this band it was partial, smaller and smaller as the points were more distant from this route of the annular eclipse. On all points of the line M M it was but 8 tenths, and along the line N N only 6 tenths. Similarly it diminished on the zones P, Q, R, S. Beyond this last line there was no eclipse at all, notwithstanding the presence of the sun above the horizon.

For each eclipse of the sun a chart similar to this is constructed.

If we expose to the sun during a partial eclipse a visiting-card pierced with a pinhole, and if we place behind it a screen to receive the solar rays which pass through the hole, we see on this screen an image of the solar disc, with the indentation produced by the interposition of the moon. The foliage of trees often permits some rays of the sun to pass, and illuminate certain parts of the ground in the midst of the shadow of the foliage. The interstices of the leaves, then, play the part we have described, and it follows that the parts of the ground illuminated are round or oval (fig. 69) During eclipses of the sun the indentations, more or less decided in the disc of the body, are reproduced in these clear spaces in the middle of the shadow, and they take the form of hollowed ellipses all on the same side and of the same size (fig. 70). This peculiarity shown by the foliage of trees during eclipses is very easily recognised. It is a real projection of the sun's image through a small opening. When solar spots are visible to the naked eye, they may be thus perceived even on a walk.

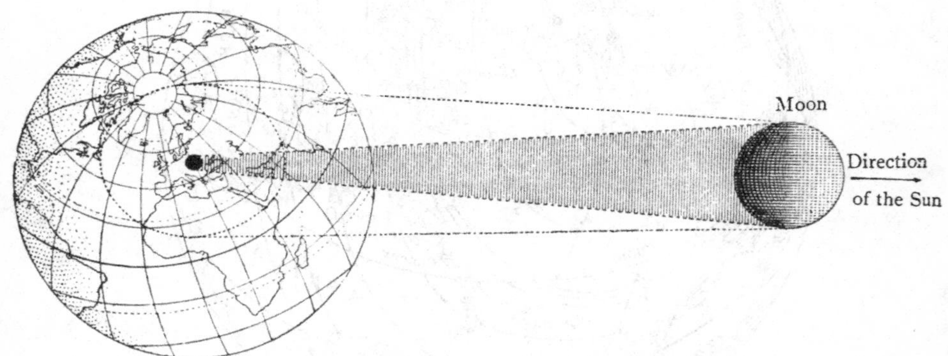

FIG. 72.—THEORY OF ECLIPSES OF THE SUN

We will now give an account of the frequency of eclipses of the sun, and we shall then have the complete theory of these interesting phenomena.

The 'tables' of the sun and moon show that, on the average, we may observe, over the whole earth, seventy eclipses in eighteen years—twenty-nine of the moon and forty-one of the sun. There can never be more than seven eclipses in one year; never less than two. When there are but two, they are both of the sun.

For the whole globe, the number of eclipses of the sun is greater than the number of eclipses of the moon, nearly in the ratio of 3 to 2. In a given place, on the contrary, for the reason we have explained of the constant visibility of eclipses of the moon in all countries in which the moon has risen, eclipses of the moon are much more frequent than those of the sun.

In every period of eighteen years there are, on the average, twenty-eight central eclipses of the sun—that is to say, capable of becoming, according to circumstances, annular or total; but as the terrestrial zone along which the eclipse can have one or other of these characters is very narrow, the total or annular eclipses at a given place are extremely rare.

Halley found, in 1715, that since 1140—that is to say, in a period of 575 years—there had not been in London a single total eclipse of the sun. Since the eclipse of 1715 London has not seen another. Montpellier, very much favoured by the combination of different elements which co-operate in the production of these phenomena, has had during five hundred years, as total eclipses, only the four following: January 1, 1386; June 7, 1415; May 12, 1706; and July 8, 1842.

At Paris, during the eighteenth century, there was but one total eclipse of the sun, that of May 22, 1724. In the nineteenth century Paris has not yet had one, and will not have any. In the twentieth century we find that of April 17, 1912, just at the limit of totality, and that of August 11, 1999, a total eclipse of short duration. In the twenty-first century *we* shall have those of August 12, 2026, and September 3, 2081, both total at Paris.

Calculation shows that the greatest possible duration of an eclipse of the sun from the beginning to the end is 4 hours 29 minutes 44 seconds for a place situated on the equator, and 3 hours 26 minutes 32 seconds at the latitude of Paris. The *total* eclipse cannot last longer than 7 minutes 58 seconds at the equator, and 6 minutes 10 seconds at Paris. In annular eclipses the moon cannot be projected wholly on the disc of the sun for more than 12 minutes 24 seconds at the equator, and 9 minutes 56 seconds at the latitude of Paris. We learn, besides, that the durations of these phenomena pass through all degrees of magnitude below the limits which have been assigned.

The maximum duration of the totality in recent great total eclipses of the sun has been :—

			M.	S.
For the eclipse of		December 22, 1870 (Algeria)	2	10
,,	,,	December 12, 1871 (Australia)	4	22
,,	,,	April 16, 1874 (Cape of Good Hope)	3	31
,,	,,	April 6, 1875 (China)	4	38
,,	,,	July 29, 1878 (United States)	3	11
,,	,,	January 11, 1880 (Pacific Ocean)	2	8
,,	,,	May 17, 1882 (Bokhara)	1	50
,,	,,	May 6, 1883 (Caroline Islands)	5	24
,,	,,	August 19, 1887 (China)	3	50

Several of the solar eclipses have been of the highest importance in the study of the solar atmosphere. It is only, in fact, in those rare and precious moments when the moon completely hides the dazzling light of the day star that we can see the marvellous vicinity of that body, the seat of unimaginable cosmical circulations, of extraordinary conflagrations, of downfalls, and tremendous eruptions, which we shall study in the next chapters, devoted to the glorious sun.

These eclipses have proved that round the sun there is an immense atmosphere of hydrogen, which burns incessantly, of which the height varies, and in which float metallic vapours—an atmosphere traversed by intermittent jets of matter thrown out from the interior of the solar body. Above this atmosphere, all round this fiery furnace, circulate corpuscles in incalculable number, swept into the solar whirlpool. We cannot form any picture of the impetuous motions which are incessantly stirred up in these tempestuous regions—motions so tremendous that masses much more voluminous than the whole earth are removed, tossed up, broken, and reproduced in a few minutes! But we will not anticipate the study of the sun.

. . . .

Without being very rare, partial eclipses are not very frequent at the same place, and are only produced at very irregular intervals.

There will not be a total eclipse entirely visible in France before May 28, 1900; even here the totality will pass along the Pyrenean frontier (see fig. 76).

It is necessary to catch them on the wing, so to say, and not to imitate the too presumptuous marquis of the time of Louis XV. when conducting to the Observatory a party of fashionable ladies, and who, having been a little delayed by the petty cares of the toilet, arrived half a minute after the end of the eclipse. As the ladies refused to alight from their coach, a little displeased by the unreasonableness of coquetry: 'Let us all go in, ladies,' cried the little dandy, with the most haughty assurance; 'M. de Cassini is one of my best friends, and he will have real pleasure in recommencing the eclipse for us!' This anecdote has been retold in our century on the authority of Arago.

Of all astronomical phenomena, there are few which have struck the human imagination so much as total eclipses of the sun. What spectacle more strange, in fact, than that of the sudden disappearance of the day star at noonday in the midst of a clear sky? In the days when humanity was ignorant of the natural causes of these effects, such a disappearance was considered as supernatural, and they saw in it with terror a manifestation of the Divine anger. Since the natural causes have been discovered, and these phenomena are seen to answer to our calculations with the most obedient fidelity, all supernatural terror has disappeared from cultivated minds, but the grand spectacle does not the less impress the beholder.

At the hour predicted by the astronomer we see the brilliant disc of the sun cut into towards the west, and a black segment slowly advancing, eating away the solar disc until it is reduced to the form of a thin luminous crescent. At the same time the daylight diminishes ; from all sides a wan and sinister gleam replaces the brilliant light in which nature rejoiced, and an infinite sadness falls upon the world. Very soon there remains nothing of the radiant star but a narrow arc of light, and hope appears disposed to wing its flight from this earth, so long illuminated by

FIG. 73.—TOTAL ECLIPSE OF THE SUN, MAY 17, 1882. SHOWING COMET SEEN NEAR THE SUN DURING TOTALITY. (Drawn from photographs by W. H. Wesley)

the paternal sun. Life seems still connected with the sky by an invisible thread, when suddenly the last ray of daylight dies out, and a darkness as profound as it is sudden spreads all around us, reducing the whole of nature to astonishment and silence. The stars shine in the sky ! The man who would still speak and communicate his impressions while attentively watching the phenomenon cries out with surprise ; then he becomes silent, struck with stupor. The singing-bird crouches under the leaf ; the dog takes refuge against the legs of his master; the hen covers the chickens with her wings. Living nature is hushed—dumb with astonishment. Night has come—a night sometimes intense and profound, but oftener incomplete, strange, and extraordinary, the earth remaining vaguely illuminated by a reddish light reflected from distant regions of the atmosphere situated outside the cone of the lunar shadow which produces the eclipse. Sometimes we see shining during the eclipse all the stars of the first and second magnitude which are above the horizon, sometimes only the

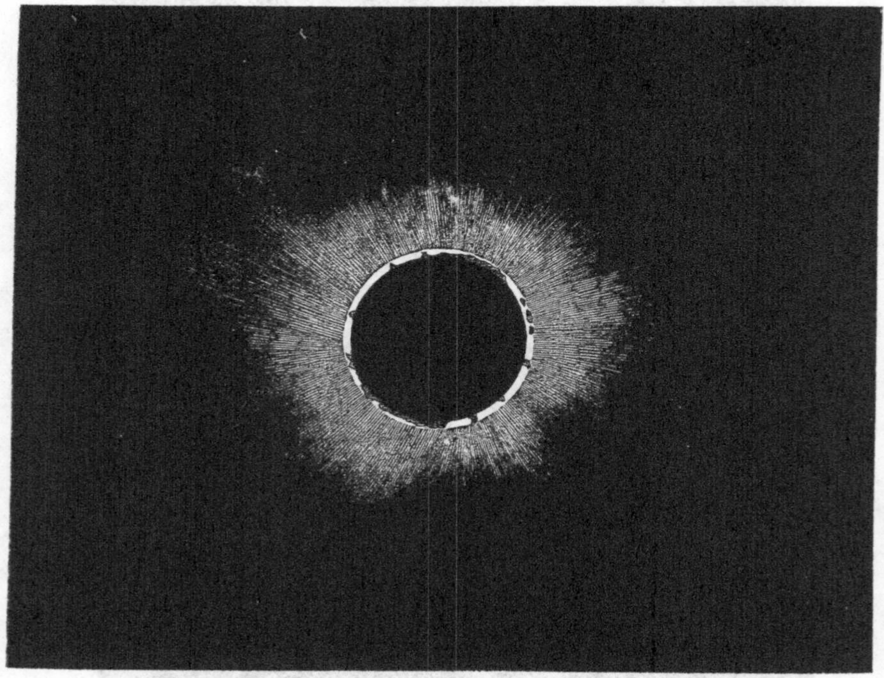

FIG. 74.—TOTAL ECLIPSE OF THE SUN, JULY 18, 1860

brightest of the planets. The temperature of the air rapidly sinks several degrees.

But what a marvellous spectacle is then afforded to all eyes directed to the same point of the sky! In place of the sun appears a black disc, surrounded by a glorious crown of light. In this ethereal crown we see immense rays diverging from the eclipsed sun. Rose-coloured flames appear to issue from the lunar screen which masks the god of day. During two minutes, three minutes, four minutes, the astronomer studies this strange frame, rendered visible by the passage of the moon before the radiant disc, while the people, surprised and still silent, seem to await with anxiety the end of a spectacle which they have never seen before and may never see again. Suddenly a jet of light, a shout of pleasure from a thousand throats, announces the return of the joyous sun, still pure, still luminous, still fiery, still faithful. We think we hear in this universal cry the very sincere expression of an undisguised satisfaction: 'It was very true: the sun, the beautiful sun, was not dead, it was only hidden; yes, see it, quite whole: what happiness! And yet it was very curious to see it thus disappear for an instant!'

The last *total* eclipse of the sun which was visible in France was that of July 8, 1842, seen as partial at Paris, but total in the South of France. I admit that I was not an eyewitness of it, firstly, because I did not live in the zone of central eclipse, secondly, and especially, on account of my extreme youth (the author was then four months and eleven days old); but he who was afterwards by his noble and powerful writings my master, François Arago, had returned to the Eastern Pyrenees, his birthplace, expressly to observe it, and the following is an extract from his account of what he saw:—

'The hour for the beginning of the eclipse approached. Nearly twenty thousand persons, with smoked glasses in hand, examined the radiant globe projected on an azure sky. Scarcely had we, armed with our powerful telescopes, begun to perceive a small indentation on the western limb of the sun, than a great cry, a mingling of twenty thousand different cries, informed us that we had anticipated only by some seconds the observation made with the naked eye by twenty thousand unprepared astronomers. A lively curiosity, emulation, and the desire of not being forestalled, would seem to have given to their natural sight unusual penetration and power.

'Between this moment and those which preceded by very little the total disappearance of the sun, we did not remark in the countenances of many of the spectators anything which deserves to be related. But when the sun, reduced to a narrow thread, commenced to throw on our horizon a much-enfeebled light, a sort of uneasiness took possession of everyone. Each felt the need of communicating his impressions to those who surrounded him: hence a murmuring sound like that of a distant sea after a storm. The noise became louder as the solar crescent was reduced. The crescent at last disappeared, darkness suddenly succeeded the light, and an absolute silence marked this phase of the eclipse, so that we clearly heard the pendulum of our astronomical clock. The phenomenon in its magnificence triumphed over the petulance of youth, over the levity which certain men take as a sign of superiority, over the noisy indifference of which soldiers usually make profession. A profound calm reigned in the air; the birds sang no more.

After a solemn waiting of about two minutes, transports of joy, frantic applause, salute, with the same accord, the same spontaneity, the reappearance of the first solar rays. A melancholy contemplation, produced by unaccountable feelings, was succeeded by a real and lively satisfaction, of which no one thought of checking or moderating the enthusiasm. For the majority of the public the phenomenon was at an end. The other phases of the eclipse had hardly any attentive spectators, apart from those devoted to the study of astronomy.'[1]

Every observation of an eclipse presents similar scenes, more or less varied. At the time of the eclipse of July 18, 1860, we saw, in Africa, men and women, some praying, others flying towards their dwellings. We also saw animals proceeding towards the villages as at the approach of night, ducks collected into crowded groups, swallows hurling themselves against the houses, butterflies hiding, flowers—and notably those of the *Hibiscus Africanus*—closing their corollas. In general, it is the birds, the insects, and the flowers which appear the most influenced by the darkness due to an eclipse.

At the eclipse of August 18, 1868, which M. Janssen went to observe in India, the natives engaged to carry out his arrangements all escaped at the moment it commenced and ran away *to bathe.* A religious rite commands them to plunge into water up to the neck in order to exorcise the influence of the evil spirit. They returned when all was over.

During that of May 15, 1877, the Turks raised a veritable riot, notwithstanding their preparations for war with Russia, and fired gunshots at the sun to deliver it from the claws of the dragon. The illustrated journals even represented this—for our age—very curious scene.

During that of July 29, 1878, which was total in the United States, a negro, suddenly seized with a paroxysm of terror, and convinced that the end of the world had come, suddenly killed his wife and children.

On December 16, 1880, the eclipse of the moon was saluted at Tashkend by an infernal clatter of teapots, stewpans, and iron pots struck by indefatigable hands, to frighten the Tchaitan devil who was devouring the moon.

On January 28, 1888, at Pekin, the same uproar occurred—this time with the drum, and by order of the mandarins—to put to flight the celestial dragon who was eclipsing the moon.

However, why should we be surprised at the terror caused among ignorant populations by an eclipse of the sun or moon? Let us consider again, for example, the eclipse of the sun of July 29, 1878 (see fig. 75). What a grand aspect! We notice round the sun, eclipsed by the moon, a luminous halo and immense rays darting into space. Three stars are visible to the left of the sun: these are Mercury, Regulus, and Mars. Two are seen to the right, Castor and Pollux; one below, Procyon; and one to the right below, Venus. Other stars were seen near the sun, and these

[1] François Arago, *Astronomie populaire*, vol. iii. p. 583.

Fig. 75.—Total Eclipse of the Sun, July 29, 1878, as observed from the Rocky Mountains (United States)

have been taken for planets in the vicinity of the radiant star ; but we shall see, further on, that there is no certainty in this observation.

In our enlightened Europe there still remain some vestiges of ancient fears, and we sometimes associate these phenomena, like the disagreeable facts of meteorology—such as storms, inundations, tempests—with the beliefs of antiquity in Divine anger. During one of the recent partial eclipses, which was well observed in our climate, that of March 6, 1867, the governesses of a girls' school in Paris took their pupils to prayers in order to avert the malediction of the Most High. I heard nothing similar said on the occasion of that of December 22, 1870. It is true that everyone had then other dominant preoccupations, and that this eclipse was itself eclipsed by that of common-sense: two intelligent and reasonable nations were mutilating each other, without anybody even knowing for what reason. Two hundred and fifty thousand men were killed, and ten thousand millions of francs were cast to the wind. Formerly they would have associated with this international slaughter either that eclipse of the 'terrible year,' or the aurora borealis which then appeared in the sky ; now everyone understands that it had no other cause but human folly.

We will conclude this long notice of eclipses with two interesting tables: (1) a list of all the eclipses of the sun and moon which will happen from now to the end of the century (those which will be visible at Paris are marked with a *) ; (2) a list of the principal total or annular eclipses of which the path will pass over France, or not far from France, during the twentieth century.

Eclipses of the Sun and Moon which will happen from now to the end of the Nineteenth Century

Date	Eclipse	Where
1894		
March 21	Partial eclipse of the moon	New Guinea
April 6	Annular and total of the sun	India, Thibet
*Sept. 15	Partial of moon	Canada
*Sept. 29	Total of the sun	Indian Ocean
1895		
[*March 11	Total of the moon]	Barbadoes
*March 26	Partial eclipse of sun	Europe
*August 20	Partial „ sun	Asia
*Sept. 4	Total „ moon	Mississippi
Sept. 18	Partial „ sun	South America
1896		
Feb. 13	Annular eclipse of sun	Southern Ocean
*Feb. 28	Partial „ moon	Persia
*August 9	Total „ sun	Norway, Lapland
August 23	Partial „ moon	Mexico

1897

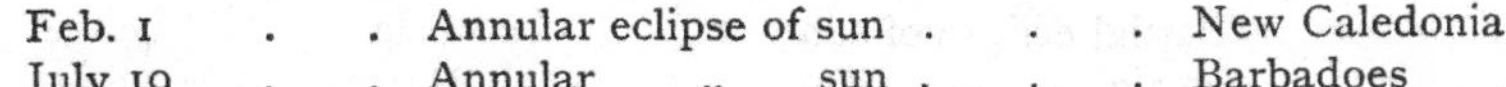

Feb. 1	Annular eclipse of sun		New Caledonia
July 19	Annular	„ sun	Barbadoes

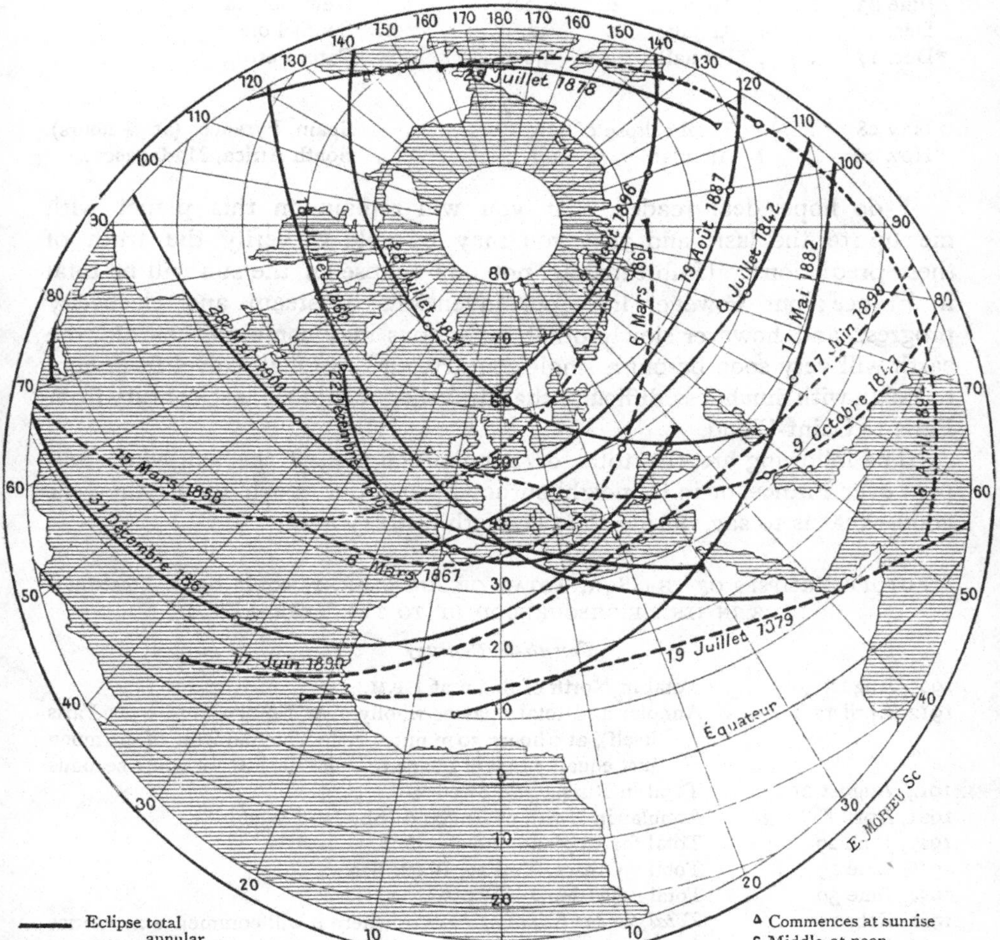

Eclipse total
„ annular
„ annular and total

△ Commences at sunrise
○ Middle at noon
▲ Ends at sunset

FIG. 76.—TOTAL OR ANNULAR ECLIPSES OF THE SUN PASSING OVER FRANCE OR ITS NEIGHBOURHOOD FROM 1842 TO 1900 [1]

1898

*Jan. 7	Partial eclipse of moon		France
*Jan. 22	Total	„ sun	India and China
*July 3	Partial	„ moon	Russia
July 18	Annular	„ sun	South America
*Dec. 27	Total	„ moon	France

[1] In figs. 76, 77, Mars is March ; Mai, May ; Juin, June ; Juillet, July ; Août, August ; Octobre, October ; and Décembre, December.

1899

Jan. 11	Partial eclipse of sun	. . .	Asia.
*June 8	Partial „ sun	. . .	Europe
June 23	Total „ moon	. . .	New Guinea
Dec. 3	Annular „ sun	. . .	South Pole
*Dec. 17	Partial „ moon	. . .	Cape Verd

1900

*May 28	Total eclipse of sun	. . .	Spain, Pyrenees (at $3\frac{1}{2}$ hours)
*Nov. 22	Annular „ sun	. . .	South Africa, Madagascar

I do hope, dear reader, that you will remain on this planet with me up to the last, and that you may be able to verify the truth of these predictions. Unfortunately, not one eclipse of the sun will be total in France; but however little our inventions of steam and electricity progress, and however slowly other inventions may come to their aid, the earth will very soon be but a single country, and we shall travel from here to Pekin with much less difficulty than travellers could go last century from Paris to Saint-Cloud.

The following are the total or annular eclipses of the sun which will pass over France, or in its neighbourhood, from our epoch up to the year 2000—that is to say, during the twentieth century:—

Future Eclipses of the Sun, Total or Annular, passing over France or in its Neighbourhood up to the Year 2000

Twentieth Century

1905, August 30	Total in North of Spain at 1 P.M.
1912, April 17	Annular and total; passes wholly over France (and over Paris itself), at 0 hours 30 minutes P.M. The diameter of the moon just equal to that of the sun; will only last for some seconds
1914, August 21	Total in Russia and Sweden
1921, April 8	Annular in North of England (Shetland Isles)
1927, June 29	Total for England and Sweden
1936, June 19	Total: Greece, Turkey, Black Sea
1954, June 30	Total: Sweden and Russia
1961, Feb. 15	*Total for the South of France*, where it will commence at sunrise
1966, May 21	Annular for Greece and the Black Sea
1976, April 29	Annular for Tunis
1984, May 30	Annular for Algeria, nearly at sunset
1999, August 11	*Total for France*; it will pass over Paris about $10\frac{1}{2}$ A.M.; great and beautiful eclipse, several minutes' duration

[The Rev. S. J. Johnson, F.R.A.S., gives the following in addition to the above:—

1908, June 28	Annular in Mexico
1916, Feb. 3	Total, Caribbean Sea
1919, Nov. 22	Annular: West Indies to Africa
1922, March 28	Annular: Guiana, North Africa

1925, Jan. 24 . . Total : United States
1929, Nov. 1 . . Annular : Sierra Leone
1945, July 9 . . Total : Canada, Greenland, Lapland, Northern Russia
1952, Feb. 25 . . Total : Guinea, Red Sea to Central Asia
1959, Oct. 2 . . Total : Atlantic, across Northern Africa to Abyssinia
1972, July 10 . . Total : Kamschatka, Greenland
1994, May 10 . . Annular, across centre of United States]

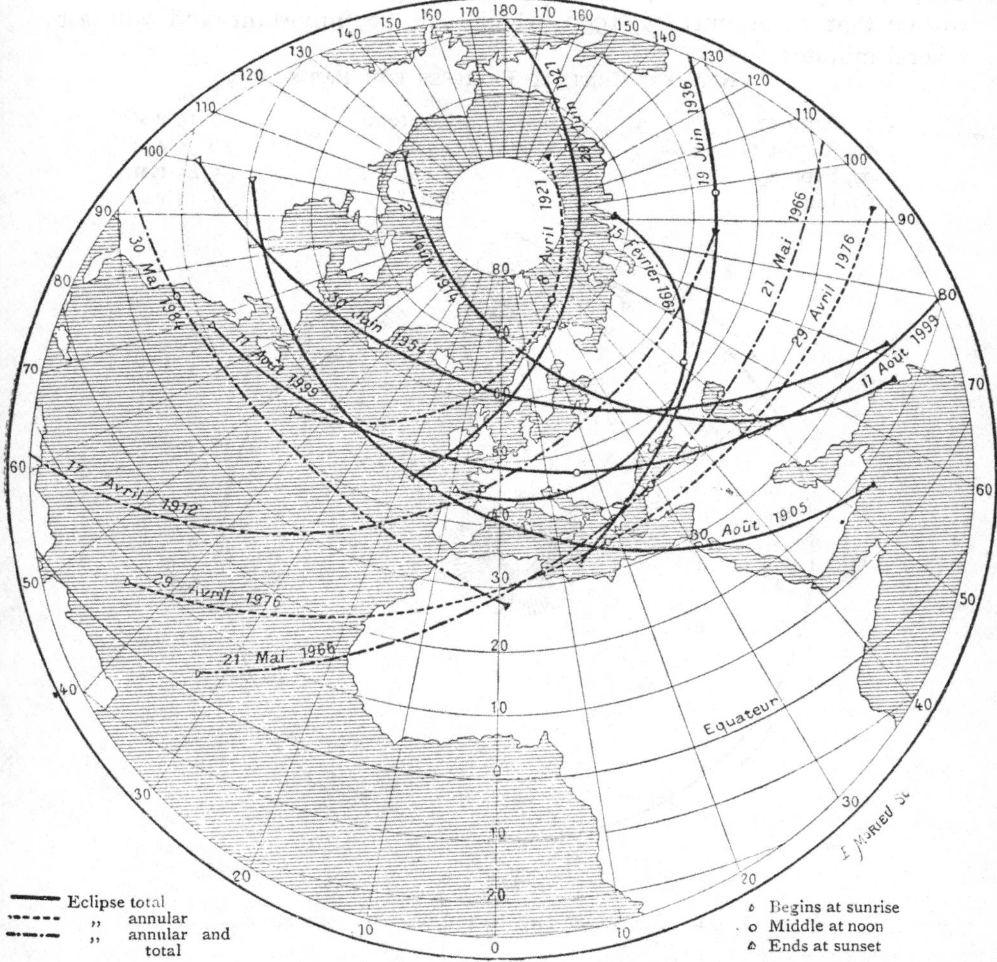

FIG. 77.—TOTAL OR ANNULAR ECLIPSES OF THE SUN PASSING OVER FRANCE OR ITS NEIGHBOURHOOD FROM 1905 TO 1999

Such are the total or annular eclipses of the sun which will happen during the next century. The preceding chart shows the course according to the 'Canon der Finsternisse of Oppolzer, corrected for certain curves. In all these tracks the eclipse moves from west to east, from left to right. In the case where an eclipse may be annular or total, it is total in the middle of the line—that is to say, for the countries which see the eclipse about noon. The nearest total eclipse which will be visible at Paris is one of this kind, that of April 17, 1912, which will occur about a quarter of an hour after noon; the totality will last but a few seconds. This scarcely deserves the name of a total eclipse. The nearest afterwards will be that of August 11, 1999, which will be important and will last several minutes.

NEAREST CENTRAL ECLIPSES FOR PARIS

		H. M.
1912, April 17	Just total	0 15 P.M.
1999, August 11	Total	10 30 A.M.
2090, Sept. 23	Total	5 45 P.M.
2160, June 4	Total	7 15 P.M.

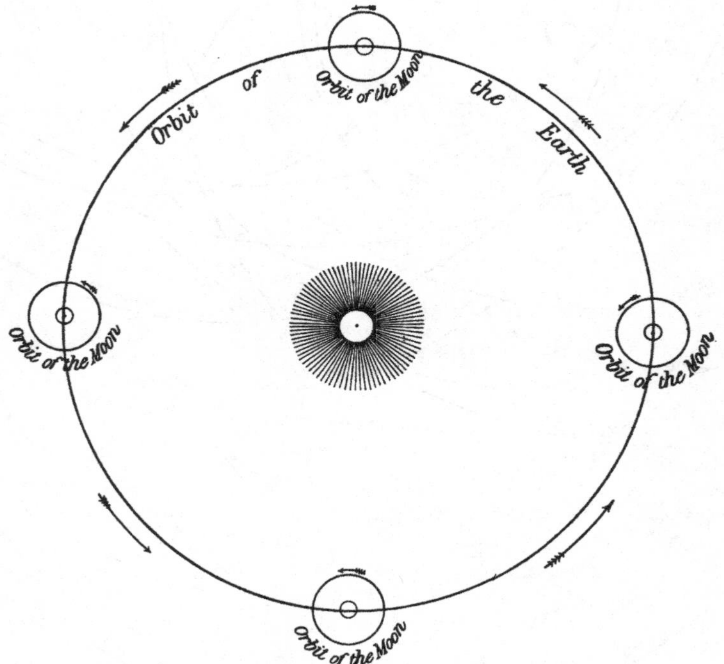

FIG. 78.—ORBITS OF THE EARTH AND OF THE MOON

Thus the celestial motions move in their perennial harmony; we cannot bear the same testimony to the fluctuations of humanity. Who can divine what the face of Europe will be in two or three centuries?

We will now leave the moon and earth and transport ourselves to the sun, the centre of the celestial system to which we belong. The logical order guides us. We wished at first to give an account of the true situation which we occupy in space, and we commenced with our own planet, the moving base of all our observations. Then we examined the situation, the motion, and the nature of the moon, our faithful satellite, and we completed this study by that of eclipses, which have already led us to refer for a moment to the sun in revealing to us the prominences and the luminous atmosphere, which the lunar screen renders conspicuous when it protects our sight from the dazzling furnace. We have also spoken of the sun with reference to the earth's annual motion of translation round him, and we already know that he reigns near the centre of the terrestrial orbit. Nothing remains, then, but to enter into complete relation with the sovereign of the world, and to estimate exactly the ratio which exists between its distance and that of the moon, our first halting-place.

And, at first, we must represent the orbit of the moon closely round the earth, while that of the earth is drawn at a great distance round the sun. While our planet turns in a year round the radiant star, it carries with it the

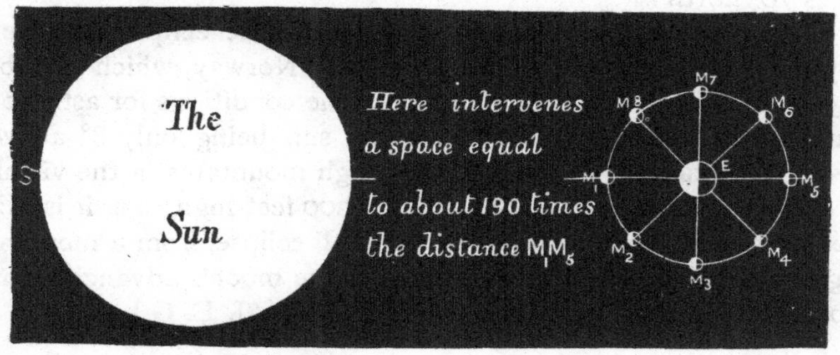

FIG. 79.—DISTANCE FROM THE SUN TO THE MOON

moon, which turns in a month round us. But the ratio between the distances is much more considerable than can well be indicated by a figure, and it is rather difficult to represent it accurately by a drawing (fig. 79). The distance of the sun is 385 times greater than that of the moon. Representing it by a line of 16 feet in length, the distance of the moon would be half an inch. The sun is, in fact, nearly twice as large in diameter as the whole orbit of the moon; and on the whole length of the line which joins the earth to the sun we could put 108 suns placed side by side, touching like the beads of a chaplet. We see, as we have said above, that the moon nearly touches the earth, and is truly but a province, an annexed island.

What do we think now as to its real proportions?

The earth measures 7,912 miles in diameter. From here to the moon is 30 terrestrial globes, and from here to the sun 11,600.

A railway train, which at the rate of thirty-seven miles an hour would reach the lunar station in thirty-eight weeks, would run on in a straight line for 266 years before it arrived at the capital of the solar empire. This is a very long time. Let us place ourselves astride a cannon-ball: we shall have crossed the lunar orbit on the morning of the ninth day, and after nine years of rapid flight we shall arrive at the threshold of the day star. The time is still a little long. Let us wing our flight on a ray of light: in a little more than a second we shall reach the moon, and in eight minutes the sun. We leave!—and arrive!

[The total eclipse of the sun which will take place on the morning of August 9, 1896, will be visible in the north of Norway, near a town named Vadso, on the Varanger Fjord, in Finmark. The central line of totality passes close to a village called Bugonaes, behind which rises a mountain 1,800 feet high, from which a good view may be obtained. An expedition to this place is being organised by Miss E. Brown, the well-known solar observer. The country south of the Varanger Fjord is said to be beautiful, and worth visiting, 'a southern flora covering hill and dale,' although the latitude is 70° north!

For those who do not care to go so far north, the eclipse may be seen from a place called Bodo, on the west coast of Norway, which lies on the ordinary steamboat route.' Here, however, the conditions for astronomical observations will not be so favourable, the sun being only 8° above the horizon during totality. There are some high mountains in the vicinity of Bodo; one of them especially, Sulitelma, 6,200 feet high, can, it is said, be ascended without danger or fatigue. A total eclipse, from a mountain of this height, would be a sight worth seeing, as the moon's advancing shadow might be detected as it approached the observer.—J. E. G.]

BOOK III

THE SUN

CHAPTER I

THE SUN, RULER OF THE WORLD

Magnitude and Proportion of the Solar System. Numbers and Harmony

DAZZLING source of light and heat, of motion, life, and beauty, the inimitable sun has in all ages received the earnest and grateful homage of mortals. The ignorant admire it because they feel the effects of its power and its value; the *savant* appreciates it because he has learned its unique importance in the system of the world; the artist salutes it because he sees in its splendour the virtual cause of all harmonies. This giant star is truly the heart of the planetary organism; each of its celestial palpitations is felt from afar by our little earth, which sails at 93 millions of miles, by distant Neptune, which revolves at 2,790 millions of miles, and by the pale deserted comets farther out still in the eternal winter, and even by the stars at millions of millions of miles; each of the throbbings of this ignited heart sheds and scatters without limit the immeasurable vital force which distributes life and happiness over all the worlds. This force incessantly emanates from the solar energy, and is precipitated into space with an unheard-of rapidity: eight minutes suffice for light to traverse the abyss which separates us from the central star; thought itself cannot see distinctly this bound of 186,000 miles cleared in each second by the luminous motion. And what energy like that of this furnace? We have already estimated the magnitude of the solar globe as 108½ times larger than the earth in diameter, 1,279,000 times more immense in volume, 324,000 times heavier in mass. How shall we picture magnitudes like these?

Representing the earth by a globe of one foot in diameter, the sun would be represented by a globe of 108½ feet. We can form an idea of a similar globe if we think that the most spacious cupola which human architecture has ever constructed, the dome of Florence, raised into the air by the genius of Brunelleschi, measures but 46 metres (151 feet); the dome of St. Peter's at Rome and that of the Pantheon of Agrippa measure less than 43 metres; the dome of the Invalides at Paris measures 24 metres (78·7 feet), and that of the Pantheon 20½ metres (67·25 feet) only. Thus, if we represent the sun by a ball of the size of the Pantheon at Paris, the earth would be reduced to its comparative dimensions by a ball of 19 centimetres (7·48 inches) in diameter.

We cannot, however, insist too strongly on the importance of the sun, nor fix too much in the mind its superiority over our globe. That is why we reproduce here the very eloquent figure of the comparative size, to which our attention will very soon be particularly directed.

If we could place the sun in the scale of a balance sufficiently gigantic to receive it, it would be necessary to pile up 324,000 earths in the other scale in order to produce equilibrium.

This enormous mass maintains in his rays the whole of his system. If the comparison were not offensive to the sun-god, we might say that he is like the spider at the centre of his web. In the net of his attraction the worlds are sustained. Relatively to his magnitude and might, the planets are but toys turning round him. We will represent at once the relation which exists between the importance of the sun and the situation of the little globes which surround him. For this purpose we will form some very interesting tables, although they *are* composed of figures! And, firstly, let us see the general form of the system.

DISTANCES OF THE PLANETS FROM THE SUN, AND PERIOD OF THEIR REVOLUTIONS

Planets	Distances from the Sun		Period of the Revolutions
	The Earth being 1	In Millions of Miles	
The Sun	—	—	—
Mercury	0·387	36	88 days
Venus	0·723	67	225 „
The Earth and Moon	1	93	365 „
Mars (two satellites) . . .	1·524	141	1 year 322 „
Jupiter (five satellites) . . .	5·203	482	11 years 315 „
Saturn (eight satellites) . . .	9·539	885	29 „ 167 „
Uranus (four satellites) . . .	19·183	1,780	84 „ 7 „
Neptune (one satellite) . . .	30·055	2,789	164 „ 281 „

This little table explains itself. We see that the last planet of the system, Neptune, is thirty times more distant than we are from the sun, and nearly eighty times farther than Mercury. As light and heat diminish in proportion to the square of the distance, this utmost province receives nearly 6,400 times less light and heat than the nearest city to the burning star. We see at the same time that the year of Neptune is nearly 165 times longer than ours, and more than 680 times superior to that of Mercury. In a Neptunian year the earth counts nearly 165 years, and Mercury 684. We will consider now the differences of size and weight of the principal globes of the system, and let us class them in decreasing progression

MAGNITUDES AND MASSES COMPARED

	Diameters	Volumes	Masses
The Sun	108·5	1,280,000	327,200
Jupiter	11·1	1,369	312
Saturn	9·3	719	92
Uranus	4·2	69	14
Neptune	3·8	55	16
The Earth	1·0	1	1
Venus	0·99	0·97	0·79
Mars	0·53	0·16	0·11
Mercury	0·37	0·05	0·07
The Moon	0·27	0·02	0·01

COMPARATIVE ELEMENTS OF THE PLANETS, THE SUN BEING TAKEN AS UNITY

	Distances, the semi-diameter of the Sun being 1	Diameters compared to that of the Sun	Masses compared to that of the Sun
The Sun	1	1	1
Mercury	83	$\frac{1}{282}$	$\frac{1}{5,310,000}$
Venus	155	$\frac{1}{115}$	$\frac{1}{409,000}$
The Earth	214	$\frac{1}{108}$	$\frac{1}{327,200}$
Mars	322	$\frac{1}{202}$	$\frac{1}{3,093,500}$
Jupiter	1,116	$\frac{1}{9·7}$	$\frac{1}{1,047}$
Saturn	2,041	$\frac{1}{11·4}$	$\frac{1}{3,502}$
Uranus	4,108	$\frac{1}{24}$	$\frac{1}{22,600}$
Neptune	6,420	$\frac{1}{25}$	$\frac{1}{18,780}$

These figures also explain themselves. We see that, representing the earth by 1, Jupiter, for example, has a diameter 11 times greater, and Mercury a diameter of only 37 hundredths, or a little less than 4 tenths of ours. The mass of the sun is represented by the number 327,200, while that of Mercury is only 7 hundredths of ours, and that of Neptune is nearly equal to 16 times that of our globe. The first of these tables shows us that, representing by 1 the distance of the earth from the sun, that of Mercury is denoted by 387 thousandths—that is to say, that Mercury is nearly at one-third of the distance from the sun to the earth, going out from the sun

Venus at about 7 tenths, Mars half as far again as we are, Jupiter 5 times farther, and so on. Now, from the absolute point of view, as it is not the earth, but the sun, which is the centre of comparison and the regulator, it will be interesting for us to represent the distances of the planets

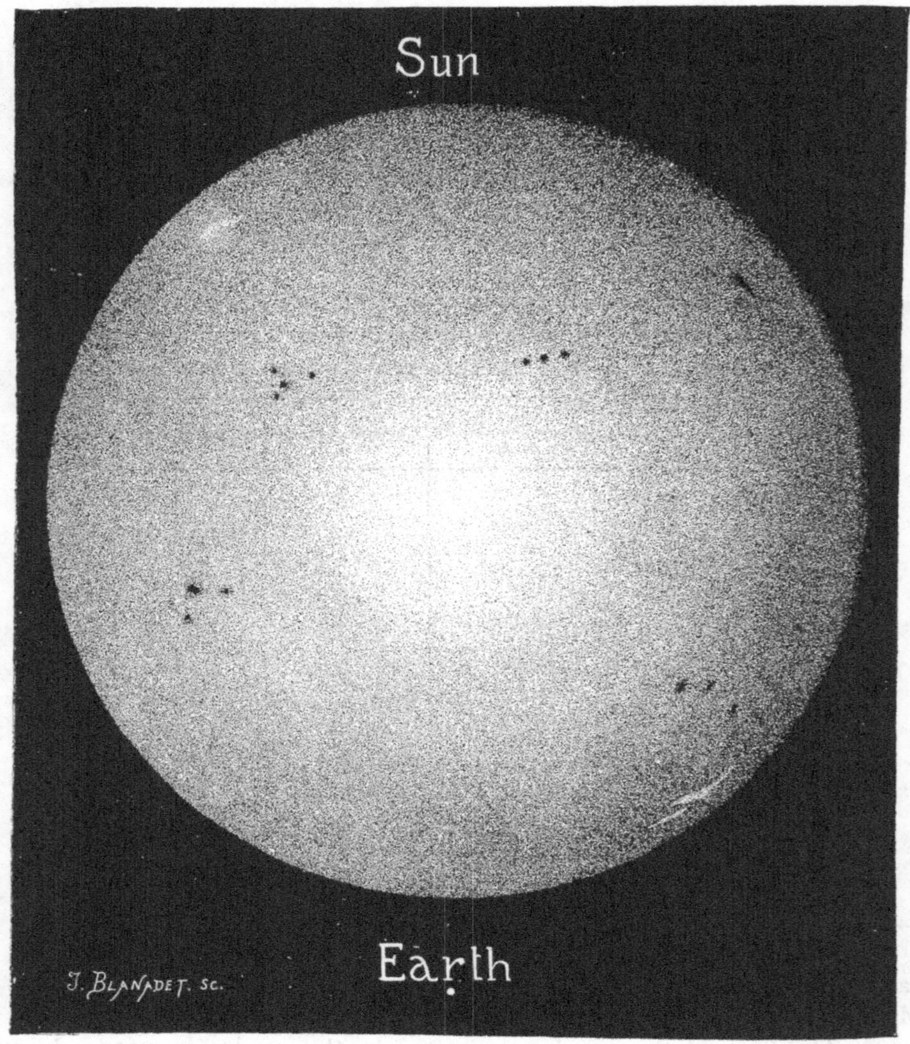

FIG. 80.—MAGNITUDE OF THE SUN. GRANULATED ASPECT OF ITS SURFACE

in proportion to the sun's diameter, the volumes and the masses in parts of the volume and mass of that body; and this new table will be more natural than the first, since the sun is the true sidereal unit of our system, to which all should be referred. (See table on page 209.)

These numbers mean, as will be easily understood, that Mercury is distant from the sun 83 times the semi-diameter of that great body, Venus 155 times, the earth 214 times, &c.; that the diameter of Mercury is but the 282nd of that of the sun—that is to say, that it would require 282 globes like Mercury side by side to pass across the solar globe, 108 globes like the earth, nearly 10 of Jupiter, &c.; and that, with respect to the masses or weights, it would require more than 5 millions of Mercuries, or 327,200 earths, or 18,780 Neptunes, to form a mass of the same weight as that of the sun. Jupiter weighs 312 times more than the earth, but 1,047 times

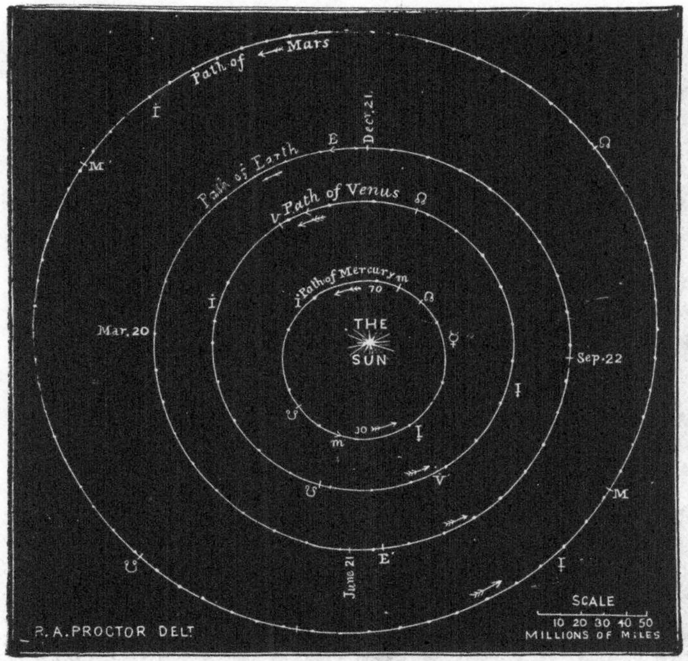

FIG. 81.—PATHS OF MERCURY, VENUS, THE EARTH, AND MARS ROUND THE SUN

less than the sun. His diameter exceeds that of the earth more than 11 times, but is inferior to that of the sun $9\frac{7}{10}$ times. This is, then, an important planet, which is, so to say, intermediate in volume and in mass between the earth and the sun. Nevertheless, the day star rules the whole, as a leviathan on the sea dominates a crowd of boats accompanying it; it alone weighs more than *seven hundred times* all the planets put together.

From the masses and volumes we infer the density of the constituent materials of each world.

Comparative Density of the Worlds of our System

Mercury	1·173	Neptune	0·300
The Earth	1	The Sun	0·253
Venus	0·807	Jupiter	0·242
Mars	0·711	Uranus	0·195
The Moon	0·615	Saturn	0·128

This little table shows that the planet of our system of which the constituent materials are the densest is Mercury, and that which is composed

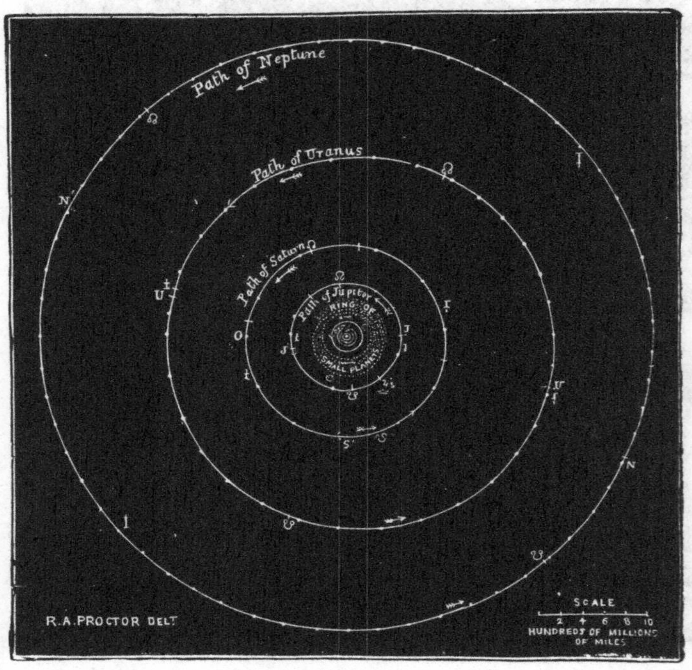

Fig. 82.—Paths of Jupiter, Saturn, Uranus, and Neptune round the Ring of Small Planets

of the lightest substances is Saturn. In the tables which precede we have not taken into account the zone of small planets which revolve between Mars and Jupiter. These are fragments—asteroids—of which a large number do not measure more than a few miles in diameter, and which have been produced either by the rupture of an original ring, or by the breaking-up of one or several planets; they occupy the greater part of the space comprised between the orbit of Mars and that of Jupiter. Already more than two hundred of these have been discovered. [Now (1893) over 380.]

Our readers will complete the exact knowledge which they may desire to have of the solar system by examining attentively figs. 81, 82, which represent the whole of the solar world. The orbits of the planets are drawn to scale in their relative order, but fig. 81 to a larger scale than fig. 82.

How interesting fig. 82 is to examine! It is there, on the third little circle, that we are, that we live and revolve—there, very near to the luminous focus. Are we not burnt, are we not blinded, like butterflies whirling round a flame? Only think that all the material destinies, moral, religious, and political, of the earth and moon take place within this little point!

The inspection of these topographical plans of the solar universe does not reveal any proportion in the distances of the orbits. Do you not think that the distance from Saturn to Uranus appears too great? It is, in fact, the same as that from Uranus to Neptune, which upsets the progression. The astronomer Titius remarked, in the last century—and Bode has published this remark, which bears his name—that we can express the successive distances of the planets from the sun by a very simple progression. Let us write, one after the other, the numbers successively doubled—

3, 6, 12, 24, 48, 96.

Putting zero for the first term, and adding 4 to all the numbers, we find—

4, 7, 10, 16, 28, 52, 100.

Now, it is found that, representing by 10 the distance of the earth, the distances of the other planets correspond approximately to these numbers, as may be seen from the following:—

Mercury	Venus	The Earth	Mars	Asteroids	Jupiter	Saturn
3·9	7·2	10	15·2	20 to 35	52	95

The planet Uranus, afterwards discovered, is placed at the distance 192, which differs very little from 196, the number obtained by continuing the series (192 + 4). But Neptune, instead of being found at 384 + 4, or 388, is at 300—that is to say, much nearer. The regularity, then, does not continue. It is a curious but not a real relation.

The progression of the *velocities* is closer. Multiplying the velocity of a planet in its course by 1·414 ($=\sqrt{2}$), we obtain a number pretty near the velocity of the inferior planet. It is possible that originally the planets may have been detached from the sun according to this law, and that subsequently several of them, from some cause, approached nearer to the sun. Perhaps, moreover, all the planets are destined to fall successively on the central star.

The solar power causes all the worlds of his system to gravitate round the sun. They all go round, like stones in a sling, with an enormous velocity. The nearer they are to the sun, the faster they revolve. Thus, as we have remarked with reference to the moon, the velocity with which the celestial globes go round gives rise to a centrifugal force which tends to increase their distance from the sun by precisely the quantity which the sun attracts them, and thus they are always maintained at the same mean distance.

We have already seen, when speaking of the motion of the moon round the earth, and of Newton's researches on the cause of celestial motions, that the attraction decreases according to the square of the distance—that is to say, according to the distance multiplied by itself. At double the distance it is four times less; at triple the distance it is nine times weaker; at four times the distance, sixteen times less, &c. It is, then, easy to show what the exact value of the solar attraction is at the distance of the different planets. The following is the distance which the planets would fall towards the sun if they were stopped in their course, or, if we like to put it so, the distance which a stone would fall towards the same attractive centre, supposing it placed at these different distances and abandoned to gravity:—

GRAVITY TOWARDS THE SUN

	Fall in the 1st second M.
At the surface of the Sun	134
At the distance of Mercury	0·0196
" " Venus	0·0056
" " the Earth	0·0029
" " Mars	0·0013
" " Jupiter	0·0001
" " Saturn	0·000032
" " Uranus	0·000008
" " Neptune	0·000003

These velocities are those of bodies falling towards the sun during the first second of the fall; after this first second, at the beginning of the second, they would be doubled, and the planets would thus fall with an increasing velocity towards the central star, at which they would arrive with the unimaginable velocity of 600,000 metres (373 miles) in the last second! And yet, during the first second the earth would fall towards the sun—would approach him—by only one-ninth of an inch, Mars by $\frac{1}{20}$th of an inch, Jupiter by $\frac{1}{250}$th of an inch, and Saturn, Uranus, and Neptune by very small fractions of an inch! Here are the times which each planet would take to fall to the sun :—

TIME OF FALL OF THE PLANETS ON THE SUN[1]

	Days		Days
Mercury	15·55	Jupiter	765·87
Venus	39·73	Saturn	1,902·03
The Earth	64·57	Uranus	5,424·57
Mars	121·44	Neptune	10,628·73

The velocity of the planets in their orbits is proportional to their distance, and combined in such a manner with the attraction of the sun that in sailing through space they develop a centrifugal force which tends to increase their distance from the sun precisely by the same quantity by which they tend to approach him by the solar attraction, from which results a perpetual equilibrium, as we have already remarked. We have seen that the earth moves round the sun with a mean velocity of 29,450 metres (18·3 miles) per second, and the moon round the earth with a velocity of 1,017 metres (0·63 mile) in the same unit of time. The following are, in

[1] There is something very curious in these times, which is that, by multiplying all by the same number, we reproduce the year of each planet :—

		Days
Mercury	15·55 × 5·656856 =	87·9692
Venus	39·73 × 5·656856 =	224·7008
The Earth	64·57 × 5·656856 =	365·2564
Mars	121·44 × 5·656856 =	686·9796
Jupiter	765·87 × 5·656856 =	4,332·5848
Saturn	1,902·03 × 5·656856 =	10,759·2198
Uranus	5,427·57 × 5·656856 =	30,686·8208
Neptune	10,628·73 × 5·656856 =	60,126·7200

The first time I remarked this (it was at the beginning of the year 1870) I was perplexed for months, and taxed my ingenuity, or searched in books, for any principle of celestial mechanics which would put me in the way of discovering the cause. What is this famous coefficient, 5·656856? It is the square root of 32. But what has this square root to do with such a curious and unexpected relation between the revolutions of the planets and their falls to the sun? Here is the explanation: If we compare the fall of the earth to the sun to the half of an extremely elongated ellipse, of which the perihelion would be nearly tangential to the sun, this ellipse would have for its longer axis the actual distance of the earth from the sun—that is to say, the half of the actual diameter of the terrestrial orbit. The squares of the times being to each other as the cubes of the distances, the revolution of the earth along this new orbit would be given by the square root of the cube of $\frac{1}{2}$ or of $\frac{1}{8}$, and consequently would be $\frac{365\cdot256 \text{ days}}{2\cdot828 \text{ days}}$, or 128 days. The half of this revolution—or, which comes to the same thing, the time of the fall to the sun—would then be given by the half of the square root of $\frac{1}{8}$, or by $\frac{365\cdot256 \text{ days}}{5\cdot657 \text{ days}}$. But the half of the square root of $\frac{1}{8}$ is the square root of $\frac{1}{32}$, or our number, 5·656856.

round numbers, the velocities with which all the planets are animated in their rapid motion round the focus of illumination :—

VELOCITY OF THE PLANETS ROUND THE SUN

	Mean velocity per second	
Mercury	29 miles, or	2,505,000 miles per day
Venus	21·7 „	1,873,000 „
The Earth	18 „	1,555,000 „
Mars	14·9 „	1,287,000 „
Jupiter	8 „	771,000 „
Saturn	6·2 „	536,000 „
Uranus	4·3 „	372,000 „
Neptune	3·1 „	268,000 „

Such are the velocities with which the planets are animated in their course round the sun. Is it possible for us to conceive this magnitude? A ball fired from a cannon leaves the mouth with a velocity of 400 metres (1,312 feet) per second ; the terrestrial globe flies 75 times quicker, Mercury 117 times faster. This is a rapidity so stupendous that, if two planets were to meet in their course, the shock would be frightful : not only would they be shattered in pieces, both reduced to powder, but, further, their motion being transformed into heat, they would be suddenly raised to such a degree of temperature that they would disappear in vapour—everything, earths, stones, waters, plants, inhabitants—and they would form an immense nebula!

On account of these different velocities the planets are constantly changing their situation with reference to each other.

This series of little tables gives us a general idea of the physiology of the planetary system.

We have seen, in treating the question of the motions of the earth, that our planet describes an ellipse (fig. 8, p. 26) round the sun ; and we have also seen how the laws of attraction were discovered by the analysis of the moon's motion. We are now sufficiently prepared to understand the laws which govern the system. Here are these laws, which it is important to remember :—

1. *The planets revolve round the sun, describing ellipses, of which that body occupies one of the foci.*

We have sufficiently studied this fact when speaking of the annual motion of the earth round the sun, and we have seen that all the planets revolve, like the earth, round the same star.

2. *The areas or spaces described by the radii vectores of the orbits are proportional to the times taken to pass over them.*

Let us consider the same planet at different epochs of its revolution, and suppose that we mark on its orbit as many arcs, A B, C D, E F, passed over by the planet in equal times, either in a month, or, more exactly, in a period of thirty days.

The velocity of the planet varies according to the position which it occupies along its orbit. It follows a mean course when it is at its mean distance, A B. When it is nearer to the sun, near the positions C D, its velocity is accelerated. When it is more distant, as in the positions E F, it moves much more slowly. Thus the motion of the earth in its orbit is not uniform: it travels much more quickly when it is at its perihelion (January) than when at its aphelion (July). The arcs described in the same time are so much smaller as the planet is more distant. But the *spaces* included between the lines drawn from the sun to the two extremities of the arcs described in equal times are *equal* to each other. Here is, then, a remarkable fact. Thus, the earth takes as long to pass from E to F as to go from C to D, although the first arc is much less than the second. We call *radii vectores* the lines, such as S E, S F, S A, S B, &c., drawn from the sun to the planet in its different positions. The spaces swept over by these *radii vectores* are proportional to the times taken in passing over them—twice, three, four times larger, if we consider an interval of time twice, three, four times as long. If we draw fig. 83 on cardboard, and if we cut out the sectors, the three pieces should have the same weight.

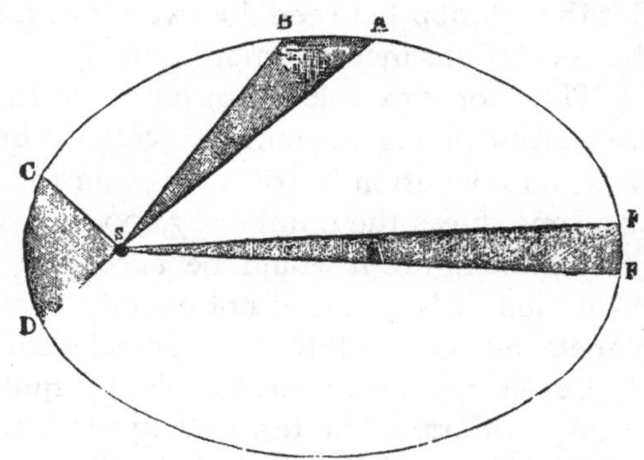

FIG. 83.—EXPLANATION OF THE MOVEMENTS OF THE PLANETS: LAW OF AREAS

The third fundamental proposition is this. It is important to know it in order to exactly represent these motions.

3. *The squares of the times of revolution of the planets round the sun are to each other as the cubes of the distances.*

This law is the most important of all, because it connects all the planets with each other.[1]

[1] This is the place to explain in a few words to those of our readers who have not studied mathematics what is a *square* and what is a cube. A square is simply any number multiplied by itself. Thus, twice 2 are 4: well, then, 4 is the square of 2; 3 times 3 make 9: 9 is the square of 3; $4 \times 4 = 16$: 16 is the square of 4.

A cube is a number multiplied twice by itself. Thus, $2 \times 2 \times 2 = 8$: 8 is the cube of 2; $3 \times 3 \times 3 = 27$: 27 is the cube of 3; $4 \times 4 \times 4 = 64$; 64 is the cube of 4, &c.

We call the *square root* of a number the number which, multiplied by itself, reproduces this number. Thus, the square root of 4 is 2, since twice 2 are 4; the square root of 9 is 3; that of 16 is 4; that of 25 is 5, &c.

We call the *cube root* of a number the number which, multiplied twice by itself, reproduces

The revolution is so much longer as the distance is greater or the orbit has a greater diameter. The order of the planets, beginning from the sun, is the same whether we arrange them according to their distances or according to the times which they take to accomplish their revolutions. But the relation between the two series is not a simple *proportional* increase: the revolutions increase more quickly.

Thus, for example, Neptune is 30 times farther from the sun than we are. Multiplying the number 30 twice by itself, we find the number 27,000. Now, its revolution is 165 years, and this number multiplied once by itself also reproduces the number 27,000 (in round numbers; in order to obtain the precise figure it would be necessary to consider the fractions, for the revolution of Neptune is not exactly 165 years). It is the same for all the planets, all the satellites, and all celestial bodies.

Let us make the same calculation, quite correctly, for another planet—for example, Mars. The terrestrial year is to a year of Mars in the proportion of 365·2564 to 686·9796, and the distances from the sun are in the ratio of 100,000 to 152,369. If we care to take the trouble, we find that

$$\frac{(365{\cdot}2564)^2}{(686{\cdot}9796)^3} = \frac{(100{,}000)^3}{(152{,}369)^3}.$$

Thus the revolutions of the planets round the sun are regulated according to their distances. The more distant the planets are, the less rapidly do they move, and this according to a mathematical proportion.

To these three laws, which justly bear the name of Kepler, who discovered them, we may add here a fourth proposition, which completes and explains them—the law of attraction or universal gravitation, discovered by Newton from the labours of Kepler.

Matter attracts matter in the direct ratio of the masses and the inverse ratio of the square of the distances.

Whether this attraction is a real property conferred on matter, or only an appearance which explains the celestial motions, the truth is that things happen *as if* matter were endowed with the occult property of attracting at

this number. Thus, the cube root of 8 is 2, that of 27 is 3, that of 64 is 4, that of 125 is 5, &c.

To indicate the square of a number we place a small 2 above it: the square of 10 is written 10^2. This signifies the second power.

The cube is indicated by a 3. The cube of ten is written 10^3.

The square root is indicated by the sign $\surd$, which comes from the letter *r*.

The sign of addition, *plus*, is written +; that of subtraction, *minus*, is written −; that of multiplication, *multiplied by*, is written ×; that of division is a line between the two numbers, written one under the other, $\frac{12}{5}$.

Many intelligent and intellectual persons are scared by mathematics: there is nothing in the world so simple, so clear, and so easy.

a distance. This attraction decreases in the inverse ratio of the square of the distance—that is to say, the more the distance increases, the more the attraction diminishes, and that not in a simple proportion, but in the ratio of the distance multiplied by itself. A body twice as far is four times less attracted ; a body three times as far is nine times less attracted, &c.

This proportion of the square of the distance will be understood at the first glance by the little figure, where we suppose the light of a candle received on a screen successively removed to a distance double, triple, and quadruple. We easily see that at the distance C, double of B, the light is scattered four times more ; at the distance D, nine times more ; and that at the distance E it spreads over sixteen equal surfaces, &c.

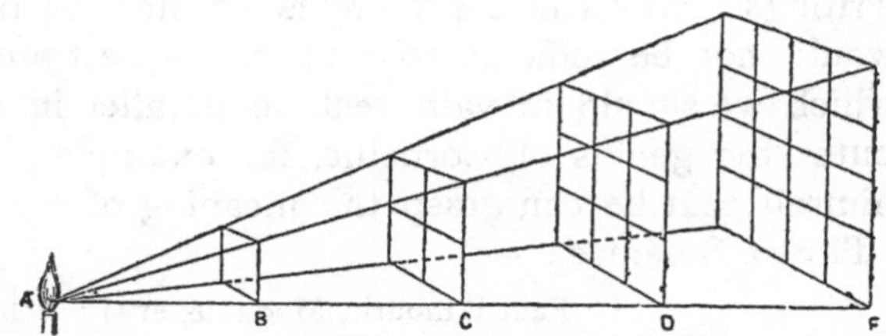

FIG. 84.—DECREASE OF INTENSITY IN PROPORTION TO THE SQUARE OF THE DISTANCE

It is possible that this attraction may be only an appearance due to the pressure of the ethereal fluid, which fills space supposed to be void. We do not know the essence of the cause of which we observe the effects. Moreover, this gravitation of celestial bodies towards each other *rules* the motion, but does not *create* it. We must at first suppose this motion of the planets in their orbits, due, doubtless, to their primordial detachment from the solar nebula.

Everything is reduced in the final analysis to two causes or two forces. One of these forces is nothing else but gravitation ; it is the tendency which two bodies have to come together, a tendency which is proportional to their respective masses, and which varies in the inverse ratio of the squares of their distances. It is gravity which makes bodies fall to the surface of the earth, and which constitutes their pressure or their weight. If gravity existed alone, the moon would join the earth, their united masses would fall with an increasing velocity into the sun itself ; and it would be the same for all the planets and all the bodies which compose the universe. Long ago the universe would have been a motionless heap of ruins.

But besides this central force of gravitation there is another force with which each planet is animated, and which alone would cause it to escape in a straight line along the tangent. It was by combining these two forces, in seeking by geometry and analysis to determine the real resultant motion of their simultaneous and constant action, that Newton showed that the laws of this motion are conformable to those which Kepler succeeded in discovering. Perhaps this *motion* itself and the forces by which we explain it, by decomposing it into two other forces, may exist only in our mind. The first thing for us is to determine the facts and to know exactly how they occur. The theory comes afterwards. Even this theory is certain and absolutely demonstrated now. But the essence itself of the force (whatever it is) which operates still remains hidden from us in the mystery of causes.

Such are the laws which rule the motions of the planets. It requires, no doubt, serious attention to understand them well, but we see that they are neither obscure nor equivocal. We often hear it said that scientific writings cannot attain the clearness or the elegance of purely literary writings. Nevertheless, there is nothing so beautiful as an equation. It would not be difficult to find in the best authors examples of nonsense which we should in vain seek to parallel in mathematics. No one disputes the genius of Corneille, for example. And yet, who can flatter himself that he can grasp the meaning of the following declaration from 'Tite et Bérénice:'—

> Faut-il mourir, Madame, et si proche du terme !
> Votre illustre inconstance est-elle encor si ferme
> Que les restes d'un feu que j'avais cru si fort
> Puissent dans quatre jours se promettre ma mort !

Read it again, if you please, in order to understand well the profound thought of the author. The actor Baron, not knowing in what manner he should deliver the end of the sentence, went to ask the advice of Molière, who, wearied with trying in vain, sent him back to Corneille himself.

'What!' said the illustrious author of the 'Cid'; 'are you sure that I wrote that?'

He then set himself to consider these four lines in every sense, and returned them, saying, 'Indeed, I no longer know exactly what I wished to say; but recite them *nobly*, so that those who do not hear them will admire them.'

It is related that the famous Bishop of Belley, Camus, being in Spain, and not being able to understand a sonnet of Lope de Vega, who was then living, requested the poet to explain it to him; but the author, having read and re-read his sonnet several times, frankly confessed *that he could not understand it himself*.

We often meet in the greatest of poets (everybody will at once name Victor Hugo) with thoughts so profound that they remain in complete obscurity. This is the infinite.

Science, on the contrary, may see its most sublime discoveries expounded with simplicity, and every eye open to the spectacle of nature can understand its grandeur.

We have considered the motions of the planets gravitating round the sun. But the solar system is not composed solely of this star, of planets, and satellites; we must not forget the comets, which likewise move according to the preceding laws, and of which a large number describe very elongated orbits, with their aphelia far beyond the orbit of Neptune.

The comet of Halley goes out to 35 times the distance of the earth (Neptune revolves at 30 times, as we have seen)—that is to say, to 3,200 millions of miles from the sun; the comets of 1532, 1661, 1862 extend their flight, like the swarm of shooting stars of August 10, to a distance of 48—that is to say, more than 4,200 millions of miles (a distance at which a trans-Neptunian planet might revolve), and there, at a distance which sound would take 668 years to pass over, the comet hears instantaneously the voice of the sun, it still submits to his magnetic influence, is stopped in the depths of the icy night of space, and returns towards the star which attracts it, describing round him this elongated and oblique flight which brings it back to his flames! Is the attractive influence of the sun arrested there? No. It extends through Infinitude, and is only humbled when it penetrates into the sphere of attraction of another sun, not milliards (1,000 millions) of miles from here, but thousands of milliards, or billions.

Thus each star, each sun in the infinite, rules round itself—in spheres of which the limits intersect—the different worlds which gravitate in its light and in its power. And the numberless suns which people immensity are mutually sustained among themselves in the net of universal gravitation.

Immense and majestic harmony of worlds! A universal motion bears along the stars, atoms of the infinite. The moon gravitates round the earth, the earth gravitates round the sun, the sun carries along all its planets and their satellites towards the constellation Hercules; and these motions are executed according to determined laws, like the hand of a watch which turns round its centre, and like the circular undulations which are developed on the surface of still water when a point has been struck. This is a universal harmony which the physical ear cannot hear, as Pythagoras supposed, but which the intellectual ear can understand. And is it not music itself which vaguely lulls us on its seraphic wings, and so easily transports our minds into those ethereal regions of the ideal where we forget the fetters of matter? Do not the sonorous modulations of the organ, the sweet quiverings of the bow on the violin, the nervous languors of the cythara, or the still more captivating charm of the human voice, unite the raptures of life with the warm colours of harmony? What is it except an undulatory motion of the air contrived to reach the mind in the depths of the brain and to impress it with emotions of a special order? When the martial tones of the spirited 'Marseillaise' are borne in the heat of the conflict to the excited battalions, or when, under the Gothic vault, the sad 'Stabat' pours out its mournful notes, it is the vibration which affects us by speaking a mysterious language. Now, all in nature is motion, vibration, and harmony. The flowers of the garden sing, and the effect which they produce depends on the number and agreement of their vibrations relatively to those which emanate from surrounding nature. In violet light the atoms of the ether oscillate with the unheard-of rapidity of 740 billions of vibrations per second; red light, slower, is produced by undulations vibrating still at the rate of 380 billions per second. The violet colour is, in the case of light, what the highest notes are in the case of sound, and the red colour represents the lowest tones. As we see an object floating in the water obeying with docility the waves which come from different sides, so the atom of the ether undulates under the influence of light and heat, the atom of air undulates under the influence of sound, and the planet and satellite circulate under the influence of gravitation.

Harmony is in everything. To the eye of a person acquainted with the principles, nothing is more interesting than the crossing of waves of water. By their interference the surface of intersection is sometimes so divided that it forms a beautiful agitated mosaic of rhythmical motions, a sort of visible music. When the waves are skilfully produced on the surface of a disc of mercury, and we illuminate this disc with a pencil of intense light, this light reflected on a screen reveals the harmonious motions of the surface. The form of the vessel determines the form of the figures produced. On a circular disc, for example, the disturbance is propagated under the form of circular waves producing the magnificent *chassé-croisé* represented in fig. 85. The light reflected by a similar surface gives a design of extraordinary beauty. When the mercury is slowly agitated with the point of a needle in a direction concentric with the circumference of the vessel, the lines of light turn round in a ring under the form of interlacing distorted threads, revealing each other in an admirable manner. The most ordinary causes produce the most exquisite effects.

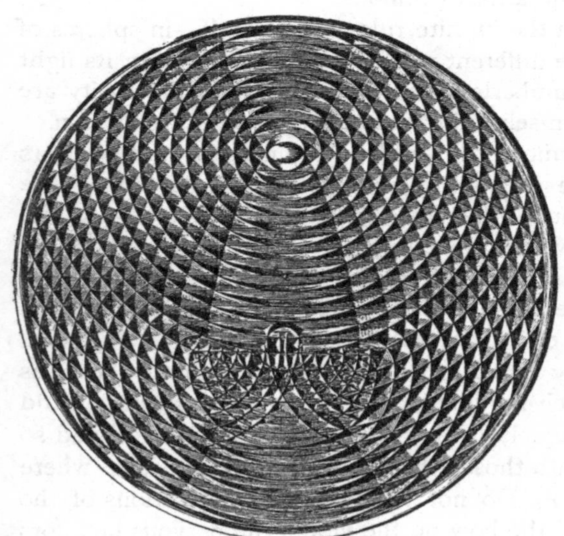

Fig. 85.—Harmony of Undulations

The undulations of sound may be expressed to the eye by figures no less harmonious, no less pleasing, than the preceding.

Let us take, in imitation of Chladni, a plate of glass or a thin plate of copper, and sprinkle it with fine sand. Let us deaden one of its edges at two points with two fingers of the left hand. and pass the bow along the middle of the opposite side (fig. 86). We shall see the sand trembling, falling back from certain parts of the surface, following the sounds obtained and designing the figures here reproduced (fig. 87) By varying the experiment we thus obtain these admirable designs, which appear at the command of the bow of the skilful experimentalist.

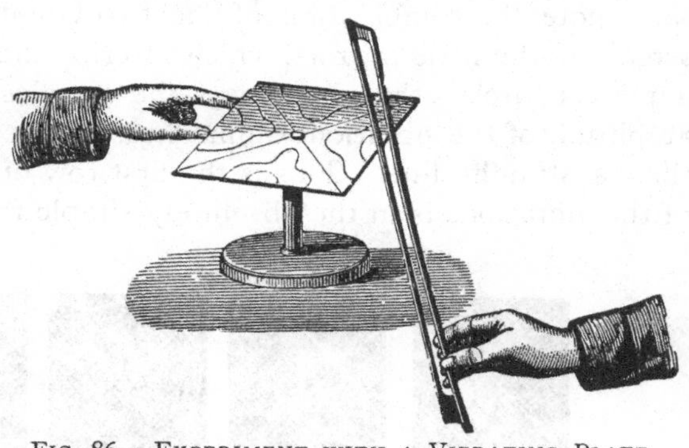

FIG. 86.—EXPERIMENT WITH A VIBRATING PLATE

The notes of the gamut are, besides, nothing else but ratios of number between the sonorous vibrations. Combined in a certain order, these numbers give perfect accord. Here the major mode rouses and enraptures us ; there the minor mode affects us, and plunges us into

FIG. 87.—HARMONY OF VIBRATIONS

melancholy reverie. And yet there is here but a matter of figures ! We can not only hear these sounds, but may even see them. Let us make two tuning-forks vibrate by the ingenious method of Lissajous, one vertical, the other horizontal, fitted with little mirrors reflecting a luminous point on a screen. If the two tuning-forks are in unison and give exactly the same note, the combination of the two vibrations rendered visible on the screen by the little mirrors, which inscribe them in lines of light, produces a perfect circle—that is to say, the simplest geometrical figure; as the amplitude of the vibrations diminishes, the circle flattens, becomes an ellipse, then a straight line. This is the first row of fig. 88, in which the number of the vibrations is in the absolutely simple ratio of 1 to 1. If, now, one of

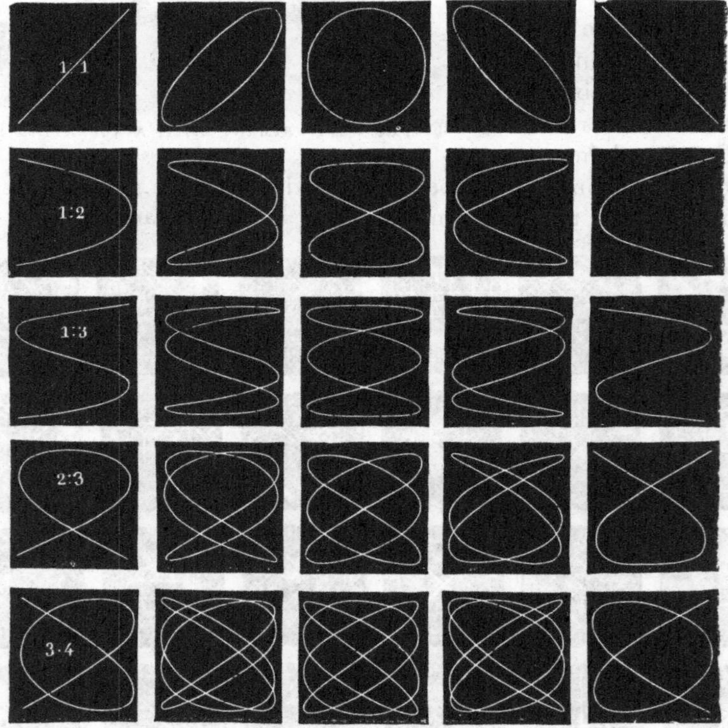

FIG. 88.—GEOMETRY OF MUSIC

the two tuning-forks is exactly an octave from the other, the vibrations are in the ratio of 1 to 2, since every note has for an octave a number of vibrations exactly double, and instead of a circle it is an 8, which is formed and modified as we see in the second row. If we take the combination of two tones of 1 to 3, say the *do* with the *sol* of the octave above, we obtain the figures of the third row. If we combine 2 to 3, as *do* and *sol* of the same octave, we produce those of the fourth row. The union of 3 with 4, of *sol* with the *do* above, gives the fifth series. What is most curious is, that in the complete figures (those of the middle of each series) the number of summits in the vertical direction and in the horizontal direction indicate

FIG. 89.—THE GEOMETRY OF SNOWFLAKES

the ratio of the vibrations of the two tuning-forks. Yes, in everything and everywhere numbers rule the world.

Why, however, seek in scientific analysis testimony to the harmony which nature has shed over all her works? Although it may be necessary for us to rise to the ideal of music, to contemplate the beautiful colours of the sky or the splendour of the setting sun, we may on a dull winter day, in the grey and monotonous hours when the snow falls in innumerable flakes, examine with the microscope some of these flakes, and the geometrical beauty of these light crystals will fill us with admiration. As Pythagoras said, 'God works everywhere by geometry:' *ἀεὶ ὁ θεὸς γεωμετρεῖ.*

CHAPTER II

MEASUREMENT OF THE SUN'S DISTANCE

Concordant Results of Six Different Methods. Transits of Venus. How we have Measured and Weighed the Sun

ALL the numbers which we have given on the magnitude and the mass of the sun, on its distance, and on the dimensions of the solar system, are found from the measurement of the distance from the sun to the earth. This distance is truly the *unit* of the system of the world and the measure of the sidereal universe itself. The relative proportions of the motions and distances stated in the preceding chapter remain the same, it is true, whatever may be the absolute distances; but these absolute distances, which are of much interest, cannot be known unless the measure which serves as a base for all the others is itself accurately determined. We *know*, for example, that the distance of the farthest planet of our system, Neptune, is 30 times greater than that of the earth from the sun, and we *know* also that that of the nearest star is 275,000 times greater than the same unit; but we should not know the absolute distance if we had not first determined this unit with the most minute precision. It is, then, quite natural that astronomers should attach the greatest importance to its precise determination.

We have seen (p. 86) by what process they have determined the distance of the moon. If we wished to find by the same mode of observation the distance of the sun, we should not succeed. This distance is too great. The entire diameter of the earth is not comparable with it, and would not form the base of a triangle. Suppose that we draw from two diametrically opposite extremities of the terrestrial globe two lines to the centre of the sun: these two lines would nearly touch all along their length, the diameter of the earth being but a point relatively to their immense length. From here to the day star is nearly 12,000 times the diameter of the earth. It is as if we attempted to construct a triangle by taking for its side a line of 1 *millimetre* ($\frac{1}{25}$ of an inch) of length only, from each extremity of which we drew two straight lines to a point placed at 12 metres (39 feet) distant. We see that these two lines would be nearly parallel, and that the two angles which they would form at the base of the triangle would be truly two right angles.

It became necessary, then, to get over the difficulty, and it was this which the astronomer Halley did in the last century by proposing to use for this measurement the transits of Venus over the solar disc. We have already seen that Venus is nearer to the sun than we are, and goes round the central star along an orbit interior to ours. This is represented exactly in fig. 90, where the two orbits are drawn on a scale of 1 millimetre to a million of leagues. Now, when Venus passes exactly between the sun and the earth, two observers placed at two extremities of our globe do not see it projected on the same point of the sun, and the difference of the two points leads to the knowledge of the angle which gives the sun's distance.

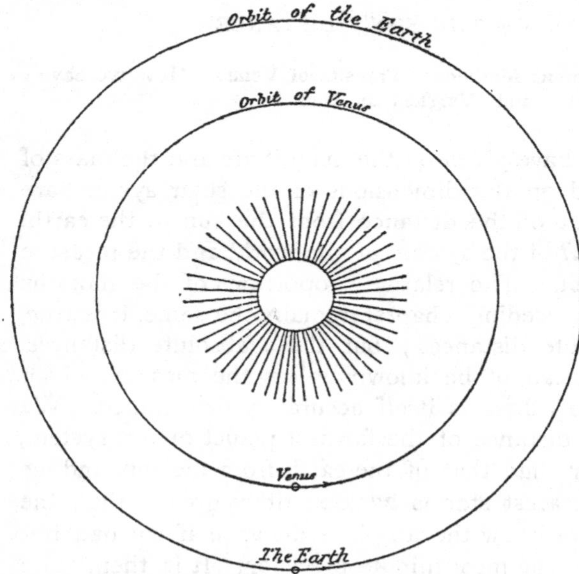

FIG. 90.—ORBITS OF VENUS AND OF THE EARTH AROUND THE SUN

Let us suppose that two observers are placed at the two extremities of a terrestrial diameter. Each of them would see Venus following a different path in front of the sun. This is a matter of perspective. Stretching out the hand and raising the forefinger vertically, it will hide from us a certain object by closing the left eye and looking at it with the right eye, and another object by closing the right eye and looking with the left eye. For the right eye it is thrown towards the left; for the left eye it is thrown towards the right. The difference of the two projections depends upon the distance at which we place our finger. In this familiar comparison, for which I humbly ask pardon of the reader, the distance which separates our two retinas represents the diameter of the earth; our two retinas are our two observers; our forefinger represents Venus herself, and the two projections of our finger represent the different places at which astronomers will see the planet on the surface of the sun. In order to make the comparison complete, it would be better, instead of holding out the finger to hold a pin with a large head at a certain distance from the eye, in such a way that its head is projected on a disc of paper placed at several yards distant, and then make this pinhead move in front of the disc by looking at it successively with each eye.[1]

[1] Let us enter into some details of this important method. Let us consider for a moment the respective positions of the sun, Venus, and the earth in space at the time of the transit. Two observers placed on the surface of the earth, as distant as possible from each other, are observing Venus; for each of them, as we have seen, it is projected at a different point, V_1 and V_2, on the sun's surface. Join these two points by a straight line. This line measures the distance which separates them from each other on the sun. Now, from these points draw a straight line which, passing through Venus, goes to meet each of the terrestrial observers. We have then constructed two triangles.

The first of these triangles has its base on the sun formed by the line of junction of the two points. Its two other sides go from these two points to Venus, the apex of the triangle.

The second triangle has likewise its apex at Venus, but in a direction opposite to that of the

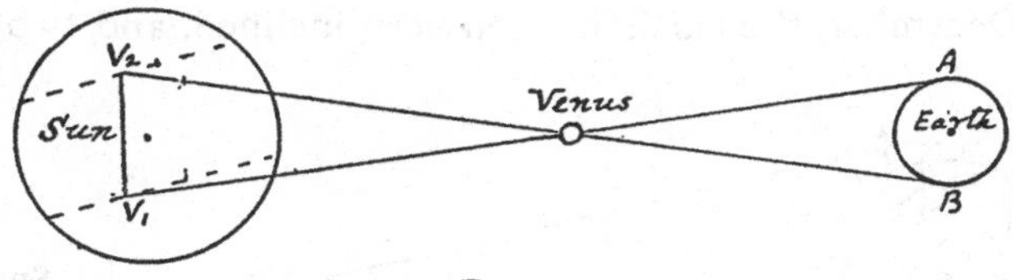

FIG. 91

preceding. Its two longer sides go from Venus to the earth, instead of going from Venus to the sun. Its third side, or base, is formed by the line which joins the two terrestrial observers, A and B.

In these two triangles the rectilinear distance which separates the two terrestrial observers is known, since we now know the dimensions of the earth. The third law of Kepler shows, on the other hand, that the sides of the two triangles are to each other in a certain determined ratio, which is equal to 0·37 for the triangle which has its base on the earth. The rectilinear distance which separates the two terrestrial observers is $\frac{37}{100}$ of the line of junction which joins the two points of projection of Venus on the disc of the sun. The problem is, then, finally reduced to measuring this line of junction as accurately as possible. Let us suppose that it has been found equal to 48 seconds of arc. This value would prove that the diameter of the earth seen at the sun's distance measures $48'' \times 0{\cdot}37$—that is to say, $17''{\cdot}76$. This is precisely the number sought.

The parallax of the sun is, then, nothing else but the angular dimension under which the earth is seen at the distance of the sun.

What is a second of arc? It is the apparent size of a metre or of any object distant at 206,265 times its length. An object which is seen under an angle of $17''{\cdot}76$ is, then, distant from the observer by a quantity equal to the number which I have given above, divided by 17·76. If, then, the earth seen from the sun subtends an angle of $17''{\cdot}76$, the distance from here to the sun is $\frac{206,265}{17{\cdot}76}$; that is to say, 11,614 times the diameter of the earth.

Instead of the whole diameter of the earth, we express the preceding values by the semi-diameter or radius, which, however, makes no change in the proportions. If the preceding number, which I have chosen for more simplicity, were exact, the parallax would then be expressed by the number $8''{\cdot}88$, the angle under which we should see the radius of the earth at the distance of the sun.

Such is the method of triangulation proposed by the English astronomer Halley to measure the sun's distance. The idea occurred to him in 1678, when he was twenty-two years old, but he did not

The combination of the earth's motion and the motion of Venus makes Venus only pass before the sun at odd intervals of 113½ years, *plus* or *minus* eight years. Thus, there was a transit in the month of December 1631; following it, one took place eight years later, in December 1639. The one which followed that occurred in the month of June, 1761—that is to say, 113½ years *plus* eight years, or 121½ years after. The next arrived eight years after, in June 1769. Now, to obtain the date of another transit, it is necessary to add to the preceding date 113½ years *minus* eight years, or 105½ years, which gives December 1874. The next transit happened eight years later, in December 1882. Now, we shall not have another before a new interval of 113½ years *plus* eight years—that is to say, before the month of June of the year 2004, which will be followed eight years afterwards by that of June of the year 2012, and so on. Another period—of 235 years—likewise brings back certain transits. As these phenomena only happen in June and December, the earth is then very inclined, and two observers,

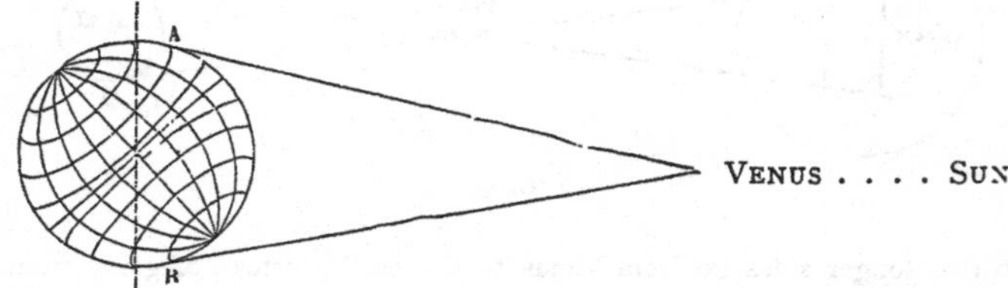

FIG. 92.—INCLINATION OF THE EARTH AT THE TIMES OF THE PASSAGE OF VENUS BEFORE THE SUN

A and B (fig. 92), can observe them from two opposite meridians, having the pole between them.

On the next page we give the dates of these transits from the invention of telescopes up to the thirtieth century of our era.

We see that astronomers do not allow themselves to be taken unawares. Astronomy is, after all, the only science which enjoys the privilege of reading the future as it does the past, and it avails itself of it.

The special details of the 'approaching' transit of June 7 of the year 2004 have already been calculated with precision, as well as those of the transit of June 5, 2012, and we might almost say that the various expeditions are arranged, with the exception of the names of the astronomers who will take part in them.

publish it till 1691. In indicating it as an excellent method of obtaining the parallax of the sun, the illustrious astronomer knew well, however, that, according to all probability, he would not himself be able to make use of it, and that he would, doubtless, have long since ceased to live when the moment for employing it should have come (1761). He recommended it, however, with pleasure, thinking more of being useful to men after he should have disappeared from among them than of expressing vain regrets on this earthly existence, which must be too short to permit him to behold the phenomenon of which he had first discovered the importance.

TRANSITS OF VENUS FROM THE SEVENTEENTH TO THE THIRTIETH CENTURY

		Central Phase, counted from Noon			Duration	
		H.	M.	S.	H.	M.
	1631, Dec. 6	17	28	49	3	10
	1639, Dec. 4	6	9	40	6	34
235 years	1761, June 5	17	44	34	6	16
	1769, June 3	10	7	54	4	0
235 years	1874, Dec. 8	16	16	6	4	11
	1882, Dec. 6	4	25	44	5	57
235 years	2004, June 7	21	0	44	5	30
	2012, June 5	13	27	0	6	42
235 years	2117, Dec. 10	15	6	37	4	46
	2125, Dec. 8	3	18	40	5	37
235 years	2247, June 11	0	50	23	4	16
	2255, June 8	16	53	56	7	12
235 years	2360, Dec. 12	13	59	9	5	25
	2368, Dec. 10	2	10	2	4	59
235 years	2490, June 12	3	58	35	2	4
	2498, June 9	20	21	2	7	33
235 years	2603, Dec. 15	12	54	16	5	53
	2611, Dec. 13	1	11	12	4	30
235 years	2733, June 15	7	23	56	Short	
	2741, June 12	23	43	59	7	46
235 years	2846, Dec. 16	11	53	15	6	14
	2854, Dec. 14	0	13	29	3	48
	2976, June 17	19	23	30	Very short	
	2984, June 14	3	2	22	7	52

The last two transits—December 8, 1874, and December 6, 1882—were observed by special scientific commissions sent out by all nations to the different points of the globe where the phenomenon would be visible. France sent expeditions to Japan, China, Cochin China, New Caledonia, to the island of St. Paul, and Campbell Island. England installed observers in India, Egypt, Persia, Syria, China, Japan, the Cape of Good Hope, Australia. The Americans were scattered in Siberia, China, New Zealand, the islands of Chatham and Kerguelen, and in Tasmania. Italy sent observers to Bengal. Germany was represented in Persia, Egypt, China, New Zealand, and in the islands of Auckland, Kerguelen, and Mauritius. Russia posted astronomers all along her immense territory up to Siberia. Thus, on these two dates, our planet was surrounded, over the whole hemisphere illuminated by the sun, by a zone of observers anxiously watching the transit of the little black disc of Venus in front of the radiant disc of the sun.

Astronomers had calculated in advance the geographical places from which observations could be made. Figs. 94 and 95 represent the result of this calculation, the first relating to the transit of 1882, and the second to that of 1874: the two projections are different, but the object is similar. The terrestrial planisphere is divided into four parts. The darkest, with horizontal hatching, represents the places for which the transit happened during the night, and was consequently invisible. The white tint indicates the places where the whole transit of the planet over the solar disc was visible. The two light tints indicate: that to the left, with horizontal hatching, the stations from which the egress of Venus could be seen, but not the ingress; that to the right, with vertical hatching, those at which the ingress was seen, but not the egress.

The weather did not favour all the expeditions, and many of the scientists had the sorrow of returning to their own country without being able even to see the sun, owing to persistent rain; while others, more favoured

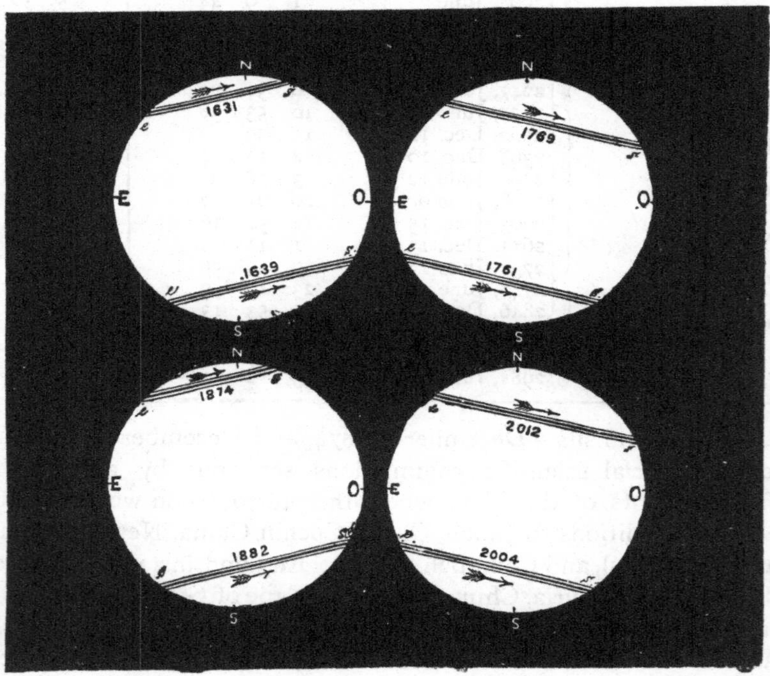

FIG. 93.—LINES FOLLOWED BY VENUS IN THE EIGHT TRANSITS FROM THE SEVENTEENTH TO THE TWENTY-FIRST CENTURY, FROM THE ENTRY TO THE EXIT

by the climate, returned with a rich collection of measurements and photographs, and received as a reward the title of Academicians.

Already, however, at the end of last century, Venus had made strange sport of astronomers who were most devoted to her. Witness the misadventure, now become legendary, of poor Le Gentil, whose name at least should have saved him from the harshness of the cruel planet, but who was, on the contrary, overwhelmed by a series of unexpected misfortunes. He went, in 1760, to observe the transit of 1761; but the English war in India prevented him from arriving; he could not land until *after* the date of the transit. Passionately fond of astronomy, he made the heroic decision to remain at Pondicherry for eight years, and wait for the next transit of 1769! As at this season (June) the weather in these latitudes is generally superb, he did not doubt of wonderful success. He built an observatory, learned the language of the country, installed excellent instruments, reached the happy year, the fortunate month of May, and the first days of June, which were illuminated with splendid sunshine. At last the day of the transit arrived; but the sky was covered, clouds eclipsed the sun, which remained obstinately hidden, Venus passed, and some minutes after the egress the sky cleared, the radiant orb shone anew, and did not cease to show itself during the following days! Not being able to wait for the next transit (of 1874), the poor astronomer decided to return to France, escaped twice from shipwreck, and arrived in Paris to ascertain that, in the absence of all news, he was believed to be dead, and was superseded in

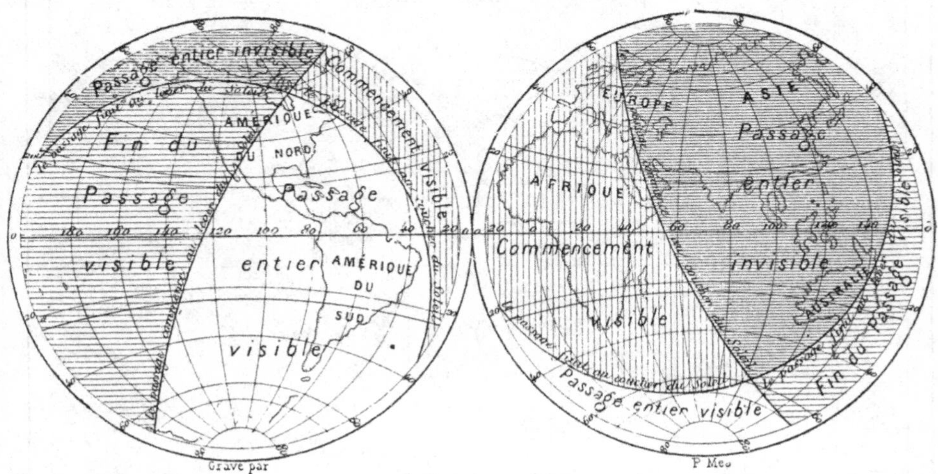

FIG. 94.—MAP (FRENCH) OF OBSERVATIONS OF THE TRANSIT OF VENUS OF DECEMBER 6, 1882

the Academy of Sciences; and, to complete his misfortunes, he was even prohibited from recovering his own inheritance, the law having decided that he was dead. He ended by actually dying!

The comparison of the whole of the observations made during the two transits of 1874 and 1882 has given as the result of the solar parallax numbers varying only between 8″·80 and 8″·86. Such is the angle which the semi-diameter of the earth measures as seen from the sun.

. . . . . .

This method—by transits of Venus over the sun—is not the only one which has been used to calculate the distance of the radiant star. Several others, absolutely different from that, and independent of each other, have been applied to the same research. Their results mutually confirm each other. Let us give a rapid sketch.

The first two are based on the velocity of light. It has been ascertained that light takes a certain time to be transmitted from one point to another, and that to come, for example, from Jupiter to the earth it takes thirty to forty minutes, according to the distance of the planet. In examining the eclipses of Jupiter's satellites we find that there is a difference of sixteen minutes twenty-six seconds between the times they happen when Jupiter is on the same side of the sun as the earth and when he is on the opposite side. Light, then, takes sixteen minutes twenty-six seconds to cross the diameter of the terrestrial orbit, and hence the half of this, or eight minutes thirteen seconds, to come from the sun situated at the centre.

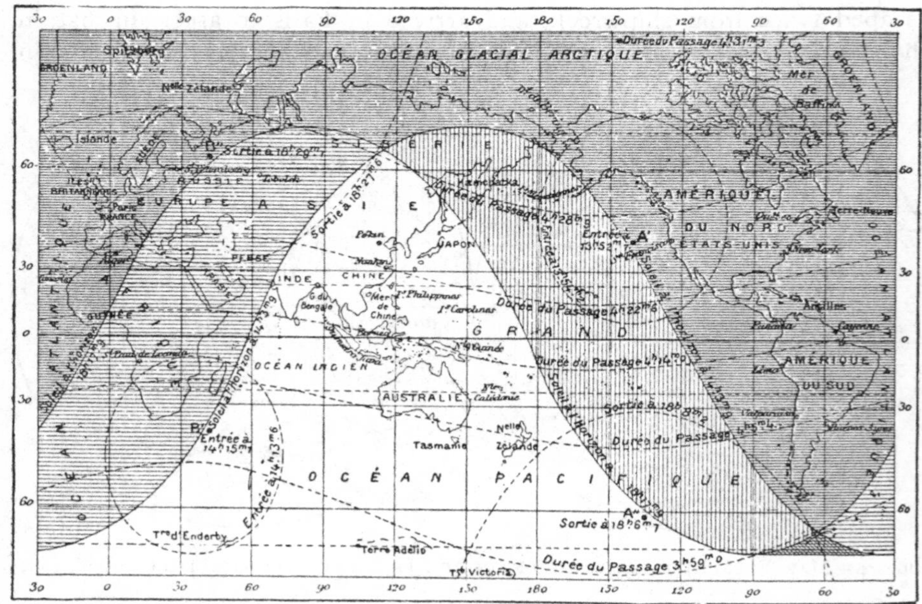

FIG. 95.—MAP (FRENCH) OF THE TRANSIT OF VENUS OF DECEMBER 8, 1874

Now, as the physicists (Foucault, Fizeau, Cornu, Newcomb) have directly measured this velocity, and have found it equal to 300,000 kilometres (186,414 miles) per second, we conclude that the distance from here to the sun is about 148 millions of kilometres (about 93 millions of miles).

Another method might likewise give this distance; it also is founded on the velocity of light. A familiar example will enable us to understand it at once. Let us suppose ourselves placed under vertical rain: if we are motionless, we hold our umbrella vertically; if we walk, we incline it before us; and if we run, we incline it still more. The degree of inclination of our umbrella will depend on the ratio of the velocity of our walk to that of the drops of rain. We observe the same effect on the railway in the slanting lines which the rain traces on the windows; of which the obliquity is the resultant of the motion of the train combined with the fall of the drops. The same effect is produced by light. The rays of light fall from the stars through space; the earth moves with a great velocity, and we are obliged to incline our telescopes in the direction in which the earth moves. This is the phenomenon of the aberration of light, which shows that the velocity of the earth equals $\frac{1}{10000}$ of that of light. We can then calculate from that the earth's velocity, which we thus find to be 30 kilometres (18·64 miles) per second; we can calculate the length of the orbit described in 365 days, and finally the diameter of this orbit, of which the half is precisely the distance of the sun.

A fourth method is furnished by the motions of the moon. The regularity of the monthly motion of our satellite is interfered with by the attraction of the sun; now, as the attraction varies in the inverse ratio of the square of the distance, we may understand that in rigorously analysing the action of the sun on the moon we can arrive at a knowledge of the sun's distance. This has been done by Laplace and Hansen.

A fifth method can be deduced from the masses of the planets, the motions of which are intimately connected with the mass of the sun and his distance. The planetary influences produce perturbations rendered perceptible by observations; when the masses have been determined by an independent method, the magnitude of the perturbations leads us to a knowledge of the distances. This calculation has been made by Le Verrier.

A sixth method is afforded by the observation of Mars and by that ot the small planets exterior to the earth; these planets pass near distant stars situated, so to say, at an infinite distance behind them, and if we observe their positions as seen from two countries on the earth very distant from each other, they are projected in two different points (like Venus on the sun); the angular separation of these two points indicates the distance from the earth to Mars, or to the other planets made use of. Mars was minutely examined from this point of view in 1832, 1862, and 1877; and also the small planets Flora in 1874, and Juno in 1877 [and Victoria in 1889].

All these measures agree with a remarkable precision, considering the difficulty of these observations. Here are the principal results:—

Transit of Venus in 1769	8″·91
" " 1874 (mean)	8″·85
" " 1882 (mean)	8″·82
Velocity of light	8″·86
Aberration of light	8″·80
[" "	8″·794 (Chandler)]
Motion of moon	8″·85
Masses of the planets	8″·86
Oppositions of Mars, Flora, &c.	8″·86
[Opposition of Victoria and Sappho, 1889	8″·799]

We see that the tenths are concordant, and that the uncertainty is only in the hundredths. The mean adopted (8·86) signifies that, seen from the sun, the semi-diameter of the earth is reduced to that angle, or its diameter to an angle of 17″·72. It is like a ball of 10 centimetres (3·937 inches) in diameter placed at 1,164 metres (3,819 feet) from the eye. Going back to the little table of angles given on page 86, we easily calculate that this parallax corresponds to a distance of 11,640 times the diameter of the earth—that is to say, in round numbers, to 148 millions of kilometres (92½ million miles).[1]

Such is the measure of the *Sun's distance.*

There is here no romance, no work of the imagination; the figures are *mathematical facts*, absolute, unquestionable for every sincere mind which chooses to examine for itself their origin and nature.

The result is not the less marvellous on that account.

Thus, if we could throw a bridge across space from here to the sun, and if we could form the arches of this viaduct of semicircles as wide as the earth, the total length of this ethereal bridge would be composed of 11,640 of these arches side by side! Or, again, it would be necessary to range 11,640 earths to form the foundation of the bridge in question.

How shall we represent this distance which separates us from the day star?

One way, perhaps, of succeeding would be to suppose that a moving body, a cannon-ball for example, was shot from here to the sun, to follow it in thought, and to *feel* the time which it would take to pass over this distance. Let us try. Propelled by a charge of thirteen pounds of powder, such a projectile moves with a volocity of 500 metres (1,640 feet) in the first second. If it preserved this uniform velocity to the sun, it would require to fly in a straight line during *nine years and eight months* in order to reach it.

We shall soon see that the sun is the seat of explosions and terrible conflagrations. If the space comprised between this body and the earth could transmit a sound with the ordinary velocity of propagation of 340 metres (1,115 feet) per second, the sonorous concussion would require 13

[1] It is probable that this parallax is a little smaller, and approaches 8″·8. A change of 0″·01 in the solar parallax corresponds to a difference of about 170,000 kilometres (105,635 miles). This parallax is certainly included between 8″·80 and 8″·86. Here are the corresponding distances:—

			Kilometres	
8″·80 = 23,439	radii of the earth		= 149,330,000	(92,790,675 miles)
8″·81 = 23,412	,,	,,	= 149,160,000	(92,685,040 ,,)
8″·82 = 23,385	,,	,,	= 148,991,000	(92,580,027 ,,)
8″·83 = 23,358	,,	,,	= 148,822,000	(92,475,000 ,,)
8″·84 = 23,332	,,	,,	= 148,653,000	(92,370,000 ,,)
8″·85 = 23,306	,,	,,	= 148,485,000	(92,265,600 ,,)
8″·86 = 23,280	,,	,,	= 148,317,000	(92,161,200 ,,)

years and 9 months to pass over this distance. Nearly 14 years would, then, elapse after a solar explosion had taken place before we could hear the sound which it produced.

A railway train will perhaps measure this distance in a still more obvious form. Let us suppose, then, in imagination a railroad going in a straight line from here to the central star. Well, an express train travelling at the constant velocity of 37 miles an hour (or a kilometre a minute), would take 148 millions of minutes to reach the sun—that is to say, 97,222 days, or 266 years. Leaving on January 1, 1895, it would not finish its run till the year 2161. In proportion to the mean duration of our life, the sidereal expedition would not attain its object till its seventh generation, and it would be the fourteenth which could report the 'news' of what the great-great-grandfather of his great-grandfather had seen! A traveller who had started with this velocity in 1628 under Louis XIII. would only arrive now (1894)! At the rate of 10 centimes per kilometre, the price of the ticket would be 14,800,000 francs (about 600,000*l.*).

We know that nervous sensation takes a certain time to transmit itself along the nerves to the brain, the seat of perception. When we burn the tip of our finger we do not feel it instantaneously, but a fraction of a second afterwards; the velocity is estimated at 28 metres (92 feet) a second. If an infant had an arm sufficiently long to touch the sun, and be burnt, he would not feel the pain until the end of 167 years! That is to say, he would be dead long before he could feel it! It would be the same if we could touch the sun by the aid of a metallic bar conducting the heat with a similar velocity of transmission.

Now that we know the distance of the sun, nothing is simpler than to calculate his real size by the aid of his apparent diameter, exactly as we have done for the moon. We have seen that the diameter of the earth seen from the sun is 17″·72. On the other hand, the diameter of the sun seen from the earth is 32′ 4″—that is to say, in seconds, 1,924″. Such is, then, very simply the proportion of the two diameters. Dividing the latter number by the former, we find that it contains it 108½ times (108·55). It is, then, *demonstrated* that the real diameter of the sun measures 108½ times 12,732 kilometres—that is to say, 1,382,000 kilometres (858,747 miles); that its circumference measures 2,690,000 miles; that its surface exceeds that of our globe nearly twelve thousand times, and presents an area of six millions of millions of square kilometres (over two millions of millions of square miles); and that its volume, 1,280,000 times greater than that of the earth, contains 333,000 billions of cubic miles, that is:

333,000,000,000,000,000.

Since the diameter of the sun is 858,000 miles, there are 429,000 miles from its centre to its surface. Now, there are 238,000 miles from here to the moon. If, then, we could place the earth at the centre of the sun, like a small kernel in the middle of a colossal fruit, the moon would revolve in the interior of the solar globe, and the distance of the moon would hardly represent more than half the way from the centre to the solar surface ; to reach this surface from the lunar orbit there would still remain 191,000 miles to pass over!

The most lively imagination cannot succeed in forming a correct idea of the difference of volume between the sun and earth. A comparison often quoted is not wanting in eloquence. It appears that in a litre of wheat there are thirty thousand grains. The measure of capacity called the decalitre contains, then, a hundred thousand grains. If, then, we pour into the same heap thirteen decalitres of wheat (28·6 gallons), and if we take one of these grains, we shall have in round numbers the prodigious difference of volume which exists between the sun and the earth. One of our globes more or less in the sun is insignificant. But we shall imagine this volume of 1,280,000 grains of wheat if we remark that each of these grains measures in reality 260,000 millions of cubic miles.

Jupiter is 1,369 times larger than the earth. Saturn, Neptune, Uranus, are also much superior in volume to our world. Nevertheless, if we united together all the planets and all the satellites, they would still form a volume six hundred times smaller than that of the sun alone.

Everyone is justly astonished at such magnitude. Well, science is not less admirable in the infinitely small than in the infinitely great. Calculation proves that there are not more cubic kilometres in the sun than there are atoms in the head of a pin. In fact, the entire body of certain infusoria, seen in the microscope, would lie between two divisions of a millimetre divided into a thousand equal parts, and consequently measure at most the thousandth of a millimetre. This little being lives, walks, feels, is provided with organs of locomotion which require muscles and nerves (some have as many as 120 stomachs !). In comparing its diameter to a metre, the most moderate supposition which we can make is that the organic molecules which constitute its body are a millimetre in diameter, and that in these molecules there are not less than ten distances of constituent atoms. We may, then, conclude with Gaudin, for the distance of the atoms, a ten millionth of a metre. It follows that the number of atoms contained in a fragment of matter of the size of the head of a pin of two millimetres would be represented by the cube of twenty millions, or by the figure 8 followed by twenty-one cyphers :—

8,000,000,000,000,000,000,000 ;

so that if we wished to count the number of metallic atoms contained in the head of a pin, detaching each second, in thought, a milliard (a thousand millions), it would require no less than *two hundred and fifty thousand years* to count them all.

Nature is immense in the little as in the great, or, to speak more correctly, for her there is neither little nor great.

But science has not only measured the sun ; it has even weighed it. Our justifiable curiosity here puts this question, not less bold than the preceding : *How can we weigh the sun ?*

This explanation is a little more difficult to 'popularise' ; it is also generally passed over in silence. It requires at least five minutes' sustained attention to understand it well. Five minutes! That is nothing in a lifetime ; it is enormous for shallow minds who prefer Offenbach to Beethoven. Ungallant physiologists assert that the feminine brain weighs 124 grammes less than the masculine brain, the first weighing but 1,210 grammes, and the second 1,334 (the question is of French women and men). It will, then, require six minutes of attention from my lady readers.

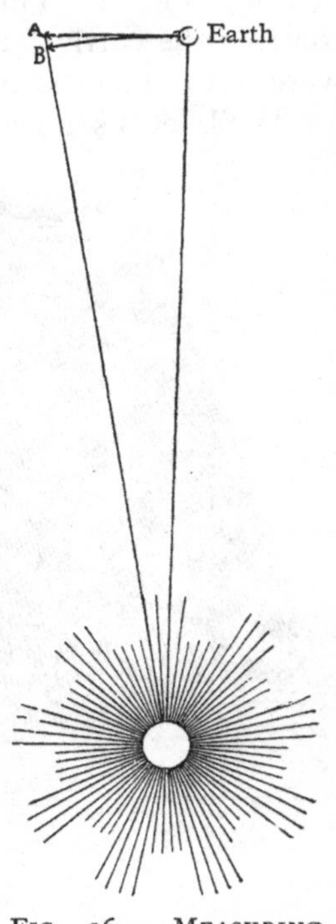

FIG. 96. — MEASURING THE SUN'S ATTRACTION

We have seen above, with reference to the moon (p. 91), that gravity and universal attraction are one and the same force, and that Newton discovered this identity by calculating the distance which exists at the end of a second between the extremity of the straight line which the moon would pass over if it were not attracted by the earth, and the extremity of the curved line which it describes in reality on account of our attractive influence. This distance, which is about $\frac{1}{20}$th of an inch, represents exactly the distance which any body would fall towards the earth in one second if it could be carried to this height and there abandoned to the influence of gravity. If, for example, an angel could seize a man by the hair and raise him to the height of the moon (they say that Mahomet had this pleasure—doubtless in a dream), then leave him there and reascend to the sky, our man would fall again towards the earth ; but in the first second of the fall he would only fall $\frac{1}{20}$th of an inch, and the fall would go on increasing by a uniformly accelerated motion.

It is by a similar process that we judge of the attractive mass of the sun.

If, instead of carrying a stone to the distance of the moon—to sixty times the radius of the earth—we could transport it to the distance of the sun—to 23,200 times this radius—how much would the intensity of terrestrial gravity be diminished at such a distance? The law is the same everywhere. The answer is, then, that it would be diminished in the ratio of the square of the distance. Now, this distance is 23,200 times the radius of the earth, the square is 538,240,000 ; instead of being $\frac{4\cdot90}{3{,}600}$ m., the fall would be $\frac{4\cdot90}{538{,}240{,}000}$—that is to say, so small that it can scarcely be expressed by a comprehensible fraction of a millimetre—it is nine millionths of a millimetre. This is the small distance which a stone would fall back towards the earth if it could be conveyed to 92 millions of miles, and if it were not influenced by the attraction of any other celestial body.

Well, let us now do for the earth what we did before for the moon.

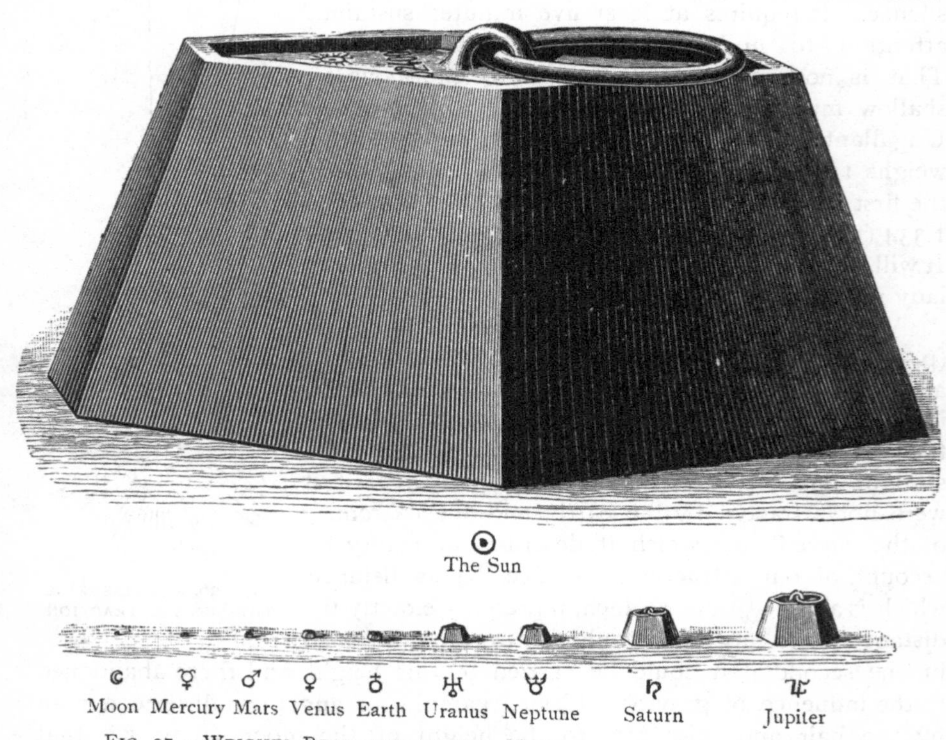

Fig. 97.—Weights Representing the Masses of the Celestial Bodies

Let us draw (fig. 96) the path passed over in a second by our planet in its annual course round the sun, and let us see what difference exists between the arc passed over and the straight line which the earth would follow if it did not experience the attractive influence of the sun : this difference shows us, as in the case of the moon, exactly the quantity which our planet falls in one second towards the sun. The precise measurement gives 2·9 millimetres (we have seen it already, p. 214) [0·114 of an inch].

Consequently, the attraction of the sun is to that of the earth in the ratio of 0 0029 m. to 0·000,000,009 m., or of 29 to 0·00009, or as 29 to 9 hundred-thousandths. Or, in other words, the attraction of the sun is 324,000 times stronger than that of the earth. We have seen that attraction is produced by the masses or by the weights of bodies. *We know, then, mathematically by this that the sun weighs* 324,000 *times more than the earth.*[1]

We see that all this is very simple. The first Bachelor of Arts we meet may now flatter himself that he can weigh the sun—when the astronomers have furnished him with the elements of the calculation. The distance of the star being the first of these elements, we can understand the importance which science attaches to its being exactly known.

Now that we have determined the volume and the mass of the sun, it is easy for us to complete these data by the determination of its density. The density of a body is the mass divided by the volume. The central star of the solar system being 1,280,000 times larger than the earth, and only 324,000 times heavier, it is much less dense than our world. This density is expressed by the figure 0·253, representing that of the terrestrial globe by 1—that is to say, that the constituent materials of the sun weigh about 25 hundredths, or a quarter, of those which compose the whole of the earth.

[1] We can arrive at the same result by another method. We have seen (p. 215) that the planets revolve so much the slower as they are more distant from the sun, and that the law of this diminution of velocity is expressed by the following formula: 'The squares of the times of revolutions are to each other as the cubes of the distances.' In other words, a body situated twice as far as another revolves in a period indicated by the square root of 8 (cube of 2), a body 3 times farther by the square root of 27 (cube of 3), and so on. Do you wish to know, for example, in what time a moon situated at double the distance of ours would revolve round us? The calculation is easy: $2 \times 2 \times 2 = 8$; the square root of 8 is 2·84; it would, then, revolve 2·84 times more slowly—that is to say, in 77 days.

In order, then, to learn the difference which exists between the attraction of the earth and that of the sun, it is simply necessary to find in what time a body situated at 93 millions of miles would revolve round us. Now 37 millions is 385 times the moon's distance. Let us make the calculation: $385 \times 385 \times 385 = 57{,}066{,}625$; the square root of this number is 7,553; this distant moon would, then, turn round us 7,553 times less quickly than the present moon—that is to say, in 206,330 days, or in 566 years.

If the values of the directing masses were judged simply by the periods of revolution, since the earth has only the power to make a satellite revolve in 566 years, and the sun has the power to make the earth revolve in one year (at the same distance of 93 millions of miles), we should conclude at once that the sun is simply 566 times stronger than the earth. But it is not the simple periods which it is necessary to compare: it is the periods multiplied by themselves.

Let us, then, multiply 566 by itself, and we find, in round numbers, 320,000 for the approximate ratio between the mass of the sun and that of the earth. If we had taken into account the decimals and fractions, we should have found the same number as before, or 324,000.

If we were dealing here with mathematical analysis, we should make a distinction between the words *mass* and *weight*, but we think that in a popular work it is unnecessary to complicate questions.

The weights represented in fig. 97 give an idea of the magnitude of the solar mass compared to that of the planets.

We may form an idea of the power of the solar attraction by the following comparison: In order to supply its place it would be necessary to suppose the earth connected with the sun by a system of steel wires of the size of the strongest telegraph wires, and forming over the whole hemisphere turned towards the sun a mass as close as the blades of grass in a meadow.

The sun weighs a little more than a globe of water of the same dimensions, and doubtless, as we shall see presently, is formed of extremely condensed gas.

A last word still on gravity at the solar surface, and we shall have studied *ex professo* all the uranographical elements of the focus of the planetary system.

The state of gravity at the surface is inferred from the mass of the globe and its volume: it depends at once on the mass of the globe on the surface of which we consider it, and on the radius of this globe—that is to say, the distance which separates the surface from the central point where the whole mass may be supposed concentrated without the total attraction which it exerts being perceptibly altered. It is not difficult to calculate the intensity of gravity at the surface of a world by taking into account these two elements. Let us make this calculation for the sun.

The intensity of gravity on the earth being represented by 1, that which exists on the sun would be represented by 324,000, if the semi-diameter of that body were equal to that of the earth. But it is 108½ times greater; the attraction exercised by the sun on its surface is, then, 11,783 times less than if its radius were equal to that of the earth (11,783 is the square of 108·55). Dividing 324,000 by 11,783, we find 27·6 for the solar gravity compared with terrestrial gravity. The sun attracts objects at its surface twenty-seven times more strongly than the earth does. This calculation would be the same for the investigation of gravity at the surface of all the planets. We have given the results above (p. 111) with reference to the moon.

A human body, if it could be transported to the sun, would be immediately flattened by gravity into a thin leaf. But, in point of fact, it would be vaporised long before it could arrive there.

CHAPTER III

LIGHT AND HEAT OF THE SUN. STUDY OF ITS SPOTS

State of its Surface. Its Rotation. Magnitudes, Aspects, Forms, and Motions of the Solar Spots

'ALREADY the star of Venus, Chasca, gives the signal of morning. Scarcely do her silvery fires sparkle on the horizon, when a gentle murmuring is heard round the temple. Soon the azure of the sky pales towards the east, waves of purple and gold inundate the celestial plains. The watchful eye of the Indians observes its gradations, and their emotion increases with each new tint. Suddenly the light rushes in great waves from the horizon; the star which sheds it rises in the sky; the temple opens, and the pontiff, in the midst of the Incas and a choir of sacred virgins, sings the solemn hymn, which at the same instant is repeated by thousands of voices from mountain to mountain.'

Thus writes Marmontel when describing the festival of the Sun—the god worshipped by primitive nations. At the return of the equinox the rising of the sun, the god of day, the king of light, was saluted by the Incas from the heights of their cyclopean terraces. The same adoration, the same worship, is met with among all the ancient peoples. Without yet taking into account the real size and the incomparable importance of the dazzling star, they already knew that he is the father of terrestrial nature; they knew that it is his heat which supports life; they knew that it is he who makes the trees in the forests to grow, the stream to flow in the valley, the flowers of the meadow to bloom, the bird to sing in the wood, the cereals and the vines to ripen; and they hailed in him their father, their friend, and their protector.

Modern science has not only confirmed, but increased tenfold, a hundred-fold, the ancient conjectures. The sun's light, heat, and power, are as much above the ancient ideas as the poetry of nature is above our interpretation. No light created by human industry can be compared with his. Interposed before his disc, the brilliant electric light appears black. The highest temperatures of our furnaces, that of the melting of gold, of silver, of platinum, of iron, are but ice compared with the solar heat. The astronomers of the school of Pythagoras, who thought they gave a grand idea of the day star by estimating its distance at 72,000 kilometres (44,740 miles), and its diameter at 618 kilometres (384 miles), were as far from the reality as an ant who believed itself the size of a horse. And yet to estimate the sun as of the size of the Peloponnesus was then such boldness in the eyes of the classical conservatives and the teaching doctors, that for having asserted this beginning of truth the philosopher Anaxagoras was outrageously persecuted and condemned to death—a sentence commuted to a decree of exile on the petition of Pericles! The trial of Galileo was, later on, a repetition of that of Anaxagoras.

Photometric measures of the solar light show that it is equivalent to 1,575,000,000,000,000,000,000 millions of wax candles, or to 157,500,000,000,000,000,000 millions of carcel lamps, supposing that one carcel lamp is equal to ten candles, or to 15,750,000,000,000,000,000 millions of gas burners each equivalent to ten carcel lamps. At the surface of the day star the intensity of the light exceeds by 5,300 times that of the incandescent metal melted in a Bessemer converter, 146 times that of calcium, and 4 times that of the electric light.

The luminous and calorific influence which we receive from the star of day being a fact of constant and universal observation, the question which is presented to us is, not to ask whether this influence is real, but to determine the intensity of a cause which at such a distance can still produce such effects. But what are our temperatures, which, after all, proceed from the sun, in comparison with that of the sun itself? The heat of boiling water appears to us enormous, and our living organism cannot bear it. Still, it represents but the ordinary scale on which our thermometers are graduated. Water boils at 100° Centigrade (212° Fahrenheit), sulphur fuses at 113° Centigrade, tin at 235°, lead at 325°, silver at 945°, gold at 1,245°, iron at 1,500°, platinum at 1,775°, iridium at 1,950°. The furnaces of our laboratories have succeeded in producing heats of 2,500 to 3,000 degrees.

What are these effects in comparison with the incandescent star, which, across a distance of 93 millions of miles, and only by a quantity of heat two milliards (2,000 millions) of times less intense than that which it radiates, is still capable of warming our planet to a point which makes it live in the fecundity of this radiation! The quantity of heat emitted by the sun was measured by Sir John Herschel at the Cape of Good Hope, and by Pouillet at Paris. The agreement between the two series of measures is very remarkable. Sir John Herschel found that the calorific effect of a vertical sun at the level of the sea is sufficient to melt a layer of ice of 0·0075 of an inch per minute; while, according to Pouillet, the quantity of ice melted would be 0·0070 inch. The mean of these two determinations cannot be far from the truth; it is 0·0072, or 0·437 inch per hour. Taking into account the thicknesses of the atmosphere traversed at different hours, we find that the quantity of solar heat absorbed by the atmosphere is four-tenths of the total radiation directed towards the earth; so that if the atmosphere were removed the illuminated hemisphere would receive nearly double the heat. If the quantity of solar heat received by the earth in a year were uniformly distributed, it would be sufficient to liquefy a layer of ice of 30 metres (98·4 feet) covering all the earth. In the same way, it would cause an ocean of cool water having a depth of 100 kilometres (62 miles) to pass from the temperature of melting ice to that of ebullition.[1]

The sun is the mighty source from which proceed all the forces which set in motion the earth and its life. It is its heat which causes the wind to blow, the clouds to ascend, the river to flow, the forest to grow, fruit to ripen, and man himself to live. The force constantly and silently expended in raising the reservoirs of rain to their mean atmospheric height, in fixing the carbon in the plants, in giving to terrestrial nature its vigour and its beauty, has been calculated from a mechanical point of view; it is equal to the work of 217 billions 316 thousand millions horse-power; 543 milliards (543,000,000,000) of steam-engines, each with an effective power of 400 horses, would have to work day and night without intermission: such is the permanent work of the sun upon the earth.

We may not think so, but everything which moves, circulates, and lives on our planet is the child of the sun. The generous wine whose transparent ruby cheers the French table, the champagne which sparkles in the crystal cup, are so many rays of the sun stored up for our taste. The most nutritious foods come from the sun. The wood which warms us in winter is, again, the sun in fragments; every cubic inch, every pound of wood, is formed by the power of the sun. The mill which turns under the impulse of wind or water revolves only by the sun. And in the black night, under the rain or snow, the blind and noisy train which darts like a flying serpent through the fields, rushes along above the valleys, is swallowed up under the mountains, goes hissing past the stations, of which the pale eyes strike silently through the mist—in the midst of night and cold, this modern animal, produced by human industry, is still a child of the sun; the coal from the earth which feeds its stomach is solar work stored up during millions of years in the geological strata of the globe. As it is certain that the force which sets the watch in motion is derived from the hand which has wound it, so it is certain that all terrestrial power proceeds

[1] Concentrating this heat by the aid of an ingenious apparatus, M. Mouchot has been able for many years past to substitute the sun's rays for the ordinary heat of our fires in order to cook beef *à la mode*, to boil coffee, to distil brandy, &c. There are climates where these operations would render kitchens unnecessary. Future industry will certainly utilise the solar rays.

from the sun. It is its heat which maintains the three states of bodies—solid, liquid, and gaseous ; the last two would vanish, there would be nothing but solids, water and air itself would be in massive blocks, if the solar heat did not maintain them in the fluid state. It is the sun which blows in the air, which flows in the water, which moans in the tempest, which sings in the unwearied throat of the nightingale. It attaches to the sides of the mountains the sources of the rivers and glaciers; and consequently the cataracts and the avalanches are precipitated with an energy which they draw directly from him. Thunder and lightning are in their turn a manifestation of his power. Every fire which burns and every flame which shines has received its life from the sun. And when two armies are hurled together with a crash, each charge of cavalry, each shock between two army corps, is nothing else but the misuse of mechanical force from the same star. The sun comes to us in the form of heat, he leaves us in the form of heat, but between his arrival and his departure he has given birth to the varied powers of our globe.

Presented to our mind under their true aspect, the discoveries and generalisations of modern science constitute, then, the most sublime poem which has ever been offered to the intelligence and the imagination of man. The physicist of our day, we may say with Tyndall, is incessantly in contact with marvels which eclipse those of Ariosto and Milton; marvels so grand and so sublime that those who study them have need of a certain force of character to preserve them from being dazed.

And still all this is nothing, or almost nothing, in comparison with the real power of the sun! The liquid state of the ocean, the gaseous state of the atmosphere; the currents of the sea; the raising of the clouds, the rains, storms, streams, rivers; the calorific value of all the forests of the globe and all the coal mines of the earth; the motion of all living beings; the heat of all humanity; the stored-up power in all human muscles, in all the manufactories, in all the guns—all that is almost nothing compared with that of which the sun is capable. Do we think that we have measured the solar power by enumerating the effects which it produces on the earth? Error! profound, tremendous, foolish error! This would be to believe still that this star has been created on purpose to illuminate terrestrial humanity. In reality, what an infinitesimal fraction of the sun's total radiation the earth receives and utilises! In order to appreciate it, let us consider the distance of 93 millions of miles which separates us from the central star, and at this distance let us see what effect our little globe produces, what heat it intercepts. Let us imagine an immense sphere *traced at this distance from the sun,* and entirely surrounding it. Well, on this gigantic sphere, the spot intercepted by our little earth is only equivalent to the fraction $\frac{1}{2,138,000,000}$; that is to say, that the dazzling solar hearth radiates all round it through immensity a quantity of light and heat two thousand one hundred and thirty-eight million times more than that which we receive, and of which we have just now estimated the stupendous effects. The earth only stops in its passage the *two thousand millionth part of the total radiation.*

It is absolutely impossible for our conception to imagine such a proportion.

All the planets of the system intercept but the 227 millionth part of the radiation emitted by the central star. The rest passes by the worlds and *appears* to be lost.

It is not impossible to *express* this marvellous power, but we can admit without shame that it is impossible to comprehend it. *The heat emitted by the sun* IN EACH SECOND *is equal to that which would result from the combustion of eleven quadrillions six hundred thousand milliards of tons of coal burning at the same time.*

This same heat *would boil* per hour *seven hundred thousand millions* of CUBIC MILES *of water at the temperature of ice.*

Attempt to understand it! As well might the ant attempt to drink the ocean!

O popes of the Aryans! O priests of the Incas! O therapeutists of Egypt! and you philosophers of Greece, alchemists of the Middle Ages, scientists of modern times! O thinkers of all ages! you should be dumb before the sublime star! What is our voice in nature? We may pile up metaphors on metaphors, we shall only lower these magnitudes to our own size. We are but pigmies pretending to scale the sky.

Scientific analysis, however, may attempt only to state the observed facts and to give an approximate idea of their immense realities. The modern physicist has sought to determine the true temperature of the sun. Berthelot and Saint-Claire Deville have estimated it at 3,000° Centigrade, Vicaire and Violle at 2,500, Pouillet at only 1,600. Zöllner has estimated at 27,000° the temperature of the surface, and at 85,000° that of the nucleus. The experiments of Rossetti would place the heat of the sun at 10,000°, those of Hirn at 2 millions, those of Soret at 5 millions, those of Waterston at 7 millions, and those of Secchi at 10 millions. The diversity of these results proves that it is still an open question, and that science does not yet possess sufficient elements to solve the problem.

We shall see further on what we should really understand by the word *heat.*

The whole surface of the solar disc does not present everywhere the same degree of light, nor of heat. We ascertain this at the first glance, when we observe the sun with a telescope. Our drawings (especially the figure on p. 210) have already given an approximate idea. Receiving the image of the sun on a screen, M N (fig. 98), Secchi has found that two holes pierced in this screen give two very different pencils of light, *a* and *b*, according to their distance from the centre of the disc. At the point *a* the light is but the fifth of that at the centre. All along the edge it is but a fourth, and reddish, which explains the tint of the horizon during eclipses.

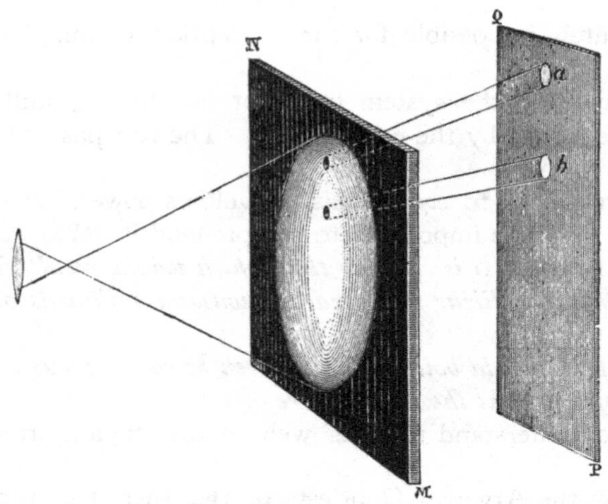

FIG. 98.—VARIATIONS OF LUMINOUS INTENSITY OF THE SOLAR DISC

According to the experiments of Pickering, the following is the variation of the light from the centre towards the edges, and according to Langley the variation of the heat:—

Distance from centre	Light	Heat
0·00	100	100
0·25	96	99
0·50	91	95
0·75	79	86
0·95	55	62
0·98	45	50

This diminution of light and heat proves that the sun is surrounded by an atmosphere. Without this atmospheric absorption this star would be, like the moon, uniformly luminous over the whole of its surface.

The temperature of the spots is inferior to that of the general luminous surface; but the difference of calorific intensity is much less than the difference of the luminous intensity.

Let us examine, however, in detail the aspect of the solar surface.

The ancients did not know any of the particulars relative to the physical constitution of the sun. They described, from time to time, black spots which they could distinguish with the naked eye when that body was near the horizon, but they took them for planets, or phenomena of which the cause was unknown. Such are the spots which were observed in 807,

840, 1096, 1588. Kepler himself thought he observed a transit of Venus over the sun: it was a spot which he had before his eyes. It was Fabricius who first, in 1610, examined the solar spots with the naked eye by projection on a screen, and discovered the rotation of the sun.[1]

We can easily observe the sun-spots, even with telescopes of rather small dimensions, taking care to place before the eyepiece a strongly-

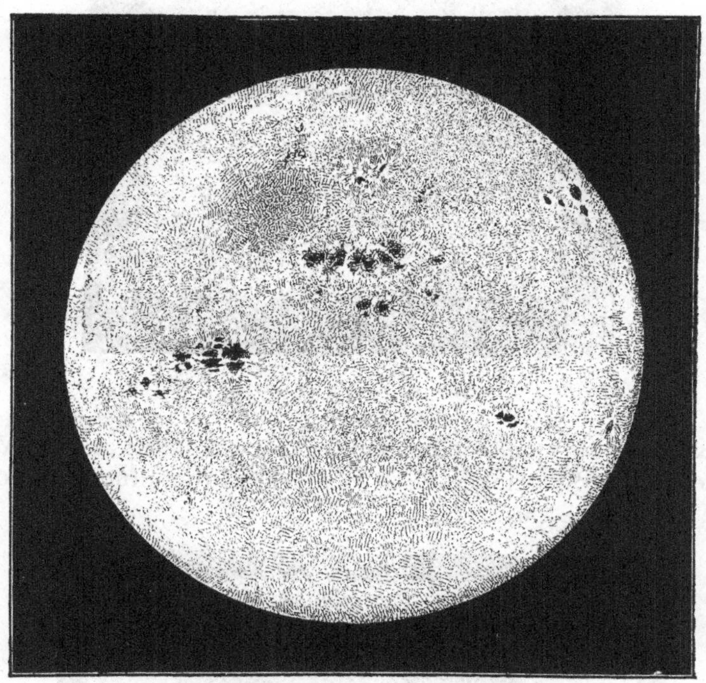

FIG. 99.—THE SUN. (From a direct Photograph)

coloured glass. They usually present themselves as nearly round black points. Very often, however, they are grouped in such a way as to form, on the whole, very irregular figures. The central part is black. It is called the *nucleus* or *umbra*. The outline is formed by a half-tint, which is called

[1] The Chinese have anticipated us in these observations. The encyclopædic work of Ma-Tuoan-Sin contains a remarkable table of forty-five observations made between the years 301 and 1205 of our era—that is to say, in an interval of 904 years. To give an idea of the relative size of the spots, the observers compare them to an egg, a date, a plum, &c. The observations often extend over several days; some have even been made during ten consecutive days. We cannot doubt the reality and the accuracy of these observations, and yet they have hitherto been unavailable to Europeans, for they have only been published in recent years. The Chinese astronomers have not made known to us the method they used in these observations, but we know that with a simple piece of smoked glass we can see the largest spots with the naked eye. Before telescopes were invented observers admitted the solar rays through a small circular hole cut in a window-shutter.

the *penumbra.* The outlines of the umbra and those of the penumbra are clearly defined, at least in most cases.

Fig. 99, engraved from a direct photograph, gives an idea of the

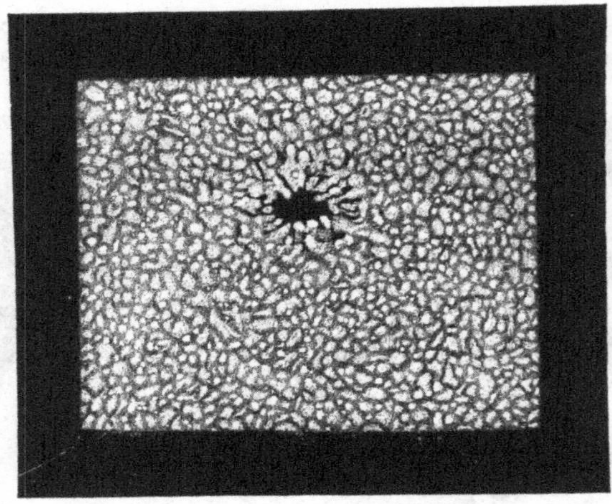

FIG. 100.—SURFACE OF THE SUN, HIGHLY MAGNIFIED

relative extent of the spots. This photograph was taken at New York, by Mr. Rutherfurd, on September 22, 1870, a year of tumultuous motions both in the sun and on the earth.

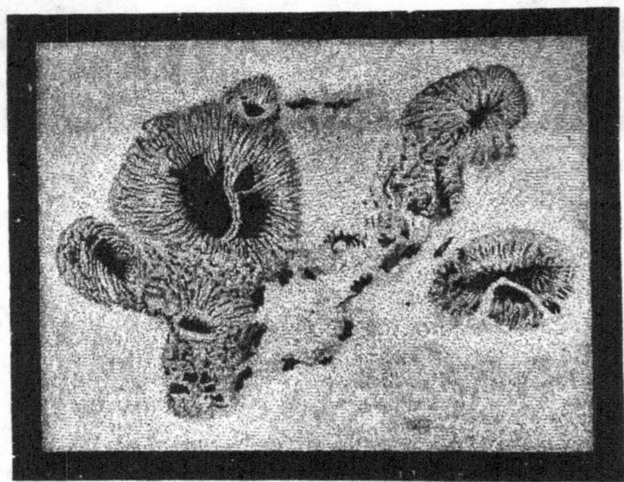

FIG. 101.—TYPICAL SUN-SPOTS

On the edges of the disc we see little white spots, to which the name of *faculæ* has been given. All these spots change in place and form.

The solar surface, far from being perfectly even, presents an irregular and granulated appearance. We recognise this aspect when we observe the sun with a powerful eyepiece, in the rather rare moments when the atmosphere is perfectly calm, and before the object-glass begins to get heated. Then we see that the surface is covered over with a multitude of little grains, having very different forms, among which the oval seems to

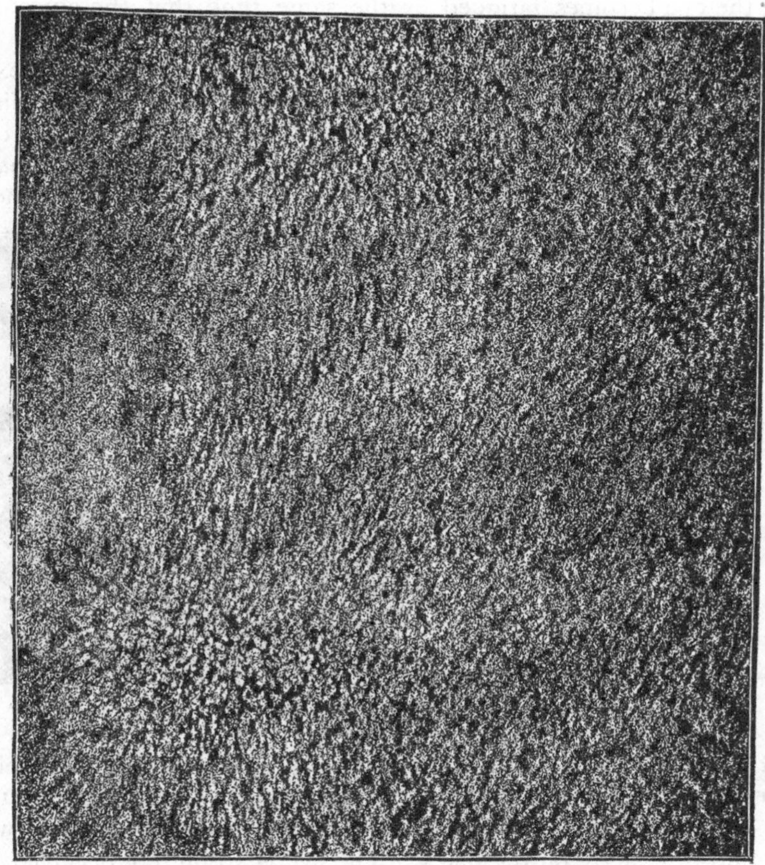

FIG. 102.—DIRECT PHOTOGRAPH OF THE SURFACE OF THE SUN, SHOWING LUMINOUS WAVES AND CURRENTS

predominate. The very thin interstices which separate these grains form a grey network. We reproduce in fig. 100 a drawing by Secchi, in which the Roman observer has attempted to make a sketch of the characteristic aspect of the surface. 'It seems to us difficult,' he says, 'to find a known object which recalls this structure. We obtain something similar in examining with a microscope milk a little dried up, of which the globules have lost the regularity of their form.' This drawing represents the grains, as well as the interstices which separate them, as we see them with a strong magnifying power in atmospherical circumstances exceptionally favourable. More often, in making use of weak magnifiers, we perceive a multitude of little white points on a black network. This structure is very apparent in the first moments of observation, but it soon becomes less distinct, because the eye becomes fatigued, at the same time that the object-glass gets heated, as well as the air which is contained in the tube.

We hasten to say that this granular structure cannot be observed except with instruments of large aperture, for the grains having very small dimensions, irradiation, by enlarging and causing them to run into each other, necessarily produces a general confusion. We know the effect of irradiation. An object appears so much the larger the more it is illuminated, and the

FIG. 103.—EFFECT OF IRRADIATION

greatest difference is presented by the contrast between white and black. See, for example, fig. 103. Which of these circles appears to you the larger?—The white. Well, *they are both rigorously equal*, and the white circle would fit exactly on the black. The eye is absolutely deceived.

The solar grains, which we can scarcely measure on account of their smallness, have, in reality, a diameter of 200 to 300 kilometres (124 to 186 miles).

The sun's surface is sometimes so covered over with these granulations, and the network is so pronounced, that we might be tempted to see everywhere pores and rudiments of spots. But this aspect is not constant, and we must seek for the cause, not only in the variations of our atmosphere, which sometimes render the observations difficult, but also in the modifications which the radiant star itself experiences.

Thus, the solar surface is not uniform, but is composed of a multitude of luminous points, scattered over a space of darker network. The knots of this network may sometimes enlarge to the extent of forming pores. The pores, enlarging further, may end by giving birth to a spot. Such is the order in which these phenomena usually succeed each other. This luminous surface of the sun has received the name of *photosphere*.

At the Observatory of Meudon, M. Janssen has succeeded in photographing all the details on plates which measure no less than thirty centimetres (11·8 inches) in diameter, with an exposure which varies between $\frac{1}{2000}$ and $\frac{1}{3000}$ of a second. These photographs show the solar surface covered with the delicate general granulation of which we have spoken. The form, the dimensions, and the arrangement of these granular elements are very varied. The sizes vary from some tenths of a second to 3″ and 4″. The forms recall those of the circle and of ellipses more or less elongated, but often these regular forms are altered. This granulation shows itself everywhere, and it does not at first appear that it presents a different constitution towards the poles of the sun. The illuminating power of the granular elements considered separately is very variable; they would appear to be situated at different depths in the photospheric layer. The most luminous occupy but a small fraction of the surface of the body. Careful examination of these photographs shows that the photosphere has not a uniform constitution in all its parts: here the grains are clear and well marked, although of very variable size; there they are half-effaced, drawn out, disturbed, or have even disappeared to make way for trains of matter which replace the granulation. Everything indicates that, in these spaces, the photospheric matter is subject to violent motions which confuse the granular elements.

It is these luminous grains which produce the light and heat which we receive from the sun. According to the American astronomer, Langley, who has made them the object of special study, they only occupy about a fifth of the solar surface. If from any cause they should be squeezed against each other by multiplying and condensing, the dark network in which they float would disappear, the sun would send us twice, three times, five times more light, and the heat which we should receive would increase in the same proportion; if, on the contrary, they should diminish in number or sink below the dark layer, farewell to the light and heat—the earth would rapidly die of cold.

But let us now fix our attention on the sun-spots. The discovery of the spots is one of those discoveries of which it may be said that they are made by an epoch, and not by a man. Several *savants* having telescopes at their disposal would sooner or later direct them towards the sun.

It was Father Scheiner, a Jesuit of Ingolstadt, who first effectually called attention to the sun-spots, and this, so to say, in spite of himself and in spite of his superior. The day star was regarded and honoured as the purest symbol of celestial incorruptibility, and the official *savants* of that age would never have dared to consent to the admission of these spots. It would have been then a crime of high treason, and dogma itself would have appeared to be compromised. After his repeated observations, which would not permit him to doubt their existence, our Jesuit went to consult the provincial Father of his Order, a zealous Peripatetic philosopher, who refused to believe it. 'I have read the whole of my "Aristotle" several times,' he replied to Scheiner, 'and I can assure you that I have found nothing similar there. Go, my son,' added he, dismissing him; 'quiet yourself, and be certain that there are defects in your glasses or in your eyes which you take for spots on the sun.' They even say that he *passed the night* in ascertaining the state of the day star! At this epoch academical routine still domineered over the study of nature. Very fortunately for science, unfettered minds would observe: what Scheiner did in Germany, Galileo did in Italy, and the solar spots were verified as *facts* by all those who wished to see them.

By his observations of 1611 Galileo determined the duration of the solar rotation. This rotation was ascertained, but not determined, by Fabricius in 1610, guessed by Galileo in 1609, and, previous to him, in 1591, by the philosopher Giordano Bruno, who was burnt alive at Rome in 1600 for his astronomical and religious opinions, and especially for his convinced affirmation of the doctrine of the plurality of worlds.

In general, the spots show themselves at the eastern edge of the sun, pass across the disc, following oblique lines with reference to the diurnal motion and the plane of the ecliptic, and after about fourteen days they disappear at the western limb. It is not rare to see the same spot, after remaining invisible during fourteen days, appear anew at the eastern edge, to make a second, sometimes a third, and even a fourth revolution; but more generally they are distorted, and end by breaking up before leaving the disc, or while they are on the opposite side.

If we note every day on the same drawing the position of the spots, we see that their apparent motion is more rapid near the centre, while it becomes very slow at the edge of the solar disc. We give in fig. 104 the paths of two spots observed by Scheiner from the 2nd to the 14th of March, 1627—that is to say, more than two and a half centuries ago. The dotted places indicate gaps due to the presence of clouds. The spots are sharply terminated, as well as the umbræ and penumbræ. This figure also suffices to show that the paths are curves, and that in approaching the edge the spots lose their rounded form, become oval, and then narrow, so as to appear almost linear. These differences are only apparent, and they follow from the fact that the motion seems to be performed on a plain, while in reality it takes place on a sphere. This is one of the first *proofs* that we have had that the sun is not a flat disc but a sphere. All these apparent variations may be very simply accounted for by pasting a small circle of black paper on a sphere, and turning it round in the hand.

These first facts of observation have proved that the spots are adherent to the sun's surface; for, if they were at a distance from it, it would be necessary to consider them as very flattened bodies, which would be contrary to what we know of the characteristic form of celestial bodies. Galileo compared them to clouds; later, Scheiner regarded them as cavities. We shall soon see what the correct view is.

In this drawing the two lines KEK, LEL represent the projection of

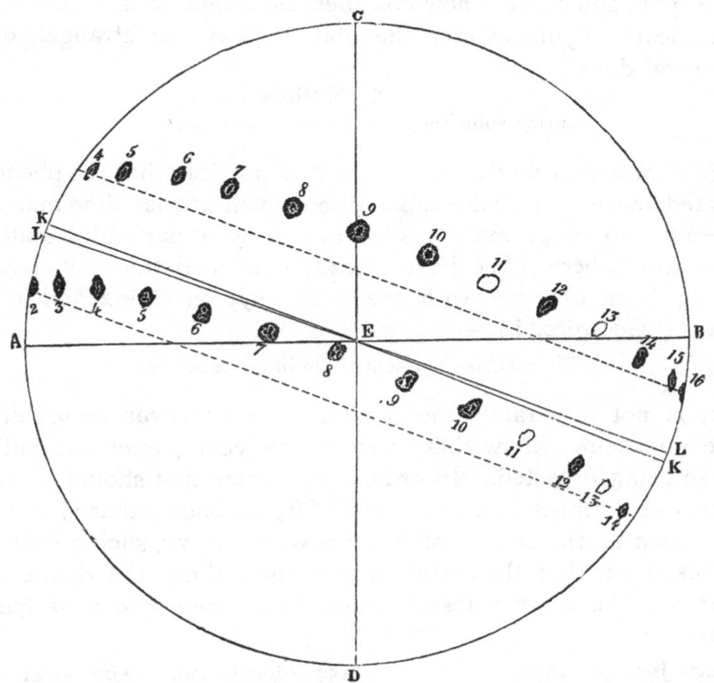

FIG. 104 — ROTATION OF THE SUN, CARRYING WITH IT TWO SPOTS
(From a Drawing made in 1627)

the ecliptic on the solar disc at the beginning and at the end of the observations.

The paths described by the spots vary with the season: in the month of March they are very elongated ellipses with their convexity turned towards the north, the major axis of the ellipse being almost parallel to the ecliptic. After this time the curvature of the ellipses gradually diminishes, at the same time that they become inclined to the ecliptic, so that in the month of June they are transformed into straight lines. From June to September the elliptic curves reappear in the inverse direction;

then they repass through the elongated curve, and again become a straight line to recommence the same series. These different aspects are due to changes in the position of the earth.

The spots do not show themselves promiscuously at all points of the disc. They are more numerous in the immediate neighbourhood of the equator, and very rare in latitudes above 35 or 40 degrees. They show themselves especially along two symmetrical zones, which, by a singular flattery, Scheiner has called *royal zones*, included between 10 and 30 degrees of latitude.

These spots sometimes show considerable dimensions. History relates that at the death of Julius Cæsar the solar star was seen strangely darkened during several days.

Phœbi tristis imago
Lurida sollicitis præbebat lumina terris,

says Ovid (' Metamorphoses,' xv.); but it is possible that the phenomenon, exaggerated, moreover, by the superstition which at that time made gods of all emperors and great men, was produced by a particular state of the terrestrial atmosphere. We have already described the spots which have occasionally been observed with the naked eye in China and in Europe. Virgil says (' Georgics,' i.):—

Sin maculæ incipient rutilo immiscerier igni.

This fact is not very rare; the readers of our 'Revue mensuelle d'Astronomie populaire' know that scarcely any year passes without several of this size being detected. In order that a solar spot should be visible to the naked eye it must measure at least fifty seconds—that is to say, since the earth seen at the same distance measures 17″·72, such a spot is about three times larger than the earth—that is, three times the diameter. The following are the most colossal which have been *measured* (penumbra included):—

Ricco, June 30, 1883 . . .	159″ (double spot, 114,000 kilometres)
Flammarion, Nov. 17, 1882 . .	140″ or 100,000 kilometres
Maunder, June 21, 1885 . . .	130″ (double spot, 93,000 kilometres)
Tacchini, Oct. 14, 1883 . . .	124″ or 89,000 kilometres
Maunder, April 21, 1882 . . .	120″ or 86,000 kilometres

[100,000 kilometres = 62,138 miles]

The diameter of the sun being 1,924″, or 866,000 miles, a second of arc measured on the sun represents 450 miles, 10 seconds represent 4,500 miles, and a minute is equal to 27,000 miles. On the sun, the imperceptible thread which crosses the eyepiece of a telescope, and is used for taking measures, covers with its thickness 240 kilometres (149 miles).

The number of spots is very variable. Sometimes (as in 1883 and 1894) they are so numerous that we can, by a single observation, recognise the zones which usually contain them. Sometimes, on the contrary, they are so rare that several months may elapse without a single one being perceived. We shall see directly that they have a curious periodicity. On the other hand, certain spots occasionally last for some days, others several weeks, and others even several months, becoming modified more or less. I followed one in 1868 which lasted through three solar rotations; Secchi

FIG. 105.—ONE OF THE LARGEST SUN-SPOTS: SEVEN TIMES THE SIZE OF THE EARTH. OBSERVED OCTOBER 14, 1883. VISIBLE TO THE NAKED EYE

followed one in 1866 which lasted four; and Schwabe saw one in 1840 which returned eight times.

We find that, on an average, a spot returns (at least in appearance) to its primitive position at the end of about twenty-seven days and a half; but there is in this estimate a source of error which it is necessary to take into account. During this time the earth does not remain motionless; it describes on its orbit an arc of about 25 degrees in the same direction as the solar rotation. At the moment when a spot completes its apparent rotation it has described a complete circle, and for nearly two days it has commenced a second revolution. This is a difference similar to that which we have noticed with reference to the revolution of the moon and the lunar month (fig. 36). Making this correction, we find for the true duration about twenty-five days and a half.

Does this number represent exactly the period of rotation of this enormous globe?

Here is a fact assuredly curious: the surface of the sun does not turn all in one piece, like that of the earth, but with a velocity decreasing from the equator to the poles. It appears with certainty from the calculation of all the observations that the velocities vary from one spot to another, so as to lead to values ranging from twenty-five to twenty-eight days for the rotation of the body. These values depend exclusively on the latitude of each spot, so that the variation of velocity from one spot to another is proportional to the latitude, like the variation of terrestrial gravity when we move from the equator towards the poles.

There is nothing more striking than the following table, where we have recorded the duration of the solar rotation deduced from the motions of corresponding spots:—

Duration of the Solar Rotation at different Parallels, Degree by Degree

Latitude	Rotation	Latitude	Rotation	Latitude	Rotation	Latitude	Rotation
Degrees	Days	Degrees	Days	Degrees	Days	Degrees	Days
0	25·187	12	25·388	24	25·975	36	26·891
1	25·188	13	25·423	25	26·040	37	26·979
2	25·193	14	25·460	26	26·107	38	27·068
3	25·200	15	25·500	27	26·176	39	27·259
4	25·210	16	25·543	28	26·248	40	27·252
5	25·222	17	25·588	29	26·322	41	27·346
6	25·238	18	25·636	30	26·398	42	27·440
7	25·256	19	25·686	31	26·475	43	27·536
8	25·277	20	25·739	32	26·555	44	27·633
9	25·300	21	25·794	33	26·636	45	27·730
10	25·327	22	25·852	34	26·717	46	27·828
11	25·356	23	25·913	35	26·804	47	27·926

Thus the solar surface rotates at the equator in about 25 days and 4 hours; in 25 days 12 hours at 15 degrees of latitude; in 26 days at the 25th degree; in 27 days at the 38th; and in 28 days about the 48th degree. We cannot follow a spot farther; but this progression should continue up to the poles. This is, then, the rotation of the *surface*, as if the earth were covered over with an ocean which turned more slowly than itself, and less and less quickly from the equator to the poles. It is probable that the solar globe itself rotates in the equatorial period. These numbers are calculated by allowing, with M. Faye, 857′6 for the diurnal motion of an equatorial spot. Carrington allowed 865′, which corresponds to 24 days 22 hours.

According to the study of the white spots or *faculæ* recently made (1888) by M. Wilsing on photographs of the sun taken at the Observatory of Potsdam, the rotation of these appears to be the same for all latitudes, and equal to that at the equator. The unequal velocity shown by the dark spots is doubtless limited to a rather thin layer of the solar surface.

Such is the first aspect presented by the telescopic image of the day star, and by the study of its spots. But what is the nature of these spots themselves?

The first attentive observer of the sun, Scheiner, at first regarded the spots as satellites—an indefensible opinion, which, however, some have attempted to revive. Galileo attributed them to clouds or vapours floating in the solar atmosphere; this was the best conclusion which could be drawn from the observations of that epoch. This opinion met for a long time with general approval; it has even been renewed in our day. Some astronomers, and among them Lalande, believed, on the contrary, that they were mountains, of which the flanks, more or less steep, might produce the aspect of the penumbra—an opinion irreconcilable with the proper motion which the spots sometimes possess in a very marked manner. It is not usual, in fact, to see mountains travelling. Derham attributed them to the smoke issuing from the volcanic craters of the sun, an opinion revived and maintained in recent times by my lamented friend Chacornac. Several *savants*, regarding the sun as a liquid and incandescent mass, have also explained the spots by immense scoriæ floating on this ocean of fire. But a century had scarcely elapsed after the epoch when the spots were observed for the first time when an English astronomer, Wilson, showed with certainty that the spots are hollow.

What events occur at the sun's surface! It is important that we should give an exact account of them.

The time necessary for the formation of a solar spot is very variable, and it is impossible to discover any law; some are formed very slowly, by the expansion of the pores or small points; others appear almost suddenly However, if we observe the sun every day with great care, we recognise that this formation is never completely instantaneous, however rapid it may be. The phenomenon is always announced some days in advance; we perceive in the photosphere a great agitation, which often manifests itself by very brilliant faculæ, giving birth to one or several pores. These pores at first move about with rapidity, perhaps disappear to be again reproduced, then one of them gets the upper hand and is transformed into a large opening. At the first moments of the formation there is no sharply defined penumbra; it develops by degrees and becomes regular as the spot itself takes a rounded form, as we see in fig. 106, which represents a regular and in some measure a typical spot.

This quiet and peaceful formation is only realised at the epochs when a calm seems to reign in the solar atmosphere; in general the development is more tumultuous and complex.

FIG. 106.—TYPE OF AN ORDINARY SUN-SPOT

We often see several spots blend into one by the dissolution of the luminous matter which separates them. The opposite sometimes happens: a spot completely formed divides into several others. I observed this especially, in 1868, in a spot which showed in succession all the aspects represented here. This spot divided into two; but the child was only separated from its mother to die, while the principal spot was seen during two solar rotations.

The penumbra varies in size according to the spots, and is far from being uniform in its structure. This penumbra is very radiated; the rays which compose it are of irregular forms. Some resemble sinuous currents, which become narrower as they withdraw from the edge; several appear formed of oval masses, like elongated clouds placed end to end. It is not difficult to verify this radiant structure of the penumbra.

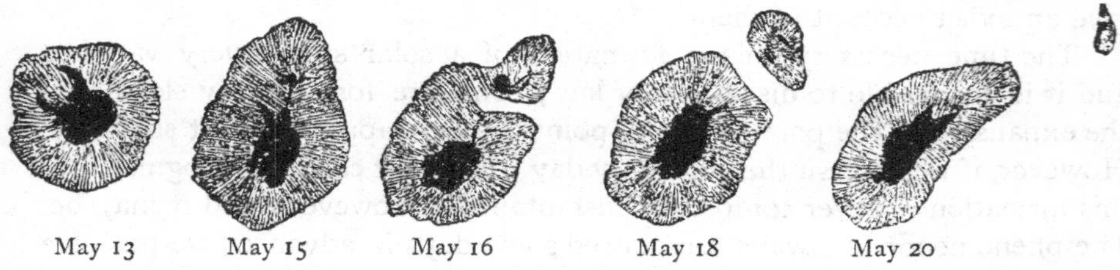

FIG. 107.—SEGMENTATION OF A SUN-SPOT

These currents are less condensed, less luminous, less clearly cut at the exterior of the penumbra, where they are detached from the photosphere;

while near the nucleus they are crowded, condensed, and become more brilliant. It thus happens sometimes that the edge of the penumbra adjoining the nucleus acquires a very strong brightness; the spot then appears composed of two brilliant concentric rings, as we have seen in fig. 106.

Sometimes the interior extremities of the currents terminate in brilliant grains projected on the dark background of the nucleus. Occasionally the currents of luminous matter throw veritable bridges across the spots, as is shown in the drawing (fig. 108), in which, particularly at the point *a*, two systems of currents superposed on each other cross at right angles. This was drawn at Rome, in 1870, by Secchi.

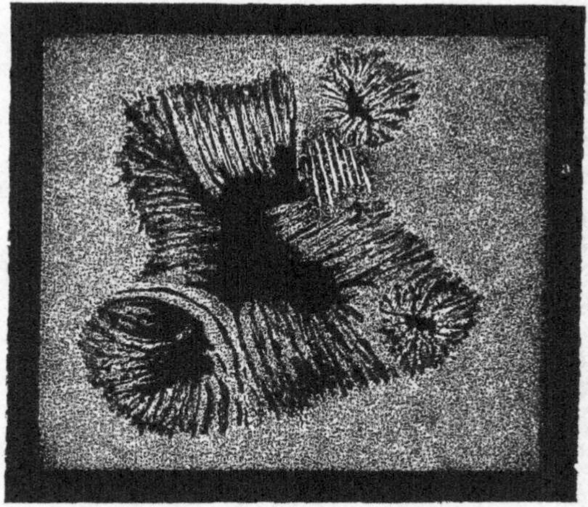

FIG. 108.—SUN-SPOT, WITH SUPERPOSED BRIDGE AND CURRENTS

We also see spots in which the luminous substance runs so evidently in threads from the exterior to the interior that we might think we were watching a whirlpool of luminous water. This is especially represented in fig. 109, drawn at Palermo, in 1873, by M. Tacchini.

It is only by an attentive examination of the spots that we can succeed in obtaining a knowledge of the nature of the solar surface. The aspect of this surface is modified up to the border of the spots, formed simply by the non-luminous gas in which float the brilliant grains constituting the photosphere. The interior heat of the solar globe radiates out externally, and it thus establishes vertical currents. The altitude at which the luminous clouds which form the solar light are condensed is comparable with that which in terrestrial meteorology we call the 'dew point.' A little more or less of height, of heat, and of condensation, and the cloud does not form. The spots would be the points where the currents come down again and scoop out the photosphere a little, bringing back the cooler elements from above. The layer in which the luminous clouds are formed may have the thickness of the earth, and the richness of the circulation is such that during millions of years there would probably be no diminution in the solar light and heat.

Still, the diameter of the sun would not appear to be constant; while the measures give regularly for the mean diameter of the moon the number 1,968″, they vary for that of the sun from 1,919″ to 1,924.″ The solar diameter must diminish slowly, but, without this diminution being yet perceptible, it may be subject to oscillations.

The places where the sun is spotted are deep relatively to the levels of the luminous clouds—that is to say, to the mean level of the visible surface: the depth appears to be one-third of the terrestrial radius, or about 1,240 miles. Occasionally it attains the semi-diameter of the earth, or 3,900 miles.

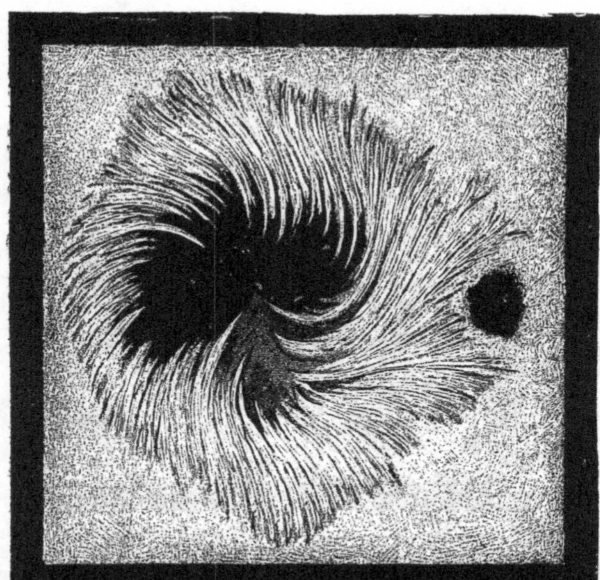

FIG. 109.—A SUN-SPOT, WITH LUMINOUS THREADS FLOWING TOWARDS ITS CENTRE

The nuclei are not absolutely black, as might be thought at first sight. Their light is only imperceptible on account of contrast; it is even five thousand times superior to that of the full moon. Dawes, in England, was the first to notice places much darker and apparently quite black; Secchi, at Rome, has observed grey and singular rose-coloured trails; Trouvelot, at Cambridge, and Ricco, at Palermo, transparent veils.

The spots are usually surrounded by very brilliant regions, to which the name of *faculæ* has been given. These are upheavals of the photosphere, and they are distinguished with clearness when a spot approaches the limb. These regions are the seat of considerable agitation, of which the extent much surpasses that of the spot itself.

Thus, the spots are the result of great convulsions, which produce differences of level, of upheavals and depressions; these depressions are formed of hollows more or less regular, surrounded by a sharp and salient swelling. These hollows are not voids; the resistance which they oppose to the course of the luminous currents proves that they are filled with vapours more or less transparent.

We now arrive at another order of phenomena, known and studied for a much shorter time than the spots, but not less important than they are in solar physics, and perhaps even more so. We refer to the solar *eruptions* which are presented to the eye and to the mind of the observer as intimately connected also with the calorific and luminous activity of the day star.

CHAPTER IV

THE ERUPTIONS OF THE SUN

Prominences. Jets of Flame. Gigantic Explosions. The Solar Atmosphere; the Corona and Halo

WE have already seen, in speaking of total eclipses of the sun, that during the very rare moments when the moon interposes itself before the day star we ascertain that the vicinity of that body is not blank and pure as it appears to the naked eye, for example, in the midst of a beautiful summer day, but that it is occupied by luminous materials, shining either by themselves, or by the reflection of the solar splendour, and appearing like a crown of glory all around the god of day.

In this halo we notice tongues of fire which emanate from the sun and are contiguous to him. It was during the eclipse of July 8, 1842, that the attention of astronomers was first attracted to these prominences, which shoot forth round the moon like gigantic flames of a rose or peach colour (they had already been seen with the naked eye, especially in 1239, in 1560, 1605, 1652, 1706, 1724, 1733, and 1766, but astronomers believed them to be optical illusions). The surprise produced by this unexpected phenomenon did not permit exact observations to be made, so that there was a complete disagreement between the different accounts. Baily noticed three enormous prominences, almost uniformly distributed on the same side.

Airy observed three, in the form of the teeth of a saw, but placed at the summit. Arago saw two at the lower part of the disc. At Verona, these flames remained visible after the appearance of the sun. These appendages have enormous dimensions; the French astronomer, Petit, of Toulon, measured the height of one of them, and found it 1′ 45″, which is equivalent to six terrestrial diameters—that is to say, to 80,000 kilometres (about 49,000 miles).

A discussion was immediately started on the nature of these prominences. They were at first taken for mountains; but this opinion was irreconcilable with the observations of Arago, some of these supposed mountains being very inclined, even overhanging considerably, so that equilibrium was impossible. Most of the *savants* regarded them as flames or clouds. They even spoke of indentations seen in the lunar disc, of flames, flashes, clouds, and storms suspended in the atmosphere of the moon.

They awaited with impatience the eclipse of 1851, which would be total in Sweden. Sir G. B. Airy, Director of the Greenwich Observatory, organised an expedition in order to take precise measures. At the moment of totality he first observed a prominence having the form of a square ending in a point; below he found a little cone, and farther off a small suspended cloud; then, at the end of a minute, a prominence and a rosy arc. Other observers noticed the same phenomena, with slight differences of form.

These observations permitted the following conclusions to be stated with certainty: (1) The prominences are not mountains; this hypothesis is irreconcilable with their forms; (2) they should be regarded as gaseous masses, of which the aspect is rather analogous to that of clouds; their curvatures somewhat remind us of the smoke which escapes from our volcanoes; (3) the variety of forms attributed to the same prominence may be due to real variations, but they may also be the result of inaccuracy in the drawings; (4) there is an evident relation between these prominences and the rose-coloured arcs already observed in 1842, but which were much better observed on this occasion; we may legitimately suppose that these arcs form the visible part of a continuous layer which completely surrounds the sun; (5) the size of the prominences was seen to increase on the side the moon was leaving, and to decrease on the side towards which she was advancing: it is certain, then, that the seat of the phenomenon is in the sun; (6) all the observers have not seen the same number of prominences; they have not assigned to them exactly the same place; this is due to the rapidity of the observations.

The eclipse of 1860, total in Spain, was observed with the same object by the Italian astronomer, Secchi, and by the English astronomer, Warren de la Rue, and they photographed it. The two following figures (110, 111) are reproductions of the photographs taken, the first at the beginning of the totality, the second at the end. We notice in the first seven principal prominences:

A. Prominence having two summits very close together and very lofty.

C. Great prominence in the form of a cloud, inclined at 45°, rounded at the base, pointed at the summit, and having a helicoïdal structure.

E. Little clouds, very slender, of which the whole forms a bent-down curve, having a height of about 2′ 40″.

H. Complicated mass of small clouds.

G. Enormous mass of brilliant matter which has over-exposed the plates, so that the interior details have disappeared. Its rounded form proves that it was not in immediate contact with the sun, but suspended in its atmosphere. Seen in the telescope it presented the appearance of a chain of mountains.

I. Gigantic flame, or rather enormous cumulus, in which were distinguished shades of yellow and red.

K. Prominence with two summits, of which one, thinner and less lively, is prolonged in the form of a horn.

In the whole of the left part we do not see any prominence. The black line X V represents a thread stretched in the telescope directed along the motion of the sun from east to west, in order to point out the position of the prominences with reference to the solar equator.

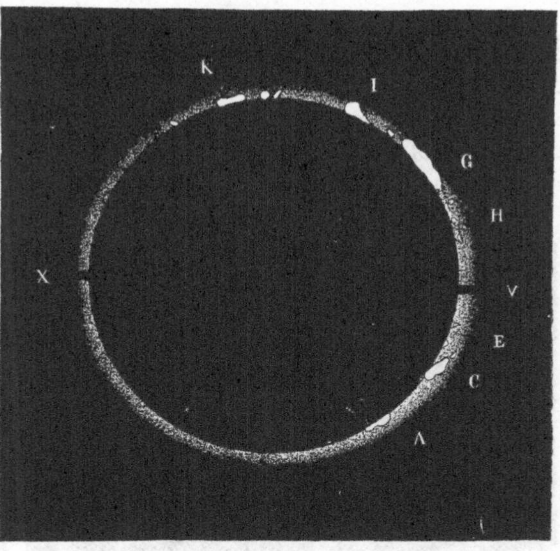

FIG. 110.—PROMINENCES OBSERVED DURING THE ECLIPSE OF 1860: FIRST HALF

We have in fig. 110 the right half of the sun, the moon coming on from the left. [The figures are inverted, as seen in a telescope. —J. E. G.] At the end of the eclipse, the moon having advanced and cleared the left half, we perceive the prominences drawn in fig. 111.

The observations just described prove that besides the prominences there exists a layer of the same substance, which surrounds the sun on all sides. The prominences proceed from this layer; they are masses which rise above the general surface, and are occasionally even detached from it. Some of them resemble the smoke which issues from our chimneys or from the craters of volcanoes, and which, reaching a certain height, yield to a current of air and bend down horizontally.

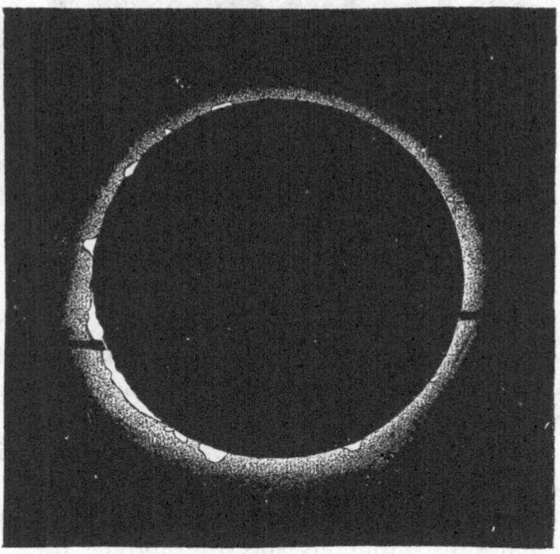
FIG. 111.—LAST PHASE OF THE ECLIPSE OF 1860: WESTERN PROMINENCES

The number of the prominences was incalculable. In the direct observation the sun seemed surrounded by flames; they were so multiplied that it appeared impossible to count them. My lamented friend Goldschmidt, whose sight was so piercing, saw them before and after the totality.

The height of the prominences was very considerable, especially if we notice that, in order to estimate it, it is necessary to take into account the part eclipsed by the moon. They were all much higher than the entire diameter of the earth—five, eight, and ten times higher.

We now know that the number of the prominences is very variable in the course of time. In 1860 the sun was at an epoch of great activity.

The phenomena observed during this eclipse have been confirmed by all subsequent observations. At the approach of the eclipse of August 18, 1868, it was proposed to study them particularly, profiting by the new discoveries of *spectrum analysis*, that marvellous study of which we shall soon explain the principles. The questions to be solved were the following:

FIG. 112.—PROMINENCES SEEN DURING THE ECLIPSE OF AUGUST 18, 1868

1. Are the prominences composed of solid matter, and should they be compared with simply incandescent clouds, or are they rather true gaseous masses?

2. What are the substances which enter into their composition?

The first of these two questions could be determined at once by directing a spectroscope towards the prominences; the question was simply to see whether the spectrum was continuous or not.

The eclipse presented very favourable circumstances; an enormous prominence, ten times larger than the earth, was immediately perceived by the observers, who directed all their instruments toward it, and instantly found a discontinuous spectrum formed of a small number of white rays. The first part of the problem was then solved; they had acquired the certainty that the prominences are gaseous masses.

The question then was to recognise the nature of the substances which

Fig. 113.—Flames of the Sun: Types of the Principal Forms

compose them ; and this second question was not so simple as the first, for it was necessary to determine the position of the rays with reference to some scale, taking as a term of comparison the spectrum of a known substance, or that of the sun. MM. Rayet and Janssen succeeded in determining these positions, and ascertained that the fundamental substance of the prominences is *hydrogen.*

This study was, however, incomplete, for it was necessary to make sure of the identity of the different rays. This determination appeared to require that we should wait for another eclipse, but M. Janssen has been able to avoid this long waiting by a very ingenious discovery. Having been greatly struck with the brilliant light of some of the rays of the prominences, he asked himself whether these rays might not be visible in full daylight. Unfortunately, the sky became covered with clouds a little after the eclipse, and it was impossible for him on that day to confirm his conjecture. On the following day he set to work, and had the signal good fortune to see the rays of the prominences in full sunshine. The slit of his spectroscope (see further on) being exactly tangential to the edge of the sun at a place where the day before he had noticed a flame, he perceived a brilliant red-coloured ray ; then in the blue another bright ray. These two rays are precisely those of hydrogen, and consequently this gas is the chief of the substances which compose the prominences.

The same day that this news arrived in Europe (October 20, 1868) Mr. Lockyer also announced that, on his part, he had been able to see, at the edge of the sun, the rays of hydrogen. We see that the fruit was ripe.

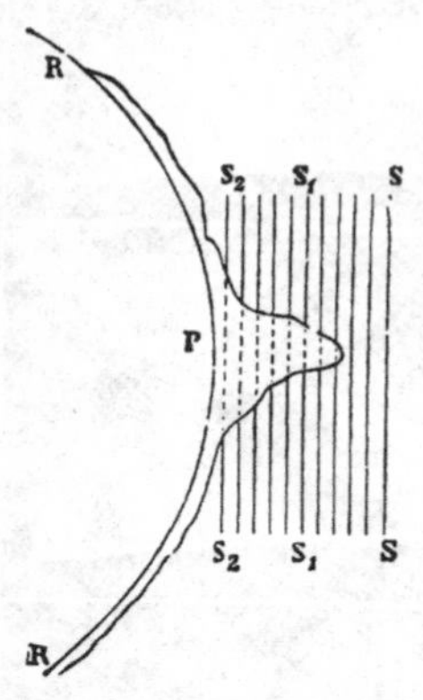

FIG. 114.—EXAMINATION OF A PROMINENCE SEEN THROUGH THE SLIT OF A SPECTROSCOPE

This method of observation permits us to recognise at all times the *prominences* of the sun, which are only visible during total eclipses. This is how we can ascertain with the spectroscope the existence of these prominences. We move this apparatus, fitted to the eye-piece of a telescope (we will give a description of it later), along the edge of the sun. This apparatus is closed by a narrow slit. This slit being placed parallel to the edge of the sun, when it meets a prominence we see, varying in length, the brilliant line of hydrogen which characterises these flames ; the variation in the length of the line indicates the form of the prominence. This process will be easily understood by fig. 114. The edge of the sun is represented by R R ; there is a prominence at P ; the lines S S, S_1 S_1, S_2 S_2 represent the successive positions of the slit of the spectroscope.

Proceeding thus, we can draw the outline of the sun, such as we might see it directly if we were not dazzled by the light of the brilliant star. See, for example, fig 115, a drawing of the whole of the sun, observed on July 23, 1871; there are seventeen prominences revealed by the spectroscope, each in its place. We can thus study their relations to the spots.

These studies have shown us that the solar globe is surrounded by an atmosphere, principally composed of rose-coloured hydrogen, from which rise these eruptions, themselves composed of this gas. This layer has received the name of *chromosphere* (from *chromos*, colour). The solar limb thus constantly shows the most varied aspects.

In certain observatories these prominences are observed and drawn every day; for example, at Rome, where I followed them, in 1872, in company with the learned Father Secchi. A special society has been founded

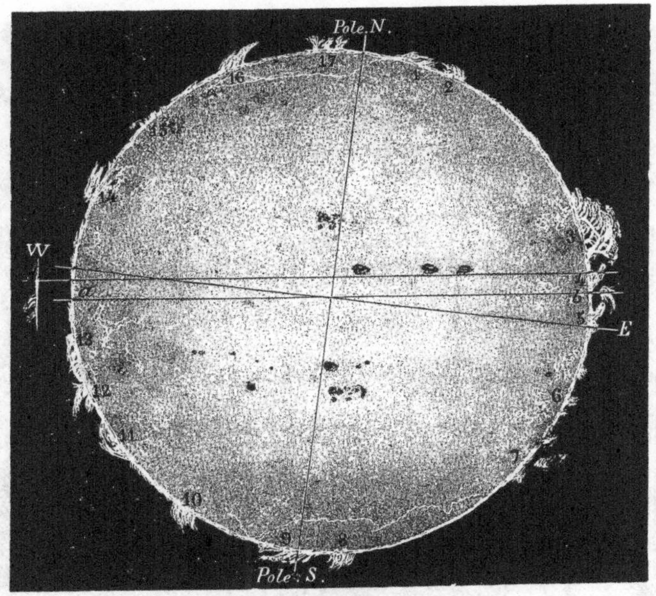

FIG. 115.—PROMINENCES OBSERVED JULY 23, 1871

in Italy for the study of this subject, the 'Society of Spectroscopists,' of which the headquarters is at Rome. It has already published a great number of drawings, of which the following engravings will give an idea. Fig. 116, drawn April 21, 1873, represents a portion of the edge of the sun; the flames in the form of jets shoot out in the atmosphere of the sun up to a height of 25,000 miles. The solar globe is surrounded with similar flames. Occasionally there is a relative calm. Sometimes, on the contrary, there are violent and tremendous eruptions.

The luminous intensity in the jets is always very great. They occasionally show magnificent forms, like the finest bouquets of fireworks which can be imagined; the branches, falling in parabolas more or less inclined, present a beauty quite artistic. Certain jets represent the heads of magnificent palm-trees with their graceful curves in branches. More usually the stem, very vivid and brilliant, appears at a certain height to divide into branches. We see the upper crest sometimes carried by the wind in the direction of the jet, and sometimes driven back in a contrary sense from the direction of the stem. These forms are always compact, fibrous at the base, and terminate in threads. Their light is so vivid that we see them through thin clouds when the chromosphere has disappeared; their spectrum indicates, besides hydrogen, the presence of several other substances. They are veritable sheafs of fire, very ephemeral; they rarely last an hour; they are often an affair of a few minutes.

FIG. 116.—AN EXPLOSION IN THE SUN

All these aspects are represented in fig. 113, p. 267: jets, sheafs, plumes, and clouds. The plumes consist especially of masses of filaments, wide at the base and narrowing to a point. We meet with them either straight or curved, evidently by the action of currents which draw them along. It is not rare to see in these plumes well-marked double inflections, as if the jet were raised in a spiral. A rather beautiful form, and one which is not rare, is the flame which clings to the chromosphere by a very thin tongue, and rises on this stalk, expanding like a flower.

These forms may attain all heights. Usually, at a certain height they open out in trains and clouds. We usually find these plumes coupled together, or convergent, or collected, but with a different inclination. It is probable that many of these forms are due to perspective, and that their bases are very distant, in the direction of the visual ray. Several cross each other oddly, standing out one before the other.

These masses reach the enormous heights of 150 to 200 seconds, sometimes 240 seconds. Their summit is, however, in general very much cut up, and similar on the whole to the masses of cirro-cumuli which we see at the extremity of stormy clouds, and which produce a *dappled* sky. Certain forms of prominences soar like clouds in the solar sky.

The study of the surface of the day star is now actively pursued, thanks to the persevering activity of a great number of observers. One of the most curious observations which have been made in this very interesting study, and one of those which may give us a better idea of the energetic forces in action at the surface of this immense body, is unquestionably that which Professor Young has made in America, and which has caught in the act a *tremendous explosion* of hydrogen in the solar atmosphere. Let us recapitulate the narrative of the observer.

On September 7, 1871, between noon and 2 P.M., there happened an explosion of solar energy remarkable for its suddenness and violence. The observer had noticed an enormous prominence or cloud of hydrogen on the eastern limb of the sun. It had been maintained with very little change since noon of the day before as a long cloud, low and quiet. It was principally formed of filaments, most of them almost horizontal, and floated above the chromosphere, the lower surface being at a height of about 24,000 kilometres (about 15,000 miles); but it was connected to it, as usually takes place, by four or five vertical columns, more brilliant and more active than the rest. It was 3′ 45″ in length and about 2′ in height at the upper surface—that is to say, about 161,000 kilometres (100,000 miles) in length and 88,000 kilometres (54,680 miles) in height.

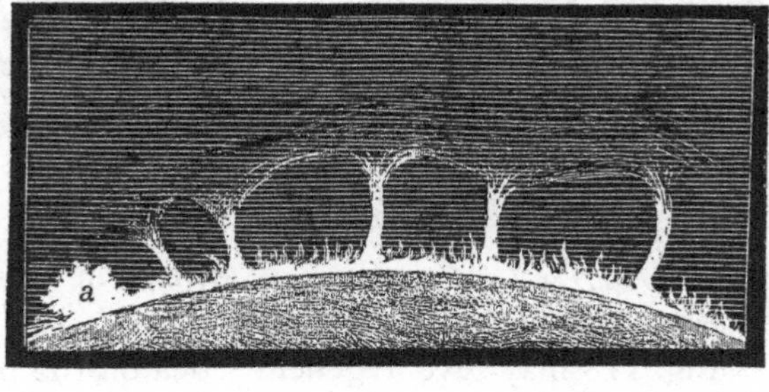

FIG. 117.—AN EXPLOSION IN THE SUN: FIRST PHASE

The column at the southern extremity of the cloud had become much more brilliant, and was curved in a curious manner at one side. Near the base of another column, at the northern extremity, there was developed a small brilliant mass, very much resembling in form the upper part of a stormy cloud of summer. Fig. 117 represents the prominence at this instant: *a* is the little stormy cloud.

At 1 o'clock the astronomer, putting his eye again to the telescope, which he had left for half an hour, was greatly surprised to find that in that interval all had been literally torn to pieces by some inconceivable explosion from below. Instead of the quiet cloud which he had left, the air, if we may use the expression, was filled with floating débris, a mass of vertical filaments, fusiform and separated, each having 16 to 30 seconds of length with 2 or 3 seconds of width, more brilliant and more drawn together where the columns at first were, and rising rapidly.

Already some of these had reached a height of 4 minutes (108,000 miles). Then, under the very eyes of the observer, they rose with a motion almost perceptible to the eye, and at the end of ten minutes most of them were more than 300,000 kilometres (186,400 miles) above the solar surface! This frightful eruption was verified by a careful measurement; the mean of three very concordant determinations gave 7′ 49″ for the extreme altitude to which the jets were shot up. This is so much the more curious as the matter of the chromosphere (red hydrogen in this case) has never been observed at an altitude higher than 5 minutes. The velocity of the ascension—

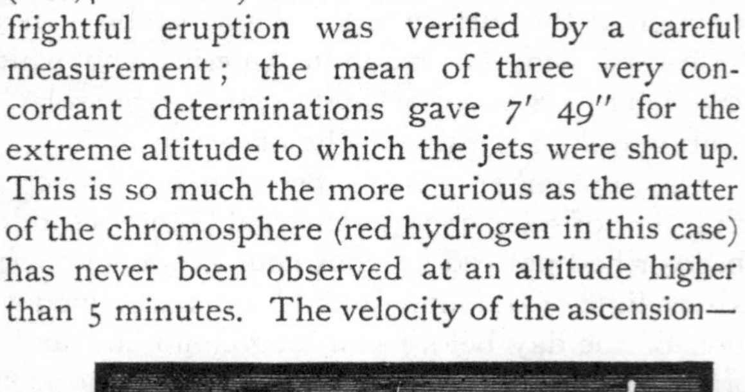

Fig. 118.—An Explosion in the Sun: at its Height

Fig. 119.—End of an Explosion

267 *kilometres* (166 miles) *per second*—is considerably greater than any other which has been observed.

Fig. 118 may give a general idea of the phenomenon at the moment when the filaments were at their greatest height. In proportion as these filaments ascended they became gradually enfeebled, like a cloud which dissolves, and at a quarter-past one there only remained to mark the place a few cloudy flakes, with some low flames more brilliant near the photosphere.

But at the same time the little mass, like a stormy cloud, had grown and developed in an astonishing manner into a mass of flames, which rolled and changed incessantly, to speak according to appearances. At first these flames were pressed together in a crowd, as if they were stretched along the solar surface. Afterwards they rose in a pyramid to a height of 80,000 kilometres (about 50,000 miles) Then their summit lengthened out in long filaments, rolled up in a curious manner before and behind, above and below, like the volutes of an Ionic capital. Finally, they became enfeebled, and at half-past two they had vanished like the rest.

The whole phenomenon forcibly suggested the idea of a vertical and violent explosion, rapidly followed by a remarkable depression.

On the same afternoon a part of the chromosphere at the opposite edge (to the west) of the sun was, during several hours, in a state of excitement and of unusual brilliancy On the evening of the same day, September 7, 1871, there was in America a beautiful aurora borealis. Was this in response to this magnificent solar explosion ?

Space fails us to describe all the varieties observed in these explosions. We may, however, notice some others which are specially curious.

On August 25, 1872, there was observed at Rome a prominence, a sort of fan-shaped sheaf of hydrogen, resembling a gilliflower detached from its calyx. This mass was suspended and isolated in space. It persisted till the following day, gradually diminishing in size (fig. 120).

FIG. 121.—A FLAME RISING TO THE HEIGHT OF ONE-FOURTH THE SUN'S DIAMETER, APRIL 1873

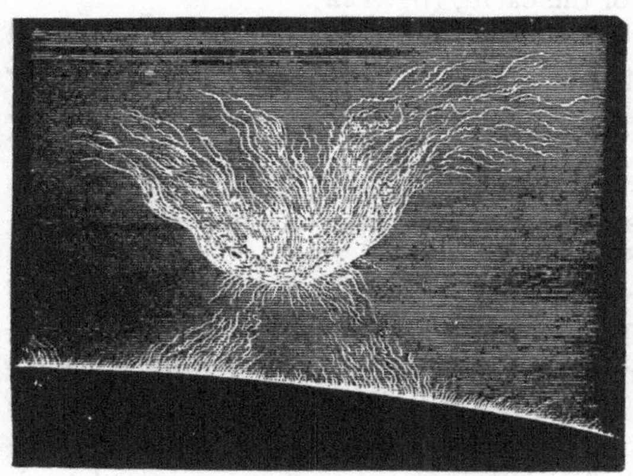

FIG. 120.—CURIOUS FORM OF SOLAR ERUPTION

On April 3, 1873, there was noticed in the morning, at a quarter to nine, above the solar border, a mass of hydrogen of enormous elevation. It appeared like a mass of light and fibrous cirri ; the convolutions were very difficult to grasp, and changed from moment to moment. At the beginning it was long and diffuse, but in twenty-five minutes it rapidly shrank, and was transformed into a sort of branched column (fig. 121), which rose to 7′ 29″, or 322,000 kilometres (200,000 miles)—that is to say, to nearly a quarter of the sun's diameter.

SUCCESSIVE MEASURES OF A PROMINENCE

Hour of Observation	Height	
	In seconds of Arc	In Miles
H. M.		
8 45	259″	115,300
8 50	345	153,700
9 0	372	165,700
9 10	449 = 7′ 29″	200,000
9 15	380	165,200

It then diminished rapidly: at 9 hours 36 minutes nothing was visible but a feeble trace of the brilliant cloud corresponding to the densest part. Taking the difference of height between 8 hours 45 minutes and 9 hours 10 minutes, we find a mean velocity of elevation of 105 kilometres (65¼ miles) per second.

On January 30, 1885, M. Tacchini, at Rome, drew one of the most remarkable of these prominences. It measured 6′ 18″, or 228,000 kilometres (18 times the diameter of the earth) (fig. 122).

On June 26, 1885, M. Trouvelot, at the Observatory of Meudon, measured two prominences still more gigantic, of a height of 10½′, or 460,000 kilometres (286,000 miles; a third of the solar diameter!), which rose just at the antipodes of each other.

It follows from these observations that the solar atmosphere must rise to eleven minutes of arc at least; for these brilliant jets are doubtless continued in a more extended dark atmosphere.

Pictures are sometimes drawn which attempt to show the colour of these flames. Some of these have been made at the Observatory of Harvard College (United States), where these phenomena are observed with the greatest care. On one of these plates two magnificent prominences of more than 60,000 miles in height are shown, the first observed on April 29, 1872, at 10 o'clock in the morning (25 minutes later it had so much changed that it was not to be recognised); the second, on April 15 of the same year, at the same hour. We may thus gain a better impression of them than by black figures. But there is something which a picture can never reproduce—the vivacity of the tints which these enormous masses present, and the rapidity of the motions with which they are animated. The best drawings will always be bodies without life, veritable corpses, if we compare them with the grand phenomena of nature. These incandescent masses are animated with an internal activity, from which life seems to breathe. They shine with a vivid light, and the colours which adorn them form a specific character, by which we can recognise, thanks to spectrum analysis, the chemical nature of the substances which compose them. Could the most perfect drawings depict this solar life?

FIG. 122.—SUN-FLAMES 142,000 MILES HIGH (EIGHTEEN TIMES THE EARTH'S DIAMETER). OBSERVED AT ROME, JANUARY 30, 1885

The solar eruptions show themselves everywhere, at the poles as well as at the equator; while the spots, as we have seen, are confined to two zones on both sides of the equator. The following diagram, drawn by

Mr. Young, represents the relative proportions of spots and prominences. The latter are exaggerated at the poles on account of the solar rotation, which allows them to be seen for a longer time than at the equator, so that they are counted several times. The dotted line represents the distribution of the prominences above 1′, or 43,000 kilometres (26,720 miles). Thus the solar globe is entirely covered with an ocean of fire, on which constantly rise rose-coloured flames of a gigantic height, and from which tremendous eruptions proceed. There is an appearance as of perpetual fireworks. The true figure of the sun should not, then, be the white disc it is generally represented, but a globe surrounded by rose-coloured flames.

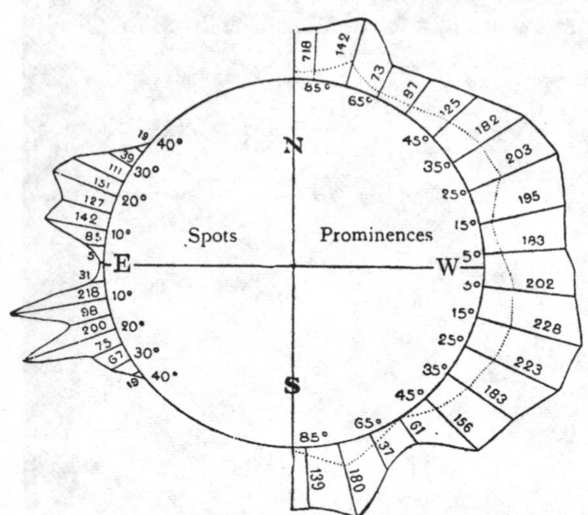

FIG. 123.—RELATIVE DISTRIBUTION OF SOLAR SPOTS AND PROMINENCES

The substances which produce the phenomenon of the prominences are generally incandescent gases raised towards the upper regions by forces of which the origin is still unknown to us. Are these motions the result of the specific lightness of the luminous matter, or should we, rather, attribute them to an impulsive force proceeding from the interior of the solar globe? The second explanation is the more probable. The substance is not simply shot up in a straight line, it is also animated with vortex motions, which give to the luminous jets the appearance of spirals, of which the axes take all positions, from the vertical to the horizontal. These whirlpool motions, especially those in which the axis is horizontal, would necessarily result from an eruptive force combined with violent currents, solar winds, and tempests.

Reaching a certain height, the luminous masses change their aspect and become nebulous, like smoke which vanishes in the air. They continue to ascend, but they are diffused by degrees, and end by fading away. We must conclude that these movements are performed in a resisting medium which is nothing but the solar atmosphere.

The rapidity with which the motions and transformations we have described are produced is truly extraordinary. We have seen just now the observed velocity, by Young, of 166 miles a second; Secchi quotes one of 230, and Respighi goes to 372, 434, and even 500! We must not, however, admit without challenge such exorbitant velocities. A body shot upwards with an initial velocity of 608 kilometres (378 miles) would indefinitely recede from the sun. Explosions capable of giving to bodies velocities of 400 to 500 miles would produce, then, a dif-

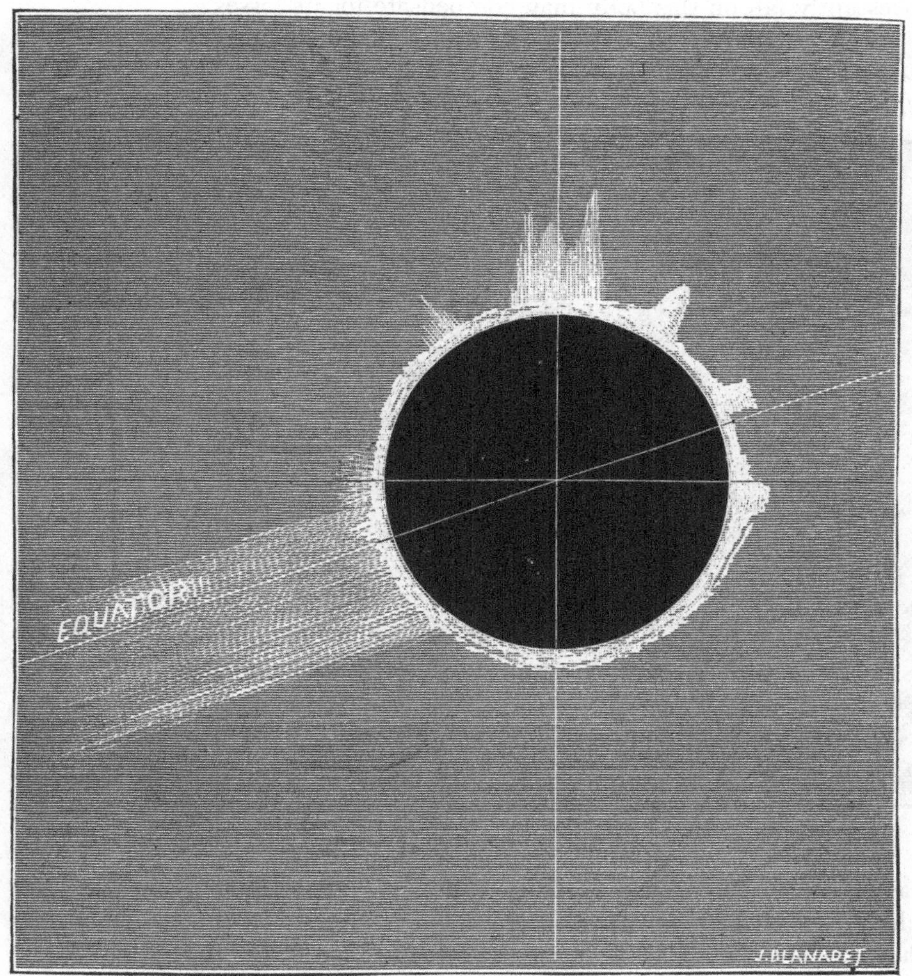

FIG. 124.—TOTAL ECLIPSE OF THE SUN. OBSERVED IN RUSSIA, AUGUST 19, 1887

fusion of solar matter in the planetary spaces. It is true that these explosions do not take place in a vacuum. The resistance of the sun's atmosphere diminishes the velocity, and may, in certain circumstances, prevent the diffusion of which we speak. But if the initial velocity were really 800 kilometres (497 miles), the resistance would not suffice to prevent the matter from passing beyond the sphere of the sun's attraction and scattering in space.

Such an effect, however, would not be inadmissible, and it would not even prove that the weight of the sun goes on diminishing, considering that the quantities of uranoliths (meteoric stones) and materials which incessantly fall on this body may compensate for its losses.

What is certain is, that the day star is really surrounded by unknown substances which extend to a distance all round him. The phenomenon which strikes us most when we observe an eclipse of the sun with the

FIG. 125.—TOTAL ECLIPSE OF THE SUN. OBSERVED IN CALIFORNIA, JANUARY 1, 1889

naked eye is the brilliant halo of light which surrounds the moon, and which has received the name of *corona*. The ancients had noticed it, and they concluded that the eclipse was never total.

The luminous intensity of the corona is difficult to estimate ; however, it is at least equal to that of the full moon.

We generally distinguish in the corona three well-defined regions, although the lines of separation are not sharply marked. The first and the most vivid of these regions is the brilliant ring which is found immediately in contact with the sun. The rose-coloured matter appears to be in suspension in this layer itself. Its light is so vivid that it may occasion doubts as to the precise moment of totality. We may estimate its width at 15 to 20 seconds. Round this first layer, and in immediate contact with it, is found another region, where the light is still rather vivid, in which the prominences are produced, and which extends to a distance of 4 or 5 minutes. Above this region begins the halo properly so called. It is often irregular, and its outline, far from being uniform, as we had supposed at first, often presents inequalities, and sometimes even very deep cavities. The name of *aigrettes* is given to those long rectilineal plumes which are detached from the halo, like the rays of light which pass

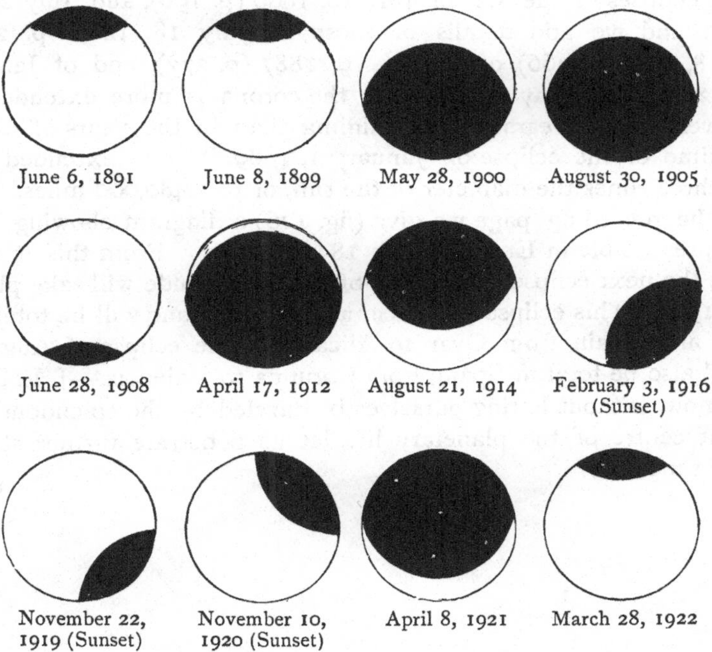

FIG. 126.—SOLAR ECLIPSES VISIBLE IN ENGLAND, 1891 TO 1922

between the clouds when the sun is near the horizon. They often extend to considerable distances.

The primary cause of the corona and the aigrettes is certainly in the sun, but their appearance may be considerably modified by the presence of the moon and by atmospherical circumstances. We do not yet know with certainty the extent of the corona. We estimate only the limit which is determined by the optical power of our instruments, by the physiological perception of our retina, and by the sensitiveness of our photographic preparations. Of the regions which pass beyond these limits we know nothing. There may exist there more rarefied matter exercising no appreciable action on our senses. Perhaps this atmosphere extends to a very considerable distance, up to the zodiacal light. We may remark, finally, that there exist masses of cosmical matter, analogous to nebulæ, circulating like the comets in the interior of the solar system, and which at the moment of perihelion are found very near the sun. This fact is, perhaps, not unconnected with certain extraordinary appearances observed during eclipses.

In order that the reader may appreciate the great variety of these aspects and forms we have given in the chapter on Eclipses drawings of the total eclipses of the sun of July 18, 1860 (p. 196), and July 29, 1878 (p. 199); and we add details of those of July 18, 1860 (p. 265), of August 18, 1868 (p. 266), of August 19, 1887 (p. 277), and of January 1, 1889 (p. 278). We may remark that the corona is more extended from east to west in the years of spot minima than in the years of maxima. At the time of the eclipse on January 1, 1889, its rays extended to the west to three times the diameter of the sun, or to 2,480,000 miles.

[On the preceding page we give (fig. 126) a diagram showing all the solar eclipses visible in England from 1891 to 1922. From this it will be seen that the next eclipse of the sun of any magnitude will take place on May 28, 1900. This eclipse will occur about 4 P.M., and will be total across Portugal and Spain, from Ovar to Alicante. The eclipse of August 30, 1905, will also be total in Spain, from Corunna to Valencia.—J. E. G.]

And now, without letting ourselves be dazzled by the splendour of the beneficent centre of the planetary life, let us penetrate further still into its magic sanctuary.

CHAPTER V

THE FLUCTUATIONS OF THE SOLAR ENERGY

Annual Variation of the Number of Spots and Eruptions Eleven-Year Period. Curious Coincidences. Terrestrial Magnetism and the Aurora Borealis

THE preceding facts have taught us that this colossal body which illuminates us is far from being calm and tranquil, and that a consuming agitation incessantly palpitates through all its being. We now come to facts more astonishing still. This prodigious energy, which appears in turn to be exhausted and revived, manifests its effects neither in a constant nor in an irregular manner, but following a fixed periodicity. As the sea rises at the flow and sinks at the ebb, to rise anew at regular intervals, like the isochronous respiration of our chest, which dilates and contracts it, like the beating of the heart of the little bird under his fine down, the solar forge shoots forth its lightnings, takes its breath and recommences, at intervals proportionate to the grandeur and energy of the gigantic furnace.

This harmonic periodicity is even perceptible here, notwithstanding the awful distance which separates us from the flaming star. Every eleven years, as we have already seen, the number of spots, of eruptions, and of solar tempests reaches its maximum; then the number diminishes during seven years and a half, sinks to its minimum, and then takes three years and six-tenths to reascend to its maximum. The period is thus eleven years and a tenth. It varies, however, sometimes shortening to nine years, occasionally extending to upwards of twelve.

But that everyone may personally verify the facts, the following are the statistics of solar spots since the year 1826, the year when the amateur astronomer, Baron Schwabe of Dessau, took it into his head to count them. Since 1878, instead of being simply counted without having regard to their dimensions, observers have measured the surface occupied by the spots, and estimated it in millionths of the solar hemisphere turned towards the sun: this estimation is more accurate than the simple enumeration of the spots. We see that the years 1828, 1837, 1848, 1860, 1870–71, 1883, have been years of maximum, while the years 1833, 1843, 1855, 1867, 1878, have been the years of minimum ; the period of decrease is longer than the period of increase (which also happens for the ebbing of the sea).

STATISTICS OF SOLAR SPOTS ACCORDING TO YEARS

Years	Number of Spots	
1826	118	
1827	161	
1828 *maximum*	225	
1829	199	
1830	190	5 years
1831	149	
1832	84	
1833 minimum	33	
1834	51	
1835	173	4 years
1836	272	
1837 *maximum*	333	
1838	282	
1839	162	
1840	152	6 years
1841	102	
1842	68	
1843 minimum	34	
1844	52	
1845	114	
1846	157	5 years
1847	257	
1848 *maximum*	330	
1849	238	
1850	186	
1851	141	
1852	125	7 years
1853	91	
1854	67	
1855 minimum	28	
1856	34	
1857	98	
1858	202	5 years
1859	205	
1860 *maximum*	211	

Years	Number of Spots	
1861	204	
1862	160	
1863	124	
1864	130	7 years
1865	93	
1866	45	
1867 minimum	25	
1868	101	
1869	198	
1870	305	4 years
1871 *maximum*	304	
1872	292	
1873	215	
1874	159	
1875	91	7 years
1876	57	
1877	48	
	Spotted Surface	
1878 minimum	24	
1879	49	
1880	416	
1881	730	5 years
1882	1,002	
1883 *maximum*	1,115	
1884	1,079	
1885	811	
1886	381	6 years
1887	179	
1888		
1889 minimum		

In general, each maximum is nearer to the preceding minimum than to the following, so that the curve presents the form drawn in fig. 127. If we draw one beside the other eleven lines, of which the highest corresponds to the number of spots in each year, and if we join the extremities of these lines by a curve, we obtain fig. 127, in which the verticals corresponding to the maxima and minima only are shown. The number increases during $3\frac{6}{10}$ years, and then diminishes during $7\frac{5}{10}$ years. The different periods are not absolutely identical.

Thanks to the old observations of Schwabe of Dessau, the state of the sun has been regularly followed since the year 1832. That observer made daily drawings (commenced in 1825) up to 1867. Carrington, in England, undertook a series of more precise measures from 1853, and this series has been continued by Mr. Warren De la Rue, and was published in 1868. They were then taken up, in 1873, at the Greenwich Observatory, and have been continued since without interruption. M. Rudolf Wolf, at Zurich, and P. Secchi, at Rome, have likewise made daily observations of the sun.[1]

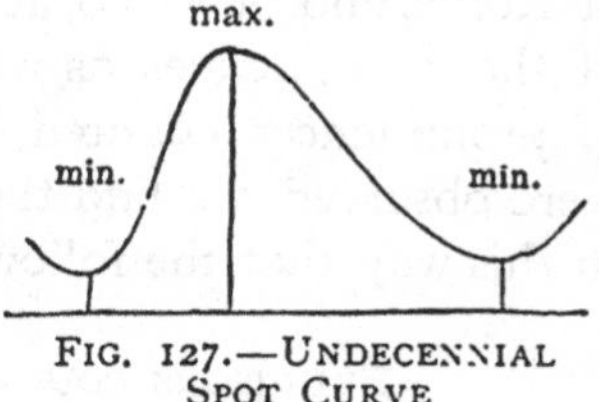

FIG. 127.—UNDECENNIAL SPOT CURVE

This periodicity was the first result of the assiduous observation of Schwabe. It was speedily adopted by M. R. Wolf, then Director of the Berne Observatory, now Director of that at Zurich, in spite of the opposition of other astronomers, and was confirmed by his personal observations, as well as by an investigation of the statements previously made on the solar spots since their discovery. That astronomer has fixed the dates of maxima and minima from the beginning of the observations up to 1878. A careful comparison of the observations, especially with reference to the work of Mr. Warren De la Rue and the Greenwich Observatory, has led us to slightly modify this table since 1847 and we have continued it up to 1889.

EPOCHS OF MAXIMA AND MINIMA OF THE SOLAR SPOTS

Maxima		Minima	
1615·0 ± 1·5	1761·5 ± 0·5	1610·8 ± 0·4	1755·7 ± 0·5
1626·0 ± 1·0	1770·0 ± 0·5	1619·0 ± 1·5	1766·5 ± 0·5
1639·5 ± 1·0	1779·5 ± 0·5	1634·0 ± 1·0	1775·8 ± 0·5
1655·0 ± 2·0	1788·5 ± 0·5	1645·0 ± 1·0	1784·8 ± 0·5
1675·0 ± 2·0	1804·0 ± 0·1	1666·0 ± 2·0	1798·5 ± 0·5
1685·5 ± 1·5	1816·8 ± 0·5	1679·5 ± 2·0	1810·5 ± 0·5
1693·0 ± 2·0	1829·5 ± 0·5	1689·5 ± 2·0	1823·2 ± 0·2
1705·0 ± 2·0	1837·2 ± 0·5	1698·0 ± 2·0	1833·8 ± 0·2
1717·5 ± 1·0	1847·8 ± 0·4	1712·0 ± 1·0	1843·7 ± 0·2
1727·5 ± 1·0	1859·7 ± 0·2	1723·0 ± 1·0	1856·2 ± 0·2
1738·5 ± 1·5	1870·9 ± 0·2	1733·0 ± 1·5	1867·0 ± 0·1
1750·0 ± 1·0	1883·9 ± 0·1	1745·0 ± 1·0	1878·9 ± 0·1
			1889·1 ± 0·1

Now, it is not only the solar spots which are subject to this periodical variation: there are also the *eruptions*, the tumultuous motions, and the

[1] The maxima and minima of sun-spots are accompanied by a rather curious circumstance. As the minimum is approached the spots are found nearer to the equator; then, when the number increases, they appear at a higher latitude. Moreover, the two solar hemispheres are not alike on both sides of the equator; sometimes it is the southern hemisphere which has most spots, and sometimes the northern; the epochs of maxima and minima are not the same for the two hemispheres. From 1883 to 1889, for example, the relative number of spots was 64 per cent. for the southern hemisphere and 36 per cent. for the northern.

surprising forms which we have described above. We have seen that by moving the spectroscope along the sun these prominences have been observed daily since the year 1871. Thanks to the labours of the 'Society of Italian Spectroscopists,' and in particular to those of Secchi and Tacchini, at Rome, and of Ricco, at Palermo, we can give an account of the variations of the prominences as we have done for the spots. Dividing the number of prominences counted on the sun by that of the days on which they were observed, we find the mean number corresponding to each day. It is in this way that the following little table has been obtained :—

Eruptions counted on an Average per Day on the Sun

Year		Year	
1871	15	1880	7
1872	12	1881	11
1873	9	1882	11
1874	7	1883	9
1875	6	1884	11
1876	5	1885	10
1877	4	1886	7·3
1878	2	1887	9·0
1879	3	1888	7·8

The observations of the faculæ give similar results for the fluctuations of the solar activity.

Thus the manifestations of the solar energy vary year by year. We do not mean to say that the number of spots always corresponds with those of the eruptions and faculæ ; no, these phenomena are both intermittent, and, up to a certain point, independent of each other; but the whole of the manifestations of solar physics present to the student of nature the curious eleven years' fluctuation which has just been explained.[1]

This periodicity of the manifestation of solar activity is a fact now proved with the most unquestionable certainty. It was discovered by him who first thought of counting the spots on the sun. What a beautiful lesson for astronomical amateurs ! How discoveries may be thus made by simple curiosity or by perseverance! What could apparently be more childish than the idea of amusing one's self by counting every day the spots on the sun ? Nevertheless, the name of Schwabe will remain inscribed in the annals of astronomy for having thus discovered this mysterious period of eleven years in the variation of the solar spots. Certain astronomers understand nothing of these delicate investigations, and Delambre, for example, whose mind was at the same time so rigid and so narrow, scarcely

[1] For a detailed discussion of the variations of solar activity, see our studies published in *Astronomie*, a monthly review of popular astronomy, February 1888, especially the general table, pp. 48–49. This table shows clearly that the fluctuations of the spots, of the faculæ, and of the eruptions are not simultaneous, although they are united to each other by a certain relation. See also the table on p. 211 (June 1888). These studies are too technical to be reproduced here.

deigned to speak of these spots; still, he took care not to compromise himself by adding this profession of faith: '*It is true that they are more curious than really useful.*' If Delambre had comprehended the grandeur of astronomy, he would have known that in this science there is nothing to be neglected.

There is no effect without a cause. What may be the cause of this motion of the solar surface?

This cause may be in the interior of the sun. It might also be exterior to him.

If it is in the interior of the solar body, it would not be easily discovered.

If it be exterior, the first idea which suggests itself is to seek for it in some combination of planetary motions.

Among the different planets of the system there is one which, from its importance, first presents itself to us, and it is found that the duration of its revolution round the sun approaches closely to the preceding period. Our readers have already named Jupiter, of which the diameter is only ten times smaller than that of the solar colossus, and of which the mass is equivalent to a thousandth of that of the central star. It revolves round the sun in 11·85 years.

During the course of its revolution its distance from the sun is subject to a perceptible variation. This distance, which is, on the average, 5·203 (that of the earth being one), sinks at the perihelion to 4·950, and rises at the aphelion to 5·456. The difference between the perihelion and aphelion distance is 0·506—that is to say, a little more than half the distance from the earth to the sun, or about 47 millions of miles. This is rather considerable. Revolving thus round the sun, Jupiter exercises on him an attraction easily calculated, and constantly displaces his centre of gravity, which can, consequently, never coincide with the centre of figure of the solar sphere, and is always found drawn eccentrically towards Jupiter. The attraction of the other planets prevents this action from being regular, but it cannot prevent it from being predominant.

It might be that this motion of the solar mass should be interpreted for us by the spots, and that it might have, for example, a maximum of spots when Jupiter attracts more, or attracts less, the solar centre. If we had here the cause of the periodicity of the spots, this periodicity should be 11·85 years.

But it is shorter. While Jupiter returns to his perihelion only after 11·85 years, the maximum of spots returns very irregularly, but on the average after 11·11 years—that is to say, 74 hundredths of a year, or 200 days,[1] sooner. This number comes from a discussion of all the observations. Does there exist in the solar system a second cause which obliges a phenomenon to advance thus on the perihelion of Jupiter? Venus revolves

[1] [This should be 270 days.—J. E. G.]

round the sun in 225 days, and about every 245 days meets the radius vector of Jupiter. The earth revolves in 365 days, and meets the radius vector of Jupiter every 399 days. These two planets certainly act on the sun in the same way as the giant planet, but with less intensity. If this common action were expressed by an increase of spots, we should see in the fluctuations of the solar spots combinations of the period of 11·85 of Jupiter with that of one year for the earth, of 0·62 for Venus, and of 0·24 for Mercury. Unfortunately, this combination does not appear to produce the observed effect.

Whether it be the perihelion or the aphelion of Jupiter which causes the maximum of solar spots, these maxima should always coincide with the same positions. But, on the contrary, each revolution of Jupiter adds the difference of 0·74 which we have just noticed, and at the end of a certain time, of thirteen to fourteen revolutions, the positions are reversed. We must, then, although with regret, give up Jupiter.

. . . .

Whatever may be the relation which exists between the two periods, the connection is, then, purely accidental, for we cannot logically admit that the same cause produces contrary effects, and that the perihelion sometimes induces a minimum and sometimes a maximum.

However, let us dismiss the idea of the variation of the distance of Jupiter and consider only its imaginary circular revolution. Let us suppose that the variation of distance does not act perceptibly. The fact still remains that the Jovian attraction makes the centre of gravity of the sun turn round his centre of figure in 11·85 years. Are the spots always on the radius vector of Jupiter? No, for the earth crosses this radius vector every thirteen months, and we do not see more spots on that solar hemisphere than on the opposite hemisphere. Moreover, the sun rotates on itself in 26 days, and would bring these spots in view of the earth, since they turn with the solar surface. Under whatever aspect we discuss the question, we are, then, led, in spite of ourselves, to eliminate the action of Jupiter. It is the same, and with much stronger reason, as regards all the other planets.

It is difficult to conceive how the planets which are so small and so distant could produce in the sun disturbances so profound and so extensive. It is scarcely possible that it should be their gravitation which acts, considering that the attractive power of Venus on the solar surface would be about $\frac{1}{750}$ of that which the sun exercises on the earth; and in the case of Mercury and Jupiter the effect would be still less, about $\frac{1}{1000}$ of the influence of the sun on the earth. The sun, considered apart from the moon, raises on the deep waters at the earth's equator a tide of a little less than 13 inches in height, so that, taking into account the rarefaction of the substances of which the photosphere is composed, it is very evident that any tide produced by a planet cannot directly explain the phenomena. If the solar spots are due in any way to planetary action, this action must be that of a different and much more subtle influence.

The cause of the periodicity of the solar spots will perhaps be found some day, after a general comparison of corresponding phenomena which may appear subject to a similar periodical motion.

While awaiting the time when we may make this discovery, let us call attention here to a truly extraordinary connection shown by terrestrial magnetism.

We know that the magnetic needle does not remain fixed in the plane of the magnetic meridian, but oscillates every day to the right and left of this plane. The greatest deviation is produced about 8 o'clock in the morning. Then the needle stops, returns towards the line of magnetic north, goes beyond it, and reaches its greater western deviation about a quarter-past one in the afternoon. This excursion from east to west is performed, then, in about 5 hours, more or less, according to the season. The needle afterwards returns towards the east, stops about 8 o'clock in the evening, turns back at 11 o'clock, and sets out again towards the east [? west] at 8 o'clock in the morning. The above figure reproduces on parallel lines these four movements, which constitute the double diurnal oscillation.

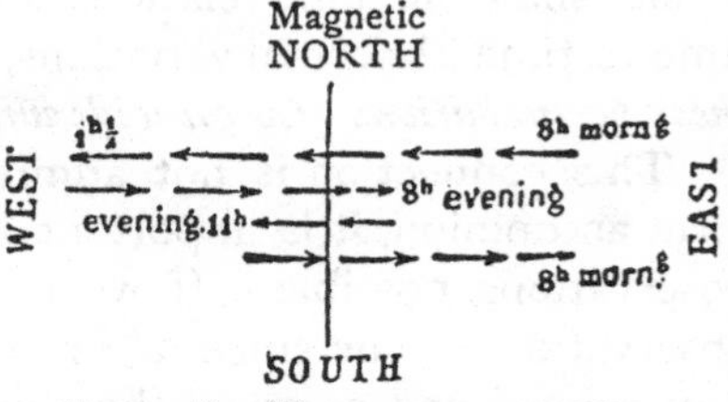

FIG. 128.—MOVEMENTS OF THE MAGNETIC NEEDLE

This phenomenon is absolutely general; it shows itself all over the earth according to the same laws; only the extent of the oscillation, which is on the average 9′ at Paris, is reduced to 1′ or 2′ between the tropics, and increases, on the contrary, towards the poles. It does not, however, increase proportionately to the latitude, for it is not greater at Christiania than at Naples. Moreover, the course of the needle, usually very regular, is sometimes accidentally disturbed by perturbations which are felt at the same moment over very great areas.

At each place, the hours at which the needle reaches the maximum of its excursion, either to the right or to the left, are so constant that the observer may almost use them to regulate his watch.

This diurnal oscillation of the magnetic needle is produced by the diurnal oscillation of the temperature, to which is superadded that of electricity, of the vapour of water, of atmospheric pressure, &c. If we examine the monthly variation, we arrive at the same conclusion. The oscillation is feebler in winter, stronger in summer. The thermometric variation is also feebler in winter, stronger in summer, &c. This same variation likewise increases from the tropical regions towards the polar regions. We can, then, affirm that this diurnal variation depends in the first place on the variation of temperature due to the sun, and acting, through the medium of atmospheric electricity, on the terrestrial magnetism, of which the magnetic needle indicates the variations.

The amplitude of these diurnal oscillations varies every day, every month, every year. If we take the mean of the observations for a whole year, we ascertain that this oscillation may lengthen from single to double in a period of about 11 years, which period—a fact eminently worthy of attention—corresponds to that of the solar spots, *the maximum of the oscillations coinciding with the maximum of the spots, and the minimum with the minimum!* All the other elements of magnetism, inclination, and intensity show the same relation. Further, the magnetic needle manifests from time to time abnormal variations, perturbations caused by magnetic storms; *these perturbations also coincide with the great agitations observed in the sun!*

This connection is not admitted by all astronomers. To enable us to form an opinion, it is important at first to compare the greatest number of observations possible. If we construct a table of the principal magnetic observations made since 1842, the year when five of the best series were commenced, and compare them with the solar spots, we see that there have been maxima in the diurnal magnetic variation in 1848, 1859, 1870–71, 1883–84, and minima in 1844, 1856, 1867, 1878, and that this oscillation corresponds to that of the solar spots. The fact is incontestable. The number of eruptions presents a connection less marked, but it is not absolute, since it depends on the number of days of observation—that is to say, days of fine weather, when only we can see the eruptions on the solar borders. But it indicates no less the state of the sun.

If we trace the curve of the number of solar spots (area of the solar surface occupied by the spots, daily mean for each year), and below the curve of magnetic variation according to a complete series, that of Prague, for example, we obtain fig. 129, which sufficiently speaks for itself, and which gives precise affirmation to a real correspondence. We see that all the maxima and minima are far from reaching the same value as well for the spots as for the magnetism.

The connection is so striking that an astronomer, M. Wolf, Director of the Zurich Observatory, has found formulæ for calculating the number of spots on the sun, or, we should rather say, the spotted area, by the examination of magnetic observations alone without the necessity of looking at the sun. He wrote to me lately that these formulæ have never been in fault more than a few months. The curve traced, so as to express on a physiological table the daily, monthly, annual state of *the sun's health*, presents the same inflexions, the same aspects as the curve traced for the daily, monthly, annual observation of the magnetic needle.

This study of the magnetism of our wandering planet is very interesting, and one which is still very little known. Here is a weak needle, a slip of magnetic iron, which with its restless and agitated finger incessantly seeks a region near the north. Carry this needle in a balloon up to the higher aërial regions, where human life begins to be extinguished, shut it up in a tomb closely separated from the light of day, take it down into the pit of a mine, to more than a thousand yards in depth, and incessantly, day and night, without fatigue and without rest, it watches, trembles, throbs, seeks the point which attracts it across the sky, through the earth, and through

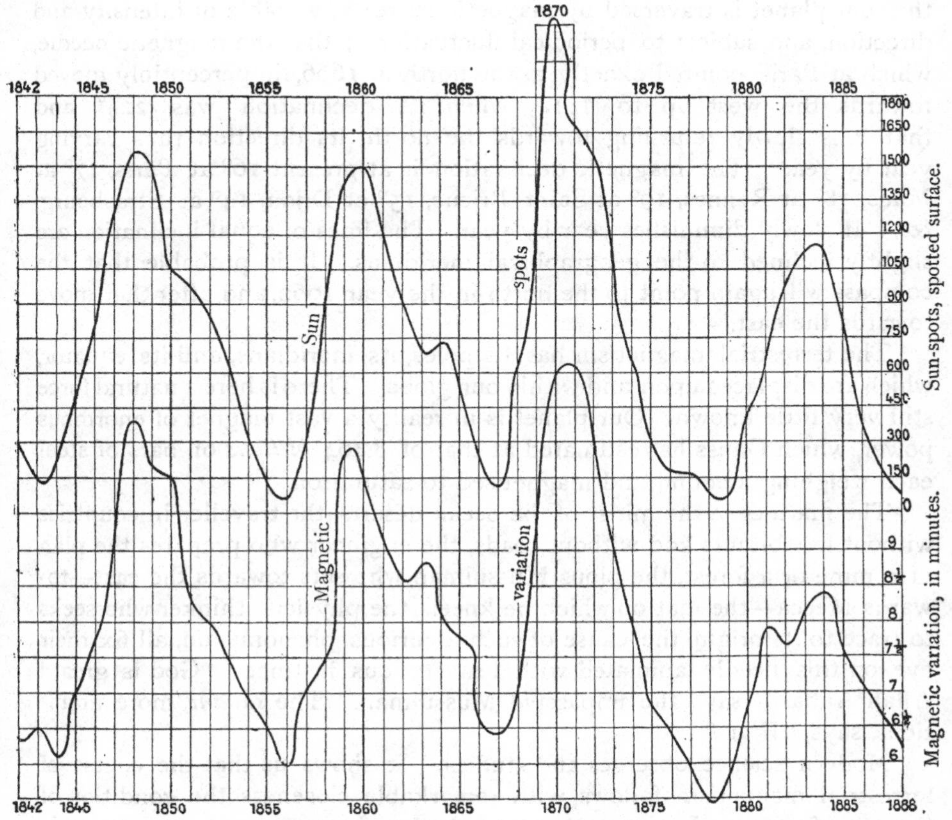

FIG. 129.—REMARKABLE CORRESPONDENCE BETWEEN THE NUMBER OF SUN-SPOTS AND MAGNETIC VARIATION

the night. Now—and here is a coincidence truly filled with notes of interrogation—the years when the oscillation of this innocent little steel wire is strongest are the years when there are more spots, more eruptions, more tempests in the sun ; and the years when its daily fluctuations are weakest are those when we see in the day star neither spots, eruptions, nor storms. Does there exist, then, a magnetic bond between the immense solar globe and our wandering abode ? Is the sun magnetic ? But the magnetic currents disappear at the temperature of redhot iron, and the incandescent focus of light is at a temperature incomparably higher still. Is it an electrical influx which is transmitted from the sun to the earth across a space of 93 millions of miles? So many questions, so many mysteries. Let us first ascertain the *facts*; we will then seek the explanation.

We have already seen, in the chapter on the life of the earth (p. 67), that our planet is traversed by magnetic currents, variable in intensity and direction, and subject to periodical fluctuations; that the magnetic needle, which at Paris pointed exactly to the north in 1666, imperceptibly moved towards the west up to 1814, when its declination was 22½°, and that it is slowly returning towards the north, its direction thus varying year by year. The magnetic declination is at present 16° at Paris, 17° at Mans, 18° at Rennes, 19° at Saint Brieuc, 15° at Dijon, 14° at Strasbourg, &c. It slowly diminishes year by year. The lines of equal inclination are slightly inclined to the geographical meridians. It is probable that the compass will again point to the north in the year 1962, and after that move towards the east.[1]

The terrestrial magnetism has its poles, its meridians, and its equator, which are displaced upon and within our globe. There is here a natural force still very little known. Our planet is in reality a vast magnet of enormous power, which Gauss has estimated at that of 8,464 *billions* of bars of steel each weighing a pound and magnetised to saturation.

The mariner in the midst of the ocean deserts, the traveller in countries without inhabitants and without roads, the engineer who prepares the plan of a mine or a forest, the pious Mussulman who sets towards the east—towards Mecca—the mat on which he kneels, the physicist thinker who seeks to trace to its origin the cause of such a curious phenomenon, all fix their eye on this needle animated with a mysterious instinct. 'God is great! Allah Akbar!' says the impassive Mussulman. The *savant*, more ambitious, says, '*Why?*'

Modern science observes and studies. It shows us that the course of terrestrial magnetism follows with remarkable closeness the condition of the solar furnace. Let us notice some further facts.

On September 1, 1859, two astronomers, Carrington and Hodgson, were observing the sun, independently of each other, the first on a screen which received the image, the second directly through a telescope, when, in a moment, a dazzling flash blazed out in the midst of a group of spots. This light sparkled for five minutes above the spots without modifying their form, as if it were completely independent, and yet it must have been the effect of a terrible conflagration occurring in the solar

[1] For details of the magnetic variation, see the chapter on 'La Terre séjour de Vie' in our work, *Les Terres du Ciel.*

atmosphere. Each observer ascertained the fact separately, and was for an instant dazzled. Now, here is a surprising coincidence: at the very moment when the sun appeared inflamed in this region the magnetic instruments of the Kew Observatory, near London, where they were observing, manifested a strange agitation; the magnetic needle jumped for more than an hour as if infatuated. Moreover, a part of the world was on that day and the following one enveloped in the fires of an aurora borealis, in Europe as well as in America. It was seen almost everywhere: at Rome, at Calcutta, in Cuba, in Australia, and in South America. Violent

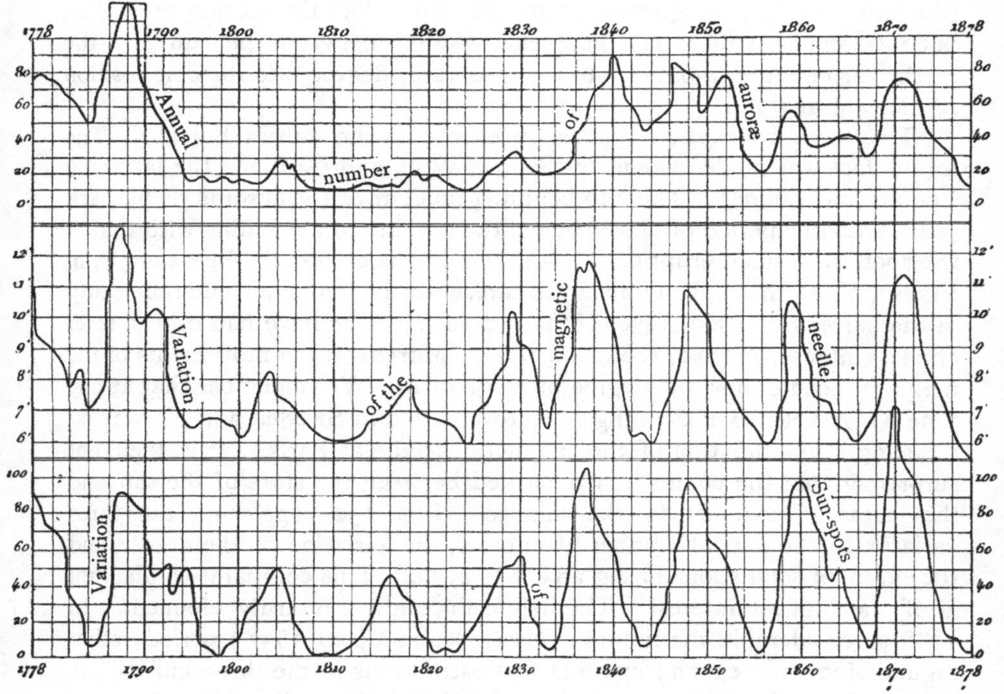

FIG. 130.—ANNUAL VARIATION OF THE AURORA BOREALIS, OF THE MAGNETIC NEEDLE, AND OF SUN-SPOTS DURING A CENTURY

magnetic perturbations were manifested, and at several points the telegraph-lines ceased to act. Why should these two curious events not be associated with each other?

A similar coincidence was observed on August 3, 1872 (by Young, in the United States): a paroxysm in the solar chromosphere, magnetic disturbances everywhere.

One of the most colossal solar spots which has ever been measured was that which crossed the disc from the 12th to the 25th of November, 1882, and which in crossing the central meridian on the 18th showed a maximum real extent equal to 2,417 millionths of the surface of the solar hemisphere. Visible to the naked eye, it measured on the 17th more than 62,000 miles in diameter. Well, on this same date of November 18 the terrestrial globe was the seat of a considerable magnetic disturbance —in the United States the telegraphs ceased to act; in France, in England, everywhere, the magnetic instruments registered fantastic oscillations.[1]

We might describe other similar examples.

The conclusion, then, is that this connection between the state of the sun and terrestrial magnetism is not fortuitous, like the motion of Jupiter, but *real*, and that there is a magnetic relation between the sun and the earth. Iron enters, moreover, to a considerable degree into the composition of the central star.

This same connection appears to extend to the aurora borealis. The former fact was made evident by Sabine, Wolf, and Gautier; the latter, by Loomis and Zöllner. The number and grandeur of the auroras visible each year varies in a period of eleven years, the maximum coinciding with that of the spots and solar eruptions. Who has not been struck, for example, in France, where these phenomena are rare, with the frequency and the beauty of the auroras of 1869, 1870, 1871, and 1872? We have had others, rarer and less intense, in 1882, 1883, and 1894. Moreover, the movements of the magnetic needle show the degree of magnetism. We remember that at one time Arago boasted of divining an aurora visible in Sweden and Norway by the simple examination of the magnetic needle at Paris. It is, then, not surprising that the connection remarked between the state of the sun and the compass extends to the auroræ. Fig. 130 represents the three variations: the annual number of auroras, the variation of the needle, and the extent of solar spots during an entire cycle. This comparative diagram is of the highest interest. The triple fluctuation is truly eloquent. In 1788 a considerable maximum; relative calm up to 1837; period rather regular since that epoch; symmetrical oscillations in the three curves. A similar connection appears to present itself with the zodiacal light.[2]

[1] See the review, *L'Astronomie*, 1883, pp. 74 and 111; 1888, p. 211.

[2] As to the influence of the solar spots on the temperature of the globe, it does not appear possible to say anything conclusive at present. The spots themselves certainly send us by radiation less heat than the general surface of the sun. According to results determined with much care by Mr. Langley, the umbra of a spot emits about 54 per cent., and the penumbra about 80 per cent. of the heat emitted by a corresponding surface of the photosphere. The spots have, then, a direct effect in cooling the earth. The total surface covered by the spots, even at the epoch of maximum, never exceeding $\frac{1}{500}$ of the total surface of the sun, it follows that they directly diminish our supply of heat by about $\frac{1}{1000}$ of the total quantity. Would this effect be perceptible? It is difficult to say.

But this direct effect may be neutralised by an action of a totally opposite character. Light and heat come to us from the photosphere, which is covered over with an atmosphere of gas endowed with a considerable absorption. Now, if the level of the photospheric surface is disturbed in such

Let us complete all these facts of observation by recapitulating here as a definite conclusion the *actual state of our knowledge of the sun.*

In order to represent as accurately as possible the physical state of the solar world, let us proceed from the exterior to the interior, considering that the external regions of the sun are better known to us than the interior regions.

I. When we approach the central star, the first material substance we meet with is the corona, which surrounds the radiant body at a height of more than 500,000 kilometres (310,000 miles), and which sometimes sends out rays up to some millions of miles (example: the eclipse of January 1, 1889). It is certain that this substance does not constitute an atmosphere, properly so called—that is to say, a continuous gaseous envelope. The two following considerations demonstrate, in fact, the impossibility of this condition.

And firstly we have seen that gravity is twenty-seven and a half times stronger on the sun than on the earth; all gas, then, is consequently twenty-seven and a half times heavier. Now, in every atmosphere each layer is compressed by the weight of the layers above it, and the density increases in geometrical progression. An atmosphere composed of the lightest gas we know, hydrogen, would, then, present in the lower layers a density incomparably greater than that which corresponds to the observed facts; indeed, it would be then no longer gaseous, but liquid or solid; it would cease to exist.

On the other hand, we have seen comets approach close to the sun. Thus, on February 23, 1843, a comet grazed, and, so to say, went out of its way through the corona. At the time of its greatest proximity it flew above the flames of the sun with a velocity of 563,000 kilometres (? 563 kilometres = 350 miles) per second, and it traversed at least two to three hundred thousand miles of the solar corona with a similar velocity without experiencing the least influence or the least delay. The same fact

a manner that it raises undulations of a considerable height, compared to the thickness of the atmosphere which covers it, then the radiation must be proportionately increased. The spots appearing to be due to an eruptive action, the interior and hotter gas rushes through the photosphere with unusual copiousness; at the epochs of maximum of spots this may produce more than a compensation.

It is worthy of attention, however, that terrestrial meteorology appears to be subject to fluctuations of the same order. Thus, in our climates, cold years, rains, and inundations appear to correspond to those when the sun is quiet, without eruptions and without spots—witness the years 1888, 1879, 1866, and 1856. The dry and warm years appear, on the contrary, to correspond to the epochs of great solar activity—examples: 1884, 1870, 1859, 1845, 1836 (and 1893). The American astronomers have noticed a similar relation in the annual number of cyclones. But it is not necessary for us to hasten to a conclusion; we should not generalise before obtaining a sufficient number of observations, and meteorology is still in its infancy.

Experiments on the solar heat seem to show that this heat varies in proportion to the spots—that is to say, that the more spots there are, the hotter is the sun. (See the review *L'Astronomie*, 1888, p. 390.)

was repeated before our eyes on September 17, 1882. In order to form an idea of what these comets would become if they passed through even the most rarefied atmosphere, it will suffice to notice that the shooting stars are instantaneously and completely reduced to vapour by the heat of friction when they reach our atmosphere, at a height of 60 to 90 miles—that is to say, at an elevation where our atmosphere has entirely ceased to reflect the light of the sun. Now, the velocity of shooting stars is but 19 to 37 miles per second. The resistance (and the heat produced by it) increasing at least as the square of the velocity, what would be the fate of any body traversing several hundreds of thousands of miles of the rarest atmosphere with a velocity of 500,000 kilometres (? metres) per second? What must, then, be the rarity of an atmosphere through which comets can pass, not only without being annihilated, but even without experiencing the least perceptible delay? The solar heat exercises there a repulsive action, which drives out in some way the tails of comets to *millions* of miles of distance away from the sun.

What, then, is the corona? It is probably a region in which is found a variable quantity of detached particles, partially or wholly vaporised by the intense heat to which they are exposed. But how can these particles be supported in these burning heights? To this question we are already able to give three replies: (1) The matter of the corona may be in a state of permanent projection, being composed of substances incessantly darted out by the sun and falling back on him; but for this, forces of projection would be necessary capable of darting out matter with a velocity of 186 miles per second, and that almost constantly, all round the sun. (2) The coronal substance may be more or less supported in the solar heights by the effect of a calorific or electrical repulsion; why should not electricity, which already plays so great a part in terrestrial meteorological phenomena, exert itself with an energy increased a hundred-fold in the stormy centre of our system? (3) Finally, the corona may be due to clouds of meteors, aerolites circulating round the sun in his immediate vicinity. All these explanations are perhaps in part true.[1]

This is the place to call attention to the existence of a still mysterious light which constantly envelops the day star to a great distance, and which we perceive after sunset or before sunrise, forming a sort of cone more or less diffused in the direction of the zodiac. This gleam has received the name of the *zodiacal light.* It stretches along the ecliptic, and is perceived in our northern latitudes, in Europe, in America, in Asia, in Japan, extending to a distance of 90° from the place occupied by the sun. Near the equator attentive observers have followed it much farther, and even to 180° from the sun—that is to say, to the point opposite to him, and

[1] Newcomb, *Popular Astronomy.*

making a complete circuit of the sky at midnight, on the one hand from the west, on the other hand from the east up to the zenith. Two explanations of this light present themselves : either it surrounds the earth, or it surrounds

FIG. 131.--ZODIACAL LIGHT, AS OBSERVED IN JAPAN

the sun. The first case is less probable, since it is not in the plane of the terrestrial equator, but in the plane of the ecliptic. It is, then, probable that it is due to an immense cloud of corpuscles surrounding the day star to beyond the distance where the earth revolves, thus marking the general plane in which the sun and all the planets turn.[1]

II. Below the corona, descending, we find the chromosphere, a sheet of fire from 6,000 to 9,000 miles in thickness, and which, here and there, is projected in immense masses which we might call flames, if this expression were not, in spite of its eloquence, very much below the reality. We call flame and fire that which burns; but the gases of the solar atmosphere are raised to such a degree of temperature that it is impossible for them to burn! Extremes meet. Hydrogen forms the upper part of the chromosphere; but as we descend we find vapours of magnesium, iron, and a great number of metals. The prominences are due to projections of hydrogen, shot up with velocities which exceed 240,000 metres (149 miles) per second. The eruption sometimes continues during several hours, and even during several days, and these immense luminous clouds remain suspended without moving until they fall back in showers of fire on the solar surface. How can we conceive, how express, these tremendous operations of solar nature! If we call the chromosphere an ocean of fire, it should be added that it is an ocean hotter than the most intense fiery furnace, and also deeper than the Atlantic is wide. If we call these movements hurricanes, it should be remarked that our hurricanes blow with a force of 100 miles an hour, while on the sun they blow with a violence of 100 miles a second! Shall we compare them to volcanic eruptions? Vesuvius buried Pompeii and Herculaneum under its lava; a solar eruption rising in a few seconds to 60,000 miles in height would swallow up the entire earth in its rain of fire, and reduce to ashes all terrestrial life in less time than you take to read these lines. If our globe could fall into the sun, it would melt and evaporate on arriving there like a flake of snow on redhot iron. When, in the Burial Service, before the catafalque

[1] The zodiacal light is rarely visible at Paris, on account of the nocturnal illumination of that capital. I observed it, however, one evening when it presented a great intensity (February 20, 1871), and I have given a description of it in a report to the Institute. It measured 86 degrees in length from the sun, and extended nearly to the Pleiades. The estimation of its intensity was so much the easier as the atmosphere of Paris was less illuminated than usual on account of the absence of gas. Calm and motionless, the zodiacal light was very different from the throbbing gleams of the aurora borealis, and dispelled rather than confirmed the idea sometimes expressed of any connection between the two phenomena. The spindle was a little more intense in the central region than at its borders, and much more so at the base than towards its summit. Its tint, about half as bright again as that of the Milky Way, was a little more yellow. The faintest stars visible to the naked eye, those of the sixth magnitude, were perceptible through this veil; in the telescope we could distinguish stars to the tenth magnitude, but the eleventh magnitude and fainter were extinguished. I have often observed it since in the sky of Nice, where it is almost constantly visible in winter and spring.

Commenced by Cassini in the seventeenth century, the study of this singular light has been much advanced in our century by the numerous observations of Jones in Japan. The theory is not yet certain, however.

lit with tapers, the priest appeals 'on the faith of David and of the Sibyl to the final conflagration of the world and the flames of hell :

> Dies iræ, dies illa !
> Solvet sæclum in favilla,
> Teste David cum Sibylla,

he does not in his conception reach the grandeur of the solar testimony in the bosom of its fiery ardour. Several theologians have, it is true, placed hell in the sun, and I have at this moment before my eyes a book entitled 'Researches on the Nature of the Fire of Hell,' by Mr. Swinden, Doctor of

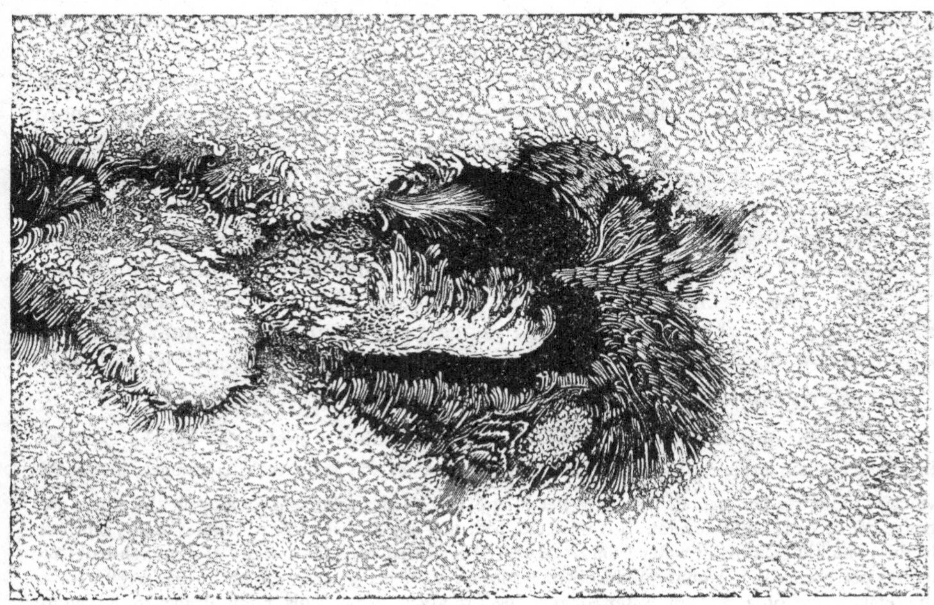

FIG. 132.—PHYSICAL CONSTITUTION OF THE SUN'S SURFACE—SPOT OF DECEMBER 1873
(From a Drawing by Langley)

Theology, of which the frontispiece is extracted from the 'Mundus Subterraneus' of Father Kircher. This drawing is remarkable, notwithstanding its exaggeration, for the solar eruptions—which were not then known, but which were guessed at.

III. The corona and the chromosphere are only visible during total eclipses or by the aid of the spectroscope. What we see of the sun with the naked eye or a telescope is the luminous surface named the *photosphere*, on which the chromosphere rests. It is this which radiates the light and heat which we receive from the brilliant star. This surface itself appears to be neither solid, liquid, nor gaseous, but composed of movable particles, nearly as the surfaces of clouds are presented when seen from the top of a balcony. No one has gone so deeply into this analysis as the American astronomer, Langley. We have reproduced (fig. 132) the drawing which he has made from nature of these solar particles caught in the act of forming a spot. It is probable that these granular elements constitute on the whole a very thick layer, like a layer of floating dust—dust by comparison, for each grain is an Alp or a Pyrenees. This embracing layer dances on an ocean of gases of prodigious weight and cohesion. The entire globe of the sun appears formed of gases enormously condensed. At the base, and in contact with the photosphere, is a sort of sheet of scarlet fire, a multitude of jets of brilliant gas gushing out from the whole surface, flames which rise and are agitated incessantly like those of a conflagration.

And now, how is this heat and light kept up? If the sun were composed of massive coal burning in pure oxygen it would not burn for more than 6,000 years without being entirely consumed: it would, then, have been burnt out since the beginning of historic times. For the maintenance of the solar heat, three principal causes appear to be at work: the contraction of the solar globe, the fall of meteors on its surface, and the liberation of heat by chemical combinations. The first cause must be the most important. We know the mechanical equivalent of heat. All bodies which fall and are stopped in their fall produce a certain quantity of heat, and the quantity of heat produced is the same whether the body is suddenly stopped or gradually 'slowed down' by resistances.[1]

[1] All bodies when their motion is stopped produce a quantity of heat which can be expressed in calories by the formula $\frac{m v^2}{8,338}$, in which m is the mass of the body in kilogrammes, and v the velocity in metres per second. A body weighing 8,338 kilogrammes and moving at the rate of one metre per second would develop, if it were stopped, just one calorie of heat—that is to say, sufficient to raise one kilogramme of water from 0° to 1°. If it moved with the velocity of a cannon-ball (500 metres per second), it would produce 500 × 500, or 250,000 times more heat, or enough to raise the temperature of a mass of water equal to itself to about 30°. This body, drawn to the sun by attraction, would reach it with a velocity of 377 miles, and would thus produce 370,000,000,000 times more heat—that is to say, it would be immediately reduced to vapour. Sir W. Thomson (now Lord Kelvin) has calculated the quantity of heat which would be produced by each planet if they fell into the sun, and has expressed as follows the number of years of maintenance of the solar heat which these falls would supply:—

Mercury	6 years	219 days
Venus	83 ,,	326 ,,
The Earth	95 ,,	19 ,,
Mars	12 ,,	259 ,,
Jupiter	32,254 ,,	
Saturn	9,652 ,,	
Uranus	1,610 ,,	
Neptune	1,890 ,,	
Total	45,604 years	

The fall of all the planets into the sun would thus produce sufficient heat to maintain its radiation during 45,600 years. A quantity of matter equal to the hundredth part of the mass of the earth

If, as is probable, the solar globe is the result of the condensation of an immense nebula which originally extended beyond the orbit of Neptune, the fall of the molecules to the present concentration has supplied about 18,000,000 times as much heat as the sun now gives per annum (Thomson). It would follow that the sun would have had but 18,000,000 years of the present radiation; but during the whole period of its condensation it was incomparably vaster, and radiated differently. On the other hand, supposing this to be the sole source of the solar heat, this body, continuing to condense, would be reduced to half its present diameter in 5,000,000 years at the latest, and as with this dimension it would have eight times its present density, it would become liquid (or solid), and its temperature would begin to decrease in such a manner that in about 10,000,000 years its heat would not be sufficient to support a state of life similar to that of the present life. The total life of the solar system would not exceed, on this hypothesis, 30,000,000 years. Young adds that the fall of meteoric matter might increase it by a nearly equivalent quantity, which would, however, not lead us to 60,000,000 years. We may add, that we do not know all the resources of nature, and that probably this amazing radiation is still sustained by other causes. But it is absolutely certain that it will become extinct, and that terrestrial life, of which it is the sole source of maintenance, will then be lulled in an eternal sleep. In all probability, the sun will be extinct before 20,000,000 years.

We see that the physical constitution of the sun is one of the most curious and important subjects of study which are presented to our attention, and every mind which takes an interest in the things of nature cannot help being impressed by these grandeurs and attracted by these problems.

Such is this immense body, on the rays of which our existence hangs. From its surface, agitated by the waves of an eternal tempest, are constantly shot forth with the velocity of light fertile vibrations which carry life to all the planets. The physical state of this gaseous globe certainly does not permit it to be inhabited at present by organised beings of the nature of those which exist on the planets; but neither our observations nor our deductions, nor even our conceptions, limit the power of nature, and there would be nothing absurd in imagining the sun inhabited by spirits of which the physical organisation should be scarcely material. But here we leave the bounds of positive science. Let us hasten to return there, remarking, however, that in the future the solar globe will be in a

falling annually on the solar surface would maintain his radiation indefinitely. This increase of the solar mass would occasion an acceleration of the motions of translation, and a shortening of the years. But as the mass of the sun is 324,000 times greater than that of the earth, the annual addition would be only a 32-millionth, and ages would be required to render the effect perceptible.

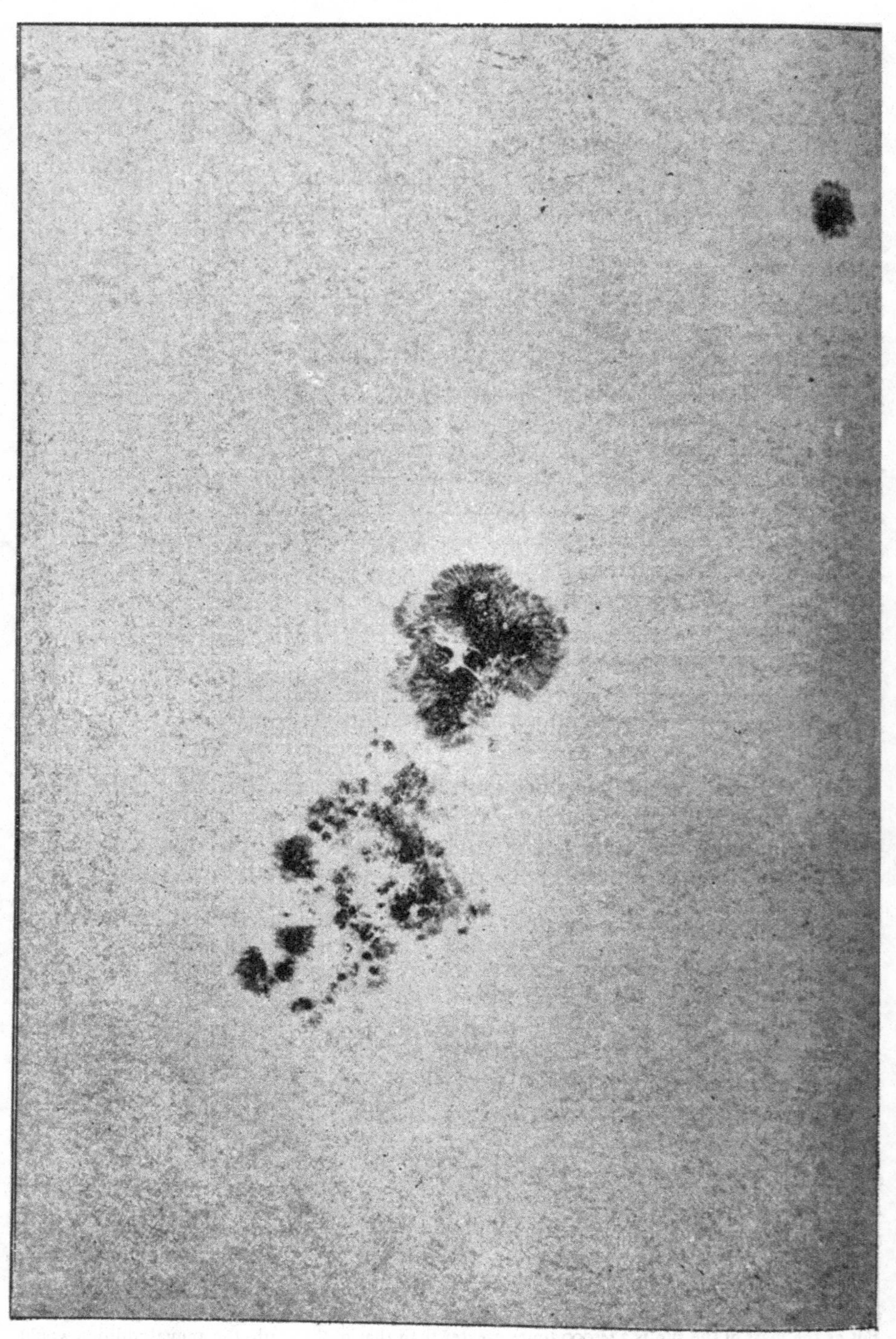

Fig. 133.—A Remarkable Sun-spot. (From a Photograph by Dr. Janssen)

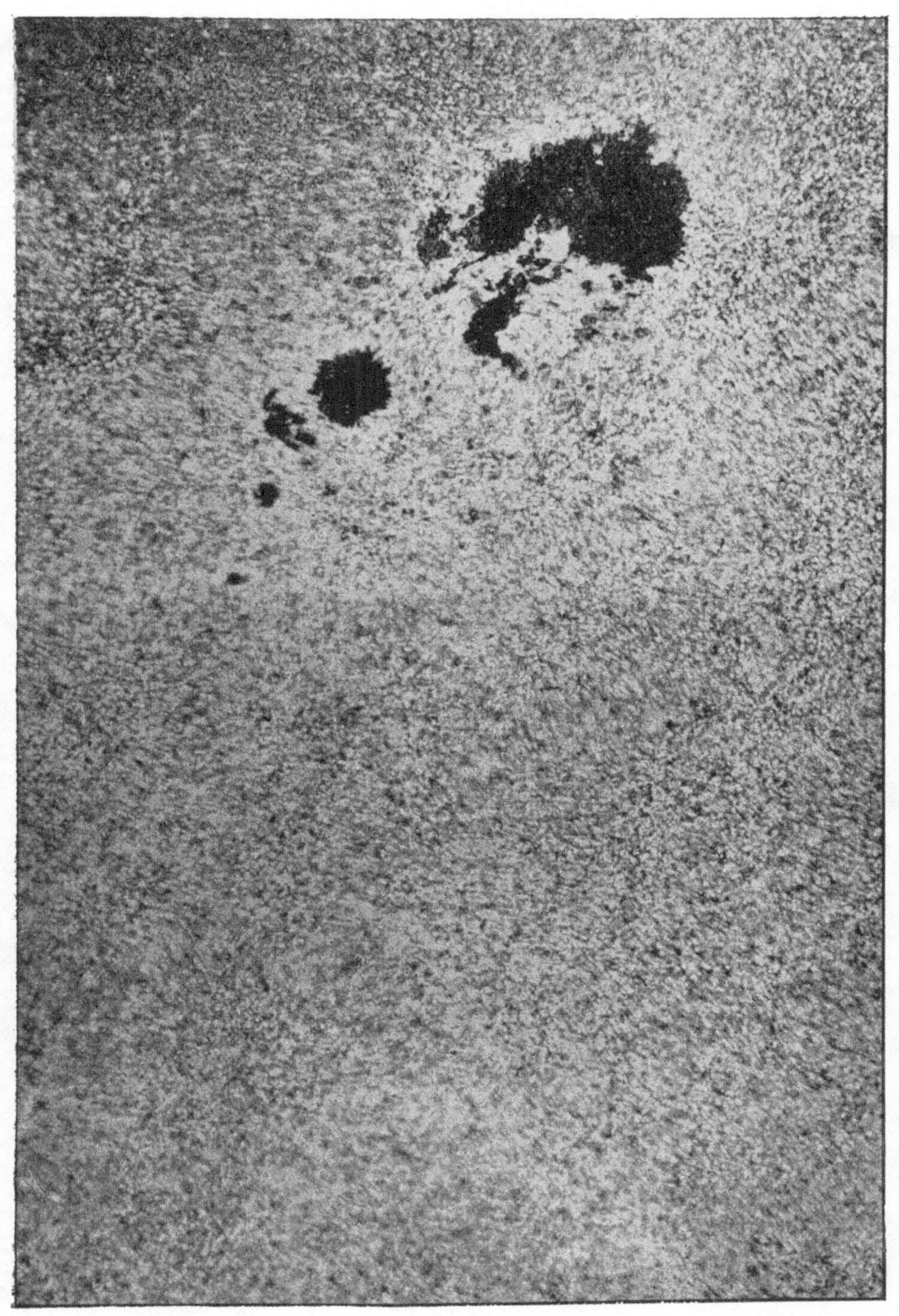

Fig. 134.—A Portion of the Surface of the Sun, showing a Group of Sun-spots and Rice-grain Structure of the Photosphere. (From a Photograph taken at Meudon, by Dr. Janssen, June 1, 1881)

planetary state, and may be inhabited by organisms as coarse as ours. But how will the sun itself be then illuminated? Perhaps by a permanent magnetic aurora. Perhaps only by the stellar light for more clear-sighted eyes than ours. But these are mysteries of the future.

[On the two preceding pages we give reproductions of some very beautiful photographs of sun-spots taken by Dr. Janssen.]

CHAPTER VI

OUR SUN IS ONLY A STAR

His Destiny

WE have contemplated the solar splendour and have estimated the stupendous forces which are at work in this immense furnace; we have greeted the sun as the father and the ruler of worlds, and we know that our life, like that of the other planets, hangs on its fertilising rays. But what is the sun in the universe? what place does it occupy in the infinite? what is its intrinsic value from a general point of view? what will be its duration in the succession of ages?

However surprising may appear to us the assertion, after the amazing wonders which we have just considered, this immense globe, more than a million times superior to the earth in volume, and more than 300,000 times heavier, is but *a point* in the universe!

When our eyes are raised to the starry skies, during those sparkling hours when the celestial vault appears constellated with veritable luminous dust, let us stop at any one of these brilliant points which scintillate in the depths of the heavens; this point is as large as our sun, and, in the universe, our sun is not more important than it. Let us remove ourselves in thought to that star, and from its distance let us look back towards the earth and search for our solar system; from there, neither the earth nor any planet is visible; from there, the entire orbit which our earth describes in a year, and which measures 186 millions of miles in diameter, would be entirely hidden behind the thickness of a hair; from there, the sun is but a point hardly perceptible.

Yes, our sun is but a star! Look at this little square taken in the sky (fig. 135). It is a reduction of one of the beautiful ecliptic charts of the Paris Observatory, which reproduces exactly, rigorously, place by place, brightness by brightness, a little region of the sky about 5° in width by 5° in height. This chart contains 4,061 stars, with their precise positions. Well, look for the sun in this cluster of stars; it would be among the largest if you were not too distant in space, among the smallest if your flight had carried you away into the ethereal depths, and it would even become wholly invisible if you plunged still farther into the abysses of Infinitude.

How do we know this? The *nearest* star looks down on us from such a distance that if we attentively follow it during the whole course of the year, the great movement which we annually make round the sun has

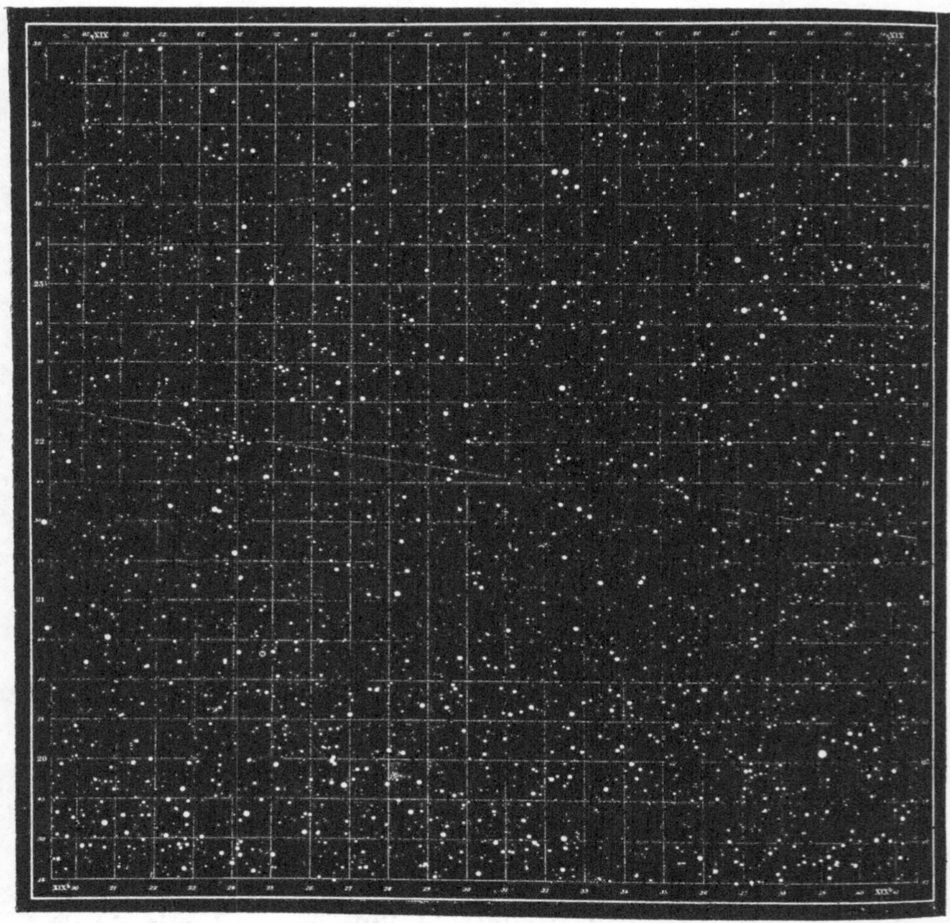

FIG. 135.—A CORNER OF THE HEAVENS CONTAINING 4,061 STARS
(From the Atlas of the Paris Observatory)

scarcely any influence in perspective on its absolute position. Now, in order that a displacement of 186 millions of miles in the path of an observer should not produce any effect on the position of an object we look at, it is necessary that this object should be tremendously far away. The entire orbit of our planet seen from this star (*Alpha Centauri*) appears quite small, and shows an angular width scarcely perceptible. We have seen (p. 86) that an angle of one degree corresponds to a distance of 57 times the size of the object, that an angle of one minute corresponds to a distance of 3,438 times, and that an angle of one second corresponds to a distance of 206,265 times. We have seen that the distances of the sun and moon have been measured by this mathematical method. Well, the entire orbit of the earth, reflected in the apparent motion of a star seen by a terrestrial observer, makes it only describe a little ellipse of less than two seconds in width (about the 900th part of the apparent diameter of the moon)—that is to say, our annual orbit seen from there is only presented under the form of a small imperceptible ellipse. The precise calculation shows that the half of this orbit—that is to say, the distance of the earth from the sun, which is, as we have seen, the unit by which we measure all the celestial distances—only appears under an angle of seven- to eight-tenths of a second (0″·75). If it were presented under an angle of a whole second, the distance of this star would be 206,265 times 93 millions of miles; as it measures but 0″·75, it is mathematically demonstrated that this distance is 275,000 times the same unit.

And this is the nearest star!

All the others are more distant still.

This single fact, now incontestable, proves: (1) that the stars are too distant to be visible if they simply received the light of the sun and did not shine of themselves; and (2) that the sun placed at similar distances would be reduced in appearance to a point so small as to appear but as a simple star.

The farthest known planet of our system, Neptune, revolves at a distance equal to thirty times the radius of the terrestrial orbit. *It would be necessary, then, to increase the length of this celestial road* 9,167 *times before arriving at the distance of the nearest star.* Then, searching through immensity in all directions round the solar system to this distance, we meet with no other sun.

In forming an idea of the immensity of the desert which surrounds our solar system some comparisons will be more easily grasped than the figures themselves. Representing by 1 yard the distance which separates us from the sun, and placing the sun at the centre of the system, that globe would be one-third of an inch in diameter, our planet would be quite a small point of $\frac{1}{300}$ of an inch in diameter, placed at 1 yard, and Neptune, the frontier of our planetary republic, would be a ball of $\frac{1}{80}$ of an inch placed at *ninety-eight feet.* To mark the distance of the nearest star, it should be removed to 170 miles, or from Paris to beyond Brussels; such is the proportion between the extent of the solar system and interstellar immensity. There, the first sun met with would be represented by a sphere of a size similar to that which we have supposed for our sun.

Let us suppose that a celestial traveller were carried out in space by a motion of such rapidity that he would in twenty-four hours pass over the distance which extends from the sun to Neptune (more than two thousand millions of miles). This velocity is so enormous that he would cross the Atlantic from Havre to New York in less than the tenth of a second. Our traveller would in forty-eight minutes pass over the space extending from the sun to the earth, and would arrive at Neptune at the end of the first day. But, having thus traversed the whole system, he would still travel in a straight line and with the same velocity for twenty-five years before reaching the first star, and he would then have the same voyage before him to arrive at the second, and so on. The earth would have disappeared from his view in the middle of the first day, and all the planets would have vanished before the end of the third day; then the sun himself, gradually diminishing in brightness, would year by year sink to the rank of a star.

We have remarked above that, if we could throw a bridge from here to the sun, this celestial bridge would be composed of *eleven thousand six hundred arches as wide as the earth.* Suppose a pillar at each extremity of this bridge. It would be necessary to repeat *this same bridge two hundred and seventy-five thousand times* to reach the nearest sun; that is to say, this marvel of imaginary architecture, more wonderful than all the fables of ancient mythology, and more fabulous than all the tales of 'The Thousand and One Nights,' would be composed of 275,000 piers, distant from each other by 93 millions of miles.

A star, a sun, may cause an explosion. If the noise of such a terrible conflagration could be transmitted to us, we should not hear it till the end of *three million seven hundred and ninety-five thousand years!*

We may add, further, that an express train which, at the constant velocity of 37 miles an hour, would pass over in 266 years the space which separates us from the sun, would not arrive at the nearest star, Alpha Centauri, until after an uninterrupted run of nearly *seventy-three millions of years!*

The sphere of the sun's attraction extends through the whole of space out to infinity. To speak accurately and minutely, there is not in the whole universe any particle of matter which does not feel to some extent the attractive influence of the sun, and even that of the earth, and of all other bodies still lighter; each atom in the universe has an influence on every other atom, and in displacing objects on the surface of the earth—in sending a ship from Marseilles to the Red Sea—we disturb the moon in its course. But, as we have seen, the action is in the direct ratio of the masses and in the inverse ratio of the square of the distances. The influence of the sun on the stars is not only excessively small with reference to the velocity of motion which it would produce in a given interval of time, but there is here only the influence of one star among its equals. On all sides, moreover, the reign of the sun is limited, for there are innumerable suns in all directions, and the sphere ruled by each star is as limited as that of our own star, so that everywhere we should find regions where his influence would be neutralised.

The sphere of the sun's attraction extends, nevertheless, out to and beyond the distance of Neptune. Strictly speaking, it extends indefinitely out to points where, in various directions, it meets with spheres of attraction of the same intensity.[1]

A planet removed to the distance of the nearest star would take 144 millions of years to travel round its orbit. This orbit would measure 1,585 billions of miles. The velocity would be 1,100,000 miles per annum—that is, 3,000 miles per day, or 125 miles an hour.

But this star is a sun like ours, of enormous volume, and of considerable mass. Since we have entered into these important considerations of celestial mechanics, and have ventured to give an account of the relations which bind our sun to the stars, let us take a step further by penetrating for a moment into the sidereal world, and take a foretaste before lingering in the flowery paths of planetary description. This will be the best way of judging the sun among his peers.

We shall see its position further on, in the section of this volume which treats of the Stars; but we can even now give an idea of its mass and weight.

[1] If, as is probable, Neptune is not the last planet of the system (there is no reason why the limits of our sight should mark the limits of nature), the planet beyond it should be situated, according to all probability, at a distance of 48, and, in this case, its year is 333 times longer than ours. Revolving thus at 10,000 times the solar semi-diameter, its gravity towards the central star would be equal to 0·0000013 m.—that is to say, its curve would only differ from a straight line by 13 ten-thousands of a millimetre per second.

The comet of 1862 and the shooting stars which cross the terrestrial orbit on August 10 go out as far as that.

The bodies, whatever they may be, which float round the sun at these enormous distances move with a slower and slower velocity. While the earth moves in its narrow orbit at the rate of 18 miles per second, Neptune only moves at the rate of 3·3 miles. From the point of view of the practical velocities we are familiar with, such as those of railway trains, even that is an enormous velocity. We may easily find the velocities which correspond to greater and greater distances. The mean velocity of a planet in its orbit may be calculated by this very simple formula:—

$$x = V\sqrt{\frac{1}{D}}$$

in which V represents the mean velocity of the earth in metres per second, and D the distance in terms of the earth's distance from the sun.

It would be necessary to go out to 110,000,000 millions of miles to find the region in which a planet would travel in a circular orbit round the sun with the velocity of an express train (37 miles an hour); but if such a body travelled in the plane of the ecliptic, it could not accomplish its circuit round the sun on account of the disturbing influence of our neighbouring sun, Alpha Centauri, which would be, in certain regions, nearer to this orbit than the sun himself.

Alpha Centauri is a double star, of which we possess nearly two centuries of observations, and of which we can calculate the orbit ; the two components of this brilliant pair revolve round each other in eighty-one years. On the other hand, the mean distance which separates the two components is eighteen seconds. Now, as at this distance from the earth the radius of the terrestrial orbit is reduced to $0''\cdot75$, 18 seconds represent about 888 millions of leagues. Such is, then, the real distance which separates these two connected suns from each other. It is a little more than the distance which separates Uranus from the sun.

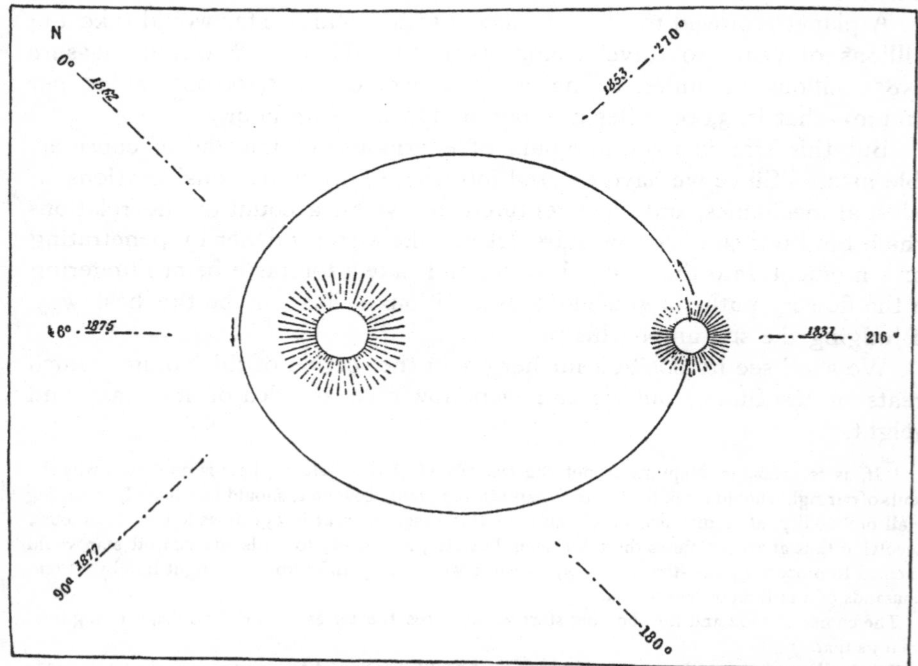

FIG. 136.—SYSTEM OF THE DOUBLE SUN ALPHA CENTAURI

As this separation cannot be measured, at such a distance, with absolute accuracy, we may without great error take for the basis of our conclusion the distance and the motion of Uranus. This planet takes precisely eighty-four years to perform its revolution ; then, according to the principles which we have explained (p. 239), the double sun Alpha Centauri turning round its centre of gravity in a period equal to that of Uranus, but its two components having between them a distance which is to the distance of Uranus from the sun in the ratio of 24 to 19, or of 126 to 100, the mass of this double sun is about twice as great as that of the sun which illuminates us.

It follows that the sun Alpha Centauri can*not* revolve round ours with the slowness which we have just now assigned to the fictitious planet which we supposed to obey our governor at this distance. This neighbouring sun exercises on ours a more powerful influence than we exercise on him. If, then, the double sun Alpha Centauri formed a system with ours, both would revolve round their common centre of gravity, situated in space at nearly a third of the distance between Alpha Centauri and the sun, in a period of 83 millions of years, if the orbit were circular.

If our sun and that of the Centaur existed alone in space and formed a system, it is thus that they would gravitate together. But it is not thus. The sun of the Centaur is carried along in space by a proper motion of 3″·67 per annum, which would cause him to make the circuit of the sky in 353,000 years, if this represented an orbital motion. We shall examine these interesting questions, however, when we take up the subject of the stars.

For the present, the important point for us is not to leave the sun without giving an account of his situation as a star, and estimating the relations which link his destiny to that of other similar centres scattered through Infinitude.

In analysing the motions of the earth we have already learned that the sun, the centre of our system, moves in space and carries us at present towards the constellation Hercules (p. 50). Is this orbit of the sun in space a closed curve? Does he himself revolve round a centre? Is this unknown centre fixed in its turn, or is it displaced from age to age, thus causing the sun and the whole planetary system to describe helices similar to those which we have found for the earth? Or, rather, does our central body, which is but a star, form part of a sidereal system, a cluster of stars animated by a common motion? Does there exist a *central sun of the universe?* Do the worlds of Infinitude gravitate as a hierarchy round a divine focus? The flight of the modern Urania's wings has not yet attained these transcendental heights. But it is certain that the sun in his course must be subject to sidereal influences, veritable perturbations which disturb his path and further complicate, in unknown ways, the motion of our little planet and that of all the others. Some day the astronomers of the planets which gravitate in the light of the suns of Hercules will see a little star appear in their sky: this will be our sun, carrying us along in its rays; perhaps at this very moment we are visible, dust of a sidereal hurricane, in a milky way, the transformer of our destinies. We are mere playthings in the immensity of Infinitude.

The proper motions with which all the stars are animated further show us that the suns of space move in all directions with considerable velocities. The analysis of their light has taught us that these distant suns are as hot and as luminous as that which illuminates us, surrounded, like him, with vaporous atmospheres, in which float molecules of the elements in combustion. The study of their masses and their motions leads us to the conclusion that these radiant foci are, like ours, the centres of planetary systems more or less analogous to that of which our abode forms part, and that in their fertile light gravitate also inhabited earths—worlds peopled like ours —planets, satellites, and comets. In the radiation of these other suns other existences throb. Some are still larger, more important, more powerful than our beautiful sun; others are different in brightness, in colour, and in character: here we see some which scintillate with an orange light; several are red like the ruby, and when they enter the field of the telescope they appear like a luminous drop of blood on the black velvet of the sky; there we contemplate the translucent brilliancy of the green emerald; here, the soft blue light of the sapphire. A large number are double, triple, multiple, so that the planets which surround them are illuminated by several suns of different colours. Some vary periodically in brightness; others are extinct, and have completely disappeared from the sky.

Our sun does not represent a privileged exception. We have already recognised this in considering our own world; we have already seen that he is destined himself to die out, like all the stars in succession; we have already foreseen what will happen to our globe and to all the other planets of the system. But we have stopped at an end which cannot be general, which can only be particular, and which does not satisfy logic.

We left the earth frozen and depopulated by the cold, the last human family lulled in its final sleep, the sun progressively obscured by the formation round him of a solid crust, the planetary system henceforth entirely deprived of the light and heat which had enabled it to live during many ages; and we left the sun, an enormous black ball, continuing its course in space, carrying with him his planets, dark deserts, wandering tombs, still gravitating round him in the eternal night. What will become of these worlds? Matter, like force, being indestructible, will it continue to gravitate eternally in space in the state of cosmic skeletons? To solve this question we are obliged to leave the domain of pure science, and to enter that of hypothesis. But, even here, let us try not to forget the rigorous principles of the method of scientific induction.

If such were the definitive end of worlds, if worlds were always dying, if suns once extinguished were not again rekindled, there would be now no stars in the sky.

And why? Because creation is so ancient that we may consider it as eternal in the past. Since the epoch of their formation the innumerable suns of space have had ample time to die out. Compared to a past eternity, they are but new suns which are now shining. The earlier stars are extinct. The idea of succession, then, forces itself on our mind.

Whatever may be the familiar belief that each of us has acquired on the nature of the universe, it is impossible to admit the ancient theory of a creation made once for all. Is not the very idea of God synonymous with the idea of a creator? As long as God exists, He creates; if He had created but once, there would be no more suns in immensity, nor planets drawing from them light, heat, electricity, and life. Creation must of necessity be perpetual. And *if God did not exist*,[1] the antiquity, the eternity of the universe, would assert itself with still greater force.

Let us, however, directly interrogate nature, and listen to her reply.

What passes around us? The same molecules of matter enter successively into the composition of different bodies. The bodies change, matter remains. In a short period our own body is almost entirely renewed. A perpetual exchange is effected between the air, water, minerals, plants, animals, and ourselves. The atom of carbon which at present burns in our lungs perhaps burned also in the candle which served Newton for his optical experiments; and perhaps you have at this moment in your hand atoms which belonged to the arm of Cleopatra or to the head of Charlemagne. The molecule of iron is the same whether it circulates in the blood which flows in the temple of an illustrious man, or whether it lies in a bit of rusty old iron. The molecule of water is the same whether it glitters in the loving glance of the *fiancée*, or intercepts the rays of the sun in a monotonous cloud, or falls in a stormy shower on the inundated earth. There is incessant exchange during life, and exchange not less rapid after the death of organisms. When war has sown its victims in the fields, life seems to rush in new waves to fill up the voids; on the carriage of the dismounted cannon, in spite of man himself, flowers bloom and birds sing: nature always asserts its rights. The matter of beings does not remain motionless, but returns into the circulation of life. What we breathe, eat, and drink has already been breathed, eaten, and drunk thousands of times. We are formed of the dust of our ancestors.

This is what passes around us. Now, there is neither great nor little in nature. The stars are the atoms of the infinite. The laws which govern the atoms also govern the worlds.

The same quantity of matter always exists. After being used in forming nebulæ, suns, planets, and beings, it does not remain inactive, but returns into a new circulation; otherwise, the world would end; otherwise, the day would come when all the worlds would be dead, swallowed up in the night

[1] This is not the place to enter into any discussion on this question, which is one of pure philosophy, and not of positive science. See our work, *Dieu dans la Nature*, or Spiritualism and Materialism confronted with Modern Science.

rolling, falling aimlessly into the black desert of space, the eternal solitude which no ray of light would ever illumine. There is here a perspective which gives no satisfaction to the most elementary logic.

But by what natural process can dead worlds again become alive? When our sun becomes extinct (and there is no doubt that he will become so in the future), how will he return into the circulation of the universal life?

The study of the constitution of the universe, which has only just begun, permits us already to give two replies to this question, and it is very probable that nature, which yields its secrets with such difficulty, may hold others still better in reserve for the science of future ages.

Two dead globes may revive and recommence a new era by uniting in virtue of the simple laws of gravitation.[1]

At the time, then, when our sun shall be extinct, and shall roll, a dark globe, through space, it will be able, like a new phœnix, to rise from its ashes

[1] Let us suppose, to fix our ideas, that a dark globe, like our earth, or even as large as the sun—it is immaterial—is shot into the void. It carries with it its living force, and, if it is alone in space, it will continue to travel in a straight line, always with the same velocity, without either slackening, accelerating, or turning aside one iota from its path, and *it would go on thus eternally*; the force which animates it would be always used by it in passing over the same number of miles per hour. But suppose, now, that exactly below, at the end towards which it travels, in a direction diametrically opposite, is found a second ball of the *same* mass, which we shoot towards the first with the *same* velocity: when they come together they will strike normally, and will be entirely stopped. What becomes of the force which animates them, since nothing is lost in nature? It will be transformed. The motion visible up to that time will become invisible motion, exactly of the same intensity as the first, which sets in vibration the constituent molecules of the two masses, separates them from each other, and the two cold and dark globes will produce a burning and dazzling sun. *Nothing is lost, nothing is created.*

The hypothesis we have just stated would realise itself, without our being obliged to shoot two globes against each other, by simply placing them in space at some distance from each other. In virtue of the laws of gravitation they would proceed slowly towards each other, and would fatally meet and be united in a dazzling collision, which would transform them into a sun or a nebula. Let us suppose, for example, that our sun and the sun Sirius were the only suns existing in the infinite, that the parallax of Sirius is half a second, and that they have the same mass, and are motionless. In virtue of the laws of gravitation they feel each other across space and attract each other: hardly are they placed in the void before they tend to approach each other. The fall at first is infinitesimal. During the first day they will fall towards each other by a small fraction of an inch. This fall is imperceptible. But the motion goes on accelerating. At the end of a year the motion is already perceptible. They are proceeding towards each other like our two balls. And after *thirty-three millions of years* of incessant fall they are precipitated on each other with such a velocity that they are joined, united, melted, and are evaporated into a single immense and brilliant nebula.

The principles of thermodynamics demonstrate that an aërolite which comes from the infinite depths of the heavens, being precipitated on the sun with the unheard-of velocity of 377 miles during the first second of its fall, the transformation of its motion produces a heat more than nine thousand times greater than would be generated by the combustion of a mass of coal equal to this aërolite. Whether the aërolite be combustible or not, the combustion would add almost nothing to the terrible heat generated by its mechanical collision.

We have seen above that if the earth fell into the sun it would augment the solar heat by a quantity sufficient to sustain the solar emission for 95 years, and that the total heat of gravitation produced by the fall of all the planets on the sun would maintain the emission during 45,600 years.

by meeting with another extinct sun, and thus light again the torch of life for new worlds, which the laws of gravitation will detach from the nebula thus formed, as our present earth and her sisters have been detached from the nebula to which we belong. At this moment our sun travels with a great velocity towards the stars of the constellation Hercules. Each star is animated with a proper motion which carries it with its system through Immensity. Several of these motions are rectilinear. There is, then, nothing impossible in two stars meeting in space, and perhaps we have here the secret of the resurrection of worlds.

Perhaps it enters into the general destinies of the universe that the sun is guided precisely towards an end which he will only attain after his death, and perhaps there is here the final cause of the proper motion of all the suns in space. But we can at the same time imagine a second process of destruction and resurrection, of which the aërolites, the shooting stars, and the comets would be evidence.[1]

Like the eagle which rises higher and higher in the upper regions where the atmosphere itself loses its density, so we sail ourselves overlooking the mysterious horizons of the future. If the earth exists a sufficient number of ages, it is possible that it may itself fall into the sun. 'Created simply,' says Tyndall, 'by the difference of position in the masses which attract it, the potential energy of gravitation was the original form of all the energy in the universe. As surely as the weights of a clock descend to their lowest position, from which they can never reascend, unless a new energy is communicated to them, in the

[1] Many of us have seen the stones or uranoliths which fall from the sky. Our museums possess them of all dimensions, from a few ounces up to several thousands of pounds in weight. Figs. 137, 138 give an idea of them: we will return to this further on in the Book on Comets.

How can a world be broken into fragments in this manner? We do not know, and the fact appears even contrary to the laws of gravitation. But what is gravitation itself in its essence? We are still ignorant. Is this force of gravitation absolute? Can bodies reach a physical or chemical state in which gravitation loses its rights? Well, let us suppose for an instant that on account of secular cooling, of solidification, of dryness, our globe happens one day to split up, and that later these constituent materials cease to obey the force of aggregation which keeps them united, our globe, stony to its centre, would be then formed of materials simply placed by side, which would be no longer retained by any central force, like the corpse which, abandoned to the work of destruction, leaves to each of the molecules which compose it the power of quitting it for ever, to henceforth obey new influences. What will happen to this dead world, to this corpse of a world? The attraction of the moon, if it still exists, will itself demolish it by producing a tide of morsels of earth in place of a liquid tide. Adding to this other planetary perturbations, we see in some centuries our poor disintegrated globe losing its spheroidal form, and becoming scattered imperceptibly along its orbit. See the planetary system in pieces! All to fall pell-mell into the sun! And if such is also the final destiny of the sun, see this black body itself disintegrated, and all the constituent particles of the solar system carried along in space and destined to be disseminated through the fields of the sky. Dust of worlds, they will float in the void till the day when, arriving in the regions of a new resurrection, they will be thrown back into the crucibles of creation, attracted by a fruitful centre, and all similar cosmical dust will reunite towards this same centre to form by their universal fall a new focus of incandescence and creation.

same way, as the ages succeed each other, the planets must fall in turn on the sun, and produce several thousand times as much heat as would be produced by burning masses of coal of the same dimensions. Whatever may be the definitive fate of this theory, it establishes the conditions which would certainly produce a sun, and shows in the force of gravity acting on dark matter the source from which all the stars might arise.'

The mathematician and physiologist, Helmholtz, supposing that the nebulous matter from which the solar system has been formed—according to the theory of Kant and Laplace—may have been in the first instance of extreme tenuity, has determined the quantity of heat which should be generated by the condensation to which we owe the existence of the sun, the earth, and the planets. Taking the specific heat of water for that of the condensing mass, the elevation of temperature produced by the mechanical

Fig. 137.—Meteorite found in Chili in 1860

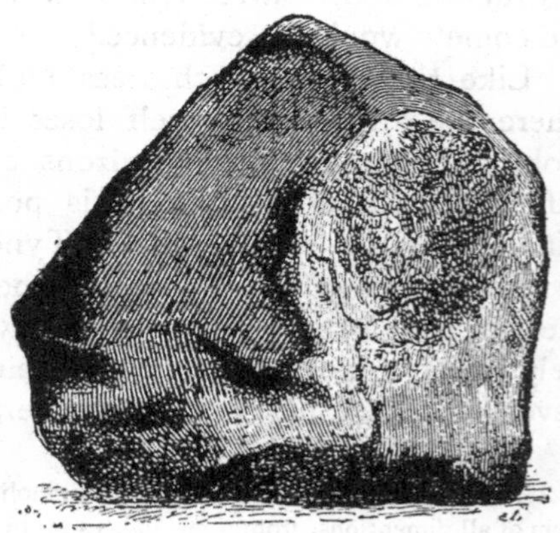

Fig. 138.—Meteorite which fell at Orgueil, May 14, 1864

formation of the sun would have been 28 millions of degrees! The subsequent condensation of cosmical dust disseminated in space would, then, amply suffice for the creation of new worlds.

We may, then, be quite certain that nature holds in reserve causes of resurrection, as it also holds in its hands the causes of destruction. For her, time is nothing. An action which requires a hundred thousand years to accomplish is as clearly determined and planned as an action which requires only a minute. Absolutely speaking, eternity alone exists, and time is but a relative form. As to our human personalities, and their immortality or resurrection, it would be of the highest interest for us to know the essence of the soul. Each of the constituent atoms of our bodies is indestructible, and incessantly travels from one incorporation to another. Logic leads us to think that our vital force, our psychic monad, our individual self, is equally indestructible—and more justly. But in what conditions does it exist? Under what forms is it re-incarnated? What were we before birth, and what shall we become after death? Astronomy gives us the first reply, worthy of the majesty of nature, and in intimate correspondence with our innate aspirations. But this reply cannot be merely the corollary of a psychological solution. Let the philosophers imitate the astronomers! Let them work at facts instead of speculating on words, and one day the veil of Isis shall be entirely raised for our souls, which so eagerly long for the Truth. Positive science, science alone will reply: *Life is universal and eternal.*

CHAPTER VII

LIGHT

Its Nature. Its Velocity. Spectrum Analysis. The Chemical Composition of the Sun and Celestial Bodies

THERE are in science few subjects so obscure as that of which we have just written the title. What is the essential nature of *light*? How do we see the universe? How does a luminous body radiate, and by what vehicle do its rays reach our eyes? What are even these rays? Man has discussed this great problem for thousands of years. The ancients believed that the rays might be shot forth from our eyes to lay hold of objects far away; Newton thought, on the contrary, that objects emitted luminous particles which pass through space and strike our retina; Young and Fresnel have since shown that luminous bodies do not emit any material particle, but cause the surrounding fluid to vibrate, as a bell makes the air vibrate. This has led us to imagine as indispensable to the propagation of light a certain fluid named *ether*, which is extremely light, and disseminated through the whole of space. It is to Young that the honour belongs of having stemmed the flood of authority which, since Newton, had opposed the progress of optics, and of having established this theory on a basis which now appears definitely assured.

Just as we see the circular waves of a piece of water succeed each other round the point where the water has been struck, as air condenses and dilates in spherical waves round the resounding tuning-fork, so the ethereal fluid which fills space gives birth to a series of spherical waves, succeeding each other all round a luminous body. The waves of water are transmitted so slowly that the eye easily follows their motion; those of air fly with the velocity of 1,100 feet per second, varying with the temperature and the density of the atmosphere; those of the ether pass through immensity with the dizzy velocity of 186,000 miles per second. The most marvellous fact is that every star, every sun in space, is the centre of constant undulations, which thus *perpetually cross each other through immensity, without ever being confused* or mutually mingled. I confess, for my part, that this fact appears to me absolutely incomprehensible.

The velocity of light has been approximately known for more than two centuries. The following is the first notification which nature gave to the human mind. The planet Jupiter sails round the sun accompanied by five satellites, which pass from time to time through the shadow which the planet forms behind it, as the moon does for us. These eclipses of Jupiter's satellites are convenient for calculating longitudes at sea, and from the time of Louis XIV. tables of their occurrence have been constructed in order to attentively observe them. But observers were not long in remarking that they did not return regularly:[1] sometimes they would be in advance of the hour indicated by calculation, and sometimes they would be delayed. The tables were corrected, but without obtaining greater precision. However, the motions of the satellites of Jupiter are regular, and these advances and delays could be in appearance only.

FIG. 139.—ECLIPSE OF ONE OF JUPITER'S SATELLITES

The classical astronomers, Cassini, Fontenelle, and Hook, searched in vain for an explanation, refusing to admit that light, of which the propagation had always been considered as instantaneous, took a certain time to come from Jupiter to the earth; but a student of nature, Olaüs Roëmer, a young Dane, then at the Observatory of Paris, set to work to discuss freely all the observations, and clearly proved (in 1675) that the eclipses are seen later when the earth is farther from Jupiter, and sooner when it is nearest, with a difference which appeared to him to rise to 22

[1] When the earth is at A (fig. 139) we see the eclipses happen sooner; when it is at B we see them happen later, by the difference of time which light takes to traverse the diameter of the terrestrial orbit. Practically, the delay increases progressively from the point A to the point B, but we do not observe the eclipses up to this last point, since the sun is then placed between Jupiter and us; we take account of this difference in making the calculation.

The experiment of Roëmer has been repeated, verified, perfected, and for a long time past predictions of these eclipses have been made by taking account of the variation of the distance from the earth to Jupiter.

minutes for the entire diameter of the terrestrial orbit; he naturally concluded that the difference proceeds simply from the distance, the light taking so much longer to come as the distance is greater.

The fact of the progressive propagation of light was confirmed in 1727 by the English astronomer Bradley, in his discovery of *aberration*, or the annual apparent motion of the stars (which we have already explained, p. 63, in the proofs of the motion of translation of the earth round the sun). This motion, which has a variation of 40½ seconds, shows that the velocity of light is 10,000 times greater than that of the earth, and that the light of the sun takes 8 minutes 13 seconds to cross the space which separates us from that body. This measure was more precise than the first.

Without making use of celestial phenomena, M. Fizeau, by the aid of a light, a telescope, a mirror, and an ingenious apparatus, measured, in 1849, this velocity between two terrestrial stations distant from each other 8,633 metres (about 5⅓ miles) only (Montmartre and Suresnes), and found it 195,000 miles per second.

New experiments, made by Foucault in 1850, and repeated in 1862, give for this velocity 185,000 miles.

Renewed in 1874 by M. Cornu, and carried out between the Observatory and the tower of Montlhéry, the experiments have given 186,660 miles. This latter value is the most certain.

Taking up the problem in another way, we find, then, that since there are 93 millions of miles from the sun to the earth, the luminous ray crosses this distance in 493 seconds, which makes precisely 8 minutes 13 seconds.

In 1882 Prof. Newcomb found by new experiments made at the Observatory of Washington 299,860 kilometres, and Mr. Michelson 299,853.

We may, then, accept, in round numbers, for this velocity 300,000 kilometres per second [186,414 miles. From a discussion of all the observations, in 1891, Professor Harkness found 186,337·00±49·722 miles. —J. E. G.].

Thus, when we see an eruption shoot out from the solar limb, eight minutes have elapsed since the event happened; when we see a satellite of Jupiter lose its light, it is at least thirty-four minutes since the eclipse took place; when we observe Neptune, we see it as it was four hours previously; when we look at a star, we see it, not as it is, but as it was at the moment the luminous ray left it—that is to say, four years ago with reference to the nearest, and ten years, twenty years, fifty years, one hundred, a thousand, ten thousand years, according to the distance. Likewise, a transcendent eye placed at these successive distances would now see the earth as it was four years, ten years, a hundred years, a thousand years ago, according to the distance. Light makes the past an eternal present.

Such is the progressive transmission of light. But how shall we represent the action of the sun in the production of this light?

Let us remark, first, that the radiant star sends us heat at the same time as light, and that very often the two species of rays are mixed up. Everyday experience shows us also that heat raised to a certain degree becomes light; on the other hand, we know that heat is nothing else but a mode of motion: *it is the motion of the molecules in rapid vibration which is felt as heat.*[1] Light is likewise but a vibration.

There is no solid matter properly so called, and this is a fact not less worthy of attention than that of astronomical magnitudes and motions. In the densest mineral, in a piece of iron, of steel, of platinum, the molecules do not touch. Cohesion, which is the attraction of the atoms, maintains them; but heat increases their distance from each other, more or less, by animating them with a vibratory motion; if this heat is sufficient the cohesion loses its power, the *solid* state disappears, and the molecules glide upon each other: this is the *liquid* state. If the heat is raised higher—that is to say, if the vibratory molecular motion is more violent—the molecules even escape altogether from cohesion, and the body becomes *vapour* or *gas*. Thus, there is no solid matter, and the heat motion makes bodies pass through the three states. It is assuredly strange to think that our own body is not more solid than the rest, but formed of molecules which do not touch and are in perpetual motion. Perhaps even the constituent atoms of bodies rotate on themselves and round each other. If you had sufficiently good sight to see exactly the materials which compose your body, you would see it no longer, because your sight would pass through it. And how small are these constituent parts! The red globules which colour the human blood have the form of microscopic lenses measuring only the hundred and thirtieth of a millimetre in diameter: it would be necessary to place 130 of these little bodies end to end to form a length

[1] Let us strike a piece of iron. The muscular motion of the arm is transmitted to the molecules of the iron in a state of invisible motion, and it is this invisible motion which we call heat. Friction produces heat, and this was the first source of fire among the ancients. Thermodynamics has estimated the mechanical equivalent of heat, and we now know that the heat necessary to raise one pound of water 1° in temperature is equivalent to a mechanical force capable of raising 772 lbs. 1 foot in height, and conversely. *Heat is a mode of motion.* A ball of lead of 1 lb. falling from 772 feet of height arrives with a velocity of 222 feet per second, and, as its calorific capacity is one-thirtieth of that of water, its collision with the ground would raise its temperature by 30° if the soil itself was not heated by the fall. Such a ball shot with a velocity five times greater, or 1,110 feet, would attain a degree of heat twenty-five times higher, or 750 degrees, in striking a target which could not be heated. That is to say, that if a supreme will were to stop at once this ball thus shot out in space, it would melt on the spot, and would flow like water. If the earth were thus suddenly stopped in its course, it would not only be melted by the transformation of motion into heat, but even reduced almost entirely to vapour.

of a millimetre. A drop of blood of a cubic millimetre contains about 5 millions of globules, a litre of normal blood contains 5,000 millions, and there flow in our arteries and our veins twenty-five to thirty thousand millions of these little organic bodies.

Let them become either reduced or multiplied, and we are dead! Let them coagulate, or become cooled or heated, and we are dead! Let them stop, and we are lost! At each throb of our heart a violent and rapid impulsion projects the blood to the extremities of the members; 100,000 times a day, 36 millions of times a year, the same pulsation recommences, until the day when the fatigued muscle stops and compels us to lull ourselves profoundly in the last sleep.

The constituent molecules of bodies do not touch. It is thus, and thus only, that the expansion and the change of state of bodies under the influence of heat can be explained. We do not doubt the energy of the atomic forces in action around us. Let us heat 1 lb. of iron from 0 to 100 degrees, it will expand about $\frac{1}{800}$, a span imperceptible to the eye, and yet the force which has produced this expansion would be capable of lifting 12,000 lbs., and raising them to the height of one yard. The power of gravitation almost vanishes in comparison with these molecular forces; the attraction exercised by the earth on the weight of half a kilogramme (about a pound) taken in a mass is nothing compared to the mutual attraction of its own molecules. In the combination of 1 lb. of hydrogen with 8 lbs. of oxygen to form water, work is performed capable of raising by 1 degree the temperature of 34,000 lbs. of water; or of lifting 15,000,000 lbs. to 1 yard high! These nine pounds of water in being formed have fallen molecularly down a precipice equal to that which would be passed over by a ton of 1,000 kilogrammes rolling down to 46,000 feet of depth!

When a bar of iron is heated and becomes sufficiently hot to be luminous, it sets the ether in vibration with the unheard-of velocity of 450 billions of undulations per second. The length of the wave of the extreme red is such that it would require 38,000 placed after each other to form a length of 1 inch. As light travels 300,000 kilometres per second, or 30,000,000 centimetres, multiplying this number by 15,000 we obtain the number given above. *All these waves* (450,000,000,000,000) *enter the eye in one second!*[1]

Let us receive a ray of light on a lens in order to produce a very pure pencil, then on a prism (a triangular piece of glass): in passing through the

[1] What comes from the sun and from all sources of light and heat is not, then, to speak accurately, either light or heat (for these are merely impressions) but *motion*—motion extremely rapid. It is not heat which is scattered through space, for the temperature of space is, and remains everywhere, glacial. It is not light, for space has constantly the darkness seen at midnight. It is motion, a rapid vibration of the ether which is transmitted to infinity, and does not produce a perceptible effect until it meets with an obstacle which transforms it.

prism this luminous ray is refracted, and in passing out, instead of forming a white point, it forms a ribbon coloured with the tints of the rainbow. In making this experiment, Newton proved that the white light gave birth to all these colours. These are arranged in the following well-known order :—*Violet, Indigo, Blue, Green, Yellow, Orange, Red.*

The colours are separated, each according to its character: the most intense, the red, does not allow itself to be turned aside from its path, and passes in a straight line; the orange submits a little to the influence of the prism, and is placed to one side; the yellow submits still more; the green, then the blue, are still milder and weaker, and continue the ribbon. It is this coloured streamer which bears the name of the *solar spectrum.* In reality there are not *seven* colours; there are an unlimited number. In the time of Newton the number VII. was still sacred.

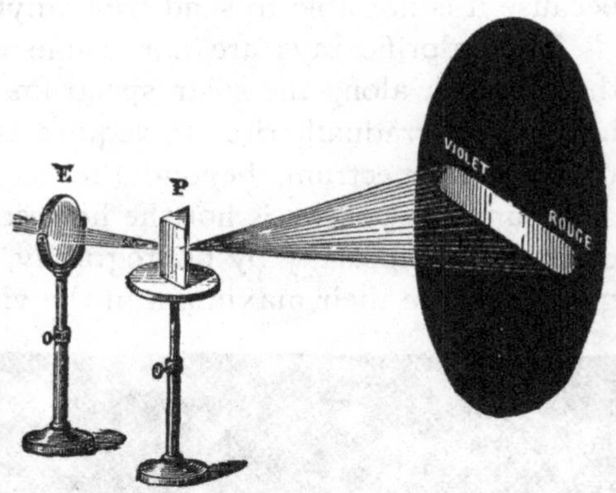

FIG. 140.—PASSAGE OF LIGHT THROUGH A PRISM

The length of the spectrum only represents the light—that is to say, the solar rays—perceptible by our retina. Our eye begins to see when the ethereal vibrations reach the number of 450 billions, and stops seeing when they exceed 700 billions (purple-violet); but beyond these limits nature still acts—unknown to us. Certain chemical substances—the photographic plate, for example—see farther than we do, beyond the violet; these are *invisible rays* for our eyes.

Our ear perceives aërial vibrations from 32 vibrations per second (low tones) up to 36,000 (high tones); beyond this we hear nothing. Thus our senses are limited, but not the facts of nature. The colours are, like the notes of the gamut, effects of number; in music, as in painting, there are *tones.*

It is the molecular arrangement of reflecting or transparent substances which gives rise to the different reflections of light—that is to say, the colours. A slight difference produces here a blue eye, pensive and thoughtful, there a brown eye with half-hidden flames, there a look dull and distasteful. The dazzling rose which blooms in the flower-garden receives the same light as the lily, the buttercup, the cornflower, or the violet; molecular reflection produces all the difference; and we might even say, without metaphor, that objects are of all colours *except that which they appear.* Why is the meadow green? Because it keeps all except the green, which it does not want, and sends back. White is formed by the reflective nature of an object which keeps nothing and returns all ; black, by a surface which keeps all and sends back nothing. Project the solar spectrum on black velvet : it is absolutely extinct ; place a band of red velvet in the blue part of the spectrum : it becomes black, because it is not able to send back anything but red,[1] &c.

The calorific rays are not visible to us. If we move the bulb of a thermometer along the solar spectrum, we find that the heat begins at the indigo, and gradually rises, to acquire its maximum intensity *near the end* of the visible spectrum, beyond the red. The most luminous part of the spectrum, the yellow, is not the hottest. On the other hand, we ascertain chemically, especially by photography, that the chemical rays begin in the green, acquire their maximum in the violet, and extend beyond it, forming also an invisible spectrum. Fig. 141 represents the relation which exists between the three species of rays. The luminous rays extend from the red to the violet (from the left of the line A to the right of the line H), and their luminous intensity is represented by the curve L, of which the maximum occurs, as we see, between the rays D and E. The curve to the left, G, represents the calorific intensity ; and the right curve, *Ch*, corresponds to the chemical action. A sixth sense is opened to the world by the calorific rays, a seventh by the chemical rays. What we *see* is nothing compared with what is constantly passing around us in nature.

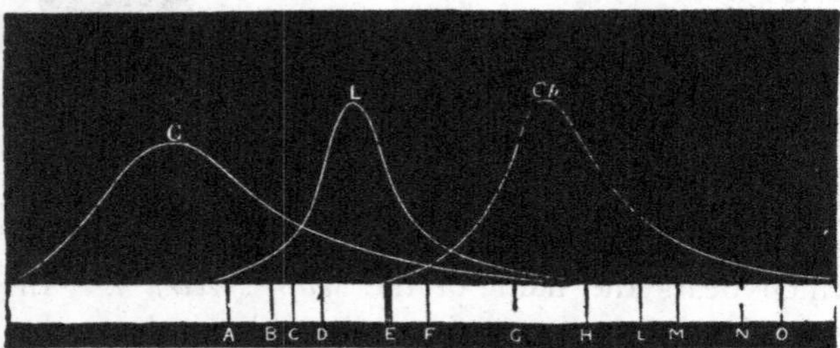

FIG. 141.—RELATIVE INTENSITIES OF HEAT, OF LIGHT, AND OF CHEMICAL ACTION IN THE RAYS WHICH REACH US FROM THE SUN (See footnote on next page)

As early as 1815 Fraünhofer, a Bavarian optician, studied with care the solar spectrum, and sought to discover some fixed points in it which might be independent of the nature of the prisms, and which could be regarded as points of reference to which the zones and colours of the spectrum might be referred ; when he perceived that, by giving the prism a certain special position, there suddenly appeared in the spectral image *dark lines* crossing the streamer transversely in the same colours. He designated the eight principal lines by the first letters of the alphabet. They are placed as follows : the first at the limit of the red, the second in

[1] With reference to this, I have noticed in my course of lectures a rather singular fact. A white ray which passes through a plate of yellow glass is projected in yellow, and a ray which traverses a plate of blue glass is projected in blue ; projecting these two colours on each other on a screen we obtain pure *white*, because these two colours are complementary. But if we place the *same* plates of yellow and blue glass in a single apparatus we obtain green.

the middle of that colour, the third near the orange, the fourth at the end of that tint, the fifth in the green, the sixth in the blue, the seventh in the indigo, the eighth at the end of the violet.[1] These are, then, the principal black lines which we distinguish in the spectrum. As to the total number of these lines, they are really amazing. Fraünhofer counted 600 with a microscope; later, Brewster carried this number to 2,000; now we count 5,000 and more (see fig. 142).

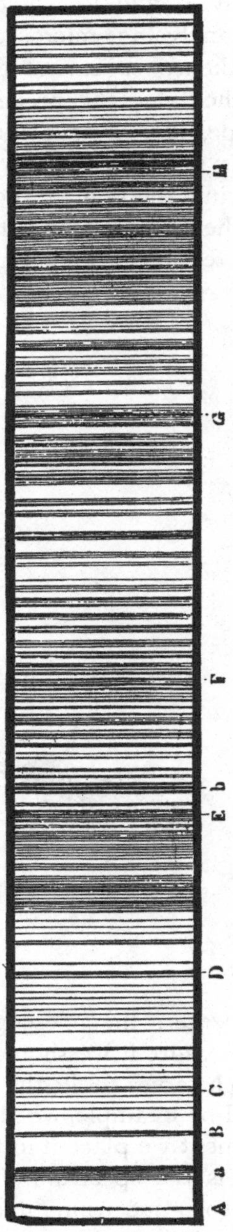

These lines of the solar spectrum are constant and invariable at all times when the spectrum studied is that of light emanating from the sun, whatever this light may be. We find them in day light, in that from the clouds, in the light reflected by mountains, buildings, and all terrestrial objects. We find them even in the light of the moon and in that of the planets, because these celestial bodies only shine by the light which they receive from the sun and reflect into space.

This discovery of microscopical lines which thus cross the solar spectrum was soon made fruitful by another not less important discovery. Admitting through a prism rays issuing from a luminous terrestrial source, such as a gas-jet, a lamp, a metal in fusion, &c., we notice at first that these artificial lights give rise to a spectrum as well as that of the sun, but that this spectrum differs from the solar spectrum by the number and arrangement of the colours; we remark in the second place—and here is the important point—that the spectrum of these lights is also crossed by lines, that the distribution of these lines differs according to the nature of the light observed, and, in short, that they *present an invariable order characteristic* of each of them.

Fig. 142.—Principal Lines of the Solar Spectrum

In order to fix our ideas, let us describe an experiment such as was made by Kirchhoff and Bunsen, the two physicists to whom we owe these brilliant researches. Let us place in a gas-jet a platinum wire, at the extremity of which we put a small fragment of the substance which we wish to analyse. Before the

[1] [These lines are universally known by the letters given in figs. 141, 142.]

flame is placed the *spectroscope*, a telescope expressly constructed for our analysis, and in which the rays from the flame pass through a prism and an analysing microscope. (The flame of our gas-jet is regulated and weakened so as not to give a spectrum itself.) The moment we place in the flame the prepared platinum wire a spectrum appears in the telescope, and an eye placed at the microscope can analyse it at its ease. This spectrum *is that of the substance which burns.* The luminous ray leaving the point L (fig. 143) is reflected from the little prism *o* at the end of the telescope, and thus appears to come from L′. Following the axis of the telescope, it is refracted successively through six prisms, A, B, G, D, E, H,

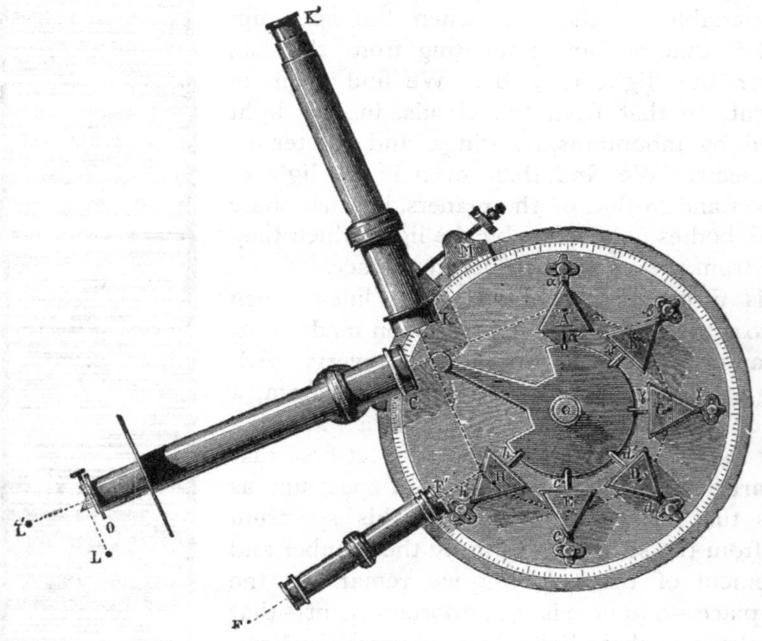

FIG. 143.—SPECTROSCOPE

and enters the telescope K, by which it is observed. In order to compare or measure it we should have in the little telescope F an image or a scale which serves to fix the positions of the rays.

For example, we dip the platinum wire in a bottle of potash. The moment we place it in the gas-jet a spectrum appears in the spectroscope; this is the spectrum of potassium. It is composed of seven colours, like the solar spectrum; in addition, it is characterised by two very brilliant red rays, situated towards each of the extremities.

Similarly, if we place small crystals of soda at the point of the platinum wire, we see a singular spectrum appear, which contains neither red, nor orange, nor green, nor blue, nor violet, and which is simply characterised by a splendid yellow ray corresponding to the position of the yellow in the solar spectrum and of the line which crosses that colour. We have here the spectrum of sodium. And so on.

This method of analysis is so marvellous and powerful that it reveals the existence of substances in quantities infinitely small, and where any other method would be completely abortive. The presence of *a millionth of a milligramme* of sodium discloses itself in the flame of a candle.

Thus, every substance analysed produces in the spectroscope an arrangement of lines which is peculiar to it; *it registers its true natural*

FIG. 144.—COMPARATIVE FIGURE OF SPECTRA

name in hieroglyphic characters; it reveals itself by itself and under an incontestable form.

The black lines which we have described above in the solar spectrum *correspond precisely to certain bright lines characteristic of the spectrum of different terrestrial substances.*

On the other hand, it has been ascertained that metallic vapours endowed with the property of emitting in abundance certain coloured rays absorb these same rays when they come from a luminous source situated behind these vapours and traversed by them. Thus, for example, if behind a flame in which sea-salt burns we kindle a brilliant Drummond light, and if we superpose the two spectra, immediately the yellow line of sodium will disappear from the spectrum of sodium, and give place to a dark line occupying precisely the same place.

It follows from this double observation that the black lines of the solar spectrum prove: (1) the existence of a burning and gaseous atmosphere round that body, and (2) the presence in that atmosphere of substances announced by the lines in question.

There have been identified, line for line, in the sun the 460 lines of the spectrum of iron,[1] the 118 of titanium, 75 of calcium, 57 of manganese, 33 of nickel, &c.; so that we now know certainly that there are at the surface of that dazzling star, and in the gaseous state, iron, titanium, calcium, manganese, nickel, cobalt, chromium, sodium, barium, magnesium, copper, potassium; but we still cannot recognise any trace of gold, silver, antimony, arsenic, or mercury. Hydrogen was discovered in 1868. Oxygen must exist in this furnace, but the oxygen lines which have been found in the solar spectrum proceed from our own atmosphere (Janssen, 1888).

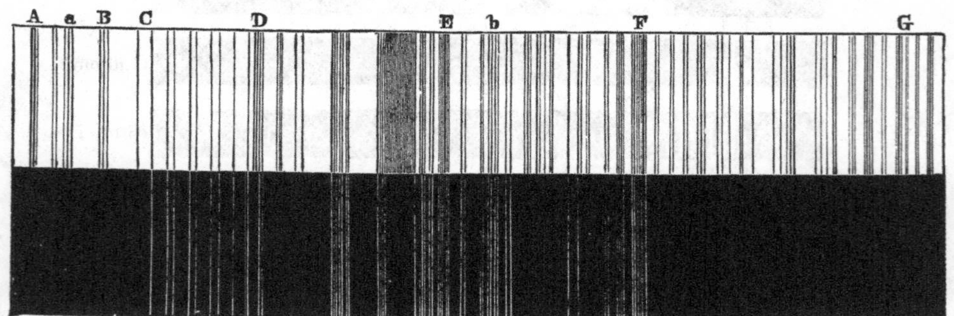

Fig. 145.—Coincidence of the Lines of Iron with those of the Solar Spectrum

We shall take up further on the applications oi spectrum analysis to the chemical knowledge of the *planets*, the *comets*, and the *stars*. It was important to give an account here of this fruitful method, and of the modern study of light.

We see that the varied horizons discovered from the height of the elevated paths which the study of astronomy has led us to follow are not less interesting than astronomy itself. The attraction, almost universal, which draws the human mind towards the most abstruse and less usual results of science is, perhaps, the most singular trait of that restless curiosity which has been given to us in order that we may observe and know. Pythagoras was asked what was the characteristic mark of man. He replied: '*The knowledge of truth for truth's sake.*' Is it not remarkable to see the human species, living on the productions of the fostering earth,

[1] [Professor Rowland has found over 2,000 lines of iron in the solar spectrum.—J. E. G.]

according to the expression of Homer, applying itself in preference to purely intellectual sciences, and giving to them the greatest part of its attention, to the exclusion of those which have for their object health, feeding, material welfare, and, in short, all the arts without which the powerful organisation of modern society cannot subsist? We feel a more lively and profound interest in studying astronomical conquests—as the distance of the stars, the nature of the sun, the planetary humanities, the destinies which await us in infinity and eternity - than in a new route opened to commerce, a new sort of eatable, or a chemical discovery which may afterwards disturb numerous interests. Thus, of the three elements which form the essence of man—his wants, his affections, and his intelligence—it is the last-named faculty which obtains the preference. It is an advantage, especially to the young, to comprehend in their totality truths the possession of which does honour to the human mind. It is thus that we learn to rise above the petty interests of life towards the higher regions to which the divine patriotism of the soul aspires.

[In his 'Handbook of Astronomy,' Dr. Lardner says: 'Nature has raised the curtain of futurity and displayed before the astronomer the succession of her decrees, so far as they affect the physical universe, for countless ages to come; and the revelations of which she has made him the instrument are supported and verified by a never-ceasing train of predictions fulfilled. He "shows us the things which will be hereafter," not obscurely shadowed out in figures and parables, as must necessarily be the case with other revelations, but attended with the most minute precision of time, place, and circumstance. He counts the hours as they roll into an ever-present miracle, in attestation of those laws which his Creator through him has unfolded; the sun cannot rise, the moon cannot wane, a star cannot twinkle in the firmament, without bearing witness to the truth of the prophetic records. It has pleased the "Lord and Governor" of the world, in His inscrutable wisdom, to baffle our inquiries into the nature and the proximate cause of that wonderful faculty of intellect—that image of His own essence which He has conferred upon us; nay, the springs and wheelwork of animal and vegetable vitality are concealed from our view by an impenetrable veil, and the pride of philosophy is humbled by the spectacle of the physiologist bending in fruitless ardour over the dissection of the human brain, and peering in equally unproductive inquiry over the gambols of an animalcule. But how nobly is the darkness which envelops metaphysical inquiries compensated by the flood of light which is shed upon the physical creation! *There* all is harmony, and order, and majesty, and beauty. From the chaos of social and political phenomena exhibited in human records—phenomena unconnected, to our imperfect vision, by any discoverable law, a war of passions and prejudices, governed by no apparent purpose, tending to no apparent end, and setting all intelligible order at defiance—how soothing, and yet how elevating, it is to turn to the splendid spectacle which offers itself to the habitual contemplation of the astronomer! How favourable to the development of all the best and highest feelings of the soul are such objects! the only passion they inspire being the love of truth, and the chiefest pleasure of their votaries arising from excursions through the imposing scenery of the universe—scenery on a scale of grandeur and magnificence compared with which whatever we are accustomed to call sublimity on our planet dwindles into ridiculous insignificancy. Most justly has it been said that nature has implanted in our bosoms a craving after the discovery of truth, and assuredly that glorious instinct is never more irresistibly awakened than when our notice is directed to what is going on in the heavens.']

BOOK IV

THE PLANETARY WORLDS

CHAPTER I

THE APPARENT AND THE REAL MOTIONS

Systems successively imagined

IN order to easily and accurately understand the true arrangement of the system of the world, the surest method is to take the path which the human mind has itself followed in its ascent to the knowledge of the truth. We do not see the universe as we are obliged to represent it in our drawings. Consider, for example, pp. 211, 212 of this volume, where the planetary system is drawn with great accuracy. On these pages we see the system in plan, and we can easily appreciate the relative distances which separate the planets from each other; but in nature we do not see it thus, since we are ourselves upon the earth, which is the third planet, and which revolves nearly in the same plane as all the others round the sun; we see it *in profile*, as if we looked at this page along the edge. Moreover, there are no real orbits drawn in space: they are ideal lines which the planets follow in their course. In reality, then, we only see with our bodily eyes the *motions* of the planets which displace them in the sky.

In the silence of a fine summer night, let us suppose ourselves in the midst of a country with a very open horizon. Thousands of stars twinkle in the sky, and we think we see millions, although in reality there are never more than 3,000 visible to the naked eye above the horizon at the same time. These stars, of different degrees of brightness, maintain the same positions with reference to each other and form figures to which the name of constellations has been given: the seven stars of the Great Bear have presented during the thousands of years they have been observed the outlined form of a waggon drawn by three horses; the six stars of Cassiopeia always form a chair turning round the pole, or the letter M with elongated strokes; Arcturus, Vega, Altair, always mark the places of Boötes, Lyra, and the Eagle. The earlier observers ascertained this fixity of the brilliant points on the celestial vault, and by uniting the principal stars by fictitious lines, by tracing the outlines, in which they were not long in finding resemblances or symbols, they peopled with objects and fantastic beings the unalterable solitude of the heavens.

If we are in the habit of observing the starry sky, we come insensibly to familiarise ourselves with the constellations and to know the principal stars by their names. This we shall do a little further on, when we reach the world of stars. At present we have not yet left the solar world. Now, it sometimes happens that, in observing the celestial vault with which we are familiar, we notice a brilliant star at a point in the sky where we know a fixed star does not exist. This new star is perhaps more brilliant than any other, even excelling Sirius, the brightest star in the sky. We find, however, that its light, although more intense, is calmer, and that it does not twinkle. Further, if we take care to examine its position well with reference to the other neighbouring stars, and observe it for several weeks, we shall not be long in recognising that it is not fixed like the others, and that it changes its place more or less slowly.

This was noticed by the first observers of the sky, the shepherds of Chaldæa and the nomadic tribes of ancient Egypt, from the earliest epochs of astronomy. These stars, sometimes visible and sometimes invisible, and moving on the celestial vault, were named planets—that is to say, *wanderers*. Here, as in all etymologies, the word expresses the first impression experienced by the observer.

Our ancestors were far from imagining that these luminous points wandering among the stars do not possess any real light of their own; that they are dark like the earth, and as large as she is; that several are even much larger and heavier than our world; that they are illuminated by the sun, like the earth and moon, neither more nor less; that their distance is small compared to that which separates us from the stars; that they form, with the earth, a family of which the sun is the father! Yes, that luminous point, for example, which shines like a star is Jupiter. It has itself no light, any more than the earth has, but it is illuminated by the sun; and as the earth shines from afar on account of this illumination, so it shines a luminous point in which is condensed all the light scattered over its immense disc. Place a stone on a black cloth in a chamber completely closed to the daylight, throw upon it the rays of the sun by means of an opening suitably arranged, and this stone will shine like the moon and like Jupiter. The planets are dark worlds like ours, and only shine by the solar light which they receive and reflect into space.

If we direct a telescope towards a star, it does not appear larger than to the naked eye; the planets, on the contrary, show themselves the more increased in size the greater the magnifying power employed. The planets are relatively near; the stars are in Infinitude, and to bring them nearer by a thousand or two thousand times signifies nothing.

What first strikes an observer of the planets is the motion which displaces them in the sky relatively to the stars which remain fixed. Follow any planet, and you will see it moving towards the east, arrested in its motion for a week or two, retrograding towards the west, again stopping and then continuing its course.

Look at Venus, the *Shepherd's star*, which appears on a fine evening in the rays of the western twilight; it elongates from the west, rises in the sky, remains after the sun has set two hours, two hours and a half, three hours and more, then again approaches imperceptibly, and plunges once more into his rays. Some weeks later the same 'Shepherd's star' precedes the day star in the morning sky and shines in the transparent dawn. See Mercury, which so rarely emerges from the solar rays; you can scarcely recognise it on two or three evenings, when it again returns towards the sun. If, on the contrary, it is Saturn which you observe, it will appear during whole months to drag its slow steps through the sky.

These motions, combined with the brightness of the planets, have suggested the names conferred upon them, the ideas which we associate with them, the influences attributed to them, and the symbolical divinities with which we identify them. Venus, white and radiant, supreme beauty, queen of the stars; Jupiter, majestically reigning over the cycle of the years; the ruddy Mars, god of war; Saturn, slowest of the inhabitants of the sky, symbol of time and of destiny; Mercury, agile and flaming, to-day following Apollo, to-morrow announcing his rising. The designations, the attributes, the influences have been so many effects produced by the same causes, until at last, in the course of ages, the symbols have been taken literally, have influenced the minds of men, and these stars have been worshipped as veritable divinities. Religions begin in the mind, but they end by the materialisation of the purest ideas; they give birth to aspirations, desires, hopes; at first they answer ideas by ideas; afterwards they manufacture idols and prostrate themselves before them.

It was by these differences of motions that the planets were at first classed. By following them attentively observers ascertained that they appear to revolve round us from west to east among the stars, with certain irregularities, and arguing, logically, that those which move the more slowly and have the longest periods are the more distant, they classed them in an order of decreasing velocity. They were thus registered for 4,000 years:—

SATURN	Revolving round us in		30 years
JUPITER	„	„	12 „
MARS	„	„	2 „
THE SUN	„	„	1 year
VENUS AND MERCURY	„	„	1 „
THE MOON	„	„	1 month

At first this was only an approximation. The motions of Mercury and

Venus especially were very difficult to unravel. As observers wished to make all the stars revolve round the motionless earth as the centre of creation, and as in reality things do not happen in this way, they could not arrive at any great precision. They were constantly obliged to correct the 'tables.' Several astronomers began to think that Mercury and Venus really revolved round the sun, and that that body carried them with him in his annual motion round us. But the majority maintained, 2,000 years ago, an harmonic regularity decided on by Hipparchus, according to the harmony of ancient observations. This is the system which has been handed down to us in that great work, the 'Almagest' of Ptolemy, written about the year 130 A.D.,[1] and which was maintained down to the eighteenth century. Cicero gives us in his 'Dream of Scipio' the following eloquent description of this ancient astronomical system :—

The universe is composed of nine circles, or rather of nine globes, which move. The external sphere is that of the sky, which includes all the others, and on which are fixed the stars. Within revolve seven globes, drawn along by a motion contrary to that of the sky. On the first circle revolves the star which men call Saturn ; on the second moves Jupiter, a star beneficent and propitious to human beings ; then comes Mars, glowing and abhorred ; below, occupying the middle region, shines the sun, chief, prince, moderator of the other stars, life of the world, whose immense globe illuminates and fills the volume of its light. After him come, like two companions, Venus and Mercury. Finally, the lower orbit is occupied by the moon, which borrows its light from the day star. Below this last celestial circle there is nothing but mortal and corruptible, with the exception of the souls given by Divine kindness to the human race. Above the moon all is eternal. Our earth, placed at the centre of the world, and separated from the sky in all directions, remains motionless, and all heavy bodies are drawn towards it by their own weight.

Formed of unequal intervals, but combined according to a correct proportion, harmony results from the motion of the sphere, which, forming grave and high tones in a common accord, makes with all these varied notes a melodious concert. Such grand motions cannot be accomplished in silence, and nature has placed a grave tone at the slow and inferior orbit of the moon, and a high tone at the superior and rapid orbit of the starry firmament ; with these two limits of the octave, the eight moving globes produce seven tones in different ways, and this number is the bond of all things in general. The ears of men filled with this harmony know not how to hear it, and mortals do not possess a more imperfect sense. It is thus that the tribes near the Cataracts of the Nile have lost the power of hearing them. The splendid concert of the whole universe in its rapid revolution is so prodigious that your ears are closed to this harmony, as your glances sink before the fires of the sun, whose piercing light dazzles and blinds you.

---

[1] *Mathematike Suntaxis*, or 'Mathematical Composition.' This is the most ancient complete treatise on astronomy which has been preserved. Several translations and editions have been made since the invention of printing, and now every learned astronomer has it in his library. It is hardly known under its true title, for it is always called the *Almagest*, a pompous name given to it by the Arabs. In the East, the admiration for this astronomical treatise went so far that the caliphs, conquerors of the emperors of Constantinople, would only consent to make peace with the latter on the condition that they were put in possession of a manuscript copy of the *Almagest*. We have a good French translation by Halma in two volumes, of which the first, printed in 1813, has for a frontispiece a medal of the Emperor Antoninus, and the second, printed in 1816, is dedicated to King Louis XVIII.

Thus speaks the eloquent Roman. Beyond the seven circles was placed the sphere of the fixed stars, which thus formed the eighth sky. The ninth was the Prime Mover, on which they installed in the Middle Ages the *Empyrean*, or Abode of the Blessed. The whole of this structure was supposed to be of rock crystal by the vulgar, and even by most philosophers. Some superior minds alone appear not to have admitted the solidity of the heavens (Plato, for example); but the majority declared that it was impossible to conceive the mechanism and the motion of the stars if the heavens were not formed of a substance solid, hard, transparent, and which will not wear out. As interesting details, we may remark that the celebrated architect, Vitruvius, asserted that the axis which passes through the terrestrial globe is solid, extends beyond the North and South Poles, rests upon pivots, and is prolonged to the sky. He also speaks of authors who thought that if the planets move less quickly when they are far from the sun, it is because they see less clearly. The ancient philosophers took the aërolites for pieces detached from the celestial vault, which, escaping from the centrifugal force, fall to the earth by their own gravity. A Roman cardinal maintained this belief to Alexander von Humboldt at the beginning of the nineteenth century.

As to the *harmony of the spheres*, Kepler believed it even in the seventeenth century. According to him, Saturn and Jupiter form the bass, Mars the tenor, Venus the contralto, and Mercury the soprano.

This system of the planets revolving round us appeared very simple. But we shall see that the agreement was only apparent; that in minutely examining the details they deviate more and more from this primitive simplicity, and that finally this structure cannot resist the attacks of discussion. In fact, in order that a universe so constructed should be able to move mechanical conditions would be necessary which do not exist; it would be necessary, for example, that the earth should be heavier than the sun, which it is not; that it should be itself more important than the whole solar system, which is still less a fact; that the stars should not be at the distance which separates them from us; in a word, in order that the universe should gravitate round us it would have to be constructed in a totally different way. As it is, the earth necessarily revolves round the sun, and obeys a stronger force than itself. We understand, then, that in proportion as astronomical observations became more numerous and precise the simplicity manifested in the preceding elementary sketch had necessarily to be corrected and the mechanism increased by indefinite additional burdens. The following are the principal complications which followed the improvement in astronomical knowledge.

Aristotle and Ptolemy had declared, in company, moreover, with all the philosophers, that the circle was the most perfect geometrical figure, and that the celestial bodies, divine and incorruptible, could only move in a circle round the central terrestrial globe.

Now, the truth is: (1) That they do not revolve round the terrestrial globe; (2) that they circulate, in company with the earth itself, round the sun, which is relatively motionless; (3) that they move, not in circles, but in ellipses.

The apparent motions of the planets which we observe from here are the result of the combination of the earth's motion round the sun with that of the planets round the same body.

Let us take, for example, Jupiter. It revolves round the sun at a distance five times greater than the distance of the earth from the same body. Its orbit, then, surrounds ours with a diameter five times greater. It takes twelve years to accomplish its revolution.

During the twelve years which Jupiter takes to make its revolution round the sun the earth has made twelve revolutions round the same body. Consequently, the motion of Jupiter seen from here is not a simple circle slowly followed during twelve years, but a combination of this motion with that of the earth. If the reader will refer to our figures on pp. 211, 212, and notice the orbit of the earth, and beyond, that of Jupiter, he will easily recognise that in revolving round the sun we cause an apparent displacement of Jupiter on the starry sphere on which it is projected. This displacement takes place during one-half of the year in one direction, and during the other half in the other direction. It is as if the orbit of Jupiter consisted of twelve rings. In order, then, to account for the apparent motion of Jupiter, the ancient astronomers could not long defend its simple circle, but were obliged to place on this circle the centre of a small circle on which the planet was set. Thus Jupiter did not follow its great circle directly, but a small one which made twelve turns while gliding along the primitive circle in a period of twelve years.

Saturn revolves round the sun in thirty years. In order to explain its apparent marches and counter-marches as seen from the earth, the ancients were obliged to add to its orbit a second circle, of which the centre followed this orbit, and of which the circumference carrying the planet turned thirty times on itself during an entire revolution. This second circle received the name of *epicycle*.

The epicycle of Mars was more rapid than the preceding. Those of Venus and Mercury were much more complicated.

Here, then, we have the first complication of the primitive circular system. We will now consider a second complication.

Since in reality the planets follow ellipses, they are nearer the sun at certain points of their course than at others. And since all the planets, including the earth, move in different periods round the sun, it follows that each planet is sometimes nearer, sometimes farther from the earth itself.

At certain points of its orbit, for example, Mars is six times farther from us than at other points. In order to explain these variations of distance the ancients supposed that the circles followed by each planet had for a centre, not exactly the terrestrial globe itself, but a point outside the earth and turning itself around us. We can easily see that by this stratagem a planet —Mars, for example—describing a circumference round a centre situated to one side of the earth, is found farther from the earth at a certain part of its course, and nearer at the opposite part. The real centre of each celestial orbit only coincides with the centre of the earth by the subterfuge of a second movable centre round which it was effected.

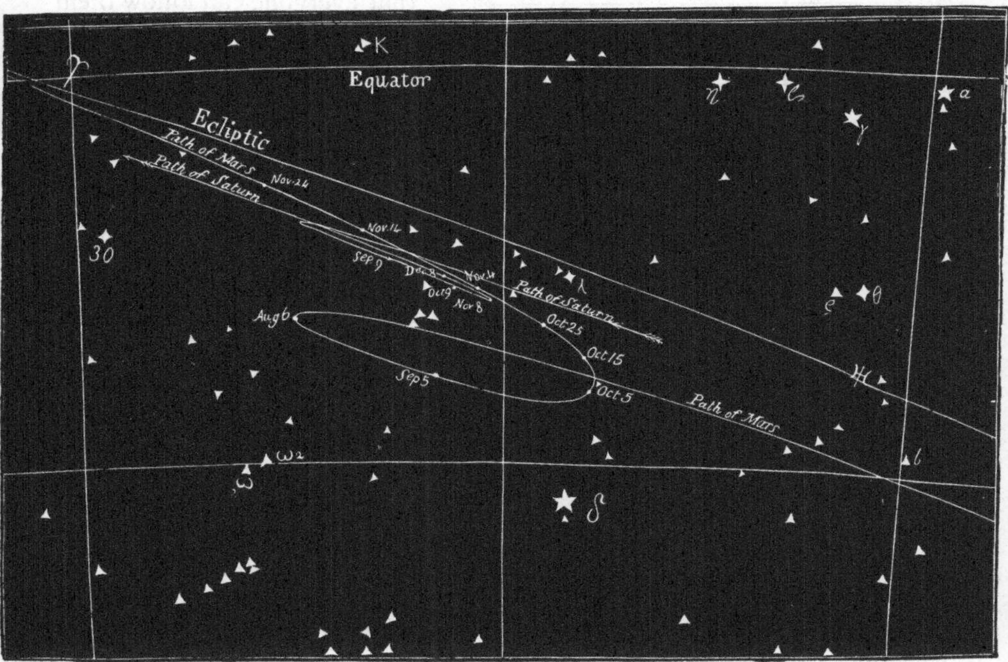

FIG. 146.—PATHS OF MARS AND SATURN DURING THE AUTUMN OF 1877

This new arrangement was designated by the name ot the system of *eccentricities*, a word which, like the first, recalls its geometrical form.

These epicycles and eccentricities were successively invented, modified, and multiplied according to the necessities of the case. As the observations became more precise it was necessary to add to their number in order to represent the facts more exactly. Each century added its new circle, its new gear to the mechanism of the universe ; so much so that in the time of Copernicus, in the sixteenth century, they had already seventy-nine arrangements piled one upon the other.

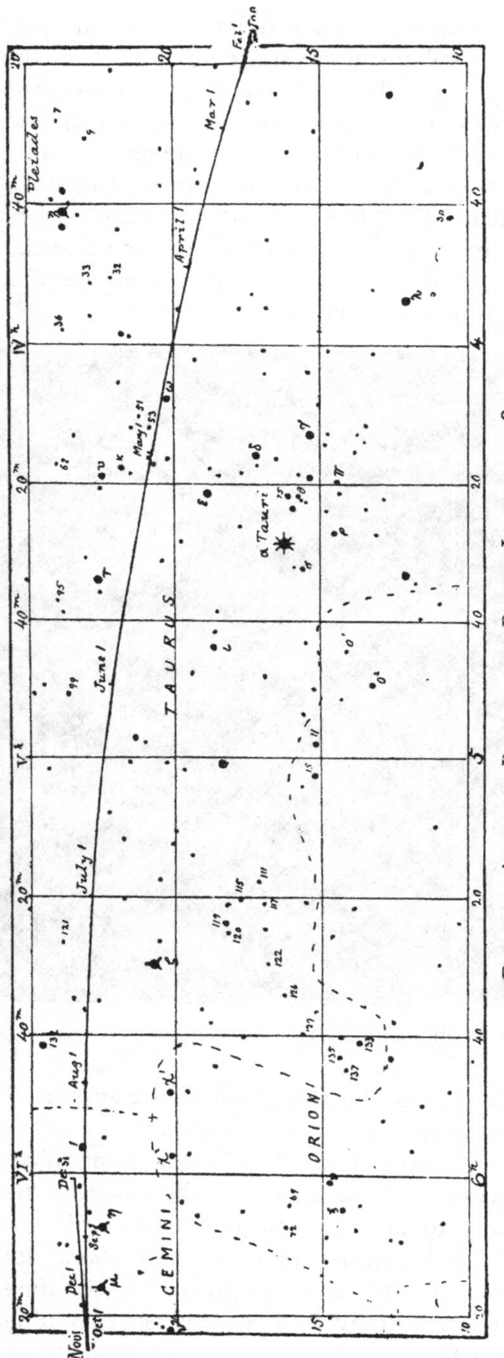

FIG. 147.—APPARENT PATH OF THE PLANET JUPITER IN 1894

We do not usually realise what singular lines the planets trace on the celestial sphere by their apparent motions as seen from the earth. In order that everyone may easily understand it we give here five small celestial maps (figs. 146–150) which show these motions, so that everyone can follow them in the sky. These paths and positions of the planets (shown here in figs. 147–150 for 1894) change perpetually.[1]

These apparent motions are the result of the combination of the motion of each planet round the sun with that of the earth. Of course, the displacements are so much the smaller and the motions so much the slower as the planets are more distant. Thus, Neptune is displaced each year but two degrees (or four times the diameter of the moon) on the average, and takes 165 years to make the circuit of the sky ; Uranus is displaced seven to eight degrees, and will return in eighty-four years to the point which he occupies at present ; Saturn makes the tour of the sky in thirty years, Jupiter in twelve years ; Mars, Venus, and Mercury move more quickly still.

We have represented each

[1] A drawing will be found each year for all the planets in our *Revue mensuelle d'Astronomie populaire.*

of these motions separately; but it sometimes happens that several planets meet in the same region of the sky, which doubles the interest of their observation. This happened especially with Jupiter and Saturn in the month of April, 1881. Mars had passed very near Saturn on July 27, 1877, June 20, 1879, July 6, 1881, &c. Neptune was also stationed in this same region, and in addition Mercury and Venus passed not far from the spot. It rarely happens that several planets are thus collected in the same region of the sky and, if the old astrologers had then been alive, they might have predicted catastrophes sufficient to make the most fearless souls shudder. For us, the scientific interest is to form an exact idea of the apparent motions of the planets in the sky, and the philosophical interest is to know that astronomy is acquainted with the future of the celestial motions as it is with their past: no miracle ever deranges them. These meetings are generally designated by the name of *conjunctions*. In astronomical language this term is especially reserved for the positions of Mercury and Venus when they pass between the sun and the earth, or behind the earth; these are their conjunctions, inferior or superior. The planets exterior to the earth are in *opposition* when the earth is found between them and the sun—that is to say, when they pass the meridian at midnight. When they pass behind the sun they are in conjunction with him.

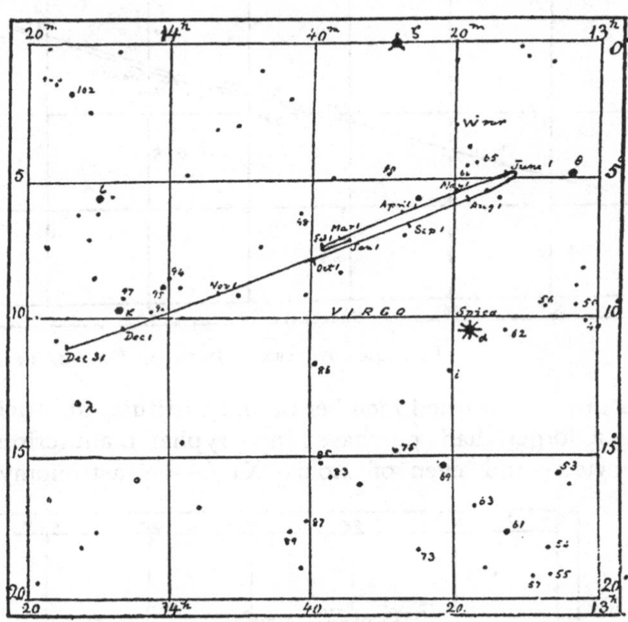

FIG. 148.—APPARENT PATH OF SATURN IN 1894.

Some *savants* think that these positions of the planets have an influence on terrestrial meteorology; the observation of facts has, however, yielded nothing positive in this respect.

Now, if we draw the plan of these motions with reference to the earth, supposed to be motionless at the centre of the system, the figures are still more singular and remarkable. I made a drawing in 1869 of Saturn's apparent motion from 1842 to 1871 (see the 'Magasin Pittoresque' for the month of April, 1870), as well as one of the secular motion of Uranus, with reference to a discussion which was raised at the Academy of Sciences on a supposed discovery of that planet made by Galileo, in 1639, in the vicinity of

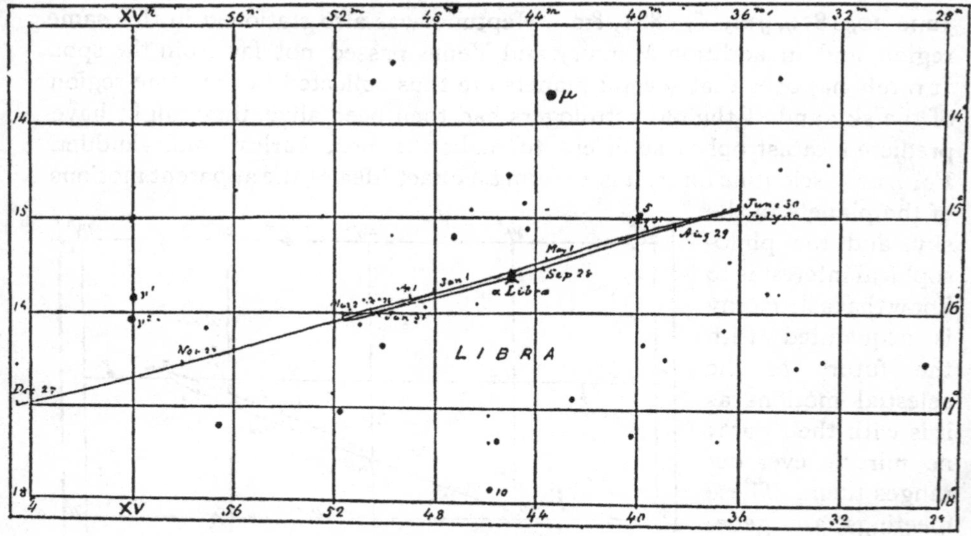

Fig. 149.—Apparent Path of Uranus in 1894

Saturn. A learned member of the Institute, M. Michel Chasles, deceived by a forger, had purchased apocryphal manuscripts of Galileo, Pascal, Newton—and even of Louis XIV.—on astronomy. The well-known

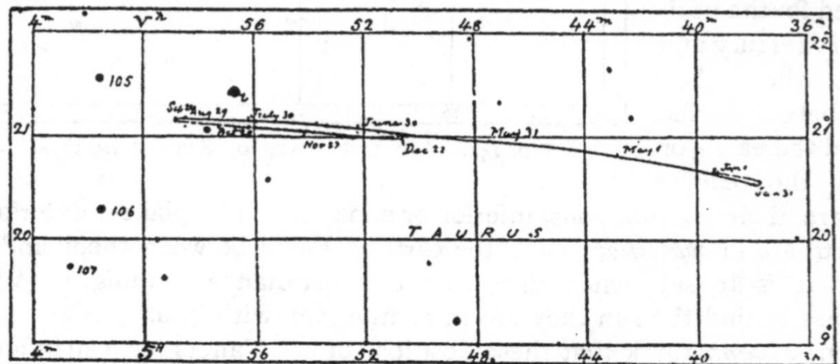

Fig. 150.—Apparent Path of Neptune in 1894

ignorance of this great king should have given warning of the falsity of these manuscripts. But the forger was so ingenious that the *savant* of whom I speak purchased for more than 1,000 francs these *dear* letters. As to the discovery of Uranus in the vicinity of Saturn in 1639, the two retrospective charts which I have constructed show with certainty that it is a story which cannot be defended, since in that year Saturn was in Capricorn and Uranus in Virgo, or more than 90° distant from each other.

Thinkers at last expressed doubts on this astronomical system, however classical it had become. A king-astronomer, who abandoned the crown for the astrolabe and forgot the earth for the sky (Alphonso X. of Castile), dared to say in a full assembly of bishops (and in the thirteenth century) that if God had asked him for his opinion when He created the universe, he would have given Him good advice as to constructing it in a less complicated manner!

But there were superior and independent minds who foresaw in the increasing complication of the system of Ptolemy testimony against its reality. The Peripatetic philosophers advanced in this discussion the singular argument, reproduced later on by the Jesuit Riccioli in his essay in refutation of Galileo's Dialogues: Let us object to the system of Ptolemy that the thousands of stars revolve round us with a regularity very difficult to understand among bodies independent of each other; that their diurnal motions should be rigorously proportional to the distance; that the size of the sun in comparison with the earth is an almost unexceptionable proof of the motion of the latter body, &c. Riccioli replied that 'there are intelligences in the stars; that the more difficult it is to explain the motion of the sky, the more the greatness of God is manifested; that the nobility of man is superior to that of the sun; that it matters little to man, for whom all has been made, whether thousands of stars turn round him, &c. . . .

Arguments of this value do not demand, in their turn, a long refutation to-day. However, they kept industrious minds in suspense, and the habit of admiring this system without discussion caused it to be preserved in the schools, in spite of all the unnatural complications with which it was surrounded.

This manner of metaphysically losing time, under the pretext of making science, endured in the schools from antiquity down to the time of Copernicus, and retarded for a long time the advent of the exact sciences. We must come to the fifteenth and sixteenth centuries to reach the establishment of the experimental method, and to find independent *savants*, unfettered by prejudices and freely seeking the truth.

By a happy coincidence, the greatest events in the historical march of humanity are met with in this same epoch. The awakening of religious liberty, the development of the most noble perception of art, and the knowledge of the true system of the world, were announced concurrently with the great maritime enterprises—the age of Columbus, of Vasco da Gama, and of Magellan. The year 1543, in which appeared the work of Copernicus, De Revolutionibus orbium celestium,' which dissected the heavens, also produced that of Vesalius, 'De Corporis humani fabrica,' which created the science of human anatomy. The terrestrial globe was revealed in all its aspects to the gaze of adventurous science, and the human mind, in verifying henceforth directly, and by experience, the sphericity of the globe and its isolation in space, acquired the most essential elements for the purpose of preparing it to understand its motion.

The system of appearances, the theory of the immobility of the terrestrial globe and the motion of the heavens, reigned, as we see, only three centuries ago—from 1500 to 1600, in the times of Francis I., the Medicis, and Henry IV.—which is not very distant from our present epoch. It was still taught under Louis XIV., and even in the time of Louis XV., in the eighteenth century. It is this idea, simple and vague, which still reigns in the ignorant mind of the present populations of Europe; for even now, among one hundred persons taken from all classes, there are but a few who understand that the earth turns, and who are certain of it, and there are not, perhaps, two who could give an exact account of the velocity of its motion of translation and of the effects of its diurnal motion.[1] Reflecting on the mechanical conditions of the system of appearances which we have sketched, Copernicus came to the conclusion that this system, so complicated and crude, could not be natural. After thirty years of study he was convinced that by attributing to the earth a double motion—one of rotation on itself in twenty-four hours, the other of revolution round the sun in three hundred and sixty-five days and a quarter—we may explain most of the celestial motions for which those innumerable crystal spheres had been imagined. The ingenious astronomer attained a knowledge of the general plan of nature, disclosed his opinion to contemporary *savants*, and published it before leaving the world. Since 1543, the epoch of the death of Copernicus, and the publication of his great work, astronomers have confirmed, proved definitely, and established for ever this theory—at first bold, but now so simple—of the motion of the earth.

The system of Copernicus is represented in fig. 151, from the work of the great astronomer himself. We see that it is essentially the basis of the system of the world as we know it to-day; that the sun is at the centre, and that the planets revolve round him; but that, nevertheless, it presents certain differences which the science of the successors of Copernicus has removed: 1. The proportions of the distances were not known. It was the genius of Kepler which found them in the seventeenth century. 2. The planets Uranus and Neptune are wanting. They were only discovered in the eighteenth and nineteenth centuries. 3. Telescopes had not been invented, and astronomers were ignorant of the existence of the

[1] And perhaps, even in France, not *one* person *in ten thousand* understands the philosophical revolution effected by modern astronomy, and knows how to estimate our planet and its humanity for what they are worth. The nations of the world live in general in the most absolute ignorance of the reality.

satellites, the form of Saturn, the relative size of the planets, &c. 4. The planets Mercury and Venus revolve in 80 days and 9 months, instead of 88 and 225 days. 5. The earth was endowed with a third motion, intended to preserve the parallelism of its axis of rotation, which the annual motion would seem to disturb. 6. The stars do not appear so distant that the sun could not illuminate them, and they reflected his light: the light of the brilliant star throned at the centre of the entire creation.

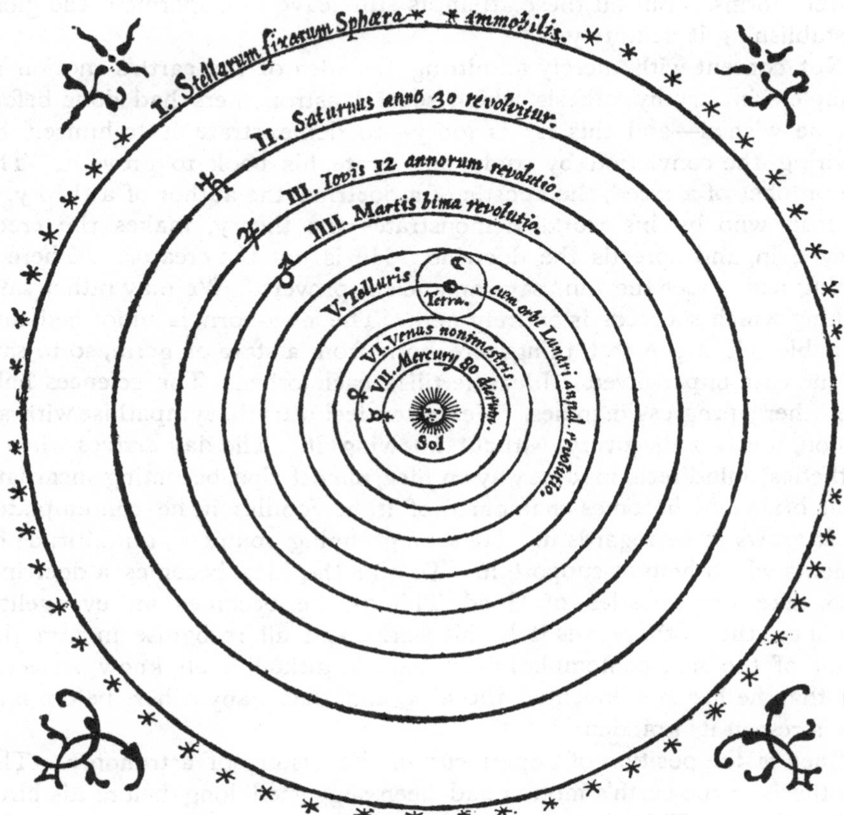

FIG. 151.—COPERNICAN SYSTEM: FACSIMILE OF THE DRAWING IN THE VOLUME BY COPERNICUS PUBLISHED IN 1543

Copernicus was not the first who thought of interpreting the celestial motions by the theory of the earth's motion. That immortal astronomer has taken care to give, with rare sincerity, the passages in the ancient writers from which he derived the first idea of the probability of this motion—especially Cicero, who attributed this opinion to Nicetas of Syracuse; Plutarch, who puts forward the names of Philolaus, Heraclides of Pontus, and Ecphantus the Pythagorean; Martianus Capella, who adopted, with the Egyptians, the motion of Mercury and Venus round the sun, &c. Even a hundred years before the publication of the work of Copernicus, Cardinal Nicolas of Cusa, in 1444, in his great theological and scientific encyclopædia, had also spoken in favour of the idea of the earth's motion and the plurality of worlds. From ancient times to the age of Copernicus the system of the earth's immobility had been doubted by clear-sighted minds, and that of the earth's motion was proposed under different forms. But all these attempts still leave to Copernicus the glory of establishing it definitely.

Not content with merely admitting the idea of the earth's motion as a simple arbitrary hypothesis, which several astronomers had done before him, he wished—and this is his glory—to demonstrate it to himself by acquiring the conviction by study, and wrote his book to prove it. The true prophet of a creed, the apostle of a doctrine, the author of a theory, is the man who by his works demonstrates the theory, makes the creed believed in, and spreads the doctrine. He is not the creator. 'There is nothing new under the sun,' says an ancient proverb. We may rather say: Nothing which succeeds is entirely new. The new-born is unformed and incapable. The greatest things are born from a state of germ, so to say, and increase unperceived. Ideas fertilise each other. The sciences help each other; progress marches. Men often feel a truth, sympathise with an opinion, touch a discovery, without knowing it. The day arrives when a synthetical mind feels in some way an idea, almost ripe, becoming incarnate in his brain: he becomes enamoured of it, he fondles it, he contemplates it. It grows as he regards it. He sees, grouping round it, a multitude of elements which help to support it. To him the idea becomes a doctrine. Then, like the apostles of Good Tidings, he becomes an evangelist, announces the truth, proves it by his works, and all recognise in him the author of the new contemplation of nature, although all know perfectly well that he has not invented the idea, and that many others before him have foreseen its grandeur.

Such is the position of Copernicus in the history of astronomy. The hypothesis of the earth's motion had been suggested long before his birth on this planet. This theory counted partisans in his time. But he—he did his work. He examined it with the patience of an astronomer, the rigour of a mathematician, the sincerity of a sage, and the mind of a philosopher. He demonstrated it in his works. Then he died without seeing it understood, and it was not till a century after his death that astronomy adopted it and popularised it by teaching it. However, Copernicus is really the author of the true system of the world, and his name will remain respected to the end of time.

This great man was neither potentate, prince, nor official personage, nor covered with titles more or less sonorous and more or less vain. He was a modest physician, the friend of humanity and the friend of science, consecrating his whole life to the study of nature, nobly indifferent as well to fortune as to glory. He was the son of a Polish baker, and became by his own labours the greatest man of his age. The physician became a priest, a physician of the soul, and the position of a canon assured to him the calm and tranquil life which he preferred. His uncle was a bishop, and was sometimes astonished that he should 'lose his time' working at astronomy.[1]

There was a slight delay in the adoption of the theory of the central sun and the motion of the earth, a delay due to the astronomer Tycho Brahé, who contrived, in 1582, a mixed system susceptible of reconciling observation with the Bible, in the name of which the teaching-schools refused to accept the theory of the earth's motion.

It was not that Tycho Brahé did not appreciate the merit of the theory of Copernicus. 'I admit,' he wrote, 'that the revolutions of the five planets are easily explained by the simple motion of the earth; that the ancient mathematicians have indeed adopted absurdities and contradictions, from which Copernicus has delivered us, and even that he has satisfied a little more exactly the celestial appearances.' But he soon stated that this system could never be reconciled with the testimony of Holy Scripture, and he believed that he would satisfy everybody by making the sun, accompanied by the planets, revolve round the earth.

Here is how the Danish astronomer himself states his theory :—

I think it is decidedly necessary, without any doubt, to place the earth motionless at the centre of the system, according to the opinion of the ancients and the testimony of Scripture. I do not admit, with Ptolemy, that the earth may be the centre of the orbits of the second 'mobile,' but I think that the celestial motions are arranged so that the moon and the sun only, with the eighth sphere, the most distant of all, and which includes all the others, have the centre of their motion near the earth. The five other planets revolve round the sun as round their chief and king, and the sun must be incessantly in the midst of their orbs, which accompany him in his annual motion. . . . Thus the sun would be the ruler and end of all these revolutions, and, like Apollo in the midst of the Muses, he would rule alone all the celestial harmony.

To the system of Tycho Brahé there exists the same serious objection which was made to that of Ptolemy, since in fixing the earth at the centre of the world system he supposed that the sun, all the planets, and the whole sky of fixed stars, described round us their immense orbits in twenty-four hours. It never enjoyed a real influence. However, we find it still, in 1651, in the curious frontispiece of the 'Almagestum Novum' of Riccioli. At the end of the seventeenth century Bossuet still declared

[1] This is like Lady Byron, who eight days after her marriage was astonished that Lord Byron persisted in writing verses, and asked him 'when he would have finished.' (I have the fact from a friend of Byron, the Marchioness de Boissy.)

imperiously that it is the sun which moves, and Fénelon placed the two theories in the same rank. The tribunal of the Inquisition, and the Congregation of the Index, presided over by the Pope, had, moreover, declared the doctrine of Copernicus to be heretical in 1616 and 1633, and condemned 'all the books which asserted the motion of the earth.' During the whole of the seventeenth and a part of the eighteenth century the University of Paris taught the motion of the earth as an *hypothesis convenient but false!* At the same epoch, under Louis XIV., the earth was still represented as seated at the centre of the system. But the consecutive labours of Tycho himself, of Galileo, Kepler, Newton, Bradley, D'Alembert, Lagrange, Laplace, Herschel, Le Verrier, and other great minds, have given to modern astronomy an absolute and immovable basis, strengthened by each new discovery, upon which the intellectual edifice of the science rises, grows, and ever mounts into Infinitude. The illusions, the errors, the shadows of night are put to flight, and the lighthouse of truth illumines the world. Only those who voluntarily close their eyes can continue to live in the illusion of the tortoise, who takes his shell for the limits of the universe.

The ancients had remarked that the planets visible to the naked eye never wandered very much from the ecliptic, the apparent annual path of the sun, and that their departure from this great circle of the celestial sphere never exceeded 8°, either to the north or south. Imagining, then, in the sky two ideal lines thus traced on both sides of the ecliptic, they described a zone of 16° in width all round the sky, which the planets never left. This zone is the *zodiac*, which derives its name from the Greek *zoon*, an animal, because the constellations which compose it are for the most part figures of animals. The ancients divided this great circle into twelve parts or signs, each of which marked the abode of the sun during each month of the year (see fig. 18, p. 44). The movements of the large planets Uranus and Neptune, discovered by modern astronomers, are also contained within the limits of the zodiac; but several of the small planets which revolve between Mars and Jupiter move at a rather high inclination, and the comets sometimes even reach the poles.

The sun, the moon, and the planets have been for a long time designated by the following signs:

The Sun	The Moon	Mercury	Venus	Mars	Jupiter	Saturn
⊙		☿	♀	♂	♃	♄

The sign of the sun represents a disc; it was used for thousands of years among the Egyptians. That of the moon represents the lunar crescent; we find it in use among all nations from the most remote antiquity. The sign of Mercury has for its origin the caduceus, that of Venus a mirror, or perhaps the mark of fecundity (union of the circle with the crossed arrow: the Egyptian signs are in favour of this origin). That of Mars has as origin a lance, that of Jupiter the first letter of Zeus, that of Saturn a scythe. We find them used by the Gnostics and alchemists from the tenth century onwards.

In the seventeenth century they began to consider the earth as a planet, and gave to it the sign ♁, a globe surmounted by a cross. In the eighteenth century the discovery of Uranus added a new planet to the system, which they designated by the sign ♅, which recalls the initial of Herschel. The discovery of Neptune, in 1846, added a new sign ♆: this is the trident of the sea-god.

But it is time to leave the history of the apparent aspects and enter directly into a description of each of the worlds of the system.

CHAPTER II

THE PLANET MERCURY

And the Suburbs of the Sun

IN the description of the planetary system we shall proceed from the centre towards the circumference. We have already contemplated the splendours of the central sun; we already know the order in which the planets follow each other; we have studied their general motions, apparent as well as real; we have examined in detail the third planet of the system (the earth) and the satellite which accompanies it. Let us, then, commence now the description of the other worlds of our solar system with the province nearest to the sun, Mercury.

Between Mercury and the sun do one or more planets exist still unknown to us? The question has been raised and very much discussed in recent years. It is interesting to examine it. Let us study it—as it is important to do in the smallest astronomical matters—from its origin, and at first hand, in order to judge it exactly and impartially.

One of the most eminent mathematicians that ever existed, the French astronomer, Le Verrier, by rigorously analysing the motions of all the planets, succeeded in constructing exact tables of the positions of Mercury, Venus, Mars, Jupiter, Saturn, Uranus, for several thousands of years. He began this immense mathematical labour about 1840, and finished it in 1877, a few months only before his death: a noble employment of a laborious life which would have been still more useful to science and humanity if he had possessed a more sociable character and a more disinterested love for the general progress.[1]

The motion of the planet Uranus showed irregularities not to be accounted

[1] But why look for a sun without spots? Did not the great Newton himself exhibit irascibility and jealousy? Was not Laplace, the French Newton, weak enough to permit himself to be decorated with the title of count by Napoleon, and with that of marquis by Louis XVIII.? Laplace, count and marquis!—did that add an iota to his value and his glory? Cuvier, the founder of palæontology, was nominated baron by the same king; did he not sacrifice the interests of pure science to official conventions? The greatest geniuses are weak. Mathematicians, whose tempers

for by the disturbing influence of the planets then known, and convinced astronomers of the existence of an unknown planet situated beyond Uranus, and causing in its path the perturbations revealed by the meridian observations of that celestial body. In 1845, Arago advised Le Verrier to solve this interesting problem of transcendent mathematics. He succeeded, with honour, and announced, as we shall see, the place which this unknown planet should occupy in the immensity of the heavens. A telescope was pointed towards this spot, and there was the planet.

Thus the unexplained perturbations in the motion of the planet Uranus revealed in theory the existence of the planet Neptune. This is one of the most admirable confirmations given by the progress of astronomy to the reality of the Newtonian theory of universal gravitation.

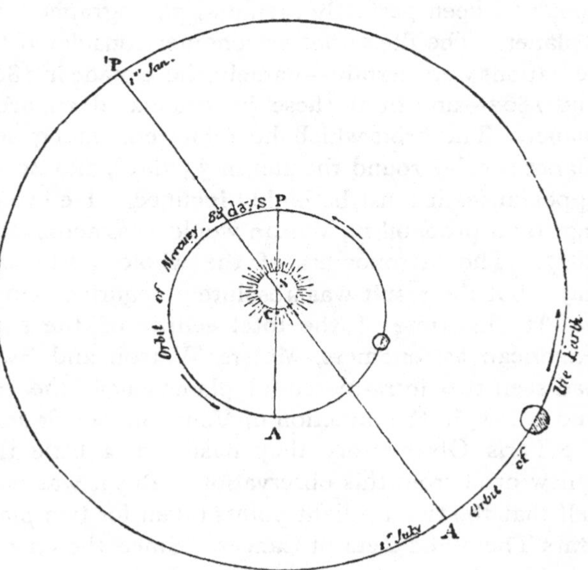

FIG. 152.—ORBIT OF MERCURY ROUND THE SUN

Now, the analysis of the motion of the planet Mercury likewise indicated to Le Verrier, in 1859, perturbations which are not explained by the action of the other planets, and which could be explained if there were between Mercury and the sun one or more planets revolving round the central body. The theory of Mercury shows a difference with the observations which produces an increase of 31″ of arc in the secular motion of the perihelion.

If this hypothesis is true, we ought from time to time to see dark bodies having a proper motion of translation passing across the solar disc. Now, only a few months had passed since the announcement of these results to the Academy of Sciences, when a country physician with a love for astronomy—one who devoted to admiration of the beauties of the sky the time which was not taken up in the alleviation of earthly miseries—my old and excellent friend, Dr. Lescarbault, announced that he had observed,

are generally intolerable, are perhaps psychologically excusable, for the constant tension of their mind is, perhaps, the cause of their bad digestion and their state of hypochondria. Let him who is without fault cast the first stone.

from his modest house at Orgères, a spot quite round and black passing across the sun on March 26, 1859; he followed it for more than an hour, and noticed its motion on the solar disc.

From 1858 to 1876 Le Verrier collected more than a hundred similar observations, most of which he rejected, because discussion showed that they were simply ordinary solar spots. Even in 1876 there was much excitement on the occasion of a spot quite round and black, apparently also endowed with proper motion, which was seen by a German observer on April 4; but it was found that on that same day the sun had been attentively observed at London and Madrid, *five hours before,* and that the said spot had been perfectly seen and photographed, and consequently was not a planet. The illustrious astronomer considered that, on the whole, six observations were certain—namely, those made in 1802, 1819, 1839, 1849, 1859, and 1862—and from these he calculated an orbit of the intra-mercurial planet. The orbit which he preferred, among several possible, made the planet revolve round the sun in 33 days, and to explain the rarity of the appearances it must be highly inclined. He even announced that, according to all probability, Vulcan would pass across the solar disc on March 22, 1877. The astronomers of the whole world watched the sun on that day; but the result was absolutely negative—no black spot appeared.

At the time of the total eclipse of the sun on July 29, 1878, two American astronomers, Messrs. Watson and Swift, announced that they had seen two intra-mercurial planets near the eclipsed sun (to the right and below, in the direction of Venus in our figure on p. 199); and even at the Paris Observatory they hastened, a little thoughtlessly, to calculate a new orbit from this observation. But it was not difficult to satisfy oneself that the two brilliant points taken for two planets were simply the two stars Theta and Zeta of Cancer. Since then a new orbit has been calculated by the German astronomer, Oppolzer, and a new transit announced. On the day predicted the sun was examined more minutely than ever, but nothing was seen. An absolutely free and impartial discussion of the subject leads us, then, to the conclusion that, *according to all probability,* there is not between Mercury and the sun a planet comparable with Mercury in size.

But, then, what becomes of the observations of black spots crossing the sun? We simply remark—not doubting, of course, the good faith and the sincerity of any observer—that there is nothing easier than being deceived in the examination of the motion of a solar spot, considering that the vertical diameter of the solar disc changes from hour to hour, and that a spot which would have been seen, for example, on the left of the disc at a certain time, would appear to have travelled if we re-observed it an hour or two later. In order to be sure of proper motion it is necessary to have followed the black point from its entry on the disc to a considerable distance from the border, or, better, to have an instrument fitted with clock-work motion. These conditions have not been complied with by any of the observers, either on account of their equipment or of the state of the sky.

But, again, what becomes of the theory of Mercury? Does not this planet show unquestionably an increase in the secular motion of its perihelion? Yes, but the cause may not be a planet. The principal reason for doubt is that for about thirty years there has not passed a single day, so to say, without the sun being examined, drawn, or photographed in Italy, England, Portugal, Spain, America, France, and elsewhere; that the supposed planet would have passed more than a hundred times across the sun, and that, nevertheless, it has never been seen. It is either well hidden, or it does not exist. Mercury was the god of thieves; his companion steals away like an anonymous assassin! The perturbations in question might be explained by a swarm of very small asteroids, too small to be visible from here on the solar disc, and by the influence of cosmical matter, which certainly exists in the vicinity of the day star—matter which is seen during total eclipses to form enormous trails on both sides of the sun, and of which the densest layers doubtless constitute the zodiacal light: these are like whirlwinds of dust illuminated in his rays.

We will, then, leave to a new race the intra-mercurial planet—already baptised by the name of Vulcan—in the domain of conjecture,[1] and we will proceed at once to the world of Mercury.

Situated, as we have seen, at 36 millions of miles from the sun, and revolving round him in 88 days, this planet describes an orbit interior to that of the earth and much smaller than ours. Fig. 152 gives an exact idea; it is drawn to a scale of 1 millimetre to a million of leagues. This orbit is not circular, but elliptical. Its eccentricity—that is to say, the distance of its centre from the focus of the ellipse, expressed in proportion to the semi-axis major or the mean distance—is two-tenths (0·2)—that is to say, 7,200,000 miles. At its perihelion the planet approaches to 28,500,000 miles from the solar focus, while at its aphelion it recedes to 43,300,000 miles. This is, relatively, the most elongated of the planetary orbits. We have drawn the orbit of the earth to the same exact scale (fig. 152).

The distance of Mercury from the earth varies, then, considerably. When it passes between the sun and us, and is at its aphelion, it may approach to less than 50 millions of miles; the apparent diameter of its disc then attains thirteen seconds: but in the most distant part of its orbit, when it passes behind the sun, its distance is increased to 136 millions of miles, and its disc is then reduced to four and a half seconds.

[1] This opinion, which we supported in the first edition of this work, in 1879, has been *absolutely confirmed* by all the researches made since that epoch during all the total eclipses of the sun. The intra-mercurial planet does not exist.

When the planet passes between the sun and us we say that it is in *inferior conjunction*; its position on the other side of the sun is called, on the contrary, *superior conjunction*.

The perihelion of Mercury is found at 76° of longitude—that is to say, at 76° from the point occupied by the sun on the ecliptic at the moment of the vernal equinox; the perihelion of the earth is found 25° farther, at 101°.

Mercury is only visible at those times when it is farthest from the sun. We then perceive it in the evening, remaining longer each day after sunset, and shining in the western sky like a star of the first magnitude. But it cannot depart more than 28° from the radiant star, nor delay more than two hours after him, so that even on the days of its greatest elongations it is lost in the glare of the twilight, or when night comes it is too low not to be hidden in the vapours of the horizon.

The author of the discovery of the true system of the world went to his grave without being able to perceive it once in Poland. In France, there hardly passes a year without its presence being verified once at least, and I, for my part, have made several observations. [I have frequently seen it in Ireland with the naked eye.—J. E. G.] One of the most interesting was that of February 17, 1868. I had then a modest observatory situated not far from the Pantheon, from which the view was very extended, and which was not long in being hidden by the encroachments of Parisian buildings. Mercury and Jupiter shone side by side on that evening—a rare conjunction; the two planets were sufficiently near each other (a degree and a half) to be visible in the field of the same telescope (in the finder). A coincidence more curious still, the planet Venus sparkled at the same time above the other two, and on January 30 had also passed near Jupiter, nearly to the point of passing over it and eclipsing it; the angular distance of the two planets was reduced to 20 minutes of arc.

FIG. 153.—ASPECT OF MERCURY NEAR QUADRATURE

The comparison of the magnitude, the brilliancy, and the colour of the three planets in conjunction was very interesting. The brilliant light of Venus, compared with that of Jupiter, produced the effect of an electric light alongside a gas jet; the beautiful planet was white and limpid, like a luminous diamond; Jupiter was, in comparison, yellowish and almost red; Mercury was still redder than Jupiter. In the telescope, Venus and Mercury showed a very marked phase.

On September 28, 1878, Mercury and Venus also came together, visible in the same field of the telescope. Venus was much brighter than Mercury, although farther from us.

To observe this planet often, the first condition is to live in a favourable climate. An amateur astronomer, Gallet, Canon at Avignon (whom Lalande called Hermophile, the friend of Mercury), observed it more than a hundred times in the last century.

Several observers follow it in France at least once or twice every year.

As to aspect and dimensions, it presents itself, in its quadratures, at its greatest elongations from the sun, nearly as shown in fig. 153; a second of arc is represented by a millimetre.

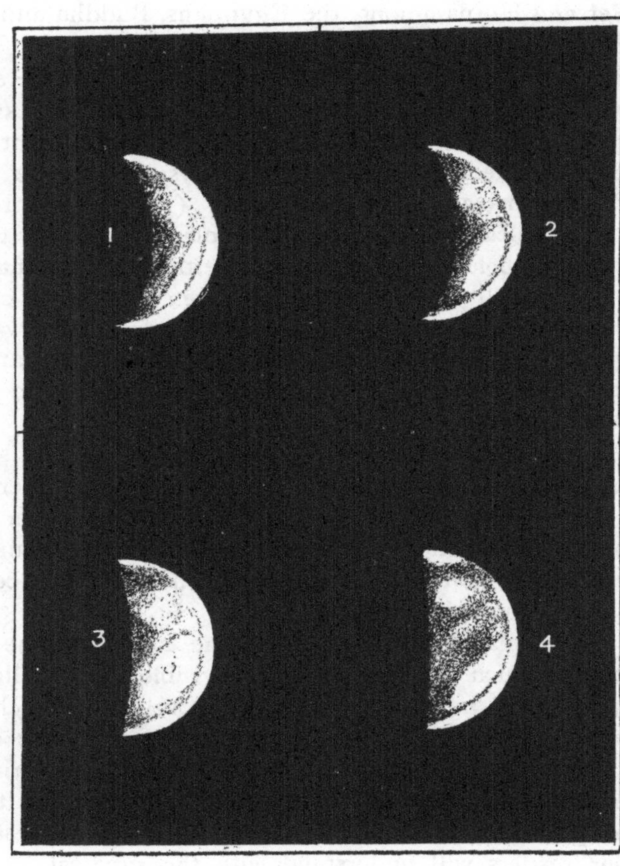

FIG. 154.—MERCURY AS A MORNING STAR, NOVEMBER 1882 (From Drawings by W. F. Denning. 1. November 5, $1^h\ 50^m$. 2. November 6, $18^h\ 55^m$. 3. November 8, $19^h\ 30^m$. 4. November 9, $19^h\ 39^m$. With 10-inch Reflector, Power 252

In rather rare circumstances we may succeed in studying its telescopic aspect and discerning some details, when it is sufficiently elevated above the horizon, in a pure atmosphere, especially if calm, either at the rising or setting of the sun. An industrious and persevering astronomer, Mr. Denning, of Bristol, especially, succeeded in 1882 (November 5, 6, 8, and 9) in obtaining the drawings we give here (fig. 154). We perceive some vague and confused spots.

By its rapid motion Mercury seems to 'play at hide-and-seek' with us. It appears only again to disappear; shines for a moment in the evening after sunset, re-plunges into the solar rays, shines in the morning in the east, preceding the sun; falls back to the flaming star, to appear anew in the evening; thus showing itself sometimes as a morning, sometimes as an evening star. This period of oscillation varies between 106 and 130 days. The ancients at first believed in the existence of two distinct stars: these were Set and Horus among the Egyptians, Buddha and Rauhineya among the Hindoos, Apollo and Mercury among the Greeks. The first shepherds who discovered Mercury in the rays of the setting sun were the ancient Egyptians, who associated the sky with all their works. Set and Horus accompanied the sun like two satellites, and, later, when the identity of the two stars was evident, the Egyptian astronomical system made Mercury revolve round the sun instead of round the earth. (It was the first to do so.) We possess astronomical observations of this planet, made by the Chaldeans, from the year 265 before our era; and since the year 118 made by the Chinese.

The agility of his motion has given to Mercury corresponding functions. He was represented with wings on his feet. He was the messenger of the gods. He was also the god of thieves, of traders, and of physicians! And even now, are not apothecaries' shops decorated with the caduceus of Mercury?

On account of our city habits we observe the stars in the evening in preference to the morning. In order to find Mercury it is necessary to search in the western sky about half an hour after sunset, at the times of his greatest elongations from the sun. These return in the evening about every four months, under various conditions, and consequently it is necessary to know these times.[1]

The planet Mercury is, like the earth and the moon, a globe of dark matter which only shines and is visible by the illumination of the solar light. Its motion round the central star, which brings it sometimes between the sun and us, sometimes in an oblique direction, sometimes at right angles, and shows us a part, incessantly variable, of its illuminated hemisphere, produces in its aspect, as seen in a telescope, a succession of phases similar to those which the moon presents to us, and which our readers will understand with the greatest ease by going back to the explanation of the phases of the moon. Fig. 155 represents the apparent variations of size and the succession of phases, visible in the evening after sunset; when the planet attains its most slender crescent it is in the region of its orbit nearest to the earth, and passes between the sun and us; then, some weeks afterwards, it emerges from the solar rays and passes again through the same series of phases in inverse order (as we see them by reversing the figure).

[1] These indications for the perpetual observation of the sky are given each month in our *Revue mensuelle d'Astronomie populaire.*

These phases are invisible to the naked eye, and this invisibility was used as an objection to the Copernican theory, on the ground that, if Mercury and Venus revolved between the sun and earth, they would present phases like the moon. 'God,' replied Copernicus, 'will cause instruments to be invented to improve the sight, and then you will see them.' Their discovery in the seventeenth century was the finishing blow to the adversaries of modern astronomy.

If this planet revolved round the sun exactly in the plane in which we ourselves revolve, it would pass across the radiant disc at each of its inferior conjunctions—that is, three times a year on the average. But it moves in a plane inclined 7° to the ecliptic, and, in order that it should pass exactly in front of the day star, it is necessary that its conjunction should happen in the line of intersection of the two planes, or the 'line of nodes,' as we have seen for eclipses of the sun by the moon and for transits

Fig. 155.—Phases of Mercury before Inferior Conjunction: 'Evening Star'

of Venus. This conjunction occurs oftener than for Venus, and the transits are much more frequent; they return at irregular intervals: 13, 7, 10, 3, 10, and 3 years. The following are their dates for two centuries:—

Transits of Mercury across the Sun

Nineteenth Century		*Twentieth Century*	
1802	Nov. 9	1907	Nov. 12
1815	Nov. 12	1914	Nov. 6
1822	Nov. 5	1924	May 7
1832	May 5	1927	Nov. 8
1835	Nov. 7	1937	May 10
1845	May 8	1940	Nov. 12
1848	Nov. 9	1953	Nov. 13
1861	Nov. 12	1960	Nov. 6
1868	Nov. 5	1970	May 9
1878	May 6	1973	Nov. 9
1881	Nov. 7	1986	Nov. 12
1891	May 10	1999	Nov. 24
1894	Nov. 10		

Fig. 156 shows each of the transits of our century in its form and magnitude. The circle represents the disc of the sun, and the lines which cross it indicate the paths followed by the planet.

We see that the length as well as the inclination of the lines differ considerably from one transit to another. The planet always enters on the left or east, and goes off on the right or west. In this apparent complication we may, however, notice a real order: all the transits which happen in the month of May are parallel to each other; all those which happen in November are also parallel among themselves.

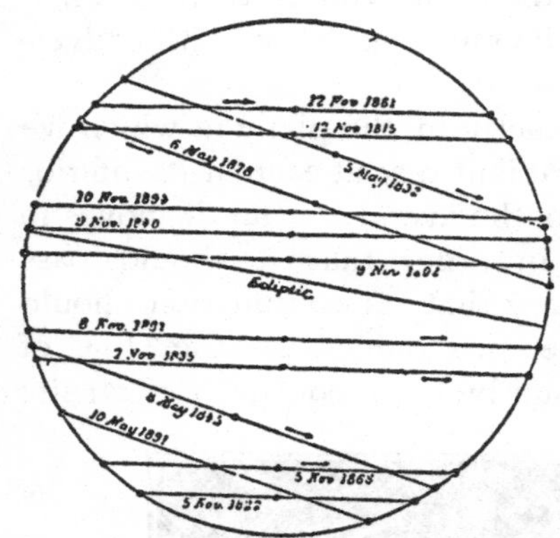

FIG. 156.—TRANSITS OF MERCURY IN THE NINETEENTH CENTURY

In 1868 the end of a transit was observable; in 1878 the sky was cloudy at Paris, with vistas through the clouds. In 1888 the phenomenon was invisible in France; in 1891 the end was visible at sunrise.

The transit of November 5, 1868, happened at sunrise. It was a spectacle very interesting and rather rare; astronomers were at their telescopes at the moment calculated for the appearance of the phenomenon. I had the pleasure of observing it, although the last scene only of this astronomical sight was favoured with a clear sky. The planet followed the line drawn on the diagram for the transit of that date. It was absolutely round and very black—much blacker than the solar spots.

During this transit several astronomers perceived a luminous point on the disc of Mercury, a point already seen in several previous transits.[1] This

[1] I did not perceive it myself, and I doubted its reality; but at the transit of 1878 it was reobserved and absolutely verified by my learned friend, Ad. de Boë, the Belgian astronomer. The most curious fact is, that during the transits of Mercury which happen in May this luminous point is found to the west of the planet's centre, while during the observations made in November it has always been seen to the east. It is not exactly at the centre, which proves that it is not an optical effect due to diffraction. Another observation not less curious is the areola with which the planet appears surrounded during its transit over the sun. Sometimes this areola is more luminous than the sun himself, and sometimes it is of a greyish or slightly violet tint. In general, the first case is presented in the month of November and the second in the month of May. (The fact is rather strange. I have observed in a balloon a phenomenon which resembles it, though it is certainly not of the same nature; several times the shadow of the balloon travelling across the meadows has been seen encircled with a luminous areola.)

We may remark now that at the time of transits in the month of May Mercury is at its greatest distance from the sun, while in the month of November it is in the neighbourhood of its perihelion—that is to say, near his least distance. There may exist a relation between this distance and the position of the luminous spot and the aspect of the areola. Doubtless the heat of the sun, four and a half times greater and hotter than that of the earth when Mercury is at its aphelion, and ten and a half times greater and more intense when it is at its perihelion, produces in the atmosphere of has been attributed to a *volcano*, and the areola (see footnote, p. 354) to an immense atmosphere. It would be singular if a colossal volcano were in eruption on Mercury, near the middle of the hemisphere turned towards the earth, exactly at the day and hour of a transit of this planet across the sun ; it would not be less strange that this planet should be surrounded by an atmospheric envelope equal to a third of its diameter ; it is as if our atmosphere were more than 2,000 miles in height. The most simple explanation is to suppose that, Mercury being but a small black point, *invisible to the naked eye* on the dazzling sun, the difficulty of observation in such a state of contrast produces phenomena purely optical. We must not imitate the inexperienced astronomer who mistook a distant fly for an elephant in the moon! The human eye is itself subject to errors and makes 'misprints.'

We may remark, however, with reference to this, that the eyes or the instruments of astronomers never commit faults so outrageous as those we often meet with even in the most carefully finished works. To mention but a few in passing, have we not read, for example, for many years, in the works of the poet Gilbert, in the midst of his pathetic adieus to nature, this singular expression :

Au *baquet* de la vie infortuné convive ! . . .

And a careless printer thus described in his newspaper an official reception by M. Guizot :

Une foule immense emplissait l'amphithéâtre. L'illustre homme d'État prend place au milieu des *gredins*, et est aussitôt accueilli par les plus *vils* applaudissements.

But the best is this. In a beautiful edition of the 'Book of Hours' of M. Affre, Archbishop of Paris, and in the part of the text relating to the order of the Mass, in place of the words, 'Ici le prêtre ôte sa calotte,' the printer put 'Ici le prêtre ôte sa culotte.' The entire edition was printed with this absurd misprint!

In all these cases, and in many others which might easily be quoted, the printer, the reader, the author, and the editor had assuredly much worse sight than any astronomer ever had at his telescope. Such typographical blunders have even sometimes led to tragi-comical consequences. Thus, lovers of bibliographical curiosities may remember that in a treatise on natural history, rather tedious, moreover, we read in the first pages, 'L'auteur est de la famille des buses.' The printer had put *auteur* for *autour*. The writer, who was somewhat irritable, believed it was malice, and challenged the printer to a duel!

But let us return to Mercury, and recapitulate the knowledge acquired of its physical condition and of its nature as a planetary world.

this planet meteorological, magnetic, and electrical phenomena quite different from those which we know on the earth. But we need not hasten to explain facts which may be purely subjective.

We have already seen that its revolution round the sun is accomplished in about 88 days. Its year is, exactly, 87 days and 97 hundredths of a day, or 2 months 27 days 23 hours 15 minutes and 46 seconds. This is less than three of our months. The inhabitants of this planet have, then, their life measured by years four times more rapid than ours. A centenarian on Mercury has lived but twenty-four of our years; in other words, a young man of twenty-four years would be a centenarian on Mercury. If biology is regulated as in our world, impressions should be more rapid and more lively, vital acts should be accomplished with greater celerity; the inhabitants should become adolescent in an interval of five terrestrial years, mature in ten years, and old in twenty terrestrial years. (It is true that we often meet Mercurians on the boulevards of Paris.) Further, the solar light and heat being much more intense than here, striking meteorological effects should be produced in rapid seasons, each of which does not last more than twenty-two days. The axis of the planet is much more highly inclined than that of ours, for this inclination appears to be 70° (the exact measurement is difficult, on account of its proximity to the sun), so that these short seasons form an enormous contrast between the Mercurian winter and summer. But still this is not all. We have seen that the orbit described by the planet is very elongated, and that it is nearly fifteen millions of miles nearer the focus at perihelion than at aphelion: fifteen millions out of thirty-six of mean distance! At the aphelion the day star shows these unknown natives a disc four and a half times larger than ours in surface, and forty-four days afterwards, at the perihelion, this enormous disc still further expands to ten and a half times greater than ours, pouring down from the torrid sky a light and a heat ten and a half times more intense. The apparent diameters of the sun are as follow:—

Seen from the perihelion of Mercury	104′
" mean distance "	83′
" aphelion "	67′
" earth	32′

Fig. 157 gives an idea of these diameters. We sometimes complain of the heat of the sun; but what is our poor luminary compared with the dazzling furnace of Mercury! It is as if ten suns together darted their rays on our heads in the month of July at noon. If the inhabitants of Mercury believe, as we did once, that this orb revolves round them, they must be puzzled to explain its periodical variations of size, its successive swellings and reductions.

We have here, then, a world ruled meteorologically by two sorts of seasons quite different from each other. Does it really endure such extremes? Yes, if it is not tempered by a sufficient atmosphere. A layer of clouds checks radiation. The Mercurian atmosphere is, perhaps, so constituted that it tempers the planet and harmonises the extremes. This atmosphere appears to be much denser and more cloudy than ours. The 'terminator' circle of the phases of Mercury is not sharp, but diffuse and shaded; there is here an atmospheric penumbra. The extent of the phases also leads us to infer the presence of an atmosphere. The spectroscope shows in the spectrum of this planet lines of absorption proving that it has a gaseous envelope thicker than ours. Whatever this atmosphere may be, it is probable that the mean temperature of Mercury is higher than that of the earth, and that a Mercurian would be frozen to death in Africa and Senegal.

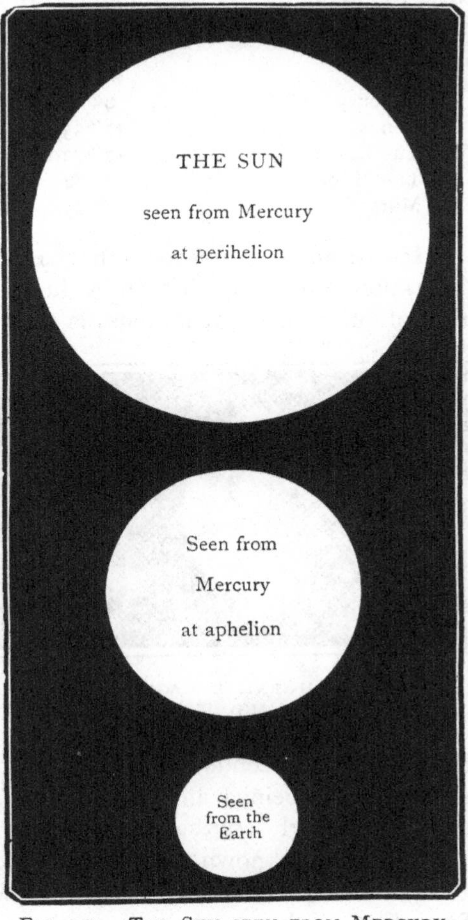

Fig. 157.—The Sun seen from Mercury

The careful observation of the terminator circle shows it to be irregular, and proves that the surface of the planet, far from being smooth, is broken by enormous irregularities, rising to the 253rd part of the diameter of this globe. Now, the globe of Mercury, much smaller than the earth (it is the smallest of the eight principal planets), is to that of our world in the ratio of 373 to 1,000 for the diameter, and measures but 4,800 kilometres, or 1,200 leagues (about 3,000 miles). The highest peaks of the Cordilleras of Mercury must rise to nearly 63,000 feet! From the periodical return of the same irregularities Schröter has found that the duration of rotation of this planet is 24 hours 0 minutes 50 seconds. This measure requires a new verification. It is probable in any case that the duration of the day and night is nearly the same as here. [From recent observations Schiaparelli finds that the period of rotation is the same as the period of revolution, or about 88 days; but this seems to require confirmation.—J. E. G.]

The earth is flattened at the poles by $\frac{1}{292}$. Mercury may have the same figure, but the proportion is so small that it is imperceptible in the best telescopes.

The diameter of this planet is but 37 hundredths of that of our globe, as we have seen. This real diameter is calculated from the apparent diameter combined with the distance. We have seen, with reference to the transits of Venus, that the conclusions relative to the solar parallax give the number 17″·72 for the diameter of the earth as seen from the sun. It is to this unit that the diameters of all the planets are referred, supposing them all to be seen at the same distance. The following are these angular diameters:—

Mercury	6″·61	Jupiter	196″·00
Venus	17″·55	Saturn	164″·77
The Earth	17″·72	Uranus	75″·02
The Moon	4″·84	Neptune	67″·29
Mars	9″·35		

It is from these numbers that the table on p. 209 has been calculated. We know from this that the volume of Mercury is but 5 hundredths of that of our globe, that its mass is only 6 hundredths, and that its density is consequently a little more than that of the materials constituting the planet which we inhabit; representing the terrestrial density by 1,000, that of Mercury is represented by the number 1,173. This is the greatest density of the whole solar system. But, as we have seen (p. 111), the intensity of gravity at the surface of this first province of the planetary archipelago is much weaker than here; a falling body descends but 8·36 feet in the first second of its fall. Thus, although the beings and things which exist on this globe are *denser* than ours, they weigh less than one-half.

FIG. 158.—COMPARATIVE MAGNITUDES OF THE EARTH AND OF MERCURY

There is no known satellite of Mercury.

Such are the positive ideas which we now possess with reference to the first planet of the system. As to the conjectures which might be suggested on the nature of the humanity which peoples it, this is not the place to enlarge on these philosophical considerations. Whatever they may be, and whatever may be the epoch of habitability of this planet in the history of the immense universe, *the inhabitants of Mercury cannot but be organised according to the special conditions of their country.* There, the eyes are constructed to support an intense light which would blind us, the blood to circulate pleasantly in a torrid heat, the muscles to move bodies endowed with an extreme lightness. It is, then, probable that, life being formed and developed on this planet under conditions altogether different from those which guide the evolutions of terrestrial life, the last animal species—that is to say, the human species, the last branch of the zoological tree—does not exactly resemble us, either in form, in size, or in suitability to the conditions of external nature. The powerful and fruitful energy of the sun, which we have already described, has doubtless developed on this tropical island a work incomparably richer than that of terrestrial nature, which is but a polar zone compared with Mercury. There especially the sun's divine rays flow in waves of gold, and waves of electricity circulate through all its beings.

The inhabitants of Mercury see us shining in their sky at midnight as a splendid star of the first magnitude! Venus and the earth are the two most brilliant objects in their starry nights. The earth and moon form to them a double star. If they have sufficiently powerful instruments, they may perhaps have already commenced to draw a geographical map of our planet—unless their religious and political principles affirm that Mercury is the only inhabited world, and prohibit the free examination of the heavens. The inhabitants of each planet have all, originally, believed themselves the centre of the universe, because they do not perceive their own motion, as the dwellers on earth do not perceive the motion of the earth, and, like the Chinese, they declare they occupy 'the central empire,' the rest of the universe being superfluous or given up to barbarians. Astronomy alone can upset these common illusions and lead the student to the mountain of Truth.

CHAPTER III

THE PLANET VENUS

The Shepherd's Star

♀

TWO worlds revolve between the earth and the sun: the first is Mercury, to which we have just directed our attention; the second is Venus, which we will now consider. The first revolves at 36 millions of miles, the second at 67, the earth at 93. We are already familiar with these ideas, and we now know the plan of the solar system as well as we know the map of England or of Europe. It was, in fact, the first knowledge to be acquired in order to travel with profit in the sky. We often meet with travellers who visit France, Switzerland, Italy, &c., without maps—that is to say, who travel without learning where they are going, and never know exactly where they are; they lose at least half their pleasure and instruction. It is true that we also meet with pretended lovers of the art who have a singular way of travelling, like the tourist who, going to visit the Museum of the Louvre, thus expressed his admiration: 'Ah! my dear, what a superb museum! Fancy, I have taken more than an hour to visit it—and you know that I walk quickly!'

We do not thus proceed in our astronomical instruction. The *method* of study is not less important than the examination of the subjects themselves; we might even say that it is more important, in that it prepares our mind to receive successively and simply all the data acquired by the science, to classify them logically, and register each in its proper place, like pieces of mosaic formed by nature itself. If the most difficult problem is well put, it is half solved.

We come, then, to the second planet of our solar system. It would be superfluous to re-draw the orbit, since we have already seen it twice, first in the general plan of the system (pp. 211, 212), and again with reference to the transits of Venus between the sun and the earth (p. 228). What we have said of the motions of Mercury applies to those of Venus also on a larger scale. As the orbit of Venus surrounds that of Mercury, Venus is much farther from the sun; she may withdraw from the sun up to 48°, and remain in the evening sky, or precede the day star in the morning by more than four hours. But she cannot move farther away, and consequently she is, like Mercury, a morning and evening star.

Revolving round the sun in 224 days, Venus has her motion combined with ours in such a manner that she passes her inferior conjunction, between the sun and us, every 584 days; but the plane in which she revolves is inclined 3° 23′ to that in which the earth itself moves, so that the transits across the solar disc only happen at the epochs already indicated. When Venus attains her greatest elongations from the sun she shines in the west in the evening, then in the morning in the east, with a splendid brightness which eclipses that of all the stars. She is, without comparison, the most magnificent star of our sky. Her light is so vivid that it casts a shadow. Sometimes, even, it pierces the azure of the sky, in spite of the presence of the sun above the horizon, and *shines in full daylight.* In ancient times Æneas, in his voyage from Troy to Italy, saw it several times shining above his head; and in modern times, in 1799, General Bonaparte, returning from the conquest of Italy, was accompanied by the same celestial diamond, to which all the Parisians directed their gaze. The great Captain was a little superstitious, something of a fatalist, like most military men: he believed himself for a long time protected by a star. Far from having large and general ideas, he attributed everything to his person and his sphere; everyone knows that he denied the power of steam and refused the offers of Fulton. One evening, leaning at a window of the Palace of the Tuileries, he appeared absorbed in contemplation, and his gaze remained fixed on a point of the starry sky, when, turning suddenly to his uncle, Cardinal Fesch: 'See, there,' said he, 'that is my star! It has never abandoned me.' Who knows?—that beautiful star was perhaps Sirius, the brightest in the sky, which shows itself in the latitudes of St. Helena and the land of the Zulus. O dynasties which believe themselves founded for eternity, and which do not exist for a single year of Uranus or Neptune!

Fig. 159.—Venus at her Greatest Brilliancy

The maximum visibility of Venus is produced by its greatest phase, by its greatest elongation from the sun, and by the clearness of our atmosphere.

The years 1716, 1750, 1797, 1849, 1857, 1889, have been remarkable in this respect.

The brilliant Venus was certainly the first planet noticed by the ancients, as much on account of its brightness as its rapid motion. Hardly is the sun set than it sparkles in the twilight; from evening to evening it removes farther from the west and increases in brightness; during several months it reigns sovereign of the skies, then plunges into the solar fires and disappears. It was pre-eminently the star of the evening, the shepherd's star, the star of sweet confidences. It was the first of celestial beauties, and the names conferred upon it correspond to the direct impression which it produced on contemplative minds. Homer called it 'Callistos,' *the Beautiful*; Cicero named it *Vesper*, the evening star, and *Lucifer*, the morning star, a name likewise given in the Bible and the ancient mythologies to the chief of the celestial army. Venus was the earliest and the most popular of ancient divinities. In the primitive ages, the hour when she lit up her limpid light was expected by the *fiancée*, who associated the beautiful planet with the sweetest sentiments of her heart.

The most ancient astronomical *observation* we have of Venus is a Babylonian record of the year 685 B.C. It is written on a brick and preserved in the British Museum.

During many centuries men believed, as in the case of Mercury, in the existence of two planets, and consequently two divinities. But when observation had shown that Lucifer and Vesper were never visible at the same time, and that the morning star appeared only when the evening star had vanished, they became convinced that it was one and the same star. Anyone may easily account for these successive appearances and observe them with interest.[1]

Like all astronomers, I make some observations every year of the planet at the epochs of its greatest visibility. Fig. 159 represents the phase of Venus at one of these epochs. This observation was made on August 16, 1879. Venus then showed a diameter of 37″ (represented by 37 millimetres).

Like Mercury, Venus shows phases corresponding to the positions which it occupies round the sun with relation to us. These phases present particular interest to beginners in the study of astronomy. A telescope of moderate power is sufficient to show them. When they are observed for the first time the observer is frequently affected by an easily explained illusion which makes him think that it is the moon which he sees. I have sometimes had considerable difficulty in persuading certain persons that

[1] We may remark here that for Venus, as for Mercury, it is necessary to know the times of the appearances of the planet if we wish to follow them with any regularity, either with the naked eye or by the aid of instruments. These epochs vary from year to year. They will be found each year in our monthly review, *L'Astronomie*. Every eight years the brightness is a maximum: 1881, 1889, 1897.

this is not the case; and nothing less than the absence of the moon from the sky would prove to them that the object visible in the field of the telescope could not be our satellite. The best hours for examining Venus in a telescope are those of daylight. In the night the irradiation produced by the brilliant light of this beautiful planet prevents us from distinguishing clearly the outlines of its phases

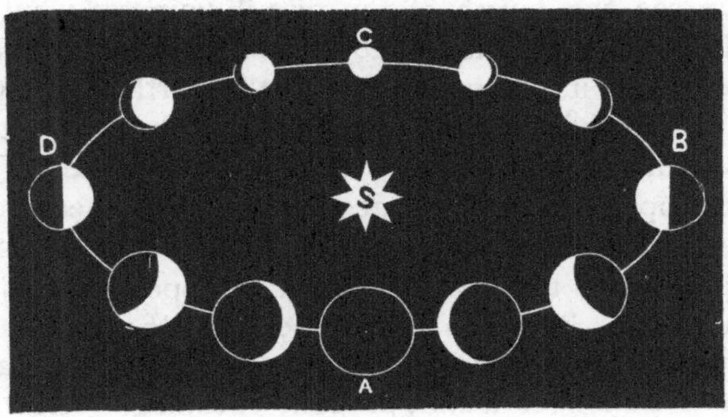

FIG. 160.—PHASES OF VENUS

Fig. 160, which shows the order of these phases, explains itself; whilst fig. 161 shows the relative magnitude of the four principal phases. When Venus occupies the region of its orbit behind the sun, with reference to us—which is called the point of superior conjunction—it is at its greatest distance, and is reduced to a disc of 9½

FIG. 161.—COMPARISON OF THE FOUR PRINCIPAL PHASES OF VENUS

seconds in diameter, represented in our drawing by a disc of 9½ millimetres. It comes imperceptibly towards us, and when it passes its quadrature, at its mean distance, it presents the aspect of a half-moon. It soon attains its most brilliant light, at the epoch when it shines at a distance of 39° from the sun, and shows the third phase 69 days before its inferior conjunction. Its apparent diameter is then 40 seconds, and the width of its illuminated part is scarcely 10 seconds. In this position we see the fourth of the disc illuminated; but this quarter emits more light than the more complete phases. Finally, when it reaches the region of its orbit nearest to the earth, it shows us nothing more than an excessively thin crescent, since it is then between the sun and us, and presents to us, so to say, its dark hemisphere. This is the position where its apparent size is greatest, and it then measures 62 seconds in diameter, which we have represented by a diameter of 62 millimetres. It is then quite near the sun, and soon disappears in his rays. Sometimes, as we have seen, it passes exactly between the sun and us, and appears a little larger (63 to 64 seconds); but it is then an absolutely black disc, and it is no longer a star, properly speaking. After passing its inferior conjunction the phases are reproduced, in inverse order, as a morning star.

The phases of Venus were seen for the first time by Galileo, towards the end of September, 1610. But his observation did not appear to him at once certain and unquestionable. In order to give himself time to verify this discovery, without running the risk of seeing it anticipated, the illustrious philosopher concealed it in this anagram:—

Hæc immatura a me jam frustra leguntur, o.y.
These things, not ripe, and hidden still to others, are read by me.

Placing the 34 letters of this Latin sentence in another order, they form these very categorical words:—

Cynthiæ figuràs æmulatur mater amorum.
The mother of love imitates the phases of Diana.

—an explicit sentence, which has no longer the vagueness of the first, and clearly asserts the existence of these phases. But who could have solved the first enigma so as to divine its true meaning?

Galileo had remarkable ingenuity of mind. On November 5, 1610, Father Castelli asked the celebrated astronomer of Florence if Venus and Mars should not present phases. Galileo replied that he had many researches to make in the sky, but that, owing to the very bad state of his health, he found himself much better in his bed than in the evening dew. It was only on December 30 that he announced the fact that he had raised the veil of Venus.

The discovery of these phases, which present on the whole exactly the same circumstances as those of the moon, overthrew one of the first objections which were raised against the system of Copernicus.

Venus is constantly visible in full daylight in astronomical instruments, even at the moment of its superior conjunction. It is then round and quite small. At the epochs of its inferior conjunction it presents itself under the form of a very thin crescent, as is shown in fig. 162, which I made at one of these epochs (September 21, 1887). It was then quite near the sun.

We sometimes notice that the interior of the crescent of Venus, the remainder of the disc, is less black than the background of the sky. This has been called the ashy light (*lumière cendrée*) of Venus, although it has no satellite to produce it. It seems to me that this visibility, rather subjective than objective, arises from clouds on the planet, which whiten its disc and vaguely reflect the stellar light scattered through space. The eye instinctively continues the outline of the crescent, and imagines, rather than sees, the rest. Besides, the aurora borealis may sometimes light up during the night the glowing sky of Venus, and its clouds may emit a certain phosphorescence, as we occasionally notice on the earth in the evenings of April and May.

FIG. 162.—CRESCENT OF VENUS ONE DAY BEFORE INFERIOR CONJUNCTION WITH THE SUN

Venus passed near Saturn (the two objects in the same field of the telescope) on August 9, 1886.

The revolution of Venus round the sun is performed in an orbit almost exactly circular, and without perceptible eccentricity (0·0068), in a period of 224 days 16 hours 49 minutes 8 seconds. Such is the calendar year of this neighbouring world. It is, therefore, about seven months and a half. In the time of Copernicus it was still believed to be nine months, as we see on the drawing from the work of Copernicus reproduced above (p. 341). While we count 100 years the inhabitants of Venus reckon 162, and those of Mercury 415! On such worlds the years pass still more quickly than here: the ladies who already fret at the rapidity of the terrestrial calendar would there be disconsolate.

The days of Venus, also, are a little more rapid than ours, but not much Since the year 1666 attentive observation of the planet led Cassini to conclude that it turns on itself in 23 hours 15 minutes. This observation is extremely difficult, on account of the brightness of the planet and the faintness of the irregularities visible on its disc. The observations of Bianchini, in 1726, indicate 23 hours 22 minutes. Those of Schröter, at the end of the last century, gave 23 hours 21 minutes. The period was definitely determined in 1841, at Rome, by De Vico, and fixed at 23 hours 21 minutes 22 seconds.

[Recent observations by Schiaparelli lead him to conclude that Venus, like Mercury, rotates on its axis in the same period that it revolves round the sun ; but the correctness of this result has been disputed by other astronomers.—J. E. G.]

This resemblance to the rotation of the earth is very curious. The year of Venus, composed of 224 terrestrial days, consequently contains 231 of its own, since the day is a little shorter there than here.

These same observations show that the axis of rotation of this planet is much more inclined than ours, and that this inclination is 55 degrees. It follows that the seasons, although each lasting but 56 terrestrial days, or 58 Venusian days, are much more intense on this world than on ours.

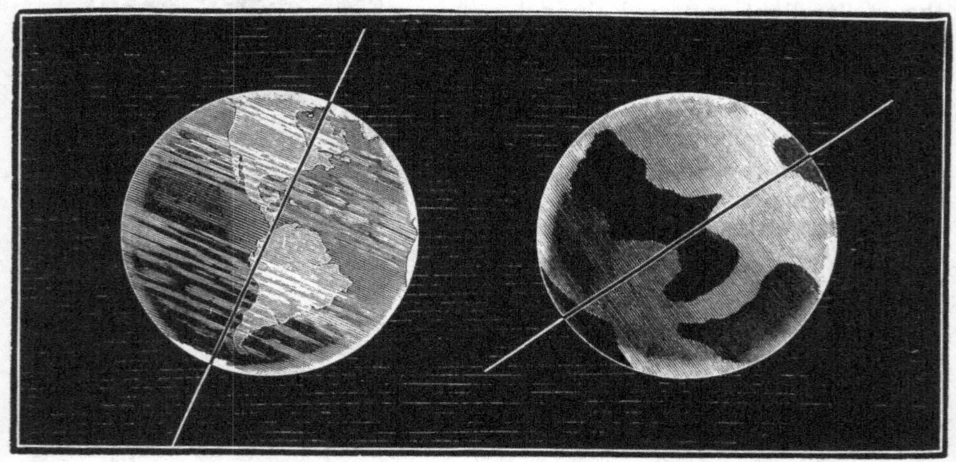

Fig. 163.—Comparative Inclinations of the Axis of the Earth and of that of Venus

They pass, without transition, from the heat of summer to the frosts of winter.

We have seen above, if we believe the traditions of which the author of 'Paradise Lost' has made himself the poetical echo, that the inclination of the axis of our planet was produced after the fall of Adam 'by angels coming in the name of the Divine wrath to chastise the disobedience of our first parents.' God, being supremely just, the chastisement was proportional to the fault. It is necessary, then, to believe that in the world of Venus the first human couple committed a much graver offence, and that on this celestial world, our neighbour, the first man and woman were much further beyond pardonable limits ; for the axis of their world has been turned aside to an inclination double that of the inheritance of Adam and Eve. It follows that this abode is far from being calm and tranquil, for it passes in turn through the extremes of heat and cold, through all the alternations of the most disordered passion.

The inclination of the world of Venus being more than twice as great as ours, we have only to take a terrestrial globe and incline it by the same quantity to understand the climates and seasons which will result. We may easily see that the torrid zone extends, in this case, up to the frigid zone, and even beyond it; and, reciprocally, the frigid zone extends to the torrid zone, and even encroaches on it; so that no place remains for a temperate zone. There is not, then, on Venus any temperate climate, but all latitudes are both tropical and arctic.

Now, in the tropics the sun darts his rays twice a year perpendicularly overhead, while in the arctic regions there are days in which he never rises at all, and others in which he never sets. What, then, must be the vicissitudes of countries which are in turn arctic and tropical? At a certain time of the year the sun remains for several days without rising, at another epoch he remains several days without setting, and between these two seasons he soars vertically overhead. The contrast between the glacial temperature of the season deprived of the sun and the intense fires of that in which the sun of Venus—*twice as great and twice as hot as ours*—pours out from the heights of heaven his burning heat certainly does not constitute a very agreeable prospect. We really do not know which region of Venus would be the least disagreeable to inhabit, and it would be almost more advantageous to choose a residence near the equator than near the poles.

It follows, then, from all these circumstances, that the seasons and climates are much more violent and more varied than ours. [In opposition to all this, however, Schiaparelli finds that the axis of rotation of Venus is nearly perpendicular to the plane of its orbit. But this is disputed by other observers, and, in fact, the inclination of the axis is still uncertain.—J. E. G.]

This neighbouring world shows nearly the same dimensions as ours. We have already seen (in the preceding chapter) the angle which it subtends as seen from the sun. This reduction to the terrestrial unit proves that, its angular diameter being 17″·55, while that of the earth is 17″·72, the two real diameters are in the ratio of 975 to 1,000. There is, then, between the two globes only a slight difference in favour of the earth. The diameter of Venus measures 12,700 kilometres (7,890 miles), and its circumference 24,700 miles. Thus this planet is truly the twin sister of ours.[1]

[1] According to a measure made during the last transit of Venus by Colonel Tennant, this planet may be slightly flattened at the poles, and even a little more so than the earth; the proportion would be $\frac{1}{280}$. From photographic measures of the same transit by M. Bouquet de la Grye it would be $\frac{1}{303}$.

It weighs a little less than our planet (79 hundredths); its density, compared to ours, is 0·81, and

The resemblance will be still more complete if we add that this world is certainly surrounded by an atmosphere. The penumbra observed along the crescent of Venus had already indicated (last century) the existence of this aërial envelope, since the dawn and the twilight on the different meridians of this globe are visible from here. Further testimony is afforded by the prolongation of the horns of the crescent beyond their geometrical limit, and again by the fact that the external outline of a phase of Venus always appears much more luminous than the interior border. These evidences have been confirmed by the revelations of spectrum analysis.

When we examine with the spectroscope the light reflected by this planet we first find the lines of the solar spectrum (and this is natural, since the planets have no light of their own, and merely reflect that of the sun); but we notice besides several absorption lines similar to those which the terrestrial atmosphere gives, and particularly those of clouds and water vapour.

The observations of Huggins, Secchi, Respighi, and Vogel are in agreement. At the time of the transit of Venus in 1874 Tacchini, stationed in Bengal, examined with care the solar spectrum at the point occupied by Venus, and also inferred the existence of an atmosphere, 'probably of the same nature as ours.' 2,500 miles from there, in Japan, and thousands of miles farther, at the island of St. Paul, and in Egypt, the missionaries of science, French and English, made a very different but confirmatory observation. At the ingress and the egress of the disc of Venus on the sun the half of Venus outside the sun was seen outlined by a faint arc of light [I saw this myself during the transit of 1874, in India, and in the transit of 1884, in Ireland.—J. E. G.], which was nothing else but the illuminated Venusian atmosphere. More complete measures still have been made in the United States. One observer, Mr. Lyman, followed Venus day by day at the epoch of inferior conjunction, and saw

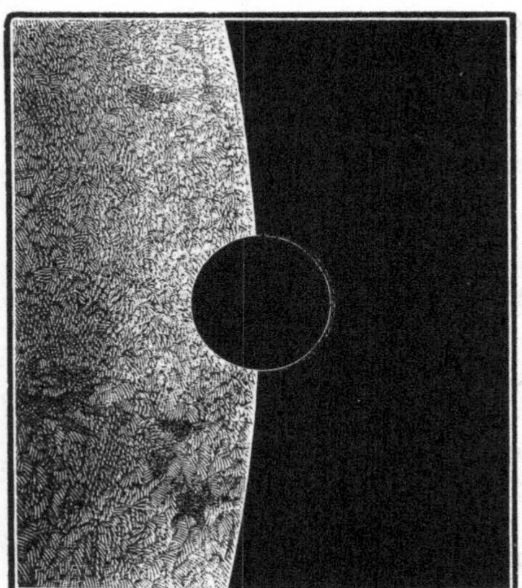

Fig. 164.—The Atmosphere of Venus Illuminated by the Sun, at the Moment of Entry upon the Edge of the Sun

gravity, at its surface, is 0·80. Under all these aspects, it is the celestial world which most resembles that which we inhabit.

its thin crescent elongating until the two points ended by passing all round the dark disc and meeting, so that the planet showed in the telescope the aspect of a luminous ring. This study led Mr. Lyman to complete all the preceding data on the atmosphere of Venus by calculating its refraction, and from that its density. This horizontal refraction is 54′. That of the terrestrial atmosphere being 33′, it follows that the density of the atmosphere at the surface of this planet is greater than ours in the ratio of 100 to 189. It has, then, an atmosphere nearly twice as dense as ours.

Fig. 165.—Venus as a Morning Star, November 1887. (From Drawings by W. F. Denning.) 1. November 2, 6h 30m A.M. 2. November 3, 7h A.M. 3. November 5, 7h A.M. 4. November 6, 7h 30m A.M. (10-inch Reflector; Power 252)

This density, vapour of water, and these clouds, appear very well adapted to temper the heat of the sun, and to give to this globe a mean temperature but little different from that which characterises our own abode.

We may also add, that attentive observation of the indentations visible on the crescent of Venus has shown that the surface of this planet is quite as uneven as that of the earth, and even more so; that there are there Andes, Cordilleras, Alps, and Pyrenees, and that the most elevated summits attain a height of 44,000 metres (27 miles). It has even been ascertained that the northern hemisphere is more mountainous than the southern.

Even the study of the geography of Venus has already been commenced. But it is extremely difficult to draw, and the hours of sufficiently pure atmosphere and possible observation are very rare. This difficulty will be easily understood if we reflect that it is exactly when Venus arrives at its nearest to us that it is least visible, since, its illuminated hemisphere being always turned towards the sun, it is its dark hemisphere which is presented to us. The nearer it approaches us, the narrower the crescent becomes. Add to this its vivid light and its clouds, and you may imagine what difficulty astronomers have in dealing with it.

However, by observing it in the daytime to avoid the glare, and not waiting till the crescent becomes too thin, by choosing the quadratures, and making use of moments of great atmospherical purity, observers succeed, from time to time, in perceiving greyish spots, which may indicate the place of its seas. On the preceding page are some drawings of Venus (fig. 165) made by Mr. W. F. Denning, the well-known English astronomer. They show the nature of the markings visible on the planet.[1]

An atmosphere and water exist there as here. From what we have seen above of the rapid and violent seasons of this planet, we might think that the agitations of the winds, the rains, and the storms would surpass everything which we see and experience here, and that its atmosphere and its seas would be subject to a continual evaporation and precipitation in torrential rains—an hypothesis confirmed by its light, due

[1] Bianchini commenced, in the last century, a rudimentary drawing, which we reproduce here (fig. 166), and which has not since been improved, nor even absolutely verified, in spite of the progress of optics. He thought he distinguished three seas near the equator and one near each pole, continents, promontories, and straits. Cassini and Schröter saw, in the last century also, spots which appeared to

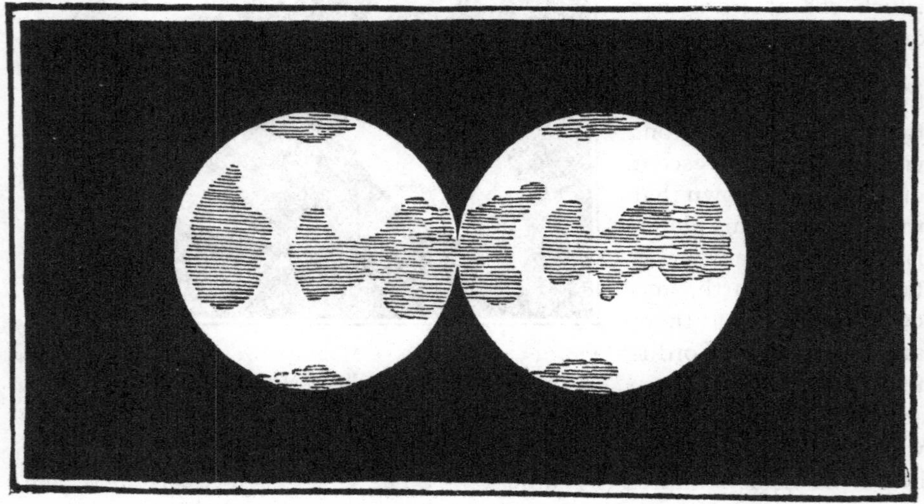

Fig. 166.—Rough Sketch of the Geographical Aspect of Venus

resemble those seen by Bianchini. In recent years Messrs. Langdon and Elger, English astronomers, have made several drawings, some of which resemble those of Cassini. I have also received one from a Belgian, M. van Ertborn. As for myself, notwithstanding all my efforts, I have never been able to clearly distinguish these spots. It would be very desirable that in Italy, or under a sky equally pure, a friend of the science should devote himself to this special observation.

doubtless to reflection from its upper clouds, and to the multiplicity of the clouds themselves. To judge by our own impressions, we should be much less pleased with this country than with our own, and it is even very probable that our physical organisation, accommodating and complaisant as it is, could not become acclimatised to such variations of temperature. But it is not necessary to conclude from this that Venus is uninhabitable and uninhabited. We may even suppose, without exaggeration, that its inhabitants, organised to live in the midst of these conditions, find themselves at their ease, like a fish in water, and think that our earth is too monotonous and too cold to serve as an abode for active and intelligent beings.

Of what nature are the inhabitants of Venus? Do they resemble us in physical form? Are they endowed with an intelligence analogous to ours? Do they pass their life in pleasure, as Bernardin de St. Pierre said, or, rather, are they so tormented by the inclemency of their seasons that they have no delicate perception, and are incapable of any scientific or artistic attention? These are interesting questions, to which we have no reply. All that we can say is, that organised life on Venus must be little different from terrestrial life,[1] and that this world is one of those which resemble ours most. We will not ask, then, with the good Father Kircher, whether the water of that world would be good for baptizing, and whether the wine would be fit for the sacrifice of the Mass; nor, with Huygens, whether the musical instruments of Venus resemble the harp or the flute; nor, with Swedenborg, whether the young girls walk about without clothing; &c. The imaginary travellers to these worlds of the sky have always carried with them their terrestrial ideas. The only scientific conclusion which we can draw from astronomical observation is, that this world differs little from ours in volume, in weight, in density, and in the duration of its days and nights; that it differs a little more in the rapidity of its years, the intensity of its climates and seasons, the extent of its atmosphere, and its greater proximity to the sun. It should, then, be inhabited by vegetable, animal, and human races but little different from those which people our planet. As to imagining it desert or sterile, this is an hypothesis which could not arise in the brain of any naturalist. The action of the divine sun must be there, as in Mercury, still more fertile than his terrestrial work, already so wonderful. We may add that Venus and Mercury, having been formed after the earth, are relatively younger than our planet.

The inhabitants of Venus see us shining in their sky like a magnificent star of the first magnitude, soaring in the zodiac, and showing motions similar to those which the planet Mars presents to us; but instead of showing a reddish brightness, the earth shines in the sky as a bluish light.

[1] See our work, *Les Terres du Ciel*, Books iv. and ix.

It is from Venus that we are most luminous.[1] The inhabitants of Venus with the naked eye see our moon shining beside the earth and revolving round it in twenty-seven days. They form a magnificent couple. Our planet seen from there measures 65″, and the moon nearly 18″; the moon seen from Venus shows the same diameter as the earth seen from the sun. Mercury is brilliant, and comes immediately after the earth in brightness. Mars, Jupiter, and Saturn are also visible as from here, but a little less luminous. The constellations of the whole sky show exactly the same aspect as seen from the earth.

Such is the second province of the solar republic. Let us cross the region occupied by the earth and moon, and approach the orbit of Mars.

[1] We are the most brilliant object in the sky of Venus, for this world has no moon, in spite of certain observations of last century which for a time produced a belief in a *satellite of Venus*. The critical examination of these observations recently made by M. Stroobant has shown definitely and without possible dispute that the supposed satellite was due to the passage of Venus near small stars. (See our *Revue*, 1887, p. 452.)

CHAPTER IV

THE PLANET MARS, MINIATURE OF THE EARTH

♂

WE now come to the best-known world of the planetary system, that which comes immediately after ours in order of distance from the sun, and which Nature seems to have placed in our neighbourhood as an eloquent example of her unity of life and action. It is the earth itself which we seem to see in space, with interesting varieties and novelties; and we would all, with pleasure, embark to-day on a voyage there if we had at our disposal a mode of locomotion certain to attain the end (going and returning included). How interesting it would be to pass half a century on another world, and then return to this! Even from a purely terrestrial point of view, how interesting and instructive it would be for us if we were enabled to return every century to see what is taking place on the earth, and to view the slow progress of humanity, inventions, sciences, arts, and industry! Mysteries of life! mysteries of death! Shall they never be unveiled?

We already know from the preceding descriptions that the planet Mars is the first we meet with outside ours; it revolves at a distance of 141 millions of miles from the solar focus (the mean distance is 141,393,900 miles), along an orbit which is exterior to that of the earth, and which it takes a year and 322 days to pass over. The combination of its motion with ours causes it to pass behind us, or opposite to the sun, about every two years, or rather every twenty-six months.

It is at these epochs that the planet crosses the meridian at midnight, and it is during these months and the three months which follow that it is most favourably situated for observation in the evening. It then shines as a star of the first magnitude, the rival of Venus and Jupiter.[1]

Its light is reddish, glowing like a flame, and gives the idea of a fire. As we see it to-day, so it shone to our ancestors. Its name in all ancient languages signifies *inflamed* (*embrasé*), and its personification is that of the God of War. Men have always attempted to partly excuse their passions

[1] For its position in the sky each year see our review, *L'Astronomie*. Mars passed near Saturn on June 30, 1879, and September 19, 1889; to the naked eye they were seen as a single star.

by attributing their most perverse acts to the fatal influence of some superior divinity or some demon ; and as war has been at all times the plaything of the great and the imbecile joy of the little, the planet of war has always been one of the most honoured and most dreaded : the temples of Mars alternate with those of Venus ; the laurel and the myrtle unite their branches ; destruction and reproduction are complementary. The burning star of Mars presided over combats ; on the field of battle of Marathon, in the midst of the carnage of the Cimbri, or in the dark pass of Thermopylæ, the imprecations of the victims accused it of barbarity, while the fact is that man has no enemy but himself, and the innocent planet soars in space without suspecting the influences of which it is accused.

The red planet varies in brightness according to its position in the sky and according to its distance. The orbit which it describes round the sun is not circular, but elliptical, the eccentricity being 0·093 :—

Perihelion distance . .	1·3826 ;	204,520,000 kilometres	(128,200,000 miles)
Mean distance . . .	1·5237 ;	225,400,000 ,,	(141,390,000 ,,)
Aphelion distance . .	1·6658 ;	246,280,000 ,,	(154,580,000 ,,)

We see that the variation of the distance is considerable, and nearly attains a fifth of the mean distance ; Mars is 26 millions of miles nearer to the sun at perihelion than at aphelion, which must produce in the temperature of this planet a very perceptible variation, independent of that of the seasons, due to the inclination of the axis. When the opposition happens at the epoch of the perihelion of Mars, the planet attains its greatest possible proximity to the earth, only 35 millions of miles, and shines with a remarkable brilliancy. Fig. 167 shows the relation which exists between the two orbits ; that of Mars is drawn outside, and that of the earth inside. The two planets revolve in the same direction, but we move more rapidly than our neighbour, and we do not meet again on the same side of the sun till after about two years and two months, and at a distance a little greater. After seven successive oppositions the two planets repass anew at their greatest proximity, which occurs nearly every fifteen years : 1830, 1846, 1862, 1877, 1892. (A rather curious coincidence is that the greatest proximities of Mars correspond with the disappearances of Saturn's ring, of which we will speak further on.) These are naturally the best epochs for observation when we apply ourselves especially to the physical study of the planet.

It was this great eccentricity which led Kepler to discover the true form of the planetary orbits, till then considered as perfectly circular ; he took no less than seventeen years of labour to attain it, and very often he despaired of success. It was the excellent observations of Tycho which proved to him the truth of the system of Copernicus, and led him to the laws which we have already recapitulated.

When Mars is in opposition his diameter may rise to 30″·4 when this happens near the perihelion of the planet and near the aphelion of the earth —that is to say, in the month of August. In 1877 this diameter very nearly approached the maximum; from August 28 to September 8 it was 29″·4. We have seen that the diameter of the moon is 31′ 24″—that is to say, since that of Mars may attain half a minute, the lunar diameter is about sixty-three times greater. In these conditions a telescope magnifying sixty-three times shows us the globe of Mars of the same size as we see the moon with the naked eye. As the planet is then illuminated full-face by the sun situated behind us at midnight at the moment we observe Mars, the observation is thus made in the best conditions, which does not happen for Venus, as we have seen. We then distinguish very clearly a circular disc, on which a white spot immediately strikes the eye, and indicates at once one of the poles of the planet. If the atmosphere is very clear, we soon notice that the red colouring of the disc is not uniform, and that there are spots. A stronger magnification shows the form of these spots.

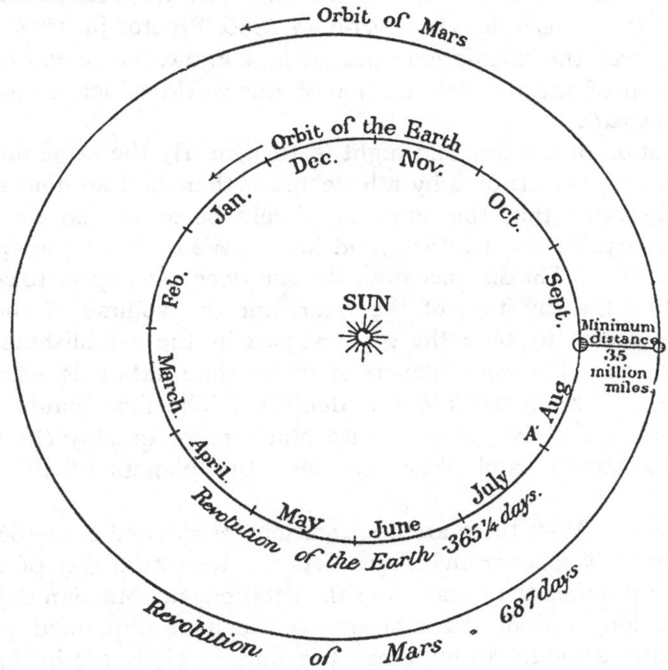

FIG. 167.—DIFFERENCE BETWEEN THE ORBIT OF MARS AND THAT OF THE EARTH

From the first telescopic observations of the planet by Galileo, in 1610, the *phases* of Mars were foreseen, but it was only in 1638 that they were confirmed by the telescope of Fontana in the sky of Naples. Our modern instruments show them easily, but they never attain the same degree as those of Venus and Mercury, for Mars always remains farther from the sun than the earth; they do not exceed the flattening of the moon three days before or after the full moon. The telescope of Galileo, it should be remembered, did not at first magnify more than eight times, and its magnifying power, afterwards increased to sixteen, never exceeded thirty-two. An examination of the motion of the spots in 1666 by Cassini gave 24 hours 40 minutes for the period of rotation. Maraldi in 1704 and 1719, William Herschel and Schröter at the end of the same century, Kunowski in 1822, Mädler in 1830, Kaiser in 1862, Wolf in 1866, Proctor in 1869, Schmidt in 1873, improved the calculation, and we now know, *to a second nearly*, the exact duration of the diurnal rotation of this world, which is 24 *hours* 37 *minutes* 23 *seconds*.[1]

The duration of the day and night is, then, nearly the same on Mars as on the earth; it exceeds ours by a little more than half an hour only. It is very remarkable that this duration should be so similar for the four planets, Mercury, Venus, the Earth, and Mars. We do not know the reason of this similitude. The distance from the sun does not appear to come into play here like the duration of the year, nor the volume of the planet. The density seems to take the greatest part in the establishment of the time of rotation. The four planets of which the rotation is effected in a period of about 24 hours are the densest. The four giants—Jupiter, Saturn, Uranus, and Neptune—rotate much more quickly (in a period of about 10 hours); and these are also the planets of the smallest density.

In the year of Mars there are 669 rotations or sidereal days (669⅔), and consequently 668⅔ solar or civil days. As the terrestrial day of 24 hours exceeds by 4 minutes the duration of the rotation, the Martian day is likewise a little longer than the rotation (as we have explained, p. 17); it lasts altogether 24 hours 39 minutes 35 seconds. There are in three years one short year of 668 days and two leap years of 669.

We see that between Mars and the earth there is but little perceptible difference with relation to the motion of rotation. The consequent phenomena there—the succession of days and nights, the rising and setting of the sun and stars, the rapid or slow flight of the hours according to the state of the mind, labours, joys, and sorrows—in a word, the daily current of life and the usual course of things, are developed under nearly the same conditions as with us.

The great difference between Mars and the earth consists in the smallness of its volume, which makes it truly a miniature of our world. As we

[1] [The latest and best value is that found by Bakhuysen, which makes it 24 hours 37 minutes 22 66 seconds.—J. E. G.]

have seen (p. 358), its angular diameter at the unit of distance is $9''\cdot35$, that of the earth being $17''\cdot72$; this is only a little more than half of ours (0·53). Expressed in terrestrial measures, this diameter is 6,753 kilometres (about 4,200 miles). The circumference of this planet is 13,000 miles. The surface of Mars is but 27 hundredths of that of the terrestrial globe, and its volume only 16 hundredths of ours. Being six and a half times smaller than the earth in volume, Mars is between seven and eight times larger than the moon, and three times larger than Mercury. Fig. 168 represents exactly these differences of volume, which are rather interesting.[1]

FIG. 168.—COMPARATIVE MAGNITUDES OF THE EARTH, MARS, MERCURY, AND THE MOON

Before the discovery of the satellites of Mars, made in 1877, it was rather difficult to determine exactly the mass of this planet. We have seen, in fact (p. 239), that the most simple proceeding to make use of in weighing a celestial body is to compare the velocity with which it causes a body submitted to its power to revolve round it with that which the earth impresses on the moon; the ratio of the velocities leads to the ratio of the masses or weights. It is thus that we have weighed the sun (p. 240). As nature did not furnish this direct way, it was necessary to use an indirect method, such as the perturbations which the planet produced on its celestial

[1] The measures made of Mars are not concordant as to its polar flattening. Herschel found $\frac{1}{16}$, Schröter $\frac{1}{80}$, Arago $\frac{1}{30}$, Hind $\frac{1}{50}$, Main $\frac{1}{39}$, Kaiser $\frac{1}{118}$, Young $\frac{1}{219}$. All these values are too great for the theory of attraction. This globe, rotating less quickly than the earth, and being smaller, develops but a feeble centrifugal force, and its flattening should be less than that of our planet, which is $\frac{1}{292}$. Perhaps the planet has been formed several times, and the layers near its surface are denser than the mean density. There is some mystery here; this planet is small, and there are several hundreds smaller behind it; we shall see further on that the nearer of its two satellites revolves more quickly than the planet itself rotates on its axis. It is the most eccentric of the principal planets. These are so many facts to be explained.

companions in their course through space, or on some wandering comet which approached sufficiently close to be subjected to a perceptible influence. It was thus that the masses of Mercury, Venus, and Mars were determined up to 1877. But when there is a satellite the operation is at once incomparably more rapid and more precise. The calculation of the mass of Mars made by Le Verrier represents a whole century of observations and several consecutive months of computation—more than a thousand hours of numeration; the satellites of Mars had hardly been discovered, on the contrary, when four nights of observation and ten minutes of calculation sufficed to prove that this planet weighs three million times less than the sun ($\frac{1}{3093000}$). It follows that, representing by 1,000 the weight of the earth, that of Mars is represented by 105; in other words, this globe weighs nine and a half times less than ours.

The density of the constituent materials of this globe is equal to 71 hundredths of the mean density of the earth, and the gravity of objects at its surface hardly exceeds one-third of that of terrestrial objects (about 37 hundredths of ours). Of the eight principal planets, this has the weakest intensity of gravity: 100 pounds carried to Mars and weighed with a dynamometer would weigh but 37 pounds.

We have seen that the revolution of this little planet round the sun is performed in 687 days. This is the length of two of our years less 43 days. As the duration of the day is a little longer on this planet than on ours, there are relatively fewer days in its year than if it turned on its axis as quickly as we do; its calendar counts 668 days in a year.

The inclination of the axis of rotation is slightly more pronounced than here. While with us the obliquity of the ecliptic is 23° 27′, it is in Mars 24° 52′; the difference is almost imperceptible, and it follows that the Martian seasons are similar to ours in intensity, although twice as long An astronomer of the earth has no need to make a voyage to Mars to learn its seasons and its climates. The considerable variation of the polar spots shows us, however, that the difference between summer and winter is more perceptible than with us. I have observed this planet with the greatest attention during all its recent oppositions since the year 1871. The extent of the polar snows always corresponds to the season. Drawings made in the month of June, 1873, show the northern polar cap reduced to a white point, and correspond precisely to the end of the summer—to the end of the melting of the snows. In 1875 I observed it in the middle of the autumn of Mars: the northern polar spot was so reduced that it could scarcely be distinguished, while the snows of the southern pole, which had just passed through a long winter of nearly twelve months, were very extended Observations in 1877 gave still more evident results. I choose from my drawings of this very favourable year four (July 30, 11 hours; August 22, 11 hours; September 14, 10 hours; and October 26, 8 hours), which show at a glance this progressive diminution (fig. 169). This polar spot is so white that, on account of irradiation, it always seems to project beyond the border of the planet's disc : its brightness is more than double that of the rest of the disc. These aspects and variations have been studied with minute care for many years, especially by Herschel at the end of the last century, and by Mädler from 1830 to 1840.

This world shows, like ours, three very distinct zones : the torrid zone, the temperate zone, and the frigid zone. The first extends on both sides of the equator to 24° 52′ ; the temperate zone extends from that latitude to 65° 8′ ; the frigid zone surrounds each pole to that distance.

Thus the duration of days and nights, their differences according to the latitudes, their variations according to the course of the year, the long days and long nights of the polar regions—in a word, everything which concerns the distribution of heat, are so many phenomena almost similar on Mars and on the earth. Between the two planets, however, there is a very notable

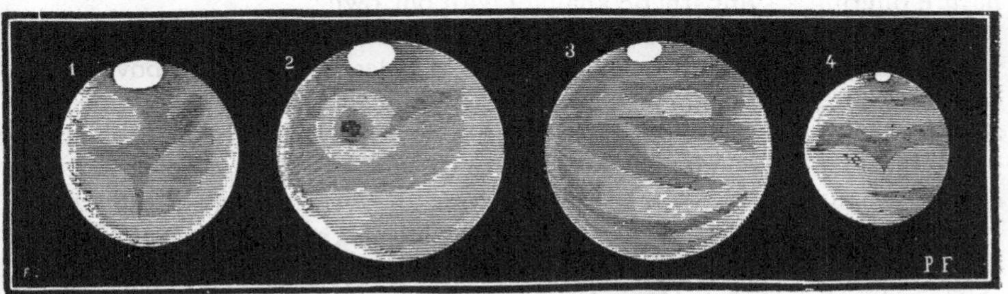

FIG. 169.—TELESCOPIC ASPECT OF MARS ON JULY 30, AUGUST 22, SEPTEMBER 14, AND OCTOBER 26, 1877

difference ; it is as regards the *duration* of the seasons. This duration is there much longer. In fact, as we have just seen, the Martian year is 687 days ; each of the four seasons is, then, nearly double as long as here. Moreover, the orbit of Mars being very elongated, the inequality in the duration of the seasons is more marked than with us. In order to make the comparison exact, let us choose the hemisphere of Mars analogous to that which we inhabit on the earth, its northern hemisphere (the southern would be symmetrical), and let us compare the durations of the seasons on the two planets :—

DURATIONS OF THE SEASONS

	On the Earth	On Mars
Spring	93 terrestrial days	191 Martian days
Summer	93 „	181 „
Autumn	90 „	149 „
Winter	89 „	147 „
	365	668

We see that the seasons of Mars are much slower and perceptibly more unequal than ours.

Thus, the spring and summer of the northern hemisphere of this planet last 372 days, while the autumn and winter last but 296. The solar heat should, then, accumulate in the northern hemisphere in a quantity considerably greater than in the southern hemisphere. But there is a compensation arising from the fact that, the orbit of Mars not being circular, the planet is much nearer to the sun at perihelion then at aphelion. It is at the summer solstice of its southern hemisphere that this planet is at present at its least distance from the sun, and consequently receives from that body the maximum of heat. It follows from this fact that the southern polar snows should vary much more in extent than those of the northern pole, and this is also what observation shows (see fig. 170). We can study from here these climatic variations, and this study is one of the most interesting which we can make, for it transports our thoughts into the midst of a physical nature offering a sympathetic analogy with our own.

Inclined as it is to its orbit, Mars is not presented to us in a direction which we may call vertical, with its two poles placed just above and below its disc, but leaning towards us. As the middle of the summer of the southern hemisphere of Mars coincides with its perihelion, it is this hemisphere which is more easily visible to us, and it is the one which we can observe when the planet is at its minimum distance; we also know this southern hemisphere much better than the northern hemisphere. Thousands of years will elapse before the northern pole of Mars is visible from the earth, at less than half the distance of the sun from the earth, or less than 45 millions of miles.

For more than a century we have observed from the earth the principal facts of Martian meteorology. We look on at the formation of the polar ice, at the melting of the snows, at the inclemencies, clouds, rains and tempests, and the return of fine weather—in a word, at all the vicissitudes of the seasons. The succession of these facts is now so well established that astronomers can predict in advance the form, the magnitude, and the position of the polar snows, as well as the probable state, cloudy or clear, of its atmosphere.

Thus, this world shows the most curious analogies with ours: the inhabitants of Venus see our planet under an appearance nearly similar to that which Mars presents to us; like the poles of Mars, ours are covered with snow and ice; it is also our southern pole which is most overrun—and for the same reasons—with the products of the congelation of water. The poles of cold do not, however, coincide with the poles of rotation. They are situated eccentrically on both sides of the geographical poles, and—a rather curious fact—they are not symmetrically placed—that is, they are not situated at the two extremities of the same diameter.

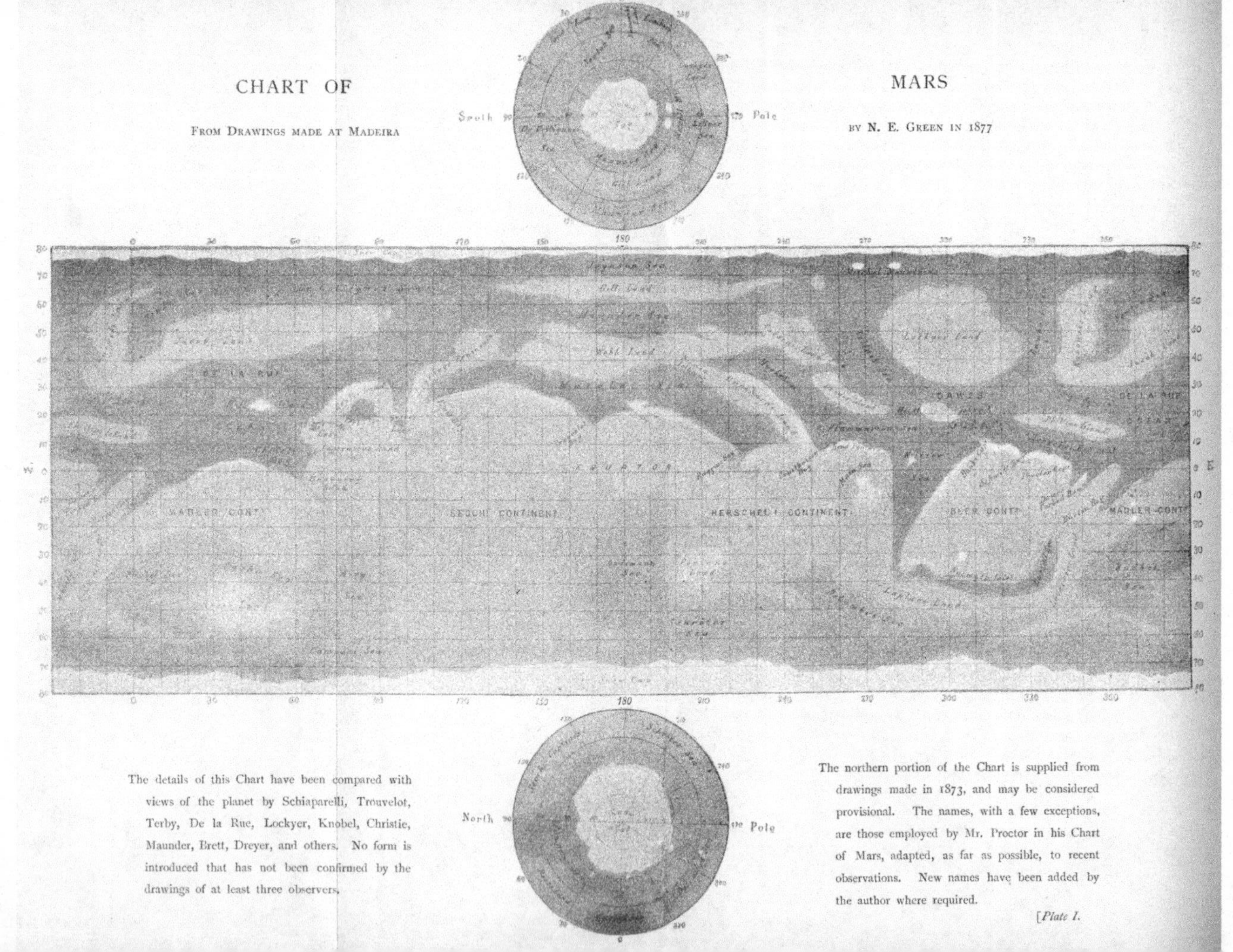

The material originally positioned here is too large for reproduction in this reissue. A PDF can be downloaded from the web address given on page iv of this book, by clicking on 'Resources Available'.

Our readers may form an idea of the aspect of Mars as seen in a telescope from the accompanying map, made at the opposition of 1877 by Mr. N. E. Green, with a 13-inch silver-on-glass reflector by With.

The geographical knowledge which we now possess of the planet Mars is sufficiently advanced to enable us to draw a general map; this several astronomers have already done. I may mention here that this neighbouring planet has always particularly interested me since the time when I wrote my first work, on the 'Plurality of Worlds' (1862), because it was the first to bear witness to the truth of that great and sublime doctrine, by the light of which life and soul are scattered through the universe, instead of the solitudes in which float the material and immaterial masses of ancient astronomy. I drew, in 1876, a geographical planisphere of the planet, constructed from a comparison of maps and drawings previously made, and for which, in addition to my own observations, I used more than a thousand drawings made since the year 1636—that is to say, since the first telescopic observations of this planet. Since 1876 the science has made further advances. In August, September, and October, 1877, the planet was at an extremely favourable proximity; it was studied with care during this advantageous period, and the knowledge of its geography has rapidly improved. We may especially mention among the most remarkable observations those of Schiaparelli in the calm and limpid sky of Milan, and those of Green in the island of Madeira.

Mars has been especially studied at each of its oppositions since 1877. Of course, the map which we have now obtained is still far from being definitive, and it is certain that a century or two will elapse before we shall know perfectly the Martian geography or 'areography'; still, this knowledge continues constantly to improve, like that of terrestrial geography itself. When shall we distinguish the great cities of this neighbouring world? Sceptics may smile, as they smiled in the time of Copernicus and Fulton; but he who has confidence in progress does not despair of such a result, which, moreover, has nothing impossible in it, and demands, that it may be obtained, only the continuance of the modern progress in optics. The general geography of Mars may already be traced with much greater certainty than that of the terrestrial latitudes that surround our two poles.

FIG. 170.—THE SOUTH POLE OF MARS

The first problem which presents itself on an examination of the geographical map of Mars is to determine whether the dark spots to which we give the name of *Seas* really indicate tracts of water. It is not impossible that we may be now, with regard to Mars, under a similar illusion to that which we laboured under up to the middle of this century with reference to the moon. That these spots *may be* seas there is no doubt, since water absorbs light instead of reflecting it like the continents; but certain purely mineral dark tints, or plains covered with a vegetable carpet, would produce the same effect, and it is this which happens in the moon, where precise observation reveals a dry and rugged soil on the vast grey plains which we have so long taken for true seas. The designation of seas applied to the dark spots on the globe of Mars might, on account of the resemblance, be retained, even if they were not true seas; if, however, it were demonstrated that there was an error here, we should be very wrong in beginning the geography of Mars by adopting it, and it would be preferable to choose an expression which would in no way prejudge the question. But we recognise that if it is still not absolutely certain that there are seas there similar to those of our planet, the fact is at least very probable.

In fact, in the first place, the existence of an atmosphere at the surface of Mars is absolutely demonstrated. It was long since revealed by the fact that the disc of this planet is more luminous at the borders than at the centre; the light reflected from Mars increases gradually from the centre towards the circumference. The most natural explanation of this fact is that which attributes it to an atmospherical absorption increasing in the ratio of the thickness traversed, with, consequently, a minimum at the centre and a maximum at the circumference. This explanation is found to be at once confirmed by a second fact of observation: the spots lose their sharpness when the rotation of the planet carries them beyond the centre towards the borders of the disc, and they disappear when they arrive at 50° or 60° distant from the central meridian, varying more or less, besides, according to the transparency of the atmosphere of Mars. This effect is never produced on the moon. A third testimony to the existence of a Martian atmosphere is afforded by the white spots which mark its poles increasing in extent during the winter, and diminishing regularly during the summer. These variable spots can only be produced by atmospherical condensation, either of snow or of clouds.[1]

[1] Their fixity eliminates the latter hypothesis and favours the first, and we may look upon them with an almost absolute certainty as masses of *snow*, similar to those which whiten the polar regions of the earth, and which, seen from Venus, must show the same aspect as those of Mars seen from here; with this difference, however, that our polar ice varies much less in extent than that of this planet. Thus, for example, measures made during the opposition of 1862 have shown a diminution of width from 20° to 7° from September 1 (day of summer solstice of the southern hemisphere) to December 1—that is to say, a diminution to one-third of the diameter in 90 days; and those of the

Further, this aërial envelope may not be composed of *air* identical with that which we breathe ; the absorbent liquids which fill the basins of the Martian seas may not be precisely *water* ; the snow may be a chemical precipitate of a nature different from ours. Well, spectrum analysis almost entirely removes these remaining doubts. Examined by Huggins, Vogel, and Secchi, the spectrum of the light reflected by Mars reproduces at first naturally the solar spectrum, but it also shows rays of absorption which correspond precisely with those of the spectrum of the terrestrial atmosphere. Certain sceptics may perhaps reply that there is nothing surprising in this fact, and that it proves nothing, since we receive the light of Mars from the bottom of our own atmosphere, which should consequently make its mark in the spectrum of this light. This is an objection to which experimentalists have themselves taken care to reply. They examined on the same day and hour as Mars the light of the moon, which also traverses our atmosphere, and have chosen for the comparison the time when the moon was lower than Mars, and should consequently be influenced in its light by a greater absorption of the terrestrial atmosphere. Now, apart from some permanent lines, the lunar spectrum was shown entirely destitute of the tell-tale lines observed in that of Mars, and the difference of the two lights proved at once the absence of a perceptible atmosphere at the surface of our satellite, and the presence on Mars of an atmosphere which cannot differ chemically from ours, and which is particularly rich in the vapour of water. We do not yet know the density of this atmosphere, but we know with certainty that *it exists*, and that it resembles that which we breathe.[1]

Thus, according to the agreement of all the evidence, the seas, the clouds, and the polar ice-caps of Mars are analogous to ours, and the study of Martian geography can be made like that of terrestrial geography. Never-

opposition of 1877 showed a diminution of width from 18° to 7° from September 18 (day of the solstice in that year) to November 1, or 43 days ; they had already diminished from 30° to 18° since August 15. In 1879 they were reduced to 4°. Their width, then, varies (taking irradiation into account) from 560 to 75 miles. We may further add that the disc of the planet shows from time to time in the telescope light spots less white than those of the poles, movable and variable, and which can only be *clouds*. All these facts unite in favour of the analogy which leads us to see on this earth an atmosphere and seas establishing there a meteorological circulation similar to that which exists on our planet.

[1] As to the thickness of this atmosphere with reference to the disc of the planet, it is necessarily too thin to be visible from here, even if it were much higher than ours. Supposing it 80 kilometres (50 miles) in height, this thickness would still be but 0″·3 when the planet is nearest to us ; the refraction would then be imperceptible.

We must not expect to easily observe the occultation of a star exactly behind Mars. The fact, however, happened in 1672. Mars passed exactly in front of the 5th magnitude star ψ of Aquarius, and, as the star disappeared 6′ from the edge of the planet, Cassini concluded the existence of an enormous atmosphere : it was simply the brightness of Mars which prevented him from seeing the star. South observed two occultations and a contact without the least variation. Mars passed exactly over Jupiter on January 9, 1591.

theless, we should not hasten to conclude an absolute identity between the geographical and meteorological systems of the two planets. Mars shows us characteristic dissimilarities. Our globe is covered with the waters of the sea over three-fourths of its surface; our great continents are, so to say, islands; the vast Atlantic, the immense Pacific, fill up deep basins with their waters. On Mars the division is more equal between the continents and the seas, and there is rather more land than water; there are veritable Mediterraneans, interior lakes, or narrow straits, which remind us of the British Channel, or Red Sea, and which constitute a geographical network altogether different from that of terrestrial geography.

Another fact not less worthy of attention is that the Martian seas show remarkable differences of intensity. On the one hand, they are darker near the equator than in adjacent latitudes; and, on the other hand, some are particularly dark, for example the sea of Hooke, the Maraldi Sea, the circular sea Terby, and the sea of Sablier [Kaiser Sea]: a comparison of old drawings shows that it was the same fifty and a hundred years ago, but that their tints vary. This gradation of intensity is, then, real. To what cause is it due? The most simple explanation is to suppose that it corresponds to a greater depth of the water.

When we pass in a balloon above a large river, lake, or sea, if the water is calm and transparent, we see the bottom sometimes so perfectly that the water seems absent (this I particularly saw one day—June 10, 1867, at 7 o'clock in the morning—when soaring at a height of 10,000 feet above the Loire); on the borders of the sea we perceive the bottom down to 30 and 50 feet in depth, out to several hundreds of yards from the shore, according to the light and the state of the sea. On this hypothesis the light-coloured seas of Mars would be those which, like the Zuyder Zee, for example, have only a depth of water of a few yards; the greyish seas would be a little deeper, and the dark seas would be the deepest of all. This is not, however, the only explanation to be given, for the tint of the water itself may be perfectly different, according to the regions; the salter the water, the darker it is, and there might result currents, such as the Gulf Stream, which flow like lighter rivers on the surface of the ocean which forms their bed. The saltness depends on the degree of evaporation, and it would be nothing surprising if the equatorial seas of Mars should be salter and deeper than the interior seas. A third explanation still presents itself to the mind. We have on the earth the Yellow Sea, the Red Sea, the White Sea, and the Black Sea. Without being absolutely correct, these names correspond more or less with the aspects of those seas. Who has not been struck with the emerald-green colour of the Rhine at Basle and of the Aar at Berne, the deep azure of the Mediterranean in the Gulf of Naples, the yellow bed of the Seine from Havre to Trouville, visible in the sea, and all the varied tints which the waters of rivers and streams present? The three explanations may then apply to the seas of the planet Mars as well as to ours. The light-coloured regions may be marshes or submerged lands. The ground of the colouring of the Martian seas is green, like that of terrestrial water. But this colouration varies, and the extent of the seas varies also. We sometimes observe aspects analogous to those presented by immense lands submerged by vast inundations. As after storms our rivers run with yellow and muddy waters, so in Mars the colouration of the waters changes according to the seasons.

The continents are yellow, and it is this which gives the planet the fiery colour by which we recognise it with the naked eye. There is here an essential difference from the earth. Seen from a distance, our planet would appear greenish, for it is green which prevails on our continents as well as on our seas ; the presence of our atmosphere slightly changes this shade to blue. With a telescope the astronomers of Venus and Mercury would see our seas tinted with a deep green, the continents shaded with a light green, more or less varied, the deserts yellow, the polar snows very white, the clouds white, and the chains of mountains marked by the snowy line of their crests. On Mars, the snows, the clouds, and the seas show nearly the same aspect as with us, but the continents are yellow, like fields of cereals—maize, wheat, barley, or oats.

This colouration is much more perceptible to the naked eye than in a telescope ; the stronger the magnifying power, the less intense it is. What is the reason of this ? It is not due to an atmosphere which might be red instead of blue, as has been sometimes supposed, for in that case this colouration should extend all over the planet, and would increase in intensity from the centre towards the circumference, on account of the increase of atmospheric thickness traversed by the luminous rays reflected by the planet. We have, then, but two suppositions to make in order to explain it : either the continents of Mars are deserts, the surface of which is covered with sand or ochreous minerals, or else the vegetation of this planet is yellow.

The first of these two hypotheses is in direct conflict with the evidence of nature on Mars, and it is surprising that several astronomers who adopt it have not perceived the contradiction. To admit that the colouration may be that of the mineral surface of this globe is to suppose that it has nothing on its surface—no species of vegetation, not the least carpet of moss, neither forests, meadows, nor fields ; for, whatever may be the vegetation which clothes this surface, it is this we see, and not the soil. This hypothesis would, then, condemn this world to a perpetual sterility.

Now, could the meteorological circulation which produces on this planet, as on ours, seasons, fogs, snows, rains, heat and humidity ; water, air, fire, earth—the four elements guessed by the ancients—act for thousands of centuries on the surface of this world without giving birth to the smallest blade of grass ? By what miracle of perpetual annihilation could the forces of nature—which, on earth, produce life multiplied to the injury of itself, and which spread with such a lavish hand thousands of millions of beings, every hour, every day, every minute over the entire surface of our globe, at the bottom of the seas as well as on the mountains—how could these forces remain unfruitful on a world placed exactly like ours, in the light of the same sun, and in the network of the same vibrations? Such an hypothesis cannot be supported for a single moment. The aspect of the continents of Mars invites us very plainly to enlarge the circle of our botanical conception, and admit that vegetation is not necessarily green in every world, that chlorophyl may be produced under different aspects, and that the varied colourings of flowers, leaves, and plants which we observe here may be repeated elsewhere a hundred-fold, or under a thousand varied conditions. We do not perceive from here the form of the Martian plants, but we may conclude that in the vegetation on the whole, from giant trees to microscopic moss, it is yellow and orange which predominate; either it has a great number of red flowers or fruits of the same colour, or in reality the vegetation itself may be not green, but yellow. An orange-tree bearing green flowers would appear to us a monstrosity, on account of our terrestrial education; but in reality it would be sufficient that chemical combinations, or even the simple arrangement of the molecules, should be accomplished otherwise than here in order that the colours should differ.

Is the vegetation of Mars persistent throughout the year, like a large number of terrestrial plants, such as the grass of the meadows, the box-tree, the rhododendron, the laurel, the cypress, the yew, the fir-tree, &c., or do the leaves fall in winter to be renewed in the spring? We do not yet know. The regions of the planet which we observe most distinctly are the equatorial and tropical, and on the earth vegetation does not vary in this zone. The other lands have been as yet too little studied to enable us to assert anything with reference to this point; but as no great differences of colouring have ever been noticed between one latitude and another, it is probable that the vegetation is not subject to the same changes as those of our northern countries. However, some variations have been already noted; thus, Hall's Land was seen in 1877 redder than the other parts of the disc.[1]

[1] This colouration of Mars is not so intense, so red, as is generally believed. In order to measure it exactly, I constructed some years since an apparatus which, on the principle of the sextant, brings into the same telescope two luminous points, whatever their distance from each other may be. We may thus bring into the field of the telescope any two stars, or a star and a source of light, such as a gas-jet, &c., for direct comparison. By repeated comparisons I have found that this planet is not red, properly speaking, nor even an intense orange, but orange-yellow, nearly the tint of a gas-light. These experiments have given me the following colours for the planets:—

1. Gas-light	Orange-coloured	5. Uranus	Light yellow
2. Mars	Orange-coloured	6. The Moon	Brass yellow
3. Mercury	Orange-yellow	7. Venus	White
4. Jupiter	Yellow	8. Saturn	Greenish yellow

These tints are written in decreasing order from red towards blue. We shall see further on that there are stars redder than Mars and greener and bluer than Saturn.

Thus red, orange, and yellow predominate on the surface of Mars.

Another difference from the earth appears to be shown by the variability of some of its geographical configurations. The continual study of the Strait of Herschel II. may lead us on this point to very curious results. In 1830 Mädler saw it several times very clearly and distinctly as it is represented at A, fig. 171. In 1862 Lockyer saw it with the same clearness as it is shown at A, and in 1877 Schiaparelli observed it as we see it in the third drawing. This point—seen round, black, and clear in 1830, so clear in reality that Mädler chose it for the origin of Martian longtitudes, as being the blackest point; already seen under the same form by Kunowsky in 1821; and also indicated, in 1798, by Schröter as a black globule—was not distinguished in 1858 by Secchi, notwithstanding a special search which he made. This same point was seen forked by Dawes in 1864, and it was certainly the same; but the region which surrounds it to the south appears covered with marshes and variable in aspect according to the year; all the

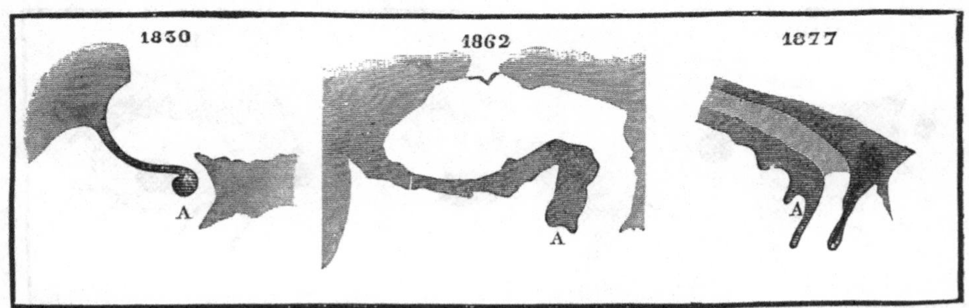

FIG. 171.—PROBABLE VARIATIONS IN THE SEAS ON MARS: HERSCHEL'S STRAIT II. IN 1830, 1862, AND 1877

drawings of 1877 show the same point, no longer like a black disc suspended by a winding thread, but with the thread widened so as no longer to bear this comparison: the gulf is as wide at the centre and at the origin as at its eastern extremity.

At present the blacker and clearer spot which we should choose preferably to mark the origin of the meridians would be the Circular Sea of Terby. We should certainly choose it in preference to the first. In 1830 the preference was given to the former, and on several drawings we see the two hanging exactly on each side of the ocean, as in fig. 172, B. These drawings are no longer correct. Here is a first variation. A second is presented by the aspect of the spot. In 1862 different observers saw it elongated from east to west. In 1877 it was seen, on the contrary, perfectly round (correction being made for perspective), and certainly not elongated in the first direction. Third variation: it appeared in 1862 united to the neighbouring ocean by a strait, and in 1877, with instruments of the same power, observers of the same skill saw nothing of this strait, but distinguished another to the north-east. Another example of variability: excellent observers perceived in 1862 and 1864, in the De la Rue Ocean, a luminous point which might be formed by an island covered with snow, and which, I believe, I indicated on the first map I made of Mars. No one has ever seen it since.

It is not necessary, doubtless, to accept as real changes all the differences which exist between the observers. Thus, for example, in 1877 several saw the seas of Hooke and Maraldi united towards the west, while the separation remained visible to others. The eye is differently impressed, and we may almost say that, for certain details, there are not two eyes which see identically the same appearance—even the two eyes of the same person. But when the attention is especially fixed on certain remarkable points which should be rendered perfectly visible by the instruments used, and when differences are thus established which appear inconsistent with

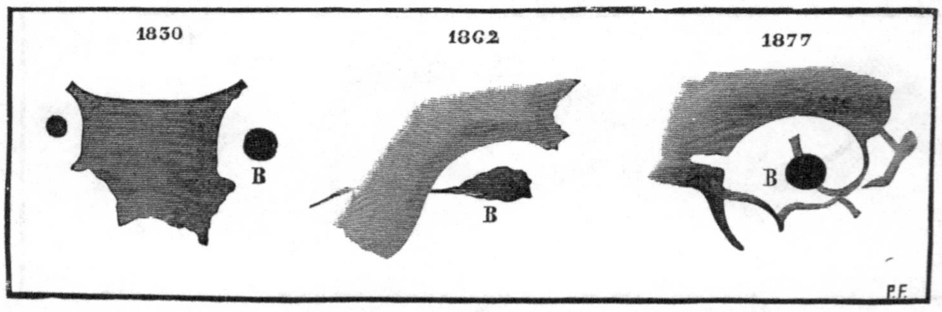

FIG. 172.—PROBABLE VARIATIONS IN THE SEAS ON MARS: THE CIRCULAR SEA OF TERBY IN 1830, 1862, AND 1877

errors of observation, the probability inclines in favour of the objective reality of the changes described.

Of what nature are these variations? This is what the future will teach us. We can now express but vague conjectures on this point. But whatever the variations may be, they do not prevent the principal configurations of Martian geography from being permanent, and consequently real, and they are seen now just as our fathers saw and drew them more than two centuries ago.

Another remark not less interesting. This neighbouring planet appears to have fewer clouds than that which we inhabit. Thus, in the month of August 1877, and in March 1878, scarcely one was seen.[1]

[1] There is here a great contrast to our globe, for there are years in which we are really deprived of fine weather. In a whole year, from August 1878 to the same month in 1879, we had at Paris 167 days during which it rained, and only 37 days of clear sky, or a little cloudy—37 days made for astronomers. It was nearly the same in 1888. On the southern hemisphere of Mars it was the contrary at the time of the observations of 1877; the planet could be observed at all times

The attentive observation of Martian meteorology, the measure of the monthly variable extent of the polar snows, and that of the annual variations, might perhaps render great service in the establishment of a basis, even for terrestrial meteorology.

Martian meteorology shows, then, the most curious analogies with that of the planet which we inhabit. On Mars, as upon the earth, in fact, the sun is the supreme agent of motion and of life, and its action causes analogous results to those which exist here. The heat vaporises the water of the seas and raises it into the heights of the atmosphere. This water vapour returns to a visible form by the same process which gives birth to our clouds—that is to say, by differences of temperature and saturation. The winds are produced by these same differences of temperature. We can follow the clouds, carried along by the aërial currents over the seas and continents, and many observations have, so to say, already photographed these meteoric variations. If we do not yet see *the rain falling* on the fields of Mars, we may at least imagine it, since the clouds are dissolved and renewed. If we cannot see the snow fall, we may also imagine it, since, as with us, the winter solstice is enveloped in frost. Thus there is there, as here, an atmospherical circulation with all its consequences. We may go still further with the induction.

In fact, the existence of continents and seas shows us that this planet has been, like ours, the seat of internal geological movements, which have given birth to elevations and depressions of the soil. There have been earthquakes and eruptions, modifying the primitively united crust of the globe. Consequently, there are mountains and valleys, plateaus and basins, steep ravines and cliffs. How do the rain-waters return to the sea? By springs, brooks, streams, and rivers. The water falling from the clouds

when it was fine with us. It is necessary to remember, in fact, that, for the observation of the Martian geography to be possible, two conditions are required above all others: it is necessary that it should be fine with us and our atmosphere clear, and it is also necessary *that it should be fine on Mars*, otherwise we can no more pierce his layer of clouds than we can in a balloon see through the clouds which hide the terrestrial villages. It is certainly remarkable that on Mars nine whole months should have elapsed almost without clouds, thus permitting us greatly to improve the geographical knowledge which we have of this neighbouring world.

Mars was, in September and October 1877, in the middle of the summer of its southern hemisphere, then much inclined towards us, and in the middle of the winter of its northern hemisphere, turned to the other side. All the clouds seemed banished to that hemisphere. On this globe, still more than on ours, the summer is the season of clear atmosphere, and the winter that of bad weather. The permanent spots show themselves divided, sharp, and clear during the summer of the hemisphere on which they are placed. When winter comes they appear vague, confused, and weak; this is doubtless because the atmosphere of Mars becomes disturbed in winter and remains very transparent in summer. We also remark a preference in the clouds to form on the marshes and shallow waters, tinted with grey, rather than on the dark and deep seas, and it is this which hinders the precise knowledge which we seek to acquire of the country situated above the Strait of Herschel II.; but we do not remark zones constantly cloudy and rainy similar to those of the terrestrial equatorial calms where it rains all the year.

penetrates, as here, the permeable lands, glides over the impervious ground, returns to the daylight in the limpid spring, purls in the brook, flows in the stream, and descends majestically in the river to its mouth. Thus it is not difficult to see in Mars scenes similar to those which constitute our terrestrial landscapes: brooks flowing in their bed of pebbles gilded by the sun; streams traversing the plains and falling in cataracts to the bottoms of the valleys, and rivers descending slowly to the sea through vast fields. The maritime rivers receive, as here, the tribute of aquatic canals, and the sea is sometimes as calm as a mirror, sometimes agitated by tempests. It is even rocked, as here, by the synchronal motion of the luni-solar tides, caused by the sun and by the two moons revolving rapidly in the sky.

It seems, however, that the continents of Mars may be flatter than ours, and formed nearly everywhere of vast plains; for, on the one hand, the seas often run to immense distances and contract in the same proportion, and, on the other hand, the straight lines or canals discovered in 1879 by M. Schiaparelli, and seen again since, not only by that astronomer, but also by many others, show us a sort of geometrical network extending all over the continents.

These straight lines place all the Martian seas in communication with each other, forming a very strange sort of geometrical network. They sometimes measure 3,000 to 4,000 miles in length, and more than 60 miles in width. Their tint seems to indicate that they are really water.

This is not the place to describe in detail these special discoveries. Most of the 'canals' consist of two parallel lines, sometimes visible and sometimes invisible. What a strange geography! The future will doubtless throw light on this mystery.

Thus we see in space, at some millions of miles from here, a world almost similar to ours, where all the elements of life are present just as they are around ourselves: atmosphere, waters, snows, heat, light, winds, clouds, brooks, fountains, valleys, and mountains. To complete the resemblance, let us remember that the seasons are of nearly the same intensity as on the earth, and that the duration of the day is only a little longer than ours. This is certainly an abode but little different from that which we inhabit.

The analogy between Mars and the earth does not cease when we come to examine this planet from the point of view of the animated beings which should people it. Its inhabitants may be considered as being those of which the conformation should approach most nearly to ours. The philosopher Kant supposed, even in the last century, that they might be ranked mentally in the category of men of the earth. He thought that the inhabitants of the inferior planets, Mercury and Venus, are too material to be reasonable, and are probably not even responsible for their actions; and he ranked the human life of the earth and of Mars in a happy moral medium, neither absolutely coarse nor absolutely spiritual. 'These two planets,' he wrote, 'are placed in the middle of our planetary system, so that we may suppose, without improbability, that their inhabitants possess an average condition, in their constitutions as well as in their morals, between the two extremes.' In order to describe the perfection and the happiness which the inhabitants of the superior planets enjoy, from Jupiter to the confines of the system, Kant quotes two verses of Haller, of which the following is a translation: 'The stars are perhaps the abode of glorified spirits; just as vice reigns here, so above virtue is sovereign.'

But these are arguments purely speculative. We have as yet no foundation from which to judge of the intellectual state of planetary humanity. All we can think is that, the moral condition being naturally in harmony with the physical, the rougher the planet is, the less must the sensibility be, so that doubtless the inhabitants of Mercury and Venus are less 'intellectual' than we are. On the other hand, humanity progresses with time, and Mars, having been formed before the earth, and having cooled down more quickly, should be more advanced, from all points of view. It has doubtless reached its apogee, while we are still but children who play seriously with politics, with rifles and cannon.

The study of modern physiology scientifically demonstrates that the human body is the product of the terrestrial planet: its weight, its size, the density of its tissues, the weight and the volume of its skeleton, the duration of its life, the periods of work and sleep, the quantity of air which it breathes and of food which it assimilates, all its organic functions, *all the elements of the human machine, are organised by the planet.* The capacity of our lungs and the form of our chest, the nature of our feeding and the length of our digestive duct, the action and the strength of our legs, sight and the construction of the eye, &c., all the details of our organism, all the functions of our being, are in intimate, absolute, permanent correlation with the world in the midst of which we live.

Now, the mean density of the materials which compose the planet Mars is less than that of the materials constituting our globe: it is 71 to 100. On the other hand, the weight of bodies is extremely light at its surface. Thus, the intensity of terrestrial gravity being represented by 100, it is but 37 at the surface of Mars—it is the *weakest* which we find on all the *planets* of the system. It follows that a terrestrial pound carried there would weigh but 6 ounces. A man of 11 stone in weight would be reduced to 4. He would not be more fatigued by walking 50 miles there than in walking 20 on the earth.

All these considerations tempt us, then, to think that the Martian population is very different from the terrestrial population. But is terrestrial life itself really so homogeneous? Do we not find in certain countries vegetables and animals absolutely different from those which we know in Europe? Does not Australia upset all our ancient ideas?

Mars and its humanity should be more advanced, and doubtless more perfect, than we are.[1] If we suppose that the celestial bodies have been formed by the condensation or consecutive agglomeration of molecules primitively scattered through an immense space, the principles of the mechanical theory of heat show that the resulting temperature was 28 millions of degrees for the sun, 9,000 degrees for the earth, and 2,000 for Mars. If we add that Mars must have been detached from the solar nebula long before the earth, we may conclude, with a great show of probability, that this world must be now cooled down to its centre, and that its surface is not subject, like that of the earth, to the influence of internal geological forces which continue to raise our lands and modify our shores. A great part of the water appears to be absorbed, and the straight and elongated form of the seas seems to indicate the bottoms of ancient beds. How interesting it would be for us to make a voyage there! In the meantime, let us improve our telescopes.

The progress of optics has already brought this world within the range of our analysis.[2] But one of the most novel and interesting facts of its study is the discovery of its two satellites.

[1] In tracing to its origin the formation of the zoological series, we may suppose that gravity will have exercised an influence of another order on the succession of species. While here the great majority of the animal races remain confined to the surface of the soil by terrestrial attraction, and a much smaller number have received the privilege of wings and flight, it is very probable that on account of the altogether peculiar disposition of things the Martian zoological series has been developed preferably by a succession of winged species. The natural conclusion is that the higher animal species may be furnished with wings. On our sublunary sphere the vulture and the condor are the kings of the aërial world; below, the great vertebrate races and the human race itself, which is the resultant of the last evolution, should acquire the privilege, much to be envied, of enjoying aërial locomotion. The fact is so much the more probable that the weakness of gravity is added to the existence of an atmosphere analogous to ours.

[2] What are the smallest objects which in the present state of optics we can perceive on the surface of Mars? This is an interesting question, which the observations of M. Schiaparelli have partly solved. His telescope of 1877, of which the object-glass measured 218 millimetres (8·58 inches) in diameter, armed with eyepieces magnifying, one 322, another 468 times, and of which the length is 10·66 feet, enabled him to distinguish: (1) luminous spots on a dark ground and dark spots on a luminous ground measuring half a second; (2) luminous lines on a dark ground measuring only a quarter of a second; and (3) dark lines on a luminous ground measuring also a quarter of a second. It follows that in excellent atmospherical conditions we may distinguish spots of which the diameter is but a fiftieth of that of the planet—that is to say, 137 kilometres (85 miles). Sicily, the great lakes of Central Africa, the island of Ceylon, and Iceland would be visible. Similarly, a line of which the width would be but the hundredth of that of the planet, or 43 miles, would be perceptible; we might distinguish, then, Italy, the Adriatic, the Red Sea, &c. Instead of continuing the duel between cannon of 80 tons, 100 tons, 150 tons, and ironclad ships, should we not be better advised in delaying for a moment this pure loss of hundreds of millions con-

We know now that this world travels round the sun accompanied by two satellites. Their discovery was made in 1877, by Professor Asaph Hall, at the Observatory of Washington, by the aid of the most powerful telescope which existed at that time. It was not due to chance, like that of a great number of small planets and comets, but it was the result of a systematic search. Most astronomers were accustomed, like ordinary mortals, to read in the standard books the usual phrase, 'Mars has no satellites'; however, some, doubting this assertion, continued to seek to surprise the secrets of nature, which always keeps more than it allows us to grasp. They had already searched the neighbourhood of Mars; but the instruments they used were much inferior to the equatorial of Washington, of which the object-glass measures no less than 66 centimetres (26 inches) in diameter, of which the focal length is 10 metres (32·8 feet), of which the optical power permits a magnification of 1,300 times, and which is moved by a mechanism of the greatest precision.[1] By the aid of this excellent apparatus the eminent American astronomer undertook the attentive examination of the neighbourhood of Mars from the beginning of the month of August 1877, in order to observe assiduously this neighbouring planet during the favourable epoch of its greatest proximity to the earth. After long evenings of barren expectation, he was about to abandon the search, when, encouraged by the entreaties of his wife, he persisted, and discovered a satellite during the night of the 11th, then a second on the night of the 17th.

This news was received by astronomers with astonishment. At least half remained incredulous till they could be more fully informed. The first care was, naturally, to seek to verify it. But eight days had not elapsed before most of the observatories of America and Europe had directed their best instruments towards the same point of the sky, and recognised the existence, if not of both satellites, at least of the more distant, which is less difficult to perceive. Now these two new worlds have been sufficiently observed to enable their astronomical elements to be determined. Here are the details:

They revolve round Mars nearly in the plane of his equator.
Their orbits are nearly circular.
The more distant satellite performs its revolution in 30 hours 17 minutes 54 seconds,
The nearer satellite in 7 hours 39 minutes 15 seconds.
The mean diameter of Mars being 9″·57,
The distance of the outer satellite from the centre of Mars is 32″·5, or 6·92 radii.

---

tributed by the taxpayers, and devoting the hundredth part to experiments capable of opening to us the divine secrets of Nature?

The instruments constructed in recent years are more powerful than the telescope of which we have just spoken. Unfortunately, the clearness of the images does not increase in proportion to the magnifying power used.

[1] This instrument has been since surpassed by those of Nice, of Mount Hamilton (California) (Lick Observatory), and Polkowa (Russia).

The distance of the inner satellite is 13″·0, or 2·77 radii.

If we express these last three values in kilometres we obtain :

Diameter of Mars	6,850 kilometres	(4,256 miles)
Distance of exterior satellite . . .	23,700 „	(14,727 „)
„ interior „ . . .	9,490 „	(5,897 „)

These distances are reckoned from the centre of Mars; but from the surface of the planet to the first moon of Mars is but 3,770 miles, and to the second 12,600, while from the earth to the moon (centre to centre) is reckoned 238,000 miles. Between the first moon of Mars and the surface of the planet there is not even the space necessary for a second globe of Mars, while twenty-nine terrestrial globes would be required to form a bridge from here to the moon.

We have represented in the accompanying figure this little system of Mars drawn to an exact scale. We may understand its difference from the terrestrial system by remarking, that if the globe of Mars represented the earth, we should on the same scale place the moon at a distance of 13½ inches.

Fig. 173.—Mars and the Path of its Satellites

We have here, then, a system very different from that of the earth and moon. But the most curious point is the rapidity with which the inner satellite of Mars revolves round its planet. This revolution is performed in 7 hours 39 minutes 15 seconds, although the world of Mars rotates on itself in 24 hours 37 minutes—that is to say, this moon turns much more quickly than the planet itself. This fact is inconsistent with all the ideas we have had up to the present on the law of formation of the celestial bodies.

Thus, while the sun appears to revolve in the Martian sky in a slow journey of more than twenty-four hours, the inner moon performs its entire revolution in a third of a day. It follows that *it rises in the west*, and *it sets in the east!* It passes the second moon, eclipses it from time to time, and goes through all its phases in eleven hours, each quarter not lasting even three hours. What a singular world!

These satellites are quite small; they are the smallest celestial bodies we know. The brightness of the planet prevents us from measuring them exactly. It seems, however, that the nearer is the larger, and shows the brightness of a star of the 10th magnitude, and that the second shines as a star of the 12th magnitude. According to the most trustworthy photometric measures, the first may have a diameter of 12 kilometres (7·45 miles), and the second a diameter of 10 (6·2 miles). *The larger of these two worlds is scarcely larger than Paris.* Should we honour them with the title of worlds? They are not even terrestrial continents, nor empires, nor kingdoms, nor provinces, nor departments. Alexander, Cæsar, Charlemagne, or Napoleon, might care but little to receive the sceptre of such worlds. Gulliver might juggle with them. Who knows, however! The vanity of men being generally in the direct ratio of their mediocrity, the microscopical reasoning mites which doubtless swarm on their surface have also, perhaps, permanent armies which mutilate each other for the possession of a grain of sand.[1]

These two little moons received from their discoverer the names of *Deimos* (Terror) and *Phobos* (Flight), suggested by the two verses of Homer's 'Iliad' (Book xv.) which represent Mars descending on the earth to avenge the death of his son Ascalaphus:

He ordered Terror and Flight to yoke his steeds,
And he himself put on his glittering arms.

Phobos is the name of the nearer satellite; *Deimos*, that of the more distant.

The existence of these little globes had already been suspected from analogy, and thinkers had frequently suggested that, since the earth has one satellite, Mars should have two, Jupiter four, Saturn eight; and this is indeed the fact.[2] But, as we experience too often in practice the insufficiency of these reasonings of purely human logic, we cannot give them more value than they really possess. We might suppose in the same way now that Uranus has sixteen satellites and Neptune thirty-two. This is possible; but we know nothing of them, and have not even the right to consider this proportion as probable. It is not the less curious to read the following passage written by Voltaire in 1750 in his masterpiece, the 'Micromégas.'

[1] Some of our readers have doubtless already asked why these moons of Mars were not discovered sooner. We might even ask whether they are not of recent creation. Without denying the possibility of an actual projection of satellites by a planet, or of planets by the sun, it is not necessary to imagine this novel formation in order to explain the recent discovery of these two satellites. They were purposely searched for, by the aid of the most powerful telescope which had yet been directed to Mars, by a careful and persevering astronomer, and at the very moment when the planet was placed in the best possible conditions for observation. Here are further reasons necessary to explain the fact. It is almost certain that they are not new formations. We have seen above that this discovery was due to an urgent feminine request. Having searched in vain during several evenings, the astronomer had given up, when Mrs. Hall begged him to search a little more. And he found.

[2] [As Jupiter is now (1893) found to have *five* satellites, this arithmetical progression is upset.—J. E. G.]

On leaving Jupiter, our travellers crossed a space of about a hundred millions of leagues and reached the planet Mars. They saw *two moons* which wait on this planet, and which have escaped the gaze of astronomers. I know well that Father Castel wrote against the existence of these two moons ; but I agree with those who reason from analogy. These good philosophers know how difficult it would be for Mars, which is so far from the sun, to get on with less than two moons. However this may be, our people found it so small that they feared they might not find anything to lie upon, and went on their way.

Here we have unquestionably a very clear prophecy, a rare quality in this kind of writing. The astronomico-philosophical romance of 'Micromégas' has been considered as an imitation of Gulliver. Let us open the masterpiece of Swift himself, composed about 1720, and we read word for word in Chapter III. of the Voyage to Laputa :—

Certain astronomers . . . spend the greatest part of their lives in observing the celestial bodies, which they do by the assistance of glasses far excelling ours in goodness. For this advantage hath enabled them to extend the discoveries much farther than our astronomers in Europe ; for they have made a catalogue of ten thousand fixed stars, whereas the largest of ours do not contain above one-third part of that number. They have likewise discovered two lesser stars, or satellites, which revolve about Mars, whereof the innermost is distant from the centre of the primary planet exactly three of his diameters, and the outermost five ; the former revolves in the space of ten hours, and the latter in twenty-one and a half ; so that the squares of their periodical times are very near in the same proportion with the cubes of their distance from the centre of Mars, which evidently shews them to be governed by the same law of gravitation that influences the other heavenly bodies.[1]

What are we to think of this double prediction of the two satellites of Mars? Indeed, the prophecies which have been made so much of in certain doctrinal arguments have not always been as clear, nor the coincidences so striking. However, it is evident that no one had ever seen these satellites before 1877, and that there was in this hit merely the capricious work of chance. We may even remark that both the English and French authors have only spoken ironically against the mathematicians, and that in 1610, Kepler, on receiving the news of the discovery of the satellites of Jupiter, wrote to his friend Wachenfals that 'not only the existence of these satellites appeared to him probable, but that doubtless there might yet be found two to Mars, six or eight to Saturn, and perhaps one to Venus and Mercury.' We cannot, assuredly, help noticing that reasoning from analogy is here found on the right road.

However this may be, this discovery truly constitutes one of the most interesting facts of contemporary astronomy.

Such is the general physiology of this neighbouring planet. The atmosphere which surrounds it, the waters which irrigate and fertilise it,

[1] [In giving this passage from Swift, quoted by M. Flammarion, I have copied *directly* from *Gulliver's Travels*, part iii. chap. iii.—J. E. G.]

the rays of the sun which warm and illuminate it, the winds which pass over it from one pole to the other, the seasons which transform it, are so many elements from which to construct for it an order of life analogous to that which has been conferred on our planet. The weakness of gravity at its surface must materially modify this order of life in adapting it to its special condition. Henceforth the globe of Mars should no longer be presented to us as a block of stone revolving in the midst of the void, in the sling of the solar attraction, like an inert, sterile, and inanimate mass; but we should see in it *a living world*, adorned with landscapes similar to those which charm us in terrestrial nature; a new world which no Columbus will ever reach, but on which, doubtless, a human race now resides, works, thinks, and meditates as we do on the great and mysterious problems of nature. These unknown brothers are not spirits without bodies, or bodies without spirits, beings supernatural or extra-natural, but active beings, thinking, reasoning as we do here. They live in society, are grouped in families, associated in nations, have raised cities, and conquered the arts. Doubtless their senses of sight and hearing do not differ essentially from ours, and if we happened to pass a day not far from their abodes, we should perhaps be surprised with their architecture, or charmed by the echo of melodious harmony reminding us of the musical inspirations of our great masters. In the midst of varieties inherent to planetary diversities and the secular metamorphoses of worlds, we should find the same vital torch kindled on all the spheres.

Seen from this neighbouring abode, the starry sky is the same as that which sparkles above our heads: the same stars attract attention and thought, and the same constellations depict their mysterious figures. But if the *stars* are the same, the *planets* differ.

Jupiter is magnificent from Mars: it appears to the Martians half as large again as it seems to us, and his satellites should be easily visible to the naked eye. Saturn is likewise very brilliant. Uranus is easily visible, and they might have discovered Neptune before we did. They must have distinguished with the naked eye a large number of the small planets which revolve between their orbit and that of Jupiter. Mercury drawn closer to the sun, and lost in his rays, is almost impossible to distinguish! Venus appears to them as Mercury appears to ourselves.

As for the *earth*, how do they see it?

The terrestrial orbit being interior to that of Mars, the earth can no longer be for Mars a night star, as it is for Mercury and Venus, but a morning and evening star only. Its greatest elongation happens when it forms a right angle with the sun, in the vicinity of its aphelion, Mars being at its perihelion. We are, then, for that planet a brilliant star, presenting an aspect similar to that which Venus presents to us, preceding the dawn and following the twilight; in a word, we are to the inhabitants of Mars the *shepherd's star.*[1]

Our natural vanity might, then, delude us with the idea that the inhabitants of Mars contemplate us in their evening sky, purpled with the last solar rays; that they admire us from afar; that they have discoverd *our phases* and those of the moon, as we have discovered those of Venus and Mercury; and that they suppose our world to be a celestial abode of peace and happiness. Perhaps, even, they raise altars to us. What a disillusion, if they could observe us a little nearer!

[1] For further details see our work, *Les Terres du Ciel.*

CHAPTER V

THE SMALL PLANETS SITUATED BETWEEN MARS AND JUPITER

ON the first day of the present century (January 1, 1801), Piazzi, an astronomer devoted to the sky, was observing at Palermo the small stars of the constellation Taurus, and noting their exact positions, when he remarked one which he had never seen before. The following evening (January 2) he directed his telescope again towards the same region of the sky, and remarked that the star was no longer at the point where he had seen it the day before, and that it had retrograded by 4′. It continued to retrograde up to the 12th, stopped, and then moved in the direct way—that is to say, from west to east. What was this moving star? The idea that it might be a planet did not immediately occur to the mind of the observer, and he took it for a comet, as William Herschel had done in 1781, when he discovered Uranus. The planetary system appeared to be completely known with respect to its essential members; to add a new planet would have been a matter of high importance, while to find one or more comets was not of great consequence.

However, the skilful Sicilian observer was a member of an association which had for its special object the search for an unknown planet between Mars and Jupiter. From the earliest times of modern astronomy Kepler had described the disproportion, the void which exists between the orbit of Mars and that of Jupiter (a void which anyone may recognise by examining the plan of the planetary system drawn at p. 212). If we omit, in fact, the orbit of the small planets or asteroids, we notice that the four planets, Mercury, Venus, the Earth, and Mars, are in some measure crowded quite close to the sun, while Jupiter, Saturn, Uranus, and Neptune extend far into immensity. We have seen (p. 213) that the law of Titius indicates a number, the number 28, as not being represented by any planet. It was in 1772 that this *savant* published this relation in a German translation which he had made of the 'Contemplation de la Nature' of Charles Bonnet. Bode, Director of the Berlin Observatory, was so astonished at the coincidence that he announced this arithmetical relation as being a real law of nature, and spoke of it in such a way that it is generally known only by his name. He even organised an association of twenty-four astronomers to explore each hour of the zodiac and search for the unknown. This systematic exploration had not yet produced any result when, by the merest chance, Piazzi saw his moving star, and at first believed it to be a comet. But on receipt of the news Bode was convinced that this was the looked-for planet. Baron de Zach, who was, by his love of the science and his activity, at the head of the movement, and kept up an astronomical correspondence over the whole of Europe, had calculated in 1784 the probable orbit of the invisible planet, and had found the number 2·82 (that of the sun being taken as unity) for its distance from the sun, and 4 years and 9 months for its period of revolution. The new planet was found to be at the distance 2·77, and to revolve within a few days of the predicted period.

Piazzi gave to the new body the name of *Ceres*, the protecting divinity of Sicily in the 'good old times' of mythology. The astronomer was an abbot of the Order of the Theatines, and owed to Pius VII. the foundation of the Observatory of Palermo; but he loved Horace and Virgil and remembered his mythology.

The gap being thus filled up at the distance 28 by the discovery of Ceres, no one thought that other planets might exist there; and if Piazzi had supposed so, he might have at once discovered a dozen of the small bodies which revolve in this region. An astronomer of Bremen, Olbers, observed this planet on the evening of March 28, 1802, when he perceived in the constellation of the Virgin a star of the 7th magnitude which was not marked on Bode's chart, which he used. The following day he found it had changed its place, and recognised by this fact that it was a second planet. But it was much more difficult to give citizenship to it than to its elder sister, because, the gap being filled up, it was not required, and it was more inconvenient than agreeable. They looked upon it, then, as a comet until its motion proved that it revolved in the same region as Ceres at the distance 2·77, and in 1,685 days (the period of Ceres is 1,681 days). They gave it the name of *Pallas*.

The unexpected discoveries of Ceres and Pallas led astronomers to revise the catalogues of stars and celestial charts. Harding was of the number of the zealous revisers. He was soon rewarded for his trouble. On September 1, 1804, at ten o'clock in the evening, he saw in the constellation of Pisces a star of the 8th magnitude which was not noted in the 'Histoire Céleste' of Lalande. On September 4 he found it had perceptibly changed its place: it was a new planet. It received the name of Juno. Its distance from the sun is expressed by the number 2·67, and its revolution is performed in 1,592 days.

After these three discoveries, Olbers, noticing that the orbits of these planets crossed each other in the constellation of the Virgin, advanced the hypothesis that they might be nothing else but fragments of a large shattered planet. The planets, in fact, are not so solid as to be proof against any accident, and it is not impossible that the earth, for example, may some day explode (if, as geology appears to indicate, the whole interior of the globe is still a glowing furnace), or that an external shock might shatter it in pieces. Mechanics show that, in this case, the fragments would again pass every year—that is to say, at each of their revolutions—through the spot where the catastrophe took place. Olbers then set himself to explore the constellation Virgo carefully, and found on March 29, 1807, a fourth small planet, to which he gave the name of *Vesta*. Its distance is but 2·36, and its revolution only 1·326 day. This is the brightest of the small planets, and it is sometimes seen with the naked eye (when we know where it is) like a star of the 6th magnitude.

It seems surprising that after these brilliant beginnings thirty-eight years should then have passed without the discovery of a single planet, for it was only in 1845 that the fifth, Astræa, was discovered by Hencke (who should not be confused with the astronomer Encke), a simple amateur astronomer, postmaster at Berlin, who amused himself by constructing charts of the stars. The principal reason for this must be attributed to the want of good star charts, for to find these little moving points the first thing necessary is to have a very precise chart of the region of the zodiac which we observe, in order to see whether one of the stars observed is in motion. The earliest good zodiacal charts are those which the Academy of Berlin commenced to publish in 1830, taking as a basis the zones of Bessel continued by Argelander. Those of the Paris Observatory, which are more perfect, were only begun in 1854. If, either in constructing these charts or in observing the stars which they contain, we notice a new star, two evenings of observation suffice to show whether this star is a planet. Consider, for example, the chart of stars which we have reproduced on p. 304. Some people in the world may, perhaps, imagine that the four thousand white points which compose it have been put there by chance. Nothing of the sort; each of these points is a distant sun, a star placed correctly in its proper position and with its exact apparent magnitude. Take a telescope, direct it towards this region of the sky, and you will find there exactly all this sidereal population. If one of these stars appears to you larger or smaller than it is marked, it is because its brightness has varied; if one is missing, it is because it has become extinct; if, at last, you notice in this region of the sky a star which is absent from the chart, this star is a planet.

These small planets are all telescopic, invisible to the naked eye, with the exception of Vesta, and sometimes Ceres, which good sight can occasionally succeed in distinguishing; they are of the 7th, 8th, 9th, 10th, and 11th magnitudes, and even still smaller, and it was for this reason also that so long an interval of time elapsed between the fourth and fifth discoveries.

It is probable that all the small planets of any importance are now known, but that a great number—several hundreds, perhaps—still remain to be discovered of which the average brightness does not exceed that of stars of the 12th magnitude, and of which the diameter is but a few miles. The diameter of the largest, Vesta, may be estimated at 400 kilometres (248 miles). [Prof. Barnard finds a diameter of 600 miles for Ceres. —J. E. G.].

Hencke found successively the 5th and the 6th in 1845 and 1847, Hind, the English astronomer, the 7th and 8th in 1847; Graham, an English observer, the 9th in 1848; Gasparis, an Italian astronomer, the 10th and 11th in 1849 and 1850, and afterwards seven others. Hind has further discovered eight others; Goldschmidt, a German painter (a naturalised Frenchman) discovered fourteen between 1852 and 1861.[1] They are now discovered by swarms; Paliser alone has found sixty-eight since 1874.

We might say, doubtless, that in order to find them we have only to look for them, and that this research merely demands a minute and persevering attention. But we should not be the less grateful to all those who, in one way or another, have contributed to the treasury of astronomical riches; it is always one step more towards the conquest of the infinite, whether this step be made in the study of the moon, in that of the planets, or in that of the double stars lost in the depths of the heavens.

In order to catch a small planet in its passage it is necessary to spread our nets well (the meshes of the net are the small squares of fig. 135), and for this work all the patience of a fisherman with a line is required. Happy, however, is he who takes anything! The principal thing is to choose the place well. . . . We know the story of that lover of fishing who arrived in a district where he found a magnificent piece of water, a true lake, apparently abounding in fish. He was confirmed in his opinion by the presence of an angler who was installed there from daybreak to sunset. However, the new-comer lost his time and art fishing during the whole day. The same total absence of gudgeon persisted for several days. What was to be done? To take the place of the fortunate angler, always so assiduous at his post. This place was necessary at any price. The next day, then, he arrived before daylight; the other was already there. Our hero, naturally,

[1] Goldschmidt passionately loved astronomy, and I have found among his papers, which his family left me, numerous observations and remarks which show how he loved the study of the sky. His greatest ambition had been, at first, to possess a small telescope, in order to make some observations, and the best day of his life was that on which he found one in the possession of a dealer in old stores. He hastened to direct it to the sky from his modest studio, situated in one of the most frequented streets of Paris (Rue de l'Ancienne-Comédie), above the Café Procope, formerly used as a rendezvous by the stars of literature. There, *from his window* he discovered, in 1852, the 21st small planet, which received from Arago the name of Lutetia; then, in 1854, the 32nd (Pomona); then, in 1855, the 36th (Atalanta); and afterwards eleven others, all from his window. Having often removed in search of a pure atmosphere, he finally retired to Fontainebleau, where the forest offered him on all sides admirable subjects for painting; and here he died in 1866.

was not more happy than on the preceding days. Stung to the quick, he made an heroic resolution, took suitable supplies of all sorts, and as soon as his rival had left the privileged place, he installed himself there, and passed the night. The morning came, and the other angler also; but the place being occupied, he went to another. However, the usurper was not happier on that account. Nothing, always nothing! The evening came, and, leaving his envied position, he went to find his fellow-angler. 'I acknowledge,' said he to him, 'that I am guilty of bad conduct towards you; but you will doubtless pardon me when you know that, in spite of all the experience which I believe I possess in our art, and especially in bait-fishing, not only have I taken nothing to-day, but I have not seen a single fish!' 'That does not surprise me at all,' gravely replied the other, 'for during the past three months I have come here every day, and I have not yet had a single bite!'

This story recalls the criticism of the good citizen who, after remaining for two whole hours watching an angler who caught absolutely nothing, became indignant with him in good earnest, and reproached him, with an air of superiority. 'How have you the patience to remain thus for two hours doing nothing? You must have nothing in your head!'

The observer of the sky thinks himself well rewarded when, after *several years* of perseverance, he detects a planet or a new star.[1]

The first thing to be done when a small planet is discovered is to ascertain its motion. When we possess three precise observations, a little separated from each other, we have three points in the unknown orbit of the new body, generally sufficient to enable us to determine the complete orbit (which requires nearly eight whole days of calculation). The most interesting element is the exact determination of the diurnal motion, which we express in seconds of arc: dividing the entire circumference—360 degrees, or 21,600 minutes, or 1,296,000 seconds—by this diurnal motion, we obtain the exact duration of the planet's revolution round the sun, expressed in terrestrial days. We have seen that the lengths of the revolutions are connected with the distances ('Third Law of Kepler,' p. 217); this length gives us, then, the distance, that of the earth from the sun being taken as unity. If we wish to obtain the distance in miles, it is sufficient to multiply the number by 93,000,000. The determination of the orbit also gives the eccentricity—that is to say, the form of the ellipse pursued by the planet in its course, and the inclination of this orbit to the plane in which the earth moves round the sun, or the ecliptic taken as a plane of comparison.

The names given to these small bodies commenced with the mythological army of divinities of the earth and ancient heaven; but even before

[1] [Numbers of the small planets are now discovered by means of photography. Of 34 discovered in 1893, all but one were found by the new method.—J. E. G.]

the list had been exhausted certain scientific, or even national or political circumstances, caused the preference to be given to more modern names. It was thus that the 11th, discovered at Naples, received the name of Parthenope; the 12th, discovered in England, that of Victoria; the 20th, that of Massilia; the 21st, that of Lutetia; the 25th, that of Phocæa, before even Urania had been restored to the skies; the 45th was named in honour of the Empress of the French; the 54th, in honour of the illustrious Alexander von Humboldt; &c. The 87th, 107th, 141st, 154th, and 169th, have been named in honour of a young astronomer who has devoted his best years to the culture of astronomy and to the apostleship of this beautiful science.

A rather curious fact is that they have put Wisdom (*Sapientia*) in the sky only at the 275th, discovered in 1888; Bellona[1] has been placed there since the 28th (1854).

Of all this number of planets, the nearest to the sun is No. 149, Medusa, of which the distance is 2·17—that is to say, about twice as far from the sun as the earth; and the most distant is No. 279, Thule,[2] of which the distance is 4·26, about $4\frac{1}{4}$ times our distance. Thus the zone which extends between the mean orbits of the two extreme planets is $4·26-2·17$, or 2·09—that is to say, $37,000,000 \times 2·09$, or 77,330,000 leagues (192,200,000 miles).

The mean distance of Mars is 1·52. There is, then, between the orbit of Mars and that of Medusa only $2·17-1·52$, or 0·65, or 24 millions of leagues. On the other hand, the mean distance of Jupiter is 5·20. There is, then, between the orbit of Thule and that of Jupiter but $5·20-4·26=0·94$, or 86 millions of miles.

A large number of these small bodies are remarkable for their great eccentricity and for their high inclination to the ecliptic, an inclination so great that some of them leave the zodiac; thus, Pallas (2) goes 34 degrees from the ecliptic; Euphrosyne (31) and Anna (265) and Istria (183), to 26 degrees. They are sometimes northern circumpolar stars, always above our horizon, sometimes southern stars, not rising above the horizon of Paris. All these orbits are so interlaced with each other that, if they were material hoops, we could by means of one or two taken by chance raise all the others.

. . . .

Fig. 174 represents some of the orbits of these small planets [they are drawn to scale, and represent in relative magnitude and position the paths of a few of the asteroids having large eccentricities and small inclinations to the plane of the ecliptic. The dotted lines show the positions of the longer axes of the elliptical orbits, and the point marked P is the position of the perihelion, or point of nearest approach to the sun. The orbits of the

[1] [The Goddess of War among the Cappadocians.—J. E. G.]

[2] [M. Flammarion gives Hilda, but the maximum distance *now* known (1893) is that of Thule, as given above.—J. E. G.]

earth, Mars, and Jupiter, are added for comparison, and the arrows show the direction of motion round the sun.—J. E. G.].[1]

The number of these small bodies increases at the rate of about ten per annum at an average, and although they have organised in several

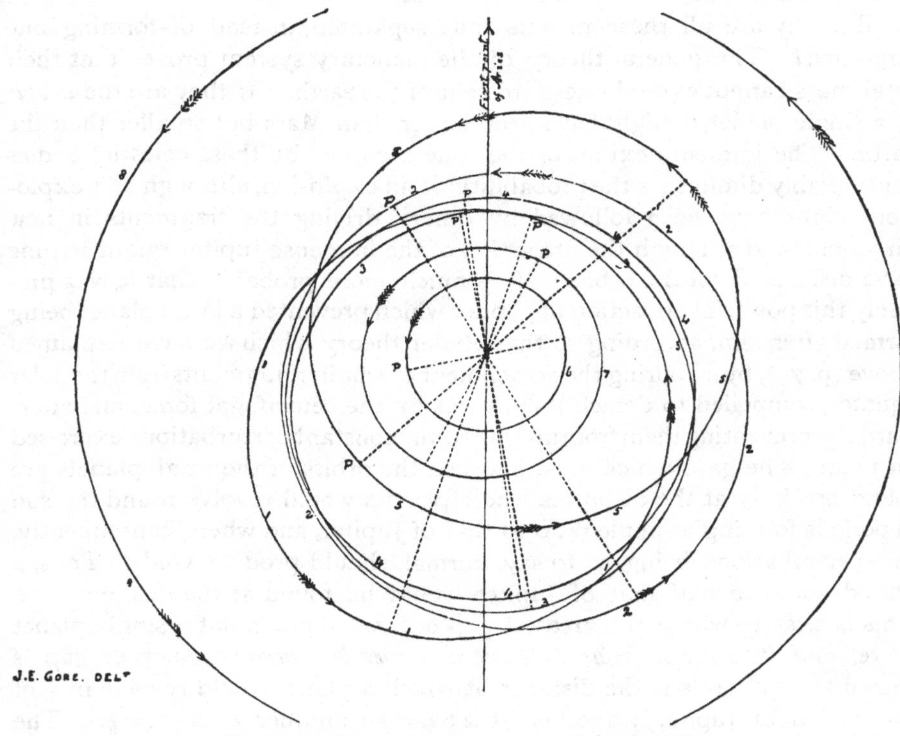

FIG. 174.—ORBITS OF SOME OF THE ASTEROIDS
1. Polyhymnia, No. 33. 2. Pales, No. 49. 3. Virginia, No. 50. 4. Vibilia, No. 144. 5. Alcmene, No. 82. 6. The Earth. 7. Mars. 8. Jupiter

observatories a special service to watch them, it is extremely difficult to follow them closely, the more so that the perturbations produced

[1] Several of these little worlds follow orbits almost identical, and there would be nothing astonishing in two of these planets approaching so near to each other as to unite their influence and form a double planet, of which the components would revolve round their common centre of gravity, both revolving together round the sun. It would not be impossible for one or more to unite into one. Examples: on the one hand, JUNO, CLOTHO, LUMEN, ADEONA, CLYTIE, EURYDICE, FRIGGA; on the other hand, FIDES, MAIA, VIRGINIA, EUNOMIA, EVA, IO, VIBILIA; or, again, THISBE, SIRONA, CERES, LÆTITIA, ALCMENE, PALLAS, GALLIA.

The orbits of Juno and Clotho approach to within 630 miles of each other. Fides and Maia follow almost in the same plane, and M. Lespiault has already proved their possible association as a double planet; Thisbe and Sirona, Astræa and Flora, present an analogy equally interesting; and we might notice a large number of others of which the orbits are very near at certain points.

by the large planets are always on the point of disturbing them and modifying the first-calculated elements. Several of these little floating islands have even been *lost*, and all the perseverance of the 'look-out men' of the sky is necessary to recover them, often at a considerable distance from the point where they are searched for.

But why are all these planets thus separated, instead of forming one large one? The general theory of the planetary system proves that their total mass cannot exceed one-third that of the earth. If they are the *débris* of a single planet, it might have been larger than Mars but smaller than the earth. The immense extent of the zone occupied by these celestial bodies considerably diminishes the probability of an explosion, although this explosion might have been followed by others, driving the fragments in new directions, and although the attraction of the immense Jupiter might in time have dislocated all the orbits. It is much more probable that it was precisely this powerful attraction of Jupiter which prevented a large planet being formed after him, according to the nebular theory which we have explained above (p. 72), by favouring the detachment of smaller fragments from the solar equator, compelled to detach themselves by the centrifugal force, and afterwards by preventing them from uniting by the constant perturbations exercised on them. The gaps which exist between the orbits of the small planets are found precisely at the distances where planets would revolve round the sun in periods forming a simple ratio to that of Jupiter, and where, consequently, the perturbations, being, so to say, normal, should produce voids. Thus, a period equal to half that of Jupiter would be found at the distance 3·28. This is exactly where the greatest gap occurs: there is not a single planet there, *and it is not probable that one will ever be found.*[1] Another gap is shown at 2·96 (this is the distance at which a planet would revolve in $\frac{3}{7}$ of the period of Jupiter); another at 2·82, $=\frac{2}{5}$; another at 2·50, $=\frac{1}{3}$. The action of Jupiter is as clear in this distribution of the orbits as that of a waterspout which passes through a forest and forms a void in its passage.

We shall see further on that it is the same with the rings of Saturn, of which the intervals correspond to zones where satellites would revolve in periods commensurable with those of the four inner satellites. We owe these interesting remarks to the American astronomer, Kirkwood. This zone of asteroids is remarkable from its situation: it separates the four smaller planets from the four larger, which, not long since, were brilliant suns at the centres of their little systems.

It may be shown, on the other hand, that the perihelia of these bodies are far from being distributed by chance uniformly all round the sun, and that they have a maximum between 26 and 72 degrees and a minimum between 153 and 293 degrees. The difference rises to treble; and what is

[1] [No. 318, discovered by M. Charlois on September 24, 1891, has, however, a mean distance of nearly 3·28 (3·27775).—J. E. G.]

more curious is, that the perihelion of Jupiter is found near the middle of this region of greater condensation of the perihelia.

To measure the diameter of these small bodies, so distant from us, is a very difficult problem. The largest does not exceed four-tenths of a second, and most of them are reduced to simple points. Combining the measures obtained with estimates founded on their light, we find the following diameters as being the most probable :—

Vesta	400 kilometres	(248 miles)	Hebe	145 kilometres	(90 miles)
Ceres	350 „	(217 „)	Lætitia	145 „	(90 „)
Pallas	270 „	(167 „)	Iris	140 „	(87 „)
Juno	200 „	(124 „)	Amphitrite	130 „	(81 „)
Hygeia	160 „	(99 „)	Calliope	125 „	(78 „)
Eunomia	150 „	(93 „)	Metis	120 „	(74 „)

There are others, on the contrary, such as Sappho, Maia, Atalanta, and Echo, which do not measure more than 30 kilometres (19 miles) in diameter. It is probable that still smaller exist, which remain absolutely invisible in the best telescopes, and which only measure a few miles, or even less, perhaps.

Are they globes ? Yes, doubtless, for the most part. But several among the smaller ones may be polyhedral, and may have proceeded from subsequent explosions ; the variations of brightness which have been sometimes observed seem to imply surfaces irregularly broken.

Are they *worlds* ? Why not ? Is not a drop of water shown in the microscope peopled with a multitude of various beings ? Does not a stone in a meadow hide a world of swarming insects ? Is not the leaf of a plant a world for the species which inhabit and prey upon it ? Doubtless among the multitude of small planets there are those which must remain desert and sterile, because the conditions of life (of any kind) are not found united. But we cannot doubt that on the majority the ever-active forces of nature have produced, as in our world, creations appropriate to these minute planets. Let us repeat, moreover, that for nature there is neither great nor little. And there is no necessity to flatter ourselves with a supreme disdain for these little worlds, for in reality the inhabitants of Jupiter would have more right to despise us than we have to despise Vesta, Ceres, Pallas, or Juno : the disparity is greater between Jupiter and the earth than between the earth and these planets. A world of two, three, or four hundred miles in diameter is still a continent worthy to satisfy the ambition of a Xerxes or a Tamerlane, and we may believe that several of them are divided into rival anthills, each of which has its king, its flag, and its soldiers, and that from time to time they go to war, and, calling on the name of the God of Armies, mutually massacre each other. An excellent design might, perhaps, be read on their mottoes and their arms, in languages special to each country : here, 'God protect France !' there, 'God protect

Belgium!'; further, 'God protect Italy!'; again, 'God protect Germany!': formulæ in which only the name of the country is changed, and which might peculiarly embarrass the Director of the solar system, if he took seriously the inscriptions on pieces of money upon which each fraction of humanity inscribes in this manner an individual invocation. But evidently all these games which seriously amuse the politicians of the great nations of the earth, might be reproduced with greater puerility still, if that were possible, in this republic of little worlds, where they may have manufactured great swords and pretty lace, and where the cavalry soldier of four inches high may scorn the foot soldier of two inches.

A good walker constituted as we are might easily make the circuit of these little worlds in a single day's journey of twenty-four hours. Gravity is inevitably very feeble on each of them, since their mass is, so to say, insensible; it is much less intense than on the moon, where an object passes over but eighty centimetres in the first second of its fall; a body would fall but a few centimetres in the first second. Suppose that the towers of Notre Dame were built in a city of these worlds, and that we hurled ourself from it into space with the feeling of terror and horrible despair which must accompany the supreme act of suicide, we should be quite surprised at remaining in the air so long; during the period of our fall, slow and gentle like that of a feather, we should have ample time to think of a thousand agreeable things, and, reaching the earth, we should feel that our attempt was by no means successful. Persons who have been nearly drowned, and by a providential hand have been brought back in good time from darkness and suffocation, relate that, in the three or four seconds which preceded their unconsciousness, they had time to see again their whole life since their earliest infancy; and those who have analysed their dreams know that a voyage of several months is easily made in less than a minute, although felt and appreciated in all its length and all its details. From this point of view an aëronaut who fell from a balloon on Vesta, or on some of its companions, would live an entire psychological life during the period of his fall.

Everything is relative. The unknown beings who inhabit these light worlds must therefore be organised altogether differently from us, in order to be appropriate to the smallness of their planet and to their special vital conditions. And how these conditions of habitability differ from ours! Let us suppose ourselves, for example, transported to one of these celestial islands. We should weigh only 1 kilogramme (2·2 lbs.). Freed from the heavy weight of matter, we should be able to run across the country with the velocity of steam, and in a step spring from a valley to a mountain; a gymnast would soar for a whole minute above our heads. We have seen, in studying the sun, that beings of large size would have the greatest difficulty in supporting their own weight on worlds of the importance of that body, and that these enormous bodies seem rather suitable for beings smaller than we are ; while small globes like the moon might be peopled with giants. But we should refrain from pushing this argument to its extreme consequences, for we might come to people these little planets with inhabitants larger than the planets themselves. We can, then, conclude absolutely nothing as to the form, the size, and the organisation of the unknown beings with which all these little worlds may be peopled ; logic alone tempts us to think that they must be smaller than we are. The intensity of gravity is so feeble that a man of power equal to ours could throw a stone into space so that it would never return ! The feeblest volcanic explosion would project from such a world materials which would be separated from it for ever !

We do not yet know the duration of rotation, nor the inclination, nor the seasons which result from them, for any of the small planets, although the variations of brightness observed in Pales by Goldschmidt led him to conclude that a rotation in twenty-four hours was probable.

We will now complete all these data by remarking that the immense vaporous atmosphere described by Herschel and Schröter round the first four small planets discovered was an illusion due to the imperfection of their instruments. We do not in reality *see* the atmosphere of any planet, any more than we *see* those of Mars or Venus: we only observe the effects. We may say, however, that these four planets, when examined with the spectroscope, show rays of absorption indicating the presence of a slight atmosphere round them. A deeper knowledge of these curious little worlds can only be obtained by a very great improvement in optics, of which we need not despair.

CHAPTER VI

JUPITER, THE GIANT OF WORLDS

♃

THE description of the system of the world brings us now to the most important planet of the whole solar family. Before the interest which the preceding provinces have been able to afford us has been effaced and disappears, and as if to prepare a surprise for us still more striking by its contrast, the natural order of things has caused the description of the giant of the worlds to be preceded by that of the minute little asteroids which we have passed through since we left the orbit of Mars.

Giant of worlds indeed! When Jupiter shines among the stars of the silent night, and when our gaze is fixed on him, who would suppose, while admiring this simple luminous point, that it is an enormous and massive globe weighing 309 times more than the planet which we inhabit, and of which the colossal volume exceeds by 1,279 times that of our earth? We have our eyes fixed on him; his light is so vivid that it casts a shadow like that of Venus, but we do not guess the marvellous grandeur of this distant body. Let us direct a small telescope towards it; it suffices to enlarge this point, to give it the round form of a disc, and to show four of the five satellites which accompany it in its celestial course. Our curiosity being aroused, let us bring the planet into the field of an ordinary astronomical telescope of which the object-glass measures 11 centimetres (4·33 inches), and its length 1·60 metres (63 inches), and suddenly we see shining in this dark field a dazzling orb, moving in majesty, and permitting us to recognise at a glance the spheroidal form of its disc very much compressed at the two poles, as well as the cloudy trails which mark its equatorial zones. How exciting is this vision! Anyone may view this spectacle, and yet no one thinks of it. It is no longer a starry point: it is a world. Our imagination can hardly make the circuit of it by thinking that if the deep waters of a vast ocean entirely surrounded it, a steamer running fourteen knots an hour, which would complete in three months the circuit of our earth, would take nearly three years to travel round the circumference of this world which shines up there. Yes, for three years the engine would keep up steam without intermission night and day, for three years the monotonous screw would beat the waves in order to make the circuit of this world, which a little leaf of a tree entirely hides from our view, and which a fly walking on the window-pane seems to swallow up.

Mercury Venus Earth Mars

Jupiter Saturn

Uranus Neptune

FIG. 175.—THE PRINCIPAL WORLDS OF THE SOLAR SYSTEM

After Venus, Jupiter is the most brilliant of all the planets. While the Evening Star became the Queen of Beauty, Jupiter was seated on the throne of the sky, and received as sovereign the homage of mortals. Weakness united to vanity associated astronomical appearances with the events of human life; each planet was endowed with an influence corresponding to its aspect. Jupiter ruled the night destinies, and in his solitary watches the astrologer of the Middle Ages, continuing the traditions of his ancestors of antiquity, still calculated the occult influences which would seem to descend from this distant and powerful light. We shall have occasion, however, further on to enter into some details of astrology, and to state that this delusive science has counted adepts even in recent centuries. Jupiter reigned at the head of celestial influences, and ruled the destinies of the 'great ones' of the earth.

The brilliant planet has preserved in modern astronomy the superiority of rank which had been assigned to it by ancient astronomy. The first telescopic observations revealed the grandeur of its sphere. Its apparent diameter is, on an average, 38″, and varies from 30″ to 47″, according to its distance. In the most favourable conditions this diameter of 47″ is only about thirty-nine times smaller than that of the moon, so that a telescope magnifying thirty-nine or forty times shows us the disc of Jupiter of the apparent size we see the moon with the naked eye,[1] and a magnifying power of eighty shows it twice as large [that is, of twice the apparent *diameter*, or four times larger in area—J. E. G.]. The real diameter of this enormous globe is, as we have seen, eleven times greater than that of our planet (11·06 exactly)—that is to say, 142,000 kilometres (88,000 miles). If this globe were spherical, its volume would exceed that of the earth 1,390 times, and it is generally so stated in works on astronomy; but the rapidity of its motion of rotation has bulged it at its equator and flattened it at the poles so considerably that it is perceptible at a glance when we observe the planet in an astronomical telescope; this flattening is $\frac{1}{17}$, while that of the earth is only $\frac{1}{292}$. It follows that the volume of Jupiter is 1,279 times greater than that of our planet. Its surface is equal to that of 114 earths.

The enormous size of Jupiter's globe will, perhaps, be still better appreciated, from what we have learned of the sun, if we remark that its diameter is about the tenth of that immense body. Its circumference being more than 270,000 miles, a band of paper as long as from here to the moon would not make the whole circuit of Jupiter. With this planet we enter the domain of the giants of the system, as we may learn by considering fig. 175, in which Jupiter and Saturn are placed in the middle, Uranus and Neptune below, and Mercury, Venus, the Earth, and Mars above.

The mass of Jupiter is known with an accuracy truly worthy of admiration. All the calculations agree in estimating it at $\frac{1}{1047}$ of that of the sun. It is determined either by the motions of the satellites or by the perturbations produced on the small planets; the determination is as accurate as if we could weigh Jupiter in the scale of a balance. It follows that this world weighs about 309 times more than ours. Taking into account the polar flattening, we find that the mean density of the sub-

[1] With reference to this size, about which everybody is deceived, see p. 88.

stances which compose it is 0·242, that of the earth being taken as unity This density is curiously similar to that of the sun. Jupiter weighs a third more than a globe of water of the same size.

We have studied the methods which are used to weigh the stars, to measure their distance, and to determine their real dimensions. We need not, then, return to this subject. We have also seen how astronomers calculate the intensity of gravity at the surface of worlds, a method which depends, on the one hand, on the mass of the globe which we are considering, and, on the other hand, on its radius, or the distance of its surface from its centre. If Jupiter were not larger than the earth, and at the same time had the weight which we have just stated, the force of gravity at its surface would be 309 times greater than it is here, and a pound would weigh 309 pounds. But as the diameter of this globe is eleven times greater than that of our planet, the intensity of gravity must be reduced in the proportion of the square of this number, or 121 to 1 (exactly, 122, for $11·06 \times 11·06 = 122$). Dividing 309 by 122, we find 2·5, or $2\frac{1}{2}$. We *know*, then, from this that the force of gravity is two and a-half times greater on Jupiter than on the earth. A man of 70 kilogrammes (154 pounds) in weight, if carried there, would weigh 342 pounds. A stone abandoned at the top of a tower to the influence of gravity would descend twelve metres (39·36 feet) in the first second of its fall.

Thus on Jupiter the constituent materials of things and beings are composed of substances lighter and less dense than those of terrestrial objects and bodies; but as the planet attracts more strongly, they are in reality heavier, fall more quickly towards the ground, and weigh more. This is the opposite of that which we have noticed in Mercury, and it is a good example of the fact that *there is nothing absolute.* Truly all is relative, and we live in the relative. Nothing seems so physically absolute to the public as a cannon-ball of forty pounds in weight. Well, this ball is only such as it is on account of its situation on the earth, Carry it to a small planet, it is a feather; imagine it on the sun, it is a massive block impossible to move. Here it kills, there it is a plaything. We live in the midst of the relative, and we would like the universe to be reduced to our size.

We have already seen that Jupiter revolves round the sun at a distance five times greater than ours—at 192 millions of leagues (477 millions of miles)—along an orbit which he takes 4,332 days to pass over, or eleven years ten months and seventeen days. We have also seen that the perspective caused by the annual revolution of the earth round the sun produces an appearance in Jupiter's motion of eleven or twelve stationary points and retrogradations along his twelve years' orbit. He returns to *opposition* relatively to the sun—that is to say, the sun, the earth, and Jupiter are found again on the same line, every 399 days, or one year and thirty-four days on an average ; 279 days are employed in the direct motion, and 121 in the retrograde direction.

This beautiful planet, then, returns to opposition every year with a delay of a month and six days. These are the epochs when it passes the meridian at midnight, soaring majestically in the south like a brilliant star. It is, consequently, very easily recognised. It moves along the zodiac, and only returns every twelve years to the same region of the sky. We may count on three months each year for its periods of favourable observation in the evening : the month in which its opposition takes place and the three months following.[1]

Its orbit round the sun is not circular, but elliptical, with an eccentricity of 0·048, which gives for its variations of distance :—

	Geometric	In Miles
Perihelion distance	4·952	459,000,000
Mean distance	5·203	482,000,000
Aphelion distance	5·454	506,000,000

There are, as we see, forty-seven millions of miles of difference between its distance from the sun (or from the earth) at its aphelion and its perihelion. These are the true seasons of Jupiter, for its axis of rotation is almost perpendicular to its orbit.

Its motion round the sun is performed in a plane but little different from that in which the earth itself moves—that is to say, the plane of the ecliptic ; the inclination of this plane to ours is only 1° 18′ (and 41″). The ancients noticed this, and they called Jupiter the ecliptic planet. Its perihelion now happens at 13° of longitude—that is to say, at 13° from the point of the Vernal Equinox ; Jupiter is then found in the constellation of Pisces, not far from the star ζ : November 16, 1868, September 25, 1880. Its aphelion of course occupies the opposite position, and was passed October 24, 1874, and April 22, 1886. Its perihelion as well as aphelion, its line of apsides, advances 57″ per annum on the ecliptic.

At this distance from the sun the disc of the day star is reduced more than five times in diameter and more than twenty-five times in surface, and in luminous and calorific intensity ; this world receives on an average twenty-seven times less light and heat than we receive from the sun. This difference of intensity would certainly have the effect of organising the life of this immense planet in a way very different from that which regulated the organisation of terrestrial life.

This globe rotates on itself, keeping its axis vertical—that is to say, instead of turning *awkwardly*, as Voltaire accused the earth of doing, it remains upright, we might say vertical, in its course round the sun. The inclination of its axis of rotation is, in fact, but 3 degrees—that is to say,

[1] The position of the planets in the sky on each day of the year is given continuously in our *Revue mensuelle d'Astronomie populaire*.

insignificant. A total absence of seasons and climates is the result. The days preserve the same length during the whole year; the sun performs his apparent diurnal motion nearly in the plane of the equator; there are neither tropical zones nor polar circles: it is in the state of a perpetual equinox, *an eternal spring* for all countries of the globe, the temperature decreasing uniformly from the equator to the poles.

But this immense world shows many other differences from that which we inhabit. Although revolving slowly round the sun, it rolls on itself with such impetuosity that its entire rotation is effected in less than ten hours. It has not even five hours of day or five hours of night![1] The diurnal rotation is performed in 9 hours 50 minutes for the equatorial region, and in 9 hours 55 minutes for the regions on both sides of the equator from 20° to 25° of northern and southern latitudes. This rotation is that of the bright or dark spots visible in the planet's atmosphere, and not that of the surface, which we do not see. There reigns constantly in the atmosphere of Jupiter an equatorial current directed from west to east, which moves with a velocity of 400 kilometres (248 miles) an hour relatively to the lateral regions.

The first thing which strikes every observer when he views Jupiter in a telescope is, that this globe is marked with bands more or less wide, more or less intense, and which show themselves principally near the equatorial region. These belts of Jupiter may be considered as characteristic of this gigantic planet. They were remarked from the first telescopic view which was given to man of this distant world, and since then they have been invisible only in extremely rare circumstances.

Sometimes, independently of these white and greyish trails, which are often tinted with a yellow and orange colouring, we notice spots, either more luminous or darker than the ground on which they are placed, or,

[1] In the year 1655 Cassini discovered this velocity of rotation by his observations made in the beautiful sky of Italy. He obtained for its duration 9 hours 56 minutes. Later, in 1672, in Paris, observations made on a spot which he believed to be similar to that which he had observed in Italy gave him 9 hours 55 minutes 51 seconds. Resuming this interesting research in 1677, he obtained 9 hours 55 minutes 50 seconds. But this close agreement vanished in 1690, for in that year he found 9 hours 51 minutes; it was the same in the following year, and even in 1692 he found it 9 hours 50 minutes. In 1713 Maraldi found 9 hours 56 minutes; and it was the same in 1773 in a determination made by Jacques de Sylvabelle. William Herschel found 9 hours 55 minutes 40 seconds in 1778, and 9 hours 50 minutes 48 seconds in 1779; Schröter, 9 hours 56 minutes 56 seconds in 1785, and 9 hours 55 minutes 18 seconds in 1786; Airy, 9 hours 55 minutes 24 seconds in 1834; Mädler, 9 hours 55 minutes 26 seconds in 1835; Schmidt, 9 hours 55 minutes 24 seconds in 1862; Lord Rosse, in 1873, 9 hours 54 minutes 55 seconds. In 1874 I found 9 hours 55 minutes 45 seconds from a spot situated near 25° of latitude. In 1888 Denning found 9 hours 55 minutes 40 seconds from the red spot, and 9 hours 50 minutes 28 seconds from the white spots at the equator.

The considerable discrepancies of these various results have led us to suppose that the spots are clouds floating in a very agitated atmosphere, which have a more rapid motion the nearer their position to the vicinity of the planet's equator. Thus, as Fontenelle said, we may compare the motions of these spots to that of the trade winds which blow near the terrestrial equator.

again, irregularities and very pronounced rents in the form of the bands. If, then, we observe with attention the position of these spots on the disc, we soon notice that they are displaced from east to west, or from left to right, if we view the planet in a telescope which does not reverse the objects. When these spots are very marked, an hour of *attentive* observation is sufficient to verify the motion.

These spots belong to Jupiter's atmosphere itself. They do not travel round the planet, like satellites, with a proper motion independent of the motion of rotation, but form part of an immense cloudy layer which surrounds this vast world. On the other hand, they are not fixed to the surface of the globe, as the continents and seas of Mars are, but are relatively movable, like the clouds of our own atmosphere.

We have here, then, a world where, instead of twenty-four hours, the

FIG. 176.—COMPARATIVE INCLINATION OF THE AXIS OF THE EARTH AND OF THAT OF JUPITER

length of day and night is not even ten hours. They do not count five hours between sunrise and sunset, and throughout the year the night is still shorter, on account of the twilights. As, on the other hand, the year is almost equal to twelve of ours, the rapidity of the days makes the inhabitants of Jupiter count 10,455 days in their year. What a strange calendar, and how rapidly the hours fly up there!

The velocity of the rotation is such that a point situated on the equator moves at the rate of 12,450 metres (7·73 miles) a second as an absolute velocity. This rotation is twenty-six times more rapid than that of a point at the terrestrial equator. It is this rapidity which has produced the flattening at the poles, and it is this, evidently, which produces the belts of Jupiter. It diminishes the gravity at the equator by $\frac{1}{12}$: an object which weighs 12 pounds at the poles would weigh but 11 at the equator.

This rapid succession of light and darkness must exercise a great influence on the manner of life of the inhabitants. Littrow has asked whether they devote their days to work and pleasure, their nights to repose

FIG. 177.—COMPARATIVE MAGNITUDES OF JUPITER AND OF THE EARTH

and sleep. 'They must,' he asserted ('Die Wunder des Himmels') 'possess a singular elasticity of mind and body. How little, in fact, should we be satisfied if the night lasted but five hours, and we were roused up so rapidly! Epicures especially would be very embarrassed if, in the space of five hours, they were obliged to take three or four meals. And then, our ladies [the author is from Vienna], how they would grudge nights so short, and balls shorter still. They would require for the preparations of their toilet almost double the time of a night of Jupiter. But, on the contrary, the official astronomers of the observatories of this world should be enchanted—if, indeed, the Jovian atmosphere permits them to work ;—they would never be fatigued !'

Thus spoke the witty Director of the Royal Observatory of Austria, at an epoch when it was possible to believe that the world of Jupiter was inhabited by beings similar to ourselves ; but attentive examination of the revolutions to which it still appears to be subject modifies our inductions in this respect.

FIG. 178.—GENERAL ASPECT OF JUPITER

In fact, we observe from here singular metamorphoses on its immense sphere. The characteristic belts which cross it do not preserve, as was so long believed, the same form, the same brightness, the same tint, the same width, or the same extent, but, on the contrary, they are subject to rapid and considerable variations.

In general, the equator is marked by a white zone. On each side of this white zone there is a dark belt, shaded with a deep reddish tint. Beyond these two dark (southern and northern) belts we usually notice parallel trails, alternately white and grey. The general tint becomes more homogeneous and greyer as we approach the poles, and the polar regions are usually bluish. This normal type is nearly that of the accompanying drawing (fig. 178).

Now, this typical aspect varies to such an extent that it is sometimes impossible to find any vestige of it. Instead of this white zone, the equator sometimes shows itself occupied by a dark belt, and we see one or more clear lines in certain latitudes more or less distant. Sometimes the belts are wide, and placed apart ; sometimes, on the contrary, they are thin, and close together. Sometimes their edges are cut up like disturbed and broken clouds; sometimes they appear under the form of a perfect straight line. Luminous white spots have been seen to float above these atmospheric belts, and sometimes quite round luminous points similar to satellites are visible; dark trails have also been seen crossing the belts obliquely, and persisting for a long time. In short, the variability of this vast world is such that it presents to the observer and thinker one of the most novel and interesting problems of planetary astronomy.

These atmospheric perturbations may, however, be performed in the immense aërial envelope of Jupiter, unless the surface of the planet itself is in a state of corresponding instability. This surface we rarely or never see through the vistas which appear dark to us.

Since the year 1868, and especially since 1872, I have followed with great assiduity the variations in the aspect of this very important world. Of all the bodies of our system, it is that which presents in the telescope the most considerable and the most extraordinary changes, not only in the markings, but also in the colouring of its disc.

The drawings on the next page (fig. 179) were made by Dr. Otto Boeddicker with Lord Rosse's 3-foot reflecting telescope. The *attentive* examination of these drawings will show better than any description the changes observed.

An elongated oval spot appeared during the year 1877, and it has always returned since, with variations of tint; it was reddish, more or less deep, and measured twenty-five thousand miles in length—more than three times the diameter of our world! It is, perhaps, a continent in formation with vapours rising from it.

This is what we see more certainly in the aspect of Jupiter: his cloudy belts and their curious variations. Sometimes, in three or four hours, a cloud lengthens along a whole latitude of the disc; sometimes the total aspect changes from day to day; at other times, on the contrary, we do not see any perceptible variation for whole weeks. Now, it is not the solar heat which can produce all this, since Jupiter has no seasons, and since in the whole length of its year its relative variation of temperature, proceeding from the central body, does not exceed that which we receive here during the fifteen days which border on the vernal and autumnal equinoxes. It is only every six years that the variation dependent on the eccentricity would show itself. How can this action, so slow and so feeble, produce the prodigious and rapid atmospheric variations observed on this planet?

It has been hitherto believed that the temperature of the surface of Jupiter is inferior to that of our atmosphere, on account of its greater distance from the sun. Now, the existence of water-vapour, which saturates the Jovian atmosphere, and the tremendous movements which we see from here, lead us, on the contrary, to think that Jupiter is warmer than the earth. It must, at least, be warmer at its surface than the sun could

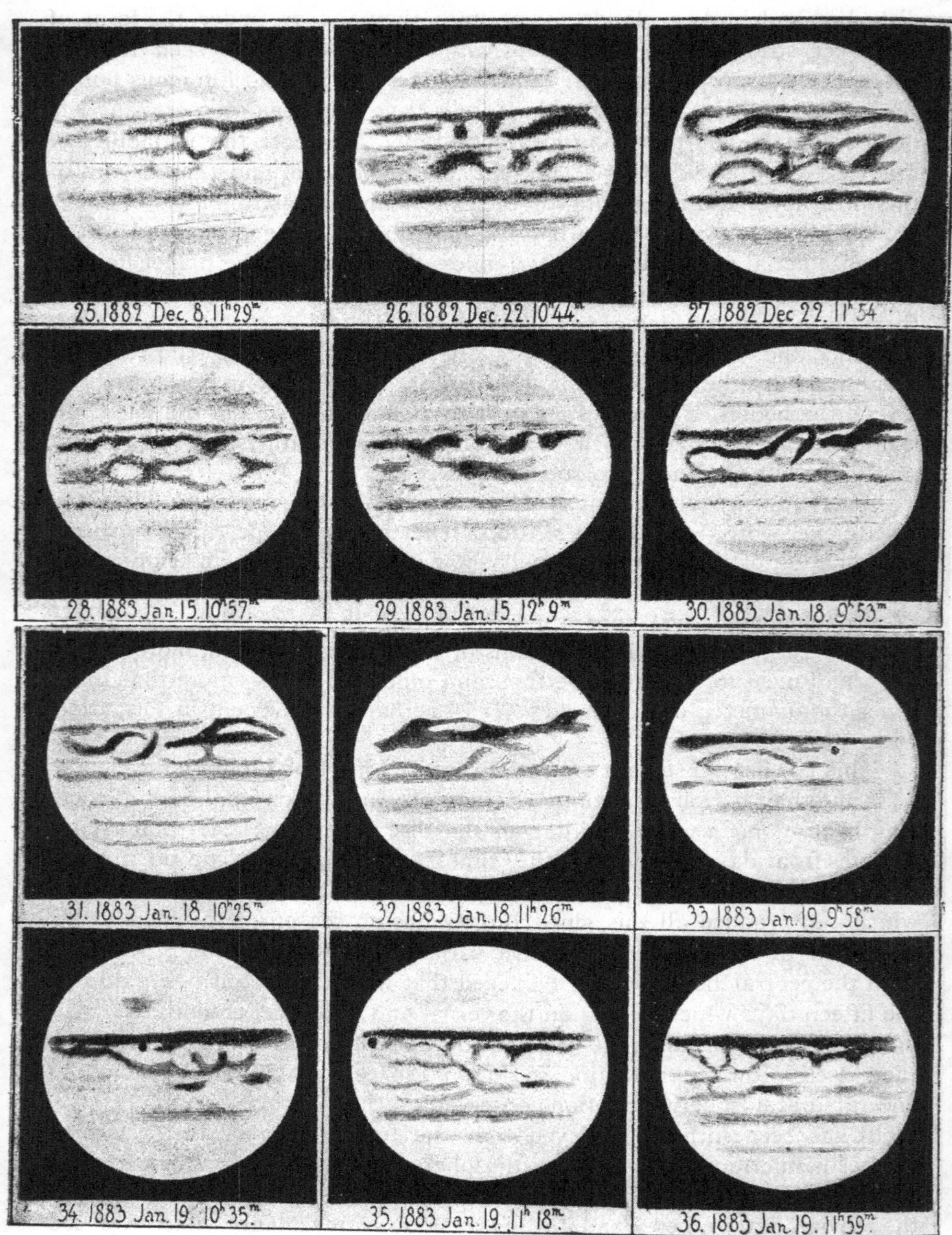

Fig. 179.—Jupiter, from Drawings by Dr. Otto Boeddicker
(Made with the aid of Lord Rosse's 3-foot Reflector)

make it. Perhaps it has volcanoes and sources of vapours ; perhaps it is the seat of revolutions capable of producing the phenomena which we observe in our atmosphere ; perhaps electricity comes into play in these variations ; and perhaps, also, the atmosphere of this planet sometimes glows with immense auroræ boreales.

With reference to the physical and chemical constitution of Jupiter's atmosphere, we may at first remark that this atmosphere must be of considerable density in its lower layers, on account of the intensity of gravity ; on the other hand, spectrum analysis shows that there is there the same vapour of water as here, except some substances which appear to be special to this world.

The white belts of Jupiter and the white spots would represent, generally, the most elevated clouds of his atmosphere. The dark regions tinted with brown maroon, and sometimes red, represent either the soil of the planet or the lower layers of the atmosphere. There are, doubtless, cloudy regions, less dense and less opaque, through which our sight can penetrate more or less to the planet's soil.

There blows constantly in the equatorial zone of the planet a violent wind at the rate of 250 miles an hour (four times the velocity of our express trains). The rotation of the equatorial clouds is performed in 9 hours 50 minutes, that of the clouds at the 25th degree of latitude in 9 hours 55 minutes.

FIG. 180.—JUPITER AND FOUR OF ITS SATELLITES AS SEEN IN A TELESCOPE

All these considerations show us that while Mars, Venus, and Mercury resemble our planet more or less, it is not so with Jupiter. There the constituent materials ; the molecular, physical, and chemical states ; the local forces, electricity, heat, are found *in conditions totally different from the four preceding worlds.*

Thus the meteorological *régime* of Jupiter, as we observe it from the earth, leads to the conclusion that the atmosphere of this planet is subject to more considerable variations than those which would be produced by the solar action alone ; that this atmosphere is very thick ; that its pressure is enormous ; and that the surface of the globe has not yet reached the state of fixity and stability which the earth has now attained. It is probable that, although born before the earth, this globe preserved its original heat long after ours, on account of its volume and mass. Is this inherent heat, which Jupiter appears still to possess, sufficiently elevated to prevent all vital manifestation? And is this globe still, not in the state of a luminous sun—for the satellites disappear in his shadow, and receive no light from him—but in the state of a dark and glowing sun, entirely liquid, or barely covered over with a coagulated crust, as the earth was before the appearance of life upon its surface? Or, again, is this colossal planet at a temperature through which our own world passed during the

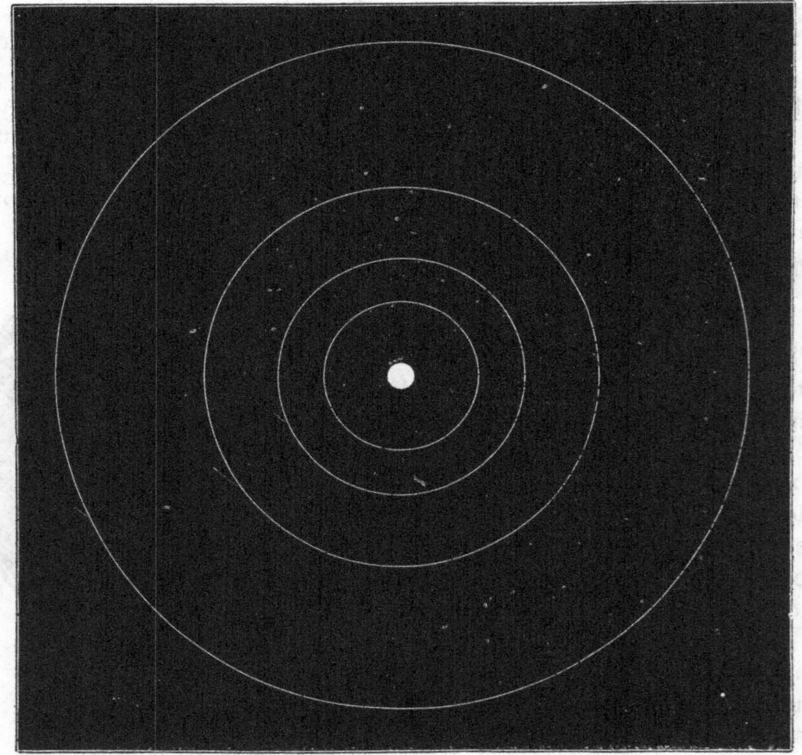

FIG. 181.—THE SYSTEM OF JUPITER
(The Orbit of the Fifth Satellite is not given)

primary period of geological epochs, when life began to be manifested under strange forms, by vegetable and animal beings of astonishing vitality, in the midst of the convulsions and storms of the infant world? This latter conclusion is unquestionably the most rational which we can deduce from the observations. However, as we do not know all the resources concealed in the fruitful power of nature, logic leads us at the same time to admit that this world, so different from ours, may well be, notwithstanding our ignorance at present, inhabited by extra-terrestrial organisms.

But, let us repeat, time is nothing; our epoch is not of importance; the present is but an open door, through which the future is hurried into the past; the time which we call the present is only the moment in which we name it, and that a world should be inhabited in the nineteenth century of the Christian era or the hundredth century before or after is the same in eternity.

The colossal globe of Jupiter moves along accompanied by a beautiful system of four satellites on duty round him [a fifth satellite of small size was discovered by Professor Barnard, at the Lick Observatory, in September 1892. It is the nearest to Jupiter of all the satellites, and revolves round him in 11 hours 57 minutes 22½ seconds, at a distance from his centre of 112,500 miles. Its probable diameter is about 100 miles—J. E. G.]. The first time that scientific curiosity directed the telescope of Galileo towards the brilliant planet (January 7, 1610), the fortunate investigator of celestial mysteries had the pleasure of discovering these four little worlds, which he took at first for stars, but which he soon recognised as belonging to Jupiter himself. He saw them alternately approaching and receding from the planet, passing behind, then in front of him, and oscillating to his right and to his left at distances limited and always the same. Galileo soon concluded that these were bodies which revolve round Jupiter in four different orbits; it was the system of Copernicus in miniature, and the views of this great man could no longer be rejected. It is related that Kepler, hearing of the observations of the Florentine astronomer, wrote, in parody of the exclamation of the Emperor Julian: '*Galilee, vicisti*!' ('Galileo, thou hast conquered!')

The nearest satellite to the planet[1] revolves round him at a distance of 430,000 kilometres, or 107,500 leagues (266,000 miles); the second, at the distance of 170,700 leagues (424,000 miles); the third, at the distance of 272,000 leagues (676,000 miles); and the fourth, along an orbit drawn at 478,500 leagues (1,189,000 miles) from the same centre.

Seen with the aid of an ordinary telescope they have the appearance of small stars arranged along a line drawn through the centre of the planet, nearly parallel to the belts, and in the prolongation of the equator. The whole system is included within a visual surface of about two-thirds of the apparent diameter of the terrestrial moon. If, then, we could place the centre of the lunar disc on that of Jupiter's disc, not only would all the Jovian satellites be covered, but the most distant of the satellites from the planet would not approach nearer to the edge of the moon than one-sixth of its apparent diameter.

The varied and ever-changing configurations of these four globes in the sky of Jupiter must present a curious spectacle. We muse sympathetically, in the midst of the profound silence of night, when our pale Phœbe

[1] That is, the nearest of the old satellites.

pours out from the height of immensity her soft and cold light, and the poetical influence of its celestial radiance slowly sinks into our soul. How would it be if, in the same sky, several moons united their light, gliding in silence through the ethereal regions, eclipsing in turn the distant constellations, which are sunk and lost in the depths of the infinite night!

These four satellites revolve round the powerful planet in orbits represented in fig. 181, which is drawn to an exact scale. The plane of these orbits is not perpendicular to our visual ray; that is to say, we do not see them full-face. On the contrary, this plane, like the equator of Jupiter, lies in the ecliptic (the plane in which we are), so that to us they only oscillate to

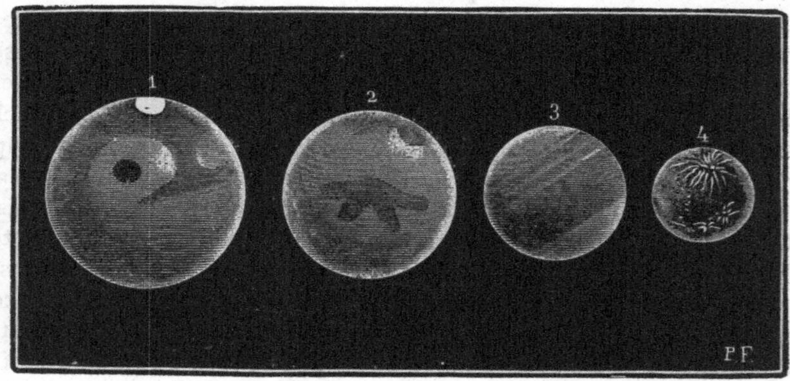

FIG. 182.—COMPARATIVE MAGNITUDES OF MARS (1), GANYMEDE (2), MERCURY (3), AND THE MOON (4).

the right and left of Jupiter: we never see them above or below the planet. It is as if in looking at fig. 181 we placed the page, not in front, but lying along our visual ray, and viewed it edgeways. The following are the astronomical elements and the relations which these four satellites have to their central world:—

	Distance from centre of Jupiter		Durations of revolutions	
	In radii of Jupiter	In miles	In terrestrial days	In Jovian days
			D. H. M. S.	
I. Io[1]	6·05	266,000	1 18 27 33	4·27
II. Europa	9·62	424,000	3 13 13 42	8·58
III. Ganymede	15·35	676,000	7 3 42 33	17·29
IV. Callisto	27·00	1,189,000	16 15 32 11	40·43
Barnard's satellite	2·50	112,500	11 57 23	1·21

and the following are their dimensions, masses, and densities:—

[1] These names are not used by English astronomers.

	Diameters				Volumes	Masses
	Apparent	Earth = 1	Jupiter = 1	In Kilometres	Jupiter = 1	Jupiter = 1
I. Io	1″·02	0·32	0·027	3,800 (2,356 miles)	0·000020	0·000017
II. Europa	0″·91	0·27	0·024	3,300 (2,046 miles)	0·000014	0·000023
III. Ganymede . .	1″·49	0·47	0·040	5,800 (3,596 miles)	0·000060	0·000088
IV. Callisto . . .	1″ 27	0·33	0·034	4,400 (2,728 miles)	0·000039	0·000042
Barnard's satellite . . .	—	—	—	100 miles	—	—

We see here a very beautiful family. The dimensions of these worlds are respectable. The third (Ganymede) has a diameter equal to $\frac{47}{100}$ of that of the earth; that is to say, nearly the half: it measures 5,800 kilometres, or 1,450 leagues (3,596 miles): in importance it is a real planet. Not only

FIG. 183.—TRANSIT OF ONE OF JUPITER'S SATELLITES, AND THE SHADOW PRODUCED BY IT

does it much exceed, like its brothers, all the small planets which revolve between Mars and Jupiter, but it even *exceeds Mercury in volume by nearly double,* and is equal to two-thirds that of Mars. It is five times larger than our moon. It is a veritable world.

Their united masses form the 6,000th part of that of Jupiter, and their volumes the 7,600th of his volume. The densities of the second and third are greater than the density of the planet. Gravity at their surface must be very feeble.

The discovery of the satellites of Jupiter was the first result of the invention of telescopes.[1] Like all discoveries, it was not accepted without criticism. An entire academy—that of Cortona—maintained that the satellites were the result of an optical illusion. There was at Pisa a philosopher named Libri who would never consent to put his eye to a telescope to see the satellites of Jupiter. He died soon after. 'I hope,' said Galileo, 'that, not desiring to see them on earth, he will have perceived them on his journey to Heaven.'

Jupiter casts, on the side opposite to the sun, a cone of shadow, into which the satellites penetrate from time to time, and which produces *eclipses* similar to eclipses of the moon. This planet being much larger than the earth, and being, moreover, more distant from the sun, the length of its shadow-cone is incomparably longer than that of the earth's shadow-cone; it is eighty-nine millions of kilometres (fifty-five millions of miles). This cone extends very far beyond the orbit of the fourth satellite. It follows that the transverse dimensions of the cone, at the points where it reaches the satellites, are almost equal to that of the planet itself. The eclipses of these satellites are much more frequent than eclipses of the moon. The first three penetrate the cone of the shadow at each of their revolutions; the fourth alone sometimes passes outside the cone, without entering it, above or below. These eclipses serve for the determination of longitudes at sea; they are phenomena which are produced in the sky, and which, being observable at the same time from a large number of points on the surface of the earth, indicate the exact hour, and consequently the longitude.

When the satellites of Jupiter pass between him and the sun, their shadow is thrown on the planet, and produces veritable eclipses of the sun, which we can observe from here.

There exists between the motions of the first three satellites a peculiar relation, from which this result follows—ascertained, moreover, by observations: that they cannot be subject to simultaneous eclipses. When the second and third are eclipsed at the same time, the first is in conjunction with the planet; if both pass in front of Jupiter, so that they produce simultaneous eclipses of the sun to him, the first satellite is found in opposition—that is to say, it is eclipsed itself.[2]

[1] To do honour to the Duke de Médicis, Galileo proposed to give to the satellites of Jupiter the name of the *Medician Stars*.

Father Rheita, of Cologne, who had mistaken five stars of Aquarius for satellites of Jupiter, proposed to give to these nine companions of the planet the names of the *Urbanoctavian* Stars, in memory of Pope Urban VIII. (Urbanus Octavus).

Hevelius, on his part, proposed for the authentic satellites of Jupiter the name of *Wladislavanian* Stars, in homage to the King of Poland (Wladislas IV.).

[2] It sometimes happens that the four satellites disappear to us, some being eclipsed or occulted, others being projected on the luminous disc of Jupiter. This observation was made, among other

These satellites vary in brightness. I have observed them with care, particularly during the years 1873, 1874, 1875, and 1876. Several interesting facts result from the comparison of these observations. The first is, that the intrinsic nature of these four worlds and of their reflecting surface is very different for each of them.

As to *dimensions*, the decreasing order has been this: III., IV., I., II.

As to *intrinsic light*, for equal surface, we have: I., II., III., IV.; sometimes II. has appeared a little more luminous than I.

As to *variability*, the decreasing order is: IV., I., II., III.[1]

dates, on March 15, 1611; November 12, 1681; May 23, 1802; April 15, 1826; September 27, 1843, from 11 hours 55 minutes to 12 hours 30 minutes; August 21, 1867, from 10 hours 13 minutes to 11 hours 58 minutes; March 22, 1874, at 1 hour 46 minutes A.M.; October 15, 1883, from 4 hours 5 minutes to 4 hours 24 minutes A.M.

I notice that between the third and fourth observation there are twenty-four years, less thirty-eight days; and that between the fifth and sixth there are twenty-four years, less thirty-seven days. The difference of a day would proceed from the hours. The period appears to be, exactly—1867·6377 − 1843·7393 = 23·8984 years, or 23 years 328 days. This period includes 523 revolutions of satellite IV., 1,220 of III., 2,458 of II., and 4,934 of I. It gives the following dates for this curious disappearance: 1819·841 (November 4), 1843·739 (September 27), 1867·638 (August 21), 1891·536 (July 16).

On August 21, 1867, satellite II. was behind Jupiter, while the three others passed in front (fig. 184); on March 22, 1874, satellite II. passed in front, and the three others behind; on October 15, 1883, satellite I. was behind, and the three others in front; on July 16, 1891, satellite II. passed behind, and the three others in front, but there was not a complete disappearance, for III. had left the disc before I. and IV. had entered it.

The dates March 15, 1611, to May 23, 1802, give 191·190 years, of which the eighth part is 23·8984 years; 1874·219 − 1826·217 = 47·932 years, of which the half is 23·966 years.

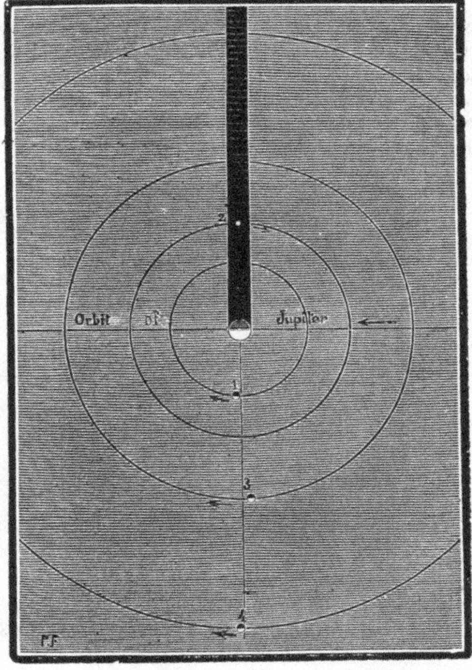

FIG. 184.—DISAPPEARANCE OF JUPITER'S SATELLITES

[1] A rare observation has confirmed me in previous conclusions on the existence of an atmosphere round these globes. On March 25, 1874, I commenced, at 8 hours 45 minutes, the study of the disc of Jupiter, when my attention was immediately arrested by the presence of a round spot, absolutely *black*, and clearly defined, situated at a little distance from the eastern edge of the planet, and admirably shown on the white ground of a wide luminous zone.

Below the round *black* spot of which I speak, and almost in contact with it, a second was distinguished, also round, but not black like the other: it was *grey*, a little smaller, and stood out clearly on the same white ground.

Observing the planet attentively, I soon distinguished a third spot, situated to the right of the

Their photometric magnitudes are respectively equal to 6·2, 6·3, 5·8, and 6·6.

These observations prove that the four satellites of Jupiter present a curious variability. They are worlds which deserve to be regularly followed. It is highly probable that they are now inhabited, and that they form the earliest abodes of life in the Jovian system. Even lately Jupiter was a sun to them.

Some words still on this vast world considered as an observatory.

The earth seen from Jupiter is a luminous point oscillating in the vicinity of the sun, from which it is never distant more than 12°; that is to say, about twenty-four times the diameter under which we see that body. It can then be perceived only in the *evening* and *morning*, as Mercury is by us, and even less so, visible with great difficulty to the naked eye, but present-

two former and more to the north, near the central meridian, visible, not on the white ground, but on the grey northern belt. It was less defined than the preceding, very difficult to see well, and scarcely darker than the grey belt on which it was placed. It appeared a little less dark than the second, on account of the ground on which it lay. After some minutes of observation it became evident that these spots were displaced on the disc. They moved towards the west, and left the disc at 10 hours 23 minutes.

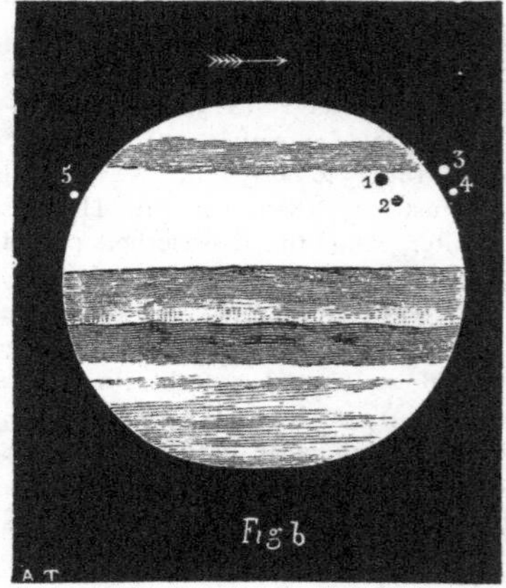

FIG. 185

In the sketches here given (fig. 185) the first of the two drawings (*a*) represents the disc of Jupiter at 8 hours 50 minutes. The spot No. 1, *black*, is the shadow of the third satellite; the spot No. 2, *grey*, is the shadow of the second; the spot No. 3 is the third satellite itself. The second drawing (*b*) represents Jupiter at 10 hours 32 minutes. The second satellite (No. 4) only became visible at the moment of egress, certainly on account of the feeble luminous intensity of the border of the planet relatively to that of the whole.

Thus the third satellite, which usually appears white, like the others, when it passes in front of the planet, was dark, and darker than the grey belt on which it was placed. *It was almost as dark as the shadow of the second satellite.*

These differences of brightness observed on these little worlds show that their soil is uneven like that of the earth, and that they are surrounded with variable atmospheres themselves.

ing in optical instruments the aspect of the moon in quadrature. If the Jovian astronomers, present or future, observe the sun with attention, it is in the transits of our little globe across his disc that it will be most easily discovered, as we might do for an intra-mercurial planet. It is thus that they see us from there. . . . . Assuredly, if it were rumoured in Jupiter that the inhabitants of this little black point maintain that the whole universe was constructed for them, it is very probable that the good Jovian citizens would burst out into such an Homeric laugh that it would be audible from here.

At night the spectacle of the sky seen from Jupiter is, with reference to the constellations, the same as that which we see from the earth. There, as here, shine Orion, the Great Bear, Pegasus, Andromeda, Gemini, and all the other constellations, as well as the diamonds of our sky: Sirius, Vega, Capella, Procyon, Rigel, and their rivals. The 390 millions of miles which separate us from Jupiter *in no way* alter the celestial perspectives. But the most curious character of this sky is unquestionably the spectacle of the four moons, each of which shows a different motion. The first moves in the firmament with an enormous velocity (and Barnard's satellite still faster), and produces almost every day total eclipses of the sun in the equatorial regions. The three inner moons are eclipsed at each revolution, just at the hours when they are at their 'full.' The fourth alone attains the full phase.

Contrary to the generally received opinion, these bodies do not give to Jupiter all the light which is supposed. We might think, in fact, as has been so often stated, that these four moons illuminate the nights four times better relatively than our single moon does in this respect, and that they supplement in some measure the feebleness of the light received from the sun. This result would be, assuredly, very agreeable, but nature has not so arranged it.[1]

Jupiter appears to be a world still in process of formation, which lately —some thousands of centuries ago—served as a sun to his own system of four [five, or perhaps more] worlds. If the central body is not at present

[1] The four satellites cover, it is true, an area of the sky greater than our moon, but they reflect the light of a sun twenty-seven times smaller than ours; indeed, the total light reflected is only equal to a sixteenth of that of our full moon, even supposing the soil of these satellites to be as white as it appears to be, especially satellite IV. It should be remarked, however, that the optic nerves of these unknown beings, being constructed for an intensity of light twenty-seven times weaker than ours, may be more sensitive than ours in the same proportion, and it is natural to think that the inhabitants of Jupiter would see as clearly as we do. Our terrestrial organisation cannot be considered as a type, for it is simply relative to our planet. Now, if the eyes of the inhabitants of Jupiter are twenty-seven times more sensitive than ours, their sun is as luminous to them as ours is to us, and it is not necessary to diminish the brightness of the satellites twenty-seven times in order to judge of its effect upon them. In reality, then, the totality of their moons gives them a maximum of light counted entirely by the area of the reflecting surface, which, consequently, exceeds by one-half that which the full moon sends us.

inhabited, his satellites may be. In this case, the magnificence of the spectacle presented by Jupiter himself to the inhabitants of the satellites is worthy of our attention. Seen from the first satellite, the Jovian globe presents an immense disc of twenty degrees in diameter, or 1,400 times larger than the full moon ! [1] What a body ! What a picture, with its belts, its cloud motions, and its glowing coloration, seen from so near ! What a nocturnal sun !—still warm, perhaps. Add to this the aspect of the satellites themselves seen from each other, and you have a spectacle of which no terrestrial night can give an idea.

Such is the world of Jupiter from the double point of view of its vital organisation and of the spectacle of external nature seen from this immense observatory. As to the nature of its inhabitants, present or future, we will not imitate the English writer, Whewell, who, on account of its small density, could only see in it 'gelatinous creatures like the medusæ which float by the seashore'; nor the German Wolf, who, on account of the feebleness of the light, supposed them to have eyes three times larger than ours, and to be of a size in the same proportion, 'which was exactly the size of Og, King of Basan, whose bed was nine cubits long'; nor an American novelist, who assures us that, the muscular force of beings varying as the square of the section of the muscles, and the weight increasing as the cube of the height, the inhabitants of Jupiter cannot exceed, on account of their weight, the size of General Tom Thumb ! . . . We should beware of measuring the inhabitants of other worlds by the conceptions, more or less incomplete, which the forms of terrestrial life may suggest to us. Nature knows how to people all the worlds at the proper time with non terrestrial beings adapted to their special situation in the universe.

[1] [Seen from Barnard's satellite, Jupiter's disc has a diameter of 47°, or more than half the distance from the horizon to the zenith.—J. E. G.]

CHAPTER VII

SATURN, THE WONDER OF THE SOLAR SYSTEM

♄

WE now come to the ancient frontier of the solar system, to the orbit of old Saturn, God of Time and of Destiny, who from the origin of planetary astronomy down to the end of last century marked for our ancestors the extreme limit of the solar realm. Even in the times of Copernicus, of Galileo, and of Newton, it was the farthest known planet; in the middle of last century the unfortunate Bailly, a learned *savant* and excellent man, who was in the revolutionary madness sacrificed to the blind passion of political parties, thought to give a grand idea of the extent of the solar system by estimating the distance of Saturn at 218,000 times the semi-diameter of the earth, or 327 millions of leagues, and by supposing that, the limit stopping there, the stars might not be much more distant. This distance for Saturn was nearly exact, since it is in reality 885 millions of miles; but in 1781 the discovery of Uranus removed the frontier to 1,780 millions; in 1846, that of Neptune extended it to more than 2,700 millions of miles; and, as we have seen, the nearest star is nine thousand times farther from us than Neptune.

The human mind has followed, step by step, in its conception of the universe the lights which the torch of Urania has thrown into Infinitude; we may say, with Laplace, that 'by the certainty of its views and the grandeur of its results astronomy is the most beautiful monument of the human mind.' In the times of Homer and Hesiod men believed that the extent of the whole universe had been measured by the myth of Vulcan's anvil, which took nine days and nine nights to fall from the sky to the earth, and as many more to descend from the earth to the infernal regions. It would have fallen, however, but 575,500 kilometres (357,600 miles); that is to say, only from a little higher than the moon.[1] It seems to us now

[1] I have found a very simple formula to make this calculation, and for all others similar. The duration of the fall of a satellite on a planet is nothing but its revolution divided by the square root of 32, or $\frac{R}{5{\cdot}656856}$.

We have here, then, revolution = 9 days × 5·656856 = 50·911704 days.

$$\text{Hence } \frac{50{\cdot}91^2}{27{\cdot}32^2} = \frac{h^3}{60{\cdot}27^3},$$

and $h = \sqrt[3]{760,200} = 91{\cdot}4$, or 581,870 kilometres, and, deducting the radius of the earth (6,370 kilometres), we have height of fall = 575,500 kilometres, as above.

that we should not be able to breathe in such a small edifice, shut in on all sides by a crystal sphere.

Saturn appears to the naked eye as a star of the first magnitude, but much less brilliant than Venus, Jupiter, Mars, and Mercury. Its light is dull and livid. The slowness of its motion and the tint of its light made it for the ancients an unlucky planet. Saturn was, indeed, considered as the gravest and slowest of stars, a god dethroned, and banished into a sort of exile. The Sabbath day was consecrated to him. During the long ages when astrology flourished, the earth and man being considered as the centre and sole end of the Creation, each planet exercised its influence proportionately to the valour of each being. The star of Saturn, with unlucky influence, was associated with the greatest griefs ; it was the voice of destiny which spoke in him. He who was born under the sign of Jupiter became celebrated, and was raised to the most brilliant positions of glory and fortune. Mars incited to war. Mercury inspired the arts. As to those who were born under the sign of Venus, they were, it appears, very happy mortals.

. . . .

The ancient opinion of Saturn has been preserved to our day, even among cultured minds. The marvellous ring which surrounds this strange world, far from effacing the legendary impression, has even further confirmed it. In the course of the year 1879 I had the honour of conversing on this subject with the great poet, Victor Hugo, and he assured me that, in his opinion, Saturn could only be a prison, or a hell.

The most ancient *observation* which we have of Saturn dates from the year 228 B.C. We speak here of *observation*—that is to say, of the precise position determined in the sky, and serving to calculate the motion of the planet, and not merely the fact of seeing Saturn in the sky ; for if our first parents inhabited the terrestrial paradise, they *saw* this planet, as they did all the stars visible to the naked eye.

The revolution of Saturn as seen from the earth is performed in twenty-nine years, and is subject to twenty-nine stationary points and retrogradations, due to the perspective of our annual motion round the sun. It comes into opposition—that is to say, behind the earth with reference to the sun—every year, with a delay of thirteen days each year. We can, then, observe it for about six months. (To learn these epochs of visibility, consult the ephemerides, such as our 'Revue mensuelle d'Astronomie populaire.') It returns every thirty years only to the same point of the sky.

Its sidereal revolution round the sun is performed in twenty-nine years five months sixteen days, in a plane which makes an angle of 2° 30′ with that of the ecliptic. The eccentricity of the orbit is 0·056, which gives for its variations of distance :—

	Geometric ☉=1	In Kilometres	In Miles
Perihelion distance . .	9·0046	1,330,000,000	826,435,000
Mean distance . . .	9·5388	1,411,000,000	876,767,000
Aphelion distance . .	10·0730	1,490,000,000	925,856,000

There is, then, a difference of more than the distance of the earth from the sun (ninety-nine millions of miles) between Saturn and the sun (or from the earth also) at his aphelion and his perihelion. The position of this perihelion is found at 91° from the point of the vernal equinox—that is to say, almost at the point of the summer solstice, near the star η of the constellation Gemini; the aphelion is, of course, found at the point diametrically opposite, at 271°, between the stars δ and λ of Sagittarius. This line revolves in the sky, advancing 1′ per annum on the ecliptic.

The apparent diameter of Saturn measures on an average 17′·5, and varies from 15″ to 20″, according to its distance from the earth. This same diameter, reduced to the distance from the earth to the sun taken as unity, is 165″, as we have seen (p. 358); that is to say, nine and a half times (9·527) larger than that of our globe. But this world is far from being spherical: it is still more compressed at the poles than Jupiter, for its polar flattening is $\frac{1}{10}$ [$\frac{1}{9\cdot18}$, Kaiser.—J. E. G.], and exceeds that of all the known planets. We may reckon its equatorial diameter at 76,000 miles. It follows that the circumference of the Saturnian world, measured along its vast equator, almost attains 240,000 miles.

The surface of this world is equal to that of eighty earths. Its volume, estimated at 864 times that of our globe when we do not take into account the polar flattening, which amounts to 6,900 kilometres (4,287 miles) in thickness at each pole, in reality exceeds that of our globe only 719 times. It is still a considerable volume; it represents three-fifths of the giant Jupiter.

On account of the velocity of the motion of rotation, gravity is diminished by a sixth at the equator, so that while in the polar regions objects weigh more than on the earth, at the equator they weigh less. A falling body passes over, on our globe, 4·90 metres in the first second of its fall, and on Saturn, 5·34 metres in the polar latitudes, and only 4·51 in the equatorial regions. If Saturn rotated only two and a half times more quickly, objects would *no longer have any weight at all* in these regions.

Again, the opposing attraction of the ring further diminishes the weights in a considerable proportion, and there is a zone between the inner ring and the planet where bodies are equally attracted upwards and downwards. No great effort of the imagination is necessary to divine that, if an intervening atmosphere permits it, the aërial inhabitants of Saturn may enjoy the power of flying on to the rings! We may remark, with reference to this, that our own globe in rotating produces a centrifugal force which is to gravity in the ratio of the fraction $\frac{1}{289}$. An object which weighs, for example, 289 lbs. at the poles would weigh but 288 at the equator. In order that this diminution should become equal to gravity it would be necessary for the earth to rotate seventeen times more quickly (for $17 \times 17 = 289$). Objects would then have no longer any weight in our equatorial regions. An inhabitant of Quito who jumped only a few inches in height would not fall back! What do I say? No one would adhere to the ground! No living being, no object, nothing would be sustained by its own weight. The least wind would sweep them all away. . . .

Like Jupiter, Saturn shows belts in the telescope, but less easily distinguished, and not straight, but arranged in curves. These belts indicate at a glance the inclination of its equator. Excellent instruments are necessary to perceive the irregularities which diversify these cloudy bands, and it is very difficult to observe them clearly. However, they supplied to W. Herschel, in 1793, the first evidence of the rotation of the planet, which he estimated at 10 hours 16 minutes. No other determination of this period of rotation was made until 1876, when Professor Hall, of Washington, while engaged in measuring the satellites with the aid of the colossal telescope of that observatory, noticed (December 7) a brilliant spot on the equator of the planet. He thought he was looking at an immense eruption of white matter violently projected from the interior of the globe; this spot extended towards the east like a long luminous trail, and it remained visible up to the month of January following, when the planet was lost in the rays of the sun. He immediately addressed a circular to a large number of astronomers, inviting them to observe the phenomenon, and from all the observations the American astronomer found for the length of the rotation 10 *hours* 14 *minutes* [24 *seconds*], a result which confirms in a truly remarkable manner the observation of Herschel

Five hours of day and five hours of night: *twenty-five thousand days in the year!* What a calendar!

The axis of rotation of Saturn is inclined 64° 18′ to the plane of its orbit, and the obliquity of the ecliptic is therefore 25° 42′. This is an inclination differing but little from that of the earth; from which we may conclude that the seasons of this distant world, although each lasting more than seven years, are, nevertheless, but little different from ours with reference to the contrast between summer and winter. The climates are divided, in the same way as those of the earth, into torrid, temperate, and frigid zones. But what a length! Seven years each. Each pole and each side of the ring remain fourteen years and eight months in sunlight.

With reference to the quantity of heat and light which this world receives from the sun, as it is nearly ten times more distant than we are from the central body, the sun is seen nearly ten times less in diameter, ninety times smaller in surface, and the planet receives also ninety times less heat and light. Here, evidently, are conditions of existence differing from those of the earth.

Hardly had the first telescopes been invented than Galileo noticed, in the year 1610, something strange in the aspect of Saturn: he seemed to see two balls, one on each side of the planet. While waiting for an explanation, he called Saturn triple-bodied, and announced the discovery in this singular logogriph:

Smaisnermilmbpoetaleumivneuvgttaviras.

Kepler sought in vain for the key to this enigma, which consisted in a transposition of the much-mixed letters. Galileo replaced them in order so as to form the following Latin sentence:—

Altissimum planetam tergeminum observavi.
I have observed that the farthest planet is triple.

'When I observe Saturn,' he wrote later on to the Ambassador of the Grand Duke of Tuscany, 'the central star appears the largest; two others, one situated to the east, the other to the west, and on a line which does not coincide with the direction of the zodiac, seem to touch it. They are like *two servants who help old Saturn on his way*, and always remain at his side. With a smaller telescope the star appears lengthened, and of the form of an olive.'

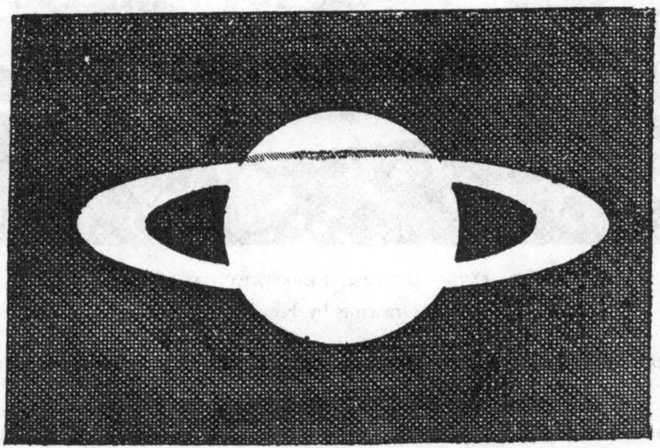

FIG. 186.—FIRST DRAWING OF SATURN'S RING, MADE BY HUYGENS IN 1657

The industrious astronomer, however, was unsuccessful; he was not favoured in his researches as he had been in those preceding. At the epoch when the rings of Saturn are presented to us edgeways they disappear on account of their thinness. This is what happened, particularly in 1612. Galileo, on a certain night, finding it absolutely impossible to distinguish anything on each side of the planet where some months before he had observed two luminous objects, was completely in despair. He began to think that the glasses of his telescopes had deceived him. Affected with deep discouragement, he took no further notice of Saturn, and died without knowing that the ring existed. Fontana drew the planet in 1645 without divining it. It was only in 1659 that Huygens, the true author of the discovery of the ring, wrote the first description and gave the first explanation. He commenced his observations in 1656, a year in which the rings were presented edgeways. His first drawing of the ring was in 1657. Still, he hid his discovery under the following mask :—

aaaaaaa, ccccc, d, eeeee, g, h, iiiiiii, llll, mm, nnnnnnnnn, oooo, pp, q, rr, s, ttttt, uuuuu.

FIG. 187.—SATURN, FEBRUARY 11, 1884
From a Drawing by Henry Pratt

Three years afterwards, only, he declared that this anagram was intended to say :—

Annulo cingitur tenui [plano], nusquam cohærente, ad eclipticam inclinato.

It is surrounded by a thin [flat] ring, not adhering to the body at any point, and inclined to the ecliptic.

These words contain the three fundamental facts of the nature of this mysterious appendage. It must be admitted that the *savants* of that epoch had singular methods of publication. It is, however, to human *curiosity* that we owe the fruitful continuance of all these efforts ; and we may say that *savants*, and especially astronomers, are the most curious of mortals.

The hypothesis of a ring surrounding the globe of Saturn on all sides without touching it was not immediately accepted; many maintained that it was only the effect of a reflection of the light from his convex surfaces. Auzout perceived, in 1662, the shadow of Saturn on the ring, an observation confirmed many times since. In 1664 Campani represented the ring as composed of two zones, the outer one dark, the inner one bright. In 1666 Hooke observed that the ring was more luminous than the planet. In 1675 Cassini saw it divided all along its length by a dark line into two parts of dissimilar intensity. 'The interior part is,' said he, 'very bright, and the exterior a little dark, the difference of tint being that of dull silver

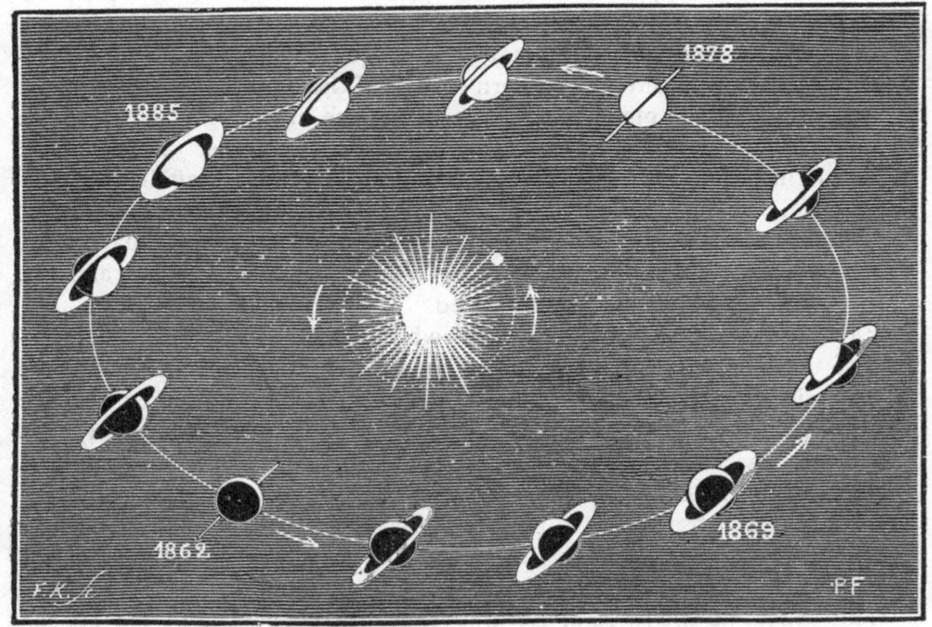

FIG. 188.—VARYING PERSPECTIVE OF SATURN'S RINGS AS SEEN FROM THE EARTH

and burnished silver.' (This discovery has been wrongly attributed to an Englishman named Ball.) There are, in fact, two perfectly distinct rings, one exterior, the other interior, separated by the black line of which we have spoken, and which bears the name of Cassini's Division. In 1837 Encke saw the exterior ring divided in two by a black line; and in 1838 Father de Vico perceived two other similar black bands on the interior ring. In 1850 Bond (and Dawes, independently) discovered a third ring, interior to the two preceding, dark, and even partly transparent.

Considered as a whole, the ring makes with the plane of the planet's orbit an angle of 28 degrees. Consequently, to an observer situated on the earth it always appears elliptical and of a variable transverse width. The shadows thrown show that the body of Saturn and his ring are illuminated, as we are, by the sun, and have no other light.

Seen full face—that is to say, from a point of space situated in the prolongation of the planet's axis—the rings would be recognised in their real form, that is to say, circular. From here we never see them except obliquely; at the epochs when they appear to us most open, the smaller apparent diameter is never equal to the half of the longer. Fig. 188 shows how these appearances are produced. Twice in the Saturnian revolution—that is to say, about every fifteen years—we see them with their maximum of opening; seven and a half years after that, and with a period of fifteen years also, they are presented to us edgeways, and disappear twice —(1) when the sun only illuminates the edge; (2) when, the sun illuminating either the northern or southern surface of the rings, the earth happens to pass through their plane, and we no longer see anything. In the most powerful instruments a thin luminous thread still remains. Thus, in the month of June 1877 the earth passed through their plane; they disappeared a first time, reappeared, then, in February 1878, disappeared again, being only illuminated on the edge. Their northern surface, which had been illuminated since 1862, then lost sight of the sun for fifteen years, and the southern surface began to be illuminated.

Fig. 189 gives a series of drawings which show at a glance the variations of these aspects according to the years. I made the drawings with a telescope of 0·20 m. (7·87 inches) not inverting the images. There is a special remark to be made on that of 1877 (September 14). It is that the ansa to the right, or west, appeared to me more brilliant and longer than the other. (In these drawings the images are erect, and not inverted.) This observation is not new; Saturn does not exactly occupy the centre of the rings, but the difference is usually so slight that it is not noticed.

This aspect of the rings of Saturn, especially when seen at the epochs of their greatest apparent width, is marvellous, and we cannot help being affected with a certain emotion when we see this astonishing system entering the field of an astronomical telescope. When we think that there is here a celestial deck on which the entire globe of the earth might roll like a ball on a road, and that the world poised in the centre is several hundred times larger than our planet, we transport ourselves easily in thought to those sublime regions where the vulgar affairs of our mortal existence vanish like an evil dream. . . . How strange it is that so small a number of human beings have ever seen this wonder otherwise than in a cold engraving, when it is so easy nowadays to possess an instrument of observation!

This celestial wreath is not homogeneous; these rings are not distributed on an absolutely plane surface, but show irregularities, which are visible

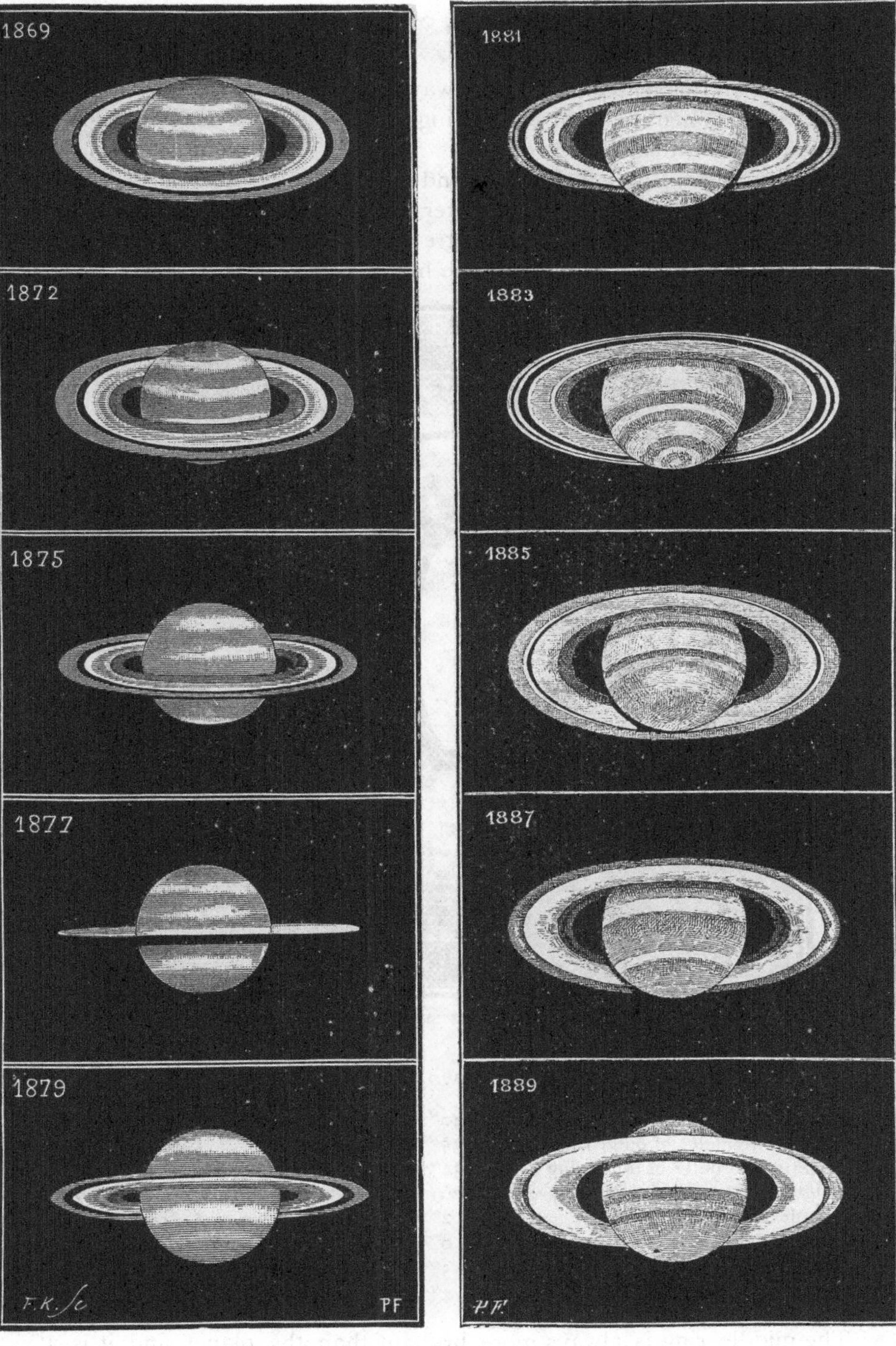

FIG. 189.—CHANGES IN APPEARANCE OF SATURN'S RINGS

when they are presented to us edgeways, and which produce shadows on the planet. When the light of the rings is reduced to a thread, we notice on this thread brilliant knots.

What an astonishing system! And these rings, which, as we see, are not less than 176,000 miles in diameter and 30,000 miles in width, are not more than sixty to seventy kilometres in thickness! [Thirty-seven to forty-three miles; perhaps fifty to one hundred miles.—J. E. G.]

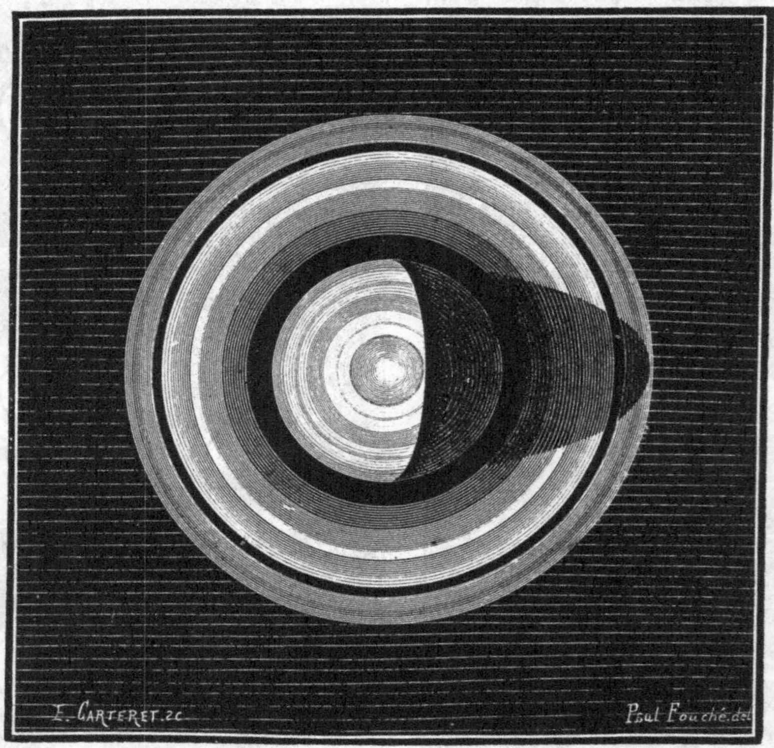

Fig. 190.—Saturn's Rings seen from the Front

Here are the measures of the two principal rings:—

Exterior diameter of outer ring . .	40″·00 or	71,000 leagues	(176,000 miles)
Interior diameter of outer ring . .	35″·29 „	62,640 „	(155,000 „)
Exterior diameter of inner ring . .	34″·47 „	61,200 „	(152,000 „)
Interior diameter of inner ring . .	26″·67 „	47,340 „	(117,000 „)
Width of exterior ring	2″·40 „	4,260 „	(10,600 „)
Width of division between rings .	0″·41 „	720 „	(1,790 „)
Width of inner ring	3″·90 „	9,306 „	(23,100 „)
Distance between the ring and planet	4″·00 „	7,000 „	(17,400 „)

The middle ring is always more brilliant than the planet, and it is at its outer edge that its brightness is most vivid. This brightness gradually diminishes to the inner border, where it has sometimes appeared so feeble as to make it difficult to distinguish it from the inner dark ring. Examined, in 1874, with the great equatorial at Washington, it showed no remarkable contrast between its inner edge and the outer edge of the transparent ring; the two borders appeared, on the contrary, to blend imperceptibly into each other. Has the dark ring increased at the expense of the bright ring?

M. Trouvelot made careful observations from 1871 to 1875, from which it would follow that the transparent inner ring has changed in aspect since its discovery in 1850. Instead of being entirely transparent, as it was seen

Fig. 191.—Telescopic Aspect of Saturn in 1874

by Bond, it is no longer so, except in its inner half; the globe of Saturn remains visible at its entry behind this veil, but is imperceptibly effaced and is no longer perceptible when it reaches the outer border. Is there here a real change, or is the fact merely due to the scrupulous attention which the author has given to his observations? It is difficult to decide as to details of such delicacy. However, it is probable that if Bond, Dawes, Lassell, Warren de la Rue, &c., had not traced the globe behind the grey ring as far as the bright ring, they would not have drawn it as sharply marked as it is shown. It would follow, besides, from a special analysis made by O. Struve in 1851, that the Saturnian system has undergone, since the epoch of its discovery, surprising changes, considering that the inner edge of the rings appears to gradually approach the planet, and that their total width has increased at the same time; the middle ring appears to extend more rapidly than the outer ring. Shall we witness some day the grand and tremendous spectacle of the disruption of Saturn's rings, and their fall on the globe? The interval between the ring and the planet appears to diminish at the rate of 1″·3 per century, at least if we are to accept the following measures as they are given :—

	Year	Distance between bright ring and the planet	Width of ring
Huygens	1657	6″·5	4″·6
Huygens and Cassini . .	1695	6″·0	5″·1
Bradley	1719	5″·4	5″·7
W. Herschel . . .	1799	5″·12	5″·98
W. Struve	1826	4″·36	6″·74
Encke and Galle . .	1838	4″·40	7″·60
O. Struve	1851	3″·67	7″·43
O. Struve	1882	3″·66	7″·54

At this rate, with this velocity of approach, the luminous ring would come into contact with the planet about the year 2150.[1]

But what is the nature of this celestial wreath?

Are these rings solid, liquid, or gaseous?

Whatever may be their number, they cannot be solid, and resemble, for example, flat hoops, more or less wide. The constant variations of the planet's central attraction, combined with that of the eight satellites, would not only have dislocated and shattered them, if they could have been formed, but would have absolutely prevented their formation.

The only system of rings which could exist is a system composed of an infinite number of *distinct particles revolving round the planet with different velocities, according to their respective distances.* No refraction being

[1] Without yet asserting the fact, we may remark that it is difficult to reconcile the descriptions of the old observers with the present aspect of the ring without admitting that changes of a certain importance may have taken place in two centuries. Anyone may notice to-day that the united width of the two bright rings is about twice as great as that of the dark space which separates the planet from the ring; while Huygens described this dark space as equal to the width of the ring, or even a little greater. The inspection of the drawings of the seventeenth century produces the same impression (see fig. 186). Can the difference be attributed to the imperfection of the instruments then in use? No, for these imperfect glasses would give, on the contrary, a preponderance to the luminous parts. Did the dark ring not exist before its discovery? This is possible, for the skilful Schröter studied especially, in 1796, the interval in question, and found it darker than the sky; the two Herschels did not perceive it from 1789 to 1830. However this may be, these rings present variable irregularities; and it may be also seen from the drawing of M. Trouvelot (fig. 191) that the shadow of the planet on the different rings indicates rather singular differences of level.

The distance from the inner edge of the interior ring to the planet was found to be 1″·61 in the measures of 1851, and 1″·49 in those of 1882. This difference of 0″·12 in thirty-one years is not insignificant either. But perhaps there are here merely variations in the luminous intensity of the rings.

observed on the edge of the planet seen through the inner ring, it follows that this ring is not gaseous, and that the rays do not pass through a gas.[1] The other rings may be of the same nature, but formed of particles sufficiently numerous to prevent them from being transparent. This vast system should rotate in the following periods :—

	Distance in radii of Saturn	Period H. M. H. M.
Interior transparent ring	1·36 to 1·57	5 50 to 7 11
Wide central ring	1·57 to 2·09	7 11 „ 11 9
Exterior ring . . .	2·14 to 2·40	11 36 „ 12 5
First satellite	3·35	22 37

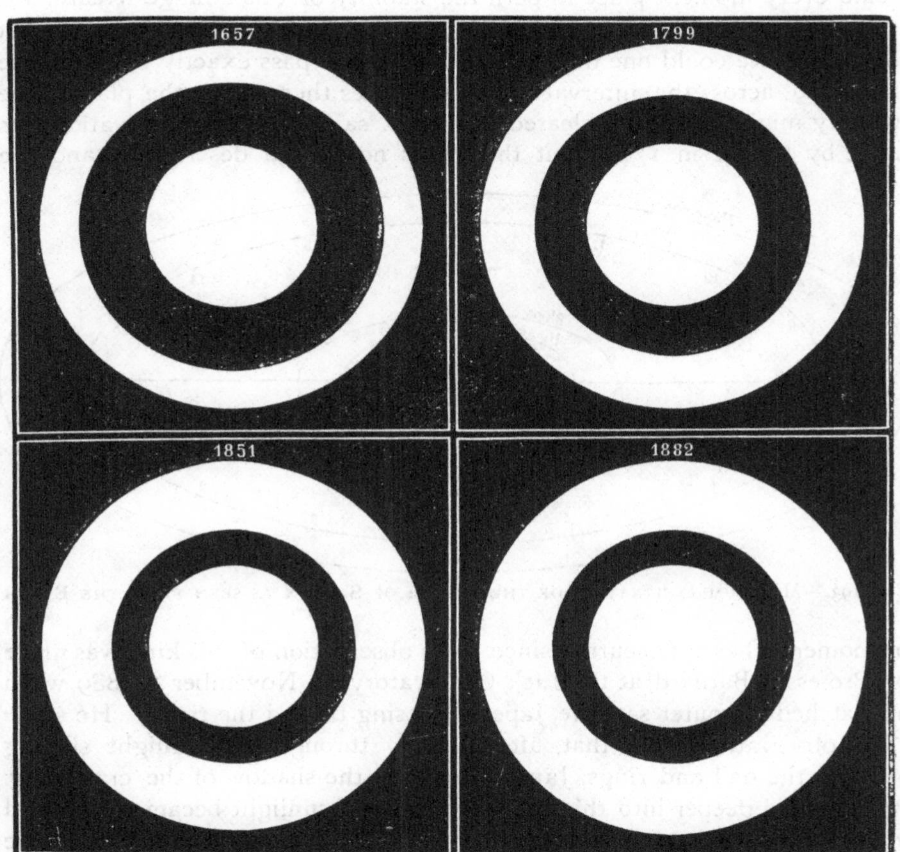

FIG. 192.—PROBABLE DRAWING TOGETHER OF SATURN'S RINGS SINCE THEIR DISCOVERY

Thus the particles forming the transparent ring should revolve in the time included between 5 hours 50 minutes and 7 hours 11 minutes, accord-

[1] [For the same reason they cannot be liquid ; and, if they had been originally liquid, they would be frozen solid at the great distance of Saturn from the sun, and would then be shattered to pieces, as explained above.—J. E. G.]

ing to their distance, the nearest zone revolving the most rapidly; those which compose the broad luminous ring should revolve in periods from 7 hours 11 minutes to 11 hours 9 minutes, also according to their distance; finally, the outer limit of this singular system should perform its revolution in 12 hours 5 minutes. But the eight satellites which gravitate outside the rings produce considerable perturbations in these motions, perturbations such that, perhaps, it is to the unstable equilibrium which they perpetuate that is due the preservation of the Saturnian appendage, for it seems that without their external support the frictions and inevitable collisions would every moment place in peril the stability of this strange wreath.

Although studied from different points of view, the problem is not yet solved. If we could one day see a brilliant star pass exactly behind these rings, and across the interval which separates them from the planet, the mystery might be partly cleared up. It is said that this observation was made by Clarke in 1707; but there was no special description, and the

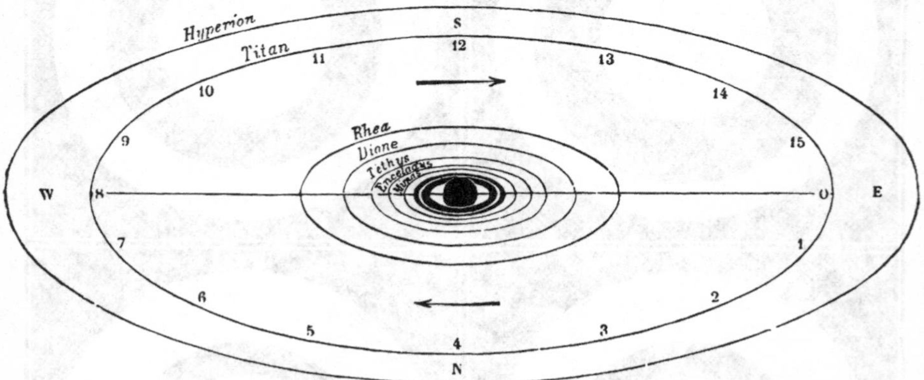

FIG. 193.—MINIMUM INCLINATION OF THE SYSTEM OF SATURN AS SEEN FROM THE EARTH

phenomenon has not occurred since. [An observation of this kind was made by Professor Barnard, at the Lick Observatory, on November 1, 1889, when he watched the outer satellite, Japetus, passing behind the rings. He says: 'The observations show that, after passing through the sunlight shining between the ball and rings, Japetus entered the shadow of the crape ring. As it passed deeper into this the absorption of sunlight became more and more pronounced, until, finally, the satellite entered the shadow of the bright rings. In a word, then, the crape ring is truly transparent, the sunlight sifting through it. The particles composing it cut off an appreciable quantity of sunlight. They cluster more thickly, or the crape ring is denser, as it approaches the bright rings.' From the total disappearance of the satellite when it passed behind the bright ring, he concludes that 'the bright ring is fully as opaque as the globe of *Saturn* itself.' [1]—J. E. G.]

[1] *Monthly Notices*, R. A. S., January 1890.

These rings are certainly variable. In 1889 a brilliant region was observed quite near the shadow of the planet (Terby, at Louvain) [possibly, however, due to an effect of contrast.—J. E. G.].

The wonderful annular system which we have been admiring has not sufficed for Saturn's ambition. He has, besides, received from Heaven the richest retinue of satellites which exists in the whole solar system: eight worlds accompany him in his destiny. It is an empire of five millions of miles in width. Saturn is, however, so distant that this width is reduced for us to a space which the moon would entirely hide! If the centre of the moon were placed on the centre of Saturn, the most distant satellite, far from going beyond the lunar disc, would not even approach its borders: it would remain within the edge by one-third of the moon's semi-diameter.

The following are the eight companions of Saturn, with their distances from the centre of the planet, estimated in miles, and their periods of revolution, estimated in terrestrial solar days:—

	Distance from centre of Saturn			Period of revolution	Order of Discovery	Discoverer
	Apparent	In radii of ♄	In miles			
				D. H. M. S.		
I. Mimas	0″·27	3·36	117,600	0 22 37 5	7	W. Herschel, 1789
II. Enceladus	0″·35	4·31	151,000	1 8 53 7	6	W. Herschel, 1789
III. Tethys	0″·43	5·34	187,000	1 21 18 26	5	Cassini, 1684
IV. Dione	0″·55	6·84	239,000	2 17 41 9	4	Cassini, 1684
V. Rhea	1″·16	9·55	336,000	4 12 25 12	3	Cassini, 1672
VI. Titan	2″·57	22·14	777,000	15 22 41 23	1	Huygens, 1655
VII. Hyperion	3″·33	26·78	951,000	21 6 39 27	8	Bond and Lassell, 1848
VIII. Japetus	8″·35	64·36	2,261,000	79 7 54 17	2	Cassini, 1671

The first three satellites are all nearer to Saturn than the moon is to the earth; and they would be still nearer if we measured their distances from the surface of the planet. Mimas is hardly, at an average, more than 80,000 miles distant; and even the fourth, Dione, is but 201,000 miles—that is to say, less than the distance of the moon. Their distances from the edge of the outer ring are less still, and Mimas approaches it within 17,450 leagues (43,000 miles).

Fig. 194 shows the system of orbits with their relative dimensions. This plane is seen full face. In reality it is always seen obliquely. At the minimum of inclination this system is presented as we see it in fig. 193, in which, however, we have not placed the outer satellite, on account of its distance.

These satellites were only discovered successively, according to their gradation of brightness and the progress of optical instruments, as we see by the last column of the preceding table. The first noticed (the largest, Titan) was discovered by Huygens in 1655. The instruments of this astronomer might have been sufficient to enable him to discover others, if

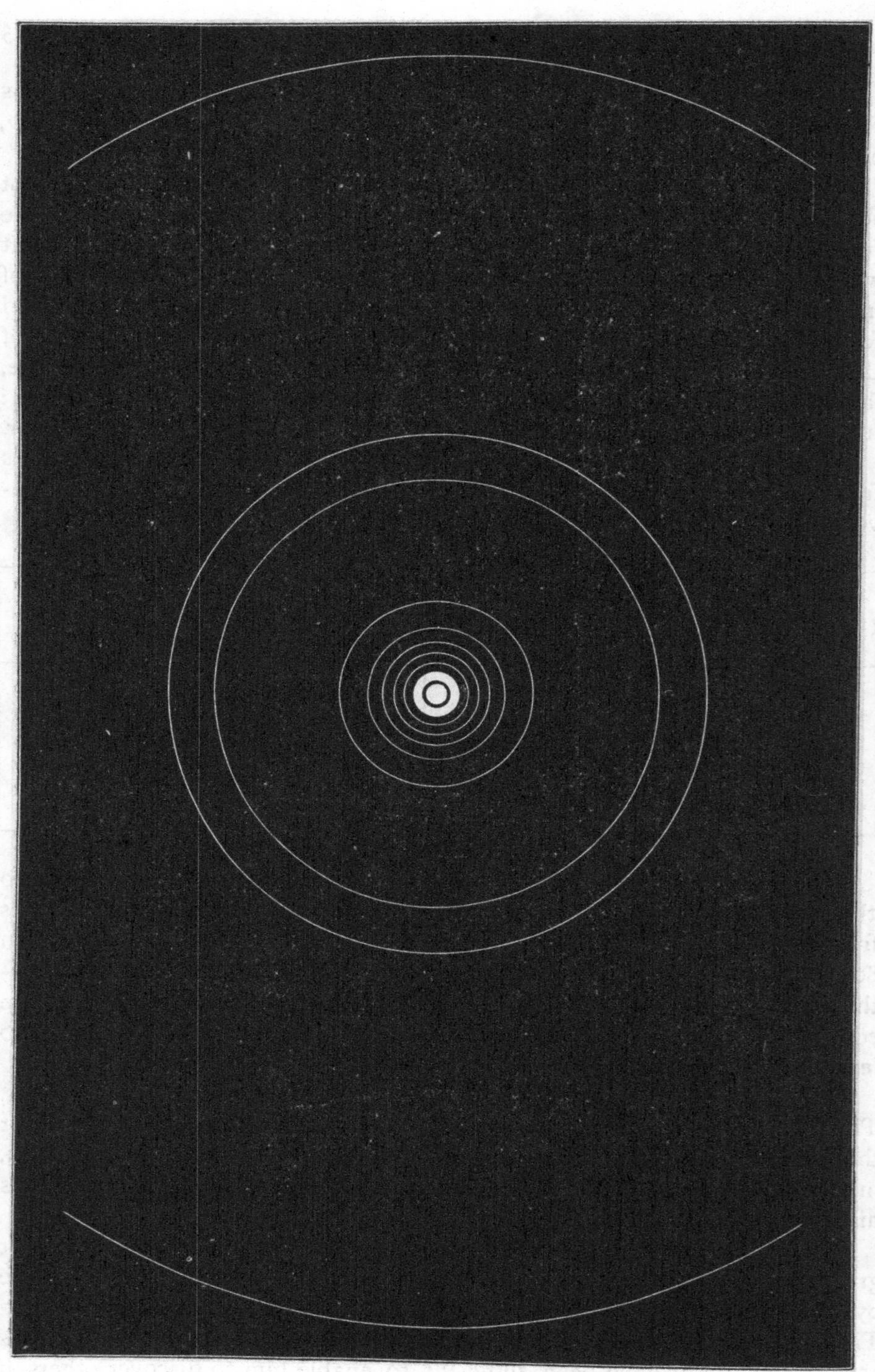

Fig. 194.—The System of Saturn

he had attentively searched for them ; but astronomers were then convinced that there could not be more satellites than planets, and they did not look for them.[1]

All these little worlds were baptized by Sir John Herschel, who gave them the names of the brothers and sisters of Saturn, the only persons to take, since this good father devoured all his children. The largest is named Titan, the most distant, Japetus ; the last-discovered received in 1848 the name of Hyperion, son of Uranus and brother of Neptune.

Variations of brightness are observed on these satellites, which show that they probably revolve round the planet, presenting to him always the same face, as the moon does with regard to the earth. Japetus is especially curious in this respect. It is almost as bright as Titan to the west of the planet, while to the east, seven degrees after opposition, it almost entirely disappears. Doubtless a part of its surface is incapable of reflecting the solar rays.

At the stupendous distance which separates us it is difficult to measure their dimensions. The principal, Titan, shows the brightness of a star of the 8th magnitude. The measures made of it do not agree ; it is probable, however, that it is not as large as our moon. The others are smaller. They are small celestial islands.

Here, then, we have a universe, a colossal system, a wonderful wreath, and eight worlds revolving in harmony. The Saturnians have assuredly a right to be proud, and to believe that the entire universe was created and brought forth expressly for them : their sky-vaults are not imaginary like ours, but real ; there the theologians have fair play, and if Voltaire came to life again he would run a great chance of being beaten.

Direct observation on the one hand, spectrum analysis on the other, prove the existence of an atmosphere similar to that of Jupiter. We distinguish in the telescope belts formed of clouds of the nature of our cirri, which are arranged in long trails in the Saturnian atmosphere, on account of the rapidity of the motion of rotation. The equatorial belt is the most permanent, on account of the attraction of the ring. This atmosphere of Saturn is so thick, moreover, and so charged with clouds, that we never see the surface of the soil, any more than on Jupiter, except, perhaps, towards the polar regions, which are usually whiter than the temperate and tropical zones, because, perhaps, they are also covered with snow, and are so much the whiter, alternately at each pole, as the winter is more advanced. But

[1] Huygens himself had the imprudence to write that this was the sixth satellite discovered, and that, 'as there are but six planets, there can exist but six satellites.' An English *savant* also said, in 1789, that if Saturn has more than six satellites (then known), they will doubtless never be discovered, 'for optics will hardly go further.' The history of the sciences shows that academical prejudices have always retarded progress. Each age has its own ; it is difficult to get free from them, and those who are sufficiently independent to do so are usually neither understood nor appreciated by their contemporaries.

we do not distinguish, as on Mars, the geographical surface, the continents, the seas, and the varied configurations which must diversify it.

The intensity of gravity at the surface of Saturn exceeds by about a tenth that which exists here; but the density of substances there is seven times less than here; and, further, the spheroidal form of the planet proves that, as in Jupiter and as in the earth, this density increases from the surface to the centre, so that the external substances must be of unimaginable lightness. On the other hand, if this atmosphere is as deep as it appears, it must be at its base of great density and enormous pressure, and heavier than objects on the surface. This is a very strange state.

Now, telescopic observations tempt us to believe, on the other hand, that there is on this planet a quantity of heat greater than that which results from its distance from the sun, for the day-star as seen from Saturn is, as we have said, ninety times smaller in surface, and its heat and light are reduced in the same proportion. Water could only exist in the solid state of ice, and the vapour of water could not be produced so as to form clouds similar to ours. Now, meteorological variations are observed similar to those which we have noticed on Jupiter, but less intense. Facts, then, combine with theory to show us that the world of Saturn is at a temperature at least as high as ours, if not higher.

But the strangest feature of the Saturnian calendar is, unquestionably, its being complicated, not only with the fabulous number of 25,060 days in a year, but, further, with eight different kinds of months, of which the length varies from 22 hours to 79 days—that is to say, from about 2 Saturnian days to 167. It is as if we had here *eight moons revolving in eight different periods.*

The inhabitants of such a world must assuredly differ strangely from us from all points of view. The specific lightness of the Saturnian substances and the density of the atmosphere will have conducted the vital organisation in an extra-terrestrial direction, and the manifestations of life will be produced and developed under unimaginable forms. To suppose that there is nothing fixed, that the planet itself is but a skeleton, that the surface is liquid, that the living beings are gelatinous—in a word, that all is unstable—would be to surpass the limits of scientific induction.

We see in Saturn a marvellous abode of habitation, and we should not doubt that nature has known how to obtain the best possible result from all these conditions, as she has done here with ordinary terrestrial conditions. Marvellous abode indeed! What would not be our admiration, our astonishment, our stupor perhaps, if it were granted us to be transported there alive, and, among all these extra-terrestrial spectacles, to contemplate the strange aspect of the rings, which stretch across the sky like a bridge suspended in the heights of the firmament! Suppose we lived on the Saturnian equator itself, these rings would appear to us as a thin line drawn across the sky above our heads, and passing exactly through the zenith, rising from the east and increasing in width, then descending to the west and diminishing according to perspective. Only there have we the rings precisely in the zenith. The traveller who journeys from the equator towards either pole leaves the plane of the rings, and these sink imperceptibly, at the same time that the two extremities cease to appear diametrically opposite, and by degrees approach each other. What an amazing effect would be produced by this gigantic arch, which springs from the horizon and spans the sky! The celestial arch diminishes in height as we approach the pole. When we reach the 63rd degree of latitude the summit of the arch has descended to the level of our horizon and the marvellous system disappears from the sky; so that the inhabitants of those regions know nothing of it, and find themselves in a less favourable position to study their own world than we, who are nearly 800 millions of miles distant.

During one-half of the Saturnian year the rings afford an admirable moonlight on one hemisphere of the planet, and during the other half they illuminate the other hemisphere; but there is always a half-year without 'ring-light,' since the sun illuminates but one face at a time. Notwithstanding their volume and number, the satellites do not give as much nocturnal light as might be supposed, for they receive, on an equal surface, only the ninetieth part of the solar light which our moon receives. All the Saturnian satellites which can be at the same time above the horizon and as near as possible to the full phase do not afford more than the hundredth part of our lunar light. But the result may be nearly the same, for the optic nerve of the Saturnians may be ninety times more sensitive than ours.

But there are further strange features in this system. The rings are so wide that their shadow extends over the greater part of the mean latitudes. During fifteen years the sun is to the south of the rings, and for fifteen years it is to the north. The countries of Saturn's world which have the latitude of Paris endure this shadow for more than five years. At the equator the eclipse is shorter, and is only renewed every fifteen years; but there are every night, so to say, eclipses of the Saturnian moons by the rings and by themselves. In the circumpolar regions the day star is never eclipsed by the rings; but the satellites revolve in a spiral, describing fantastic rounds, and the sun himself disappears at the pole during a long night of fifteen years.

From this distant abode the earth is, as it is for Jupiter, and, indeed, more so, a luminous *little point*, which does not depart more than six degrees from the sun—that is, about twelve times the apparent width which the sun appears to us. It will be still more difficult to discover than it is from Jupiter, for it is but an imperceptible point, and it is very doubtful whether they have been able to notice it even when it passes in front of the sun which happens every fifteen years, even supposing—which is, of course, possible—that the Saturnians enjoy transcendent visual powers. However this may be, *this planet is the last* from which our little world could be distinguished, and for the rest of this universe, for the whole of Infinitude, we are as if we did not exist. It is evident, moreover, that if the Saturnians have discovered our globe, they do not think of *us* for all that, and probably this little globule has been declared by their academies to be unimportant, scorched, desert, and uninhabitable.

CHAPTER VIII

THE PLANET URANUS

♅

ABOUT the year 1765 there was living at Bath a German organist, born in 1738 in the Duchy of Hanover, who had emigrated to England to earn his living. An indefatigable worker, the study of music had led him to the study of mathematics, and the latter to that of optics. One day a telescope of two feet in length fell into his hands; he directed it to the sky, was astonished, and wondered at splendours which he had not suspected: the fixed stars increased in number, and presented the most vivid colouring; the planets acquired considerable dimensions and varied forms. His imagination had often dreamt of the sky, but it remained powerless to picture the splendours of so dazzling a spectacle. The musician was transported with enthusiasm.

From that day he had no rest till he became possessed of an instrument capable of revealing to him the sublime marvels of the heavens. Not having the means to pay the price which a London optician asked to supply him with one, he immediately set to work to construct a telescope with his own hands. Launching, then, into a multitude of ingenious attempts, he arrived, in the year 1774, at being able to contemplate the sky with a Newtonian telescope of five-foot focus executed entirely with his own hands. Encouraged by this first success, the German musician soon obtained telescopes of seven, eight, ten, and even twenty feet of focal length. Later he constructed a truly gigantic one of four feet in diameter and forty feet in length, surpassing by his own efforts all the opticians of Europe and all astronomical observers of the time.

The enthusiastic astronomer was engaged on March 13, 1781, in observing with a telescope of seven feet in diameter and a magnifying power of 227 times a small group of stars situated in the constellation Gemini, when he found one of these stars with an unusual diameter. Substituting eyepieces magnifying 460 and even 932 times for that which the telescope carried at first, he saw that the apparent diameter of the star increased in proportion to the magnifying power, while this was not so with the neighbouring stars which served for comparison. This little star showed to the naked eye the aspect of a star of the 6th magnitude—that is to say, was hardly visible. The magnifying powers which could be applied to the little star had, however, a limit, because beyond a certain magnification its disc was obscured and became badly defined at the borders, which did not happen with the other stars; these latter preserving their brightness and sharpness.

This new body became displaced among the stars. It has been remarked, with reason, that if Herschel had directed his telescope to the constellation Gemini eleven days earlier—that is, on March 2, instead of the 13th—the proper motion of the little star would have escaped him, for it was then at one of its stationary points.

What could this new star be? It would be very extraordinary if there still existed in the sky an unknown planet. It seemed that astronomers had for a long time had the right to consider that the planets had all been discovered, and to assert that their number was irrevocably fixed at six, since from historical times, and especially since the invention of the telescope, they had not found any new ones.[1] The author of the discovery was not so rash as to think that his little star might be a planet, and, although it had neither tail nor apparent coma, he did not hesitate to call it a comet. It was under this designation that he described it to the Royal Society of London in a memoir on April 26, 1781: 'Account of a Comet.'

The name of the musician-astronomer became known over Europe together with the news of this discovery. The newspapers and the scientific periodicals of that time vied with each other in repeating his name, but nearly all wrote it in a different way: thus, the Germans, his fellow-countrymen, spelled it, in 1781, *Merthel*, *Hersthel*, *Hermstel*, &c.; the French astronomers

[1] Uranus had already been seen nineteen times as a star. It might have been discovered to be a planet in 1690 if the instruments used had given it a perceptible disc, or if it had been followed several days in succession; and in 1750, if Lemonnier had copied his observations on the same sheet of paper, the motion would have manifested itself.

In his *Histoire de l'Astronomie* (1785) Bailly speaks of this discovery, which he attributes to a *German named Hartchell*; he describes the star as a comet, but remarks that in France and England they begin to believe that it is rather a planet. Pingré, in his *Cométographie*, published in 1784, classes Uranus under the title of the *first comet* of 1781. 'This comet or planet,' says he, '(for it is not yet decided if it is one or the other), was discovered in England by M. Herschel, ASTROPHILE, they say, rather than ASTRONOMER.'

The official French astronomers still called it *Horochelle* in the *Connaissance des Temps* for 1784. Bernardin de St. Pierre remarks, in his *Harmonies*, that the astronomical editors of that publication have intentionally thrown much doubt and obscurity on the discoveries of this great man. The *savants* who are supported by the State often do very little that is useful; but the worst of it is that they criticise those who work, and who devote themselves with disinterestedness and zeal to the progress of science.

called it *Horochelle.* The illustrious man who made his appearance in so brilliant a manner signed his name *William Herschel.*

From that day the reputation of Herschel—no longer in the character of a musician, but rather in the capacity of a constructor of telescopes and an astronomer—rapidly increased. King George III., who loved the sciences and patronised them, had the astronomer presented to him; charmed with the simple and modest account of his efforts and his labours, he secured him a life pension and a residence at Slough, in the neighbourhood of Windsor Castle. His sister Caroline assisted him as secretary, copied all his observations, and made all his calculations; the King gave him the title and salary of Assistant Astronomer. Before long the Observatory at Slough surpassed in celebrity the principal observatories of Europe; we may say that, in the whole world, this is the place where the most discoveries have been made.

Astronomers soon applied themselves to the observation of the new body. They supposed that this 'comet' would describe, as usually happens, a very elongated ellipse, and that it would approach considerably to the sun at its perihelion. But all the calculations made on this supposition had to be constantly recommenced. They could never succeed in representing all its positions, although the star moved very slowly: the observations of one month would utterly upset the calculations of the preceding month.

Several months elapsed without a suspicion that a veritable planet was under observation, and it was not till after recognising that all the imaginary orbits for the supposed comet were contradicted by the observations, and that it had probably a circular orbit much farther from the sun than Saturn, till then the frontier of the system, that astronomers came to consider it as a planet. Still, this was at first but a provisional consent.

It was, in fact, more difficult than we may think to increase without scruple the family of the sun. Indeed, for reasons of expediency this idea was opposed. Ancient ideas are tyrannical. Men had so long been accustomed to consider old Saturn as the guardian of the frontiers, that it required a rare boldness of spirit to decide on extending these frontiers and marking them by a new world.[1]

William Herschel proposed the name of *Georgium Sidus*, 'the star of George,' just as Galileo had given the name of 'Medician stars' to the satellites of Jupiter discovered by him, and as Horace had said *Julium Sidus.*

[1] This was a case like that of the discovery of the small planets situated between Mars and Jupiter. When, two centuries before this discovery, Kepler had imagined from the harmony of the world a large planet in this interval, he was opposed by considerations the most frivolous and devoid of sense. For example, reasonings like this were advanced: 'There are but seven openings in the head—two eyes, two ears, two nostrils, and a mouth; there are, then, but seven planets,' &c. Considerations of this kind, and others no less imaginary, often delay the progress of astronomy.

Others proposed the name of *Neptune*, in order to maintain the mythological character, and to give to the new body the trident of the English maritime power; others, *Uranus*, the most ancient deity of all, and the father of Saturn, to whom reparation was due for so many centuries of neglect. Lalande proposed the name of *Herschel*, to immortalise the name of its discoverer. These last two denominations prevailed. For a long time the planet bore the name of Herschel; but custom has since declared for the mythological appellation, and Jupiter, Saturn, and Uranus succeed each other in order of descent—son, father, and grandfather.

The discovery of Uranus has extended the radius of the solar system from 885 millions to 1,765 millions of miles.

The apparent brightness of this planet is that of a star of the 6th magnitude; observers whose sight is very piercing may succeed in recognising it with the naked eye when they know where to look for it. Uranus moves slowly from west to east, and takes no less than eighty-four years to make the complete circuit of the sky. In its annual motion round the sun the earth passes between the sun and Uranus every 369 days—that is to say, once in a year and four days. It is at these times that this planet crosses the meridian at midnight. We can observe it in the evening sky for about six months of every year.[1]

On June 5, 1872, Jupiter and Uranus met in perspective in the field of the sky at a distance of only one and a half times the diameter of Jupiter. I had announced this curious conjunction some years before, and I was doubly interested in verifying it myself. The drawing here given represents the observation which I made. The diameter of Jupiter was 33″·4, that of Uranus 3″·8; the minimum distance between the centres took place at 6^h 29^m 53^{sec} at 1′ 9″·8, and from the edge of Jupiter's disc to that of Uranus was 51″·2. What an approach! The first satellite of Jupiter revolves at six times the semi diameter of the planet. At half-past five the daylight prevented observation, and the more so as the phenomenon passed to the west. At nine o'clock Jupiter was admirably shown in the field of the telescope, accompanied by five satellites, of which one

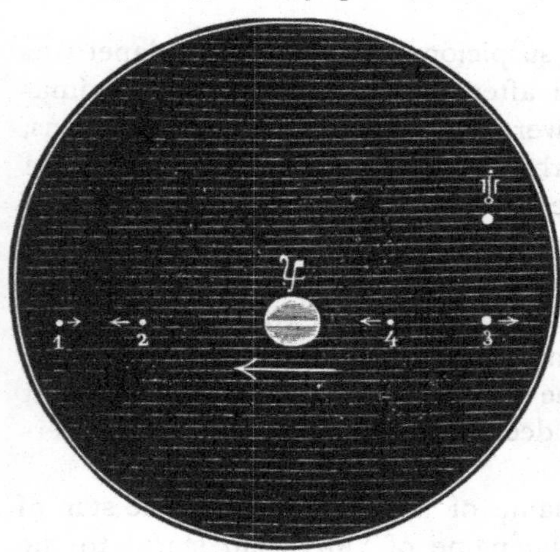

FIG. 195.—URANUS AND THE SATELLITES OF JUPITER, JUNE 5, 1872

[1] For its positions each year, see our *Revue mensuelle d'Astronomie populaire.*

was Uranus; this planet appeared perceptibly brighter than the largest satellite (the third). This observation permitted me to determine that the brightness of Uranus exceeds that of the brightest satellite of Jupiter (the third), and that its magnitude should be noted as 5·7.

The orbit of Uranus round the sun is placed at a mean distance of 1,765 millions of miles from the central star, at about nineteen times (19·18) that at which the earth revolves. This elliptical orbit has for its eccentricity the ratio 0·0463, so that its distance varies as follows:

	Geometric	In Kilometres	In Miles
Perihelion distance	18·295	2,700,000,000	1,678,000,000
Mean distance	19·183	2,840,000,000	1,765,000,000
Aphelion distance	22·071	2,968,000,000	1,844,000,000

Thus this planet is 67 millions of leagues (166 millions of miles) nearer to the sun at its perihelion than at its aphelion. Its distance from the earth varies in the same way from 1,585 to 1,750 millions of miles. The perihelion of Uranus is situated at 171° from the equinox; the planet passed it in 1799 and in 1883; it will return to it in 1967. Its orbit lies almost exactly in the plane of the ecliptic. The period of its revolution, recently calculated from all the observations made since its discovery, is 30,688 days, or 84·022 years, or 84 years 8 days: it is two days longer than it was thought to be some years since. Herschel's planet returned on March 21, 1865, to the point of the sky where it was discovered on March 13, 1781.

The calendar of this distant world must, in all probability, reckon 60,000 days in a year, if we may judge by the velocity of rotation of the large and light exterior planets of which we have already been able to observe the motion. The smallness of the disc of Uranus has not yet permitted us to discover a spot sufficiently plain to render this observation certain; however, we have an index of the probable velocity of the rotation of this globe by that of its satellites and by its polar flattening.

The diameter of Uranus measures 4″. Combining this with the distance, we find that it corresponds to a line of 13,400 leagues (33,300 miles)—that is to say, more than four times greater than the diameter of our globe. It follows that the volume of this planet is sixty-nine times greater than that of the earth. It is much larger than the four interior planets Mercury, Venus, the Earth, and Mars) put together. Its mass has been determined, according to the principles already explained, from the velocity of its satellites and by its influence on Neptune, and it has been found that this mass exceeds by thirteen and a half times that of our planet. It follows that the matter of which Neptune is composed is much lighter than that of which our world is constituted: its density is but the fifth of ours (0·195).

The atmosphere of Uranus has been ascertained by spectrum analysis. It differs from ours by its powers of absorption, resembles those of Saturn and Jupiter rather than that which we breathe, and contains a gas *which does not exist on our planet.*

This distant world moves in the sky accompanied by a system of four planets, of which the elements are:

	Distance in radii of Uranus	Distance in Miles	Period of Revolution				Discoverer
			D.	H.	M.	S.	
I. Ariel . .	7·72	127,000	2	12	29	21	Lassell, 1851
II. Umbriel . .	10·76	177,000	4	3	27	37	Lassell, 1851
III. Titania . .	17·65	291,000	8	16	56	29	W. Herschel, 1787
IV. Oberon . .	23·60	389,000	13	11	7	6	W. Herschel, 1787

This gives to the Uranians four species of months, of two, four, eight, and thirteen days, without reference to other satellites which may remain undiscovered.

There is in this system a surprising peculiarity: the satellites of Uranus do not revolve like those of the other planets. When we consider Mars, the Earth, Jupiter, and Saturn, their moons revolve from west to east in the plane of the equators of those planets, or approximately so, and this plane does not make a considerable angle with that of their orbits round the sun. The satellites of Uranus revolve, on the contrary, in a plane nearly perpendicular to that in which the planet moves, and from east to west, as we see in fig. 196. This figure, drawn to a scale of 1 millimetre to 2″, shows points which, reckoning from the bottom of the longer axis, indicate the positions of each satellite from day to day.

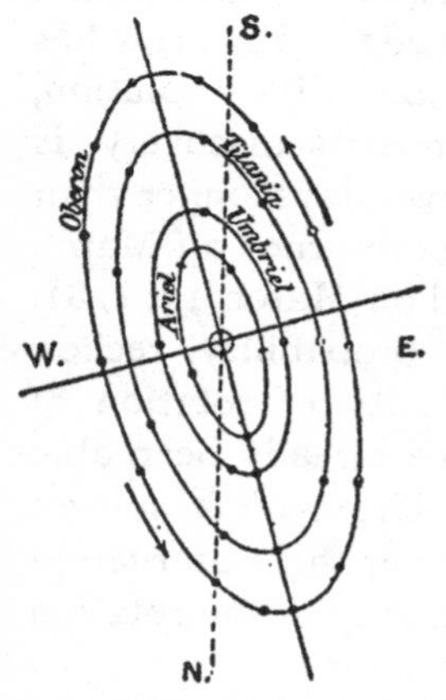

FIG. 196.—INCLINATION OF THE SYSTEM OF URANUS, SEEN FROM THE EARTH

Observations made in 1884 at the Paris Observatory by M. Henry indicate that there are on its surface belts analogous to those of Jupiter, as is shown in fig. 197; the direction of these bands does not coincide with the projection of the longer axis of the apparent orbits of the satellites, but makes with it an angle of 40°. Supposing that the equator of Uranus is parallel to these bands, we find about 41° for the angle contained between the plane of the equator of Uranus and that of the orbits of the satellites. The equator would be inclined 58° to the plane of the ecliptic, and the satellites would be inclined 98°.

Measures made, in 1883, by M. Schiaparelli at Milan gave $\frac{1}{11}$ for the polar flattening, and those of Mr. Young at Princeton (United States), made in the same year, gave $\frac{1}{14}$. The rotation of this planet must, then, be very rapid. It is doubtless not far from 11 hours.

We might almost say that we have here an inverted world. But this is not all. The equator of this singular world being inclined 58°, the Uranian sun wanders during the course of its long year up to this latitude: it is as if our sun forsook the sky of Central Africa and the tropics to soar over Siberia, or as if at Paris we should see the day star in summer turning round the pole without setting, even at midnight, during a summer of 21 years, and remaining invisible in winter for 21 years also. The

FIG. 197.—BANDS OBSERVED ON URANUS, AND PROBABLE INCLINATION OF ITS EQUATOR

seasons there are incomparably stranger than those which we have noticed in Venus.

Seen from Uranus, the starry sky is the same as seen from here, but it is not so with the solar system. Mercury and Venus are absolutely unknown there, and we may, notwithstanding the regrets which such a conclusion may occasion, say the same of the earth. In fact, our little planet, besides being completely invisible from its smallness, is, moreover, lost in the glare of the sun, from which it does not depart more than three degrees. Thus, to the inhabitants of this world we do not exist, the whole earth itself *does not exist*, and it is the same for all the rest of the universe. Mars, and

FIG. 198.—PROPORTIONAL SIZE OF THE SUN AS SEEN FROM DIFFERENT PLANETS

Jupiter himself, are invisible; Saturn appears as a small morning and evening star; Neptune as a small night star.

At this distance the day star shows a diameter nineteen times smaller than that which it presents to us, and a surface 368 times (19·18 × 19·18)

less in area. Thus, this world receives from the sun 368 times less light and heat than we do: to judge from our terrestrial impressions, it must be a desert of ice compared with which the polar solitudes or the snows and storms of Mont Blanc would be a Senegal or a Sahara. The diameter of the Uranian sun measures 1′ 40″, but its light is still equal to that of 1,584 full moons. Fig. 198 shows the comparative size of the sun as seen from the different planets. We see that, arriving at the distant regions of Uranus and Neptune, the day star is reduced to dimensions which would hardly tempt us to transfer our household gods to these northern latitudes.

But ought we to judge of the infinite universe by the particular aspect of our little floating island? We generally make the mistake of considering as utterly uninhabitable those regions which beings of our species could not inhabit. This is to have a very poor opinion of the power of nature, which would on this hypothesis have the power of constituting enormous globes at immeasurable distances, but not that of organising them with appropriate beings. If we judged of the temperature of distant planets from our view of things, we should not hesitate for a moment to declare them for ever uninhabited on account of the excessive cold which must reign there. We cannot imagine that men can exist who have not the same conformation and the same wants as we have.

The population of worlds depends so much on different causes that it would be puerile even to ask whether a large world is more thickly populated than a little one. On the earth the human population is constantly increasing over the whole world, although decreasing at several places; our planet might easily support ten times as many human beings as at present; 14,000 millions might live here with no greater difficulty than 1,400 millions.[1]

The conditions of life on these planets do not, however, appear more different from those of the earth than the condition of a terrestrial animal differs from that of a fish. 'The inhabitants of Saturn,' said Huygens even in his day, 'have no more to complain of than the owls and bats of the little light which they receive from the sun, for it is more advantageous and agreeable to them to sport in the twilight, or that which remains during the night, than that which shines in the daytime.'

Fontenelle, always so ingenious in the determination of the conditions of existence in planetary worlds, advances considerations with reference to Saturn which we may also apply to Uranus. 'We should be very astonished, said he, 'if we were in the world of Saturn, to see over our heads during

[1] The proportion of increase of each family is subject to considerable fluctuations. A large number of families become absolutely extinct; others develop like the foliage of an oak. The most curious example of human fecundity which the annals of anthropology mention is that which is related by Derham, of an Englishwoman who died at the age of ninety-three years, having had *two hundred and fifty-eight* children, grandchildren, and great-grandchildren.

the night this great ring, which would extend from one side of the horizon to the other, and which, sending us light, would have the effect of a continuous moon! Nevertheless, these people are sufficiently miserable even with the help of the ring. It gives them light, but how little light, at the distance it is from the sun! The sun itself, which they see a hundred times smaller than we see it, is to them but a little white and pale star with a very feeble brightness and heat; and if you could place them in our coldest

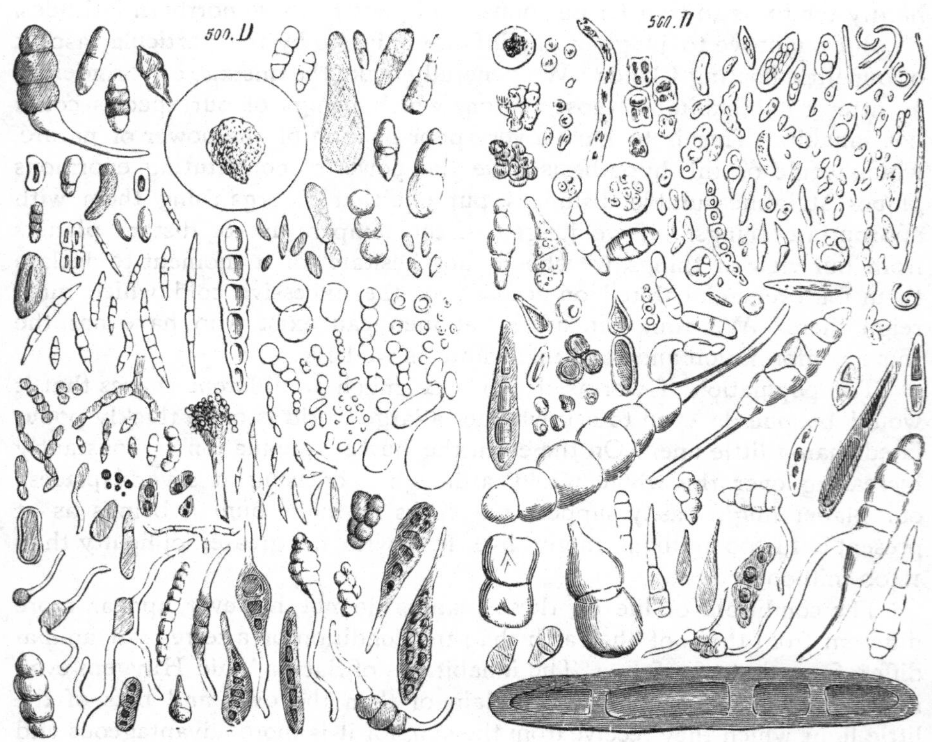

FIG. 199.—WHAT WE BREATHE: MICROSCOPIC ANIMAL AND VEGETABLE OBJECTS FLOATING IN THE AIR

countries, in Greenland or Lapland, you would see them perspire in great drops and die of the heat. If they had water, it would not be water to them, but a polished stone, a slab of marble; and spirits of wine, which never freezes here, would be as hard as our diamonds.'

Having accused the men of Mercury of folly on account of their vivacity, owing to their proximity to the sun, Fontenelle treats as phlegmatic those of Uranus, for the contrary reason: 'They are people,' said he, 'who know not what it is to laugh, who often take a day to reply to the smallest question put to them, and who might have found Cato of Utica too jocular and playful.'

Without prejudging anything as to the character of the Uranians, the study of nature and the variety of its manifestations absolutely convinces us that distance from the sun cannot be an absolute hindrance to the manifestation of life. The new worlds discovered by the telescope in the infinite depths have coincided with the great discoveries of the microscope

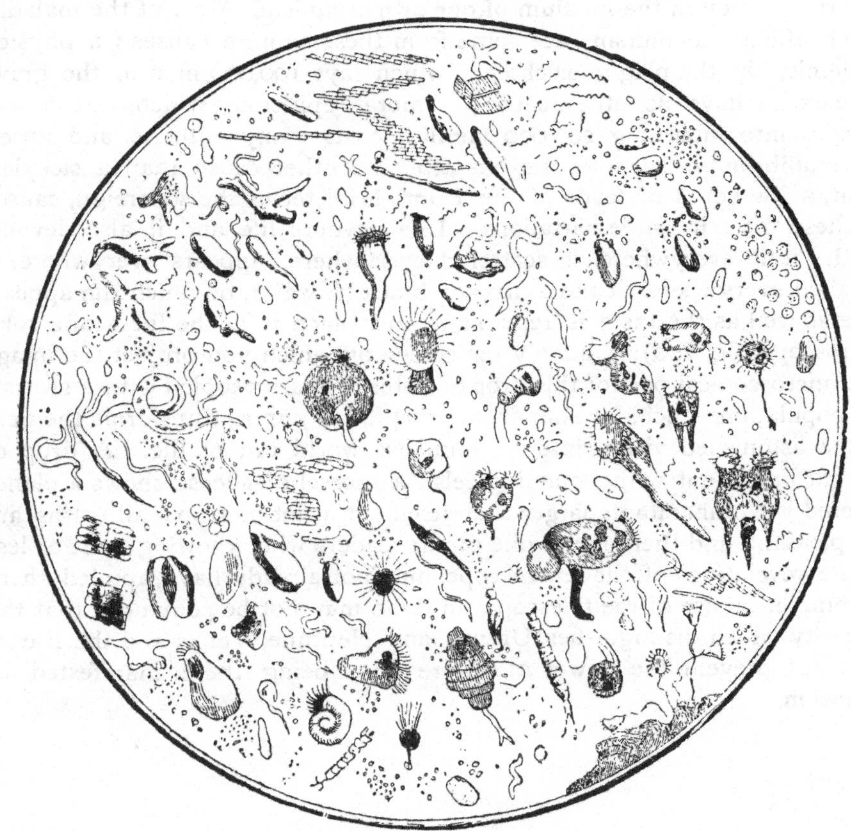

FIG. 200.—POPULATION OF A DROP OF IMPURE WATER

in the universe invisible to our eyes, although present all around us. The air which we breathe is filled with germs, and our lungs constantly absorb an enormous quantity of organisms and vegetable and animal *débris*. We open the mouth, we breathe—what do I say?—on the contrary, we hardly breathe, for, notwithstanding all possible precautions to breathe but the purest air, we inhale incessantly, unknown to us, innumerable corpuscles suspended in the air, spores of cryptogams, grains of pollen, ferments, vibriones, bacteria, eggs, organised cells, various microbes, living and dead bodies in *débris*, of which scientists have counted by thousands in a cubic yard, and of which fig. 199, due to the analysis of M. Miquel, may give an idea. These microscopical beings are here magnified 500 times in diameter; several of their forms are very curious. Who knows? perhaps they are in their turn the receptacles of beings infinitely small relatively to themselves. Where does life stop? And these beings are not insignificant: they are those which govern us through the medium of our own organism. Most of the maladies which afflict the human race come from these minute causes; a physical epidemic, like the plague or cholera, which lays 100,000 men in the grave, appears to have no other cause; a moral epidemic, which, like a war, plunges into mourning 200,000 families, costs many millions, and upsets the equilibrium of all interests, has often no other cause than a sleepless night, a few hours of fever of the Prime Minister or the sovereign, caused by these little invisible battalions. Life devours life, and it also devours death; it is everywhere, is scattered everywhere, appears everywhere, is installed everywhere. Take a drop of brackish water, of which the appearance as well as the taste is repugnant, and place it in the focus of a solar microscope (fig. 200): suddenly the screen on which you project the image of a microscopical part of this drop of water appears peopled with a swarming population, which, by bounds and magnified jumps, transforms the field of the astonished vision into an immense world full of life. A drop of vinegar bursts out with bounding eels; a morsel of cheese shows a planet covered with inhabitants larger than itself. But let us stop: all truths are not pleasant, and there is not one of our readers who, knowing more or less of the revelations of the microscope, may not already have applied them to complete those of the telescope, and who may not be convinced that the diversity which distinguishes Uranus and Neptune, Venus and the Earth, does not prevent the power of nature from being there manifested in profusion.

CHAPTER IX

THE PLANET NEPTUNE AND THE FRONTIERS OF THE SOLAR DOMAIN

♆

IT has been said, with reason, that the labours of astronomy are those which give the highest measure of the powers of the human mind. The discovery of Neptune, due to the sole power of numbers, is one of the most eloquent witnesses of this truth. The existence of this planet in the sky was revealed by mathematics. This world, distant more than 2,700 millions of miles from our terrestrial station, is absolutely invisible to the naked eye. The perturbations manifested by the motion of the planet Uranus permitted the mathematician to say that the cause of these perturbations was an unknown planet which revolved beyond Uranus at about such a distance, and which, to produce the effect observed, should be found at a certain point of the starry sky. A telescope was directed towards the point indicated, the unknown was searched for, and in less than an hour it was found!

If the planets only obeyed the action of the sun they would describe round him the elliptic orbits which we have studied in Chapter I. of 'The Sun' (p. 211). But they act on each other; they likewise act on the central star, and from these various attractions perturbations result.

Astronomers construct in advance tables of the positions of the stars in the sky, in order to know where they should be found exactly, and to observe them according to the interest presented by their situations, either from the point of view of their physical constitution, or to verify their motions, or for the numerous applications of astronomy to geography and navigation. An astronomer of Paris, Bouvard, calculating in 1820 the tables of Jupiter, Saturn, and Uranus, ascertained that the theoretical positions given by his tables agreed perfectly with modern observations for the first two planets, while for Uranus there were inexplicable differences. From 1820 to 1840 these discrepancies struck all astronomers; several (Bouvard himself, Mädler, Bessel, Valz, Arago) expressed the opinion that these perturbations must proceed from an unknown planet, and Bessel himself had commenced the mathematical research when he was seized with an illness which carried him to the grave. However, the difference between the calculated and observed positions of Uranus went on increasing: it was 20″ in 1830, 90″ in 1840, 120″ in 1844, and 128″ in 1846. To a man of the world, an artist or a merchant, this would have been, in the affairs which interest him, a difference so slight that it might not have struck him—it is like a comma in music; and if there were two adjacent stars in the sky at this distance from each other, excellent sight would be required to separate them clearly. But to an astronomer such a difference becomes altogether intolerable and a veritable cause of insomnia.

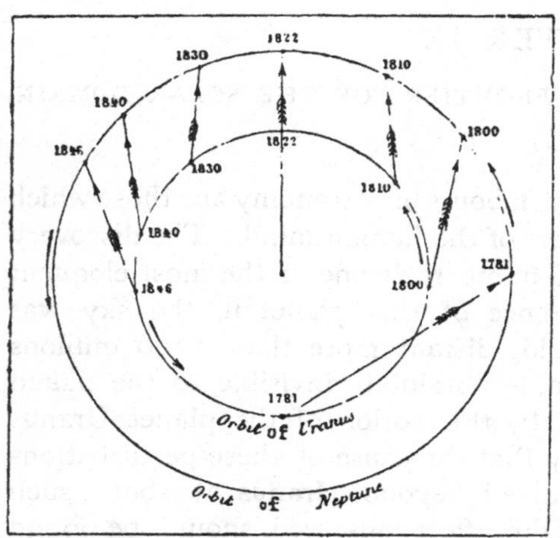

Fig. 201.—Disturbances of Uranus by Neptune

We may very easily understand the disturbing action of an exterior planet on the positions of Uranus by an examination of fig. 201, which shows the positions of the two planets from the discovery of Uranus to that of Neptune. We see that from 1781 to 1822 the influence of Neptune tends to draw Uranus in advance—that is to say, to accelerate his motion; while from 1822 to 1846, on the contrary, it remained behind and tended to retard it by diminishing its longitude.

This problem was the topic of the day, and Arago, always in the vanguard of progress, advised a young and skilful calculator, a stranger to the Observatory of Paris, the young mathematician, Le Verrier, to solve this magnificent problem. Already accustomed by his researches on comets to difficulties in the calculation of perturbations, the young *savant* set to work immediately. He began by verifying the tables of Bouvard, in which he corrected several errors; but these errors did not account for the differences found. Recommencing the whole of the calculations of the perturbations of Saturn on Uranus, he added those of Jupiter, recalculated the orbit of Uranus from nineteen old observations of the planet observed as a star before 1781, and 179 observations made from 1781 to 1845, and proved that the discrepancy between the observed and calculated positions could not be explained by the perturbations produced by Saturn and Jupiter. 'I have demonstrated,' said he, 'that there is a formal inconsistency between the observations of Uranus and the hypothesis that this planet is subject only to the actions of the sun and the other planets acting according to the principle of universal gravitation. We can never succeed in representing, by this hypothesis, the observed motions.' In presence of this well-established inconsistency Le Verrier did not doubt for a single moment the accuracy of the law of universal gravitation; he recollected that over and over again, in order to explain inequalities which could not be accounted for, fault had been found with this law, which, however, always came out victorious after a more searching examination of the facts. He boldly advanced the hypothesis of a planet acting in a continuous manner on Uranus and changing his motion very slowly. The fact of the existence of an exterior planet being henceforth certain, he supposed, according to the law of Titius, explained above (p. 213), that this planet should be at the distance 36, and consequently revolve round the sun in 217 years; and on this hypothesis he calculated what position it should have in the sky behind Uranus in order to produce by its attraction the observed differences, and what its mass should be to explain the amount of the deviation. He then re-calculated the orbit of Uranus, taking into account the perturbations thus produced by the disturbing planet, and found that all the positions agreed with the theory (the greatest differences between the observed and calculated positions not exceeding $5''.4$). From that time the problem was practically solved. On August 31, 1846, Le Verrier announced to the Academy of Sciences that the planet should be found at the longitude 326°, which placed it 5° to the east of the star δ of Capricorn.

On September 18 he wrote to Dr. Galle, of the Observatory of Berlin, where they were constructing star charts of the ecliptic zone, to request him to search for the planet. That astronomer received the letter on the 23rd. It was a fine evening. He directed his telescope towards the point indicated, and perceived a star which was not on the chart, and which showed a perceptible planetary disc. Its position in the sky was 327° 24′; calculation had indicated 326° 32′; the longitude had therefore been precisely stated to within one degree!

Here we have the discovery of Neptune in its simple grandeur. It calls to memory the beautiful apostrophe of the poet Schiller, who, representing Christopher Columbus sailing to the discovery of a new hemisphere, said to him: 'Pursue thy course towards the west, bold navigator; the land which you seek will rise, even if it should not exist, from the depths of the waters to meet thee, for Nature goes hand in hand with Genius!' There is here, under the form of a great picture and a proud exaggeration, the expression of one of the most real conditions of true genius in the sciences, to which discoveries do not come by chance, but which anticipates them by a sort of presentiment.

This discovery is splendid, and of the highest order from a philosophical point of view, for it proves the security and the precision of the data of modern astronomy. Considered from the point of view of practical astronomy, it was but a simple exercise of calculation, and the most eminent astronomers saw in it nothing else! It was only after its verification, its public demonstration—it was only after the visual discovery of Neptune—that they had their eyes opened, and felt for a moment the dizziness of the infinite in view of the horizon revealed by the Neptunian perspective. The author of the calculation himself, the transcendent mathematician, did not even give himself the trouble to take a telescope and look at the sky to see whether a planet was really there! I even believe that he never saw it. For him, however, then and always, to the end of his life, astronomy was entirely enclosed in formulæ—the stars were but centres of force. Very often I submitted to him the doubts of an anxious mind on the great problems of Infinitude: I asked him if he thought that the other planets might be inhabited like ours, what might be especially the strange vital conditions of a world separated from the sun by the distance of Neptune, what might be the retinue of innumerable suns scattered in immensity, what astonishing coloured lights the double stars should shed on the unknown planets which gravitate in these distant systems. His replies always showed me that these questions had no interest for him, and that, in his opinion, the essential knowledge of the universe consisted in equations, formulæ, and logarithmic series having for their object the mathematical theory of velocities and forces.

But it is not the less surprising that he had not the *curiosity* to verify the position of his planet, which would have been easy, since it shows a planetary disc; and, besides, he might have had the aid of a chart, because he had only to ask for these charts from the Berlin Observatory, where they had just been finished and *published*. It is not the less surprising that Arago, who was more of a physicist than a mathematician, more of a naturalist than a calculator, and whose mind had so remarkable a synthetical character, had not himself directed one of the telescopes of the Observatory towards this point of the sky, and that no other French astronomer had this idea. But what surprises us still more is to know that *nearly a year before*, in October, 1845, a young student of the University of Cambridge, Mr. Adams, had sought the solution of the *same* problem, obtained the same results, and communicated these results to the Director of the Greenwich Observatory, and that the astronomer to whom these results were confided had said nothing, and had not himself searched in the sky for the optical verification of his compatriot's solution!

We have said just now that the mysterious disturbing planet was supposed to be placed at the distance 36, as the series of Titius indicated. But in reality it is much nearer. The theoretical elements of Le Verrier are not, then, those of Neptune, as we may see from the following:

—	Elements of Le Verrier	Elements of Neptune
Distance from the sun	36·154	30·055
Period of revolution	217 years 140 days	164 years 281 days
Eccentricity of the orbit	0·10761	0·00896
Longitude of the perihelion	284° 45′	46° 0′
Mass compared to that of the sun . .	$\frac{1}{9300}$	$\frac{1}{19700}$

These two series of elements are as different one from the other as if it were a question of two planets having no relation to each other. Must we, then, believe that Le Verrier did not discover Neptune? No, assuredly. The principal cause of the difference proceeds from the distance 36, instead of 30; but in this problem, as in many others where there are several unknown quantities, there are several solutions possible. It is necessary either to assume a distance and calculate the mass, or to assume a mass and calculate the distance. The more distant the planet was supposed to be, the greater should be the disturbing mass, and conversely. The problem is none the less solved; for, as we remarked just now, it was merely a mathematical problem, and it is its verification which is of such importance to the thinker. But then, it may be said: How does it happen that, with such a difference between his results and the reality, he came so near the real position occupied by the looked-for star? The answer is, that this *position* was comparatively independent of the calculated orbit. In fact, it is sufficient to look at fig. 201 to see that, whatever may be this orbit, whatever may be the distance, and whatever may be the mass of Neptune, this planet was in 1822 exactly beyond Uranus, that it was before it from 1781 to 1822, and from 1822 to 1845 it was behind it; the acceleration and the slackening of the motion of Uranus indicated this position. The analysis of the perturbations gave, then, the longitude to a close approximation.

Arago wished to give to this planet the name of the learned mathematician who had discovered it 'at the end of his pen'; but mythological memories prevailed again, as they had done for the Herschel planet, and the name of Neptune, son of Saturn, god of the sea, already proposed for Uranus, was given by common consent to the star of Le Verrier.[1]

Neptune presents the aspect of a star of the 8th magnitude. An astronomical telescope of moderate power is sufficient to find it when we know where it is. A magnifying power of 300 times gives it a perceptible disc. This disc measures but 3″ in diameter, and appears, in powerful telescopes, slightly tinted with blue. Lalande had observed it as a star on

[1] Le Verrier succeeded François Arago in 1854 as Director of the Paris Observatory, where he died on September 23, 1877, the anniversary day of the optical discovery of Neptune, and only two months after having finished the complete theory of the planetary motions, including the theory of the motion of Uranus, which he had investigated in 1845.

May 8 and 10, 1795, and Lamont on October 25, and even on September 11, 1846, without suspecting it. Lalande had even noticed a difference between his two observations, but, attributing this to an error, he suppressed the first one. If he had thought of following the star, he would have discovered Neptune half a century before Le Verrier. With an *if* we might go to the moon!

According to the latest calculated elements, the real distance of Neptune from the sun is 30·055, that of the earth being taken as unity—that is to say, 1,112 millions of leagues (2,764 millions of miles). The diameter of its orbit is, then, 2,224 millions of leagues (5,528 millions of miles), and the whole circumference measures 2,224 × 3·1416, or 17,000 millions of miles. We have, then, 17,000,000,000 miles described in 60,151 days, which gives a velocity of 299,000 miles a day, or 12,460 an hour, 200 per minute, or 3·33 miles per second. This is, of course, the smallest of the planetary velocities which we know, since this planet is the most distant from the sun.

This distant planet takes more than a century and a half to make the circuit of the sky. The long and slow revolution of Neptune round the sun requires 60,181 of our days for its performance—that is to say, *one hundred and sixty-four years and two hundred and eighty-one days.* Such is the year of the Neptunians.

The real diameter of Neptune is nearly four times greater than that of the earth (3·8), and its volume 55 times superior to ours. Its density is hardly a third of ours[1] (=0·300), but gravity at its surface is almost identical with terrestrial gravity (=1·14).

We do not yet know the duration of the diurnal rotation of this distant planet; it may be very rapid, like those of Jupiter, Saturn, and Uranus. Great improvements in optics will still be necessary before we can succeed in magnifying this pale disc so as to discover the aspects of its surface, and thus reveal its motion of rotation.

Spectrum analysis has succeeded, however, notwithstanding the weakness of Neptune's light, in ascertaining the certain existence of an absorbent atmosphere, in which are found gases which do not exist on the earth, and showing a remarkable resemblance in chemical composition to the atmosphere of Uranus.

At the distance of Neptune from the sun the day star, if it still deserves this title, is reduced 30 times in diameter, 900 times in surface and in luminous and calorific intensity; it does not measure more than 64″ in diameter. What is this light and heat? Doubtless the sun is not exactly a star, for the diameter of the most brilliant star, Sirius, is not even a hundredth of a second, and consequently the Neptunian sun still shines

[[1] About $\frac{1}{5·4}$ of that of the earth.—J. E. G.]

with a light of more than 40 millions of stars of the 1st magnitude. But to leave the earth and go to Neptune would be to forsake heat and light and plunge into ice and darkness.

Is this equivalent to saying that this world is condemned to remain eternally in the state of a sterile and uninhabited desert? Nature herself replies that such a supposition would be entirely contrary to her acts and her views. Short-sighted naturalists, who think they know everything, would teach dogmatically that a pressure of so many atmospheres prevents life from being produced; that a certain amount of light is indispensable to life, and that the ocean depths are absolutely destitute of all vital manifestation. A ship starts[1] on an immense liquid plain to visit the equatorial and polar zones, casts the sounding-line at 2,000 fathoms, at 10,000, feet in depth, in eternal night—a black darkness, where the pressure is such that could a man descend there he would have to support a weight equal to that of twenty locomotives, each accompanied by a train of waggons loaded with bars of iron. Evidently there is nothing there! The sounding-line is drawn in, however, and brings up charming delicate beings which the lightest touch of the finger of Psyche would kill: they live there tranquil, happy, 'like the fish in water,' and, since there is no light there, they make it! If they could understand you, you should not speak to them of your castles, your parks, and venerable trees, nor of the Paris worldling and the boulevards which you love so much; they prefer their abode, their dark abode in the depths of the sea, scarcely illuminated with the light of their own phosphorescence, and to them there is the true medium, there is real happiness. And when you cast these living *dèbris* on the deck of the ship, and when these marvellous beings with variegated embroideries die before your eyes, overwhelmed by the light of the sky, suffocated by the rarefaction of the air which nourishes your lungs, do you not think of Neptune? Do you not see that the god of the ocean has down there an empire as vast as the one we see? And as they have there 900 times less light and heat than on the deck of your ship, you imagine that nature has been unable to produce anything there! Error, foolish, insane error, excusable perhaps in the time of Aristotle, but absolutely unpardonable now.

But doubtless the Neptunians differ very much from us. They have neither our heads, nor our bodies, nor our limbs. The brain is but an expansion of the spinal marrow; it is this which has made the skull, and it is the skull which has formed the head. Our legs and arms are but the transformed limbs of the quadruped; it is the position, gradually becoming vertical, which has made the feet, and it is exercise, gradually perfected, which has made the hands. The stomach is but the cover of the intestines; the form and length of the intestines result from the mode of feeding; and there is not upon or in our whole body a square inch which is not due to our vital

[1] See our *Contemplations scientifiques.*

working in the midst of the planet which we inhabit. Now, do you think that they eat in every world? That would be an error. Where they do not eat, the digestive duct is useless, and consequently does not exist, very fortunately. An infinite unimaginable variety exists, then, between the different worlds; upon each of them beings from the first to the last are intimately organised by the forces in action at the surface of each globe. Man is everywhere but an animal more or less rational, and our terrestrial species would appear to be less favoured from this point of view. Our life is half lost in the time devoted to sleep and eating. There may exist worlds where they never sleep, as there may exist worlds where they sleep for ever. This is perhaps the case in Neptune.

There a single year lasts 164 of ours; a child of ten years has lived 1,648 terrestrial years; a young girl of eighteen years marries, at the age of 2,950 terrestrial years, the young man of her dreams, aged himself more than 3,000 years; and a retired general should have been born 13,000 years ago if things are organised there as here—which is not probable.

The slowness of this distant and gloomy world recalls the *shades* (ghosts) which Scarron speaks of in his visit to the infernal regions:—

> I saw the *shade* of a coachman
> Who with the *shade* of a brush
> Was rubbing the *shade* of a carriage.

I need not say that from Neptune the earth is completely invisible, as well as Mercury, Venus, and Jupiter. Saturn is a little star which departs from the sun up to 18°. For the Neptunians the solar system appears to be composed of the Sun, Saturn, Uranus, their own world, and the planet which doubtless gravitates beyond Neptune. These beings should have excellent sight, for it is formed in a medium 900 times less illuminated than ours. They should perceive the stars by day as well as by night, if the state of their atmosphere permits them; and their enormous base of operation, thirty times greater than ours, should have enabled them to calculate long before us, and much better, the parallaxes and the distances of the stars.

Hardly had Neptune been revealed to the inhabitants of the earth when, on October 10, 1846, a satellite was discovered by an English astronomer, Mr. Lassell. It showed the feeble light of an imperceptible star of the 14th magnitude. Its distance from Neptune is fourteen times the semi-diameter of the planet, which corresponds to about 260,000 miles. It revolves round Neptune in five days twenty-one hours. A circumstance worthy of attention is that the motion of this satellite is retrograde, like that of Uranus. This body has not yet received a name. The god Neptune, however, was not without a son; would not the name of *Triton*, one of the most diligent companions of his father on the ocean, be suitable for it? It is probable that this distant planet is accompanied by a large number of satellites.

From the fact that Neptune is the farthest planet we know we have by no means the right to conclude that there are not others beyond it.

> Croire tout découvert est une erreur profonde :
> C'est prendre l'horizon pour les bornes du monde.

We should not even despair of soon finding the first, when the observations of Neptune extend over a space sufficiently large, so that, its orbit being rigorously calculated, the perturbations produced by the exterior planet may be manifested in a perceptible manner. This research may be undertaken next century, unless the observers who pass their nights in the search for small planets find it by chance by the displacement of a small star on their celestial charts ; but, on the one hand, it may be a star below the 12th magnitude, and, on the other hand, it could only move with extreme slowness. The mean diurnal motion of Saturn is 120″, that of Uranus 42″, that of Neptune 21″, that of the exterior planet may not exceed 10″.

We are about to treat of the comets, and we shall see that, in all probability, the periodical comets owe their presence in our solar system to the influence of the planets. In fact, every comet which, coming from the outside, passed sufficiently near a planet to be subject to its attraction and be captured by it, would continue its voyage to the vicinity of the sun, and would afterwards return to the point where it had been diverted from its primitive course, and would thus continue to revolve round the sun. All the periodical comets have their aphelia near the orbit of a planet. Now, the third comet of 1862 and the swarm of shooting stars of August 10 follow an orbit of which the aphelion is at the distance 48. There should exist there a large planet, sailing at 4,000 millions of miles from the sun in a revolution of about 330 years. This revolution would be double that of Neptune, as Neptune's is about double that of Uranus.

Such is the last halting-place in our planetary voyage ; such is the last station of the sun's vast empire. We have recognised from the beginning of our description that the same force, the same motion, and the same law rules the harmony of all these worlds ; we may notice in conclusion that they are all constituted of a substance originally the same : the primitive cosmical nebulous matter. The idea of the unity of substance urges itself on the mind like that of the unity of force. The variety of the successive conditions of organisation has induced a correlative variety in the definitive products. The bodies called simple in chemistry may not be so in reality. Oxygen, nitrogen, carbon, mercury, gold, silver, iron, and all other bodies, may be but different molecular arrangements of primitive atoms, of mineral species, as it is produced afterwards from vegetable and animal species, of which the constituent substances are likewise derived from previous mineral substances. The unity of origin is not doubtful, and for a long time past we have been warranted in thinking that hydrogen is the body which ap-

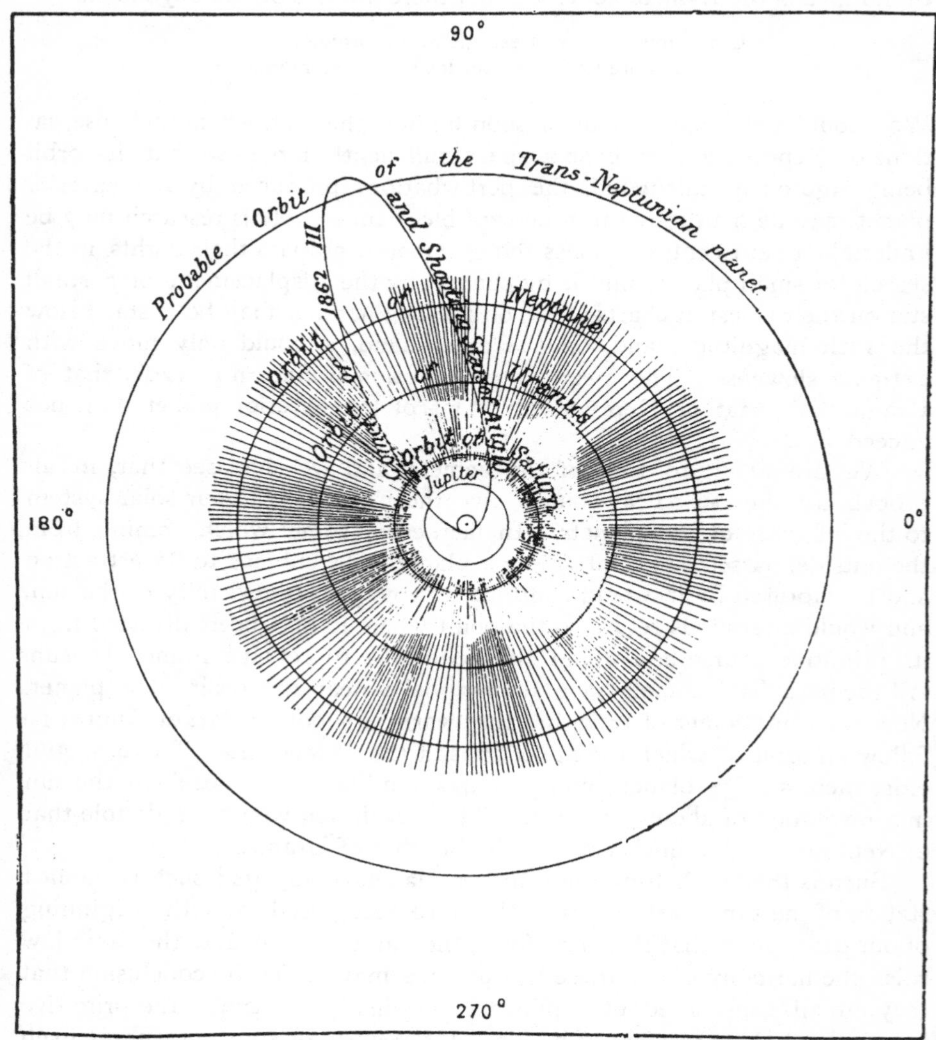

FIG. 202.—DIAGRAM SHOWING THE PROBABLE EXISTENCE OF A TRANS-NEPTUNIAN PLANET AT FORTY-EIGHT TIMES THE DISTANCE OF THE EARTH FROM THE SUN

proaches nearest to the essentially simple primitive substance. Spectrum analysis now confirms these views. The differences between the planets of our great solar family are not, then, material ; they are especially differences of-degree. The spectrum analysis of the stars confirms this way of looking at it. The stars, suns of space, are themselves sisters of our sun. Unity of origin, unity of force, unity of substance, unity of light, unity of life in the immense universe, through an infinite variety of aspects and generations!

[Professor Nichol says in his 'Views of the Architecture of the Heavens':

'Apart from all speculation, surely the view of an actual order whose beginnings are hid in what seems in our eye nothing less than Eternity, cannot but elevate our thoughts of that BEING who, amid change, alone is unchangeable—whose glance reaches from the beginning to the end—and whose presence occupies all things! If uneasy feelings are suggested—and I have heard of such—by the idea of a process which may appear to substitute *progress* for *creation*, and place *law* in the room of *providence*, their origin lies in the misconception of a name. LAW of itself is no substantive or independent power; no causal influence sprung of blind necessity, which carries on events of its own will and energizes without command. Separated from connexion with an ARRANGER in reference to whose mind alone, and as an expression of the Creative Idea it can be connected with the notion of control—Law is a mere name for a long order—an order unoriginated, unupheld, unsubstantial, whose floor sounds hollow beneath the tread, and whose spaces are all void; an order hanging tremblingly over nothingness, and of which every continent, every thing, and every creature fails not to beseech incessantly for a substance and substratum in the idea of ONE—*who liveth for ever*!'

And again: 'We are not rendered incredulous by the *nature*, but overwhelmed by the magnitude of the works; our minds will not stretch out to embrace the periods of this stupendous change. But Time, as we conceive it, has nothing to do with the question—we are speaking of the operations and tracing the footsteps of ONE who is above all time—we are speaking of the energies of that ALMIGHTY MIND, with regard to whose infinite capacity a day is as a thousand years, and the lifetime of the entire Human Race but as the moment which dies with the tick of the clock that marks it—which is heard and passes!'

The famous writer, Richard Hooker, has well said: 'Of Law there can be no less acknowledged than that her seat is in the bosom of God, her voice the harmony of the world: all things in heaven and earth do her homage—the very least as feeling her care, and the greatest as not exempted from her power.']

But we have tarried long enough in the vicinity of our terrestrial abode, already remote in the invisible, already lost to our view since we left Saturn, already forgotten in its mediocrity. Let us unfold our wings. By a rapid flight let us clear the abyss and reach the stars! But no; we must delay a little longer, we who have resolved to make an instructive voyage and to let nothing pass unnoticed. Between the solar world and the stars strange forms rush dishevelled through celestial space, appearing to throw a bridge for our mind across the fathomless abyss, and to place us in communication with other universes. Let us observe these comets in passing, but let us take care not to be delayed too long by these fantastic creatures, the sirens of the sidereal ocean, whose revelations concerning immensity are full of charms, and whose hands, extended towards the inaccessible horizons, seem to show us from afar the mysterious dreams of Infinitude.

BOOK V

COMETS AND SHOOTING STARS

CHAPTER I

COMETS IN THE HISTORY OF HUMANITY

OF all the heavenly bodies, comets are assuredly those whose appearance strikes most forcibly the attention of mortals. Their rarity, their singularity, their mysterious aspect astonish the most indifferent minds. The things which we see every day, the phenomena which are constantly or regularly reproduced before our eyes, no longer strike us, and excite neither our attention nor our curiosity. 'It is not without reason that philosophers are astonished to see a stone fall,' wrote D'Alembert, 'and the people who laugh at their astonishment soon share it themselves if they reflect a little.' Yes, it is necessary to be a philosopher, it is necessary to reflect, in order to seek the why and the how of facts which we see daily, or, at least, the occurrence of which is frequent and regular. The most admirable phenomena remain unnoticed; custom, deadening the impression, leaves us but indifference. It is a rather curious fact that the unforeseen and the extraordinary always give rise to fear, and never to joy or hope. So in all countries, at all epochs, the strange aspect of a comet, the pale gleam of its nebulous coma, its sudden appearance in the firmament, have produced on the minds of men the effect of a formidable power, menacing the order anciently established at the Creation; and as the phenomenon is limited to a short duration, there results the belief that its action should be immediate, or at least at hand. Now, the events of this world always show in their course some fact which may be considered as the fulfilment of a fatal omen.

With some exceptions, the ancient astronomers regarded comets either as atmospheric meteors or as quite transient celestial phenomena. To some, these bodies were *terrestrial exhalations*, kindling in the region of fire; to others, they were the *spirits of great men*, which were mounting to the sky, and which, in leaving it, were handing over our poor planet to the plagues with which it is so often attacked. The Romans appear to have

FIG. 203.—A COMET SEEN OVER THE SEA

seriously believed that the great comet which appeared at the death of Cæsar, in the year 43 B.C., was really the spirit of the Dictator.[1] In the seventeenth

[1] It is by this Metamorphosis that Ovid concludes his great work dedicated to Augustus himself. 'Venus,' said he, 'descends from the ethereal vaults, invisible to all eyes, and stops in the midst of the senate. From the body of Cæsar she takes his spirit, prevents it from evaporating,

century, Hevelius and Kepler himself were still disposed to see in them emanations from the earth and the other planets. We can understand that, with such ideas, the determination of cometary motions would be rather neglected. It is due to the efforts of Tycho Brahé at first, then of Newton, of Halley, and especially of more modern astronomers, that it has been raised to the rank of the theory of planetary motions.

Unquestionably, at first sight the majestic uniformity of the celestial motions would appear to be deranged by the sudden apparition of a dishevelled comet, whose extraordinary aspect seems to give it the form of a supernatural visitor. Thus, ancient writers always depicted them under the most frightful images: they were javelins, sabres, swords, horses' manes, decapitated heads with hair and bristling beard; they shone with the red light of blood, yellow or livid, like that, of which the historian Josephus speaks, which showed itself during the terrible siege of Jerusalem. Pliny found in this same comet 'a whiteness so brilliant that one could hardly look at it; men saw there the image of God under a human form.'

The historian Suetonius ascribes to the influence of one of these bodies the horrors committed by Nero, who had attached to himself the astrologer Babilus, and asserts that a comet announced the death of Claudius. We also read in Dion Cassius: 'Several prodigies preceded the death of Vespasian: a comet appeared for a long time; the tomb of Augustus opened of itself. As the physicians blamed the Emperor, who, attacked by a serious illness, continued to live and attend to the affairs of state, "It is necessary," replied he, "that an Emperor should die standing." Seeing some courtiers conversing in a whisper about the comet, "This hairy star does not concern me," he said, laughing; "it menaces rather the King of the Parthians, since it is hairy and I am bald."' This reply is as good as that of Hannibal to the King of Bithynia, who refused to fight on account of omens read in the entrails of the victims: 'So you prefer the advice of a sheep's liver to that of an old general.' Each epoch has its prejudices, and we have in our time things equally ridiculous.

The same beliefs showed themselves among the Greeks. A comet which appeared in 371 B.C., and was described by Aristotle, announced, according to Diodorus of Sicily, the downfall of the Lacedæmonians, and according to Ephorus, the destruction by the waters of the sea of the towns of Helice and Bura, in Achaia. Plutarch relates that the comet of B.C. 344 was to Timoleon of Corinth an omen of the success of the expedition which he led the same year against Sicily. The historians Sazoncenes and Socrates relate that in the year 400 of our era a comet in the form of a sword

and bears it to the region of the stars. In rising, the goddess feels it transformed into a divine and glowing substance. She allows it to escape from her bosom. The spirit flies away beyond the moon and becomes a brilliant star, which draws through a long space its ignited hair.'

shone above Constantinople, and appeared to touch the city at the moment of the great misfortunes with which the perfidy of the Gainas threatened it.

The Middle Ages outdid, if this were possible, the foolish ideas of antiquity, and gave fantastic descriptions of certain comets which exceed anything which can be imagined.[1] Paracelsus asserted that they were sent by the angels to warn us. The sanguinary madman who was called Alphonsus VI., King of Portugal, hearing of the arrival of the comet of 1664, threw himself to the ground, loaded the comet with abuse, and threatened it with his pistol. The comet majestically pursued its course.

We shall see further on that, of the periodical comets, the most famous in history is that which now bears the name of Halley, in memory of the astronomer who calculated and predicted the first of its returns. This comet has, in fact, already appeared to the earth twenty-four times since the year 12 B.C., the date of the earliest appearance of which the record has been preserved. Its first memorable appearance in the history of France is that of the year 837, in the reign of Louis I., le Débonnaire. An anonymous chronicler of the time, surnamed the Astronomer, speaks of it in the following terms: 'In the midst of the holy days of Easter a phenomenon, always fatal and of sad omen, appeared in the sky. From the time that the Emperor, who gave much attention to such phenomena, had perceived it he gave himself no rest. "A change of reign and the death of a prince are announced by this sign," he said to me.' He took counsel of the

[1] Comets appeared at the deaths of Constantine (336), of Attila (453), of the Emperor Valentinian (455), of Mérovée (577), of Chilpéric (584), of the Emperor Maurice (602), of Mahomet (632), of Louis le Débonnaire (837), of the Emperor Louis II. (875), of Boleslas I., King of Poland (1024), of Robert, King of France (1033), of Casimir, King of Poland (1058), of Henry I., King of France (1060), of Pope Alexander III. (1181), of Richard I., King of England (1198), of Philip Augustus (1223), of the Emperor Frederic (1250), of Popes Innocent IV. (1254) and Urban IV. (1264), and of John Galeas Visconti, Duke de Milan. This tyrant was ill when the comet of 1402 appeared. From the time he perceived the fatal star he despaired of life; 'for,' said he, our father on his deathbed revealed to us that, according to the testimony of all the astrologers, at the time of our death a similar star would appear during eight days. I thank my God that He has willed that my death should be announced to men by this celestial sign.' (What monk-like humility! There are, however, people who seriously imagine themselves to be of a different constitution from that of their dependents.) His malady growing worse, he died shortly after at Marignan, on September 3. It was also thought that cometary apparitions coincided with the deaths of Charles the Bold (1476), of Phillippe le Beau, father of Charles V. (1505), of Francis II., King of France (1560), &c. The list might be easily extended. Comets were even invented for the occasion; for example, at the death of Charlemagne (814). And what descriptions! Here, for example, is the account of the historian Nicetas of the horrible aspect of the comet of 1182:

'After the Latins had been driven from Constantinople, they saw a prognostication of the madness and the crimes to which Andronicus was to deliver himself. A comet appeared in the sky; like a winding serpent, it soon extended, coiled on itself, and, to the great terror of the spectators, *it opened a vast mouth*; they might have said that, thirsty for human blood, it was about to satisfy itself.'

bishops, and they advised him to pray, build churches, and found monasteries—which he did. But he died three years later.

The comet of Halley appeared again in 1066, at the time when William the Conqueror invaded England. The chroniclers unanimously write: 'The Normans, guided by a comet, invaded England.' The Duchess-Queen Matilda, wife of William, has represented this comet and the amazement of her subjects on the tapestry (230 feet long) which may be seen at Bayeux. Queen Victoria has in her crown a jewel the design of which was suggested by the tail of this comet, which had the greatest influence on the victory at Hastings.

But the most celebrated of its appearances is that of 1456, three years after the capture of Constantinople by the Turks. Europe was still a prey to the emotion produced by this terrible news; it was said that the church of Saint Sophia was converted into a mosque, and that all the Christian people were either killed or reduced to captivity. Men trembled for the safety of Christianity. The comet appeared in June 1456; it was large and terrible, say the historians of the time; its tail covered two celestial signs—that is to say, 60°; it had a brilliant gold colour, and presented the aspect of a waving flame. They considered it a certain sign of Divine wrath; the Mahomedans saw in it a cross, the Christians a yataghan. In so great a danger Pope Callixtus III. ordered that the bells in all the churches should be rung every day at noon, and he invited the faithful to say a prayer in order to exorcise the comet and the Turks. This custom is still kept up among all Catholic nations, although we have no longer any fear of comets, and still less of Turks. From that time dates the *Angelus*.

This comet, however, formed no exception to the general rule, for these mysterious bodies have had the gift of exercising on the imagination a power which plunged it into ecstasy or terror. *Swords of fire, bleeding crosses, flaming daggers, spears, dragons, mouths*, and other names of the same kind were lavished on them in the Middle Ages and the Renaissance. Comets like that of 1577 appear, however, to justify, by their strange form, the titles by which they were usually saluted. The most serious writers were not free from this terror. Thus, in a chapter on 'Celestial Monsters,' the celebrated surgeon, Ambroise Paré, describes in the most vivid and frightful colours the comet of 1528. 'This comet was so horrible and frightful, and produced such great terror among the populace, that some died of fear; others fell sick. It appeared as a star of excessive length and of the colour of blood; at its summit was seen the figure of a *bent arm* holding a great sword in its hand, *as if about to strike*. At the point there were three stars. On both sides of the rays of this comet were seen a great number of axes, knives, spaces coloured with blood, among which were a great number of hideous *human faces* with beards and bristling hair.'

We see that the imagination has good eyes when it sets to work.

Several well-known personages believed so much in the end of the world in 1528 and 1577 that they bequeathed their goods to the monasteries; without, however, sufficiently reflecting, for doubtless the catastrophe would have been universal. The monks showed themselves to be better philosophers, and accepted the goods of the earth while awaiting the caprices of the sky.

However, the astrological ideas began to be sharply attacked. 'Yes,' said Gassendi, at the beginning of the reign of Louis XIV.—'yes, the comets are really frightful, but through our own folly. We gratuitously fabricate for ourselves objects of panic, and, not satisfied with our real evils, we keep up imaginary ones.'

'Would to God,' said Erasmus, a century later, 'that wars might have no other cause than the spleen of sovereigns excited by some comet! A skilful physician with a dose of rhubarb would soon restore to us the pleasures of peace!'

In 1661 Madame de Sévigné wrote to her daughter:—

> We have here a comet which is of great extent; it has the most beautiful tail which it is possible to see. All the great personages are alarmed, and believe that the sky, much occupied with their ruin, gives them warnings by this comet. They say that Cardinal Mazarin, being despaired of by the physicians, his courtiers believed that it was necessary to honour his death agony by a prodigy, and told him that there appeared a great comet which made them fear. He had the strength to laugh at them, and told them pleasantly that the comet did him too much honour. Indeed, they should say it as much as him, and human pride did itself too much honour in believing that men may have business in the stars when they die.

Twenty years later, however, the nobles of the Court of Louis XIV. were not so wise as Mazarin. We read in the 'Chroniques de l'Œil-de Bœuf,' at the date of 1580:—

> All the telescopes have been pointed for three days to the firmament; a comet such as has not been seen in modern times occupies day and night the learned of our Academy of Sciences. The terror is great in the town; timorous minds see in this the sign of a new deluge, considering, say they, that water is always announced by fire, which will not appear to me a demonstrative reason unless M. Cassini takes the trouble of confirming it. While the timorous make their wills, and, foreseeing the end of the world, bequeath all their goods to the monks, the Court vigorously discusses the question whether the wandering star does not announce the death of some great personage, as it announced, they say, that of the Roman dictator. Some freethinking courtiers laughed yesterday at this opinion; the brother of Louis XIV., who apparently believes that he has become all at once a Cæsar, exclaimed with a very sharp voice: 'Ah, sirs, you and others speak at your ease; you are not princes!'

The learned Bernouilli himself was not free from prejudice, and he perpetuated it by saying that, if the body of the comet is not a visible sign of the wrath of God, *the tail might well be one.*[1] It was to this comet that

[1] This famous comet (that of 1680) deeply impressed all men—Catholics, Protestants, Turks, and Jews were afraid. It even made an impression on the fowls. I have found in the cartoons

Whiston attributed the Deluge, relying on mathematical calculations so abstract that they had little connection with their point of departure.

A contemporary of Newton, at once a theologian and astronomer, this author published in 1696 a 'New Theory of the Earth,' in which he proposed to explain by the action of a comet the geological revolutions and the events in the narrative of Genesis. His theory was at first entirely hypothetical, not applying to any particular comet; but when Halley had assigned to the famous comet of 1680 an elliptical orbit described in 575 years, and when Whiston, finding among the dates of its ancient appearances one of the epochs fixed by chronologists for that of the Deluge, the theologian-astronomer hesitated no longer, but stated his theory precisely, and gave to this comet not only the part of destroyer of the human race by water, but, further, that of an incendiary in the future.

When man had sinned, he said, a small comet passed very near the earth, and, cutting obliquely the plane of its orbit, gave it a motion of rotation. God had foreseen that man would sin, and that his crimes, reaching their consummation, would demand a terrible punishment; consequently He had prepared from the moment of the Creation a comet which should be the instrument of His vengeance. This comet was that of 1680. How was the catastrophe caused? In this way:

Either on Friday, November 28 of the year of sin 2349, or on December 2, 2926, the comet cut the plane of the earth's orbit at a point from which our globe was distant but 9,000 miles. The conjunction happened when they reckon midnight at the meridian of Pekin, where Noah, it appears, lived before the Deluge. Now, what would be the effect of this meeting? A stupendous tide, produced not only in the waters of the seas, but also in those which may be found below the solid crust. The mountain chains of Armenia, which would be nearest to the comet at the moment of conjunction, were disturbed and opened out. And thus 'the fountains of the great deep were broken up.' The disaster did not stop there. The atmosphere and the tail of the comet, reaching the earth and its atmosphere, were precipitated in torrents which fell for forty days; and thus 'the flood-gates of heaven were opened.' The depth of the waters of the Deluge was, according to Whiston, about 6¼ miles.

Now, how will this comet, which the first time drowned the human race, be able to set fire to the earth at a second encounter? Whiston had no difficulty in explaining this: it will arrive behind us, retard the motion of our globe, and change its orbit. 'The earth will be brought nearer to the sun; it will experience a heat of great intensity; it will be consumed.

of the National Library of Paris a print of the epoch with this title: 'Extraordinary Prodigy: How at Rome a Hen laid an Egg on which was engraved an Image of the Comet.' The engraving represents the egg in question under different aspects, and there is an inscription explaining that the fact has been 'certified by the Pope and by the Queen of Sweden.'

Finally, after the saints shall have reigned a thousand years on the earth, regenerated by fire, and rendered again inhabitable by the Divine will, another comet will strike the earth, the terrestrial orbit will be excessively elongated, and the earth, becoming a comet, will cease to be habitable.'

After this we cannot say that comets serve no purpose!

The ignorance of astronomical questions was still so general in the last century, that there was no absurdity so gross that it was not repeated once it had been said, and, above all, once it had been printed. Did they not maintain in 1736 that the sun had retrograded? Did they not add, in 1768, that the planet Saturn, with its rings and satellites, was lost? Everyone believed it, the most respectable periodicals spread this singular news, and sensible men, whose knowledge should have guarded them against such a rumour, echoed the report. Some years after a panic was produced in Paris of which, perhaps, we have never had another example; this reached such a point that the Government was obliged to interfere to put a stop to it, although at that time the indefatigable Messier[1] was discovering comet after comet, and causing these hairy stars to lose the importance attached to their ancient rarity.

Lalande, one of our most illustrious astronomers, had just published a memoir entitled 'Réflexions sur les Comètes.' As he relates himself, he had only spoken of those which, in certain cases, might approach the earth; but people imagined that he had predicted an extraordinary comet, and that this comet would bring about the end of the world. From the highest ranks of society the panic descended to the multitude, and it was generally believed that the fatal comet was on its way, and that our globe would cease to exist. The general alarm assumed such proportions that, by order of the King, Lalande was requested to explain his views in a memoir intended for the public. It was necessary to reassure timorous minds and to make the world resume its plans for the future, which had been for the moment neglected.

We might easily find similar examples in the present century. The fear of comets is a periodical malady, which never fails to return according to circumstances when the appearance of one of these bodies is announced and talked about. An event happened in our time when the fear seemed, so to say, scientifically justified; we refer to the return of the small comet of Biela in 1832.

[1] Messier discovered sixteen comets. His ardour for this kind of research was such that, having lost his wife at the moment when the astronomer of Limoges, Montagne, had in his turn discovered a new comet, he received the condolences of his friends by saying, 'I have already discovered eleven; was it necessary that this Montagne should rob me of the twelfth?' Then, perceiving that they spoke to him, not of the comet, but of his wife, he added, 'Ah, yes, she was a very good woman!' Then he continued to weep for his comet.

This famous discoverer of comets published in 1808 a pamphlet on a comet which he had observed in 1769, with the title: *The Great Comet which appeared at the Birth of Napoleon the Great.*

In calculating the epoch of the reappearance of the new body, Damoiseau had found that the comet would, on October 29, 1832, before midnight, cross the plane in which the earth moves at the only place where a comet would be likely to encounter the earth. The passage of the body would, according to calculation, take place in the plane, but a little inside the earth's orbit, and at a distance equal to four and two-thirds terrestrial radii. As the length of the comet's radius was equal to five and one-third terrestrial radii, it was probable from all the evidence that on October 29, 1832, before midnight, a part of the terrestrial *orbit* would be occupied by the comet.

These results, supported by all desirable scientific authority, were brought by the newspapers to the notice of the public; we may imagine the profound sensation which they produced. It was a fact! the end of time was near! the earth was about to be shattered, pulverised, annihilated by the shock of the comet! Such was the subject of all conversation. The strongest minds were for a moment disturbed.

But a question remained to be asked, and the newspapers had neither stated it nor even anticipated it. At what place in its immense orbit would the earth be found on October 29, 1832, before midnight, at the moment when the comet would cross this orbit at one of its nodes? Calculation very quickly settled this difficulty. Arago wrote in the 'Annuaire' for 1832: 'The passage of the comet will take place very near *a certain point* of the terrestrial orbit on October 29, before midnight; well, the earth will not reach the *same point* till the morning of November 30—that is to say, *more than a month after*. We have now only to recollect that the mean velocity of the earth in its orbit is 1,670,000 miles a day, and a very simple calculation will prove that *the comet will pass at fifty millions of miles from the earth.*'

It happened as had been predicted, and the earth was again released from fear.

The history of the past, we must admit, is always the history of the present. Although the general level of intelligence may be raised, there still remains in the heart of society a sufficiently deep stratum of ignorance in which absurdity, with all the ridiculous and often fatal consequences which it entails, has always a chance of germinating. Thoughtless fear, fear of which the motive is not stated, is one of these consequences. Many of our readers may remember that the return of the comet of Charles V. had been announced for June 13, 1857. On that very day the comet would encounter the earth, and the end of the world would follow. The people of the provinces were veritably seized with panic, and even at Paris we incessantly heard the comet spoken of with dread.[1]

[1] Here is a fact which bears witness to the fear which it inspired. At the same time the planet Venus was in the position where it shines with its greatest light. It was so brilliant that it was

The destruction of the earth by a comet was announced more recently still for August 12, 1872, on the pretended authority of M. Plantamour, of Geneva, who was certainly quite unaware of such an announcement. Some were frightened; but they did not the less live as they had been accustomed, and the fatal date passed without a catastrophe.

We will examine further on, no longer from the legendary point of view of the end of the world, but under an exclusively scientific aspect, what would be the result of the encounter of a comet with our globe.

Eighteen centuries ago, Seneca was more advanced than many of his successors. Alone, or almost alone, that philosopher opposed his powerful logic to the superstitious ideas of his contemporaries and to those of Aristotle, who attributed these bodies to exhalations from the earth. 'Comets,' he says, move regularly in the paths prescribed by Nature;' and, casting a prophetic look into the future, he asserts that posterity will be astonished that his age should have disregarded truths so palpable. He was right against the whole human race, which is nearly equivalent to being in the wrong; and during sixteen centuries more the question made no progress, even in the sixteenth century, which was bold enough to shake off the yoke of authorities otherwise very powerful. Kepler himself after 1600—Kepler the freethinker, the astronomical innovator, the discoverer of the laws which rule the celestial motions—admitted prognostics and cometary influences. However, we cannot reproach with superstitious weakness the man who dared to say to the theologians who attacked the doctrine of Copernicus and Galileo: 'We do not compromise mathematical truths: the axe with which they try to cut iron cannot afterwards make an incision even in wood.'

Observers of the sky, accustomed to the great regularity of the motions of the heavenly bodies, to the calm, the peace which characterise the celestial regions, could not see without surprise and terror bodies which seem to suddenly come to light in all regions of the sky, of which the form and the appendages differ in aspect from other stars, which seem to be followed or preceded by luminous and often immense trains, and of which the course, contrary to that of all other moving celestial bodies, ends by a disappearance as abrupt as their arrival has been sudden. It is not surprising that fear was born from the union of astonishment and ignorance, so natural is it to see prodigies in things which appear extraordinary and inexplicable.

It must be confessed, however, that the apparition of an immense comet like that of 1811 strikes with astonishment all those who behold it Without having recourse to the figures, more or less whimsical, attributed to comets

perceived even in full daylight before sunset. In the fine evenings of February numerous groups were seen in places occupied in looking at Venus, which they took for the comet; some of them, who had doubtless more piercing sight than the others, were even heard to maintain that they could distinguish the tail.

which appeared in the ages when credulity was so intense and the critical faculty so little developed, the simply grand aspect of a celestial visitor of the size of that comet explains and excuses the exaggeration of fears suggesting celestial anger or the demons of the infernal regions. Let us judge each epoch by its own light. We shall see further on that this comet, which our fathers may still remember, did not measure less than 110 *millions* of miles in length.

In order to upset the theory of prodigies it was necessary to find the LAWS of the motion of comets. This is what Newton did in the case of the

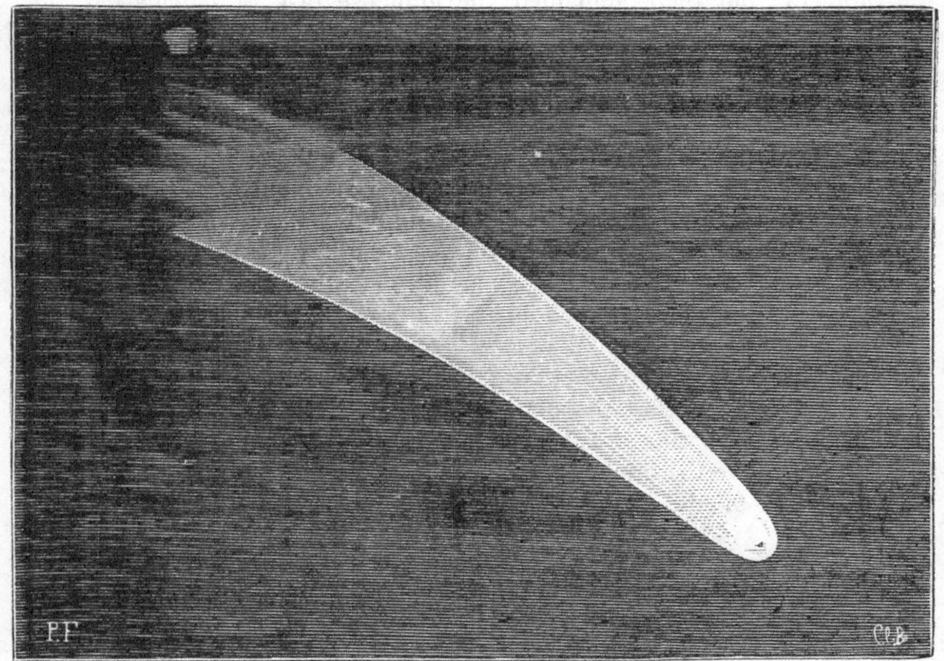

FIG. 204.—THE GREAT COMET OF 1811

comet of 1680. Having ascertained that, according to the laws of universal gravitation, the path of the comet should be a very elongated curve, he attempted, assisted by Halley, his coadjutor and friend, to represent mathematically the course of the new body, and completely succeeded. Halley energetically took up this branch of astronomy, and finding later on that the comet of 1682 was similar in its path round the sun to two comets previously observed, in 1531 and 1607, recognised it as undoubtedly the same comet, which should, therefore, reappear about 1758.

By the theoretical labours of Newton and by the calculations of Halley the prediction of Seneca was fulfilled ; comets, or at least some of them, follow regular orbits. Their return could be foreseen; they ceased to be accidental apparitions; they were true celestial bodies, with a fixed and regular course. The marvellous disappeared, or, to speak more correctly, it was transformed.

Halley calculated, with great pains, that the influence of the planets would delay the next return of the comet, and he predicted it for the end of 1758 or the beginning of 1759. It was necessary, with the improved mathematical formulæ, to calculate exactly the epoch of this return. Clairaut undertook this, and performed in a masterly manner the algebraical part of the problem; but there remained the immense task of calculating the formulæ numerically. Two computers had the courage to do this—the astronomer Lalande and Madame Hortense Lepaute. During six months the two calculators, hardly taking time to eat, put into numbers the algebraical formulæ of Clairaut. Lalande finished the calculation, and found that Saturn would delay its return by 100 days and Jupiter by 518 days, in all 618 days' delay—that is to say, that its revolution would be a year and eight months longer than its previous revolution; and that, in fact, its perihelion passage would take place within a month of the middle of April 1759.

Never did scientific prediction excite more lively curiosity from one end of Europe to the other. *The comet reappeared*; it traversed the course announced among the constellations; it passed its perihelion on March 12, 1759, just a month before the day indicated. 'We have all observed it,' wrote Lalande,[1] 'so that it is beyond doubt that comets are truly planets which revolve, like the others, round the sun.' Halley's comet, in fulfilling the prediction of the astronomers, opened a new era in cometary astronomy.

This prediction was truly worthy of admiration. When we recollect that at this epoch the orbit of Saturn marked the limit of the solar system, we may imagine the boldness which computers then had to suppose a comet going out to such a distance (beyond the orbit of Neptune). The orbit of

[1] 'The universe,' again wrote Lalande in 1759, 'sees this year the most satisfactory phenomenon which astronomy has ever presented to us; an event unique till to-day, it changes our doubts into certainty, and our hypotheses into demonstrations. In fact, although at all times intelligent physicists have hoped for the return of comets, although Newton asserted it and Halley dared to fix the time, all, down to Halley himself, would have appealed to the event and to posterity. What a difference between his position and ours, between the pleasure which this happy conjecture gave him and the advantage we have now in seeing it verified! To combine all the facts which history presents and to deduce the consequences was the work of Halley. To see these consequences justified after more than fifty years by a complete fulfilment is a satisfaction which was reserved for us, and which in more remote times philosophers would have grudged to posterity.

'M. Clairaut,' adds Lalande, 'demanded a month's grace in favour of his theory; the month was found to be exact, and the comet came back after a period of 586 days longer than the last, thirty-two days before the time which he had fixed; but what is thirty-two days in an interval of more than 150 years, of which astronomers had observed roughly hardly the two-hundredth part, and of which all the rest remained beyond our view?'

this comet is now completely determined. It performed one revolution from 1759 to 1835. Its last perihelion passage took place on November 15 of the latter year, which gives 28,006 days for the revolution from 1759 to 1835, instead of the 27,937 days which elapsed between 1682 and 1759; there was an increase of 135 days due to the action of Jupiter, and a diminution of 66 due to Saturn, Uranus, and the earth. The next return should happen on May 24, 1910 this revolution being shorter than the preceding, or 27,217 days only, or 74 years and 6 months, according to the calculations of Pontécoulant. From 1835 to 1873 the comet was increasing its distance from the sun; in the latter year it attained the icy gloom of its aphelion, and since that time *it has commenced its return voyage* towards the brilliant regions of the earth and sun. May we all see it with pleasure in 1910!

Thus have comets passed from the domain of legend into that of reality.

CHAPTER II

MOTIONS OF COMETS IN SPACE

Cometary Orbits. Periodical Comets now known

THE first result of the mathematical analysis of the orbits described by comets in space was, as we have seen, to show that some of the comets at least revolve round the sun like the planets, but follow a much more elongated ellipse; and that all those which pass sufficiently near us to be visible, either to the naked eye or in a telescope, pass round the sun in the part of their orbit which we can observe, and afterwards pass away from the sun to distances greater or less, and perhaps infinite for some of them.

The fine comets which attract public attention by their brightness and the grandeur of their form are rather rare. Thus, in our century there have been but twenty-five which have been visible to the naked eye [1]—those of 1807, 1811, 1812, 1819, 1823, 1830, 1835, 1843, 1844, 1845, 1847, 1850, 1853, 1858, 1860, 1861, 1862, 1863, 1864, 1874, 1881, 1882, 1883–1884, 1886, 1887 (1892 *a*)—and of this number the only really fine and striking comets have been those of 1811, 1843, 1858, 1861, 1862, 1881, 1882, 1887; and, again, those of 1844 and 1887 were visible to the naked eye only in the southern hemisphere.

Further on we will study in detail the most important. Their celebrity depends especially on the effect which they produce when a clear sky coincides with the epoch of their greatest beauty, and when they appear in the evening, attracting all eyes to their mysterious form. A comet which shines before sunrise has but few beholders. They may be admirable, like those of 1861 and 1862, but if they follow a splendid apparition like that of 1858 (Donati's), and cease to be rare, public attention does not grant them more than a look of politeness. A child thinks wonderful an ordinary comet which for the first time gives him an idea of these celestial apparitions, this was how the comet of 1853 struck myself, if I may be permitted to recall a personal recollection, when in the month

[1] We do not include in this number the small comets which just reached the limit of visibility to the naked eye, and which have been only observed by astronomers or persons acquainted with celestial positions.

of August of that year I viewed it from the top of the ramparts of the ancient city of Lingons, shining with its calm light in the northern sky, still illuminated with the warm brightness of the summer twilight. I even made a drawing of its appearance, without suspecting that in the future this little drawing would have the honour of publicity. (My teacher of the sixth class informed me on that evening that the word comet is derived from the Latin *coma*, hair.)

Four principal characteristics distinguish comets from planets: (1) their nebulous aspect and their tails, often considerable; (2) the length of the elliptical orbits which they describe; (3) the inclination of these orbits, which, instead of lying in the plane of the ecliptic, or at least in the zodiac, like those of the planets in general, are inclined at all degrees up to a right angle, and sometimes carry the comets to the polar constellations; (4) the directions of their motions, which, instead of being performed in the same direction as those of planets, are, some direct, others retrograde, and appear to be strangers to any unity of plan. From these circumstances the certain conclusion follows that comets have not the same origin as the planets, that they did not originally belong to the solar system, that they travel through immensity, that they may be transported from one sun to another, from star to star, and that those which revolve round our sun have been caught in their passage by his attraction, having had their course curved and closed by the influence of the planets of our system.

FIG. 205.—THE COMET OF 1853

Comets usually consist of a point more or less brilliant, surrounded by a nebulosity which extends, in the form of a luminous train, in a particular direction. This brilliant point is named the *nucleus* of the comet; the luminous train which accompanies this nucleus is called the tail, and that part of the nebulosity which immediately surrounds the nucleus, setting aside the tail, is named the *coma*. The name of *head* is given to the nucleus and coma combined.

All comets do not appear in the form which we have just indicated.

Some are accompanied by several tails. There are others which have a nucleus and coma without a tail. There are even some which are completely wanting in coma; so that they present the same aspect as the planets, with which they may be confounded. The planet Uranus, discovered in 1781, and Ceres, discovered in 1801, were for some time, as we have seen, taken for comets. Finally, some are seen which are formed solely of nebulosity, without any appearance of a nucleus.

Certain comets have appeared accompanied by tails which extended over a length equal to a quarter, a third, or even a half of the sky; such were those of 1843, 1680,. 1769, and 1861. In 1744 there was seen a comet with six tails; each of these tails had a length of thirty to forty degrees. The whole of the tails occupied in width a space of about 44°. But the different examples which we have mentioned are but exceptions; more often comets have much smaller dimensions.

These cometary trains generally appear straight, or at least, by the effect of perspective, they seem to be directed along the arcs of great circles of the celestial sphere. Some are recorded, however, which presented a different appearance. Thus, in 1689 a comet was seen whose tail, according to the historians, was curved like a Turkish sabre; this tail had a total length of 68°. It was the same with the beautiful comet of Donati, which we admired in 1858, and of which the tail had a very decided curvature.

A comet can only be observed in the sky for a limited time. We perceive it at first in a region where nothing was visible on preceding days. The next day, the third day, we see it again; but it has changed its place considerably among the constellations. We can thus follow it in the sky for a certain number of days, often during several months; then we cease to perceive it. Often the comet is lost to view because it approaches the sun, and the vivid light of that body hides it completely; but soon we observe it again on the other side of the sun, and it is not till some time afterwards that it definitely disappears. Some have been visible at noonday, quite near the sun, like that of 1882, on September 17.

In order that we may exactly understand the motion of comets, it is necessary to form a clear idea of the curve which is called a parabola. We have previously given the definition of an ellipse (p. 26). Suppose that, leaving the focus to the left (F′, fig. 8) and the neighbouring vertex A′ unmoved, we remove the focus F towards the right along the axis prolonged, we shall draw ellipses more and more elongated, which will include the first, and extend farther and farther towards the right. Let us suppose this second focus at an infinite distance (this is an abstraction which calculation permits us to realise); in this case our ellipse has but one focus, its two branches open out without closing again, it ceases to exist as a closed ellipse, and becomes a *parabola*. This definition is not more difficult to understand than others explained in this volume.

Thus the parabola is a curve with one focus, of which the branches separate indefinitely from each other. A comet which follows a parabola passes but once over the path which it describes round the sun ; it comes from the infinite, and returns there. We have seen that the term *eccentricity* is given to the distance from the centre of the ellipse to one of its foci expressed as a fraction of the semi-axis major, or the mean distance. In the circle the eccentricity is nil. In the orbit of Mercury it is equal to two-tenths (fig. 152, p. 347). In the orbit of the small planet Æthra it is over three-tenths. In that of Halley's comet it exceeds nine-tenths. When it reaches unity, the ellipse thus prolonged to infinity becomes a parabola. In the parabola the eccentricity is equal to 1. There is still another curve more open than the parabola : it is called the *hyperbola* ; in this the eccentricity is a number greater than unity.

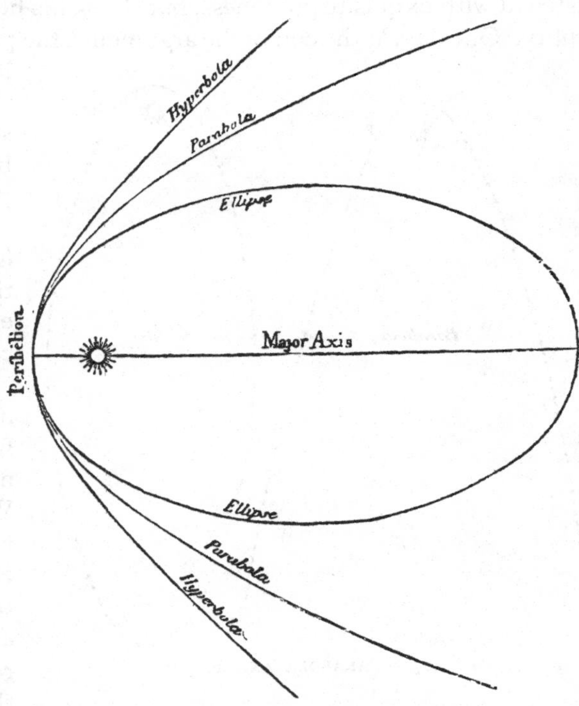

FIG. 206.—FORMS OF COMETARY ORBITS

The curve followed by all celestial bodies round the focus which attracts them depends on the velocity of the motion with which they are animated. A circular velocity is that which causes a body to describe a circle with a uniform motion. A greater velocity makes it describe an ellipse, which is the more elongated the greater the velocity ; if the velocity exceeds the circular velocity in the ratio of 1,000 to 1,414 (which is expressed by $V \times \sqrt{2}$), the ellipse becomes a parabola. A body which is animated with a parabolic velocity at the moment when, reaching its least distance from the sun, it passes the perihelion, is a body which comes from infinity and returns there. A still greater velocity produces an hyperbola.

These explanations are *indispensable* for a real knowledge of the motions of comets. I can hardly imitate here that Academician who, in order to prove a mathematical truth, was contented to give his word of honour, because the intelligence of his pupil was not equal to comprehending the demonstration. This pupil was the Duc d'Angoulême, and I venture to hope that my readers are somewhat superior to him in that respect. We know that when he was nominated Chief Minister of Marine it was perceived with dismay that he could hardly count up to a hundred. The most celebrated geometer of France was at once sent for to instruct him *in the mathematics*, as they said in old times. But it was in vain that he tried to prove the most elementary principles to his august pupil. The latter listened with exquisite politeness, but shook his head with a mild air of incredulity. One day, at the end of the arguments, the poor master exclaimed, 'My lord, I give you my word!' 'Why did you not say so sooner, sir?' said the Duke, bowing; 'I shall never permit myself to doubt it.'

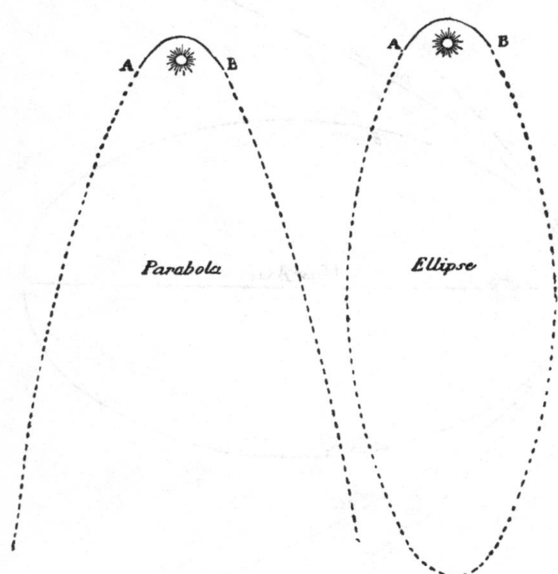

FIG. 207.—PARABOLA AND ELLIPSE

For us, a demonstration is better than an affirmation. In the present case especially, the motion of comets being rather difficult to grasp well, it is important to understand it from a geometrical. and mechanical point of view. We have seen the difference which exists between the ellipse, the parabola, and the hyperbola. We may add that, in the celestial region where we observe the comets—that is to say, in the vicinity of the earth—the part of the curve which we can draw from direct observation of the positions of the comet is exactly that which can be interpreted in two ways, as we may see by fig. 207, in which the full line (A B) of each curve represents the visible part of the comet's orbit, the comet being invisible all along the dotted curve. It is by the *velocity* of the motion that we can determine the nature of the orbit. We have already seen (p. 142) that a projectile discharged from the earth with a velocity of 11,300 metres (7 miles) per second (neglecting the resistance of the air) would never fall back to the earth, because this would be a parabolic velocity relatively to the circular velocity (26,300 feet) which would cause the body to revolve round us like a satellite. The projectile would leave the earth for ever. We have also seen that if the velocity of our planet in its orbit were increased in the same proportion of 1,000 to 1,414, and were 136,500 feet instead of 96,600, we should forsake parabolically for ever the beneficent focus of heat and light round which we revolve. When, therefore, we see in our regions a comet which moves through space with this velocity, we know that it describes a parabola. Now, in the majority of cases, the velocity of comets is precisely of this order, so that very often we may conclude that they will not return again, or that they describe ellipses so elongated that they will only return after thousands of years.

We may consider the comets as small nebulæ, wandering from system to system, and formed by the condensation of the nebulous matter scattered with such profusion in space. When these bodies become visible to us they present so perfect a resemblance to the nebulæ that they are often confounded with them, and it is only by comparing them with a chart of the nebulæ which exist in the region of the sky where they appear, and by following their motion, that we succeed in recognising them. Let us imagine a light mass of nebulous matter travelling in space beyond the solar system ; either by the direction of its motion, or in consequence of the translation of the solar system through immensity, this nebula feels the attraction of our sun and proceeds towards him. If the sun were not surrounded with planets, or if it were motionless, our comet would move regularly, increasing its velocity by degrees, would turn round our focus, following a parabolic orbit, and the velocity which it acquired in approaching the sun would be just sufficient to send it back to infinity along a second branch of the parabola symmetrical with the first. But, on account of the motions of the planets in their orbits, the comet experiences a change of velocity in passing at a certain distance from them, accelerating it or retarding it according to the path which it travels. If the total accelerations produced by all the planets exceed the delays, the comet will leave our system with a velocity greater than the parabolic velocity, and will *never* return, even if it should exist during the whole of eternity. If the retarding forces are, on the contrary, in excess, the orbit will be changed into an ellipse more or less elongated according to the amount of this excess. A planet such as Jupiter, for example, which seems to watch comets like a prey, may transform the orbit into an ellipse of short period and make the comet a permanent member of our system.

When a comet suddenly appears in the sky, we often hear excellent citizens accuse astronomers of having failed in their duty in not announcing it, or they make use of this pretext to express well-felt doubts on the value of astronomical theories. These brave people (of whom many are journalists) only prove by that one thing very clearly that they do not consider what they say or what they write. From what we have just seen, comets being strangers to our system, we can only predict the return of those which go round the sun in a closed orbit of which the elements have been calculated by the aid of one or several previous passages within sight of the earth.

Of the total number of observed comets, we know but thirteen of which the periodicity has been confirmed. These are, in order of date, Halley's comet, of which the periodicity was announced in 1704, and confirmed in 1759 and 1835; that of Encke, of which the periodicity was announced in 1819, and has been verified every three years since, for its revolution is very short; that of Biela, calculated in 1826, and which returned every six and a half years up to 1852; that of Faye, discovered and calculated in 1843, and which regularly returns every seven years; that of Brorsen, calculated in 1846, and which returns every five years; that of D'Arrest, calculated in 1851, and which returns every six and a half years; that of Winnecke, calculated in 1858, and which returns every five and a half years; that of Tuttle, calculated in the same year, and which returned in 1872 and 1885; two comets discovered by Tempel in 1867 and 1873, calculated in those years, and of which the period is five and six years; a comet discovered in 1869 by Tempel in Europe, and in 1880 by Swift in America, and which was found to be the same, with a revolution of five and a half years; a comet discovered by Pons in 1812 and which returned in 1883; and, lastly, a comet discovered by Olbers in 1815, which returned in 1887. The following are the elements of these comets in order of their aphelion distance:—

Table of Periodical Comets of which the return has been observed [1]

No.	Name	Aphelion distance	Period	Perihelion distance	Eccentricity	Inclination	Perihelion passage	Longitude of perihelion
1	Encke	4·097	3·308	0·343	0·845	13	1888, June 27	159°
2	Tempel, 1873	4·665	5·211	1·346	0·552	13	1889, Feb. 2	306°
3	Tempel, 1867	4·897	6·507	2·073	0·405	11	1885, Sept. 25	241°
4	Tempel-Swift	5·163	5·535	1·073	0·656	5	1886, May 9	43°
5	Winnecke	5·582	5·812	0·883	0·727	14	1886, Sept. 4	276°
6	Brorsen	5·613	5·462	0·590	0·810	29	1879, March 30	116°
7	D'Arrest	5·772	6·686	1·326	0·626	16	1884, Jan. 13	319°
8	Faye	5·970	7·566	1·738	0·549	11	1881, Jan. 22	51°
9	Biela	6·182	6·608	0·860	0·755	13	1832, Sept. 23	109°
	Distance of Jupiter: 4·9	to 5·5						
10	Tuttle	10·460	13·760	1·025	0·822	55	1885, Sept. 11	116°
	Distance of Saturn: 9·0	to 10·1						
11	Olbers	33·616	72·63	1·200	0·931	45	1887, Oct. 8	150°
12	Pons-Brooks	33·616	71·48	0·775	0·955	74	1884, Jan. 25	93°
13	Halley	35·411	76·37	0·589	0·967	162	1835, Nov. 15	166°
	Distance of Neptune:	29·8 to 30·3						

All the above have direct motion, with the exception of Halley's comet, which is retrograde.

The examination of this little table is not devoid of interest. We notice at first that the periodical comets now known are distributed in three groups, of which the first has an aphelion distance of 4 to 6 (the radius of

[1] [To the above should be added Wolf's comet (1891 II), of which the aphelion distance is 5·600 and period 6·821 years, and Finlay's, with an aphelion distance of 6·063 and period 6·622 years.]

the terrestrial orbit being taken as unity), the second of 10·46, and the third of 33 to 35. Now, these distances correspond precisely with the orbits of three planets: Jupiter, whose distance from the sun varies from 4·9 to 5·5; Saturn, whose distance varies from 9·0 to 10·1; and Neptune, which revolves between 29·8 and 30·3.

This grouping cannot be considered as accidental. It shows us that the attraction of the planets has played a preponderant part in the form of the cometary orbits, and it is on this very remarkable agreement

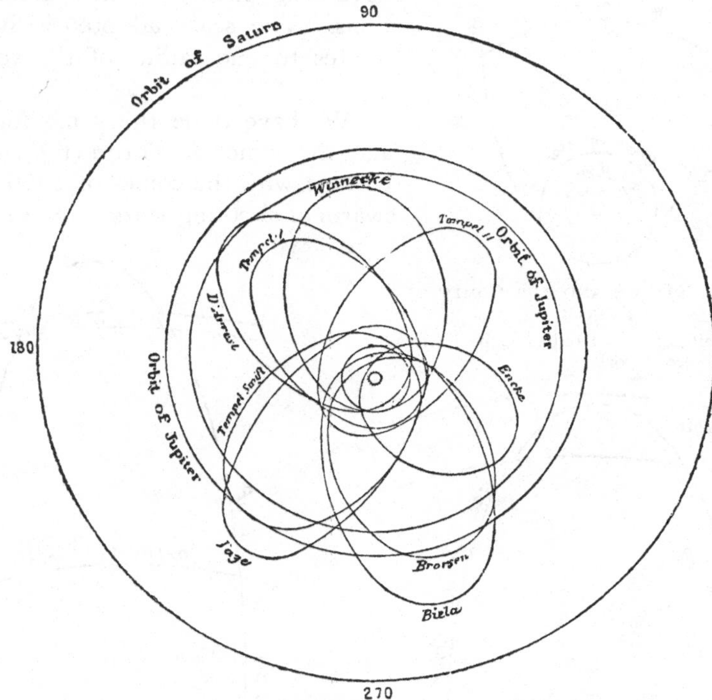

FIG. 208.—ORBITS OF THE NINE COMETS CAPTURED BY JUPITER
Scale: 5 millimetres = 1 radius of the Earth's orbit

that we based our opinion in the preceding chapter, when considering the existence of a trans-Neptunian planet at the distance 48, which borders the orbits of the third comet of 1862 and the swarm of shooting stars of August 10, as well as those of the comets of 1532 and 1661. Uranus has captured two at least: (1) the comet of 1866 and the swarm of shooting stars of November 13; (2) Comet I. 1867.

In order to give a still better account of this state of things we have drawn (fig. 208) on the same diagram: (1) The present orbit of Jupiter, marked by two circles, one at the perihelion distance, the other at that of the aphelion (the longer axis is displaced from age to age in space, and the value of the eccentricity itself varies, so that the zone swept over by the body of Jupiter is larger still); (2) the present orbits of the nine comets captured by this giant of the solar system (we say present, because these orbits themselves vary from age to age on account of planetary perturbations); the aphelia have been placed in their present positions. The scale adopted is five millimetres to one radius of the terrestrial orbit.

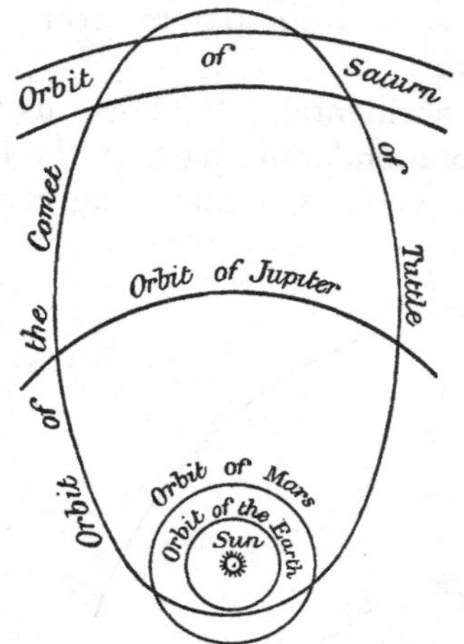

FIG. 209.—SATURN AND THE COMET OF TUTTLE
Scale: 5 millimetres = 1 radius of the Earth's orbit

We have done the same for Saturn and the comet of Tuttle (fig. 209); for Uranus, with the comet of 1866 and the swarm of shooting stars of November 13

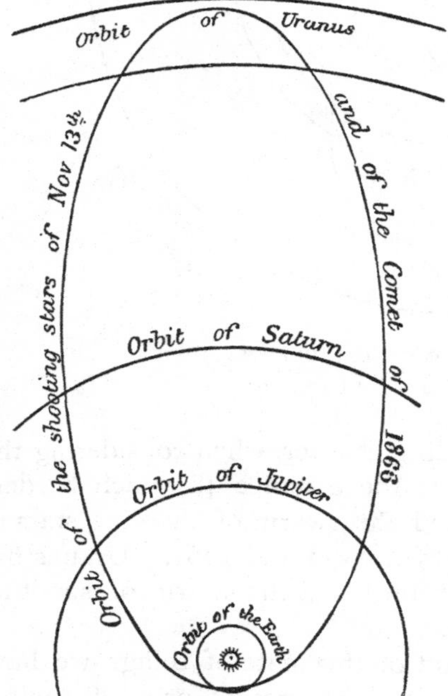

FIG. 210.—URANUS, THE COMET OF 1866, AND THE SWARM OF SHOOTING STARS OF NOVEMBER 13
Scale: 4 millimetres = 1 radius of the Earth's orbit

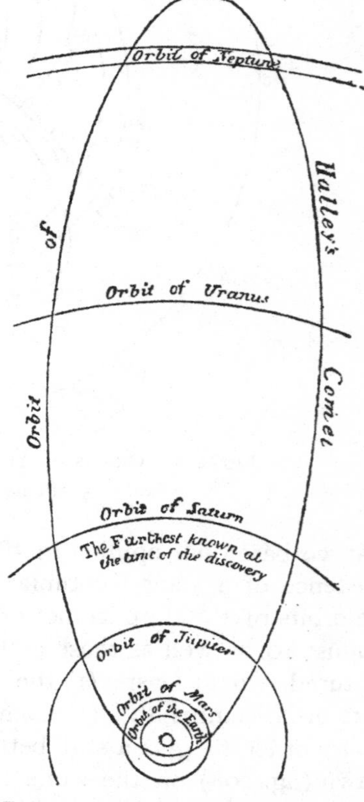

FIG. 211.—HALLEY'S COMET, PONS'S COMET, AND THE ORBIT OF NEPTUNE
Scale: 3 millimetres = 1 radius of the Earth's orbit

(fig. 210); for Neptune, with the comets of Pons and Halley united for greater convenience in the same section of the orbit (fig. 211).

The preceding table suggests other remarks still. All these periodical comets, with the exception of Halley's comet, revolve in the direct sense—that is to say, from west to east—like all the planetary motions. Further, those which are connected with Jupiter have in general an inclination of about 13°, which can hardly be merely an accident.

These periodical comets are in general, like most comets, invisible to the naked eye. However, there is among them a fine comet, that of Halley, of which we have already spoken above in connection with the conquest of England by William Duke of Normandy; and a rather fine one, that of Pons of 1812 (which should not be confounded with that of 1811), which

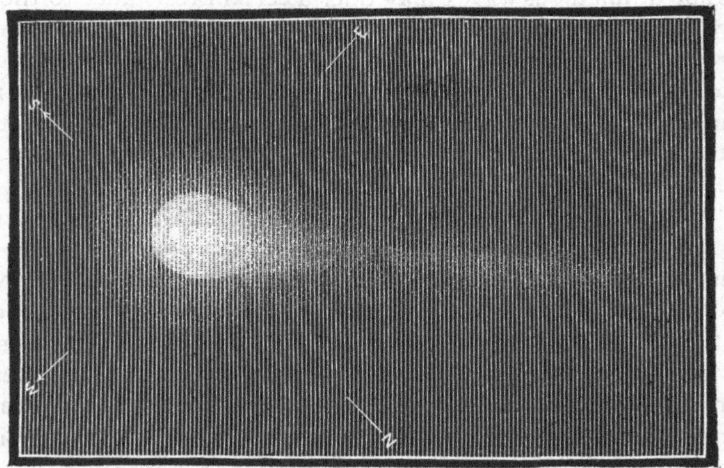

FIG. 212.—THE COMET OF PONS 1812, WHICH RETURNED IN 1883

was visible to the naked eye at its last return, from December 1883 to March 1884. At its maximum brightness it resembled a star of the 2nd magnitude provided with a tail of eight degrees in length. It underwent several singular transformations of brightness. Fig. 212 shows its appearance on December 17, 1883.

Halley's comet, much more brilliant, showed an aspect similar to that of the comet represented in fig. 203, p. 476.

Several of these comets have an interesting astronomical history. The first, discovered by Pons, porter of the Marseilles Observatory, on November 26, 1818, was found to be identical with those of 1786, 1795, and 1815. Calculations made by the astronomer Encke, of Berlin, showed that they were one and the same comet, of which the period of revolution was only 3 years and 106 days, or about 1,212 days. This period varies by several days, according to the perturbations of the planets. Since 1818 this telescopic body has always returned punctually to the rendezvous; but--a very curious fact—at each of its revolutions there is a slight diminution of the period, amounting to a tenth of a day, or about two hours and a half. At the end of the last century this period was nearly 1,213 days; it was 1,212 in 1818, 1,211 in 1838, 1,210 in 1858, and it is now 1,209, a correction being made for the perturbations of the large planets. To what cause is this diminution due? If it continues, the comet would go on approaching the sun by degrees, following a slowly contracting spiral, and it would end by falling into it and being consumed. That is, however, perhaps the fate of a large number of comets. It has been supposed that the interplanetary medium, the ether, which serves as a vehicle for the transmission of luminous waves, may offer a certain resistance to the motion of cloudy bodies so light as comets, and thus cause the observed diminution. But there is one objection to this, which is, that the other comets of short period do not show indications of a similar diminution. It seems to me that the influence of the small planets in the zone in which the comet moves during two years out of three may not be insignificant, and may explain this acceleration of its motion.

This comet presents the aspect of a feeble nebulosity accompanied by a slight tail; it was visible to the naked eye as a star of the 6th magnitude, vague and diffuse, in 1828, 1838, and 1847. Attempts have been made to estimate its mass, but it exercises no effect on the planets; on the contrary, it is the comet alone which is disturbed. It is so light that stars can be distinguished through it. However, from the intensity of the solar light which it reflects, we can believe that this mass is not altogether insignificant, and M. Roche has even estimated it at one-thousandth of that of the earth. This is an open question.

The ninth of our periodical comets is still more curious. Discovered on February 27, 1827, by Biela, and ten days later at Marseilles by Gambart, who calculated the elements and recognised that it was the same as that of 1772 and 1805 (it should, then, bear his name, if we follow the precedent of the principle which determined the name of Halley's comet), it returned six and a half years later, in 1832, and it was the announcement of this return which caused the public fear of which we have spoken in the preceding chapter. In fact it crossed, as we have seen, the plane of the terrestrial orbit at the respectable distance of 50 millions of miles from the earth; but if there was any danger in this meeting, it was rather for it than for us, for it was certainly strongly disturbed in its course. It returned in 1839, but under conditions too unfavourable to enable it to be observed—in the month of July, in the long days, and too near the sun. It was seen again in 1845 on November 25, near the place assigned to it by calculation, and its course was duly followed. Everything went on to the general satisfaction, when—unexpected spectacle!—on January 13, 1846, *the comet split into two!* What had passed in its bosom? Why this separation? What was the cause of such a celestial cataclysm? We do not know; but the fact is that, instead of one comet, two were henceforth seen, which continued to move in space like two twin sisters—two veritable comets, each having its nucleus, its head, its coma, and its tail, slowly separating from each other; on February 10 there was already 60,000 leagues (about 149,000 miles) of space between the two. They would seem, however, to have parted with regret, and during several days a sort of bridge was seen thrown from one to the other. The cometary couple, departing from the earth, soon disappeared in the infinite night.

They returned within view of the earth in the month of September

FIG. 213.—THE DOUBLE COMET, NOW LOST

1852; on the 26th of this month the twins reappeared, but much farther apart, separated by an interval of 500,000 leagues (about 1,250,000 miles).

But this is not the strangest peculiarity which this curious body presented to the attention of astronomers. The catastrophe which was observed in 1846 was only a presage of the fate which awaited it; for now its existence is merely imagined, the truth being that *this comet is lost.* Since 1852 all attempts to find it again have been unavailing; according to its elliptical motion, it should have returned within sight of the earth in 1859, 1866, 1872, 1877, and 1885. It has certainly not returned. The observer on the look-out along the course of the comet now finds himself in the same perplexity as the station-master who does not see the expected train arrive; here, the pointsman may make a mistake and the chief at the starting-point may delay; but a comet should not, cannot, go astray. A serious accident must, then, have happened to it—a very serious accident assuredly, since it no longer exists.

A similar accident happened in 1779 to the comet of Lexell; but here the cause is known, for it was led astray by Jupiter, like a bat which strikes its head against a wall. This comet, observed in 1770, was moving in an ellipse, and should have returned in 1781, but it approached so near to Jupiter that its fate seemed imminent. These fears were not exaggerated: the attraction of the immense planet, having opened very much the branch of the ellipse which it followed, performed precisely the office of a pointsman on the railway—it shunted it on another road, and it has been, not exactly lost, but led astray. This is less serious than the preceding catastrophe. This indiscreet or awkward comet was destined, indeed, sooner or later, to a similar fate; on June 28, 1770, it approached the earth to within six times the distance of the moon, and twice it passed near Jupiter's system of satellites, in 1767 and 1779.

Another comet certainly appeared to be periodical—that discovered at Rome, in 1844, by De Vico, which should have returned in 1850, 1855, 1861, 1866, &c., but has never been seen since.

But the comet of Biela could not have encountered in its course either Jupiter or any large planet; at the most it might have been caught in passing near one of the small planets, but this is almost impossible; and besides, these little planets are themselves so light, as we have seen, that they could not have prevented it from pursuing its course.

To be lost is interesting, especially for a comet; but this, doubtless, was not enough, for it reserved for us a still more complete surprise. Its orbit intersects the terrestrial orbit at a point which the earth passes on November 27. Well, nothing more was thought about it, it was given up as hopeless, when, on the evening of November 27, 1872, there fell from the sky a veritable *rain of shooting stars.* The expression is not exaggerated; they fell in great flakes; lines of fire glided almost vertically in swarms and showers, here with dazzling globes of light, there with silent explosions recalling to mind those of rockets; and this rain lasted from seven o'clock in the evening till one o'clock next morning, the maximum being attained about nine o'clock. At the Observatory of the Roman College 13,892 were counted; at Montcalieri, 33,400; in England a single observer counted 10,579, &c. The total number has been estimated at *a hundred and sixty thousand.* They all came from the same point of the sky, situated near the beautiful star Gamma of Andromeda.

On that evening I happened to be at Rome, in the quarter of the Villa Medicis, and was favoured with a balcony looking towards the south. This wonderful rain of stars fell almost before my eyes, so to say, and I shall never cease regretting not having seen it. Convalescent from a fever caught in the Pontine Marshes, I was obliged to go into the house immediately after the setting of the sun, which on that evening appeared from the top of the Coliseum to sleep in a bed of purple and gold. My readers will understand what disappointment I felt next morning when, on going to the Observatory, Father Secchi informed me of that event! How had he observed it himself? By the most fortunate chance: a friend of his, seeing the stars fall, went to him to ask an explanation of such a phenomenon. It was then half-past seven. The spectacle had commenced, but it was far from being finished, and the illustrious astronomer was enabled to view the marvellous shower of nearly *fourteen thousand* meteors.

This event made a considerable stir in Rome, and the Pope himself did not remain indifferent; for, some days afterwards, having had the honour of being received at the Vatican, the first words that Pius IX. addressed to me were these: 'Have you seen the shower of Danaë?' I had admired some days before, in Rome, some admirable 'Danaës' painted by the great masters of the Italian school in a manner which left nothing to be desired; but I had not had the privilege of finding myself under the cupola of the sky during this new celestial shower, more beautiful even than that of Jupiter.

What was this shower of stars? Evidently—and this is not doubtful—the encounter with the earth of myriads of corpuscles moving in space along the orbit of Biela's comet. The comet itself, if it still existed, would have passed twelve weeks before. It was not, then, to speak correctly, the comet itself which we encountered, but perhaps a fraction of its decomposed parts, which, since the breaking-up of the comet in 1846, would be dispersed along its orbit behind the head of the comet.[1]

Such is the history of this singular body. The fact of the division of a comet into two or more parts is not unique in history. The Greek historian Ephorus relates that in the year 371 B.C. a comet was divided into two bodies, each following its own course. Seneca attributed this account to an error; but Kepler, a better judge in such matters, remarked that there was nothing impossible in it, and that a similar division had taken place in the second comet of 1618. Here, then, are several analogous facts. The Chinese astronomers have recorded in their annals three comets united, which appeared in the year 896, and travelled along their orbits in consort. The nucleus of the comet of 1652 divided into five or six parts, which

[1] A German astronomer, Klinkerfues, believed that the comet itself had encountered the earth, and sent to the other side of the globe, to Madras, a telegram thus worded, puzzling to the telegraph clerks: '*Biela touched earth on the 27th; search near Theta Centauri.*' The Madras astronomer, Pogson, looked in the place indicated, and actually saw a comet, but bad weather prevented him from observing it; so that nothing could be learned from it in order to complete the preceding history.

No doubt can remain of the identity of this swarm of shooting stars with the comet of Biela. On November 27, 1885, the same encounter occurred; a magnificent shower of stars was observed all over Europe just at the moment when the earth crossed the comet's orbit. (See the details of this observation in our *Revue d'Astronomie*, 1886, pp. 19-35.)

showed a density a little greater than the rest of the comet; and similar observations were made on the comets of 1661 and 1664. Four nuclei were distinguished, four points of condensation, in Brorsen's comet (the sixth of our periodical comets) on May 14, 1868. The comet of 1860 was seen perfectly in Brazil by Mr. Liais, and at the time it disappeared the principal nucleus showed three centres of condensation. In 1889 a small telescopic comet showed itself divided into three. Thus it is certain that comets may divide into several parts, that they may even be broken up into small fragments, and that shooting stars may represent the *débris*.

To those comets of which the return has been observed we may add as periodical comets those for which calculation gives the orbits as ellipses of more or less length, but of which the return has not been observed. These are more numerous than the preceding; but their future course is not certain; for if a regular comet can become lost and disappear, a similar fate is, with stronger reason, to be feared for those of which the orbit has only been guessed at by a single examination of the small portion of the ellipse near the sun, which may so easily be confounded with a parabola. Then, too, the attraction of Jupiter, among others, may entirely transform the orbits of those which approach too near. Thus the orbit of Brorsen's comet was considerably modified in 1760 and 1842 by Jupiter, and it will be again in 1937; it may even, in the future, become a parabola. It would, then, be superfluous to give here the periodical comets whose return has not been observed.

It is not necessary to exaggerate the value of observations. To affirm, as has been done, for example, that the second comet of 1864 goes out to exactly 80,485 times the distance of the earth from the sun, and that it will return to this region of space in the year of grace 2801864, would be to surpass, not, assuredly, the powers of the intrepid calculator, but those of simple common-sense. We have seen above what snares, known and unknown, are laid for comets.

Observers expected in 1848 the return of the comet of 1556, named after Charles V., of which the period had been fixed at 292 years, and which appeared to be the same as that of 1264. It coincided the first time with the death of Pope Urban IV., and the second time with the abdication of Charles V., and it would have coincided in 1848 with the last days of the French monarchy. It did not return, and, notwithstanding all the delays and all the admissible excuses, we now no longer expect it.

Among the great comets observed, the most important were those of 1680, 1811, 1843, 1858, 1861, 1874, 1880, 1881, and 1882.

The first of these, according to the earlier calculations of Halley, should have previously appeared in the year 1106 of our era, in the year 531 B.C., and in the year 43 B.C., the year of the death of Julius Cæsar, whose deified spirit it represented; at the epoch of the Trojan War; and, going back still farther, at the epoch of the Biblical Deluge, of which it would have been the direct cause, according to Whiston, as we have seen. But it was not the romance of this astro-theologian that made this comet celebrated: it was the calculations of Newton, thanks to whom the theory of comets was worked out; and, above all, it was the astonishing, unheard-of, extraordinary—I might almost say incomprehensible—fact, that it passed close to the sun without being consumed, and without being caught in its passage by the fiery focus of attraction of our system. Indeed, it passed round the solar star on December 8, 1680, at the small perihelion distance of 0·0062—at only six-thousandths of the earth's distance, or 230,000 leagues (about 570,000 miles)—moving with a velocity of 480,000 leagues an hour, or more than 500,000 metres (310 miles) a second! At this distance from the radiant star—equal to only the 160th part of that which separates us—it had to endure a heat equal to that which we should receive if we had over our head at noon on a summer's day not only 160 suns, but 160 × 160, or 25,600! This is a heat 2,000 times greater than that of red-hot iron. A globe of iron equal to the earth in volume and raised to this temperature would take 50,000 years to cool, and some theorists who imagine comets habitable might suppose that, in thus passing the sun's vicinity, stores of heat would be laid up for their long and rigorous winter. But, in reality, they move so fast that they have not time to receive a very intense heat. This immense comet of 1680, whose tail extended over a length of 149 millions of miles, goes out to 855 times the distance of the earth from the sun, or to 79,400 millions of miles, and its probable period is eighty-eight centuries—forty-four centuries for *going*, and as many for the *return*. Here we are very far from even the long calendar of Neptune.

But the comet of 1843 was more astonishing, more incomprehensible still, in its course. Its perihelion distance, determined with an absolutely certain precision, is but 0·0055—that is to say, 201,250 leagues (500,000 miles) from the centre of the solar sphere; so that the comet passed at only 31,000 leagues (77,000 miles) from the fiery surface of the day star, thus certainly passing through the hydrogen atmosphere of which the corona in total eclipses has revealed the existence. From surface to surface there was a distance of not more than 13,000 leagues (32,300 miles). We have seen above that the solar furnace darts out all round it explosions, several of which have been measured up to 200,000 miles in height. How was it that this imprudent celestial butterfly was not burnt, consumed in these flames, of which the inconceivable heat rises to several hundreds of thousands of degrees, and which, united to the tremendous power of the solar attraction, ought to have seized, broken up, and annihilated the poor celestial adventurer? There is in that region a temperature at least 30,000 times greater than that which we receive from the blazing star.

However, the celestial visitor came out safe and sound, without being at all disturbed in its majestic flight.

Le vrai peut quelquefois n'être pas vraisemblable.

This event, of which the consequences might have been so dramatic from the point of view of the unalterable order and harmony of the heavens, occurred on February 27, 1843, at $10^h\ 29^m$ (Paris mean time). Borne along in its rapid flight, the comet took but two hours—from half-past nine to half-past eleven—to pass round the whole solar hemisphere turned towards its perihelion. It flew, then, with a velocity of more than 550,000 metres (342 miles) a second. This is the greatest velocity which we have measured in the whole universe. Behind it, relatively to the sun, stretched a tail of 80 millions of leagues (198 millions of miles), thus exceeding more than double the distance of the earth from the sun. As to the velocity of the extremity of the brandished tail—always remaining opposite to the sun, owing to the course of the comet in space—it exceeds everything which can be imagined,[1] and it seems to me to lead to the conclusion that these cometary tails are not substantial, but merely represent a state of the ether set in a particular undulatory motion by the influence of the comet. We shall discuss this interesting subject further on.

This marvellous and glowing child of space showed itself for the first time at noonday on February 28, by the side of the sun, and in spite of its light. (A century before, the comet of 1743 was likewise visible at noonday; it was the same with those of 1547, 1500, 1402, and 1106.) No one had seen it coming; it was perceived for the first time, on February 28, at Parma, Bologna, in Mexico, at Portland (United States), in full sunshine, at 1° 23′ to the east of the sun's centre, with a tail of 4° to 5° in length, which lost itself in the atmospheric light. The following day (March 1) the brilliant comet was seen from Copiapo (Chili), with a tail of 30° in length, shortened, of course, by the brightness of the twilight. On March 4, on the equator, a ship captain measured the tail and found it 69°. At Paris it was seen for the first time on the 17th, and the tail was not measured till the 18th; it was 43° long and only 1° 2′ wide, measurements

[1] Reducing by more than one-half the observed dimensions of the tail, and supposing that at the moment of the perihelion passage it ended at the distance of the earth, this extremity would have passed over at least half the circumference of the terrestrial orbit, or 288 millions of miles, in two hours; the velocity would, then, have been 16,111 leagues, or more than 64 millions of metres (40,000 miles) a second! This is difficult to believe. Calculation proves besides that, even in the vicinity of the sun, a moving body animated with a velocity of 608,000 metres (377 miles) a second would be driven into a parabola, and would pass away indefinitely from the sun. The particles of the tail of the comet of 1843 would, therefore, cease from that time to belong to that body, of which the motion is elliptical—that is to say, that almost the whole of the tail would escape and be scattered in space. Now, we observe nothing of the sort; these tails are not thus broken off; they present a rigid aspect, especially in the two famous comets of 1680 and 1843. How can we admit such a rotation?

which correspond to 6,000,000 leagues (15,000,000 miles) by 130,000 (322,000 miles). Arago, who measured it on that date, supposes that it may have had the same length on the day of perihelion ; and this is probable, since in the region of their perihelia comets present in general the longest trains ; but how is it that he did not remark the practical impossibility which has occurred to us ?

No one having seen the comet before its perihelion passage, it is almost certain that its grandeur dates only from its point of nearest approach to

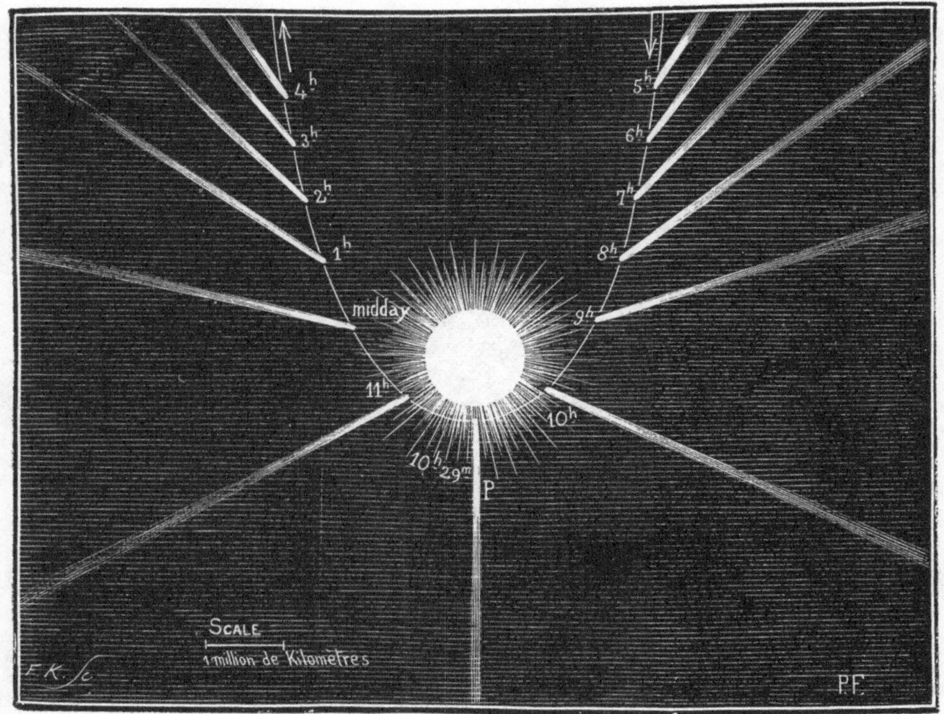

Fig. 214.—Passage of the Comet of 1843 close to the Sun, February 27, $10^h\ 29^m$

the radiant star. (We may even remark that, if it had been shot from the sun on February 27, at $10^h\ 29^m$, with a velocity of about 600,000 metres (372 miles), its appearance and its course would nearly correspond with all the observations made. The right half of fig. 214 is purely theoretical.

The comet of 1680 was seen before its perihelion passage, on November 14, at Coburg. Its tail was straight, like that of 1843. But evidently it was not the same tail which was seen before and after the perihelion passage.

After having passed without accident from the midst of the scorching heat of its perihelion, the enormous comet of 1843, slackening its course took its flight into space; in a single day (February 27) its distance from the sun's centre varied in the ratio of 1 to 10; it passed in view of the inhabitants of Mercury, Venus, and the Earth, disappeared from our sight, and withdrew to the distances of Mars, Jupiter, and Saturn. If, as is probable, notwithstanding the preceding hypothesis, the comet arrived incognito and follows an orbit described in 376 years, it continues to depart, and will arrive in the year 2031 at the extremity of its course, at 104 times the distance we are from the sun—that is to say, more than three times the distance of Neptune; it will then resume its return voyage, to rush again

FIG. 215.—THE GREAT COMET OF 1843

round the sun about the year 2219, perhaps to be entirely consumed this time. Three months and a half after its perihelion passage, in the month of June 1843, a year of minimum of solar spots (see p. 282), there was noticed on the sun, with the naked eye, one of the largest and most surprising spots which had ever been seen; its diameter was 119,000 kilometres (74,000 miles), so that its surface much exceeded that of the earth; it remained visible to the naked eye for a whole week. According to all probability, this spot did not belong to the regular cycle of solar spots, and it may have been produced by the fall into the sun of an enormous meteorite (forming part of a train of shooting stars following the orbit of the comet of 1843) which, having passed a little nearer to the sun than the head of the comet, had been caught in its passage.

To this curious comet of 1843 we may add those of 1880, 1882, and 1887, which have singularly resembled it in their aspects and their orbits, and which must be very near relations. The great southern comet of 1880, of which the tail reached a length of 40°, grazed the flames of the sun on January 27; it was discovered on the 31st, as it left the sun's

FIG. 216.—APPEARANCE OF THE GREAT COMET OF OCTOBER 9, 1882, AT 4h A.M.

vicinity. The great comet of 1882, seen at noonday near the sun on September 16, 17, 18, and 19, passed in front of the solar disc on the 17th, but in an invisible manner, like a transparent flame; it passed through the solar atmosphere with a velocity of 480,000 metres (298 miles) a second. It was observed up to the month of June. Fig. 216 represents a drawing which I made on October 9, and fig. 217 shows its apparent course in the sky; we see how, on September 17, it went round the sun.

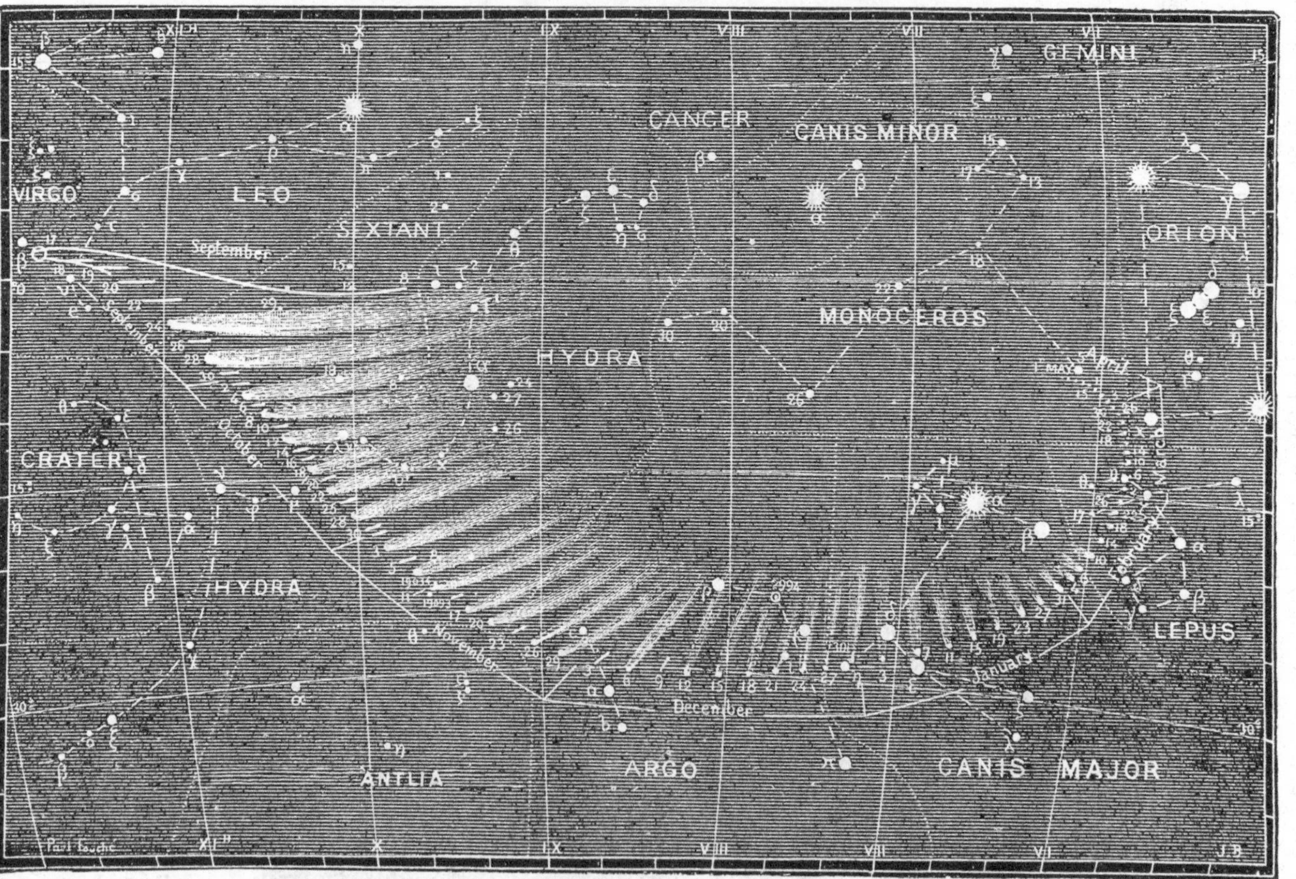

Fig. 217.—Path on the Celestial Sphere of the Great Comet of 1882

Lastly, the singular southern comet of 1887 (fig. 218) likewise passed very near the sun on January 11. It had no nucleus.

We see what unexpected interest is presented by the study of these hairy stars, which were formerly the terror of humanity, and which, in the

FIG. 218.—THE GREAT SOUTHERN COMET OF 1887

eyes of some modern astronomers, have all at once fallen below zero, some calling them 'visible nothings,' others 'nihilités chevelues.' They are perhaps destined to reveal to us many mysteries with regard to the origin and the end of things.

Two other comets of the preceding list are still particularly interesting —those of 1811 and 1858. If the former could relate its history, it would remind us that at its last appearance Europe was covered with soldiers of different colours, with bayonets, cannon, and tents, a skilful strategist being then occupied in the conscientious extermination of five millions of men; that at its penultimate passage, in the year 1254 B.C., the civilised world was then in arms, on fire with battles, in the famous Trojan War, under the pretext of the abduction of a young lady; that at its ante-penultimate voyage, in the year 4320 B.C., it saw Egypt bristling with men armed with knives, javelins, spears, and swords, tearing each other in the defiles, while armies of slaves, ruled by the lash, raised the Pyramids; that previously, about the year 7400 B.C., Asia appeared to it covered with wild hordes, fighting with clubs and slinging-stones in the conquest of China, under the direction of princes mounted on elephants; that still farther back, about the year 10450, it had seen companies of savage men, holding in their hands slings, stone axes and hammers, and spears pointed with stone, fighting in the forests for the possession of a deer; that some cometary years before, twenty or thirty thousand terrestrial years previously, the great apes, which it had observed till then at the head of terrestrial animal life, appeared to be slightly transformed, a little more upright, a little larger, a little less hairy, and a little more sociable, but still vicious and barbarous; and thus, going farther and farther up the stream of time, the comet of 1811 might relate a history of our planet full of glory for those who love battles. This great comet, in which the Russians saw with terror the omen of the terrible war which was to bring such disasters upon France, measured no less than 450,000 leagues (1,118,000 miles) in diameter at the head. The nucleus, more luminous, surrounded by this colossal nebulosity, measured 1,089 leagues (2,700 miles). The tail extended to a length of 44 millions of leagues (109 millions of miles). The appearance of this comet is represented in fig. 204.

Let us further add that one of the finest comets of our century was the great comet of 1858, discovered on June 2 of that year by my lamented friend Donati, and visible to the naked eye in September and October. Its tail attained an angular length of 64°, which was equivalent to 22 millions of leagues (54½ millions of miles). Comets have been seen to exceed 90° in length (those of 1680, 1769, 1264, 1618, and 1861), so that the head might be below the horizon, the extremity of their phosphorescent tail being still in the zenith. Its nucleus attained 900 kilometres (559 miles) in diameter, and showed rapid variations in size. Its period seems to be 1,950 years. We have reproduced in fig. 219 a drawing made by Mr. R. A. Proctor of this comet as it appeared on September 24, 1858.

The comet of 1861, which may be considered a rival of the preceding, suddenly appeared to the eyes of all Europe on Sunday, June 30, above the place where the sun had set. Its tail attained 118°! It was then not far from the earth, and its real length was but 17 millions of leagues (42 millions of miles). Its head changed wonderfully in aspect, as we shall see in the following chapter, and its study has permitted us to penetrate further into the physical examination of these strange bodies. Its period of revolution appears to be 422 years.

The most recent fine comets which have been visible in Europe were those of 1874, 1881, and 1882. We have already spoken of the last. That

FIG. 219.—DONATI'S COMET ON SEPTEMBER 24, 1858

of 1881 was visible to the naked eye from June 23 to August 12, and with an opera-glass till September 4.

Perhaps we may also refer to the little comet of 1886, visible to the naked eye in the morning sky from April 18 to 28. But it was only visible to zealous observers. The conditions were unfavourable: three o'clock in the morning and moonlight.

This general account of comets, their principal characteristics and their motions in space, will be more complete if we add that the comets of short period, of which the return has been observed, and the comets of long period, which also describe closed orbits round the sun placed at one of the foci of their immense ellipses, constitute but a part, doubtless the most interesting, but numerically the smallest, of observed comets. In fact, in the annals of astronomy, from the Chinese, Chaldean, and Greek astronomers, who, several centuries before our era, kept a register of cometary apparitions, down to the time when we write these lines, there have been seen, either with the naked eye or in astronomical instruments since their invention (1609), 806 comets, of which the following table is a curious list :—

STATISTICS OF COMETS SEEN

—	Comets observed	Recognised identical Comets or reappearances	Different Comets	Calculated Orbits
Before the present era	79	1	78	4
1st century	22	1	21	1
2nd "	22	1	21	2
3rd "	39	2	37	3
4th "	22	1	21	—
5th "	19	1	18	1
6th "	25	1	24	4
7th "	29	2	27	—
8th "	17	1	16	2
9th "	41	—	41	1
10th "	30	3	27	2
11th "	37	2	35	4
12th "	28	1	27	—
13th "	29	3	26	3
14th "	34	3	31	7
15th "	43	1	42	12
16th "	39	4	35	13
17th "	32	5	27	20
18th "	72	8	66	64
19th " (1801–1885) . .	270	68	202	249
Totals . .	929	109	822	392

If we deduct from the total number of observed comets (929) those which have returned several times, we obtain the number of 822 distinct comets. But it must be remembered, in order to interpret this number, that, up to the sixteenth century, all the comets were observed with the naked eye, while since the invention of the telescope a large number have been found with the aid of that instrument. The preceding table gives to the year 1600 only the brightest comets ; now it has been ascertained that the telescopic comets, either too small or too distant to be visible to the naked eye, are much more numerous ; thus, for example, in our century, of 270 observed comets, only 25 could have been discovered with the naked eye, while 270 have been observed with the telescope. On an average, six or seven are observed each year : 1881, seven new and one periodical ; 1882, five new ; 1883, two new and one periodical ; 1884, four and two ; 1885, seven and one ; 1886, four and one ; 1887, five and one ; 1888, four and two; 1889, seven; [1890, five new and one periodical; 1891, two new and three periodical; 1892, six new and one periodical.—J. E. G.][1]

These general statistics bring us to the question so often asked: 'How many comets are there in the sky?' 'As many as the fishes in the ocean,'

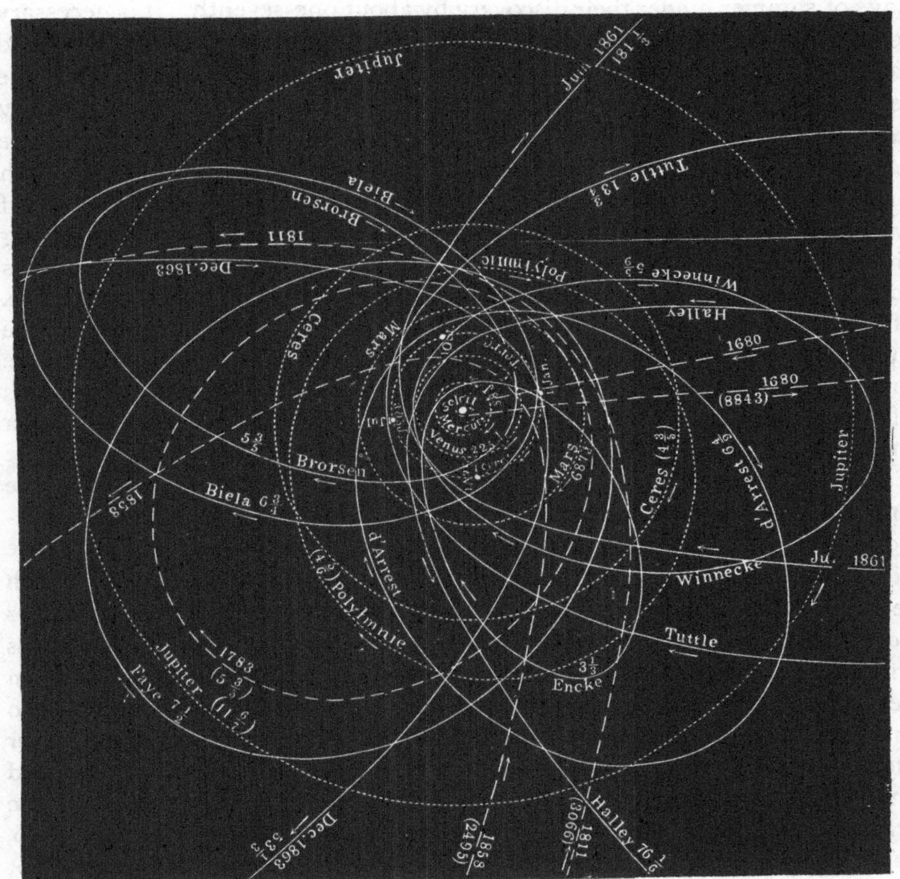

FIG. 220.—ORBITS OF SOME COMETS

replied Kepler. This reply was not at all exaggerated. In fact, we have just seen that already more than 800 different comets have been observed.

[1] Notwithstanding the extreme condensation of these pages, the abundance of matter, which overwhelms us more and more as we proceed, obliges us to reserve for another volume (*Les Etoiles*) technical papers, a knowledge of which is indispensable for 'popular astronomy,' properly so called. Readers who wish to penetrate farther into the science will find in this work, among other things, the *Complete Catalogue of all the Comets Observed and Calculated up to the Present.* It is this Catalogue which it is necessary to consult, when a comet appears in the sky, in order to know whether we are witnessing the return of a comet already seen, or have before our eyes a new visitor.

But this number includes, down to the seventeenth century, only those which were visible to the naked eye, and it is probable that if observers had searched for comets with a telescope during the last twenty centuries they would have seen more than 8,000. We may add, that the long days of summer hinder their discovery by about one-seventh. It is necessary to add this fraction to the preceding number to render it exact, which would raise it to more than 9,000. Now, we are very far from seeing all the telescopic comets which pass in sight of the earth ; all regions of the sky are not constantly scrutinised by astronomers, and more than half the comets must pass unperceived, without allowing for cloudy nights which prevent all search. We shall, then, certainly be below the actual number if we admit that 20,000 comets have passed within sight of our planet during the last 2,000 years. But still this is not all. Our calculation only applies to the comets which approach sufficiently near the earth to become visible. If these bodies are uniformly distributed in the inter-planetary spaces, the number should increase, orbit by orbit, as the cube of the distance, and we find that from the sun to Neptune there would be more than 20 millions of cometary orbits. It is probable, however, that there are more perihelia in the regions near the sun enclosed by the orbits of Mercury, Venus, the Earth, and Mars than in the external regions of less attraction, and that the progression should be less than the cube. But, on the one hand, it cannot be doubted that those which approach sufficiently near the sun to be visible from here form but a very small fraction of the total number which revolve round the sun ; and, on the other hand, the orbit of Neptune does not limit the sphere of the solar attraction. Comets may gravitate towards the sun, not only from thirty times the distance of the earth, but from very much farther, as we have already seen. They fly from one sun to another in all directions of Infinitude ; they gravitate also round other stellar foci, and perhaps it is their fall on certain stars which has produced the frightful conflagrations which we have observed from here in the stars which suddenly blazed out in 1572, 1604, 1670, 1848, 1866, 1876, [and 1892], only to mention those best observed. Thus, finally, it is not only by millions, nor even by hundreds of millions, but by *thousands of millions*, that we must estimate the real number of comets. If we are already stupefied by the wonderful interlacing indicated in fig. 220, where, however, the orbits of only thirteen comets and seven planets are drawn, how would it be if we attempted to represent the real interlacing of the thousands of cometary orbits which intersect the region where we sail in space? These mysterious bodies are the couriers of the sky. What do they teach us?

CHAPTER III

CONSTITUTION OF COMETS

Mode of Communication between the Worlds. Possible Encounters with the Earth. Whence come the Comets?

WHAT, then, is a comet? It is an extremely light nebulous mass, of which the nucleus may be solid or formed of solid aërolites, raised to incandescence at the perihelion, but of which the larger volume is formed of gas, in the chemical composition of which the vapour of carbon predominates.

Isolated in the depths of space, these masses naturally take the spherical form, and are destitute of tails, tufts, and irregular coma. When they arrive in the solar regions they are more sensitive than the massive planets to the calorific, luminous, electrical, and magnetic action of the sun: the comet expands, its vapours are developed and escape in jets towards the radiant star; then we see them driven back on each side of the head and the caudal train commencing. The jets often cover the head, and sometimes form a multiple veil composed of a series of successive envelopes. These gases are afterwards driven back behind, while the comet rapidly advances on its course. It is electricity which appears to play the principal part in these effects. From that time the comet ceases to be spherical and becomes oval, elongated in the direction of the sun.

The sun acts on the comet: (1) by its attraction, producing a double atmospherical tide similar to that of the earth, but so much the more intense as the cometary atmosphere is vaster and the comet approaches nearer; (2) by its heat, warming the nucleus, expanding the gas, producing new vapours, operating by physical and chemical transformations; (3) by electricity and magnetism, with opposite currents, attractions and repulsions being the inevitable consequence; (4) by a repulsive force, of which the nature is still unknown.

An idea generally diffused among the public leads them to believe that the tails of comets *follow them* in their course, like a train of phosphorescent matter. This opinion is incorrect. These appendages are always opposite to the sun, as if they were the luminous shadow of the comet, showing often a slight inclination in the direction opposite to the motion. Let us trace the curve of any cometary orbit (fig. 221), and we shall see that, if the tail appears to follow the comet before its perihelion, it precedes it, on the contrary, after that epoch. This train is extended in the *plane of the orbit* of the comet, and does not descend below nor rise above it. Its apparent length, its width, even its form, depend on the perspective under which we see it. This has been already understood from the cometary courses which we have traced.

Among the largest comets which have appeared to the eyes of the earth's inhabitants, that of 1843 had a tail absolutely straight and exactly opposite to the sun. We have seen that this comet closely approached

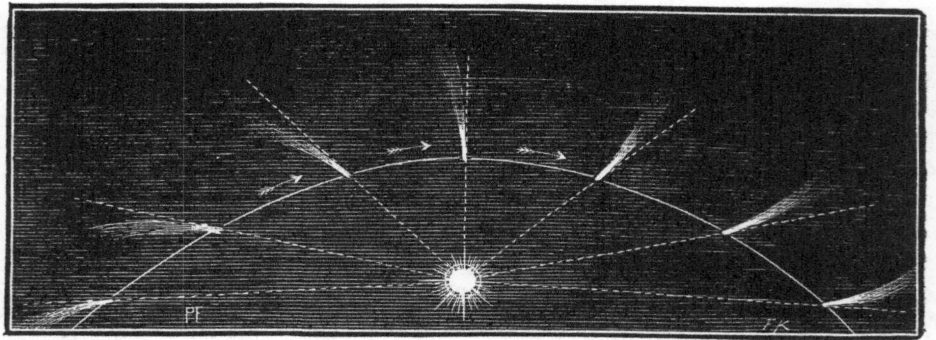

FIG. 221.—THE TAILS OF COMETS ARE ALWAYS TURNED AWAY FROM THE SUN

the radiant star, that its tail measured 200 millions of miles in length, and that it is impossible to admit its materiality on account of the fabulous velocity with which it would be animated. The same reasoning applies to those of 1680, 1880, 1882, and 1887, and to all the great comets at their perihelion passage. We are, then, led to think that the tails of large comets are not material. The tail is in some measure a luminous shadow of the comet, travelling along and bending slightly : it is comparable to a cloud which forms and evaporates incessantly in the track of this shadow. It is an electrical or other ray. We cannot say exactly that it illuminates space, for space is constantly lit up by the sun, and yet it is not visible on that account. The comet must, then, produce an altogether special effect on the ether, an *ethereal motion*. The ether exists, since the waves of light could not be transmitted without it ; we are forced to admit that space is filled with a fluid of extreme tenuity ; but whatever may be its tenuity, this fluid is not the less *real.* In order to explain the immense tails of comets, which always appear opposite to the sun, and to avoid the impossible motions which we computed with reference to the great comets of 1680, 1843, and 1882, it is necessary and it is sufficient that the comet should act on the ether in the manner of a lens, not exactly by refracting the luminous rays, but rather by producing an electrical undulation still lighter than that of the aurora borealis, which is formed near the limits of our own atmosphere. It is not matter which travels in a telegraphic message sent from Paris to New York : it is *motion.* A wave travels through a piece of water without the water travelling on that account. These immense tails cannot be substantial, cannot travel themselves, but represent a particular and momentary state of the ether set in motion by the interposition of the comet before the sun. The light of cometary tails, especially those of 1843, 1860, 1874, has been seen to undulate like that of the aurora borealis. We do not know the substance of the ether ; why should it not be luminous when electrified or traversed by a motion of a certain order? A mystery, doubtless ; but it is better to admit it than to believe the accepted theory. The objection to the immense tails at perihelion is fatal.[1]

We may form an idea of the transforming power exercised by the sun on comets by examining figs. 222 and 223, which represent the luminous jets projected from the head of the comet of 1861, as it was observed at Rome by Secchi, at an interval of twenty-four hours only, on June 30 to

[1] Having studied with the most serious attention the theories successively advanced by Kepler, Newton, Laplace, Olbers, Bessel, Liais, Secchi, as well as the recent works of Messrs. Faye, Roche, and Brédichin, as each of these theories assumes the materiality of the tails, it seems to us impossible to accept any of them. Cardan proposed, in the sixteenth century, to explain these tails by the refraction of the solar light passing through the globe of the comet ; but evidently this is not sufficient, for the comet is not a lens, and space cannot be illuminated by the solar light alone. Gergonne and Saigey have revived this hypothesis by supposing that it is the atmosphere of the comet which is thus illuminated by refraction ; but for this it would be necessary to admit that comets carry with them atmospheres measuring 60, 80, and 100 millions of leagues in diameter ! Tyndall explains the phenomenon by saying that the comet is formed of a vapour decomposable by the light of the sun, and that the head and the tail are a chemical cloud resulting from this decomposition. The enormous size of the cometary atmospheres supposed in this hypothesis also prevents us from adopting it. How can we admit that a comet can be a vaporous mass measuring 100, 50, 30, 20, or even only 10 millions of leagues in diameter ? How can we suppose, on the other hand, that the extremity of a vaporous train moves in space with a velocity of 370,000 miles a second, or even more ? To the impossible no one is bound. The astronomer has not the resource of the fatalist Mussulman, who simply believes that 'God has willed it so,' nor that of the devotee, still more humble, who goes as far as saying *Credo quia absurdum.* The study of the universe should be, above all things, rational.

May not comets be formed of that kind of rarefied substance which has received the name of 'fourth state,' and which Mr. Crookes calls 'radiant matter?' Their light would thus be of electrical origin. This constitution would not prevent repulsive forces, so well studied by M. Brédichin, from acting on these substances, which we may almost call immaterial.

July 1. Reaching a certain height, these jets form a halo or brilliant arc extending backwards into the tail.

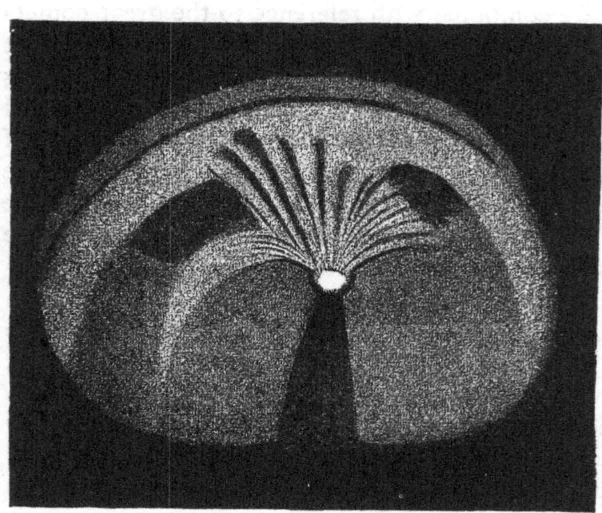

FIG. 222.—CHANGES OBSERVED IN THE HEAD OF THE COMET OF 1861

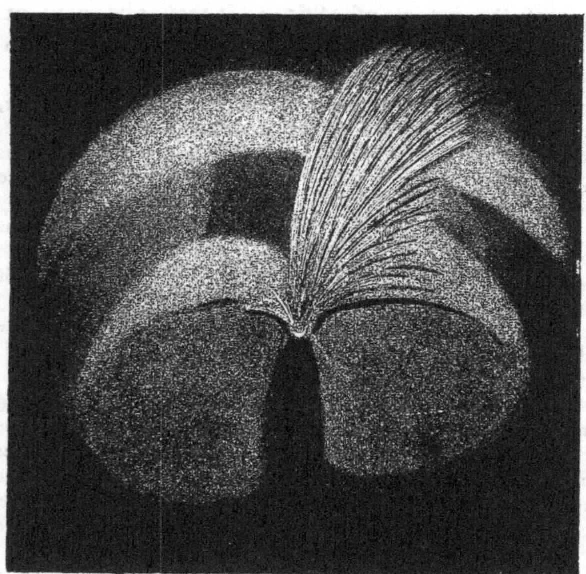

FIG. 223.—FURTHER CHANGES OBSERVED IN THE HEAD OF THE COMET OF 1861

In general, when a comet first appears in the depths of space, proceeding towards the sun, it resembles a feeble nebulosity, round or oval. Approaching the fiery focus, it appears to increase, and develops a more brilliant interior part, which is called the *nucleus*. This nucleus is surrounded by a vaporous atmosphere, usually elongated and unsymmetrical, of which the narrowest side is turned towards the sun. Such is the definitive form of small comets; but in approaching the perihelion the large comets give birth to luminous jets, which seem to dart from the nucleus towards the sun, afterwards bending round to begin the *tail* of the comet behind. The maximum brightness is attained some days after the perihelion; from this time the body becomes less luminous, the jets disappear, the tail disperses, and the comet resumes again the aspect of a simple nebulosity, which it presented at the commencement of its appearance. Such is the history of all comets.

The sun, then, acts on these bodies when they approach him, produces in them important physical and chemical transformations, and exercises on their developed atmosphere a repulsive force, the nature of which is still unknown to us, but the effects of which coincide with the formation and development of the tails. The tails are thus, in the extension of the cometary atmosphere, driven back, either by the solar heat, by the light, by electricity, or by other forces; and this extension is rather a motion in the ether than a real transport of matter, at least in the great comets which approach very near the sun, and in their immense luminous appendages. The effects produced and observed are not the same in all comets, which proves that they differ from each other in several respects. The tails have sometimes been seen to diminish before the perihelion passage, as in 1835; luminous envelopes have also been seen succeeding each other round the head, concentrating themselves on the side opposite to the sun, and

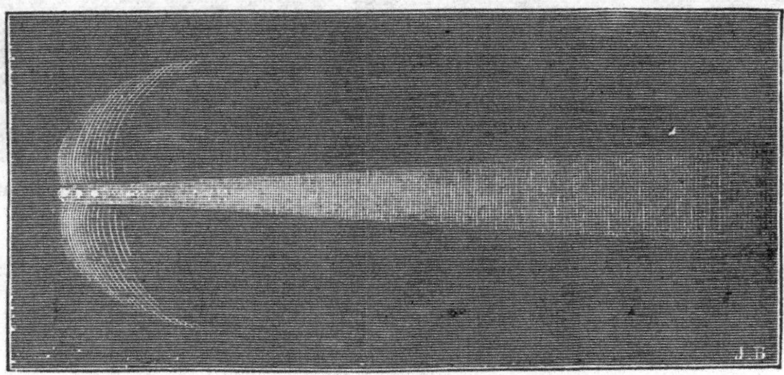

FIG. 224.—THE FIRST COMET OF 1888: JUNE 4

leaving the median line of the tail darker than the two sides. This is what happened in the Donati comet and in that of 1861. Sometimes a secondary tail has been seen projected towards the sun, as in 1824, 1850, 1851, and 1880. Comets have been seen with the head enveloped in phosphorescence, surrounding them with a sort of luminous atmosphere.[1] Comets have also been seen with three, four, five, and *six tails*, like that of 1744, for example, which appeared like a splendid aurora borealis rising majestically in the sky, until, the celestial fan being raised to its full height, it was perceived that the six jets of light all proceeded from the same point, which was nothing else but the nucleus of a comet. On the other hand, the nuclei themselves show great variations; some appear simply nebulous, and permit the faintest stars to be visible through them; others seem to be formed of one or more solid masses surrounded by an enormous atmosphere; in others, again,

[1] For an account of all these curious cometary aspects, see our *Revue d'Astronomie*, years 1882, 1883, and 1884.

a nucleus does not exist, as we have seen above (fig. 218) in the Southern comet of 1887. One of the comets of 1888 showed a triple nucleus and a bristling coma, as we see in fig. 224. We may, then, consider that the wandering bodies collected under the name of comets are of several origins and *several different species.*

The following are the real lengths of the longest cometary tails which

FIG. 225.—PASSAGE OF THE EARTH AND THE MOON THROUGH THE TAIL OF A COMET, JUNE 30, 1861

have been measured. The first three are very remarkable in this respect, and also for their near approach to the sun at their perihelia :—

Comets	Length of Tail		Distance at perihelion (Earth's distance=1)
	Leagues	Miles	
1843, I.	80,000,000	198,000,000	0·0055
1680	60,000,000	149,000,000	0·0062
1847, I.	53,000,000	131,000,000	0·0426
1811, I.	44,000,000	109,000,000	1·0355
1858, IV. . . .	22,000,000	55,000,000	0·578
1618	20,000,000	50,000,000	0·389
1861	17,000,000	42,000,000	0·822
1769	16,000,000	40,000,000	0·123
1860	9,000,000	22,000,000	0·292
1744	7,000,000	17,000,000	0·222

We may imagine what a wonderful length would have been attained by the comet of 1811 if, instead of stopping at the earth's distance from the sun, it had approached the radiant star like its sisters of 1843 and 1680.

What is the real weight of these strange bodies ? We may at once remark that in general these nebulosities are very light. In fact, when they pass near the planets they do not cause any perturbation in the motions of these planets, nor even in those of their satellites ; thus, the comet of

Lexell passed near Jupiter in 1769 and 1779—not through his system, as had been for some time believed, but at a distance of about 150,000 leagues (372,000 miles)—and in 1770 it passed at 610,000 leagues (150,000 miles) from the earth. None of the satellites of Jupiter nor the moon were in any way disturbed. Again, the comet of 1861 passed at 110,000 leagues (273,000 miles) from us on June 30, and it is almost certain, according to the most trustworthy calculations and observations of M. Liais, that the earth and moon passed through its tail at six o'clock that morning. In fact, neither the earth nor the moon perceived it: only a slight aurora borealis was seen, as if the tail itself were simply an aurora; the encounter was only really known and calculated after the passage. If we may now judge of the density of comets by their transparency, not to speak of the tails which are *absolutely transparent*, in many cases stars of the 5th, 6th, 7th, 8th, 10th, and 12th magnitude have been seen occulted by the *heads* of comets without being obscured or refracted; in 1828 the thickness of the nebulosity traversed was not less than 500,000 kilometres (310,000 miles), and the star was only of the 10th magnitude. A third method, which is based neither on planetary perturbations nor on the transparency of comets, but on the transformations of cometary atmospheres, has given to M. Roche for the mass of Donati's comet a twenty-thousandth of that of the earth, and for the mass of Encke's comet a thousandth. This result appears exaggerated.

It seems that the passage of a comet in front of the sun should be very valuable for the analysis of the physical constitution of the nucleus. There have been attributed to an event of this kind several darkenings of the day star described in history or in legend, especially the supposed eclipse of the sun which happened at the death of Christ (at the epoch of full moon). Although evidently very rare, the transit of a comet across the sun *has happened*, especially on November 18, 1826, and September 17, 1882. On the first date the bad weather caused universal disappointment to all astronomers; a very dense fog prevailed over the whole of Europe all that day; Gambart (who had announced the transit) at Marseilles, and Flaugergue at Viviers, succeeded, however, in observing the sun through an opening in the clouds, but it was pale, and they perceived no unusual form. The transit of the comet of 1882 across the sun was observed on September 17 by Mr. Gill and his assistants at the Cape of Good Hope; they saw it come into contact with the sun and pass off, but it was completely invisible during the passage.

The heads of comets are, then, certainly constituted of extremely light matter, of which the density varies, moreover, from one comet to another, and even in the same comet after some days' interval; thus, for example, while, on November 7, 1828, Struve perceived a star of the 11th magnitude through Encke's comet, twenty-one days later Wartman saw, on the contrary, a star of the 8th magnitude disappear behind the same comet, which was therefore eight times more condensed than on the first date. In general, stars are not diminished in light even when the nucleus of a comet passes over them ; the fact is not very rare. Several comets appear to be entirely nebulous ; others seem formed of a solid or liquid nucleus surrounded by a nebulosity ; some have shown luminous granulations indicating the presence of several nuclei. From experiments in polarisation commenced by Arago on the comet of 1819, we know that comets shine partly from the solar light reflected by them, as the planets do. But they add to this reflection a light which is their own.

In general, the spectra of comets show three brilliant bands, which do

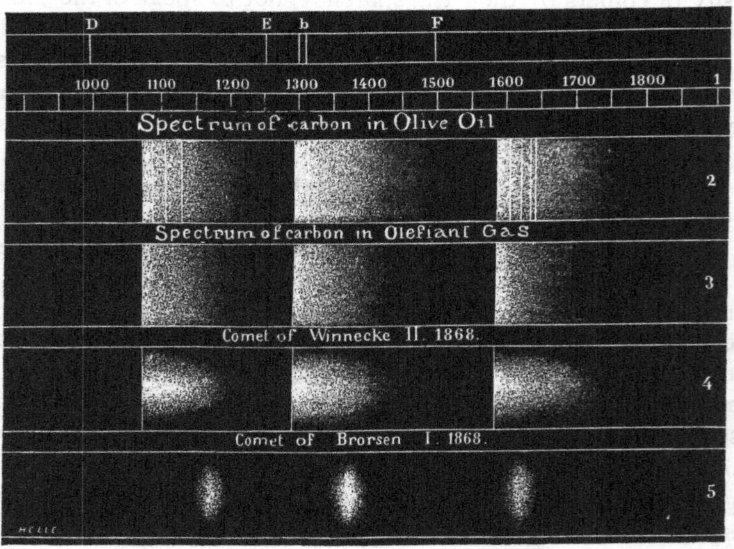

FIG. 226.—SPECTROSCOPIC ANALYSIS OF COMETS

not coincide with any of the principal rays of the solar spectrum (the second of these bands occupies the position of the little double line *b* in fig. 226) The second band is always the most luminous ; it is green ; the first to the left is yellow, and the third is found in the blue. In 1868 Secchi represented in the drawing which we reproduce here (fig. 226) the spectra of the two periodical comets of Winnecke and Brorsen, observed in that year. The first shows a resemblance, or, we should rather say, a very curious identity, with the spectrum of *carbon* observed in the light of a lamp. The coincidence of the lines is truly striking. The analysis of Huggins led exactly to the same result. In the drawing, the part of the solar spectrum which corresponds with this spectrum is given at the top, and also the scale of length of the lines. We see that the three zones begin near the lines 1070, 1295, and

1590. Could this comet, then, be *volatilised coal*? Is it carbon combined either with oxygen or with hydrogen? How can such combinations be found in the state of vapour in comets? A new mystery!

We must not hasten to conclude, as some have done, that 'all comets are carbon,' either hydrocarbon, or oxide of carbon, or carbonic acid, for the spectra of these different bodies, although showing the analogy of three bands similarly placed, are not absolutely identical; and besides, the conditions of diffuse matter in the interplanetary void are so different from those of our terrestrial laboratories and of the solar furnace, that similar aspects may very well be produced by substances altogether different.

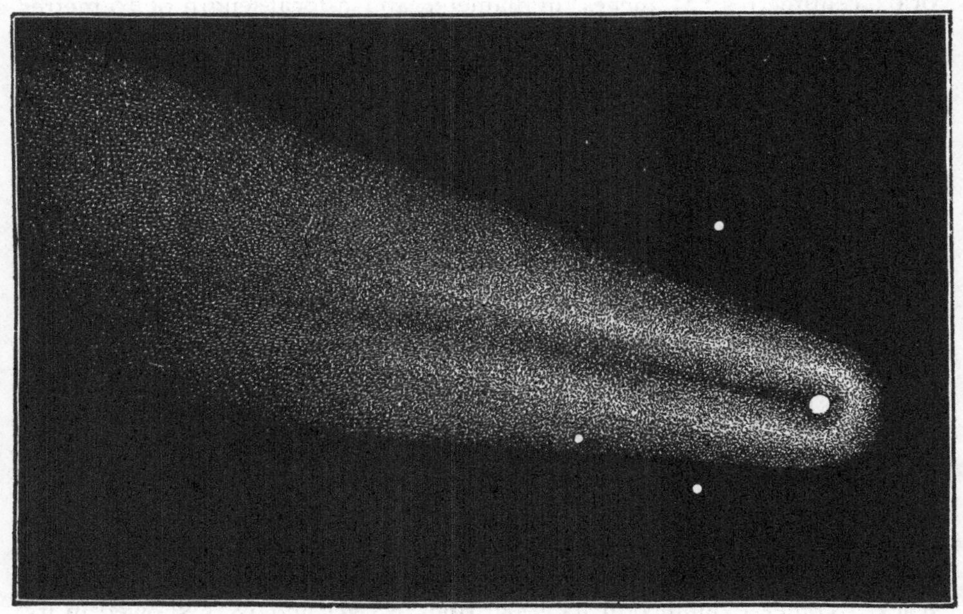

FIG. 227.—COMET III. OF 1874, ON JULY 6

There has also been found in the spectra ot certain comets a rather close resemblance to the spectrum of the gases which are developed by heat in meteoric stones, to that of shooting stars, of the aurora borealis, and of the electric spark. We notice in fig. 226 that the three bands in Brorsen's comet resemble much less the spectrum of carbon than do the bands of the other comet. [Wells's comet of 1882 showed the spectrum of sodium.—J. E. G.]

In 1874 a comet which seemed about to rival in brightness those of 1858 and 1861—that discovered by M. Coggia at Marseilles on April 17, 1874—was made the object of numerous observations. I made a drawing at the telescope on June 11 which shows a small brilliant nucleus, surrounded by a vaporous nebulosity. The general light resembled that of a ray of sunlight which penetrates into a dark room and illuminates the dust in the air. It diminished in intensity with its distance from the nucleus, but extended sufficiently far to announce the presence of the comet, even when the head was not in the field of the telescope. The *tone* of the light was a pale greenish white, which contrasted singularly with the warm and yellower tint of the neighbouring stars. In the month of July the comet became visible to the naked eye and presented a rather fine appearance. The best drawing of it which was made is that which is reproduced in fig. 227, which is due to Mr. Newall, of Newcastle, an English amateur, who at that time possessed the most powerful telescope in Europe, with an object-glass of 63 centimetres (25 inches) in diameter, and a focal length of 10 metres (32·8 feet). It cost £10,000.

A large number of observers, especially M. Rayet in Paris, have studied the spectrum of this comet; the three brilliant bands were almost united by a continuous horizontal train (fig. 228). A considerable portion of the light of this comet was derived from the sun by means of reflection. In

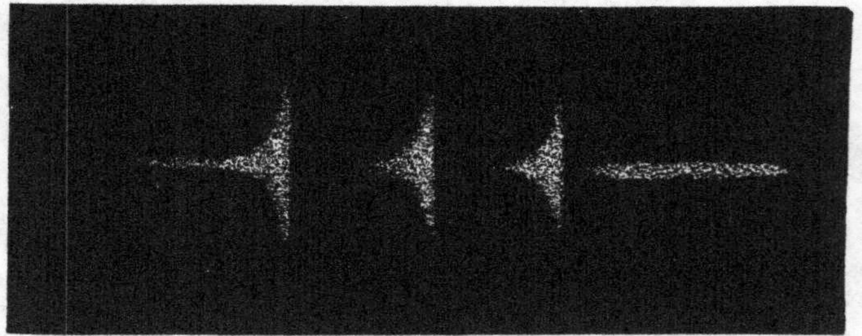

FIG. 228.—SPECTRUM OF COMET III. OF 1874

the tail there was no solid incandescent matter in perceptible quantity. The three bands coincided *nearly* with those of carbon. We may add that Encke's comet, examined at its apparitions in 1871 and 1875, showed in its spectrum three bands which coincide *nearly* with those of hydro-carbon. It is also believed that in several comets the brilliant ray of nitrogen has been recognised.

Comets also emit a light of their own, which often varies considerably from day to day, as was seen in the comets II. and III. of 1887. The maximum brightness generally coincides with the perihelion; the brightness varies according to the distance of the comet from the sun and earth.

M. Brédichin, Director of the Moscow Observatory, has been specially occupied for more than a quarter of a century in the study of cometary tails, and has obtained on this point results which, if they are not definitive as to the explanation of the motion of the immense straight tails at the perihelion, should nevertheless be recorded here, as representing the most important work which has been done with regard to these mysterious appendages. His examination of cometary tails has shown him that they may be divided into three very distinct types: (1) almost straight lines, thin and very long, represented by type I. of fig. 229; (2) multiple fan-shaped tails, more curved and shorter, represented by type II.; (3) tails more curved still and shorter, of the third type. All cometary tails can be classed in one or other of these three types. Spectrum analysis has shown that hydrogen predominates in the chemical composition of the first, and hydro-carbons in those of the second. In the third type the presence of iron, chlorine, and other elements having a great atomic weight, has been noticed.

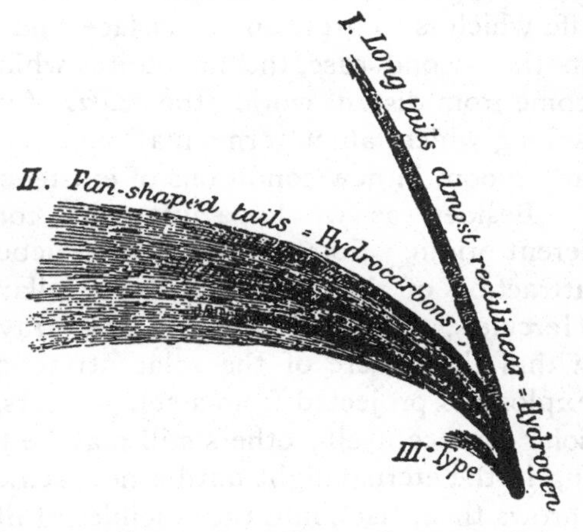

FIG. 229.—THEORY OF VARIOUS COMETARY TAILS

The great comet of 1858 had tails of the first two types, for it showed by the side of the principal tail, curved and fan-like, two curious, straight, and very thin trains. The comet of 1744, with six tails, showed the finest example of the second type. The compounds of carbon being very varied, the tails of the second type are much more diverse than those of the first.

The repulsive force necessary to produce tails of the first type must be twelve times stronger than the attraction; the tails of the second type could be produced by a force simply equal to gravitation, and those of the third type by a force equal to a quarter of that of gravitation. Is not this repulsive force electricity? That the sun is a great focus of electricity is not doubtful, as we have seen.

Such is the present state of our knowledge of the physical and chemical constitution of comets. It is important to state all that we know, but nothing more. The analysis of these singular bodies is far from being finished. Are they really composed of carbon? Everyone now knows that the diamond is pure carbon, and there is nothing so easy as to reduce one of these precious stones to charcoal. In the eyes of the chemist the most sparkling brilliants, the Regent or the Koh-i-noor, are simply little pieces of admirably formed charcoal. May not the comets, then, be the diamonds of the sky? After terrifying the planetary populations by their always strange and often sinister aspect, shall they now be admired as the precious stones of the celestial casket? Their importance would be much greater still if they should be found to carry in them the first combinations of carbon, for it is probable that it was by these combinations that vegetable and animal life commenced on the earth and the other planets and thus these vagrant bodies might be the sowers of life on all the worlds! Whence came the first seed, the first germ of terrestrial life? It was either a spontaneous generation or it descended from the sky. In the first case, which is the more probable, each world bears in itself, from its glowing genesis, the principles of its future development, and of the tree of life which is to grow on its surface and cover it with its fruitful branches; in the second case, the meteorites which follow the cometary trains may come from distant worlds, the *débris* of which they transport through space, among which latent germs may survive, ready to fall on a prepared earth and bloom in new conditions of existence.

Besides, as we have just said, comets may be bodies of very different origin. Some may be small nebulæ caught in their passage by the attraction of the sun in its interstellar course towards the constellation Hercules; others, cosmical masses travelling through space, and coming within the sphere of the solar attraction; others may be the results of explosions projected from a star; others, again, may be shot out from our solar furnace itself; others still may be the remains of ruined worlds, moving in the eternal night until a new attraction seizes them on their way and throws them back into the crucibles of life. Everything tempts us to think that there exist here and there, disseminated upon the planetary shores, floating on the ethereal waves, broken-up comets, remains of the shipwrecks which may have happened among so many millions of worlds; they are the waifs of ships powerless, for the most part, to accomplish their voyage without damage. Such fragments, more or less disintegrated, do not, however, wander at random through space: they move in orbits of which the form depends on the modifications which the disturbing actions have produced on their original velocity. The number of comets which penetrate into our system is, according to all probability, so immensely great that, during the hundreds of millions of years which it is permissible to assign to the past duration of this system, the interplanetary spaces must have been ploughed by a wonderful multitude of streams of matter, of disintegrated comets, and of fragments of comets, which the planets cannot fail frequently to encounter.

We shall see directly, with reference to shooting stars, how the periodical comets and meteorites have been incorporated into our system by the attraction of the planets. If the comet arrives near the planet in such a way that its motion is accelerated, its orbit becomes hyperbolic, and it is for ever lost to us; it will return no more. If, on the contrary, it arrives in such a manner as to be retarded, its orbit becomes elliptical, and the comet is captured by our system.

Planetary explosions projecting substances with a velocity superior to attraction may also have given birth to periodical comets, of which all the aphelia are near a planet.

We have already seen that a body projected from the sun with a velocity of 608,000 metres (377 miles) will depart from him indefinitely; this would be a veritable comet. It appears certain that these velocities really exist, and that the solar projectiles, being cooled down, may come to us under the form of meteoric stones. Since the other stars are suns like ours, we may presume that the projectiles shot from their bosom can also reach us. The attraction of the sun cannot give a body coming from infinity and meeting the earth a velocity greater than 72,000 metres (45 miles) a second; every bolide which may be recognised as reaching us with a velocity superior to that carries with it its certificate of birth, and proves that it has been projected by a star or a stellar explosion; such was probably the case with that of September 5, 1868, which passed over Austria and France with a velocity of 1,493 kilometres in 17 seconds (or 54½ miles a second).

Every comet or meteoric stream which follows a parabolic orbit possesses a velocity greater than that which the sun's attraction could give it, and it certainly enters the sphere of the solar attraction with a considerable original velocity. There is, then, no other way of explaining the interstellar velocities of comets and hyperbolic bolides but by tracing back their course to the time when their substance was projected from a star with a velocity exceeding by several miles per second that with which a body would reach that star if it had been drawn by gravity alone from an infinite distance. Although the influence of a planet, such as Jupiter or Saturn, may, strictly, transform a parabola into a hyperbola, this effect can only be produced very exceptionally, and the hyperbolic orbits of comets and bolides indicate the origin of forces superior to simple stellar attractions.

The aphelion of every comet points to the celestial region from which it has been sent to us. There are systems of comets which appear to have travelled together in space, and which have been separated by the attraction of the sun and planets. Thus, the astronomer Hoek has shown that the comets of 1860 III., 1861 I., and 1863 VI., formed a group before their entry into the solar system; the same fact has been recognised with regard to other comets. According to the researches of Kirkwood, the comets of 1812 I. and 846 IV. were introduced into our system by the attraction of Neptune, near which they passed about the year 695 before our era, forming in that region their first aphelion at 272° of longitude.

We notice a preponderance of aphelia in the direction of Arcturus and the constellation Hercules, arising from our general translation towards that region of the sky; if a few more comets come to us from that side, it is because we go to meet them.

Do we ever think what an immense voyage they must have made to come from there to here? Do we imagine for how many years they must have flown through the dark immensity to plunge themselves into the fires of our sun? If we take into account the directions from which certain comets come to us, and if we assign to the stars situated in that region the least distances consistent with known facts, we find that these comets certainly left their last star more than 20,000,000 *years ago*.

In thus putting to us from the height of their celestial apparitions so many notes of interrogation on the grandest problems of creation, comets assume to our eyes an interest incomparably greater than that with which superstition blindly surrounded them in past ages. When we reflect for a moment that a certain comet which shines before us in the sky came originally from the depths of the heavens, that it has travelled during millions of years to arrive here, and that, consequently, it is by millions of years that we must reckon its age if we wish to form any idea of it, we cannot refrain from respecting this strange visitor as a witness of vanished eras, as an echo of the past, as the most ancient testimony which we have of the existence of matter. But what do we say? These bodies are neither old nor young; there is nothing old, nothing new; all is present: the ages of the past contemplate the ages of the future, which all work, all gravitate, all circulate in the eternal plan. Musing, you look at the river which flows so gently at your feet, and you believe you see again the river of your childhood; but the water of to-day is not that of yesterday, it is not the same substance which you have before your eyes, and never, never shall this union of molecules, which you behold at this moment, come back there, never till the consummation of the ages!

If the appearance of comets forebodes absolutely nothing as to the microscopical events of our ephemeral human history, it is not the same with the effects which might be produced by their encounter with our wandering planet. In such an encounter there is nothing impossible; no law of celestial mechanics forbids that two bodies should come into collision in their course, be broken up, pulverised, and mutually reduced to vapour.[1]

[1] Is the approach of a comet ever attended by any astronomical or meteorological effect? The coincidence of the comet of 1811 with the great heat and fruitful vintage of that year caused it to be supposed that comets might exercise an influence on terrestrial temperatures; it must be admitted, also, that the fine comet of 1858 seems to have confirmed this supposition. But it is very necessary to guard ourselves against generalising in this way; besides, we do not see how or why the apparitions of great comets should occasion particularly warm years or coincide with them, and observation proves that splendid comets have likewise coincided with years of great cold. This happened especially in 1305, the year in which Halley's comet frightened the people, and coincided with one of the most frosty years which have been registered in the annals of meteorology. The year 1882, also remarkable from a cometary point of view, was particularly rainy and cold.

What would be the effects of such an event? Can we believe, with Whiston, that such an encounter brought on a universal deluge? or, with Maupertuis, that it would place the equator at the poles and the poles at the equator, and scald us 'like a swarm of ants in the boiling water which the husbandman pours on them?' or, with Pingré, that it would lift us to the moon? or, with Lambert, that it would even raise the earth and carry us 'into a winter of several centuries which neither men nor animals would be capable of resisting?' or, with Laplace, that it would raise the seas from their ancient beds and pour them over the continents, annihilating entire species, and placing humanity on the brink of ruin? Can we believe in such curious catastrophes? No; the knowledge which we now have of the smallness of cometary masses is expressly opposed to such ideas.

Ought we, then, to laugh at them, and, with Sir John Herschel and Babinet, treat them as *visible nothings*? No; that would be the other extreme.

Several comets seem to have solid nuclei. Solid bodies have already encountered the earth, have fallen on its surface, have killed men and set fire to houses, as we shall see in the following chapter. Most of the meteorites collected are, it is true, but small fragments of some few pounds in weight; but some have been met with which weigh several thousands of pounds. This is not a question of principle, but only a relation of the little to the great. Now, bolides have been measured which have, so to say, grazed the earth, and which have been several miles in diameter. The nucleus of the comet of 1811 was 690 kilometres (428 miles) in diameter; that of the great comet of 1843 measured 8,000 kilometres (4,970 miles); that of the comet of 1858, 9,000 kilometres (5,580 miles); that of the comet of 1769 measured 44,000 kilometres (27,000 miles, 11,000 leagues) in diameter! Whatever may be the intrinsic nature of these nuclei, it is not doubtful that, if one of them were to encounter our globe in its passage, both moving with a velocity of more than 60,000 miles an hour, we should certainly perceive the shock.[1]

What might happen more easily would be the passage of the earth through a comet's atmosphere. In fact, while the probably solid nucleus of the comet of 1811 measured but 428 miles in diameter, the atmosphere

[1] The encounter of these two lightning trains would probably not be harmless. A continent broken up, a kingdom crushed, Paris, London, New York, or Pekin annihilated, would be one of the least effects of the celestial catastrophe. Such an event would evidently be of the highest interest to astronomers placed sufficiently far from the point of encounter, especially if they could approach the fatal spot and examine the cometary remains left on the ground. The comet would bring, doubtless, neither gold nor silver, but mineralogical specimens, perhaps diamonds, and perhaps, also, certain vegetable *débris* or fossil animals, much more precious than an ingot of gold of the size of the earth. Such an encounter would, then, be eminently desirable from a purely scientific point of view; but we can scarcely hope for it, for we must admit, with Arago, that there are 280 millions to one against such an occurrence. However, although the probability against it is so great, we need not entirely despair.

which surrounded it (the largest which has been observed) attained 1,118,000 miles! We have seen above that the sun is 866,000 miles in diameter; this comet was, then, *larger than the sun*; it was twice the volume. If such a comet were to pass us at only 500,000 miles, we should be in its head.[1]

There is nothing, then, to prevent a comet some day encountering our planet in its course; but the effect which would be produced can hardly be determined in advance, since it would depend on the mass, on the density and constitution of the portion of the comet traversed. A chemical combination, a mixture of carbonic acid or of some other deleterious gas in our respirable atmosphere, a general poisoning of the human species, a universal asphyxia, an unexpected explosion, a sudden electrisation, a transformation of motion into heat, a shock partially or universally mortal, are so many possible effects. These bodies are not, then, absolutely inoffensive. But let us hasten to add that, notwithstanding the considerable number of comets and the variety of their irregular courses round the sun, it is probable that such a catastrophe will never happen till the natural death of the earth itself, because space is immense, because our floating island moves with amazing rapidity, and because the point of the infinite which we occupy at each instant of its duration is imperceptible in immensity.

[1] Such an event has not yet happened in the memory of humanity; but a comet has already touched us with its tail in passing, without counting the shower of shooting stars from the comet of Biela, of which we have spoken above. We have seen, in fact, that on June 30, 1861, the great comet of that year probably grazed us with its tail, of which the length then exceeded 2,000,000 miles. From what we have said of the tails of great comets, it is not surprising that the inhabitants of the earth should have slept that night as usual, and that they noticed nothing strange on awakening in the morning. However, an English astronomer, awaking at an early hour and observing the sky, wrote in his diary: 'Strange yellow, phosphorescent light, which I should have taken for an aurora borealis if there was not so much daylight.'

CHAPTER IV

SHOOTING STARS, BOLIDES, URANOLITHS OR METEORIC STONES

Orbits of Shooting Stars in Space. Stones fallen from the Sky

In the clear and transparent night a distant star seems to detach itself from the heavens, glide in silence on the nocturnal vault, shoot along, and disappear. The heart tried by terrestrial sorrows believes that the heavens are concerned with our destinies, and that the shooting star marks the departure of a spirit to another life; the young girl whose pensive gaze is attracted for a moment to the meteor hastens to form a wish, with the hope of seeing it speedily granted; the poet dreams that the stars, the flowers of the sky, bloom in the celestial fields, and thinks he sees their luminous petals swept away by upper winds through the infinite night; the astronomer knows that this ephemeral body is neither a star nor a spirit, but a molecule, a cosmical atom, a fragment more or less minute itself, but of which the lesson may be great, if we can learn whence it comes and how it thus encounters the earth in its course.

The apparition of a shooting star is an occurrence so frequent that there is not one of our readers who may not have observed it several times. Perhaps some have had the much rarer privilege of seeing not only a *shooting star*, but a more brilliant and rather exciting phenomenon: the passage of a flaming *bolide*, rapidly traversing space, and shedding in every direction a sparkling light—a globe of fire, leaving a luminous train behind it, and sometimes bursting with an explosion like that of a colossal rocket, and with a thunder resounding like the dull discharges of artillery. Perhaps some have also been able, by a chance still more fortunate and rare, to pick up a fragment of the exploded bolide—a *uranolith*, or mineral fallen from the sky.

Here are three distinct phenomena, which appear, however, to be connected by relations of origin.

The first point to be examined in the study of shooting stars is the measurement of the height at which they appear. Two observers placed at points distant from each other ascertain the path of a shooting star among the constellations. The line is not absolutely the same for both, on account of perspective. Calculating the difference, we obtain the height. In general this height is 120 *kilometres* (74 miles) at the beginning of the apparition, and 80 kilometres (50 miles) at the end of the visible path.

All nights of the year are not the same with reference to the number of shooting stars. There are a number of periods—*annual*, *monthly*, and *diurnal*—ascertained by persevering examiners of the sky, among whom we should particularly mention the French observer, M. Coulvier-Gravier [and the English observer, Mr. W. F. Denning.—J. E. G.]

The most remarkable epochs are the night of August 10 and the morning of November 14. These fixed dates prelcude all theories which attempt to attribute this phenomenon to a meteorological cause. The apparition in the month of August lasts several days, and it has its maximum on the 10th; that of November takes place in the morning of the 14th. On the latter date the meteors have sometimes been so numerous that they have been compared to showers of fire. Since 1833 the narratives of the ancient chroniclers have been studied, and the American astronomer Newton has recognised that the showers of fire which have at certain epochs struck terror among the nations were nothing else but the apparition of the shooting stars of November. This apparition is not equally remarkable every year, but its brightness varies periodically: the maximum returns every thirty-three years nearly; the shower is then renewed for several years, but gradually diminishes; and at last ceases to be noticed during a long period, to be reproduced later on, and pass again through a maximum at the end of thirty-three years. Moreover, the swarm of asteroids of the month of November having a small thickness, the earth takes but a few hours to traverse it; the maximum also is only visible in some circumscribed regions, which vary each year. The apparition of the month of August is more constant, but it is never so brilliant; it also is subject to curious fluctuations of intensity.

It has been ascertained that the paths of the different meteors diverge from the same point of the sky, which is called the point of emanation, or *radiant*. This point is found in the constellations of Perseus and Cassiopeia for the meteors of August 10, and for those of November 14 it is found in that of the Lion. This is why the shooting stars of August 10 are sometimes called the Perseids, and those of November 14 the Leonids. A large number of radiant points have been determined for different epochs of the year. Fig. 230 represents that of November 27, observed in 1872 and 1885—the *débris* of Biela's comet. It is not necessary to believe that all the shooting stars start in reality from the radiant; only that their paths, prolonged backward, all meet in the same point, except a small number which are designated by the name of *sporadic* stars. This convergence is an effect of perspective: the true paths are practically parallel, but they appear to diverge according to the same law which shows us as divergent the rays of the setting sun passing between the clouds, an avenue of trees, &c.

The shooting stars must be small solid bodies, for if they were gaseous they would not have the strength to penetrate so deeply into our atmosphere, and would be dispersed before being ignited. We occasionally see a mass divide into two or three parts, sometimes more, each of them preserving a clearly defined form; they are, therefore, composed of solid substances capable of flying in pieces during their combustion. It has also been often ascertained that in the spot where the meteors appear, they form small clouds, which persist for some time after the disappearance of the meteors, and which are drawn along by the atmospheric currents.

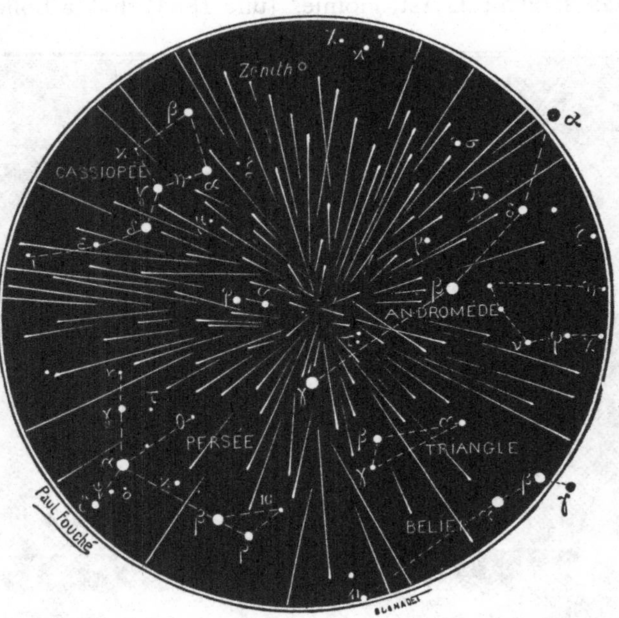

Fig. 230.—Point of Emanation of the Shooting Stars of November 27

In all the apparitions we find a diurnal and an annual period. In the diurnal period the maximum takes place from three to six o'clock in the morning. In the annual period the meteors are more numerous in the second part of the year than in the first. These two remarkable circumstances are due to the fact that the earth encounters the swarms of meteoric matter more directly in the morning than in the evening, and during the second half-year than during the first. We can, in fact, compare the earth passing through a swarm of these corpuscles to a cannon-ball passing through a swarm of gnats; it will encounter a much larger number in its anterior part, and will leave a veritable void behind it. And if the cannon-ball rotates on itself, the points situated in front, and which on that account are more exposed to collisions, will vary in the same way. The hourly number of shooting stars will depend, then, on the point towards which the earth's motion is directed at each instant with reference to the vertical of the observer: it will be a maximum when this point is as near as possible to the zenith.

If we see shooting stars in that part of the earth which is opposite to where the maximum takes place, it is because their velocity is greater than that of the terrestrial globe.

When they reach the higher regions of our atmosphere, these cosmical particles of dust—doubtless quite small and of the size of heads of pins, grains of shot, bullets perhaps—are ignited by friction. In a remarkable analysis, like everything which emanates from his judicial intellect, M. Hirn has shown ('L'Astronomie,' June 1883) that a bolide which penetrates the

FIG. 231.—THE GREAT SHOWER OF SHOOTING STARS OF NOVEMBER 27, 1872

upper regions of our atmosphere with a relative velocity of 30 kilometres (18·64 miles) per second compresses the air in front of its path to the point of increasing the pressure from the hundredth of an atmosphere at 37 kilometres (23 miles) high, to 56 times that which it has at the surface of the ground—that is to say, that the normal atmospherical pressure, which is 10,333 kilogrammes (22,780 lbs.) per square metre, would be raised in front of a bolide of 1 square metre of surface to 582,000 kilogrammes (1,283,000 lbs.). This increase of pressure is followed by a considerable increase of heat. The temperature of space being 273° below zero (on the Centigrade scale), our bolide, endowed with this ultra-glacial temperature before touching our atmosphere, becomes in a few seconds heated to 3,340°, a heat which our most intense furnaces cannot produce. This increase of heat would be attained even if the bolide traversed only the most rarefied layers of the aërial heights, where the pressure does not even reach the thousandth of an atmosphere. At 100,000 metres (62 miles) of height a shooting star becomes visible on account of this transformation of its motion into heat and light.

It follows as a consequence that shooting stars cannot reach the earth: they are inevitably vaporised before penetrating to the lower layers of our atmosphere. At first they do not come straight towards us. The earth, crossing a swarm of shooting stars, always cuts it more or less obliquely. These corpuscles glide in some measure on the exterior rotundity of our atmosphere, however traversable and rarefied this limit may be, and pass out after following tangents rather than secants. The projectiles which arrive full-face penetrate more, and remain, but are vapourised, and their velocity is destroyed before the resistance of the air permits them to reach the ground. They are met with in the state of microscopical ferruginous dust on the surface of the soil.

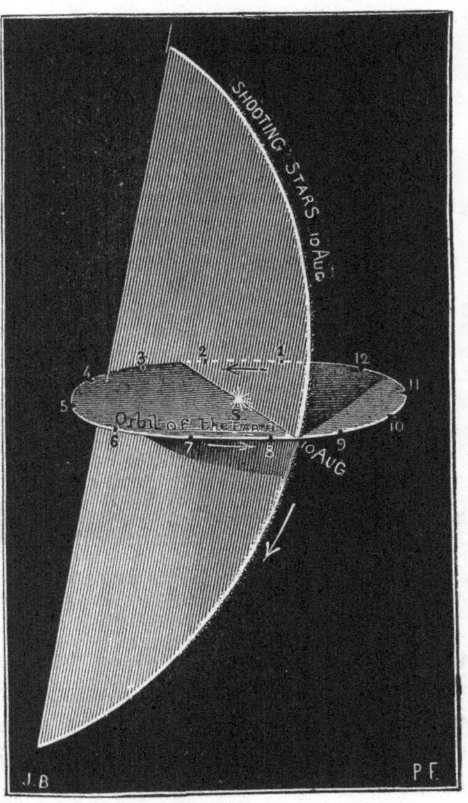

FIG. 232.—INTERSECTION OF THE ORBIT OF THE SHOOTING STARS OF AUGUST 10 WITH THE PLANE OF THE EARTH'S ORBIT

We may now remark that these meteors play a much more important part than we were formerly disposed to believe. A single night, a single hour, a single minute, does not pass without the fall of a star. The terrestrial globe sails in the midst of a space full of diverse corpuscles circulating in all directions—some in elliptical streams of various inclinations, others even in the plane of the ecliptic, as we see by the zodiacal light which extends from the sun to beyond the terrestrial orbit. By enumerating the number of shooting stars which are seen above a given horizon during the different nights of the year, calculating the number of similar horizons which would comprise the whole surface of the globe, and taking into account the directions of the shooting stars, the monthly variations, &c., an eminent American astronomer, Mr. Simon Newcomb, has demonstrated that no fewer than *one hundred and forty-six thousand millions* (146,000,000,000) of shooting stars fall per annum on the earth.[1]

We have seen above what a splendid shower of shooting stars was seen on November 27, 1872, as well as on November 27, 1885. That of the night of November 12 to 13, 1833, was more marvellous still. The stars

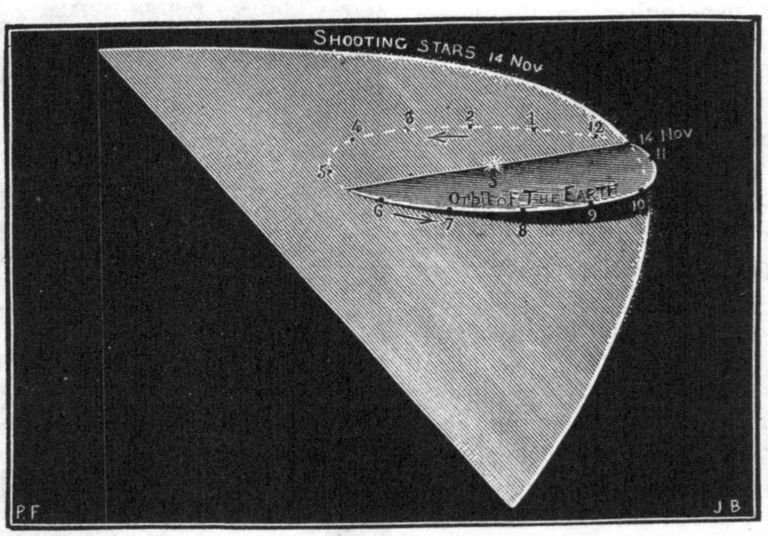

FIG. 233.—INTERSECTION OF THE ORBIT OF THE SHOOTING STARS OF NOVEMBER 14 WITH THE PLANE OF THE EARTH'S ORBIT

were so numerous and showed themselves in so many regions of the sky at the same time that in attempting to count them no one could hope to arrive at anything but rough approximations. The Boston observer, Olmsted, compared them, at the moment of maximum, to half the number of flakes which we perceive in the air during an ordinary shower of snow.

[1] The velocity with which these dust particles of worlds encounter our planet in space may, and often does, attain the order of parabolic velocity, of which the simple formula is $V\sqrt{2}$. It is equal to the velocity of translation of the earth multiplied by the square root of 2, or 1·414, and since the mean orbital velocity of our planet is 29,460 metres (18·3 miles) a second, that of a star is 42,570 metres (26·45 miles). If the shooting star reaches us directly in a direction contrary to our motion, the two velocities are added, and the collision is at the rate of 72,000 metres (44·74 miles) in the first second of the encounter. If the body arrives behind us, its velocity may be reduced to 16,500 metres (10·25 miles) a second. With the exception of massive uranoliths, which have, as has been ascertained, a weight ranging from some ounces to thousands of pounds, every shooting star encountering the earth must, then, be dissipated by the simple transformation of its motion into heat in penetrating our atmosphere, become absorbed, and then arrive slowly under the form of a deposit on the surface of the globe.

When the shower had considerably diminished he counted 650 stars in fifteen minutes, although he confined his observations to a zone which was not a tenth of the visible horizon; and he estimated at 8,660 the total number visible for the whole hemisphere. This latter number would give 34,640 stars per hour. Now, the phenomenon lasted more than seven hours, and hence the number of those visible at Boston would probably exceed 240,000!

Reaching the terrestrial atmosphere, these little bodies are heated by friction, and their motion is transformed into heat. If a shooting star weighs only a few grains, or even less, it is entirely volatilised and evaporated

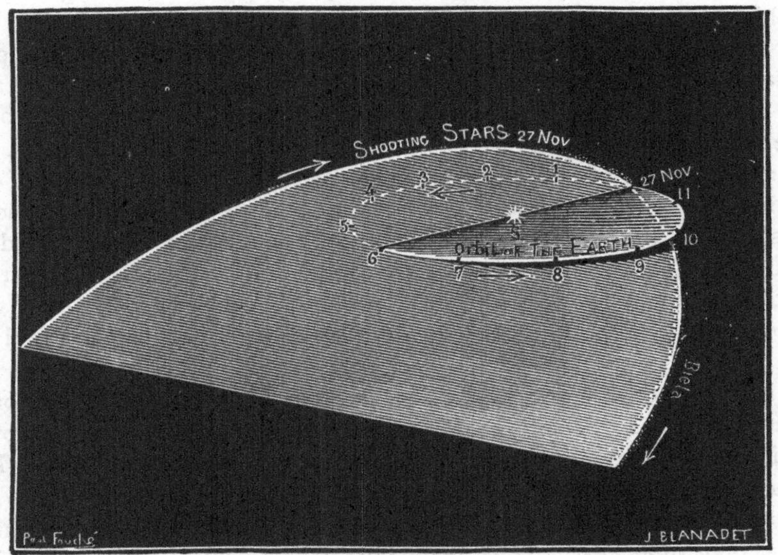

FIG. 234.—INTERSECTION OF THE ORBIT OF THE SHOOTING STARS OF NOVEMBER 27 WITH THE PLANE OF THE EARTH'S ORBIT

in the air; if it be a heavier bolide, it resists, but the whole of its external surface melts and becomes covered with a glazed layer. Supposing a bolide of 4 inches in radius, and of a density equal to 3·5 (that of water being 1), to enter the atmosphere with a velocity of 50,000 metres (31 miles) a second, we find that it suddenly develops a heat equal to 4,397,000 calories, and should lose 49,000 metres (30·4 miles) of velocity when it arrives at a height of 15,000 metres (9·32 miles); so that it reaches the surface of the ground with the small velocity of 5 metres, which explains the small depth of the holes which uranoliths make on reaching the ground. It is important, in fact, to distinguish between the sidereal velocity of bolides on their arrival, and that of their fall after their explosion.

We see now how and why these apparitions return periodically at fixed dates, and are subject to the intermissions which we have described.

Formerly, astronomers regarded the shooting stars as circulating round the sun in elliptical orbits, nearly circular, with a velocity analogous to that of the earth. Professor Schiaparelli, of Milan, struck with their velocity—which indicates a parabolic orbit, as we have remarked—suspected, in 1866, that they might have, like the comets, an origin foreign to our system, and advanced the following theory:—

Suppose a mass, nebulous or formed of corpuscles, situated at the limit of the sphere of action of our sun, which, endowed with a feeble relative motion, commences to feel the solar attraction; its volume being very considerable, its various portions are situated at very different distances. From this it follows that, when it commences to fall towards the sun, the points unequally distant acquire in time unequal velocities. Notwithstanding these differences, calculation proves that the perihelion distances of the different corpuscles will be very little modified, and the orbits will be so similar that the molecules will follow each other, forming a species of chain or stream which will take an extremely long time to pass round the sun. A mass of which the diameter would be equal to that of the sun would take several centuries to perform this movement. This stream will represent physically and visibly the orbit of the meteoric corpuscles, as a jet of water represents the parabolic trajectory of each molecule as an isolated projectile.

If in its motion of translation the earth encounters this procession of corpuscles, it will pass through, and a certain number of them will encounter it, their own velocity combining with that of the terrestrial globe. If the stream is very long, the earth will thus traverse it every year at the same point, encountering at each passage corpuscles different from those which it met the preceding year. It is then easy to calculate the position of this stream.

M. Schiaparelli has made these calculations for the two streams of August and November, and, by a fortunate circumstance, he has found that two well-known comets have their orbits precisely coincident with these two streams of meteors. The first is the great comet of 1862, which passed through perihelion on August 23 of that year, and of which the period of revolution is 121 years. Its orbit coincides with that of the meteors of August 10. The second is that which appeared in 1866, of which the period is thirty-three years, and which forms part of the November meteors.

This unexpected result throws a great light on the nature of shooting stars and their connection with cometary orbits. We may conclude at once that comets, like shooting stars, must be swarms of meteors derived from nebulous masses strangers to our planetary system.

It might be objected to this identity that the spectrum analysis of comets shows that they are formed, at least partly, of gaseous matter, while the shooting stars must be solid. But the spectroscope has solved even this difficulty. In fact, apart from the supposition that these stony masses may be surrounded by a gaseous and nebulous atmosphere to which we may attribute the cometary spectrum, spectrum analys s proves that their mass contains a large quantity of cometary gases in their pores, gas which is developed by the simple application of even a very moderate heat. Indeed, it has been ascertained that several meteorites contain carbon, as that of the Cape and that of Orgueil. Now, this substance may be vaporised at the time of the passage of the comet through perihelion and thus give the spectrum observed. The multiplicity of nuclei in certain comets is favourable to this hypothesis.

Besides the two comets mentioned above, several others have been found of which the orbits coincide with streams of meteors ; thus, the swarm of shooting stars of April 20, of which the centre of emanation is found in the constellation Hercules, is connected with Comet I. of 1861. It will be remembered that on the day on which the earth should have crossed the orbit of Biela's comet (November 27, 1872), the famous shower of shooting stars took place of which we have spoken, thus proving that, if we did not encounter the head of the comet, a little delayed, we at least traversed the stream which follows it. It was the same in 1885.

But we must not expect to find a comet for each shower of shooting-stars. The perturbations caused by the large planets on bodies so light are very considerable, and during so many ages, since the meteoric streams entered our solar system, they must have modified their primitive state.

The repulsive force exercised by the sun on the coma of a comet, which drives out the particles to begin the tail, exceeds that of the solar attraction, and at a relatively small distance from the nucleus of the comet the attraction of this nucleus should not be capable of preserving its substance. What becomes of it? It must be scattered in space. At each of its passages through the perihelion a comet must, therefore, lose a part of its substance, and the fact is that all the comets of short period are faint and, so to say, telescopic. According to the fantastic descriptions of the ancient chroniclers, it is certain that the old apparitions of Halley's comet must have been incomparably finer, more brilliant, and more astonishing than at the two last returns (1759 and 1835). Thus it is almost certain that comets diminish in size at each of their voyages round the sun.

Our readers will understand the orbits of the shooting stars and of those of the comets with which they are associated by figs. 232, 233, and 234, which represent the form and the position of these orbits relatively to that of the earth.

We may, then, admit that comets and shooting stars come from infinite space, and have been incorporated into our system—when they describe closed orbits—by the influence of the planets. Thus, for the swarm of shooting stars of November 14 Le Verrier has calculated that it entered for the first time into our system in the year 126 of our era, at a point near where the planet Uranus was then situated, and that it is this planet which has transformed the parabolic into an elliptic orbit. If the planet had not been there, the meteors would have continued their course along the dotted

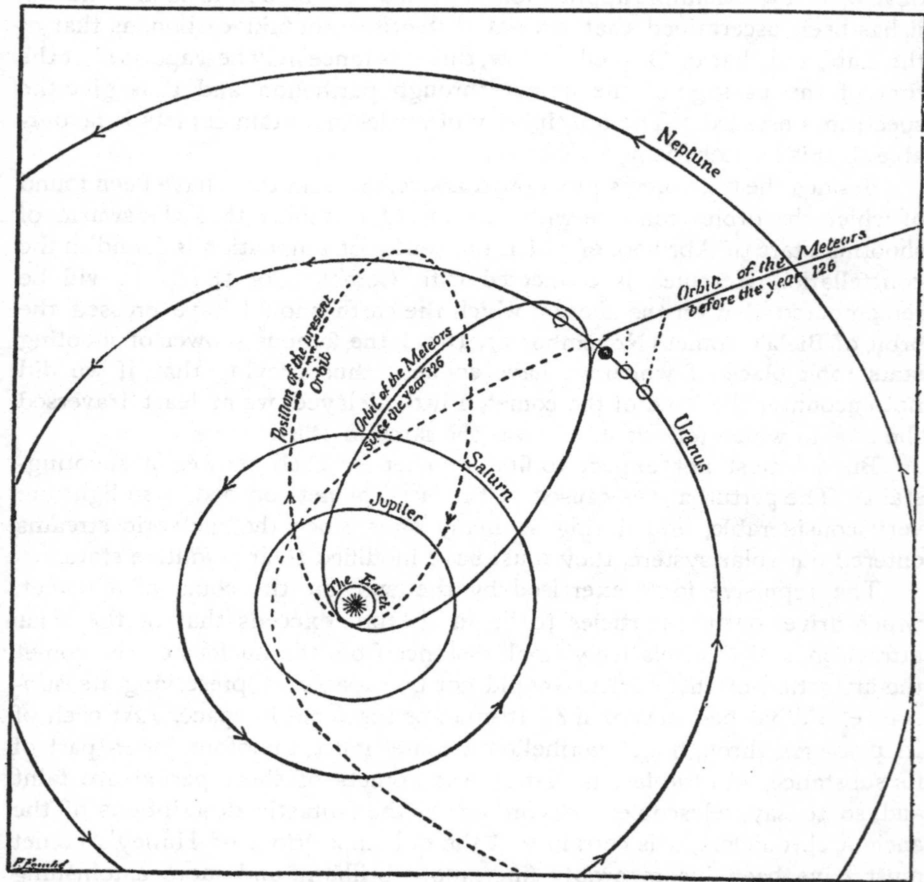

FIG. 235.—SHOWING HOW THE SHOOTING STARS OF NOVEMBER 14 HAVE BEEN DRAWN INTO OUR SYSTEM BY THE ACTION OF URANUS

line shown in fig. 235. The influence of Uranus, continuing to act, has displaced the orbit from the right towards the left, as we see in the figure.

Certain comets and shooting stars may, however, also be the volcanic products of the planets, as we shall see shortly with reference to uranoliths.

Such is the course of these minute shooting stars, a course now perfectly determined, as we see. A lesson as profound as unexpected, the shooting star itself does not glide by chance, borne along by an arbitrary wind; it describes a mathematical orbit as well as the earth or the colossal Jupiter. All is ruled, decreed by the supreme Law; and—who knows?—perhaps each of our frail existences, each of our ephemeral actions, is also determined by the invisible nature which places the star in the sky, the infant in the cradle, the old man in the tomb.

This perpetual addition of shooting stars is not without importance to our planet; it must slowly increase the volume and mass of the earth.[1]

[1] Assuming for their average size about a cubic millimetre, the annual number of shooting stars would represent a volume of 146 cubic metres and a weight of 876,000 kilogrammes (about 862 tons). In a hundred centuries this increase of volume would be 1,898,000 cubic yards, and the increase of weight would be 8,760 millions of kilogrammes (about 86,200 tons).

The superficial area of our planet measuring about 200 millions of square miles, if we suppose this cosmical dust uniformly spread, we see that in about 34,900 years the globe would increase by a layer of one centimetre in thickness, its diameter being increased by two centimetres. Doubtless this increase is of the order of the infinitely little, but it is precisely this order which acts most efficaciously in the whole of nature.

The weight of the terrestrial globe being estimated at 13,000 sextillions of pounds, the increase of mass in 100,000 years is but the 15-trillionth of the total mass of the earth. This is small, doubtless; but our planet has existed for many hundreds of thousands of years.

Thus *the earth sails in the midst of a space filled with cosmical materials, and increases gradually in weight and volume.* It has been calculated that in passing through these materials across a cylinder of a diameter equal to its own the earth encounters 13,000 bolides, or shooting stars, visible to the naked eye, and 40,000 telescopic shooting stars. Among these elliptical streams of shooting stars, that which the earth encounters on November 14 extends along a length of more than 1,600 millions of kilometres (994 millions of miles), revolves in thirty-three years round the sun, and contains a number of objects represented by the figure 1,000,000 × 100 × 100, or 100,000 millions of corpuscles!

Everywhere on the earth ferruginous dust is found of which the origin is due to myriads of meteoric corpuscles, which are ignited in our atmosphere, or vaporised, and slowly fall down. My industrious friend Silbermann, who has observed the shooting stars for so many years from the top of the College de France, and sees in them one of the most important springs of celestial mechanics, has often remarked undulatory motions, motions of recoil, and persistent trains of luminous dust. On October 5, 1879, a magnificent shooting star left behind it a spiral cloud, which remained visible for more than half an hour.

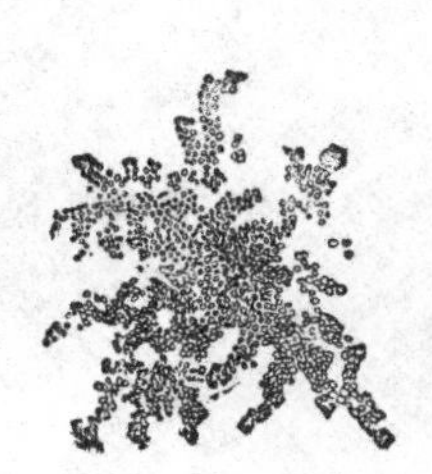

FIG. 236.—DUST FROM A SHOOTING STAR PICKED UP ON A SHIP. (Natural size)

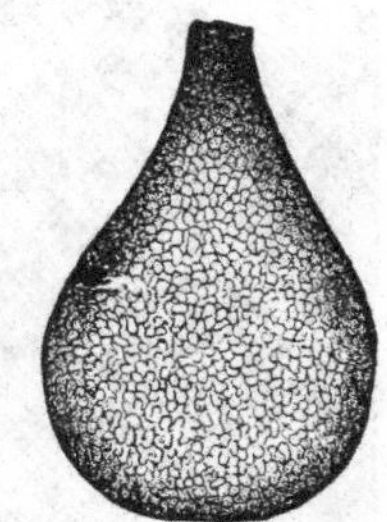

FIG. 237.—A GRAIN OF THE SAME DUST UNDER THE MICROSCOPE

More than half a century ago Mlle. Ehrenberg drew the dust of meteorites (here reproduced of the natural size, fig. 236) which fell on the deck of a ship crossing the Indian Ocean, and found that these fragments showed in the microscope blown forms similar to the residue of combustion of steel wire. The chemist Reichenbach found on isolated mountains, which probably the foot of man had never trod, traces of iron, cobalt, and nickel, which could only have fallen there from the

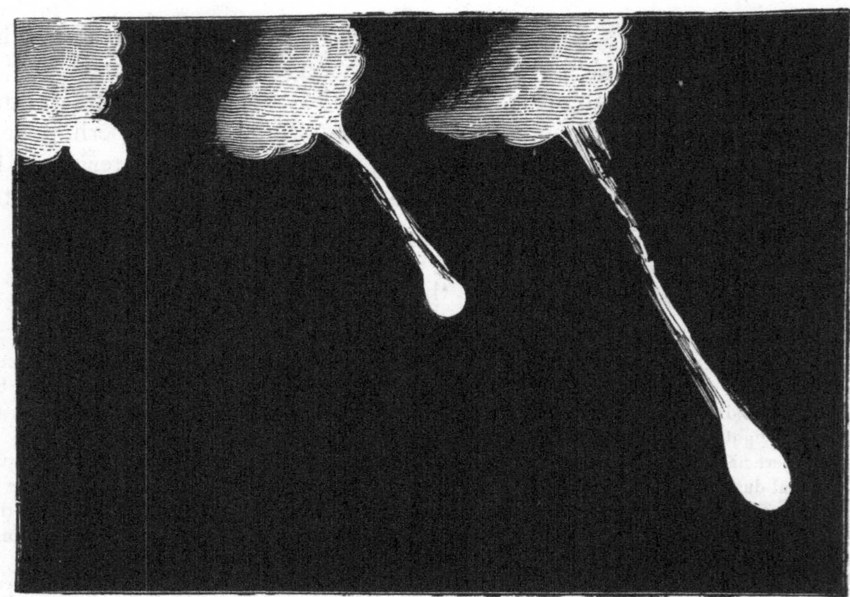

FIG. 238.—FIREBALL EMERGING FROM BEHIND A CLOUD, NOVEMBER 23, 1877, 8^h 24^m. (Drawn by J. Plant)

sky. M. G. Tissandier collected from the snows of Mont Blanc, on the towers of Notre Dame, and from rain water, the ferruginous particles here represented (fig. 239). We see that these celestial visitors—shooting stars, bolides, and aërolites—everywhere leave their traces.

The causes which act now have always so acted in variable proportions ; and as in geology we

We now come to the bolides and uranoliths.

A luminous body of sensible dimensions rapidly traverses space, diffusing on all sides a vivid light: like a globe of fire of which the apparent size is often comparable with that of the moon. This body usually leaves behind it a perceptible luminous train. On or immediately after its appearance it often produces an explosion, and sometimes even several successive explosions, which are heard at great distances. These explosions are also often accompanied by the division of the globe of fire into luminous fragments, more or less numerous, which seem to be projected in different directions. This phenomenon constitutes what is called a *meteor*, properly speaking, or a *bolide*. It is produced during the day as well as the night. The light which it causes in the former case is greatly enfeebled by the presence of the solar light, and it is only when it is developed with sufficient intensity that it can be perceived.

On the other hand, we sometimes find on the ground solid bodies of a stony or metallic nature, which appear to have nothing in common with the earth or rocks on which they lie. From time immemorial the vulgar have attributed to these bodies an extra-terrestrial origin; they believed them to be stones fallen from the sky. More than 2,000 years ago the Greeks venerated the famous stone which fell from heaven into the river Ægos; the chroniclers of the Middle Ages have preserved for us rough drawings of these inexplicable falls. Some naturalists designated them by the names of *lightning-stones, stones of thunder*, because they regarded them as matter shot forth by the lightning. They have also, it is true, confounded under the same name the mineral pyrites which is found in such great quantities in chalk lands; but this confusion of old times does not preclude the real existence of stony or ferruginous fragments which have authentically fallen from the sky. A rather curious fact is that, although the ancient traditions,

now explain the modifications of the terrestrial surface in the past and those of living species themselves by the slow action of causes which at present act before our eyes, so we may conclude that this slow and secular shower of shooting stars through the solar system *has increased the volume and mass of all the planets.* We may mention here a very remarkable consequence of the increase in volume of

FIG. 239.—DUST OF SHOOTING STARS GATHERED ON MONT BLANC

our planet. It is that *the earth's motion of rotation must be slackened*, and the length of the day increased; while the revolution of the moon must be accelerated, and appear to increase more rapidly than it does in reality. We have seen, in speaking of the moon, that this acceleration is, in fact, very probable.

the histories of antiquity and of the Middle Ages, and the popular beliefs, had distinctly spoken of *stones fallen from the sky*, stones of the air, aërolites, the *savants* would not believe in them. Either they denied the fact itself, or they interpreted it quite otherwise, regarding the stones fallen on the earth as shot out by volcanic eruptions, raised from the ground by waterspouts, or even produced by certain condensations of matter in the midst of the atmosphere. In 1790 the illustrious Lavoisier, and in 1800 the whole Academy of Sciences, declared these facts to be absolutely apocryphal. In 1794 Chladni proved the extra-terrestrial origin of these mysterious objects.

This almost general incredulity of the *savants* gave way when Biot read to the Academy of Sciences his report on the memorable fall which took place at Laigle, in the Department of the Orne, on April 26, 1803. After a minute inquiry made on the spot, the perfect accuracy of the circumstances related by public rumour of this very remarkable fall was verified. Numerous witnesses affirmed that some minutes after the appearance of a great bolide moving from south-east to north-east, and which had been perceived at Alençon, Caen, and Falaise, a fearful explosion, followed by detonations like the report of cannon and the fire of musketry, proceeded from an isolated black cloud in a very clear sky. A great number of meteoric stones were then precipitated on the surface of the ground, where they were collected, still smoking, over an extent of country which measured no less than 7 miles in length. The largest of these stones weighed less than 10 kilogrammes (22 lbs.).

Since then numerous falls have been no less authentically established. A single year does not pass without several being met with and several pieces being collected; some are shattered on the rocks, some buried in the soil to a depth of several feet. On July 23, 1872, on a fine summer's day, a stone fell at Lancé, near Blois, after an explosion which was heard for 80 kilometres (50 miles) round. It weighed 57 kilogrammes (126 lbs.), fell at a distance of 50 feet from a shepherd (who was naturally terrified), and buried itself 5¼ feet in a field. On April 31 following one fell near Rome with such a noise that the peasants believed 'that the vault of heaven had fallen down.' Its velocity was 59,500 metres (37 miles) a second on its arrival in the terrestrial atmosphere, and the explosion shattered it into fragments. This bolide appeared at $5^h\ 15^m$ in the morning at a vertical height of 184 kilometres (114 miles) above Rome; and what is more curious is, that an hour and a half before there was seen on the sea, in the direction from which the bolide came, a luminous mass, large and motionless. On May 14, 1864, a bolide which fell at Orgueil (Tarn-et-Garonne) was seen at a height of 65 kilometres (40 miles), and was perceived from Gisors (Eure) 500 kilometres (310 miles) distant. On January 31, 1879, one fell at Dun-le-Poëlier (Indre) near a husbandman, who gave himself up for dead. On April 6, 1885, at Chandpoor, India, a fall accompanied by a thunderclap and lightning frightened the natives, who saw an ignited object falling from the sky, and found it buried in the ground and quite hot. On July 7 of the same year a small uranolith fell in the prison of Valle, in Spain, and was picked up by one of the prisoners. On November 22, 1886, there fell in Russia, at Nowa-Urei, government of Penza, uranoliths containing diamonds! (See 'L'Astronomie,' 1888, p. 311.) We might easily multiply these examples. The names aërolite and meteorite are generally given to these minerals fallen from the sky. The term uranolith would suit them better, etymologically.

These masses are not insignificant, as we may judge by the following examples:—

1. A ferruginous uranolith found on a plain of sand in Chili, and weighing 104 kilogrammes (229 lbs.), was sent to the Paris Exhibition of 1867, and is now in the Paris Museum. Height, 48 centimetres (19 inches).

2. Stony uranolith fell at Murcia (Spain) on December 24, 1858, weighing 114 kilogrammes (251 lbs.); also sent to the Exhibition of 1867, and taken back to the Madrid Museum.

3. Uranolith which fell on November 7, 1492, at Ensischeim (Haut-Rhin), before the Emperor Maximilian at the head of his army (historical miracle, omen of victory: it would have been more curious still if it had fallen exactly on the head of the Emperor), weighing 158 kilogrammes (348 lbs.). It was placed at first in the church as a relic, and it is now in the Mineralogical Museum of Vienna.

4. Several thousands of stones fell on June 9, 1866, at Kniahynia (Hungary), with a terrible noise; the largest fragment, which is to be seen at Vienna alongside the preceding one, weighs 293 kilogrammes (646 lbs.).

5. Block of meteoric iron which had served from time immemorial as a seat at the door of the church of Caille (Alpes-Maritimes). Its weight is 625 kilogrammes (1,378 lbs.). It has been conveyed to Paris.

6. The Museum of Practical Geology, London, possesses a mass of iron, found in 1788 at Tucuman (Argentine Republic), which weighs 635 kilogrammes (1,400 lbs.).

7. A mass of meteoric iron found by Pallas in Siberia in 1749. This was one of the first aërolites recognised. It weighed 700 kilogrammes (1,543 lbs.), but the detaching of fragments from it has reduced it to 519 kilogrammes (1,142 lbs.). It forms part of the Paris Collection.

8. Uranolith of 750 kilogrammes (1,653 lbs.), fell in 1810 at Santa Rosa, New Granada. Its volume is nearly the tenth of a cubic metre.

9. Uranolith of 780 kilogrammes (1,720 lbs.), which served as an idol in the church of Charcas, Mexico; rescued by the care of the Commander-in-Chief of the expedition to Mexico, and at present in Paris. Height, 1 metre.

10. Uranolith of 1 metre in diameter fell on December 25, 1869, at Mouzouk, near a group of frightened Arabs.

11. Uranolith discovered in 1861 near Melbourne, Australia. Two fragments weigh together 3,000 kilogrammes (nearly 3 tons); one of these is at Melbourne, and the other in the British Museum, London.

12. The heaviest authentic aërolite which any collection possesses is that which was discovered in 1816 at Bendego, near Bahia, in Brazil, and which was conveyed in 1886 to Rio de Janeiro; it weighs 5,360 kilogrammes (about 5¼ tons). There was a model of it at the Exhibition of 1889.

To these uranoliths, weighed, analysed, and classed, we may add two other planetary fragments which are larger still, one weighing more than 10,000 kilogrammes (about 9½ tons), which fell in China near the source of the Yellow River, and measures 15 metres (50 feet) in height. The Mongols, who call it the *Rock of the North*, say that this mass fell after a great fire in the sky. The second lies in the plain of Tucuman (South America), and weighs about 15,000 kilogrammes (about 15 tons).

Savants have associated with these masses the enormous blocks of native iron of 10,000, 15,000, and 20,000 kilogrammes, found in 1870 by Professor Nordenskiöld at Ovifalk, Greenland, on the shore of the sea; but these blocks of native iron are of terrestrial origin. More authentic are the 25,000 kilogrammes of iron found in 1875 on a mountain in the province of St. Catherine, in Brazil, divided into fourteen blocks lying east and west in a straight line.

We see that, having begun with fragments of a few grains, we have arrived at respectable masses. Moreover, there must have fallen from time immemorial quantities of celestial iron, for the first instruments of iron forged by man were made from meteoric iron, and the ancient word by which they designated this metal, the word *sideros*, signifies a *star* as well as *iron*.

From several hundreds of analyses due to the most eminent chemists, it seems that meteors do not show any simple substance foreign to our globe. The elements which have been recognised with certainty up to the present are twenty-two in number. They are as follows, nearly in order of their quantity:—

Iron constitutes the predominant part; then come magnesium, silicon, oxygen, nickel (which is the principal companion of iron), cobalt, chromium, manganese, titanium, tin, copper, aluminium, potassium, sodium, calcium, arsenic, phosphorus, nitrogen, sulphur, traces of chlorine, and finally carbon and hydrogen. On the other hand, M. de Konkoly has analysed with the spectroscope several hundreds of shooting stars, and found in their nuclei a continuous spectrum with the lines of sodium, magnesium, strontium, lithium, and iron.

The density of uranoliths varies from 3 to 8, that of water being taken as unity; it is higher than that of the materials which form the exterior layers of the terrestrial globe, as far as we know, and reaches that of the lower layers. M. Daubrée, who has collected in the Paris Museum specimens of 260 falls, has classed these bodies in different types, according to the quantity of iron which they contain: (1) *holosidères*, entirely composed of pure iron, which can be forged directly (nickel is not always associated with it; such pure native iron has never been found on the earth)—specimens rare; (2) *syssidères*, composed of an iron paste in which there are stony parts, usually of olivine, resembling slag; (3) *sporadosidères*, composed of a stony paste in which iron, instead of being continuous, is disseminated in granules—very frequent; (4) *asidères*, in which there is no iron at all, like the aërolite of Orgueil—very rare.

Whence come the stones which fall from the sky? Their identity with the bolides is not doubtful, since every fall of an aërolite comes from a bolide. Should we go further, and identify the aërolites and bolides with the shooting stars? It seems not, for in the showers of shooting stars we do not remark enormous bolides nor falls of stones corresponding with these showers.[1] This fact shows us that, if the shooting stars move in space along elliptical orbits of the cometary order, the bolides and aërolites may have a different origin and course.

In studying the composition of the uranoliths we cannot help noticing the fact, assuredly well worthy of attention, that the constituent materials of these objects are identically the same as those which exist in the interior of our globe, below the sedimentary strata and the granitic layers. And not only are these bodies absolutely of the same nature as our minerals, but many of these fragments show mineral species associated in the same manner as in certain terrestrial rocks. Thus, for example, the stone which fell at Juvinas (Ardèche) in 1821 is almost exactly identical with certain lavas of Iceland; that which fell at Chassigny (Haute-Marne) in 1815 shows all the characters of terrestrial olivine, with the grains of chromic iron disseminated exactly as in the rock called *dunite*, discovered in New Zealand. The aërolite which fell at Soko-Banja (Servia), in 1877, presents a conglomerate or trachytic tufa similar to that which is found in the ancient volcanoes of Auvergne and on the banks of the Rhine. Olivine, pyroxene enstatite, anorthite felspar, chromic iron, magnetic pyrites, oxide of iron, graphite, and probably water, may be quoted among the minerals common to meteorites and the terrestrial globe. It is extremely remarkable that the three elements which predominate in meteorites generally—iron, silicon, and oxygen—are also those which predominate in our globe.

On the other hand, we never find in the uranoliths any fragment belonging to strata analogous to those of our stratified rocks, those which constitute the external and, so to say, vital casing of our globe. Never limestone, never the least sand, the smallest shell, the least trace of a fossil. All the silicious rocks which form the crust of our globe for a considerable thickness are wanting in the uranoliths.

It follows from these certain facts that the meteoric fragments which reach us proceed only from the interior parts of planetary bodies which must be constituted like our globe; and that, if they are the *débris* of the rupture of a globe, they must proceed from a world not arrived at the decay which the earth may reach some day, but from a globe unfinished, so to say, on which there have never been traces of life, and which must have followed an evolution less complete than the planet which we inhabit.

Thus, uranoliths are identical with the mineral rocks which exist at several miles below our feet, and which constitute the interior mass of our globe, materials denser than we ever find on the surface of our planet, unless in those which may have been projected by volcanic eruptions, or, on account of energetic pressure, have been driven into faults in the upper rocks [like metallic ores found in lodes.—J. E. G.]

The identity is so great between these materials and ours that M. Daubrée, having made his beautiful synthetical experiments to produce artificially minerals similar to those of uranoliths by imitating the processes of nature, adds, to complete the resemblance between the deep constitution of our globe and the composition of the stones fallen from the sky: 'There is nothing to prove that below these geological strata, which have furnished in Iceland, for example, lavas so similar to the type of the meteorites of Juvinas—that below our peridotic rocks, which so closely resemble the meteorite of Chassigny—there are not masses in which native iron begins to appear—that is to say, like the meteors of the common type; then, descending lower, types richer and richer in iron, of

[1] There has been one exception, however—on November 27, 1885. During the shower of shooting stars from the comet of Biela a magnificent uranolith, weighing 3,950 grammes (8·7 lbs.), fell at Mazapil, in Mexico; a description of this fall and a fragment of stone were sent to us by our learned Mexican colleague, M. Bonilla, Director of the Observatory of Zacatecas. This uranolith establishes for the first time a bond of union between the celestial bodies and comets. But perhaps there is here merely a coincidence.

which the meteorites show us a series of increasing density, from those where the density of iron represents nearly half the weight of the rock to massive iron itself.'

Let us now state another fact not less remarkable than this analogy between the uranoliths and the deep rocks of our own planet—*the identity of structure* of stones which have fallen from the sky *at very different epochs.* Meteorites distant from each other from the double point of view of geography and chronology sometimes present the most complete identity, so much so that *it is impossible to distinguish the different specimens.* The conclusion is that almost all these fragments, at least a very large number, proceed from the same bed, which belonged to a planet like that which we inhabit, or, we should rather say, similar to the earth in early ages, to the earth anterior to the ocean and the sedimentary deposits—that is, to a protozoic earth.

It happens also, on the other hand, that the same fall of uranoliths brings down stones of different natures. Thus, on November 17, 1773, at Sigena, Spain, and on November 12, 1856, at Trenzano, Italy, falls of stones brought two very different types at the same time; one of these types is similar to that of a meteorite which fell at Bustu, India, on December 2, 1852, the other to that of a meteorite which fell at Parnally in 1857; these were different rocks which arrived together, and this twice under the same conditions.

We might imagine an ideal globe from which the uranoliths would proceed, an ideal globe of which the density would increase from the surface towards the centre. At the outside there would be the aluminous rocks, then would come olivine stones, those of the common type, the *polysidères*, the *syssidères*, and finally the *holosidères*. This theoretical section is not without analogy with an ideal section of the terrestrial globe below the sedimentary strata and the granitic layer. In this section the lavas would correspond with the aluminous meteorites; below, the olivine would be analogous to the stone of Chassigny.

From this it is but a step to see in the uranoliths the *débris* of one or more ruined worlds. 'All the *débris of celestial bodies* scattered with profusion through space, and which fall on our planet,' again writes M. Daubrée, 'are products certainly formed under the action of a strong heat, and thus confirm the universality of origin, by the igneous method, of cosmical bodies. Those fragments which come into contact with our planet show us one of the modes of change which are produced in the universe by the division of the *débris* of demolition of certain stars or asteroids into other stars. These encounters do not constitute an accidental fact or exception, but rather a rule, a sort of evolution.'

The same idea had already been expressed by Chladni in 1794. 'Nature,' he wrote ('Journal des Mines,' year XII., p. 479), 'has the power of forming celestial bodies, of destroying them, and of re-forming others from their *débris*.' And the ideal globe of which we have just spoken had already been imagined, in 1855, by the American mineralogist, Lawrence Smith, in 1850 by Boisse de Rodez, and even in 1840 by Angelot.

M. S. Meunier has attempted to reconstitute theoretically this imaginary globe from which the uranoliths might proceed, and has advanced the idea that it might have been an ancient satellite of the earth which was shattered in pieces.

We have now explained the course of ideas relating to the search for the origin of uranoliths. It seems that the imagination may have gone a little too far, and perhaps to no purpose, in this very interesting research.

Celestial mechanics prove that those bodies which, at the distance of the moon, would revolve round the earth with the velocity of only 1,000 metres (3,280 feet) per second would, in reaching our atmosphere and continuing to revolve round us, attain a maximum velocity of only 8,000 metres (5 miles). Now, there are numerous examples of velocities considearlby greater than this.

Thus, the bolide which passed over England on November 6, 1869, from north-east to south-west, travelled in five seconds from the zenith of Somersetshire to St. Ives, in Cornwall, describing 273 kilometres (170 miles) in this interval, or 54,000 metres (33½ miles) a second; its height was 145,000 metres (90 miles) above the first point, and 43,000 metres (26¾ miles) above the second, when the meteor disappeared at the sea horizon. That which traversed Austria and France from east to west on September 5, 1868, passed in seventeen seconds from the zenith of Belgrade to that of Mettray (Indre-et-Loire), having described 1,493 kilometres, which gives 88 kilometres (54½ miles) per second. That of June 14, 1877, which burst between Bordeaux and Angoulême at 252 kilometres (156 miles) high, arrived with a velocity of 68 kilometres (42 miles) a second; &c.

In order to admit that this imaginary world ever existed, and that we now receive its *débris*, it would be necessary to suppose that it was destroyed before being finished—that it never passed through the phases through which our own planet has passed, and through which we see all the other planets also passing; that it never had any condensation of water, any seas, any sedimentary strata, since these strata are absolutely foreign to all the *débris* which reach us, but that it had the density and the mineralogical composition of our own rocks.

If only these fragments which arrive from space had anything celestial, were it only the form! if only they were large enough! But no. It is but to exaggerate their importance to give them the title of '*débris* of worlds in ruins.' If they were really pieces of planets or satellites, would they not show, sometimes at least, dimensions in harmony with their origin? Should we not see Alps or Pyrenees arrive, or at least some mountain or hill? But nothing of the sort. The largest of the uranoliths collected does not measure more than a cubic yard. In general they are of the size of a stone which we can hold in the hand—often that of an egg, more often that of walnuts, or even hazel-nuts.

Well, these stones from the sky being *of the same composition* as the minerals of which our own planet is formed, is it not natural to ask simply whether *they may not have had the earth itself for their origin?*

But how? May not the violent volcanoes of geological times, the eruptions, the tremendous conflagrations, the fierce fires of the ancient Pandemonium, have shot into space lava, scoria, stones, with such a force of projection that these objects would be despatched to thousands, millions, hundreds of millions of miles, in orbits which would not take less than a thousand, ten thousand, a hundred thousand years or more to describe?

The question is worth the trouble of examination.

What velocity would it be necessary to give a terrestrial projectile in order to make it a uranolith?

The reply to this problem is not very difficult. The velocity which it would be necessary to give to a body to send it to infinity is exactly the same as that with which a body reaching the earth from infinity would be animated, in virtue of the earth's attraction alone. It is also the same as that which would animate a body falling from the surface of the earth to its centre, supposing gravity constant. The formula simplified is reduced to—

$$v = \sqrt{2gR},$$

in which g is the acceleration of gravity at the surface of the earth, supposed spherical (9·80 metres), and R the mean radius of the earth (3,956 miles).

This velocity is 11,000 metres (6·82 miles per second).

Every body shot from the earth with a velocity *superior* to that would be freed for ever from the attraction of our planet: *it would travel eternally in the infinite, and would never return.* It would fly along the trajectory of a *parabola*.

Every body shot with a velocity less than this limit, but greater than 8,000 metres (5 miles), would recede from the earth describing the first half of an ellipse; then it would stop, close its curve, *and return towards the point whence it started.* It would follow in space an ellipse of which the eccentricity would depend on the force of projection.

If the earth existed alone, motionless in Immensity, our projectile, not being subject to any foreign influence, any perturbation, would describe a regular ellipse, and would return at the end of a certain time into the mouth of the very volcano from which it started. This time might be very long; for a velocity very near the limiting velocity it would extend to *several millions of years.* Its velocity on its return to the earth would be likewise 11,000 metres (6·82 miles).

But our planet is not alone in the universe. The fate of each projectile shot from its bosom would be different, according to the direction of departure. A body shot towards the sun would simply go on and fall on him. A body shot in a direction opposite to the sun would penetrate far into the depths of space, if—a very rare chance, however—Jupiter, Saturn, Uranus, or Neptune were not exactly on its road to modify its course. But on its return, instead of being subject to the attraction of the earth only, it would be subject to that of the whole solar system; it would make the circuit of the sun like a comet, and it would return towards the point of the terrestrial orbit from which it started; and that at each of its revolutions, so long as it did not encounter the earth itself, and thus finish its career. If our planet formerly had volcanoes capable of sending out such projectiles, all those shot out, whatever their number, would return at each of their revolutions to cross the terrestrial orbit, and arrive with the velocity of comets and shooting stars, with the velocity due to the sun's attraction added to their own velocity—that is to say, not only with the circular velocity of the earth's annual translation round the sun (29,500 metres (18·3 miles) per second), but with an elliptical velocity, of which the maximum limit is 29,500 $\sqrt{2}$, or 41,700 metres (26 miles). A uranolith coming towards the earth in a direction contrary to our motion of translation might then, from this fact even, be animated on its part with a velocity of 41,700 metres (26 miles), and encounter the earth coming with a velocity of 29,500 metres (18·3 miles), which gives a total of 71,200 metres (44 miles) as a maximum for the arrival of an aërolite in our atmosphere. (That of the diurnal rotation of our globe should be added or deducted according to the direction.) Now, these numbers are precisely those of the velocities ascertained by observation.

Thus, every stone shot from a terrestrial volcano with a sufficient initial velocity, included between 5 miles and 6·82 miles per second, neglecting the resistance of the air, would describe in space a closed curve, an extremely elongated ellipse—would become by this fact an asteroid of the solar system, and would cross the terrestrial orbit at each of its revolutions. Is such a force admissible? What is the power of our present volcanoes? During the eruption of Vesuvius in 1822, an enormous mass of lava, weighing several thousand kilogrammes, was shot by the volcano to a distance of nearly 5 kilometres (over 3 miles) into the garden of Prince Ottajana. The force which raised this mass and projected it thus was evidently much superior to that of our most powerful cannon, which already reach 2,300 feet for the initial velocity of certain projectiles. The fantastic eruption of Krakatoa, August 26, 1883, shot dust to 70 kilometres (43½ miles) in height, disturbed the whole terrestrial atmosphere, and was felt even at its antipodes! Given the elements in activity in volcanic eruptions, especially in the ancient convulsions of the planet, and the force required to produce the effects of which we speak is in harmony with the conditions of its production.

Consequently, the chemical constitution of the stones fallen from the sky showing us in themselves specimens of the constituent materials of our globe, it is possible, it is even probable, that these stones originally came from the earth; that they were formerly shot into space; that, freed from the attraction of our planet, they revolve round the sun along very elongated ellipses, and that the earth encounters them and recovers possession of them anew by this fact alone—that they ***inevitably return*** to the terrestrial orbit from whence they started. The translation of the solar system in space does not prevent these stones from returning to us, on account of the well-known principle of the independence of simultaneous motions.

Nature even seems to have wished to put us in the way of this very simple explanation by a recent proof. There are on the surface of the earth blocks of native iron absolutely identical with uranoliths. During the year 1870 M. Nordenskiöld discovered in Greenland, on the island of Disco, near Ovifalk, 15 blocks of iron, the largest of which, weighing 20,000 kilogrammes (about 20 tons), exceeds the greatest masses of the same nature which have been described. These blocks were found one beside the other on a surface which did not exceed 50 metres (165 square feet). They were considered as having fallen from the sky, and such was the opinion formed and adopted, especially by our Academy of Sciences.

Well, these blocks are not meteorites at all. It has been shown that they proceed from a terrestrial bed, and that they are associated with eruptions of basalt belonging to Greenland itself. On the other hand, the phenomena of terrestrial magnetism unite with geology to lead us to conclude that native iron exists at the base of the whole crust of our globe, at some miles in depth below the inhabited surface.

To complete this account we may add now that, if we consider as very likely this terrestrial origin of stones fallen from the sky, we do not on that account decline to admit that a certain number of these stones—those, for example, which are not exactly identical with terrestrial materials—may come from elsewhere. If our planet has been able to give birth to such projectiles, it does not form an exception in the universe, and the other celestial bodies may be in the same case. Thus, the sun itself is seen to be almost constantly surrounded with tremendous metallic gaseous eruptions, which are shot out to thousands and even hundreds of thousands of miles above its surface. On September 7, 1871, the American astronomer, Young, measured an explosion which rose in ten minutes to 300,000 kilometres (186,400 miles), with a velocity of ascension of 267,000 metres (166 miles) a second. Secchi, at Rome, measured one day a velocity of 370,000 metres (230 miles), and Respighi mentions 600,000 (373 miles), 700,000 (435 miles), and even 800,000 metres (497 miles). Now, according to our formula, we have for the sun—

$$v = \sqrt{2g\,\mathrm{R}} = \sqrt{2 \times 268^{\mathrm{m}} \times 691{,}000{,}000^{\mathrm{m}}} = 608{,}000 \text{ metres};$$

and if we divide this last number by $1{\cdot}414 = \sqrt{2}$, we find for the minimum limit 430,000 metres (267 miles). Hence, every body shot from the sun (neglecting the resistance of the solar atmosphere) with an initial velocity included between 430,000 and 608,000 metres would escape from the sun, would circulate in the solar system like a periodical comet, would be liable to encounter the planets, and would be, for the planet encountered, a bolide and uranolith.

We have for the moon—

$$v = \sqrt{2 \times 1^{\mathrm{m}}{\cdot}60 \times 1{,}742{,}000^{\mathrm{m}}} = 2{,}360 \text{ metres}.$$

Every body shot from the moon with this maximum velocity and down to the minimum velocity of 1,668 metres (5,472 feet), would not fall back to the moon, and might either fall to the earth, if its direction were suitable for this end, or revolve round the earth like a satellite. Some uranoliths may, then, have been sent from the moon, supposing that there are on its surface volcanoes endowed with this power—a rather feeble one, moreover. Laplace and Berzelius adopted this origin as a general thesis.

Similar projectiles might also come from the small planets, from Mars, Jupiter, Uranus, and Neptune, and even from the stars; but all these causes united amount to but an extremely small probability of an encounter with our globe, whilst the terrestrial origin necessarily brings back their battalions towards the path annually described by our planet round the sun. The results of calculation, then, united to the identity of uranoliths with our terrestrial volcanic rocks, renders very admissible, very likely, the theory of the terrestrial origin of the stones which come to us from the sky. This theory is not, however, as novel as it appears. It is, perhaps, the most ancient of all the explanations which have been given of the aërolites since the fall of the famous stone of Ægos Potamos, which fell into that river in Thrace in the year 465 before our era. But the ancients did not take into account either the intimate structure of these minerals or the volcanic force necessary to project the materials into space—they even believed them for a long time to be atmospherical concretions or thunderbolts—or the nature of the orbits which the volcanic products should describe in space in order to fall again on the earth in the state of celestial bodies.

To recapitulate. The following hypotheses are advanced to explain the origin of stones which fall from the sky, authentic aërolites, which we see arriving with a tremendous velocity from the heights of the atmosphere, and which are collected, analysed, and preserved.

1. *Volcanoes of the Moon.*—The great velocities observed, the densities, the directions of arrival, the curves which these projectiles describe, and the probable present state of the lunar volcanoes, unite to show that this origin can only be rare and exceptional.

2. *Ruins of Worlds scattered in Space which the Earth would encounter in its Passage.*—The annual number of falls of uranoliths is so large that the quantity of this *débris* should be considerable, and should add to the solar system a real increase of mass, perceptible to astronomical observations—which is not the case.

3. *Swarms of Matter analogous to the Elliptical Streams of Shooting Stars.*—In this case the falls would show characters of annual periodicity which is not remarked.

4. *Débris of a Vanished World.*—If this world were a stranger to our system, an encounter would be very improbable. If it had been a planet of our system, the earth could not cross its orbit. If it had been a satellite, the directions and velocities would be different. In these three cases the fragments would be in harmony with the origin. And then, such a world would have been destroyed before being finished, which is improbable.

5. *Remains of Primitive Cosmical Matter.*—An hypothesis still less admissible. These bodies are not spherical, are not whole—they are pieces, and, moreover, they are true minerals formed in the interior of a planet under conditions of temperature and pressure which accompany geological formations.

6. *Results of Volcanic Eruptions.*—This is the most rational hypothesis. Such eruptions may take place on all worlds. However, the terrestrial eruptions would make the products return to us, whereas the others would be sent in all directions. Moreover, the identity of structure of most of the uranoliths with terrestrial minerals presents itself as an eloquent witness in favour of this hypothesis, which may be summed up thus :

Most of the stones which fall from the sky may be natives of the earth itself, having been projected into space by the volcanic eruptions of geological times.

Having seen that these enormous masses have fallen from the sky, we may ask whether their fall on the earth may not produce accidents, not only to human life, but to the planet itself. Observation has already ascertained certain historical facts with reference to this point. Such are, for example, the fall of the year 616, which shattered chariots, the Chinese annals say, and killed ten men ; that of 944, which, according to the chronicle of Froissart, ignited houses ; that of March 7, 1618, which set fire to the Palais de Justice in Paris ; those of 1647 and 1654, which killed, the first, two men at sea, the second, a Franciscan at Milan. In 1879, a peasant of Kansas City (United States) was killed by an aërolite, which shattered a tree, reaching the ground with an amazing velocity. After the preceding descriptions there is nothing surprising in these facts. However, aërolites, whatever may be their number, are incomparably less destructive than lightning, for this kills every year ninety persons in France alone. Human folly destroys on an average, over the whole globe, 400,000 per century by perpetual international war, which amuses everybody.

BOOK VI

THE STARS AND THE SIDEREAL UNIVERSE

CHAPTER I

THE CONTEMPLATION OF THE HEAVENS

THE earth is forgotten, with its small and ephemeral history. The sun himself, with all his immense system, has sunk in the infinite night. On the wings of inter-sidereal comets we have taken our flight towards the stars, the suns of space. Have we exactly measured, have we worthily realised the road passed over by our thoughts? The nearest star to us reigns at a distance of 275,000 times 37 millions of leagues—that is to say, at ten trillions [1] of leagues (about twenty-five billions of miles); out to that star an immense desert surrounds us, the most profound, the darkest, and the most silent of solitudes.

The solar system seems to us very vast, the abyss which separates our world from Mars, Jupiter, Saturn, and Neptune appears to us immense; relatively to the fixed stars, however, our whole system represents but an isolated family immediately surrounding us: a sphere as vast as the whole solar system would be reduced to the size of a simple point if it were transported to the distance of the nearest star. The space which extends between the solar system and the stars, and which separates the stars from each other, appears to be entirely void of visible matter, with the exception of nebulous fragments, cometary or meteoric, which circulate here and there in the immense voids. Nine thousand two hundred and fifty systems like ours (bounded by Neptune) would be contained in the space which isolates us from the nearest star!

If a terrible explosion occurred in this star, and if the sound could traverse the void which separates it from us, this sound would take more than three millions of years to reach us.

It is marvellous that we can perceive the stars at such a distance. What

[1] [The French trillion is equivalent to the English billion, or a million times a million (1,000,000,000,000).—J. E. G.]

an admirable transparency in these immense spaces to permit the light to pass, without being wasted, to thousands of billions of miles! Around us, in the thick air which envelops us, the mountains are already darkened and difficult to see at seventy miles; the least fog hides from us objects on the horizon. What must be the tenuity, the rarefaction, the extreme transparency of the ethereal medium which fills the celestial spaces!

Let us suppose ourselves, then, on the sun nearest to ours. From there our dazzling furnace is already lost like a little star, hardly recognisable among the constellations: earth, planets, comets sail in the invisible. We are in a new system. If we thus approach each star we find a sun, while all the other suns of space are reduced to the rank of stars. Strange reality!—the normal state of the universe is night. What we call day only exists for us because we are near a star.

The immense distance which isolates us from all the stars reduces them to the state of motionless lights apparently fixed on the vault of the firmament. All human eyes, since humanity freed its wings from the animal chrysalis, all minds since minds have been, have contemplated these distant stars lost in the ethereal depths; our ancestors of Central Asia, the Chaldeans of Babylon, the Egyptians of the Pyramids, the Argonauts of the Golden Fleece, the Hebrews sung by Job, the Greeks sung by Homer, the Romans sung by Virgil—all these earthly eyes, for so long dull and closed, have been fixed from age to age on these eyes of the sky, always open, animated, and living. Terrestrial generations, nations and their glories, thrones and altars have vanished: the sky of Homer is always there. Is it astonishing that the heavens were contemplated, loved, venerated, questioned, and admired, even before anything was known of their true beauties and their unfathomable grandeur?

Better than the spectacle of the sea calm or agitated, grander than the spectacle of mountains adorned with forests or crowned with perpetual snow, the spectacle of the sky attracts us, envelops us, speaks to us of the infinite, gives us the dizziness of the abyss; for more than any other, it seizes the contemplative mind and appeals to it, being the truth, the infinite, the eternal, the all. Writers who know nothing of the true poetry of modern science have supposed that the perception of the sublime is born of ignorance, and that to admire it is necessary not to know. This is assuredly a strange error, and the best proof of it is found in the captivating charm and the passionate admiration which divine science now inspires, not in some rare minds only, but in thousands of intellects, in a hundred thousand readers impassioned in the search for truth, surprised, almost ashamed at having lived in ignorance of and indifference to these splendid realities, anxious to incessantly enlarge their conception of things eternal, and feeling admiration increasing in their dazzled minds in proportion as they penetrate farther into Infinitude. What was the universe of Moses, of Job, of

Hesiod, or of Cicero, compared to ours! Search through all the religious mysteries, in all the surprises of art, painting, music, the theatre, or romance, search for an intellectual contemplation which produces in the mind the impression of truth, of grandeur, of the sublime, like astronomical contemplation! The smallest shooting star puts to us a question which it is difficult not to hear; it seems to say to us, What are we in the universe? The comet opens its wings to carry us into the profundities of space: the star which shines in the depths of the heavens shows us a distant sun surrounded with unknown humanities who warm themselves in his rays. Wonderful, immense, fantastic spectacles, they charm by their captivating beauty and transport into the majesty of the unfathomable the man who permits himself to soar and wing his flight to Infinitude.

Nel ciel che più della sua luce prende
Fu' io, e vidi cose che redire
Nè sa, nè può qual di lassù discende.

'I have ascended into the heavens, which receive most of His light, and I have seen things which he who descends from on high knows not, neither can repeat,' wrote Dante in the first canto of his poem on 'Paradise.' Let us, like him, rise towards the celestial heights, no longer on the trembling wings of faith, but on the stronger wings of science. What the stars would teach us is incomparably more beautiful, more marvellous, and more splendid than anything we can dream of.

Among the innumerable army of stars which sparkle in the infinite night, the gaze is especially arrested by the most brilliant lights and by certain groups which vaguely present a mysterious bond between the worlds of space. These groups have been noticed at all epochs, even among the rudest races of men, and from the earliest ages of humanity they have received names, usually derived from the organic kingdom, which give a fantastic life to the solitude and the silence of the skies. Thus were early distinguished the seven stars of the North, or the Chariot, of which Homer speaks; the *Pleiades*, or the 'Poussinière;' the giant *Orion*; the Hyades in the head of Taurus; *Boötes*, near the Chariot or Great Bear. These five groups were already named more than 3,000 years ago, and so were the brightest stars of the sky, *Sirius* and *Arcturus*, &c.

The epoch of the formation of the constellations is unknown, but we know that they were established successively. The centaur Chiron, Jason's tutor, has the reputation of having first divided the sky on the sphere of the Argonauts. But this is mythology; and, besides, Job lived before the epoch at which Chiron is supposed to have flourished, and Job had already spoken of Orion, the Pleiades, and the Hyades 3,000 years ago. Homer also speaks of these constellations in describing the famous shield of Vulcan. 'On its surface,' says he, 'Vulcan, with a divine intelligence traces a thousand varied pictures. He represents the earth, the heavens, the sea, the indefatigable sun, the moon at its full, and all the stars which wreath the sky: the Pleiades, the Hyades, the brilliant Orion, the Bear, which they also call the Chariot, and which revolves round the pole; this is the only constellation which does not dip into the ocean waves' ('Iliad,' chapter xviii.).

Several theologians have affirmed that it was Adam himself, in the terrestrial paradise, who gave their names to the stars; the historian Josephus assures us that it was not Adam, but his son Seth, and that in any case astronomy was cultivated long before the Deluge. This nobility is sufficient for us.

Attentive observation of the sky also noticed from the beginning the beautiful stars *Vega* of the Lyre, *Capella* of Auriga, *Procyon* of the Little Dog, *Antares* of the Scorpion, *Altair* of the Eagle, *Spica* of the Virgin, the *Twins*, the *Chair* of Cassiopeia, the Cross of the White *Swan*, stretched in the midst of the *Milky Way*. Although noticed at the epoch of Hesiod and Homer, these constellations and stars were probably not yet named, because doubtless men had not yet felt the necessity of registering them for any application to the calendar, to navigation, or to voyages.[1]

At the epoch when the maritime power of the Phœnicians was at its apogee, about 3,000 years ago, or twelve centuries before our era, it was the star β of the Little Bear (see fig. 17, p. 40) which was the nearest bright star to the pole, and the skilful navigators of Tyre and Sidon (O purpled kings of former times! what remains of your pride?) had recognised the seven stars of the Little Bear, which they named the Tail of the Dog, *Cynosura*; they guided themselves by the pivot of the diurnal motion, and during several centuries they surpassed in precision all the mariners of the Mediterranean. The Dog has given place to a Bear, doubtless on account of the resemblance of the configuration of these seven stars to the seven of the Great Bear, but the tail remains long and curled up, in spite of the nature of the new animal.

Thus the stars of the North at first served as points of reference for the first men who dared to venture on the seas. But they served at the same

[1] The Chinese had designated them all, it is true, at the same epoch, but their groups as well as their denominations are absolutely different from ours, and do not appear to have exercised any influence on the foundations of astronomical history. It was another world, other methods, other inspirations, as if Asia and Europe formed two distinct planets. A distinguished author, M. Schlegel, published in 1875 a Chinese Uranography, which is composed of 670 asterisms, and of which he believes he can trace back the origin to 17,000 years before our era. His argument is not convincing, and it seems to me that the origin of the astronomy of the Celestial Empire cannot be very much anterior to the reign of the Emperor Hoang-Ti—that is to say, to the twenty-seventh century before our era—and would go back at farthest to the time of Fou-Hi—that is to say, to the twenty-ninth century. It was about the same epoch—the twenty-eighth century before our era—that the Egyptians, observing *Sirius*, the early rising of which announced the inundation of the Nile, formed their canicular year of 365 days.

time as guides on the mainland for the nomadic tribes who carried their tents from country to country. In the midst of savage nature, the first warriors themselves had nothing but the Little Bear to guide their steps.

Imperceptibly, successively, the constellations were formed. Some groups resemble the names which they still bear, and suggested their denomination to the men of ancient times, who lived in the midst of nature and sought everywhere for relations with their daily observations. The Chariot; the Chair; the Three Kings, also named the Rake; Jacob's Staff and the Belt of Orion; the Pleiades, or the Hen and Chickens; the Arrow (Sagitta); the Crown; the Triangle; the Twins; the Dragon; the Serpent; and even the Bull, the Swan, the Giant Orion, the Dolphin, the Fishes, the Lion, Water and Aquarius (the Water-bearer), &c., have given rise to the analogy. These resemblances are sometimes vague and far-fetched, like those we find in the clouds; but it appears much more natural to admit this origin than to suppose, with the classic authors, that these names were suggested by the concordance between the seasons or the labours of the fields and the presence of the stars above the horizon. That the name of the Balance (Libra) was given to the constellation of the equinox because then the days are equal, seems to us more than questionable; that Cancer (the Crab) signifies that the sun goes back to the solstice, and that the Lion has for its object to symbolise the heat of summer, and Aquarius the rain and inundations, appears to us no less imaginary. However, they have also had other origins. Thus, the Great Dog Sirius certainly announced the rising of the Nile and the dog-days (which remain in our calendar as a fine type of anachronism). Poetry, gratitude, the deification of heroes, mythology, afterwards transferred to the sky the names of personages and sovereigns—Hercules, Perseus, Andromeda, Cepheus, Cassiopeia, Pegasus; later, in the Roman epoch, they added the Hair of Berenice and Antinous; later still, in modern times, they added the Southern Cross, the Indian, the Sculptor's Workshop (Cœlum), the Lynx, the Giraffe (Cameleopardus), the Greyhounds (Canes Venatici), the Shield of Sobieski, and the little Fox (Vulpecula). They even placed in the sky a Mountain, an Oak, a Peacock, a Swordfish, a Goose, a Cat, a Crane, a Lizard, and a Fly, for which there was no necessity.

This is not the place to describe and draw in detail all these constellations, with their more or less strange figures. The important point for us here is to form a general idea.

The sky remains divided into provinces, each of which continues to bear the name of the primitive constellation. But it is important to understand that the positions of the stars themselves, as we see them, are not absolute, and that the different configurations which they may show us are only a matter of perspective. We already know that the sky is not a concave sphere on which brilliant nails could be attached ; that it is not a species of vault ; that an immense infinite void envelops the earth on all sides, in all directions. We know also that the stars, the suns of space, are scattered at all distances in the vast immensity. When, therefore, we remark in the sky several stars near each other, that does not imply that these stars form the same constellation, that they are on the same plane, and at an equal distance from the earth. By no means ; the arrangement which they assume to our eyes is but an appearance caused by the position of the earth relatively to them. This is a mere matter of perspective. If we could leave our world, and transport ourselves to a point in space sufficiently distant, we should see a variation in the apparent arrangement of the stars so much the greater as our station of observation were more distant from where we are at present. A moment's reflection is sufficient to convince us of this fact, and save us from insisting further on this point.

Once these illusions are appreciated at their true value, we can begin the description of the figures with which the ancient mythology has constellated the sphere. A knowledge of the constellations is necessary for the observation of the heavens and for the researches which a love of the sciences and curiosity may suggest ; without it we find ourselves in an unknown country, of which the geography has not been made, and where it would be impossible to know our exact position. Let us make, then, this celestial geography ; let us see how to find our way, in order to read readily in the great book of the heavens.

CHAPTER II

GENERAL DESCRIPTION OF THE CONSTELLATIONS

How the Principal Stars are recognised

THERE is a constellation which everybody knows; for greater simplicity we will begin with it. It will serve us well as a point of departure from which to go to the others, and as a point of reference to find its companions. This constellation is the *Great Bear*, which has also been named the *Chariot of David.*

It may well boast of being celebrated. If, however, notwithstanding its universal notoriety, some of our youngest readers have not yet made its acquaintance, the following is a description by which they may always recognise it.

Turn yourself towards the north—that is to say, opposite to the point

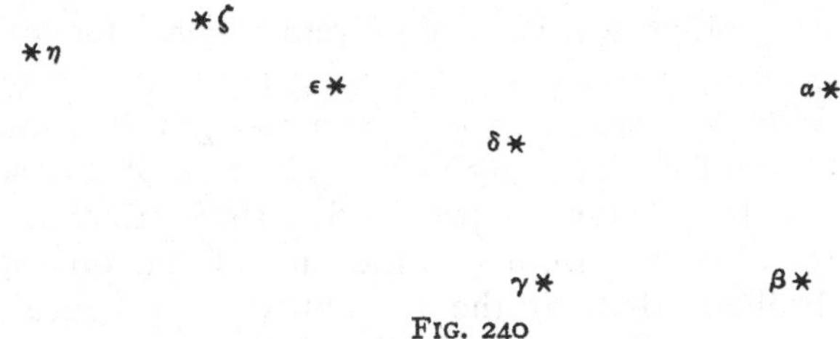

FIG. 240

where the sun is found at noon. Whatever may be the season of the year, the day of the month, or the hour of the night, you will always see there a large constellation formed of seven fine stars, of which four are in a quadrilateral, and three at an angle with one side; all are arranged as we see in fig. 240.

You have all seen it, have you not? It never sets. Night and day it watches above the northern horizon, *turning slowly* in twenty-four hours round a star of which we shall speak directly. In the figure of the Great Bear, the three stars of the extremity form the tail, and the four in the quadrilateral lie in the body. In the Chariot, the four stars of the quadrilateral form the wheels, and the other three the pole, the horses, or the oxen. Above the second of these latter stars, ζ, good sight distinguishes quite a little star named Alcor, which is also called the Cavalier. It serves to test the power of the sight. Each star is designated by a letter of the Greek alphabet: α and β mark the first two stars of the quadrilateral, γ and δ the two following, ε, ζ, η, the three of the pole. Arabic names have also been given to these stars, which we will pass in silence, because they are generally obsolete, with the exception, however, of that of the second horse—Mizar. With reference to the Greek letters (of which we have given a list on p. 33), many persons think that it would be preferable to suppress them and to replace them by numbers. But this would be impossible in the practice of astronomy; and, moreover, inevitable confusion would result, on account of the numbers which the stars bear in the catalogues.

The Latins gave to ploughing oxen the name of *triones*; instead of speaking of a chariot and three oxen, they came to call them the seven oxen (*septemtriones*). From this is derived the word septentrion, and there are now doubtless but few persons who, in writing this word, know that they are speaking of seven oxen. It is the same, however, with many other words. Who remembers, for example, in using the word *tragedy*, that he speaks of a song of a goat: *tragôs-odè*?

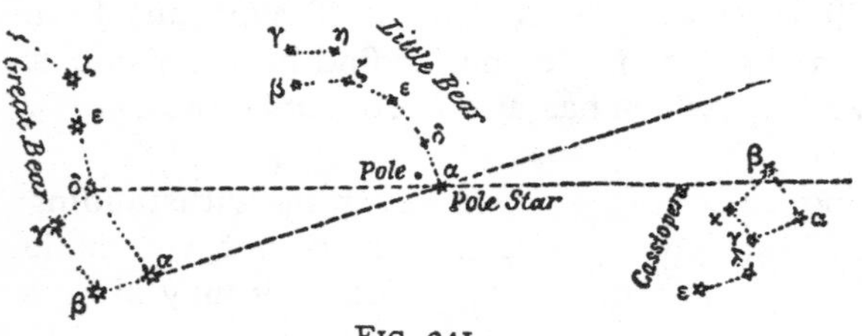

FIG. 241

Let us go back to fig. 240. If we draw a straight line through the two stars marked α and β which form the right side of the square, and produce it beyond α to a distance equal to five times that from β to α, or to a distance equalling that from α to the end of the tail, η, we find a star a little less brilliant than at the extremity of a figure similar to the Great Bear, but smaller and pointing in the opposite direction. This is the *Little Bear*, or the *Little Chariot*, also formed of seven stars. The star to which our line leads us—that which is at the tip of the tail of the Little Bear, or at the end of the pole of the Little Chariot—is the *polar star*.

The polar star enjoys a certain fame, like all persons who are distinguished from the common, because, among all the bodies which scintillate in the starry night, it alone remains motionless in the heavens. At any moment of the year, by day or by night, when you observe the sky, you will always find it. All the other stars, on the contrary, turn in twenty-four hours round it, taken as the centre of this immense vortex. The pole star remains motionless at the pole of the world, from whence it serves as a fixed point to navigators on the trackless ocean, as well as to travellers in the unexplored desert.

In looking at the pole star, motionless in the midst of the northern region of the sky, we have the south behind us, the east to the right, the west to the left. All the stars turn round the pole star in a direction contrary to that of the hands of a watch; they should, then, be recognised according to their mutual relations rather than by reference to the cardinal points.

On the other side of the pole star, with reference to the Great Bear, is found another constellation which we can also recognise at once. If from the middle star, δ, we draw a line to the pole, and produce this line by the same distance (see fig. 241), we arrive at Cassiopeia, formed of five principal stars arranged somewhat like the strokes of the letter **M**.

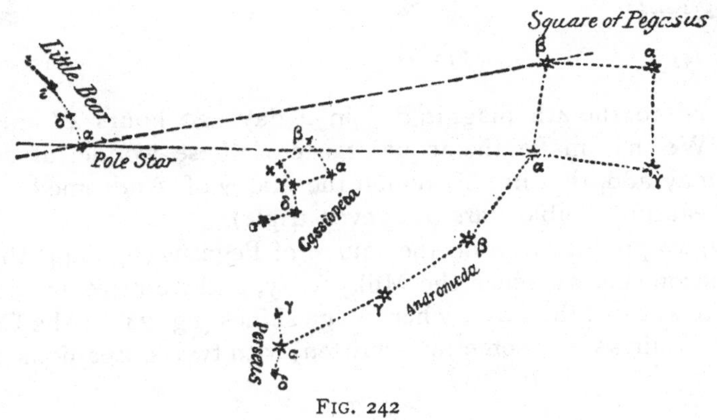

FIG. 242

The little star κ, which completes the square, gives the constellation the form of a *chair*. This group assumes all possible positions in turning round the pole; it is found sometimes above, sometimes below, sometimes to the right, and sometimes to the left; but it is always easily recognised, for, like the preceding group, it never sets, and is always opposite to the Great Bear. The pole star is the axle round which both these constellations turn.

If, now, we draw from the stars *a* and δ of the Great Bear two lines though the pole, and produce them beyond Cassiopeia, we come to the square of Pegasus (see fig. 242), which shows a line of three stars somewhat similar to the tail of the Great Bear. These three stars belong to *Andromeda*, and lead to another constellation, *Perseus*. The last star of the square of Pegasus is, as we see, the first (*a*) of Andromeda; the three others are named γ, *a*, and β. To the north of β of Andromeda is found, near a little star, ν, an oblong nebula, which can be distinguished with the naked eye. In Perseus, *a*, the brightest—on the prolongation of the three principal stars of Andromeda—appears between two others less brilliant, which form with it a concave arc very easy to distinguish. This arc serves us for a new alignment. Producing it in the direction of δ, we find a very brilliant star of the 1st magnitude; this is *Capella* (the Goat). Forming a right angle with this prolongation towards the south we come to the *Pleiades* (fig. 243). Not far from that is a variable star, *Algol*, or the *Head of Medusa*, which varies

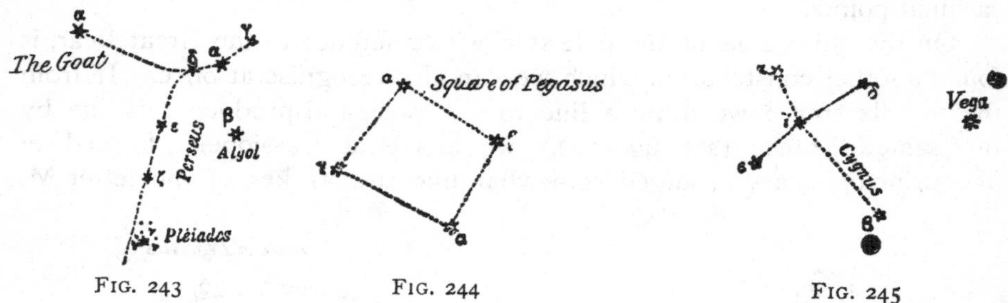

FIG. 243 FIG. 244 FIG. 245

from the 2nd to the 4th magnitude[1] in 2 days 20 hours 48 minutes 51 seconds. We shall make the acquaintance of these wonderful stars later on. We may add, that in this region the star γ of Andromeda is one of the most beautiful double stars (it is even triple).

If, now, we produce beyond the square of Pegasus (fig. 244) the curved line of Andromeda, we reach the Milky Way, and we meet in these parts Cygnus, like a cross; the Lyre, where Vega shines (fig. 245); the Eagle, and Altair (not Atair, as it is sometimes written) with two companions (fig. 246).

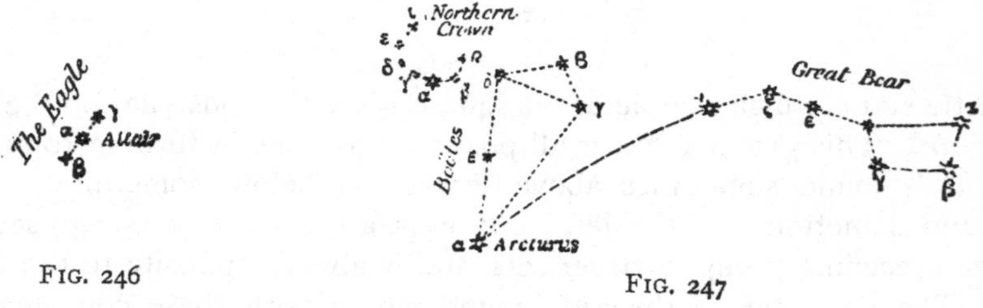

FIG. 246 FIG. 247

Such are the principal constellations visible in the circumpolar regions on one side; we shall make a fuller acquaintance with them directly. While we are tracing the lines of reference let us still have a little patience, and finish our summary review of this part of the sky.

Look now at the side opposite to that of which we have just spoken. Let us return to the Great Bear. Producing the tail along its curve, we find at some distance from that a star of the 1st magnitude, *Arcturus* (fig. 247), or

[1] [More correctly, from 2·3 magnitude to 3·5 magnitude.—J. E. G.]

a of Boötes. A little circle of stars which we see to the left of Boötes constitutes the Northern Crown (Corona Borealis). In the month of May 1866 there was seen shining there a fine star, the brightness of which lasted only fifteen days. The constellation of Boötes is traced in the form of a pentagon. The stars which compose it are of the 3rd magnitude, with the exception of Arcturus, which is of the 1st. This is one of the nearest to the earth; at least, it is one of a small number whose distance has been measured. It shines with a beautiful golden yellow colour. The star ε, which we see above it, is *double*—that is to say, the telescope resolves it into two distinct stars, one yellow, the other blue.

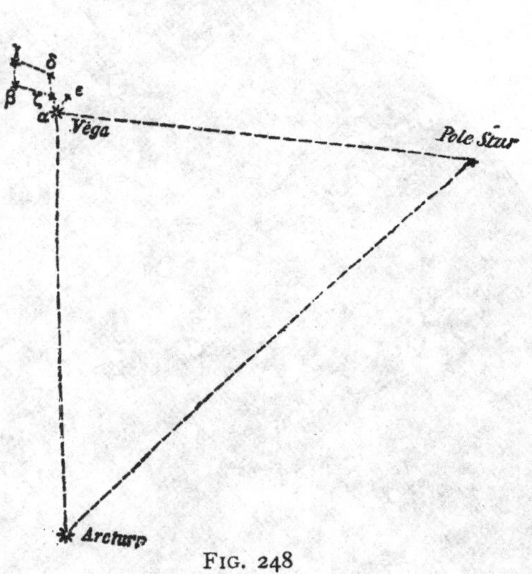

FIG. 248

This technical description is far from the poetry of Nature; but it is especially important here to be clear and precise. Let us suppose ourselves, however, under the starry vault on a beautiful summer's night, splendid and silent, and let us consider that each of these points which we seek to recognise is a world, or rather a system of worlds. Look at this equilateral triangle (fig. 248); it permits us to cast our eyes successively on three important suns: Vega of the Lyre, Arcturus of Boötes, and the pole star, which watches above the solitudes of our mysterious North Pole. Many martyrs of science have died looking at it! In twelve thousand years our descendants will see the Lyre at the pole, ruling the harmony of the heavens.

The stars which are near the pole, and which have for that reason received the name of circumpolar stars, are distributed in the groups which have just been indicated. I earnestly invite my readers to profit by fine evenings, and try and find for themselves these constellations in the sky. The best method is to make use of the preceding alignments and the drawing of the whole reproduced in fig. 249.

We have here the principal stars and constellations of the northern hemisphere, the North Pole being at the centre of the circle. We come now in the order of our description to the twelve constellations of the zodiacal belt, which makes the circuit of the sky, inclined at 23° to the equator, and of which the ecliptic, the apparent path of the sun, forms the centre line. The name of zodiac, given to the zone of stars which the sun traverses during the course of the year, comes from ζώδια, *animals*, an etymology which is due to the species of figures traced on this belt of stars. Animals, in fact, predominate in these figures. The entire circumference of the sky has been divided into twelve parts, which have been named the twelve

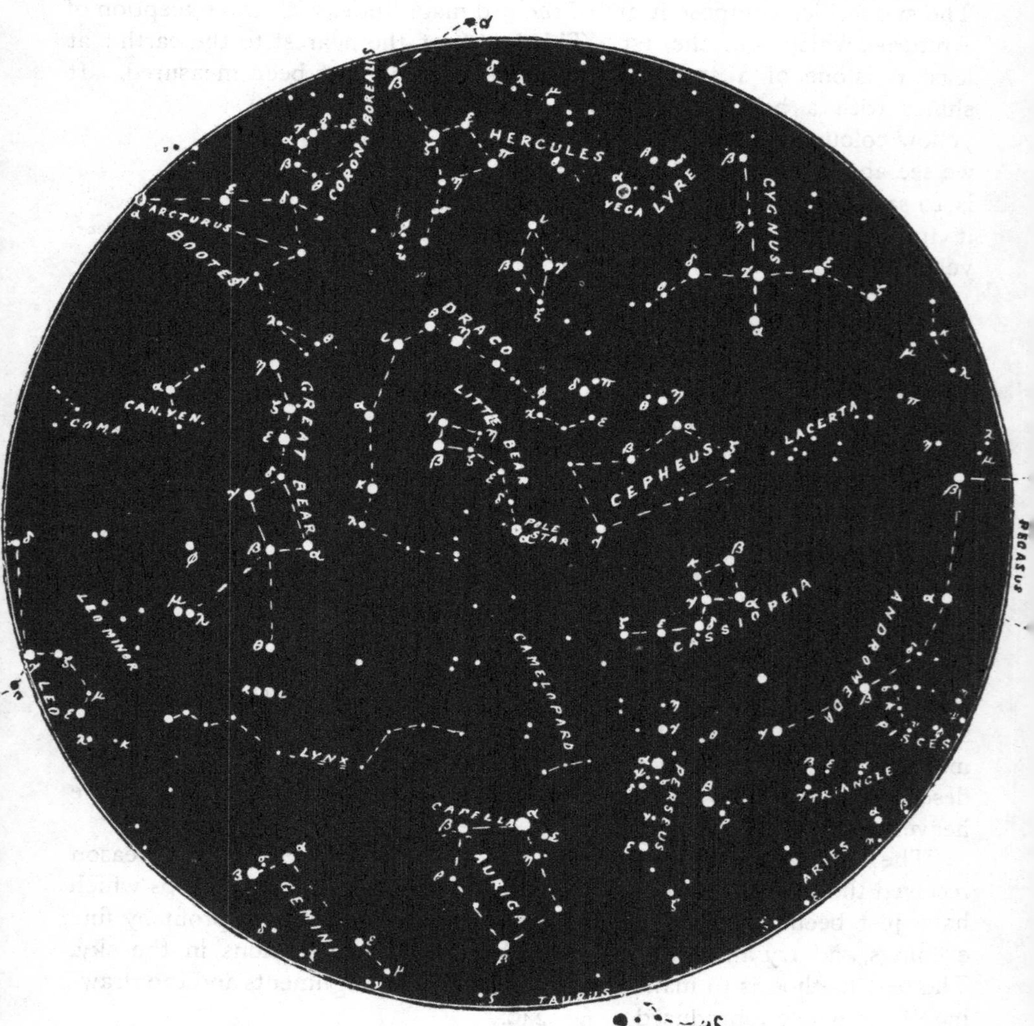

FIG. 249.—PRINCIPAL STARS OF THE NORTHERN HEMISPHERE

signs of the zodiac; our ancestors called them the 'houses of the sun,' or 'the monthly abodes of Apollo,' because the day star visits them each month, and returns every spring to the beginning of the zodiacal city. Two memorable Latin verses of the poet Ausonius present to us these twelve signs in the order in which the sun travels through them, and this still appears the easiest method of learning them by heart.

> Sunt *Aries*, *Taurus*, *Gemini*, *Cancer*, *Leo*, *Virgo*,
> *Libraque*, *Scorpius*, *Arciteneus*, *Caper*, *Amphora*, *Pisces* ;

or, in English, the Ram ♈, the Bull ♉, the Twins ♊, the Crab ♋, the Lion ♌, the Virgin ♍, the Balance ♎, the Scorpion ♏, the Archer ♐, Capricornus ♑, Aquarius ♒, and the Fishes ♓. The signs placed beside these names are a vestige of the primitive hieroglyphics which described them : ♈ represents the horns of the Ram, ♉ the head of the Bull ; ♒ is a stream of water, &c.

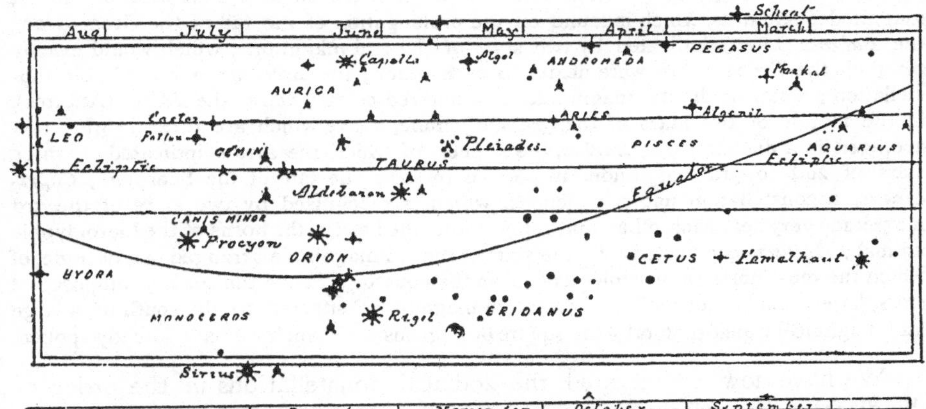

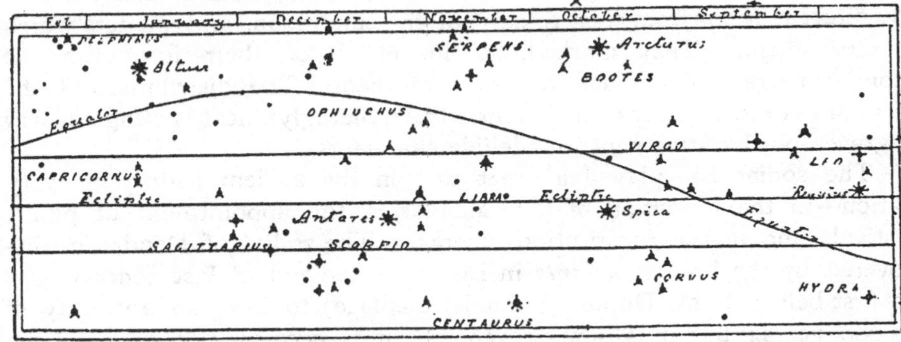

FIG. 250.—PRINCIPAL STARS AND CONSTELLATIONS OF THE ZODIAC

If we now know our northern sky, if its most important stars are sufficiently noted down in our mind, with the reciprocal relations which they preserve among themselves, we have no more confusion to fear, and it will be easy to recognise the zodiacal constellations. This zone may be of use to us as a line of division between the north and the south. Here is a description of it :—

The Ram, which, moving in front of the herd, and regulates, so to say, the march, opens the series. This constellation has in itself nothing remarkable; the brightness of its stars indicates the base of one of the horns of the leader of the sheep; it is but of the 2nd magnitude. After the *Ram* comes the Bull. Admire on a fine winter's night the charming Pleiades which scintillate in the ether; not far from them shines a fine red star—this is the *eye* of the Bull—Aldebaran, a star of the 1st magnitude and one of the finest of our sky. (Follow in this description our plan of the zodiac, fig. 250.) We now arrive at the Twins, whose heads are marked by two fine stars of the 2nd magnitude, situated a little above a star of the 1st magnitude—*Procyon*, or the Little Dog; *Cancer*, or the Crab, a constellation very little conspicuous (its most visible stars are but of the 4th magnitude, and occupy the body of the animal); the *Lion*, a fine constellation, marked by a star of the 1st magnitude, *Regulus*, by one of the second, β, and by several others of the 2nd and 3rd magnitudes arranged in a trapezium; the *Virgin*, indicated by a very brilliant star of the 1st magnitude, *Spica*, situated in the neighbourhood of a star, also of the 1st magnitude, Arcturus, which is found on the prolongation of the tail of the Great Bear; the *Balance* (Libra), indicated by two stars of the 2nd magnitude, which would exactly resemble the Twins if they were nearer to each other; the *Scorpion*, a remarkable constellation; a star of the 1st magnitude, of a fine red colour, marks the *Heart* (Antares), in the middle of two stars of the 3rd magnitude, above which are three bright stars arranged in a diadem; *Sagittarius*, the Archer, of which the arrow, indicated by three stars of 2nd to 3rd magnitude, is pointed towards the tail of the Scorpion; *Capricornus*, a constellation not conspicuous, which is recognised by two stars of the 3rd magnitude very near each other, and representing the base of the horns of the hieroglyphic animal; *Aquarius*, indicated by three stars of the 3rd magnitude arranged in a triangle, of which the most northern occupies a point on the equator; *Pisces*, the *Fishes*, composed of stars, barely conspicuous, of the 3rd to 4th magnitudes, situated to the south of a large and magnificent quadrilateral—the square of Pegasus—of which we have already spoken.

We have now enumerated the zodiacal constellations in the order of the direct motion (from west to east) of the sun, moon, and planets which traverse them. They marked, at the epoch of their formation, the monthly passage of the sun into each of them. The distribution of the stars in figurative groups was the first truly hieroglyphical writing; it was engraved on the firmament in indelible characters.

The zodiac has played a great part in the ancient history of every nation—in the formation of the calendar, in the appointment of public festivals, and in the constitution of eras. The zodiac of Denderah, discovered by the French *savants* in Egypt at the end of last century, was at first believed (*see* Dupuis, Lalande, Laplace) to have an antiquity of 15,000 years; but it is now proved that it is necessary to deduct from that number of years half the cycle of precession—that is to say, nearly 13,000 years—which brings down the date of this sculpture to 2,000 years before our epoch; and this in fact corresponds with the evidence of archæology. It is remarkable that all the ancient zodiacs and calendars which have been preserved to us begin the year with the constellation of the Bull, as we have already noticed (p. 43). The zodiac of the Elephanta Pagoda (Salsette) has at the head of the procession the sign of the sacred Bull, the ox Apis, Mithra—of which the promenade of the fat ox, which is still performed in the environs of Paris, is a vestige. The ceiling of a sepulchral chamber at Thebes shows the Bull at the head of the procession. The zodiac of Esne, the astronomical picture discovered by Champollion in the Ramesseum of Thebes, carries us back to the same origin, between two and three thousand years before our era; Biot supposes the date of this to be the year 3285, the vernal equinox passing through the Hyades on the forehead of Taurus. Father Gaubil has proved that from ancient times the Chinese have referred the beginning of the apparent motion of

FIG. 251.—ANCIENT ARABIAN ZODIAC

the sun to the stars of Taurus; and we have a Chinese observation of the star η of the Pleiades as marking the vernal equinox in the year 2357 before our era. Hesiod sings of the Pleiades as ruling the labours of the year, and the name of Vergilia, which the ancient Romans gave them, associate them with the beginning of the year in spring.

Without entering into any details of the different zodiacs which have been preserved to us from the most ancient and diverse nations, a glance at those which are reproduced here will lead us to appreciate the part which they have played in ancient religions. Several zodiacal signs have become veritable gods. The zodiac represented by fig. 251 was engraved, in the thirteenth century, on an Arabic magic mirror, and dedicated to the sovereign prince Aboulfald, 'Victorious Sultan, Light of the World,' if we are to believe the bombastic inscription which encircles it. Fig. 252 shows an ancient Hindoo zodiac. Fig. 253 shows a Chinese zodiac stamped upon a talisman, even now in use. The twelve signs differ from ours; they are: the Mouse, the Cow, the Tiger, the Rabbit, the Dragon, the Serpent, the Horse, the Ram, the Ape, the Hen, the Dog, and the Pig. Fig. 254 represents a Chinese medal, on which we see the constellation *Teou*, the Great Bear[1] (which they call the Bushel), the Serpent, the Sword, and the Tortoise. This is a talisman intended to give courage; it appears that it is in great demand among the Chinese, and is as well circulated as the medals of the Immaculate Conception are in France.

FIG. 252.—ANCIENT HINDOO ZODIAC

Of all the zodiacal constellations, that of the Bull has played the principal *rôle* in ancient myths; and in this constellation it was the sparkling cluster of the Pleiades which appears to have regulated the year and the calendar among all the ancient nations. The Mosaic deluge itself, referred to 17 Athir (November), in commemoration of an important inundation, had its date coincident with the appearance of the Pleiades.[2]

But we forget the stars. If our descriptions and maps have been carefully followed, the reader will now know the zodiacal constellations as well as those of the north. There remains but little to do to know the entire sky. But there is an indispensable addition to be made to what precedes. The circumpolar stars are perpetually visible above the London horizon;

[1] The author possesses in the Museum of the Observatory at Juvisy a Japanese executioner's sword, on the guard of which this constellation is engraved. Was it believed that the souls of executed criminals were sent there?

[2] See *Astronomical Myths, based on Flammarion's History of the Heavens.* By J. F. Blake. London, 1876.

at any time of the year when we wish to observe them it is sufficient to turn towards the north, and we shall always find them, either above the pole star or below it, to one side or the other, and always maintaining among themselves the relations which we have employed to find them. The stars of the zodiac do not resemble them from this point of view, for they are sometimes above the horizon, sometimes below. It is necessary, then, to know at what epoch they are visible. For this purpose it will be sufficient to remember the constellation which is found in the middle of the sky at *nine o'clock in the evening* on the first day of each month—that, for example, which crosses at that moment a line descending from the zenith to the south. This line is the *meridian*, of which we have already spoken ; all the stars cross it once

FIG. 253.—CHINESE ZODIAC, FROM A TALISMAN

FIG. 254.—CHINESE MEDAL, SHOWING THE GREAT BEAR

a day, moving from east to west—that is to say, from left to right. In indicating each of the constellations which pass at the hour indicated, we also give the centre of the visible constellations.

On January 1 Taurus passes the meridian at 9 o'clock in the evening ; notice Aldebaran, the Pleiades. On February 1 the Twins (Gemini) are not yet there ; we see them a little to the left. March 1, Castor and Pollux have passed ; Procyon to the south, the little stars of the Crab (Cancer) to the left. April 1, the Lion, Regulus. May 1, β of the Lion, Berenice's Hair. June 1, Spica of the Virgin, Arcturus. July 1, the Balance (Libra), the Scorpion. August 1, Antares, Ophiuchus. September 1, Sagittarius, Aquila. October 1, Capricornus, Aquarius. November 1, Pisces, Pegasus. December 1, Aries, the Ram.

Our general review of the starry sky must now be completed by the stars of the southern heavens.

Look at our zodiacal map (fig. 250) : below Taurus and Gemini, to the south of the zodiac, you notice the giant Orion, who raises his club towards the forehead of the Bull. Seven brilliant stars are here distinguished ; two of them, α and β, are of the 1st magnitude; the five others are of the 2nd magnitude. α and γ mark the shoulders, κ the right knee, β the left knee; δ, ϵ, ζ mark the belt or girdle. Below this line is a luminous train of three stars, very near each other; this is the Sword. Between the western shoulder and Taurus is seen the Shield, composed of a row of small stars. The head is marked by a little star (λ) of the 4th magnitude.

On a fine winter's night turn towards the south, and you will immediately recognise this giant constellation. The four stars, α, γ, β, κ, occupy the angles of a great quadrilateral. The three others, δ, ϵ, ζ, are crowded in an oblique line in the middle of this quadrilateral; α, at the north-east angle, is named *Betelgeuse* (not Beteigeuse, as some books print it); β, at the south-east angle, is called *Rigel.*

The line of the Belt, produced both ways, passes to the north-east near *Aldebaran*, the Eye of the Bull, which we know already, and to the south-east near *Sirius*, the finest star of the sky, which we shall soon consider.

This fine constellation is easily recognised: (1) on the zodiacal plan, fig. 250; (2) on our celestial planisphere (plate ii.), on which the stars in the sky are shown to the 5th magnitude.

It is during the fine nights of winter that this constellation shines in the evening above our heads. No other season is so magnificently constellated as the months of winter. While nature deprives us of certain enjoyments in one way, it offers us in exchange others no less precious. The marvels of the heavens present themselves from Taurus and Orion in the east to Virgo and Boötes on the west. Of eighteen stars of the 1st magnitude which are counted in the whole extent of the firmament, a dozen are visible from nine o'clock to midnight, not to mention some fine stars of the 2nd magnitude, remarkable nebulæ, and celestial objects well worthy of the attention of mortals. It is thus that nature establishes an harmonious compensation, and while it darkens our short and frosty days of winter, it gives us long nights enriched with the most opulent creations of the sky.

The constellation of Orion is not only the richest in brilliant stars, but it conceals for the initiated treasures which no other is known to afford. We might almost call it the California of the sky.

To the south-east of Orion, on the line of the Three Kings, shines the most magnificent of all the stars, *Sirius*, or α of the constellation of the Great Dog. This star of the 1st magnitude marks the upper eastern angle of a great quadrilateral, of which the base, near the horizon of London, is adjacent to a triangle. This constellation rises in the evening at the end of November, passes the meridian at midnight at the end of January, and sets at the end of March. It played the greatest part in Egyptian astronomy, for it regulated the ancient calendar. It was the famous Dog Star; it predicted the inundation of the Nile, the summer solstice, great heats and fevers; but the precession of the equinoxes has in 3,000 years

CELESTIAL P

Containing the principal st

London: Cassell & Company Limited.

F.S.Weller

LANISPHERE.

ars visible to the naked eye.

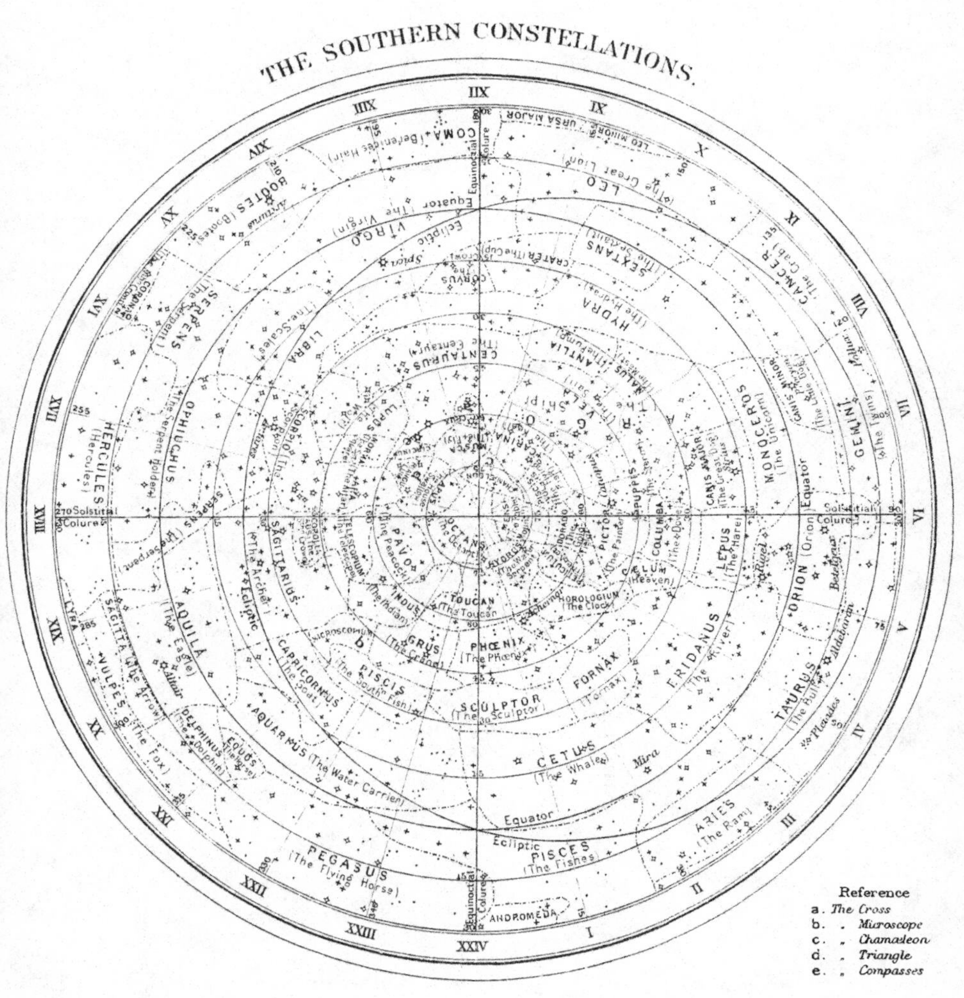

F.R.G.S.

London: Chatto & Windus, Publishers.

Plate II.

moved back the time of its appearance by a month and a half, and now this fine star announces nothing, either to the Egyptians who are dead or to their successors. But we shall see farther on what it teaches us of the grandeur of the sidereal universe.

The *Little Dog*, or Procyon, which we have already seen on our zodiacal maps, is found above the Great Dog and below the Twins (Castor and Pollux), to the east of Orion. With the exception of α Procyon, no brilliant star distinguishes it.

Hydra is a long constellation, which occupies a quarter of the horizon, under Cancer, the Lion, and the Virgin. The head, formed of four stars of the 4th magnitude, is to the left of Procyon, on the prolongation of a line drawn from that star to Betelgeuse. The western side of the great trapezium of the Lion, like the line from Castor and Pollux, points to α, of the 2nd magnitude. This is the Heart of Hydra; we remark the asterisms of the second class, Corvus the Crow, and Crater the Cup.

Eridanus, *Cetus*, *Piscis Australis*, and the *Centaur*, are the only important constellations which remain to be described. We find them, in the order which we have indicated, to the right of Orion. Eridanus is a river composed of a train of stars winding from the left foot of Orion and losing itself below the horizon. After following long windings, it ends with a fine star of the 1st magnitude, α Eridani, or Achernar. This is the river into which Phaeton fell when he unskilfully directed the Chariot of the Sun. It was placed in the sky to console Apollo for the death of his son.

To find the Whale (Cetus), we may notice below the Ram a star of the 2nd magnitude which forms an equilateral triangle with the Ram and the Pleiades; this is α of Cetus, or the Jaw; α, μ, ξ, and γ form a parallelogram which represents the head. The base, α, γ, may be produced to a star of the 3rd magnitude, δ, and to a star of the neck marked o. This star is one of the most curious in the heavens. It is named the Wonderful, *Mira Ceti*. It belongs to the class of variable stars. Sometimes it equals in brightness stars of the 2nd magnitude, sometimes it becomes completely invisible.[1] Its variations have been followed since the end of the sixteenth century, and it has been found that they are reproduced periodically every 331 days on the average. The study of these singular stars presents us with curious phenomena.

Lastly, the constellation of the Centaur is situated below Spica of the Virgin. The star θ, of the 2nd magnitude, and the star ι, of the 3rd, mark the head and the shoulder. This is the only part of this figure which rises above our horizon. The Centaur contains *the nearest star* to us (α) of the 1st magnitude, the distance of which is about 25 billions of miles. The feet of the Centaur touch the *Southern Cross*, formed of four stars of

[1] [That is, to the naked eye; it never descends below the 10th magnitude, and always remains visible in a 3-inch telescope.—J. E. G.]

the 2nd magnitude, always hidden below our horizon. It reigns in silence above the icy solitudes of the Southern Pole, where ships proceed only with difficulty. Farther on, at the centre of the other hemisphere, is the southern celestial pole, which is not marked by any remarkable star. It was from this region, Dante relates, that, having visited Hell, enclosed in the centre of the earth, he went to the Mountain of Purgatory, and from there to the Heights of Paradise. These beautiful dreams have disappeared in the sunshine of modern astronomy.

We will complete these descriptions by a little astronomical chronology, which is not without interest. From a careful examination of the most ancient historical sources of classical astronomy, the following is the order in which the constellations appear to have been noticed, formed, and named, beginning with the most ancient :—

	Most ancient reference
The Great Bear	*Job* (ch. xxxviii. ver. 32) (seventeenth century before our era), *Homer* (ninth century).
Orion	*Job* (ch. ix. ver. 9), *Homer*, *Hesiod*.
The Pleiades (the Hyades)	*Job* (ch. xxxviii. ver. 31), *Homer*, *Hesiod*.
Sirius and the Great Dog	*Hesiod* mentions it. *Homer* calls Sirius the Star of Autumn.
Aldebaran (Taurus)	*Homer*, *Hesiod*.
Boötes, Arcturus	*Job* (ch. xxxviii. ver. 32), *Homer*, *Hesiod*.
The Little Bear	*Thales* (seventh century), *Eudoxus*, *Aratus*.
Draco (the Dragon)	*Eudoxus* (fourth century), *Aratus* (third century).
The Man on his Knees, or Hercules	*Id.*
The Branch and Cerberus [1]	*Id.*
Corona Borealis	*Id.*
Ophiuchus or Serpentarius	*Id.*
The Scorpion	*Id.*
Virgo and Spica	*Id.*
Gemini (the Twins)	*Id.*
Procyon	*Id.*
Cancer (the Crab)	*Id.*
Leo (the Lion)	*Id.*
Auriga (the Charioteer)	*Id.*
Capella (the Goat, the Kids)	*Id.*
Cepheus	*Id.*
Cassiopeia	*Id.*
Andromeda	*Id.*
Pegasus (the Horse)	*Id.*
Aries (the Ram)	*Id.*
The Triangle	*Id.*
Pisces (the Fishes)	*Id.*
Perseus	*Id.*
Lyra	*Id.*
The Bird, or Cygnus (the Swan)	*Id.*

[1] A constellation wrongly attributed by Arago and others to Hevelius. It is found on the sphere of Eudoxus.

	Most ancient reference
Aquila (the Eagle)	*Eudoxus* (fourth century), *Aratus* (third century).
Aquarius	*Id.*
Capricornus	*Id.*
Sagittarius	*Id.*
Sagitta (the Arrow)	*Id.*
Delphinus (the Dolphin)	*Id.*
Lepus (the Hare)	*Id.*
Argo (the Ship)	*Id.*
Canobus (afterwards written Canopus)	*Id.*
Eridanus	*Id.*
Cetus (the Whale)	*Id.*
Piscis Australis (the Southern Fish)	*Id.*
Corona Australis	*Id.*
The Altar	*Id.*
The Centaur	*Id.*
The Wolf (Lupus)	*Id.*
Hydra	*Id.*
Crater (the Cup)	*Id.*
Corvus (the Crow)	*Id.*
Libra (the Balance)	*Manetho* (third century B.C.), *Geminus* (first century B.C.).
The Hair of Berenice [1]	*Callimachus*, *Eratosthenes* (third century).
Feet of the Centaur	*Hipparchus* (first century B.C.).
Propus (η of Gemini)	*Hipparchus.*
The Manger and Donkeys	*Id.*
The Little Horse (Equuleus)	*Id.*
The Head of Medusa	*Id.*
Antinous [1]	Under the Emperor Adrian (130 A.D.).
The Peacock (Pavo)	*John Bayer*, 1603.
Toucan	*Id.*
Grus (the Crane)	*Id.*
Phœnix	*Id.*
Doradus	*Id.*
The Flying Fish	*Id.*
Hydrus	*Id.*
Chamæleon	*Id.*
The Bee (Musca)	*Id.*
The Bird of Paradise (Apus)	*Id.*
Triangulum Australis	*Id.*
The Indian (Indus)	*Id.*
The Giraffe (Camelopardus)	*Bartschius*, 1624.
The Fly (Musca)	*Id.*
The Unicorn (Monoceros)	*Id.*
Noah's Dove (Columba)	*Id.*
The Oak of Charles II.	*Halley*, 1679.
The Southern Cross (already seen by the ancients)	*Augustine Royer*, 1677.

[1] Constellations incorrectly attributed to Tycho Brahé. The first is given by Eratosthenes, the second dates from the Emperor Adrian.

	Most ancient reference
The Great and Little Cloud (Magellanic Clouds)	*Hevelius*, 1690.
The Fleur de Lys	*Id.*
The Greyhounds (Canes Venatici)	*Id.*
The Fox and Goose (Vulpecula et Anser)	*Id.*
The Lizard (Lacerta)	*Id.*
The Sextant of Urania (Sextans)	*Id.*
The Little Lion (Leo Minor)	*Id.*
The Lynx	*Id.*
The Shield of Sobieski	*Id.*
The Little Triangle	*Id.*
Mount Mænalus	*Flamsteed*, 1725.
The Heart of Charles II. (α Canum Venaticorum)	*Id.*
The Sculptor's Workshop (Sculptor)	*Lacaille*, 1752.
The Chemical Furnace (Fornax)	*Id.*
The Clock (Horologium)	*Id.*
The Rhomboid Reticule (Reticulum)	*Id.*
The Engraver's Pen	*Id.*
The Painter's Easel (Pictor)	*Id.*
The Compass (Circinus)	*Id.*
The Air Pump (Antlia)	*Id.*
The Octant (Octans)	*Id.*
The Compass and Square	*Id.*
The Telescope (Telescopium)	*Id.*
The Microscope (Microscopium)	*Id.*
The Table Mountain (Mensa)	*Id.*
The Reindeer	*Lemonnier*, 1774.
The Solitaire (Indian Bird)	*Id.*
Le Messier	*Lalande*, 1776.
The Bull of Poniatowski	*Poczobut*, 1877.
The Honours of Frederick	*Bode*, 1786.
The Harp of the Georges	*Hell*, 1789.
The Telescope of Herschel	*Bode*, 1787.
The Electrical Machine	*Id.*, 1790.
The Printer's Workshop	*Id.*
The Mural Quadrant	*Lalande*, 1795.
The Air Balloon	*Id.*, 1798.
The Cat	*Id.*, 1799.

Such are the constellations, ancient and modern, venerable or recent, into which the celestial sphere has been divided. The ancient names are respectable and respected, on account of their relations, known or unknown, with the origins of history and religion ; the new ones must be ephemeral ; and the double celestial chart given in our volume 'Les Étoiles' is the only one which contains them all. It is useful to know them, because several stars celebrated under different titles have for their principal designation their position in these asterisms ; but what we should wish would be to see them disappear.[1]

[1] Especially those which are absolutely superfluous, and occupy places stolen from the ancient constellations, like the Heart of Charles II., the Fox and Goose, the Lizard, the Sextant, the

Many other substitutions have, however, been attempted. I have in my library a splendid folio of the year 1661, containing twenty-nine engraved plates, illuminated in gold and silver, among which are two which represent the sky delivered from the pagans and peopled with Christians. Instead of divinities more or less virtuous, in place of animals of forms more or less fantastic, we behold the elect—apostles, saints, popes, martyrs, sacred persons of the Old and New Testament—seated in the celestial vault, clothed in rich costumes of all colours, embroidered with gold, and carefully installed in the place of all the pagan heroes who for so many ages reigned in the sky.

The author of this metamorphosis was named Jules Schiller, and it was in the year 1627 that he introduced it, coupling his name with that of John Bayer. He began his dissertation by showing how the pagan constellations are opposed to Christian opinion and even to common sense. He quoted the Fathers of the Church who expressly disapprove of them: Isodorus, who treats them as diabolical; Lactantius, who condemns the corruption of the human race; Augustine, who sends their heroes to hell, &c.

These constellations formed by chance, in the course of ages, without a fixed object; their inconvenient size, the uncertainty of their boundaries; the complicated designations, for which it was sometimes necessary to exhaust whole alphabets; the bad taste with which observers have introduced into the southern sky the frigid nomenclature of instruments used in science alongside mythological allegories—all these accumulated defects have often suggested plans of reform for the stellar divisions, and even the banishing of all configuration. But ancient customs are difficult to overcome, and it is very probable that, except the recently-named groups, which we may now suppress, the venerable constellations will always reign.

Such are the provinces of the sky. But these provinces are of no intrinsic value; the important point for us is to make acquaintance with the inhabitants.

Shield of Sobieski, Mount Mænalus, the Reindeer, the Solitaire, the Messier, the Bull of Poniatowski, the Honours of Frederick, the Harp, the Telescope, the Mural Circle, the Air Balloon, the Electrical Machine, the Printer's Workshop, and the Cat. I know, however, with reference to this last animal, that Lalande wrote: 'I love cats! I adore cats! I may be pardoned for having placed one in the sky after my sixty years of assiduous labours.' But the illustrious astronomer had no necessity for this plea in order that his name should remain inscribed in letters of gold on the tablets of Urania. The Heart of Charles II. is but the flattery of a courtier; the Shield of Sobieski, the Bull of Poniatowski, should fall from the sky; the Messier is but a play on words which makes the celestial flocks guarded by a pastor whose name is the same as that of the prolific hunter of comets, Messier. As for the Honours of Frederick, they usurp an unmerited place, for, in order to make room for them, Andromeda has been obliged *to draw in her arm, which she had stretched out there for three thousand years.*

CHAPTER III

POSITIONS OF THE STARS IN THE SKY

Right Ascensions and Declinations. Observations and Catalogues

IT was formerly thought sufficient to indicate the stars by their positions in the constellation to which they belong. Thus, Regulus was called the Heart of the Lion (Cor Leonis); Antares, the Heart of the Scorpion; Aldebaran, the Eye of the Bull; Rigel, the Foot of Orion, &c.; later, the designation by letters, made by Bayer in 1603, was extended to a large number of stars, and was more precise. But in practical astronomy we cannot be contented with approximate positions; it is important to have absolutely precise places, and the following is the method by which these are obtained.

As we have seen, the constellations play the part in astronomy of the divisions into kingdoms and provinces in geography; the proper names of the principal stars are like the names of towns. Now, these names are not sufficient to fix their precise position on the terrestrial globe; it is necessary also to have recourse to geographical co-ordinates, the *longitude* and *latitude*. Astronomers use a similar system for the stars.

The position of a star, said Herschel, once well ascertained, constitutes a fixed point of immense importance in the constitution of the universe. The instrument which has determined it may moulder; it will be the same with the astronomer and his generation: but this point remains as a fixed term of eternal stability, more unalterable than monuments of bronze or pyramids of marble.

In astronomy, the fundamental circle to which the positions of the stars are referred is the celestial equator; this has been chosen because it is always easily determined. The distance of a star from the equator is called the *declination*; it is north or south according as the star is to the north or to the south of the equator. We see that this co-ordinate corresponds to geographical *latitude*. The other co-ordinate is analogous to the *longitude*; this, in geography, denotes the arc of the equator included between the meridian of the place and that of another place (for example, Paris, London, Rome), taken at will as the *first meridian*. In astronomy, the origin of right ascensions is not arbitrary; it is defined by nature, and is placed at the point of intersection of the ecliptic with the equator.[1]

Thus the position of every star in the heavens is exactly determined by the knowledge of its right ascension and declination. We should add, that it is important to denote whether the declination is north or south by placing after it the letter N or S, or putting before it the sign + or −. To avoid the possibility of an error of sign, the declination is frequently replaced by the distance from the North Pole, which cannot be equivocal, and which comes to exactly the same thing, since this polar distance is nothing else but the complement of the northern declination to form 90° if the star is between the equator and the North Pole, and the southern declination increased by 90° if the star is beyond the equator. An example will make these definitions clear:—

Let A (fig. 255) be any star on the celestial sphere. The distance A E, which separates the star from the equator, measured on the circle P Q perpendicular to the equator, is called the *declination*. Here it is north, since the star is between the North Pole and the equator. Let us suppose it to be 40 degrees: we write it thus—

Declination = + 40°,

or, if we prefer to express the same position in polar distance, a simple subtraction shows us the distance PA = PE − AE; that is to say, 90° − 40°. It is, then, 50°, and we may write it—

Polar distance = 50°.

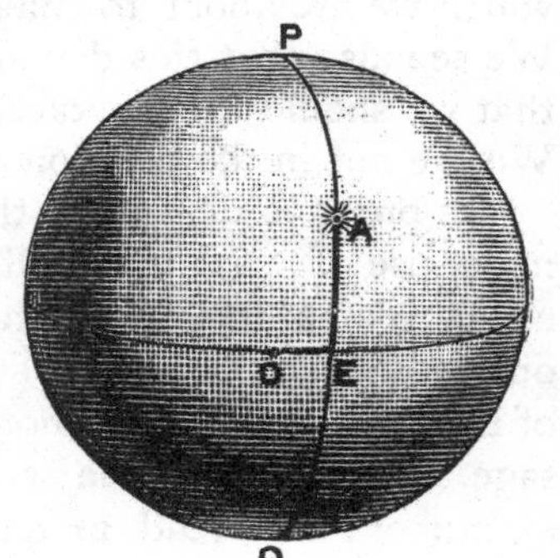

FIG. 255

If our star were beyond the equator, at the same distance, it would be necessary to add its declination to 90° instead of deducting it, and the polar distance would be 90° + 40°, or 130°.

But this determination is not sufficient to make known the position of our star, because it might be found at any point along a circle drawn at 40° above the equator; it is still necessary to know its position on this circle, which will be obtained by determining the

[1] Thus right ascension is reckoned from the point which the sun passes at the moment of the vernal equinox, from 0° to 360°, always in the direction of the sun's *annual* motion—that is to say, from west to east, following the order of the signs. It is expressed either in *degrees* or in *time*, exactly as terrestrial longitudes are. Thus, to express the distance in longitude from Paris to Vienna, we may at pleasure say that there are 15° or 1 hour of difference; in one case, as in the other, it is the twenty-fourth part of the circumference of the world, as well as of the sky. Each hour of right ascension represents 15°, so that one degree represents four minutes of time. Generally it is expressed in hours, because the instrument which is used to determine it is the meridian telescope (of which we are about to speak), which remains fixed in the plane of the meridian, and by which is determined the precise hour at which a star passes behind the wire where the point chosen for the origin of right ascensions has itself passed previously.

distance E O, which separates its vertical circle from the point O, chosen as the origin of right ascensions. Let us suppose that the interval E O is an hour and a half; we write—

Right ascension = $1^h\ 30^m$.

And thus the position of the star is completely determined. It is written—

Right ascension = $1^h\ 30^m$; declination = $+40°$.

These two important elements of celestial geography are written briefly: right ascension by Æ, which comes from the Latin *ascensio recta*; and the declination by the sign δ. It is indispensable that we should know all these details, for we shall have to use them in the following pages.

There are also celestial latitudes and longitudes. The latitude is the distance of a star from the ecliptic, and the longitude is the distance from the vernal equinox reckoned on the ecliptic. These are calculated on the same system as the declinations and right ascensions, with this difference, that the positions are referred to the ecliptic instead of the equator. These co-ordinates are but rarely used.

A knowledge of the principle of right ascensions and declinations is very important, for without it the geography of the sky and the studies which we are about to make in sidereal astronomy would be impossible. We see also that this demands but a moment of *attention*. It is necessary that we should know exactly, clearly, and unequivocally of what we speak. We are not in the position, intellectually, of a soldier who mounts guard.[1]

In order to determine the positions of the stars in the sky astronomers make use of what they call the meridian circle, or transit instrument. It consists of a telescope which moves exactly in the meridian, and they observe the passage of the stars behind cross-wires placed in the optical field of the telescope. They note by the clock the precise moment of the passage, and on the circle which is carried by the axis of rotation of the instrument are read in degrees, minutes, and seconds, the distance of the star from the equator or the North Pole. The right ascension and the declination are thus obtained with the greatest ease and precision.

[1] One evening in December of the year 1871, passing near the foot of the Vendôme Column, then demolished, I was astonished to see a half-frozen sentry mounted there as a guard of honour, as in the time when the Emperor looked down on the bronze of the converted cannon. There was nothing left but the railing and the base of the demolished column. I approached him quietly, and asked him what he was guarding. 'Keep clear!' said he. 'But,' I added, 'there is no longer a column.' 'Keep clear!' he again replied. 'Why do you not tell your sergeant that there is no longer a column?' The sentry came to the charge, and I had no choice but to turn away. However, some days after this useless post was abandoned.

One day a French diplomatist, walking with the Czar in the Summer Garden of St. Petersburg, noticed in the middle of a grass plot a motionless sentry, and asked the Emperor what the man was doing there. 'I forget,' replied the Czar, and he turned to an adjutant to ask him the same question. He in his turn went for information, and everywhere received the same reply, which told him nothing except 'It is the order.' The archives were consulted, but nothing was found. At last an old lackey recollected that his father, also an old lackey, had related to him that in the last century the Empress Catherine had discovered one fine morning in this spot a snow-drop, and had protected it from being picked. A soldier had been sent to keep his eye on the flower, and a sentry had been stationed there ever since.

This instrument is, properly speaking, the fundamental instrument of every observatory. We may say that the essential object of the establishment of great national observatories, such as those of Paris, London, Washington, Berlin, and Vienna, is not to make discoveries, but to ascertain slowly and laboriously the precise positions of the stars in the sky. These patient and silent labours are distinct from the discoveries which dazzle the world and give to their authors glorious and popular renown. The unknown astronomer instals himself at the meridian telescope,[1] watches a star which crosses the field, notes its precise passage behind the vertical wires which cross this field, determines the infinitesimal moment when the star crossed the centre wire, which represents the meridian, reads the declination circle, with microscopes which accurately indicate the altitude of the star, corrects the deviations which may result from the levelling of the feet of the telescope and its slight flexure, corrects the observed position for the apparent elevation caused by atmospheric refraction, which raises all stars above their real position, takes into account the effect of temperature, which disturbs the images (for astronomers observe in the frosty nights of winter as well as in the warm summer nights), corrects the moment of transit for his own personal equation (for every eye does not see nor every ear hear at the same moment the beat of the second which indicates the sidereal time or right ascension) and, after a series of corrections and verifications, completes *one* observation of a star intended to be entered in a catalogue which contains thousands. After thirty years of such labour the modest observer is generally knighted and elected a member of the Institute; this is his reward!

No work which proceeds from human hands is comparable in precision with that of the instruments by the aid of which astronomers determine the exact positions of the celestial bodies. Suffice it to remark that the thickness of a spider's thread is considered as enormous in the micrometrical measure of a star; it is sufficient to glance at the machine for dividing the circles of these instruments to understand what minute care is taken with all the details.

The most ancient catalogue of stars which has been preserved dates back only 2,000 years. It contains 1,025 stars observed at Rhodes by Hipparchus about the year 127 B.C. According to Pliny, this was the first catalogue which man ventured to undertake, and the work was due to the

[1] Sometimes an astronomer devotes himself to such work, and, a martyr to the science, loses his sight, his health, and even his life. On the day I wrote these lines (October 20, 1879) I received from the other side of the globe thirty volumes of astronomical observations, and among these volumes I notice especially a magnificent *Catalogue of Stars observed at the United States Naval Observatory, Washington*, 1845–1871, containing the positions of 10,658 stars, each observed seven or eight times on an average (some more than three hundred times). The author of this catalogue, Mr. Yarnall, worked at it for twenty-six years, finished, printed, and published it; then he died suddenly, *an hour* after receiving the first copy!—*Note to the First Edition.*

curiosity aroused by the phenomenon, then rather rare (and quite miraculous), of the appearance of a new star in the sky. The stars of this catalogue, which have been preserved for us in the 'Almagest' of Ptolemy, were re-observed a thousand years later, about the year 960 of our era, at Bagdad, by the Persian astronomer, Abd-al-Rahman-al-Sufi; then again, nearly five centuries later, about the year 1430, at Samarkand, by Prince Ulugh Beigh (grandson of the monster Tamerlane) who died, the victim of his good-nature, assassinated by his own son, who coveted his throne; then again, about 1590, at Uraniberg, by Tycho Brahé, who had received from the King of Denmark the principality of the island of Huen, where he had established his magnificent observatory. In 1676, the English astronomer Halley, being at the island of St. Helena, formed a first catalogue of southern stars invisible in our latitudes which preceding astronomers had observed. In 1712, Flamsteed, first Director of the National Observatory of England, published his catalogue of 2,866 stars observed at London. In 1742, Lacaille constructed his catalogue of 9,766 stars of the southern hemisphere. We may, further, mention among the best works of this kind the catalogue of Bradley (1760) and that of Piazzi (1800). The catalogue of Lalande gives the number, the magnitude, and positions of 47,390 stars observed at Paris (Observatory of the Military School, since pulled down) from 1789 to 1800. The immense atlas of Argelander presents to our wondering view 324,000 stars observed at Bonn, placed exactly in their precise positions and drawn of their exact magnitude. We now know more than a million stars separately observed, catalogued, and registered on celestial charts. It is thus that sidereal astronomy has by degrees developed and increased, by the number and the precision of observations, and that it henceforth presents to us a subject of study incomparably vaster than planetary and cometary astronomy.

In this long and careful series of observations it has been remarked that the stars are not fixed or unalterable, as they appear to be. There are some which since the time of Hipparchus have slowly diminished in brightness, and have even ended by becoming completely extinct. There are others whose light has gradually increased, and which are now much brighter than they were formerly. Others, again, have changed in tint, and have become more or less coloured [on the other hand, Algol, the well-known variable, which was formerly red, is now white.—J. E. G.] There are some, also, which have suddenly appeared, have shone with a dazzling brightness for several weeks or months, and have then relapsed into obscurity. In a large number a periodical variation of light has been established, in virtue of which certain stars, at first invisible to the naked eye, appear, increase progressively in brightness, then gradually diminish, and disappear, to again reappear after a certain number of days has elapsed : their periodicity is, sometimes, so exact that they are now calculated in advance. Stars have also been noticed which, instead of showing a white or golden light, as is generally the case, are coloured with the most vivid tints, such as those of the emerald, sapphire, ruby, topaz, garnet, and the finest of our precious stones. The telescope has discovered a large number which, instead of being single, as they appear to the naked eye, are double, composed of two stars close together which turn round each other in revolutions which we have already been able to calculate, and which include the most varied periods, from a few years to several centuries and even thousands of years. Sometimes the system is triple : a bright star is seen accompanied by two little companions, and while these two revolve round each other, they move together and revolve round the large one. It is among these multiple systems that we find the most wonderful contrasts of colours. The science is already so far advanced in this respect that I have been able to form a catalogue of nearly 1,000 double stars in certain motion, and to construct a chart of more than 10,000 double stars which have been discovered.

The minute examination of the positions of the stars has also led us to recognise remarkable motions in these little luminous points, which appear to be fixed, attached to the firmament, and which we now know to be veritable suns immensely distant from each other. One of these motions makes the entire sky turn slowly in a secular period which requires no less than 25,735 years to accomplish; this is the general motion of the precession of the equinoxes. But this motion does not belong to the stars : it belongs to the earth ; and it is only an appearance, like the diurnal motion of the sky and the annual motion of the sun ; however, it obliges astronomers to correct annually the geometrical divisions of the celestial maps, because these divisions are gradually displaced before the stars ; such a motion, moreover, belonging to the earth and not to the stars, does not modify their relative positions; the sky appears to turn all in one piece round an ideal axis passing through the poles of the ecliptic. But the careful measurement of the absolute positions of the stars gives evidence of other motions which belong to the stars themselves. Thus, for example, the fine star Arcturus, which anyone may admire every evening on the prolongation of the tail of the Great Bear, is slowly withdrawing from the fixed point where the celestial charts placed it 2,000 years ago, and is moving towards the south-west. It takes 800 years to describe a space in the sky equal to the apparent diameter of the moon ; nevertheless, this displacement was sufficiently perceptible to attract attention more than a century and a half ago, for Halley had noticed it in 1718, as well as those of Sirius and Aldebaran. However slow it may appear at the distance we are from Arcturus, this motion is, at a minimum, 1,637 millions of miles a year. Sirius takes 1,338 years to pass over the same angular space in the sky; at the distance of that star this is, at a minimum, 397 millions of miles per annum. The study of the proper motions of the stars has made the greatest progress in the last half-century, and especially in recent years. All the stars visible to the naked eye and a large number of telescopic stars show displacements of this kind.

We must study here in detail all the facts revealed by the minute analysis of modern science. The determination of the precise positions of the stars constitutes, in fact, the fundamental and classical work of the official observatories. It constitutes the *foundations of astronomy.* It is by

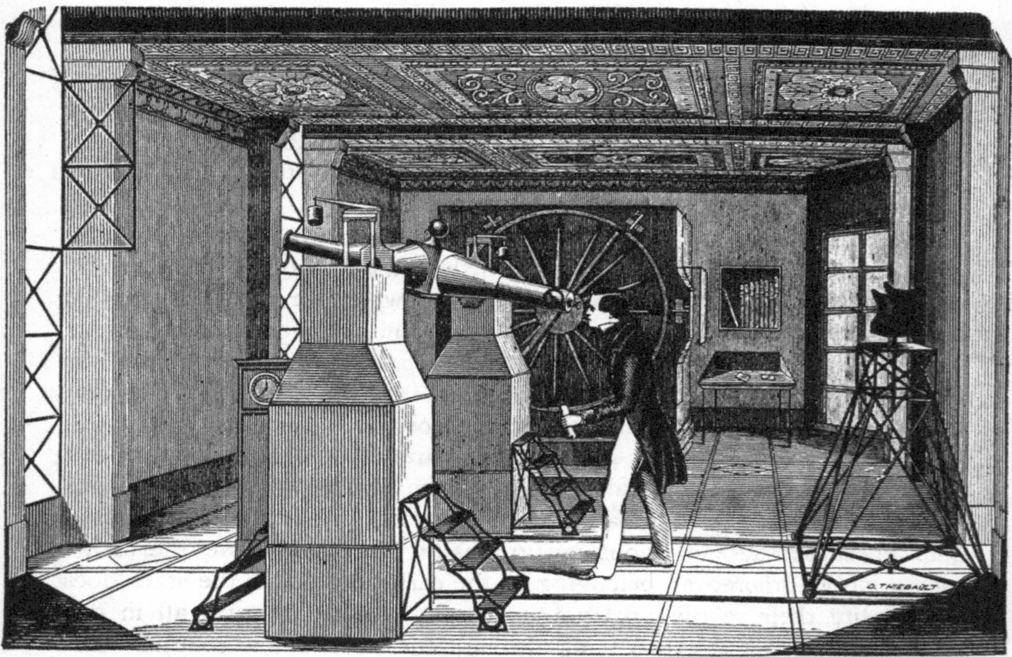

FIG. 256.—MERIDIAN TELESCOPE AND MURAL CIRCLE USED IN THE PARIS OBSERVATORY

this that we know how the earth turns and what variations its motions are subject to, like those of all worlds. The work will never be finished, and is always being recommenced; for, as we shall soon see, no star remains absolutely fixed in the midst of immensity, and from age to age its position perceptibly changes. The knowledge of the proper motions of the stars, and the deductions relating to the transport of the solar system in space, are entirely due to meridian observations.

The meridian circle of the Paris Observatory is installed in the Meridian Room in the pavilion on the side towards the garden of the Luxembourg, as we see the building when we come from the centre of Paris.

The flat roof of this pavilion opens by three trap-doors corresponding to the three windows to the north and the three other windows symmetrically situated to the south; it is as if it were slit by three enormous strokes of a saw. Here are installed three telescopes, which can be turned from north to south in the plane of the meridian, across which all the stars pass once a day. The first of these telescopes is the meridian circle. The two other instruments are an ancient meridian telescope and a mural circle, which we see in fig. 256. These two instruments performed separately the work now united in the function of the meridian circle. the first was used to take the time of the passage, or right ascension; the second, the altitude of the star observed, or its distance from the equator (declination).

As we are now describing the Observatory,[1] we may add that in the centre of the large building is the spacious hall, traversed from north to south by the meridian of Paris, inlaid in copper on the floor; it is furnished with ancient instruments and souvenirs. This is a museum.[2]

[1] Permission may be obtained to visit the Observatory by addressing a request to the Director; the visits take place on the first Saturday of each month, but by day; never by night. This is almost as if we were to go and see a play at the theatre before the arrival of the actors: the scenery would assuredly give but a very imperfect idea of it.

[2] On the upper terrace we see in the centre a little observatory with three small cupolas (the middle one contains a small equatorial by Gambey), and at each side two large cupolas shelter two powerful telescopes, one of five metres, the other of nine metres. All these instruments are moved by clockwork, in a direction opposite to the motion of the earth. They may be directed to all points of the sky, and once fixed on a star and put in communication with the clockwork motion, the star remains in the middle of the field of the telescope, permitting the observer to study it at his ease, as if the earth had ceased to turn; whilst in motionless telescopes the star observed rapidly crosses the field, thus showing the rapidity of the earth's motion, a motion magnified according to the power of the eyepiece used. In placing the instrument at the altitude of its declination on the one hand, and at the precise hour of its right ascension on the other, it is unnecessary to see the star in advance to know that it is in the field of the telescope. The dome may be hermetically closed, or clouds may obscure the sky; let us open the dome and wait for a vista through the clouds: the desired star shines exactly in the middle of the field of the telescope.

From the terrace of the Observatory let us descend to the garden. We admire the southern façade, which bears the impress of the age of Louis XIV.; it is the façade of a palace, while that of the north side reminds us rather of the ancient castles of feudalism. The garden is spacious, but already crowded. Besides the pavilion of the great meridian circle, of which we have spoken, and a magnetical pavilion, it is further occupied by two cupolas, each sheltering an equatorial, by a new meridian circle, and by the colossal telescope of 1·20 metre (4 feet) in diameter and 7·30 metres (24 feet) in height, finished in 1876. This is one of the largest telescopes which exist, but it is not one of the best. Here are also installed two great equatorials of the Coudé system, which permits the observer to use the eyeglass as a microscope whilst quietly seated in an armchair, whatever may be the direction which the tube of the instrument may take when directed towards the star to be observed.

The Observatory was erected in 1667, by the influence of Colbert, under the auspices of the Academy, and by the architect Perrault, designer of the Colonnade of the Louvre. Its height is 27 metres (88 feet), and its depth below the ground is also 88 feet. Here are the cellars, with a constant temperature (11°·7) (53° Fahr.), where, since September 25, 1871, thermometers have been observed as standards of graduation. The influence of the solar heat does not traverse the soil beyond 25 metres (82 feet); the maximum of the annual temperature happens in July at the

All the official observatories have, like ours, for their principal object this minute and permanent verifying of the precise positions of the stars. The search for new planets and comets; the study of the physical constitution of the sun, of the moon, or the planets; the investigations of spectrum analysis, the measurement of double stars, the observation of the variable stars—in a word, all the innumerable researches to be made in the inexhaustible fields of Infinitude, are labours outside the province of these observatories —'extra-meridian' labours, which require special services in these observatories, or, better still, independent observatories.

The Observatory of Juvisy, which I founded in 1882, represents one of these independent observatories, rare in France, but numerous in England and the United States.

Thus have laboured for hundreds and thousands of years the astronomers who have devoted their lives to the patient and laborious search for the secrets of the constitution of the universe. It is to these labours that we owe the knowledge of the true place which we occupy in Nature.

surface of the ground, in August at 25 centimetres, in September at 50 centimetres, in October at 1 metre, in November at 3 metres, in December at 7 metres, in January at 10 metres, and in February at 15 metres. Afterwards the curve is hardly perceptible, and the constant temperature is exactly that of the mean of the year at the place observed, increasing according to the depth at the rate of 1° for 30 metres. (The mean temperature of Paris is 10°·7.) A sepulchral silence reigns in these solitudes, which correspond with the bone caves of the Catacombs: dark passages lead to the gallery of thermometers, where hovers the memory of the *savants* who once read them—Cassini, Réaumur, Lavoisier, Laplace, and Arago. The storms of the atmosphere and of humanity do not penetrate into these lonely depths.

CHAPTER IV

MAGNITUDE OR BRIGHTNESS OF THE STARS

Their Distribution in the Sky. Their Number. Their Distances

A SINGLE glance at the sky is sufficient to show that the stars are not all of equal brilliancy: while some are endowed with a very vivid light, others are so faint that we can hardly distinguish them. The greater part of the stars visible to the naked eye are included between these two limits, and present, so to say, all degrees of brightness which can be imagined, passing imperceptibly into each other between these limits. There are, moreover, a considerable number of stars (by far the greater number) which can only be seen by the aid of opera-glasses or telescopes, and which have also very different magnitudes, from stars which observers gifted with excellent sight can perceive with the naked eye, to those which can with difficulty be seen as faint points in the dark field of the most powerful instruments.

FIG. 257.—RELATIVE BRILLIANCY OF STARS OF THE FIRST SIX MAGNITUDES. (THE SURFACE OF THE DISCS IS PROPORTIONATE TO THEIR BRILLIANCY.)

In order to facilitate the indication of the brightness of a star, all stars have been classed in order of magnitude. This word *magnitude* is incorrect, as it has no connection with the dimensions of the stars, since these are still unknown; it dates from the time when it was believed that the brightest stars were the largest, and this was the origin of the designation; but it is important to know that this is not its real meaning. It simply corresponds to the *apparent light* of the stars. Thus, stars of the 1st magnitude are those which shine with the greatest brilliancy in the dark night; those of the 2nd magnitude are those which are less brilliant, &c. Now, this apparent brightness depends, at the same time, on the real size of the star, or its intrinsic light, and on its distance from the earth; it possesses, consequently, an essentially relative meaning, although we can say assuredly that in general the most brilliant stars are the nearest, while those whose faint light is scarcely perceptible in the field of the telescope are the most distant. When we speak of the magnitude of stars, it is understood that it is simply a question of their apparent light.

The stars of the 1st magnitude are nineteen in number. In reality, the nineteenth—that is to say, the least brilliant of the series—may well be entered in the first rank of stars of the 2nd magnitude, or the first of this second series might in the same way be added to the stars of the 1st magnitude. These divisions, which our classifications require, do not exist in nature. But as we must limit ourselves to a particular star if we wish to make a series, the following list of stars of the 1st magnitude has been agreed upon :—

STARS OF THE FIRST MAGNITUDE IN DECREASING ORDER OF BRIGHTNESS

1. *Sirius*, or α of the Great Dog.
2. *Canopus*, or α of Argo.
3. α of the Centaur.
4. *Arcturus*, or α of Boötes.
5. *Vega*, or α of the Lyre.
6. *Rigel*, or β of Orion.
7. *The Goat*, or α of Auriga (*Capella*).
8. *Procyon*, or α of the Little Dog.
9. *Betelgeuse*, or α of Orion (slightly variable).
10. β of the Centaur.
11. *Achernar*, or α of Eridanus.
12. *Aldebaran*, or α of Taurus.
13. *Antares*, or α of Scorpio.
14. α of the Southern Cross.
15. *Altair* of the Eagle (Aquila).
16. *Spica*, or α of Virgo.
17. *Fomalhaut*, or α of Piscis Australis.
18. β of the Southern Cross.
19. *Regulus*, or α of the Lion.

These are the nineteen brightest stars of the whole sky ; they are given in order of brightness. Then come the stars of the 2nd magnitude, and all the others successively, showing the following total :—

19	stars	of the	1st	magnitude
59	"	"	2nd	"
182	"	"	3rd	"
530	"	"	4th	"
1,600	"	"	5th	"
and 4,800	"	"	6th	"

It will be observed that each class after the first is about three times more numerous than the one which precedes it, so that multiplying the number of stars which compose any series by three, we get nearly the number of those composing the following series. According to this estimation, the number of stars of the first six magnitudes, or, in other words, the total number visible to the naked eye, is about 7,000. Excellent sight distinguishes 8,000, average sight about 5,700. Generally, we think we see many more ; we believe we can count them by myriads, by millions ; in this, as in other things, we are always led into exaggeration. As a matter of fact, however, the number of stars visible to the naked eye in both hemispheres all over the earth does not exceed the above figures. The stars visible to the naked eye for ordinary sight are in reality so few in number that we might easily show them in an illustration of the size of these pages, and count them: the southern hemisphere has 3,307, and the northern 2,478; total, 5,785, without counting, of course, the star dust of the Milky Way. Thus we see with the naked eye fewer stars in the sky than there are inhabitants in a small town. It is, then, not so difficult to make their acquaintance as might be imagined. It is but an hour's amusement.

But when our feeble sight fails us, the telescope, that giant eye which grows from age to age, piercing the depths of the heavens, constantly discovers new stars. Exceptional sight penetrates beyond the 6th magnitude. An ordinary opera-glass shows the stars of the 7th magnitude, which number 13,000. A terrestrial telescope shows those of the 8th magnitude, which number 40,000. Thus the number of stars increases as we penetrate further beyond the sphere of natural vision. A small astronomical telescope discloses stars of the 9th magnitude, the number of which exceeds 100,000. And so on. A telescope of moderate power shows the stars of the 10th magnitude, which number nearly 400,000. Here already the spectacle is immense and dazzling.

The progression continues. We may estimate at a million the number of stars of the 11th magnitude, and at three millions those of the 12th. According to astronomical gauges made for the purpose of sounding space, the number of stars of the 13th magnitude rises to no fewer than 10 millions, and those of the 14th to nearly 30 millions. If we add these numbers, we find for the total of stars down to the 14th magnitude inclusive the number, already difficult to imagine, of *forty-four millions.*

But these are not *all* the stars. Already even the powerful telescopes constructed in recent years have penetrated the depths of immensity so far as to discover stars of the 15th magnitude, and the stellar statistics have now risen to *one hundred millions*! Celestial photography penetrates further still, and the numbers become so enormous that we are overwhelmed by their weight without understanding them.

One hundred millions of stars! This gives 17,000 stars for each of those which we see with the naked eye—seventeen times more than we can count in both hemispheres. We shall shortly estimate the distances which separate them, and the incomparable space over which their empire extends.

One hundred millions of suns similar to ours, and surrounded by worlds counted by thousands of millions! these are, unquestionably, very amazing numbers, and it would not be surprising if they should not be at once realised in their prodigious magnitude by our brains, unaccustomed to such enormous figures. We may remark, however, in passing, that a number *well understood* tells more than the finest phrases.

Thus, for example (to digress for a moment: *similia similibus curantur*), what idea does your imagination give you of the most enormous sum of money which has ever been calculated? This sum—a rather surprising one—is that which would be produced by the compound interest on five *centimes* (a halfpenny) invested at the birth of Christ. A mathematician will in vain assure you that this sum would be so enormous that all the waggons of all the railways in the world could not carry it; it would be in vain for him to tell you that if the Alps and Pyrenees were mines of diamonds they would not represent its value; but if a calculator proves that it could only be written by the following row of thirty-nine figures—

342,653,248,699,000,000,000,000,000,000,000,000,000 of francs—

we are absolutely stunned by the force of such a number. This number, then, enlightens, illuminates, transfigures itself, when we reflect that the entire globe of the earth weighs but 5,875 sextillions of kilogrammes, and that, if it were formed of solid gold, it would be three and a half times heavier, would weigh 20,562 sextillions, and would be worth only 2,796,400,000 milliards of milliards[1] of pounds! If, then, our planet were of massive gold, it would still require 4,900 millions of globes like the earth to pay this precious capital. *Supposing there fell from the sky every minute an ingot of gold as large as the earth, it would be necessary that this fall should continue for* 9,300 *years to arrive at the payment of the total sum.*[2]

Who will now maintain that figures are not eloquent! Here is a numerical result incomparably superior to all those of astronomy. The population of the heavens does not yet lead us to rows of thirty-nine figures!

FIG. 258.—A REGION OF THE HEAVENS SEEN WITH THE NAKED EYE

But the sky is rapidly transformed in the field of progressive optics. Already we distinguish neither constellations nor divisions; a fine dust shines where the eye, left to its own power, sees but a black darkness which contains but two or three stars. In proportion as the marvellous discoveries of optics increase the visual power, all the regions of the heavens will become covered with this fine golden sand; and the day will come when the astonished gaze, rising towards these unknown profundities, and finding itself arrested by the accumulation of stars which succeed each other to Infinitude, will no longer see before it anything but a delicate tissue of light.

[1] A milliard is 1,000 millions.

[2] We made this calculation in 1879 for the first edition of this work, and it refers to the sum produced up to the year 1880. Now, this sum goes on doubling every fourteen years (exactly, 14·21).

Each of these points is a sun, the centre of force, of activity, of motion, and of life. Every telescopic increase of power brings millions before the eye of the astronomer.

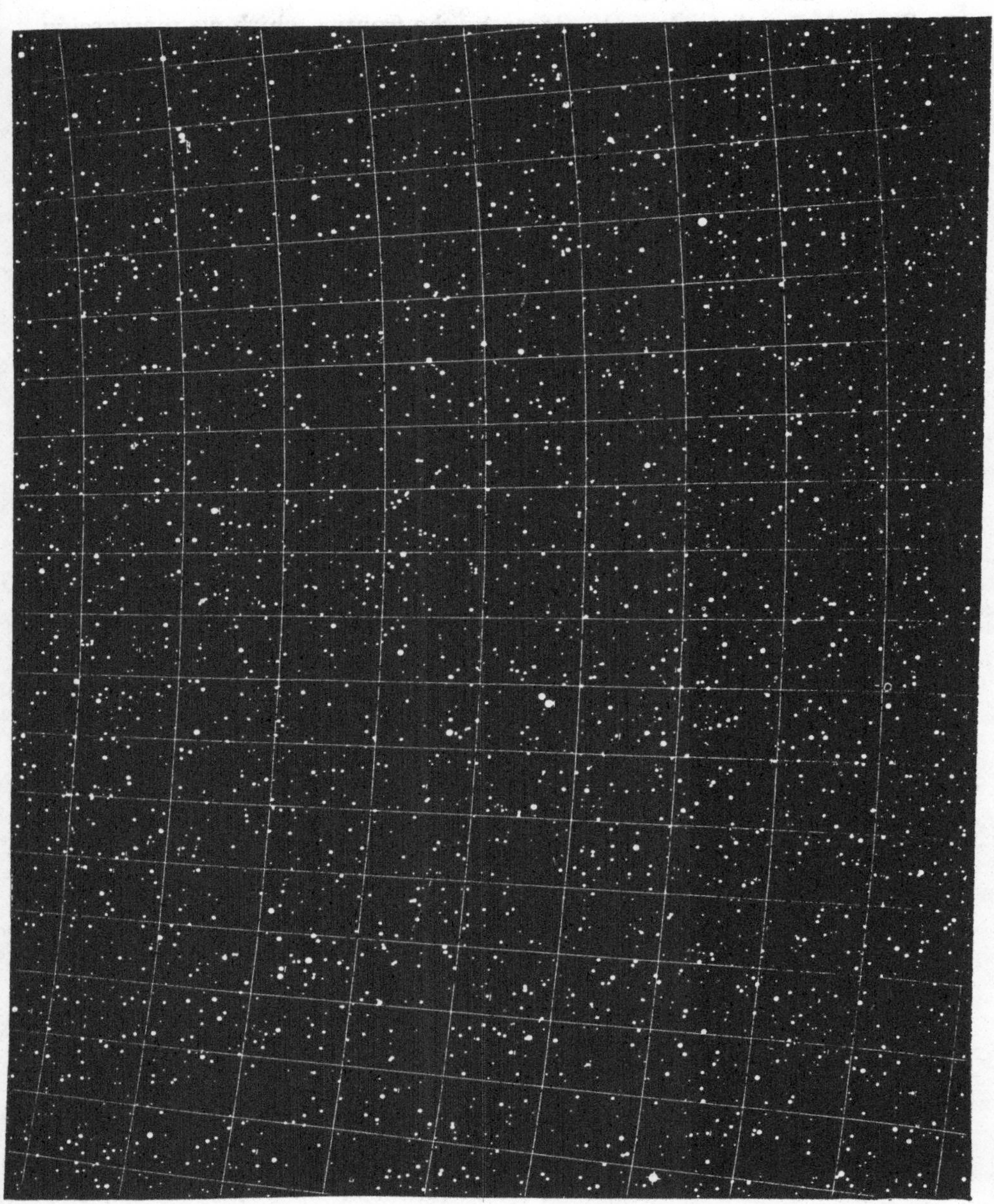

FIG. 259.—TELESCOPIC VIEW OF THE REGION OF THE HEAVENS SHOWN IN FIG. 258

But this is still only our visible universe. There, where the telescopic power stops, there, where the flight of our tired conceptions flags, immense and universal nature continues her work: the telescope carries us into Infinitude—*and leaves us there.*

Space has no bounds. Whatever be the frontier which we may assign to it in thought, our imagination immediately flies across this frontier, and, looking beyond, still finds space. And although we cannot comprehend the infinite, each of us feels that it is easier to conceive space as unlimited than to imagine it limited, and that it is impossible that space should not exist *everywhere.*[1]

Let us attempt now to gauge these depths. The first method which presents itself is to examine the ratio according to which the light of the stars diminishes with their distance.

The estimate of distances by photometry rests on two principles, the truth of which cannot be disputed—(1) the stars may not all be placed at the same distance from us; (2) the most distant should, from this fact alone, appear to us the smallest. These principles would even lead us to a direct and certain estimate of their relative distances, if we could further assert that all the stars have an equal intrinsic light. But this equality is neither proved nor probable.

The problem must, then, be treated by the calculation of probabilities.

Given a star of a determined magnitude, how much should its distance be increased in order that its brightness should be diminished by one unit?

For the brightest stars the luminous intensity is more than doubled when we pass from one order of magnitude to that which immediately precedes it, but for the fainter the ratio between the intensities approaches nearer to the number 2. Thus, setting aside the exceptional Sirius, we find that from the 1st magnitude to the 2nd the ratio is 3·75; from the 2nd to the 3rd, 2·25; from the 3rd to the 4th, 2·20. When we arrive at telescopic stars the proportion follows nearly the same law, although it may be interrupted in the passage from the 6th to the 7th magnitude—that is to say, at the limit of stars visible to the naked eye. On the whole, we find as a general mean the ratio 2·42.[2] [The ratio now

[1] The contemplation of the immensity of the heavens unavoidably gives us the idea of the infinite. Theologians and scholastic philosophers have in vain piled up quibbles on the points of needles, to make us believe that they know the attributes of the Creator, and that 'space cannot be infinite, because it would be God'; these are the arguments of preachers, the value of which is no longer in dispute since the time of Erasmus, the author of *The Praise of Folly* (*L'Éloge de la Folie*): the most timid of astronomers can now affirm that space is necessary, infinite, and eternal—three qualifications theologically reserved to God alone.

[2] Notwithstanding 'the obscure light which falls from the stars,' their total light is not so 'obscure' as it appears. At midnight in a very pure atmosphere one can always see the hour by a watch and read large characters by their aid. Sailors do not like artificial lights; they prefer to remain with the light of the stars alone, which is sufficient for all their operations; to read the compass

generally adopted by astronomers for stars of *all* magnitudes is 2·512, of which the logarithm is 0·4.—J. E. G.]

We can, then, by this ratio calculate the distance at which it would be necessary to place successively an average star of the 1st magnitude in order that it should become equal to a 2nd, a 3rd, &c. Here is the result of this calculation:

Magnitudes	Distances[1]	Magnitudes	Distances
1	1·00	9	34
	1·55	10	53
3	2·42	11	83
4	3·76	12	129
5	5·86	13	200
6	9·11	14	312
7	14·17	15	486
8	22·01	16	735

Thus, the stars of the 6th magnitude, the faintest which we can perceive with the naked eye, would be nine times more distant than those of the 1st magnitude; those of the 13th magnitude 200 times farther, &c. As we descend in the scale of magnitudes, the quantity of light emitted diminishes according to a geometrical ratio, the stars of each order being in general about two and a half times brighter than those of the order immediately inferior. Supposing that this proportion represents the general course, we find that it requires

2½ stars of the 2nd magnitude to equal the brightness of one of the 1st
6 " " 3rd " "
16 " " 4th " "
42 " " 5th " "
109 " " 6th " "
278 " " 7th " "
712 " " 8th " "
1,822 " " 9th " "

4,656 stars of the 10th magnitude to equal the brightness of one of the 1st
11,900 " " 11th "
30,420 " " 12th "
77,750 " " 13th "
199,000 " " 14th "
500,000 " " 15th "
1,280,000 " " 16th "

they use a feeble light illuminating the card by transparency. At great heights, in a balloon, I have always been able to see the position of objects (I have never, however, been able to read the degrees of the barometer without light). All our artificial lights, and even the dazzling electric light, are soon effaced by distance. Even at the distances which separate us we divine that each of the stars is a true sun.

[1] [For a ratio of 2·512 these distances are as follow:—

Magnitudes	Distances	Magnitudes	Distance
1	1·0	9	40
2	1·58	10	63
3	2·51	11	100
4	3·98	12	158
5	6·31	13	251
6	10·00	14	398
7	15·48	15	631
8	25·12	16	1000

—J. E. G.]

The number of stars of the several orders of magnitude varies in a ratio which is not very different from the inverse of that of their brightness. I need not say, however, that these magnitudes merge imperceptibly into each other, and if we wish to express the light of any star with greater exactness, we must go to fractions; thus, a star is noted of magnitude 2½, another of 3¼. Astronomers usually extend the approximation to a tenth, and in photometric measures they now go even to the hundredth.

In general, the marvellous difference between the telescopic and the natural vision is not realised. Argelander has observed and catalogued in their precise positions the stars of our northern hemisphere down to the 10th magnitude. There are 324,000 in this half of the sky alone. Let us look at any point: we count a dozen stars; take a telescope of only 7 centimetres (2¾ inches)—the telescope of Argelander—and let us compare this telescopic vision with the natural vision: we obtain the eloquent comparison shown by figs. 258 and 259 (pp. 588, 589). Here we see the beginning of the telescopic revelation.

These considerations give us a first idea of the scale of the sidereal universe. But it is important to make here some restrictive remarks.

Taking a star of any magnitude, there is nothing to prove that it is at the distance indicated by the preceding considerations. We cannot, then, apply these rules to any given star. Some star, invisible to the naked eye, of the 7th, 8th, or 9th magnitude, may be nearer than one of the 1st magnitude. We shall soon have proof of this.

On the other hand, it may be that the preceding determinations should not have even the average value which we attribute to them; if, for example, nature had decreed that we should be surrounded by small stars, and that the most magnificent suns should be very distant from our position in space. But this is a chance which the theory of probabilities indicates as very improbable. That does not, however, prevent it from being partly realised to a greater or less degree. Conclusion: we do not grant to the preceding determinations all the value which eminent astronomers, such as Sir William Herschel, William Struve, and Secchi, have attributed to them, and we consider them only as a first *gauge*, destined to make us search with attention through the population of the heavens.

What are the real distances of the stars?

CHAPTER V

MEASUREMENT OF CELESTIAL DISTANCES

Stars of which the Distance is known. Relation of our Sun to his nearest Peers

WHAT method can the microscopical inhabitant of this little terrestrial globule make use of to measure the distance which separates it from the enormous suns burning in the midst of the infinite depths? Does not such an attempt exceed the limits of his power? Will not the contrast between celestial immensity and terrestrial littleness overwhelm the audacious pigmy who attempts to scale the sky? No! Human hope is infinite, and, like it, the power of genius rises to the highest summits which it contemplates in the ethereal splendour. Where shall the human mind be stopped in the conquest of eternal realities? When will it be satisfied with the present, and no longer spread the wings of its desire towards horizons always fading into the future? It will never be satisfied, it will always aspire towards a higher progress. This is its nature, this is its destiny, this is its grandeur, and this is its true happiness. Forward! excelsior! always higher!

To measure such distances the diameter of the terrestrial globe can no longer be used as the base of a triangle, as in the measurement of the distance of the moon, and the difficulty can no longer be surmounted, as in the case of the sun, by the aid of another planet. But, fortunately for our knowledge of the dimensions of the universe, the construction of the solar system affords a means of survey for these distant perspectives, and this method, at the same time that it demonstrates once more the motion of the earth round the sun, is utilised for the solution of the greatest of astronomical problems.

In fact, the earth, in revolving round the sun at 93 millions of miles distance, describes yearly a circumference (in reality it is an ellipse) of 580 millions of miles. The diameter of this orbit is, then, 186 millions of miles. Since the revolution of the earth is completed in a year, our planet is found at any particular moment opposite to the point where it was six months before, and exactly where it will be six months later. In other words, the distance of any point of the terrestrial orbit from the point where it will pass after six months' interval is 186 millions of miles. This is, then, a respectable length, which may serve as the base of a triangle of which the vertex would be a star.

The method of measuring the distance of a star consists, then, in observing minutely this little brilliant point at six months' interval, or rather during the whole year, and seeing whether the star remains fixed, or is subject to a small apparent displacement of perspective on account of the annual displacement of the earth round the sun. If it remains fixed, it is because it is at an infinite (or rather immeasurable) distance from us—on the horizon of the sky, so to say—and because 74 millions of leagues (186 millions of miles) are as zero compared with this distance. If it is displaced, we find that it describes during the year a little ellipse, the reflection of the earth's annual motion. Everyone has noticed, in travelling on a railway, that the trees and the nearest objects seem to move in a direction opposite to our motion, and so much the more rapidly the nearer they are, while distant objects situated near the horizon remain fixed. It is absolutely the same effect which is produced in space on account of our annual motion round the sun. Although, however, we move incomparably faster than an express train (eleven hundred times quicker), and travel 1,600,000 miles a day, the stars are all so distant that they appear hardly aware of our displacement. Our 186 millions of miles is almost nothing for even the nearest. How unfortunate that we do not inhabit Jupiter, Saturn, Uranus, or especially Neptune! With their orbits, five, nine, nineteen, and thirty times wider than ours, the inhabitants of these planets have been able to determine the distance of a much larger number of stars than we have yet been able to measure.

This method of measuring the distance of the stars by the effect of perspective due to the annual displacement of the earth had already been divined by the astronomers of the last century, and in particular by Bradley, who, in attempting to measure the distance of the stars by observations made at intervals of six months, found another thing. Instead of discovering the distance of the stars, which was the object of his observations, he discovered a very important optical phenomenon—the *aberration of light*, an effect produced by the combination of the velocity of light with the earth's motion in space. His case was similar to that of William Herschel, who, in searching for the parallax of stars by comparisons between bright stars and their near companions, found the double-star systems; or like Fraunhofer, who, in seeking the limits of the colours of the solar spectrum, found the rays of absorption, the study of which founded spectrum analysis. The history of the sciences shows us that discoveries have very often been made by researches which are only indirectly related to them. In attempting to reach by the West the eastern frontiers of Asia, Christopher Columbus discovered the New World. He might not have discovered it, and he might not have looked for it, if he had known the true distance from Portugal to Kamschatka.

We did not know the distance of any stars till the year 1840. This shows how recent this discovery is; indeed, we have hardly now begun to form an approximate idea of the real distances which separate the stars from each other. The parallax of 61 Cygni, the first which was known, was determined by Bessel, and resulted from observations made at Königsberg from 1837 to 1840. Since then, the first figure obtained has been corrected by a series of more recent observations.

We can very easily understand the relation which connects the distance of a star with its parallax by an examination of the accompanying figure (260). The angle under which the diameter of the terrestrial orbit is seen is so much the smaller as the star is more distant, and the apparent motion of the star, which reflects in perspective the real motion of the earth, diminishes in the same proportion. Thus, the lowest star of this little figure shows an annual motion performed under an angular width of 20°; the next gives an angle of 15°, and the highest an angle of 11°. The geometrical relation which we learned in the first chapters of this work, from the distance of the moon (p. 86), immediately gives the distance. In fig. 260 the proportions are very exaggerated, since a parallax of 1° corresponds to fifty-seven times the length of the base. Now, the angular motion of the nearest star is not 2″; on the scale adopted for this figure, the nearest star to us should be placed, at least, at 100,000 times the base of the triangle, which is 2 centimetres—that is to say, at 2 kilometres (1¼ mile). It would assuredly be difficult to put such a figure in any book.

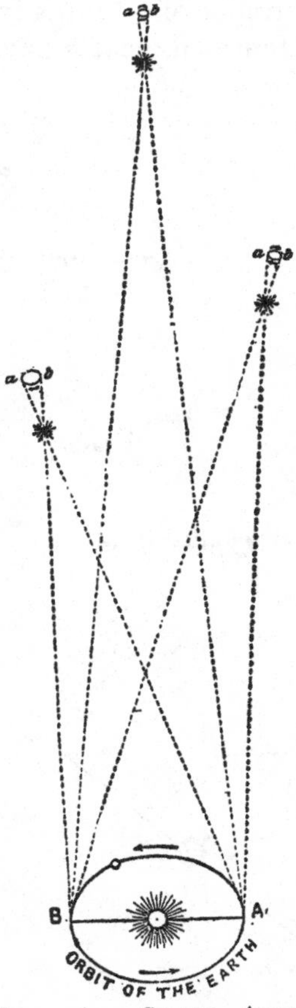

FIG. 260.—SMALL APPARENT ELLIPSES DESCRIBED BY THE STARS AS A RESULT OF THE ANNUAL DISPLACEMENT OF THE EARTH

Let us continue here the little table of p. 86 from the point of view of the smallest parallaxes.

			Units				Units
An angle of 10″ corresponds to a distance of		.	20,626	An angle of 0″·6 corresponds to a distance of		.	343,750
„	5″	„ . .	41,253	„	0″·5	„ .	412,530
„	2″	„ . .	103,132	„	0″·4	„ .	515,660
„	1″	„ . .	206,265	„	0″·3	„ .	687,500
„	0″·9	„ . .	229,183	„	0″·2	„ .	1,031,320
„	0″·8	„ . .	257,830	„	0″·1	„ .	2,062,650
„	0″·7	„ . .	294,664	„	0″·0	„	Immeasurable

The parallax of a star is usually expressed by *the angle under which the radius or semi-diameter of the terrestrial orbit is seen from the star.* Consequently, a star whose parallax should be 1″ would show by that that it is distant 206,265 times 93 millions of miles (19 billions of miles); a parallax of 9 tenths indicates a distance of 229,183 times the same unit; 8 tenths indicates 257,830 times, and so on.[1] We had good reason, in the

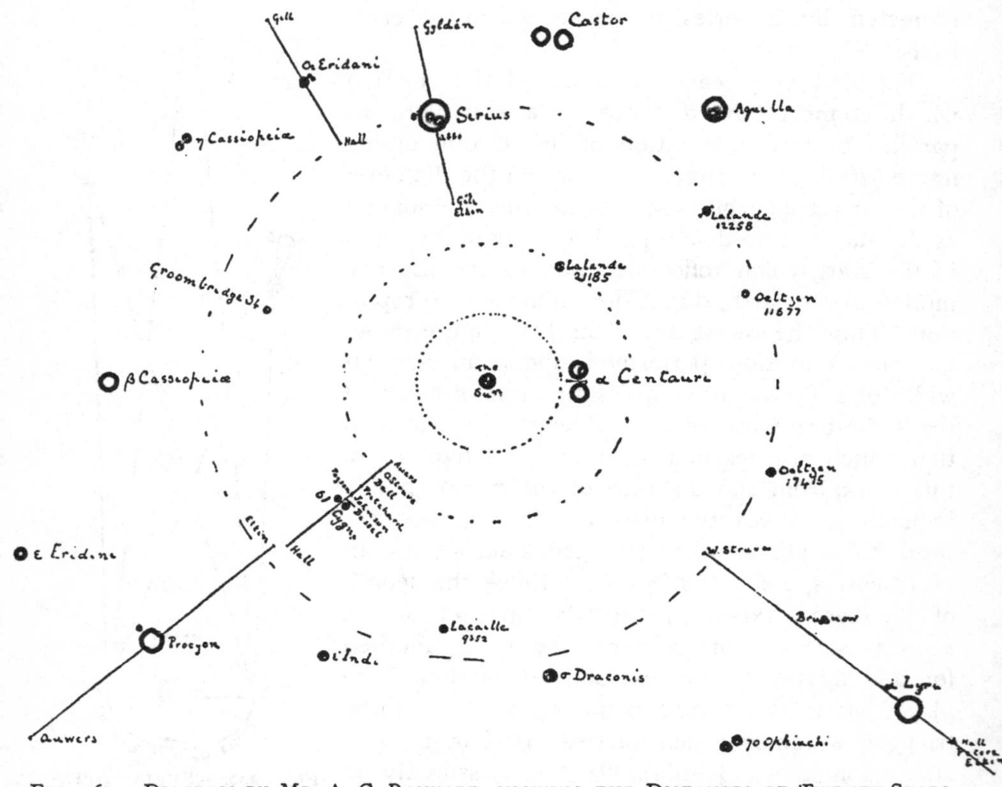

FIG. 261.—DIAGRAM BY MR. A. C. RANYARD, SHOWING THE DISTANCES OF TWENTY STARS FOR WHICH LARGE PARALLAXES HAVE BEEN FOUND. (From *Knowledge*, June 1, 1889.)

Book on the Sun, to especially dwell on its distance, because it is the unit by which we must measure everything in the universe.

In the whole of sidereal astronomy there is, perhaps, nothing more difficult to determine than the parallax of a star. To think that among all the stars in the sky there is not one which shows a parallax of one second—that is to say, an annual motion of two seconds! Now, two seconds is a millimetre seen at a hundred metres, it is a hair of a tenth of a milli-

[1] Here is the formula. The distance of any star is given by $\frac{206265}{p}$ R, R being the radius of the terrestrial orbit, and p the parallax.

metre seen at 10 metres (32·8 feet)! Well, it is in this width that the annual motion of a star is performed. The telescope magnifies it, of course; without this it would be absolutely imperceptible; but how easily it can be concealed by the imperceptible motions of the telescope, by the influences of temperature, by refraction, precession, nutation, aberration, and by the proper motion of the star itself in space! All these united influences amount to several seconds, and are themselves subject to some uncertainties, and instrumental errors must still be added to them. How, then, shall we extricate trustworthy indications of the minute displacement due to the effect of the earth's motion? Astronomers have, however, succeeded in doing so for *some* stars.

When the parallax is obtained, nothing is simpler than to translate it into the distance by using the little table on p. 595 and the formula on p. 596. If this parallax is 1″, we know that the distance is 206,265 times 93 millions of miles; if it is 0″·9, the result is 229,183 times the same unit (the sun's distance from the earth), and so on. This is a mathematical and unquestionable result, however marvellous it may appear, and however obstinate certain minds may be in not accepting it; there is here neither miracle nor mystery.

We recently calculated ('Revue d'Astronomie,' December 1889) the results inferred from all the measures of parallaxes attempted. Without entering into too technical details, we give here a list of the parallaxes most certainly determined (from several series of concordant observations):—

TABLE OF STARS OF WHICH THE DISTANCES ARE BEST KNOWN

Name of Star	Magnitude	Parallax	Distance in radii of terrestrial orbit	Distance in billions of miles	Duration of light journey in years
α Centauri	1·0	″ 0·75	275,000	25 billions	4·35
61 Cygni	5·1	0·45	458,000	43 „	7·2
Σ 2,398, Draco	8·2	0·35	589,000	55 „	9·32
Sirius.	1·0	0·39	525,000	58 „	8·36
9,352, Lacaille	7·5	0·29	711,000	66 „	11·24
Procyon	1·3	0·27	761,000	71 „	12·0
Lalande, 21,258	8·5	0·26	793,000	74 „	12·5
Œltzen, 11,677	9·0	0·26	793,000	74 „	12·5
σ Draconis	4·7	0·25	838,000	78 „	13·2
Aldebaran	1·5	0·24	874,000	81 „	13·8
ε Indi.	5·2	0·22	937,000	87 „	14·4
Œltzen, 17,415	9·0	0·20	1,010,000	94 „	16·3
Σ 1,516, Draco	7·0	0·19	1,086,000	101 „	17·1
o² Eridani	4·4	0·19	1,086,000	101 „	17·1
Altair.	1·6	0·19	1,086,000	101 „	17·1
Bradley, 3,077	5·5	0·19	1,086,000	101 „	17·1
η Cassiopeiæ	3·6	0·16	1,272,000	118 „	20·1
Vega	1·0	0·15	1,375,000	128 „	21·7
Capella	1·2	0·11	1,875,000	174 „	29·6
Arcturus	1·0	0·094	2,194,000	204 „	34·7
Pole Star	2·1	0·089	2,318,000	215 „	36·6
μ Cassiopeiæ	5·2	0·060	3,438,000	320 „	54·4
1830, Groombridge . . .	6·5	0·045	4,583,000	426 „	72·5

We see that there is no star which shows a parallax of a whole second. The immense majority of stars give zero for the result.

The numbers which express the stellar distances in kilometres or in leagues (or miles) are so vast that they indicate nothing to our mind. It is, perhaps, a little less difficult to conceive the extent of these interstellar abysses by attempting to follow a ray of light, which, rapid as the lightning,

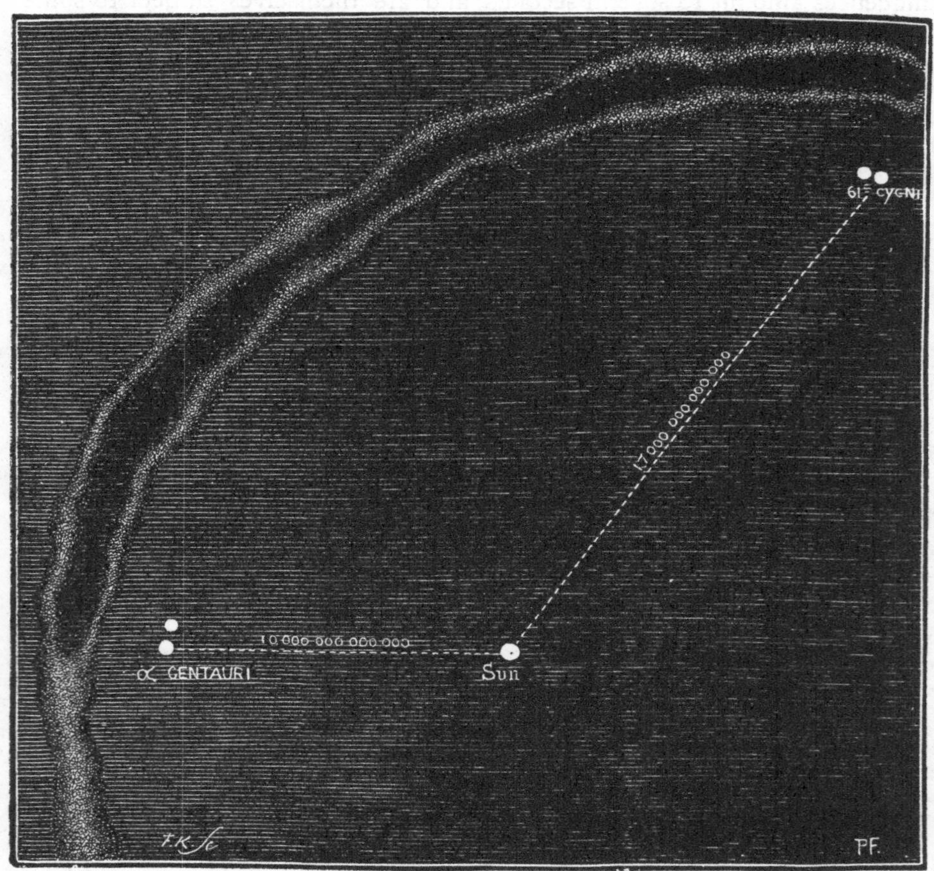

FIG. 262.—THE TWO SUNS THAT ARE NEAREST OURS

shoots through immensity with the velocity of 75,000 leagues (186,400 miles) a second, and crosses in 8 minutes and 13 seconds the 148 millions of kilometres (93 millions of miles) which separate us from the sun. We can easily calculate that for a distance corresponding to a parallax of 1″ the duration of the journey of a luminous ray is 3·262 years, which gives 6·524 years for a parallax of half a second, 13·048 years for a parallax of a quarter second, &c. It is in this proportion that the last column of the preceding table has been calculated.[1]

This table presents the most trustworthy data which we have yet obtained with reference to stellar distances. As a great number of attempts have been made on stars which, by their brightness or by the magnitude of their proper motion, would appear to be the nearest to us, we may believe that the star now considered as the nearest is really so, and that there is no other less distant. Thus our sun, a star in the immensity, is isolated in Infinitude, and *the nearest* sun reigns at 10 trillions of leagues (25 billions of miles) from our terrestrial abode. Notwithstanding its unimaginable velocity of 186,400 miles a second, light moves, flies, during four years and 128 days to come from this sun to us. Sound would take more than 3 millions of years to cross the same abyss. At the constant velocity of 60 kilometres (37 miles) an hour, *an express train starting from the sun Alpha Centauri would not arrive here till after an uninterrupted course of nearly 75 millions of years.*

We have already remarked that a bridge thrown from here to the sun would be composed of 16,600 arches of the width of the earth, and to reach the nearest sun it would be necessary to join 275,000 such bridges end to end.

This is our *neighbour* star. The second, the nearest after it, is nearly double as far, and is found in quite another region of space, in the constellation of Cygnus, the Swan, always visible in our northern hemisphere. If we wish to understand the relative situation of our sun and the nearest two, let us take a celestial globe, and draw a plane through the centre of the globe and through α Centauri and 61 Cygni ; we shall thus have before us the relation which exists between our position in Infinitude and those of these two suns. The angular distance which separates them on the celestial sphere is 125°. Let us make this drawing, and we shall discover certain rather curious particulars : in the first place, these two nearest stars are in the plane of the Milky Way, so that we can also represent the Milky Way on our drawing ; again, this celestial river is divided into two branches, precisely in the positions occupied by these two nearest stars, the division remaining marked along the whole interval which separates them. This drawing shows us, further, that if we wish to trace the curve of the Milky Way with reference to the distance of our two stars, it will be nearer to us in the constellation of the Centaur than in that of the Swan ; and, in fact, it is probable that the stars of that region of the sky are nearer than those of the opposite region.[2] Another very curious fact is that both the nearest stars are double.

[1] The ormula is : light years = $\frac{3{\cdot}262}{p}$, p = parallax.

[2] The plane of the solar system and the direction of the sun's motion in space do not correspond

Thus our sun and the neighbouring suns are isolated. Each is an independent king in its own province, and if they feel each other across the infinite, and are subject to the influence of their reciprocal attraction, it is but a suzerainty of little effect. The motions which animate them are of an order superior to their respective attractions.

Here, then, are the nearest suns to us. These stars, twenty-three in number, are almost the only ones which have shown a perceptible parallax; still, the result is very doubtful for the last four, of which the parallax is less than a tenth of a second. Attempts have been made to ascertain the parallax of all the stars of the 1st magnitude, and the result has been negative for those which are not entered in this list. Canopus, Rigel, Betelgeuse, Achernar, Alpha of the Cross, Antares, Spica, and Fomalhaut, do not show a perceptible parallax. The fine star Alpha Cygni, which shines near 61 Cygni, does not present to the most accurate researches any trace of fluctuation: it is, then, incomparably more distant than its modest neighbour—at least five times, and perhaps twenty times, fifty times, one hundred times beyond that. What must be the colossal size and amazing light of these suns, of which the distance is greater than 300 to 400 billions of miles, and which nevertheless still shine with so splendid a brightness!

Copernicus supposed the sphere of the fixed stars to be at an immense distance beyond Saturn, 'as the annual motion of the earth round the sun does not produce any parallax.' Tycho Brahé, not being able or not daring to conceive such a distance, used, on the contrary, this absence of parallax to infer the absence of motion in the earth. 'Copernicus,' said he, 'supposes an incredible and absurd distance. There must be a proportion in everything: *the Creator loves order*, and not confusion. Such a space would be void of stars and of planets, and would serve no purpose. Placing the orb of Saturn at 12,300 semi-diameters of the earth, the new star of 1572 would be at 13,000, and the distance of all the stars would be at 14,000. We can in this case measure them all. Those of the 1st magnitude would appear to have a diameter of 2′, which would be equivalent to 68 times the volume of the earth; those of the 2nd have $1\frac{1}{2}'$, or 28 times the same volume; those of the 3rd have $1\frac{1}{12}'$, or 11 times our globe; those of the 4th have 45″,

either the one or the other, with the plane which we have just drawn. Neither one nor the other of these stars can be considered as revolving round the sun in imitation of the planets or comets. If the nearest, Alpha Centauri, had no perceptible mass, and revolved round our central star, the period of its revolution would be 144 millions of years, and supposing its mass to be double that of our sun, the period of mutual revolution would be 83 millions of years. Now, its proper motion is 3″·67 per annum. At this rate it passes over 1° in 981 years, and it would not require more than 353,000 years to make the whole circuit of the sky. This is a period much shorter than that which the united action of the two masses of our sun and the star would produce. The examination of the proper motion observed in 61 Cygni leads to a similar result; its extremely rapid motion of 5·10″ a year causes the star to pass over 1° in 706 years, and would make it describe the whole circuit of the sky in 254,000 years, which gives a period a thousand times more rapid than that which would be produced by the solar gravitation alone.

or $4\frac{1}{2}$ times the earth ; those of the 5th, 30″, or $1\frac{1}{18}$; and those of the 6th, 20″, so that they are three times smaller than the earth.'

How telescopic discoveries and the micrometrical studies of sidereal astronomy have transformed the idea of the universe in the three centuries which have elapsed since the epoch of the great Danish observer!

If the neighbouring stars are placed at tens and hundreds of billions of miles from us, it is at quadrillions, at quintillions of miles that most of the stars lie which are visible in the sky in telescopic fields. What suns! what splendours! Their light comes from such distances! And it is these distant suns which human pride would like to make revolve round our atom; and it was for our eyes that ancient theology declared these lights, invisible without a telescope, were created! It was because the philosophical astronomer, Giordano Bruno, asserted these distant suns to be centres of other systems that the Inquisition caused him to be burned alive at Rome before the terrified people; and it was because Galileo persisted in maintaining that our planet is subject to the sun, and that that body is itself but a star lost in Infinitude, that this same Inquisition ordered him under pain of death to kneel before the Gospels (Church of Minerva at Rome, June 22, 1633) and abjure the truth which his conscience believed!. Was he wrong, the poor septuagenarian, in thus denying his faith? No. All the formulæ which the teachers of the day compelled him to pronounce did not prevent the earth from turning; and if it were not a terrible drama in the history of progress, it would be a veritable comedy. Pope Urban VIII. and the Cardinals laboured in vain:

> La Terre nuit et jour à sa marche fidèle
> Emporte Galilée et son juge avec elle.

No contemplation expands the thought, elevates the mind, and spreads the wings of the soul like that of the sidereal immensities illuminated by the suns of Infinitude. We are already learning that there is in the stellar world a diversity no less great than that which we noticed in the planetary world. As in our own solar system the globes already studied range from 10 kilometres (6 miles) (satellites of Mars) up to 142,000 (88,000 miles) (Jupiter)—that is to say, in the proportion of 1 to 14,000—so in the sidereal system the suns present the most enormous differences of volume and brightness: 61 Cygni, the stars 2,398 of the catalogue of Lalande, 9,352 of the catalogue of Lacaille, and others of the 8th or 9th magnitude in the preceding table, are incomparably smaller or less luminous than Sirius, Arcturus, Capella, Canopus, Rigel, and the other brilliants of the firmament. This fact proves to us that we must not take literally the layers of successive magnitudes supposed by Herschel and Struve on the hypothesis of an equality of brightness among the stars, but that all varieties of dimensions, masses, brightness, heat, and power exist simultaneously in infinite space.

Adopting and developing the views of W. Herschel on the distances of the stars, supposing, with him, that the stars of the faintest magnitudes are as large as the brightest, and that their apparent smallness is chiefly due to the distance which separates us, William Struve estimated that 'the faintest stars visible to the naked eye are 9 times more distant than the mean distance of stars of the 1st magnitude, that the faintest stars of Bessel's zones (9·5) are 38 times more distant, and that the smallest stars observed by Herschel are 228 times more distant.' He even calculated a series of parallaxes diminishing with the magnitudes, of which the principal are as follows :—

Magnitude	Parallax	Distance Units	Magnitude	Parallax	Distance Units
	"			"	
1·0	0·209	986,000	6·0	0·027	7,616,000
2·0	0·116	1,178,000	7·5	0·014	14,230,000
3·0	0·076	2,725,000	8·5	0·008	24,490,000
4·0	0·054	3,850,000	9·5	0·006	37,200,000
5·0	0·037	5,378,000			

This theory prevails even now in works on astronomy. We have already remarked that it is far from being established, and we may now recapitulate the facts we have studied in the series of arguments which follows :

(1) The distances determined up to the present show that the nearest stars are of all orders of magnitude. With the exception of *α* Centauri, the parallaxes hitherto obtained indicate as the nearest stars 61 Cygni of the 5th magnitude, Σ 2,398 of the 8th ; Sirius comes next, but we immediately see others of the 7th, 8th, and 9th magnitude. On the whole, of 23 stars measured up to the present, 13 are of the 4th to the 8th magnitude, and only 10 belong to the first three orders. On the contrary, the brilliant stars of the 1st, 2nd, and 3rd magnitude show no perceptible parallax.

(2) From the 7th magnitude the number of stars increases in a much more rapid proportion than for the preceding magnitudes. This fact may be explained by supposing that there may be a large number of small stars in the zones of space near where we in general imagine only bright stars to be.

(3) On the chart of proper motions which I constructed in 1877 (see Plate III.) one cannot help noticing groups of stars in which the smallest are incomparably nearer to us than the larger. Such, among others, is the star *μ* Cassiopeiæ, of 5½ magnitude, which is near *θ* of 4½ magnitude ; while the latter remains almost fixed in the sky, the former rushes towards the east with an enormous velocity. Again, while *ψ* of the Great Bear, of the 3rd magnitude, remains nearly fixed, a star quite near it, of 8½ magnitude, Lalande 21,258, moves rapidly towards the west, &c.

(4) A fact independent of the preceding presents itself from an examination of the comparative number of stars of all magnitudes per square degree of the celestial sphere ; it is that, far from being disseminated in space according to a homogeneous distribution, they are more abundant in certain regions, and more thinly scattered in others. There are regions completely destitute of stars, and others where all the magnitudes are found associated.

(5) The rectilinear motions, which I have inferred from an analysis of the double stars, present a certain number of perspective groups formed of two stars of similar brightness. In these groups one star passes before another without experiencing attraction ; the latter is therefore situated far beyond the former, and is perhaps much more distant from the former than this is from the earth, for it remains fixed in the sky. It is, however, as bright in appearance. There are even cases where it is the smaller which, by the amount of its proper motion, appears to be the nearer.

(6) If the distance corresponded to decrease of brightness, the angular distances of physical double stars (that is, revolving double stars, or binaries) should, on the average, diminish with the magnitudes. But this is not what we observe. We notice among stars of the 6th to the 9th magnitude binary systems of which the components are quite as wide apart as those which belong to bright stars. These systems are not, then, immensely distant from us.

(7) The proper motions of stars arising from the perspective due to our translation on the one hand, and on the other hand from a real displacement of the stars, the most rapid motions should indicate the nearest stars. It seems that the value of these motions might furnish a more certain basis than the brightness for the estimation of distances. Now, the largest proper motions, far from belonging to the brightest stars, belong for the most part to the smallest stars. On the contrary, brilliant stars, such as Canopus, Rigel, Betelgeuse, Achernar, Antares, Spica, α Cygni, show hardly any perceptible motion.

It seems, then, that if, on the one hand—and this is quite unquestionable—the light of the stars diminishes in the ratio of the square of the distance (and perhaps even more rapidly, if the ether is not absolutely transparent)—it seems, I say, that we should no longer base all estimates of distance on differences of brightness. On the other hand, photometric measures, the revelations of spectrum analysis, as well as the masses determined, unite with the preceding considerations to convince us that the greatest differences of intrinsic brightness, of dimensions, and of masses, exist among the stars. There are perhaps as many differences among the stars as between the planets of our system.

Indeed, when we see Jupiter shining not far from Sirius in the southern sky, and when we compare these two brilliant celestial lights, we do not generally remember that Jupiter is more than a thousand times larger than the earth, that the sun is more than a thousand times larger than Jupiter, and that Sirius is more than a thousand times larger than the sun.[1] This fine star, which in the most powerful instruments made by human hands appears still but a simple bright point, is in reality a globe endowed with such a luminous and calorific power, that if it were to take the place of our sun every terrestrial creature would be immediately consumed under the fiery action of this dazzling furnace. Moreover, when Sirius enters the field of a large telescope its arrival is announced like that of the rising sun, the astronomer is dazzled by its splendour, and the most experienced eye can only look at it with pain. To reduce the brightness of our sun to that of Sirius it would be necessary to remove it to a distance of about 13 billions of miles!

To measure directly the diameter of a star is impossible, since the brightest, even Sirius itself, are reduced by distance to simple luminous points. If, then, we wish to ascertain the real dimensions of a sun such as Sirius, we must attack the problem by another method—by photometry. Let us suppose that each square foot of this sun emits the same quantity

[1] [Recent researches seem to show that Sirius is not quite so large as here stated. The mass ot Sirius is probably about three times the mass of the sun.—J. E. G.]

of intrinsic light as that which illuminates us; the result of the calculation indicates that, supposing it to be at the distance given in the table on p. 597, Sirius should be fourteen times larger in diameter—that is to say, 2,800 times more voluminous than the sun in the rays of which we gravitate.[1]

Thus, from distances to distances, from billions to billions of miles, from immensities to immensities, suns succeed each other; brilliant foci, enormous globes, centres of planetary families, of varied sizes, of varied power, soaring, reigning in all directions of the infinite! What are they? What is their nature, their intrinsic value, and their importance in the constitution of the heavens? This is what you will learn from the recent progress in the inexhaustible field of sidereal astronomy.

[1] [Recent researches, however, seem to show that the great brightness of Sirius is due rather to its greater intrinsic brilliancy of surface than to its large diameter.—J. E. G.]

CHAPTER VI

THE LIGHT OF THE STARS

Scintillation. Spectrum Analysis.—Physical and Chemical Composition. Application of Photography. Measurement of the Heat of the Stars

THE soft and pleasing light which the stars send us, without which we should be condemned to live in the midst of a darkness as black as that of a tomb (if, indeed, life could have appeared on our planet under such conditions)—this soft and charming light is the only mode of communication which enables us to know of the existence of the universe, and which places us in relation with its constituent parts. It is only by sight that we know of the existence of nature, even when the question is of invisible stars revealed by calculation, for the calculation itself is based on observations due to the sense of sight. We may ask what would happen to our world if we were destitute of that slight and delicate fibre which is called the optic nerve. The reply is not difficult. None of the other senses, neither hearing, nor smelling nor taste, nor touch, play an important part in the general classification of human acquirements; astronomy, especially, would not have existed, and we should live like the blind who grope about in the dark. But perhaps—who knows?—we might be endowed with a sense which would perceive things which are now unknown, and which may exist around us in space without our being able to perceive their existence. Who knows what a sixth sense would reveal to us? The half-informed, who imagines he holds the universe in his narrow brain, smiles at this imaginary question, and supposes nothing to be invisible or unknown. Well, we have neither the credulity of the kisser of relics, nor the scepticism of the young doctor who takes the programme of his examinations for a scientific encyclopædia; if we had in our organism some means of feeling as the magnetic needle feels when it vibrates at the approach of a magnetic storm or when it is agitated at the time when a solar tempest bursts out on the surface of the day star, should we not be endowed with a sixth sense? and would not this sense reveal to us wonders which remain unknown?[1]

[1] Doubtless, among persons who read these lines, several may have observed certain surprising facts of intelligence in dogs. Let them analyse these facts, and they will recognise that it is not by

May not these luminous rays, which already place us in communication with the stellar bodies, be able to teach us more of the nature even of the bodies from which they emanate? They come to us from such an enormous distance, from stars so stupendous, and they have travelled so long to reach us—can they teach us nothing of these distant and inaccessible regions? This is a question which remained without reply even a few years ago, and which now, on the contrary, having been answered permits us to imperceptibly remove the veils which hid from us the depths of the universe.

And, in the first place, where is the contemplator of the heavens who has not been struck with the scintillation of the stars? While the planets, even the brightest, radiate a calm and motionless light, the stars, even the least brilliant, appear more or less agitated by a wavering and variable light. This light, which glimmers sometimes vividly, sometimes feebly, in intermittent gleams, sometimes white, green, or red, like the flashing fires of a limpid diamond, seems to animate the interstellar solitudes, and makes us think of eyes opened in the heavens. It is like a calm and transparent sea on which flit lamps lighted by other mortals; the silence is as profound, but the desert is less void, and it seems that we divine better the distant life which is in motion round each of these brilliant fires burning in Infinitude.

Studied by a large number of observers, especially by Arago at Paris, Respighi at Rome, and Dufour at Lausanne, the scintillation of the stars has only in recent years become an exact science through the continuous and persevering labours of M. Montigny, of the Academy of Sciences of Belgium, who, from the year 1870 up to the moment when we write these lines, has made thousands of observations on the intensity and vivacity of the scintillation of different stars in the sky. The results obtained may be recapitulated as follows:—

Scintillation is a phenomenon caused partly by the intrinsic light and partly by the state of our atmosphere.

The stars which scintillate most are the white stars, like Sirius, Vega, Procyon, Altair, Regulus, Castor, β, γ, ϵ, ζ, η of the Great Bear, α Andromedæ, and α Ophiuchi. We shall see further on that these stars, examined with the spectroscope, present a spectrum formed of the ordinary total of seven colours, crossed by four principal black lines (those of hydrogen); the spectral rays are few in number. The degree of scintillation of these

---

sight that these creatures guide themselves, but by scent, and if they could make any classification of their acquaintances, it would not be by the external form, the size, the colour, that they are classed, but by the *odour* which they emit. A dog smelling the footstep of his master several days, weeks, even several months after his passing, succeeds in finding him. For example, the dog of Beresina returned alone from Russia to Florence two months after the return of his wounded master. There are certainly worlds where scientific acquirements are classed quite otherwise than with us.

stars, or the number of variations of colour per second, is on an average 86, all the stars observed being at the same altitude of 30° above the horizon.

The stars which scintillate least are the orange or red stars, like Antares, α Herculis, Aldebaran, Arcturus, Betelgeuse, α Hydræ, ε Pegasi, ο Ceti, and β Andromedæ. The stars of this type show a spectrum crossed by large dark nebulous bands, which form a kind of colonnade; most of them are variables. The average variations of colour per second are 56.

Between these two extreme groups are ranged stars with an average scintillation (69 per second) of which the light is yellow, like Capella, Rigel, Pollux, α Cygni, γ Orionis, α Arietis, β Tauri, β Leonis, and α of the Great Bear. The spectrum of these stars is similar to that of the sun, crossed by very fine and very close black lines.

There is thus a certain correspondence between the scintillation of a star and its physical constitution; the stars whose spectrum presents a double system of dark bands and dark rays, and to which correspond consequently the most numerous and most marked gaps between their rays, separated by dispersion in our atmosphere, scintillate less than the stars with fine spectral lines, and much less than those of which the spectrum presents only four black lines, and which thus show but a very small number of gaps between their pencils of rays dispersed by the air.

Our atmosphere plays a considerable part in the scintillation: the lower a star is, the more it scintillates; the scintillation is proportional to the product which we obtain by multiplying the thickness of the layer of air traversed by the luminous ray emanating from the star by the astronomical refraction for the altitude at which it is observed.

The scintillation is more pronounced as the cold is greater; it is stronger in winter than in summer—a fact which may be noticed by everybody.

There is another fact of common observation, and now scientifically established: strong scintillations foretell rain. It is the presence of water in greater or less quantity in the atmosphere which exercises the most marked influence on the scintillation, and which modifies its character according to the quantity, either when the water is dissolved in the air, or when it falls to the level of the ground in the liquid state, or in the solid state in the form of snow.

Thus, the light which reaches us from the stars is subject, in traversing our atmosphere, to slight variations of aspect, according to its original intensity, its vivacity, its tint, in a word, according to its own nature. The higher we rise in the air, the more the scintillation diminishes. At the tops of mountains it appears very feeble. During the nights which I have had the pleasure of passing in a balloon I have been surprised at the calm and majestic tranquillity of the celestial torches, which seemed to correspond with the silence and the profound solitude by which I was surrounded.

We now come to the revelations given by the light of the stars themselves as to their own physical constitution.

Until recently astronomy had always exclusively occupied itself with the magnitude and the distance of the stars, and with a small number of physical particulars. The claim to know the nature of their substance and their chemical composition would have passed as an absurdity even a few years ago; but now the astronomer can analyse the stellar materials with the same ease that the chemist analyses terrestrial substances in his laboratory.

We have already explained (Book III., Chap. VII.) in what the spectrum analysis of light essentially consists.

The first person who scientifically studied thes pectrum of a star was the German optician, Fraunhofer. After determining with great accuracy and precision the solar spectrum with its numerous lines (1814–15), he undertook the study of other lights, and especially of some stellar lights. He thus found that the Moon, Venus, and Jupiter show a spectrum identical with that of the sun, as we might have expected; but that the stars in general present very different spectra. He even began to study under this special aspect Sirius, Castor, Pollux, and Capella; but the faintness of the light rendered observation very difficult with the prisms used.

This study remained nearly stationary up to 1860, when the astronomer Donati revived stellar spectroscopy. By the aid of a lens of forty-one centimetres (16·14 inches) he determined with precision the position of the principal lines in thirteen stars: Sirius, Vega, Procyon, Regulus, Fomalhaut, Castor, Altair, Capella, Arcturus, Pollux, Aldebaran, Betelgeuse, and Antares.

Two English astronomers, Messrs. Huggins and Miller, afterwards began to apply to the stars the method of spectrum analysis which Kirchhoff had discovered and so brilliantly inaugurated by the study of the chemical nature of the sun. These two *savants* and Lockyer in England, Secchi at Rome, Janssen, Wolf, and Rayet in France, Vogel in Germany, d'Arrest in Denmark, Rutherfurd and Langley in America, and Dunér in Sweden, are the astronomers to whom we owe, in this branch of research, the most important labours.

Let us briefly describe the principal revelations due to this ingenious examination of the chemical constitution of the stars—*celestial chemistry.* The following are the first results obtained by the English observers:

Aldebaran.—The light of this star is of a pale red. Seen in the spectroscope it presents at a glance a great number of strong lines, particularly in the orange, green, and blue. The positions of about seventy of these lines have been measured, and coincidences have been found with the spectra of *sodium*, *magnesium*, *hydrogen*, *calcium*, *iron*, *bismuth*, *tellurium*, *antimony*, and *mercury*. Seven other elements have been compared with this star, namely—*nitrogen*, *cobalt*, *tin*, *lead*, *cadmium*, *lithium*, and *barium*; but no coincidence has been observed.

α Orionis.—The light of this star is of a pronounced orange tint. Its spectrum is complex and remarkable. The position of about eighty lines has been measured, and those of *sodium*, *magnesium*, *calcium*, *iron*, and *bismuth* have been found.

β Pegasi.—The colour of this star is a fine yellow; its spectrum has a great analogy to that of α Orionis, but is much fainter. Nine elements have been compared. Two among them, *sodium* and *magnesium*, and perhaps a third, *barium*, show spectra in which we see lines coinciding with certain lines in the spectrum of the star.

The absence of any lines corresponding to those of *hydrogen*, ascertained in the spectrum of α Orionis and also in that of β Pegasi, which so much resembles it, is a fact of considerable interest.

Sirius.—The spectrum of this brilliant white star is very intense; but seen at its small altitude above the horizon, even when it is most favourably situated, the observation of the finest lines is rendered very difficult by the motions of the atmosphere. Three, if not four, elementary bodies show spectra in which the lines coincide with those of Sirius; these are *sodium*, *magnesium*, *hydrogen*, and probably *iron*. The lines of hydrogen are abnormally strong compared to those which exist in the solar spectrum.

Vega, α Lyræ.—This white star has a spectrum of the same class as Sirius, and as full of fine lines as the solar spectrum. *Hydrogen*, *sodium*, and *magnesium* are visible.

The increasing interest which similar results produced in the study of the spectra of different stars induced Secchi to undertake a general review of the starry sky, in order to lay a foundation for the complete study of all these stars, and he began by establishing among them a methodical classification destined to serve as a guide in subsequent researches. Profiting by the fine sky of Rome, and using a powerful instrument specially adapted for this kind of observation, this astronomer compared with each other the spectra of more than 300 stars. His researches led him to divide these distant suns into three principal types.

The first type is that of stars commonly called *white*, and even a little bluish, like Sirius, Vega, *α* Aquilæ, and many others, which include about half the stars of the firmament, with a composition of light especially uniform. They have generally two strong lines: one in the blue, at the limit of the green, which coincides with the solar line F; the other in the violet, which is very near the solar line H, but nearer to the red end. A third line is found in the extreme violet, but it is only visible in the brightest stars.

The second type is that of stars with fine lines similar to our sun: yellow stars, such as Arcturus, Capella, Pollux, and most of the fine stars of the 2nd magnitude. The rays are seen very distinctly, notwithstanding their fineness and feebleness.

The third type, which is totally different in character from the first, is the type with bright zones, wide and strong, to the number of six or seven, separated by black rays and semi-obscure or nebulous intervals. The principal representatives of this class are *α* of the Scorpion, *α* Herculis, β Pegasi, &c. These stars are generally of a yellow or red colour. One of the most singular stars of this family is *a* Herculis: its spectrum presents itself as a series of columns illuminated from the side, a true architectural colonnade; the stereoscopic effect is surprising. The stars of this type are not so numerous as those of the other two, and in many cases the type approaches more to the second, of which it seems to mark an extreme limit. Aldebaran is found at the common limit.

The first fact which strikes one in the spectrum analysis of the stars is their great uniformity and the small number of types. When we see different terrestrial substances giving spectra so different, according to their state and their temperature, we naturally expect to find in the stars a diversity still more considerable; the fact is, however, quite otherwise. The fundamental differences are very few, and are reduced to three only.[1]

The following figure (263) represents the types to which nearly all the stellar spectra may be referred: (1) the white stars (Sirius, Vega); (2) the yellow stars (the Sun, Arcturus); (3) the orange-coloured stars (*a* Orionis, *a* Herculis), which are divided into two sections; (4) the reddish stars (*o* Ceti, *μ* Cephei).

The fundamental lines of the first type seem to be those of hydrogen at a high temperature. This gas burns in these distant suns as it burns in our apparatus, and it is the *same gas*. We also see in ignition sodium and magnesium.

The structure of the second type seems more susceptible of variety: nevertheless, we find a rather remarkable constancy, an almost complete chemical identity with that of our sun—iron, titanium, calcium, manganese, sodium, magnesium, potassium, and hydrogen.

The third type is the least numerous of all, but not the least important. It is distinguished from the other two by the large faint and nebulous gaps which divide the spectrum into zones. These spectra have a special characteristic, which seems to indicate the presence of gaseous bodies at a low temperature. They present the aspect of spectra of the first and second type, the light of which might have passed through the absorbing atmosphere of planets. *Hydrogen is absent.* It is very probable that the yellow and red suns which present this spectrum are the oldest and least intense, and that they are casting into immensity their last gleams.[2]

[1] Another fact not less important is that the different types predominate preferably in certain regions of the sky. Thus, in the constellations of the Lyre, the Great Bear, Taurus, and particularly in the groups of the Pleiades and Hyades, the type of Vega predominates. In Cetus, Cepheus, Draco, the solar type predominates. The spacious constellation of Orion is singular in that it contains a special modification of the first type, which renders it very different from the others; we see the rays of this type, but they are remarkably narrow, and are accompanied by a large number of very fine lines scattered throughout the whole spectrum; moreover, the green colour predominates in all these stars, while the red is deficient. We cannot suppose that these coincidences are accidental; they must be connected with the original distribution of matter in space

[2] [Recent researches seem to render this very doubtful.—J. E. G.]

The planets which revolve round suns destitute of hydrogen very probably resemble them, and doubtless do not possess this element, which is of

Fig. 263.—Principal Types of Stellar Spectra

such high importance here. To what forms of life can such planets be suitable? 'Worlds without water,' remarks Dr. Huggins: 'we should require the powerful imagination of Dante to suppose such planets peopled with living creatures. Apart from these exceptions, it is worthy of notice that the terrestrial elements which are most widely diffused in the vast army of stars are precisely the elements essential to life as it exists on the earth—hydrogen, sodium, magnesium, and iron. Hydrogen, sodium, and magnesium represent, moreover, the ocean, which is an essential part of a world constituted as the earth is.'

Spectrum analysis applied to the double stars has proved that the beautiful colours presented by these pairs are not due to the simple effect of contrast, but are real. The two suns which compose the double star *β* of Cygnus, one coloured yellow and the other blue, show two spectra absolutely different. A similar observation, made on the two components of *α* Herculis, of which one is orange-coloured and the other bluish green, has also shown spectra totally different. In each of these two cases the special colour of each star agrees with the way in which the light is distributed in the different regions of its spectrum.

We shall see presently that a certain number of stars exist of which the brightness varies periodically, and this with a degree of regularity which is not the same for all. Different conjectures have been advanced to explain this variability; but they do not rest on any solid foundation. As soon as spectrum analysis was applied to the stars, observers naturally sought in this new method of examination for indications capable of suggesting the causes of so curious a phenomenon.

The most celebrated variable star, Algol, or *β* Persei, examined many times at the epoch of its minimum brightness, has always shown the type of Vega (first type); from which we may conclude that the variation of the star is not due to a chemical phenomenon, that the star does not change, and is doubtless eclipsed by a planet of its system which passes in front of it. This idea, previously suggested, of attributing the periodical diminution of the brightness of Algol to an eclipse produced by an opaque body revolving round the star agrees, moreover very well with the regularity of the phenomenon and with the short duration of the phase of light-diminution. We shall return to this further on.

Another variable star, with which we have already made acquaintance, Mira, or *o* of Ceti, presents a magnificent spectrum of the third type, comparable in beauty with *β* Pegasi and *α* Orionis, and also easily resolved. This is one of the most curious spectra which the observation of the sky affords, and it proves that the variability of this star, like that of almost all the variables (Algol excepted), is due, not to eclipses produced by opaque bodies, but to crises, and to photospheric motions analogous to those which we observe in the sun.

The temporary stars which show themselves more or less suddenly in the sky, then gradually diminish in brightness, to afterwards disappear completely, are worthy of the greatest attention ; one of these mysterious appearances took place in 1866 in the constellation Corona Borealis, and it was speedily subjected to the test of spectrum analysis. The light of this new star, examined by Messrs. Huggins and Miller, gave a very peculiar spectrum (fig. 264), proving that it emanated from two different sources. It had two spectra, each analogous to that of the sun (dark lines in great number), superposed—one evidently formed by the light of an incandescent solid or liquid atmosphere, which had been subjected to an absorption of part of the vapours by an envelope cooler than itself ; the other composed of a small number of bright lines indicating matter in the state of luminous gas. The character of this star's spectrum, compared with the sudden outburst of its light and the rapid decrease of its brightness, leads us to suppose

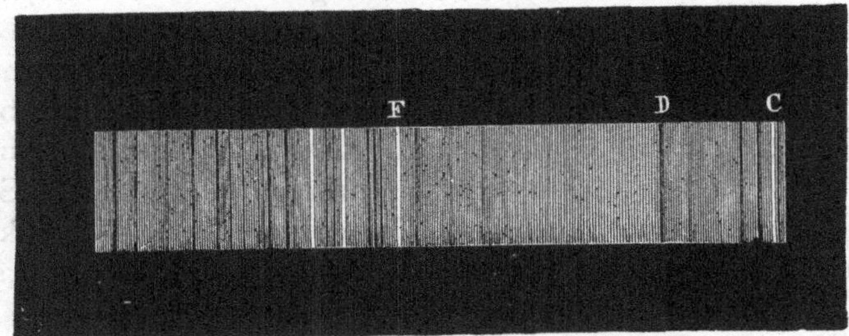

Fig. 264.—Spectrum of the New Star in Corona Borealis, 1866

that on account of some great internal convulsion immense quantities of gas were liberated ; that the hydrogen which formed part of it was set on fire by combining with some other element, thus furnishing the light represented by the bright lines, and that at last the flames heated the solid matter of the photosphere to a vivid incandescence. When the hydrogen had been exhausted, the star was rapidly extinguished. Have we not here all the characters of a veritable *conflagration* which we have been permitted to perceive in the depths of the celestial spaces ?

We shall return to this star further on, as well as to another similar one which suddenly increased in brightness in 1876 in the constellation Cygnus, and the spectrum of which, examined by M. Cornu, showed the lines of hydrogen and the chromosphere of our sun. Another curious variable which appeared in 1885 in the nebula of Andromeda has also shown in its spectrum the line 1,474 K of the solar chromosphere.

Thus do revelations descend from the sky to-day. We analyse the constituent substances of the stars as if we could touch them, and subject them to the tests of our laboratories.

Attempts have also been made to photograph the stars, and have met with complete success. Admirable photographs of the sun have been obtained for more than forty years past—and this is natural; but when we come to the moon, the operation is more difficult, on account of the feeble intensity of the lunar light compared to that of the sun, and the difference of tint of the various regions of our satellite, as we have already seen. Skill and perseverance have, however, surmounted the greatest difficulties, and now we have photographs of the moon, enlarged to more than a yard in diameter, which show the smallest details with a truly admirable clearness. The planets Jupiter, Venus, Saturn, Mars, then presented themselves to the growing ambition of the photographic astronomer, and impressed on the sensitive plate the general configuration of their physical aspect. But when the stars were in question the difficulties were still greater. However, in the year 1857, Bond photographed the beautiful star Mizar, accompanied by Alcor and all the stars visible in the field of the telescope to the 8th magnitude. This fine double star (Mizar and its near companion) was admirably photographed, and with a precision so perfect that we can use the photograph to measure the angle and distance of the companion (at 14″). Since that time celestial photography has made immense progress, thanks to the ingenious and fruitful labours of a large number of astronomers, among whom we may mention as in the first rank MM. Henry in France, Huggins, Common, Roberts in England, Pickering in the United States, and Gill at the Cape of Good Hope. A congress was convened in 1887, at the Paris Observatory, with the object of preparing a photographic chart of the entire heavens. It is probable that our century will not close before this fine work is accomplished.

We may form an idea of the perfection already obtained from the plate which we reproduce in fig. 265, which was taken in 1887, by MM. Henry, at the Paris Observatory.

A rather curious fact, and one, besides, very easily explained, was presented during these operations: it is that stars of the same luminous intensity are not equally easy to photograph. Thus, for example, although Vega and Arcturus are nearly of the same magnitude, the former is seven times more actinic: the second requires seven times longer exposure to be photographed. The red and yellow stars are rebels, the white or bluish stars yield with a good grace.[1]

[1] Everyone knows that photography has not the same sensibility for all colours. Red gives black, orange a very dark tone, while blue gives white, because it is in this part of the spectrum that the chemical rays are the most active; this is why a child whose cheeks are of a fine cherry-red colour becomes a little negro in a photograph, while a lady dressed in a blue gown sometimes appears to be in white. A rather curious fact is that in the twilight, red colour, which is so striking during the day, no longer attracts attention, while blue or violet shows a preponderant intensity.

In general the stars are white, and the length of exposure necessary to photograph them is

Fig. 265.—Direct Photograph of a Region in the Constellation Gemini
(Three exposures of an hour each. Untouched heliogravure reproduction)

The stars, which send us so feeble a light, transmit a quantity of heat still less perceptible to our senses, and yet attempts have been made to measure it. According to the experiments of Stone, Arcturus sends us much more heat than Vega. At 25° of altitude at Greenwich the first appears to emit heat equal to that of a Leslie tube of boiling water at a distance of 115 metres (377 feet), while at 60° of altitude Vega gives but a quantity of heat equal to that of the same tube at a distance of 260 metres (853 feet). It would be almost impossible to denote this feeble heat as a fraction of a thermometer degree. Thus, the sun Arcturus is hotter than Vega, and it is the rays of the red end of the spectrum which act in its light; while the sun Vega is less ardent, and it is the chemical rays at the blue end which have most action. Here, then, is the 'new astronomy' which now penetrates by its universality all the other sciences, and which draws them all into its sphere to receive still more unexpected and truly marvellous developments in the future; chemistry and physics take possession of the sky; the universe becomes for man an immense laboratory.

The rays of light which fall in silence from the distant splendours of the starry night bring to us, then, the most curious revelations on the state of creation in these inaccessible universes, and prove to us that the substances and forces which we see in activity around us exist there as well as here, producing effects analogous to those which surround our field of view; developing the sphere of our conceptions at the same time as that of our observations, and permitting us to divine the things, the beings, the populations, the unknown works which reproduce in Infinitude the spectacles of life, the sports of nature, and the varied operations of which our solar system presents but an ordinary and incomplete scene. Light transports us into the *infinite life*. It transports us also into the *eternal life*.

We have seen that light is not transmitted instantaneously from one point to another but gradually, like everything movable; that it flies at the rate of 186,000 miles a second, or 11 millions of miles in one minute; that it takes more than eight minutes to pass over the distance which separates us from the sun, four hours to come from Neptune, and four years and four months to come from the nearest star, &c.

There is here, then, a surprising *transformation of the past into the*

proportional to their brightness. The following is the time of exposure necessary for the different magnitudes:—

Magnitude	Exposure	Magnitude	Exposure
1st magnitude	0·005 sec.	9th magnitude	8·0 sec.
2nd ,,	0·01 ,,	10th ,,	20 ,,
3rd ,,	0·03 ,,	11th ,,	50 ,,
4th ,,	0·1 ,,	12th ,,	2 min.
5th ,,	0·2 ,,	13th ,,	5 ,,
6th ,,	0·5 ,,	14th ,,	13 ,,
7th ,,	1·3 ,,	15th ,,	33 ,,
8th ,,	3·0 ,,	16th ,,	1 hour 20 ,,

present. For the star observed, it is the past—already vanished. For the observer, it is the present, the now. The past of a star is strictly and positively the present of the observer. As the aspect of worlds changes from year to year, from one season to another, and almost from one day to the next, we can represent this aspect as escaping into space and advancing in Infinitude to reveal itself to the eyes of distant beholders. Each aspect is followed by another, and so on successively; and it is as if a series of waves bearing from afar the past of worlds should become present to observers ranged along its passage! What we believe we see now in the stars is already past; and what is now being accomplished we do not yet see.

We do not see any one of the stars as it is, but as it was at the moment when the luminous ray which reaches us left it. *It is not the present state of the sky which is visible, but its past history.* There may even be stars which have not existed for the last 10,000 years, but which we still see, because the ray which reaches us left them long before their destruction. A double star, of which we seek with much care and toil to determine the nature and motions, may have ceased to exist before astronomers appeared on the earth. If the visible heavens were annihilated to-day, we should still see the stars to-morrow, and even next year, and perhaps for one hundred, one thousand, fifty thousand, one hundred thousand years, or more, with the exception only of the nearest stars, which would be extinguished successively when the time necessary for the luminous rays which emanate from them to pass over the distance which separates us had elapsed; *α* of the Centaur would be the first extinguished; &c.

If from the earth we see a star, not as it is at the moment we observe it, but such as it was a hundred years previously, so from this star its inhabitants see the earth after a delay of a hundred years. Light takes the same time to accomplish the same journey.

A man, a spirit, leaving the earth, either by death or otherwise, this year, and transported in some hours or days to a great distance, would see the earth of former times, and would see himself again a child, for the aspect of the earth would not arrive where he was till after a long delay.

There is here neither a vision, nor a phenomenon of memory, nor a marvellous or supernatural action, but an actual, positive, natural, and incontestable fact; what has been for a long time the past for the earth is only the present for a distant observer in space. This vision is, none the less, very astonishing. Indeed, it is a singular fact that it is impossible to see the stars as they are at the moment when we examine them, and that we are only able to see their past! [1]

Thus the progressive motion of light carries with it through Infinitude

[1] See our *Récits de l'Infini: Lumen* ('History of a Soul').

the ancient history of all the suns and all the worlds expressed in an *eternal present.*

The metaphysical reality of this vast problem is such that we can now conceive the omnipresence of the world in all its duration. Events vanish for the place which brings them forth, but they remain in space. This successive and endless projection of all the facts accomplished on each of the worlds is performed in the bosom of the *Infinite Being* whose omnipresence thus maintains everything in an eternal permanence.

CHAPTER VII

CHANGES OBSERVED IN THE HEAVENS

Brightness of the Stars. Temporary Stars which have suddenly Appeared in the Sky. Variable Stars. Periodical Stars. Stars which have Disappeared from the Sky

DOES the above title truly agree with the reality? The stars being not merely brilliant points attached to the vault of the firmament, but each star being a veritable sun similar to ours, is it possible that a sun can increase and diminish in brightness? May our own sun, then, some day increase in light and heat, dazzle us, blind us, scorch us, consume the vegetation of the globe, cause animal life to perish in a stifling desert, and lay panting humanity in the burning sands of a perpetual Sahara? Or again, on the contrary, might the beneficent focus of our natural heat become enveloped in a veil, suspend its radiation, keep back its golden rays, the arrows of flame shot out since the beautiful days of Apollo, deny the spring and the flowers, the summer and the harvest, the autumn and the vine, spread over the globe the frosts of an eternal winter, congeal the blood in our veins, cause all beings to shiver in a foggy atmosphere, and lay the whole of humanity under a thick and increasing shroud of snow? Yes, our good and beautiful sun might grow dim and light up again; it might in a few weeks permit death to overrun the world; it might shine in the grey sky like a wan spectre reigning over a vast cemetery; it might rise again from its ashes and revive the life momentarily suspended during months, years, or centuries—it could do so, and doubtless it has before now done so.

Yes, the earth has already been buried in a shroud of snow and ice, and all living species have been plunged into a silent catalepsy. The world was already old, however. For many centuries, for many thousands of centuries, it gravitated in cadence in the fruitful light and heat of the celestial star. Its living population had been many times transformed and renewed; the splendid and impenetrable forests of ferns had given place to sunny oases full of light, of perfumes, and birds with glittering plumage; the monstrous and ferocious saurians of the secondary epoch had made way for the superior species of the tertiary period; then the mammoth led his herds, then the rhinoceros with open nostrils hunted in the woods; then the gigantic stag bounded like an arrow across the valleys and ravines, the bear installed his family in the caverns, the monkeys gambolled in the fruit-trees, the horse galloped in the fields, and groves where the bubbling brook overflowed with love and song: when the temperature fell to such a point as no longer to leave a single drop of water in the liquid state. A gloomy sky overspread the world. Then, again, nature was arrested like a man who staggers, and life was extinguished; the birds sang no more, the plants bloomed no longer, the stream flowed no more, and the sun ceased to shine. This glacial epoch, of which geology now finds traces everywhere visible, extended over the entire globe. France, Switzerland, Italy, the different countries of Europe, of Asia, of Africa, as well as those of the American continent, still bear the marks. Did man already exist? Was he a witness of this immense catastrophe? Did he find to shelter and save his infant race a beneficent volcano, an equatorial island, a refuge forgotten in the universal cataclysm? This time is already so far from us that we remember it no longer. But the glacial period is written in full in the great book of nature; only its explanation is still floating in the doubt of theories; and among the hypotheses advanced to explain it, that of comparing our sun with the other variable suns of the universe, and supposing that the variation of heat has been sufficient to deliver up our planet to the ice which enveloped it, is not the least worthy of attention.

We see similar examples produced before us in the heavens. One of the most remarkable is that which is shown by a star in the constellation Argo, the star η situated in the midst of a singular nebula. In 1837 this star was of the 1st magnitude, and up to 1854 it surpassed in brightness the finest stars in the heavens, only yielding the palm to Sirius, which it almost equalled in 1843, exceeding Vega, Arcturus, Rigel, α Centauri, and Canopus. Now, in 1856 it began to decrease, and descended below all the stars of the 1st magnitude, coming to rank among those of the 2nd. Continuing to decrease, it became in 1859 equal to stars of the 3rd magnitude, in 1862 to those of the 4th, in 1864 those of the 5th, in 1867 those of the 6th, and in 1870 it disappeared to the naked eye. From 1871 it slowly descended the degrees which separate the 6th magnitude from the 7th, and in 1886 it reached a minimum of 7·6. Since then it has seemed to revive. Thus, from the year 1856 to the year 1886, under our eyes, so to say, this distant sun, of which the parallax is imperceptible, whose distance is enormous, whose volume is stupendous—this colossal hearth of an unknown system fell seven magnitudes in brightness, and gave 600 times less light than it had previously radiated. What opinions can we form on such variations with reference to the conditions of habitability of a planetary system subject to the irregularities of such a sun! If there is near it any inhabited earth similar to ours, a glacial period might be produced on its surface by the gradual extinction of its sun.

Will it awake, this sun of Argo? Is it about to revive completely, and project anew around its brightening sphere the radiation of light and heat which seemed to have departed from it for ever? We may, we ought to hope for it, and this hope would be partly justified by its acts and doings during the two hundred years that it has been observed. Halley saw it of the 4th magnitude in 1677, Lacaille of the 2nd in 1751, Burchell of the 4th in 1811, Brisbane of the 2nd in 1822, Burchell of the 1st in 1827, Johnson of the 2nd in 1830, and Herschel of the 1st in 1837. It is, then, a sun which varies rapidly and to a great extent, and we may expect that it will soon re-ascend through all the degrees of light through which it has descended.

To what cause can this enormous variation of light be due? Could the star withdraw from us with an extreme rapidity, and could it approach us when its brightness increases? No; for, on the one hand, we do not perceive any motion (it would be necessary that the motion should be performed exactly in the direction of the visual ray, which is improbable, and even impossible if we consider the large number of variable stars now known); and on the other hand, it would be necessary to suppose that the star had been removed, from 1856 to 1867, the whole distance which would reduce a star of the 1st magnitude to one of the 6th—that is to say, to at least ten times the distance of a star of the 1st magnitude, or nine million times the radius of the terrestrial orbit—which would imply an extravagant velocity, and, moreover, one impossible to admit, considering that the ray of light which would take fifteen years to come from the first distance would require 150 years to reach us from the second. The variation of light is not, then, due to variation in the distance of the star.

Could it be produced by an eclipse? For this it would be necessary to suppose that a dark globe as large as the star itself passed exactly between it and us, and took several years to mask its light. The very nature of celestial motions is opposed to this hypothesis.

Could this surprising variation be due to a rotation of this distant sun on itself, supposing a part of its surface to be incandescent, and another part covered with spots crusted and almost dark? It is improbable that a star takes thirty years at least to perform a semi-revolution on itself, and, on the other hand, the phenomenon does not show the regularity which would correspond to this hypothesis.

The most natural explanation is to suppose that these periods of superabundance of brightness correspond to an over-excitement in the luminous photosphere of these distant suns. We have seen, in studying our own sun, that its light is due to clouds of solid or liquid particles burning in its fiery atmosphere, as carbon, lime, or magnesia in our artificial flames. As M.

Faye has established, especially with reference to variable stars, the *solar* phase, the period of brightness and activity of a star, begins when the surface of the incandescent gaseous mass is cooled enough to admit of a precipitation of the liquid or solid clouds capable of emitting a vivid light. It is thus that the photosphere of a new sun is formed. From a certain moment the phenomena of the photosphere may assume an oscillatory character. The equilibrium of the gaseous mass is at first disturbed by the showers of scoria which descend and by the vapours which rise, exactly as the equilibrium of our atmosphere is disturbed by the circulation of water in its three states; then, when this exchange between the surface and the interior begins to be obstructed by the encroachment of scoria, we see produced eruptive phenomena, periodical cataclysms, of which the consequence is a recrudescence of light, rapid but transient. To each collapse of the thickened photosphere corresponds a sudden outburst of incandescent gas issuing from the interior. Finally, these alternations are only presented by fits and starts, and at last completely cease.

Of all the stars which have changed in brightness, the most remarkable is that which in the sixteenth century (in 1572) suddenly acquired such a light that it eclipsed all its sisters of the firmament and became visible at noonday. It was observed by Tycho Brahé, and Humboldt has preserved for us the following curious account:—

When I left Germany to return to the Danish shores (says Tycho), I stayed at the ancient and admirably situated residence of Herritzwaldt, belonging to my uncle, Stenon Bille, and I was in the habit of remaining in my chemical laboratory until nightfall. One evening, when I was contemplating, as usual, the celestial vault, whose aspect was so familiar to me, I saw, with inexpressible astonishment, near the zenith, in Cassiopeia, a radiant star of extraordinary magnitude. Struck with surprise, I could hardly believe my eyes. To convince myself that it was not an illusion, and to obtain the testimony of other persons, I called out the workmen employed in my laboratory and asked them, as well as all passers-by, if they could see, as I did, the star, which had appeared all at once. I learned later on that in Germany carriers and other people had anticipated the astronomers in regard to a great apparition in the sky, which gave occasion to renew the usual railleries against men of science (as with comets whose coming had not been predicted).

The new star (continues Tycho) was destitute of a tail; no nebulosity surrounded it; it resembled in every way other stars of the 1st magnitude. Its brightness exceeded that of Sirius, of Lyra, and of Jupiter. It could only be compared with that of Venus when it is at its nearest possible to the earth. Persons gifted with good sight could distinguish this star in daylight, even at noonday, when the sky was clear. At night, with a cloudy sky, when other stars were veiled, the new star often remained visible through tolerably thick clouds. The distances of this star from the other stars of Cassiopeia, which I measured the following year with the greatest care, has convinced me of its complete immobility. From the month of December 1572 its brightness began to diminish; it was then equal to Jupiter. In January 1573 it became less brilliant than Jupiter; in February and March, equal to stars of the 1st order; in April and May, of the brightness of stars of the 2nd order. The passage from the 5th to the 6th magnitude took place between December 1573 and February 1574. The following month the new star disappeared without leaving a trace visible to the naked eye, having shone for seventeen months.

These circumstantial details permit us to imagine the influence which such a phenomenon must have exercised on the minds of men. Few historical events have caused so much excitement as this mysterious envoy of the sky. It first appeared on November 11, 1572, a few months after the massacre of St. Bartholomew. General uneasiness, popular superstition, the fear of comets, the dread of the end of the world, long since announced by the astrologers, were an excellent *mise en scène* for such an apparition. It was soon announced that the new star was the same which had led the Wise Men to Bethlehem, and that its arrival foretold the return of the Messiah and the Last Judgment. For the hundredth time, perhaps, this sort of prognostication was recognised as absurd. It did not, however, prevent the astrologers from being believed twelve years later, when they announced anew the end of the world for the year 1588; these predictions exercised the same influence on the public mind.

After the star of 1572, the most celebrated is that which appeared in October 1604 in Serpentarius, and which was observed by two illustrious astronomers, Kepler and Galileo. As happened with the preceding, its light imperceptibly faded; it remained visible for fifteen months, and disappeared without leaving any traces. In 1670 another temporary star, blazing out in the head of the Fox (Vulpecula), showed the singular phenomenon of being extinguished and reviving several times before it completely vanished. We know of *twenty-four* stars which during the last 2,000 years have presented a sudden increase of light, have been visible to the naked eye, often brilliant, and have then again become invisible. The last apparitions of this kind happened before our eyes in 1866 and 1876 [and in 1885 and 1892.—J. E. G.], and spectrum analysis enabled us to ascertain, as we have seen, that they were due to veritable combustion—a fire caused by a tremendous expansion of incandescent hydrogen, and to phenomena analogous to those which take place in the solar photosphere. A rather curious fact about these stars is that they do not blaze out indifferently in any point of the sky, but in rather restricted regions, chiefly in the neighbourhood of the Milky Way, as we see in fig. 267.

How can stars, *suns*, thus suddenly shine out in space? The idea that these temporary stars may perhaps be new creations cannot now be accepted. Their ephemeral apparition affords a striking contrast to the permanent brightness of the stars in general; they are evidently variable stars, irregular and not periodic; they existed in the sky before undergoing these extraordinary exaltations, and have relapsed to their primitive rank, as has been ascertained from those which we have been able to follow. There is an essential difference between these tremendous outbursts and the regular variations of periodical stars, which we shall study directly. Nevertheless, it is necessary to say that between the former and the latter we find, so to say, all degrees of irregularity. Thus, for example, the star η of Argo may be considered as intermediate between the two classes.

It is probable that these variations of light proceed from effects produced in the suns themselves, and analogous to those which we observe in our own sun. We have seen that the number of solar spots varies in a period of eleven years. This variation is indeed considerable, since there are over ten times as many spots in the years of maximum as in the years of minimum; only, as they intercept but a small part of the solar light, a distant observer, who watched our sun with attention, would hardly perceive any variation. It is sufficient to suppose the phenomenon of our solar spots reproduced in other suns on a much vaster scale in order to obtain an explanation of the variable stars in connection with what we know of the physical constitution of suns. We have only to suppose a general ex-

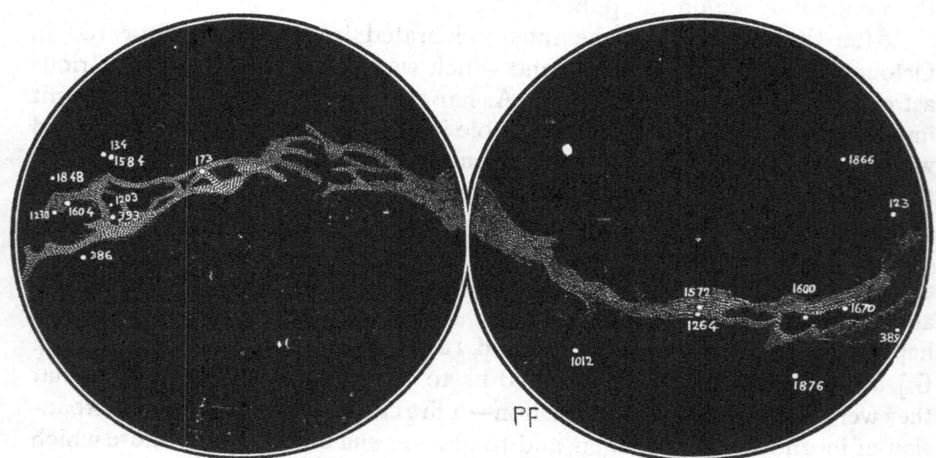

FIG. 266.—POSITION IN THE HEAVENS OF TEMPORARY STARS

plosion produced in one of these suns, which should all at once be enveloped in ignited hydrogen prominences, and that the dark network in which we have seen floating the luminous granules of the solar photosphere should disappear under the condensation of brilliant faculæ; or that a sun which has begun to cool down and become covered with a solid crust is torn by eruptions from the interior furnace. Again, we might suppose that the fall of an enormous bolide or an encounter with a celestial body breaks up a continent newly formed on a crusted sun; or that two tremendous meteoric streams collide in space—and we have an explanation of our temporary stars, which shine out all at once with a brilliant light, to afterwards relapse into their original state, or into nothingness.

Thus the knowledge of our sun itself may teach us very much with reference to the most distant phenomena which take place in the heavens, and by a natural result of the same reasoning we see that the sun may find itself some day a victim to similar disturbances.

More curious still, perhaps, than these sudden conflagrations are the regular, rapid, and periodical variations which we observe in certain stars.

One of the most famous of these variable stars is Omicron (*o*) of the Whale (Cetus), also called *Mira Ceti*, the 'Wonderful of the Whale,' and it fully merits this title. This star rises to the 2nd magnitude, and becomes as bright as the finest stars of the Great Bear, and so everybody can observe it for fifteen days; then it diminishes imperceptibly, and becomes absolutely invisible to the naked eye. It may be looked for in vain during five whole months. Afterwards it reappears, and increases slowly in brightness during three months, to reascend to the 2nd magnitude.[1] This extraordinary variation of light is performed in 331 days.

At its maximum this star is yellow; when it is faint, it is reddish. Spectrum analysis shows in it a striped spectrum of the third type, and when its light diminishes it preserves all the principal bright rays reduced to very fine threads. The most plausible explanation of this variability is to suppose that it periodically emits vapours similar to the eruptions observed in the solar photosphere. Instead of its periodicity being like that of the sun—eleven years and hardly perceptible, this variation of the sun of the Whale is 331 days and very considerable. It is subject to oscillations and irregularities similar to those which we remark in our sun.

Of all the variable stars, this is the easiest to observe, and it has been known for nearly 300 years. Another, no less curious, and incomparably more rapid, is β of Perseus, or *Algol*, which in the short period of 2 days 20 hours 48 minutes 51 seconds descends from the 2nd to the 4th magnitude [more correctly, from 2·3 to 3·5 magnitude—J. E. G.], and returns to its former brightness. During 2 days 13 hours the brightness is constant, of the 2nd magnitude; then it begins to fade, and in 2 hours 30 minutes is reduced to below the 4th magnitude;[2] it remains in this state five or six minutes [fifteen or twenty—J. E. G.], and returns to its primitive brightness in 3 hours 30 minutes.[3]

[1] It is generally stated as varying from the 2nd to the 12th magnitude. This is an error. There is near it a little star of 9½ magnitude with which I have often compared it. Mira has never become fainter than this. Its minimum is, then, 9½. [The brightness at maximum varies from 2nd to 5th magnitude.—J. E. G.]

[2] [This is incorrect. Algol is never so faint as 4th magnitude. Its usual minimum magnitude is 3·5.—J. E. G.]

[3] A variation so rapid and so characteristic cannot be analogous to that of the eleven-year period of the solar spots, and the most plausible explanation which presents itself is to attribute it either to a motion of rotation of this sun upon itself, supposing that its two hemispheres are of very different brightness, as a sun would be, for example, on which a dark continent existed, or to an eclipse produced by an enormous satellite revolving round Algol in the plane of our visual ray. The first

In Sagittarius there are three variable stars, of which the period is about 7 days.

Several other stars—δ Libræ, U Coronæ, λ[1] Tauri, δ Cephei—also show this curious rapidity in their period (2 to 5 days). We can follow these variations with the naked eye. Others take several weeks; the great majority, several months; no period exactly determined exceeds two years; in general, the longer the period, the greater the variation.[2]

In another, R Hydræ, we see the period diminishing rather rapidly. According to the researches of Schönfeld this period was—

500 days in 1708,
487 „ 1785,
437 „ 1870.

Analogy leads us to believe that for a large number of stars the variation is produced by a rotation of the star on itself. Several explanations are, then, presented: (1) Real variation produced in the photosphere, analogous to the period of the solar spots, but more intense and more rapid; (2) rotation of a globe of which the different meridians maintain for many years enormous differences of luminous intensity; (3) circulation of a luminous ring round a sun; (4) eclipses produced by the transits of dark planets. We may add that nature, which only opens a finger at a time to let out truths, of which her hands are full, certainly holds in reserve other explanations, which the progress of science will reveal later on. As to the temporary stars with which we began this study, they may be subject to veritable conflagrations—conflagrations seen at billions of miles distant!

And what variations of light! We have R of Sobieski's Shield, which varies from the 4th to the 9th magnitudes, or five orders of brightness, in 72 days; Mira Ceti, from 2 to 9·5, or seven and a half magnitudes; χ Cygni, from the 4th to the 13th, or nine magnitudes! Here, then, is a sun which radiates 4,600 times more light and heat at the epoch of its maxi-

hypothesis is open to the objection that it is rather difficult to suppose such a spot remaining for years and centuries motionless on a star; the second has against it the rapidity of the planet's motion; and, in fact, the double star of which the period is the shortest known still shows a revolution of about eleven and a half years. However, all things considered, it is to this last hypothesis that we must look for the cause of this curious variability, and so much the more that, as we have seen above, spectrum analysis shows in it not a star of the variable type, but a spectrum of the first type, which it preserves without change. Thus, the sun of Algol presents itself to our view as the centre of gravitation of a planetary system of which the larger world doubtless revolves in 2 days 21 hours. This is nearly the revolution of the fourth satellite of Saturn. We notice in this period even small irregularities, which may arise from planetary perturbations; further, the revolution has appeared to be slowly diminishing during the two centuries in which it has been observed. It has diminished 6 seconds since last century. [The truth of the above theory has been verified with the spectroscope by Professor Vogel. See p. 651.—J. E. G.]

[1] [γ in original, but incorrect.—J. E. G.]

[2] A catalogue of all the variable stars now known will be found in my work, *Les Étoiles*, together with their position in the sky, and the precise determinations of these astonishing periodicities.

mum than at that of its minimum! What imagination can picture the work of nature in such systems!

These are rapid variations. Do the stars also show evidence of slow secular variations? Do we now see identically the same sky which our forefathers saw? Have certain stars diminished in brightness since the origin of astronomical annals? Have some even entirely disappeared from the sky? Have others increased in brightness? Yes. The heavens appear to us as an immense laboratory, from which inertia and immobility are excluded; it is not activity, it is not death, which reigns in its depths; it is life, immense, universal, varied, always reviving. Generations succeed each other in the heavens as on the earth; worlds, suns, and systems are born and die like living beings, and if the general aspect of the universe is that of permanence and immutability, it is because our life is too ephemeral and our sphere of observation too contracted to permit us to appreciate things in their reality. The dragon-fly which floats above the waters in the warm hours of July knows not that winter exists, nor imagines that the sun will set.

Secular changes are accomplished in the heavens.[1] If we could embrace a period ten times, a hundred times longer than our 2,000 years of observations, we should witness metamorphoses much more profound. The suns themselves are not eternal. Although a space of 2,000 years, or sixty human generations, represents but a short moment of universal history, several suns have diminished in brightness in this interval, several have increased in light, and among the faintest some have even entirely disappeared. Doubtless there exist in space a large number of extinct suns, enormous black balls, round which gravitate other dark masses in the invisibility of the infinite night. The population of the heavens shows us a field diversified by a thousand productions incomparably more varied than all those of terrestrial nature.

The diversity of beings which people the immense universe must be infinite. It is not philosophical to maintain that we know from our own planet the absolute conditions of life. Here the extreme limits of life appear to be 40° below zero [−40° Fahrenheit] and 61° above (Centigrade scale) [142° Fahrenheit]. The vital terrestrial temperature depends on the state of water. If that element were not the only one necessary, and

[1] This is not the place to describe all these changes in detail; they have, however, their interest, so much the more that several astronomers, especially Cassini, have published lists of variations of brightness, and of stars that have disappeared, without sufficient proofs. In order to learn the truth of this matter I have constructed a comparative catalogue of all the stars observed for 2,000 years, by combining the catalogues of Hipparchus (127 B.C.), Abd-al-Rahman-al-Sufi (960 A.D.), Ulugh Beigh (1430), Tycho Brahé (1590), Hevelius (1660), Flamsteed (1700), Piazzi and Lalande (1800), Argelander (1840), and Heis (1870), and verifying directly in the sky the present state of the stars which have shown evidence of variation. This comparative work, which includes exactly 2,000 years, has led me to form a list of stars which have changed in brightness during 2,000 years; it is published in my work *Les Étoiles*, with the indications relative to each epoch.

if life could be attached to other elements, who could fix a limit to it—who could assert that it does not exist on the suns?

We do not thoroughly understand either mind, life, or matter. Perhaps spheres exist where the simplest, the most ignorant, and the coarsest men feel, divine, and see directly and intuitively the solutions of transcendent mathematical problems which no terrestrial genius has yet been able to solve, notwithstanding the differential and integral calculus.

It would only be by removing from the celestial region where we are that we could form a sound idea of the extent of creation; even in our longest stellar excursions we have but a confused idea of the unknown reality. But in diverting our attention for some moments from the earth and its limitations, we learn at least to judge better of its value and its relation to the whole. This is the one necessary condition for our progress in the science of the world. It is then that our conception rises above false appearances, that it takes its flight into the fields of the sky, and develops as it advances in the boundless creation. It is by the fulness of examination, the height of the eye, the extent of the horizon, that we judge of the natural value of a country; the ant knows neither the heavens nor the earth, and has never seen anything but the grains of dust which he piles up on the ground; the eagle soars in the heights of the air, and measures with his glance the lofty mountains as well as the immense plains.

Distant universes whose riches and beauty are unfolded in the depths of inaccessible space, who shall tell us of the wonders of your unknown nature? The luminous ray, more rapid than the lightning, takes centuries to reach us; the immeasurable separates us. Under what form do the universal laws of heat act in you? under what aspect are they manifested? what is the mode and extent of their power? With what properties are the elements which compose you endowed? For you, we know that the earth, the field of human observation, is not the book of nature, of which it forms but a chapter, a page.

Beautiful summer evenings which slowly descend from the heavens on the bright day, still come to bathe the earth with your golden halo! Open still to the perfumed breeze the gates of the winding valleys; allow still to fall, like dew of the air, the mist of the twilights; let the harmonious tints which imperceptibly fade away from the rosy west to the azure zenith still adorn this superb vault, that our delighted gaze may always wander in this floating depth! Sweet hours of evening, do not flee away! We love this universal calm which surrounds nature before it sleeps; we love this unchangeable peace which descends from the rising stars! Be present still at this profound meditation, in which all beings participate as if they had consciousness; listen still to the last rustling of the quivering foliage! The starry sky which lights up, the earth which falls asleep—these are spectacles which draw us away from a world of clamorous passions, pleasures of the soul which we enjoy in peace. But whatever O beautiful evenings, may be the pleasures of your contemplation, however delightful may be the moments which you give us, the first stars which you light up in infinity will always be there to attract us more powerfully still, to enrapture more closely our attention and our thoughts. They tell us that if the earth is beautiful, and if man can draw precious gratifications from his abode, the heavens are more magnificent still, and should be for us an inexhaustible source of study, of contemplations always new, and of intellectual pleasures incessantly reviving.

But as yet we have visited only single suns like our own ; the time has now come to penetrate into the still more magnificent field of multiple and coloured stars.

CHAPTER VIII

THE DOUBLE AND MULTIPLE STARS

Coloured Suns. Worlds illuminated by several Suns of different Colours

IN the depths of the heavens, among the varied stars which shed their silent light from the high regions of the starry night, the searching eye of the telescope has discovered stars of a character which differ from ordinary stars by their aspect as well as by their *rôle* in the universe. Instead of being single, like the great majority of stars in the sky, they are double, triple, quadruple, and multiple. Instead of being white, they often shine with a coloured light, showing in their strange couples admirable associations of contrast—where the astonished eye sees the fires of the emerald united with those of the ruby, of the topaz with those of the sapphire, of the diamond with the turquoise, or the opal with the amethyst, thus sparkling with all the tints of the rainbow. Sometimes the wonderful stars which form these celestial couples lie in the midst of the infinite, fixed and unchangeable, and for more than a century, since the attentive astronomer has contemplated and observed them, they have not varied in their relative position to each other. As the scrutinising gaze of the patient William Herschel observed them a hundred years ago, so we find them to-day. Sometimes, on the contrary, the two associated stars revolve one round the other, the fainter round the brighter, rocked on the wing of attraction, like the moon round the earth and the earth round the sun. A certain number of these pairs have already described several complete revolutions before the eyes of observers, the period of these revolutions differing from one couple to another, and showing the greatest variety, from a few years up to thousands. Our little terrestrial calendar does not extend its empire to these distant universes ; our ephemeral periods, our ant-like measures, are strangers to these grandeurs ; the earth is no longer the standard of creation ; our most sacred eras are unknown in the sky.

The study of these stellar systems constitutes one of the greatest and grandest problems of contemporary astronomy. Every star being a gigantic sun shining by its own light, a focus of attraction, of heat, and of light, the problem presented to the human mind by these systems of multiple suns is unquestionably one of those which may most perplex the imagination, impassion the thought, and move even the heart of a philosopher. What part does attraction play in these solar families, so different from ours? What is the numerical importance of these systems in the sidereal world? What is their mode of distribution in the universe? What bonds may they have with single suns like ours? What is the nature of their strange and fantastic light? At what respective distances can stars be associated and carried along by a common proper motion in space? What is the condition of the planetary systems which may revolve round these double suns? What can be the physiology of these planets, ruled, illuminated, warmed, alternately or simultaneously, by suns of different masses, different distances, and different lights? And, finally, what are the astonishing and extraordinary conditions under which life may exist on these unknown worlds, lost in the depths of the unfathomable heavens? Such are the questions which are now presented to our curiosity and study.

We have said that a large number of stars which appear single to the naked eye become double when we observe them with a telescope. We then distinguish two stars instead of one. If the telescope used is only endowed with a weak magnifying power, the two stars appear to touch each other, but become wider apart in proportion to the magnifying power employed. This double star, then, becomes for the contemplative mind a system of two neighbouring suns, separated from each other by millions of miles, and revolving round each other in periods which vary for each system according to the laws of universal gravitation. The immense distance which separates us from these celestial couples is the only cause of their invisibility to the unaided eye. The stars, divided into two in the telescope, still appear to touch, notwithstanding the millions of miles which really separate them, because they are distant from us by billions of miles.[1]

Several double stars were discovered so long since, and form systems so

[1] When we direct a telescope to a star, and, instead of a single star, see another star near it, it is not always certain that it is truly a double star. In fact, infinite space is peopled with numberless stars scattered in all the depths of immensity. There is, then, nothing astonishing in the fact that in pointing a telescope towards any star we discover one or more small stars situated behind it, farther off, and at a distance as great and even greater beyond it than that which separates the nearer one from us. As in a vast plain two trees may appear to touch because they are placed one before the other in our perspective, although they may be very distant from each other in reality, so in celestial space two stars may be found on the same visual ray and appear to touch, although separated from each other by an abyss. These are pairs of stars which are purely optical, and due to the position of the two stars on the same visual ray. To ascertain if this union is not only apparent but real, we must study it with attention. The probability that a pair of stars are really united is so much the greater the nearer they are to each other. But still, this would not be a sufficient reason for admitting its reality. It must be observed attentively and during several years. If the two stars are truly associated, if they form a system, we recognise that they move through space together, and that in general they revolve round each other. They are connected by the bonds of universal attraction; they have the same destiny. If the union is only apparent, we shall recognise in time that the two stars, thus fortuitously united by perspective, have nothing in common with each other; their proper motions, being different, will end by separating them completely.

rapid, as to have accomplished one or even several revolutions before our eyes; others have only traced in the sky a part of their orbits, but with an angular motion sufficient to permit us to calculate likewise the elements of their orbits; others, very numerous, have described but a small arc of their curve, insufficient for calculation of the whole orbit, but sufficient to vouch for the orbital nature of the motion; in certain pairs the components move in a straight line in virtue of parallactic displacement, proving that they are not physically associated, and that they are only temporarily connected by the chance of perspective.

There are still other more singular systems, of which the components describe straight lines in space, being animated by a common proper motion, which has led me to correct orbits prematurely computed (like that of 61 Cygni), and even to conclude that these suns cannot revolve round each other, but follow straight lines, obeying a force which sways them and leads them together through space. Several very distinct causes thus act on double stars to give them a real or apparent motion: the gravitation of the components, of a binary, ternary, or multiple system round their centre of gravity; the gravitation of two or more stars borne along together in space, under the influence of unknown sidereal attractions; the different proper motion of two distant stars fortuitously placed on our visual ray—causes to which we must add the secular translation of our solar system in space, which is revealed by the least distant stars having an apparent displacement in the opposite direction.[1]

In the present state of science we know 819 groups in certain relative motion. There are 558 certain or probable orbital systems, 317 perspective groups, 17 physical systems the components of which are displaced in a straight line, 23 ternary systems, 32 triples not ternary, formed of a binary system and an optical companion, &c. In order that the reader may understand the nature and variety of these orbits, I have collected here the systems which have been calculated up to the present, writing them in the order of increasing periods; there are over 70 which have, since the year of their discovery, described a part of their orbit sufficiently large to enable their orbit to be calculated and the period determined.

We see that the periods of revolution already calculated range from a few years to sixteen centuries. I might add others, almost as certain, which require no less than 2,000 years for their performance, and others still of which the period extends to 4,000, 5,000, and 6,000 years; but the

[1] I published in 1878 a first *Catalogue of Double and Multiple Stars in Certain Motion*, the result of a comparison which I made (1873–1877) of 200,000 observations made on the 10,000 double stars known in the sky, and a careful discussion of the motion of each star. From this work results a Catalogue of 819 groups in certain motion, of which I have measured micrometrically 133, chosen from among the most doubtful. This Catalogue contains 28,000 measures and the history of each star. It is absolutely impossible to enlarge here on this vast and important subject, and we can only recapitulate it from a descriptive point of view.

observation of these distant systems was begun so recently (the most ancient *measure* dates from 1709) that their long periods hardly commence to be revealed. When one of the two associated suns is endowed with a much greater mass than the other, it appears to be the centre of motion, as our sun appears to be the centre of motion of the translation of the earth and planets, although in reality the planets and the sun himself revolve together round their common centre of gravity: the smaller of the two stars revolves round the larger. Although a little technical, this comparative table is of the highest interest:—

DOUBLE STARS OF WHICH THE PERIODS HAVE BEEN DETERMINED [1]

Stars	Magnitudes	Colours	Period in Years	Stars	Magnitudes	Colours	Period in Years
κ Pegasi	—	—	11·4	ι Leonis	3·9, 7·1	Yellow, blue	116
δ Equulei	4·1, 5·0	White	11·5	37 Pegasi	6·0, 7·5	—	117
β 883	—	—	16·3	OΣ 285	7·1, 7·6	White	119
ζ Sagittarii	3·5, 4·1	—	19	25 Can. Venat.	5·0, 7·6	White, blue	120
85 Pegasi	6·0, 11·0	—	22	Σ 1,785	7·2, 7·5	Pale yellow, bluish	126
β 524	—	—	22	ξ Boötis	4·7, 6·6	Yellow, red	127
β Delphini	3·5, 11·0	—	24	4 Aquarii	5·9, 7·2	Yellow	130
42 Comæ	6, 6	White	26	OΣ 4	7·4, 8·1	—	135
β 612	6, 6	—	30	OΣ 20	5·9, 7·0	Yellowish-white, bluish-white	136
ζ Herculis	3, 6·5	Yellow and reddish	34				
β 416	—	—	35	Σ 73	6·2, 6·9	Golden	138
Σ 3,121	7·2, 7·5	White and yellow	35	Σ 2,525	7·2, 7·3	—	139
β 101	5·6, 6·7	—	41	40 Eridani	9·0, 11·0	Yellow	139
η Coronæ Bor.	5·5, 6·0	Golden yellow	42	*h* 5014	—	—	141
Σ 2,173	5·8, 6·1	Yellow	45	Σ 186	7·2, 7·2	White	151
OΣ 269	6·5, 7	—	48	OΣ 400	7·2, 8·2	—	170
μ′ Herculis	9·5, 10·5	—	49	γ Virginis	3·0, 3·0	White, yellowish	180
Sirius	1·0, 10·0	White	53	Σ 2107	6·5, 8·0	Yellowish, bluish	186
99 Herculis	6·0, 11·5	—	54	14 (*i*) Orionis	6·0, 6·8	—	190
OΣ 298	7·0, 7·3	—	57	π Cephii	5·2, 7·5	Yellow, purple	198
ζ Cancri	5·0, 5·7	Yellow	59	Dembowski 15	8·0, 8·0	—	205
ξ Ursæ Maj.	4·0, 5·5	Yellow, ashy	61	τ Ophiuchi	5·0, 5·7	White	218
γ Centauri	4·0, 4·0	—	62	η Cassiopeiæ	4·0, 7·6	Yellow, purple	222
OΣ 234	7·0, 7·4	White	63	*p* Eridani	6·0, 6·0	—	224
α Centauri	1·0, 2·5	White, yellow	81	35 Comæ	5·0, 7·8	Yellow, blue	228
γ Coronæ Bor.	4·0, 7·0	Yellow, purple	85	44 Boötis	5·2, 6·1	Yellow, bluish	261
OΣ 149	6·5, 9·0	—	86	Σ 1,757	7·8, 8·9	White, yellow	277
70 Ophiuchi	4, 6	Yellow, rose	88	μ² Boötis	6·7, 7·3	White	280
Σ 228	6·7, 7·6	White	89	36 Andromedæ	6·2, 6·9	Orange, yellow	349
γ Coronæ Aust.	5·5, 5·5	Golden yellow	93	λ Ophiuchi	4·0, 6·0	Yellow, bluish	374
λ Cygni	5·0, 6·3	Pale yellow	93	δ Cygni	3·0, 8·0	White, blue	377
OΣ 235	6·0, 7·3	Yellow, red	94	γ Leonis	2·0, 3·5	Golden yellow	407
8 Sextantis	—	—	94	12 Lyncis	5·2, 6·1	Yellowish, bluish	486
Σ 3,062	6·9, 8·0	Yellow, olive	104	μ Draconis	5 5	White	648
ξ Scorpii	4·9, 5·2	Yellow	105	61 Cygni	5·3, 5·9	Golden yellow	782
OΣ 215	6·7, 7·2	—	108	σ Coronæ Bor.	5·0, 6·1	Yellowish-white	846
ω Leonis	6·2, 7·0	White, blue	110	Castor	2·5, 3·5	White	1,001
OΣ 387	7·2, 8·2	—	110	ζ Aquarii	4·0, 4·1	White, green	1,624
OΣ 208	5·0, 5·6	—	115				

No spectacle is more imposing than that of these sidereal revolutions. In some systems the revolution is described in less than half a century Example: the star η of Corona Borealis, composed of two golden-coloured suns, of which the period is forty-two years. In other systems the period approaches a century, as in that of 70 Ophiuchi, composed of a clear yellow sun and a rose-coloured one, which revolve round each other in a period of ninety-three years [eighty-eight years.—J. E. G.]. The brilliant

[1] [For the figures given by M. Flammarion I have substituted the most recent results, and have added orbits recently computed.—J. E. G.]

pair, γ of the Virgin, is composed of two equal suns which slowly revolve round their common centre of gravity in a period of 175 years [180 years.—J. E. G.]. The ternary system, ζ of Cancer, is composed of three suns; the second revolves round the first in a period of fifty-nine years, and the third round both stars in 600 years, describing epicycloids which I discovered in the beginning of the year 1874, and which had puzzled me very much, as well as other astronomers to whom I had communicated them at that epoch.

Finally, we know orbital systems, such as those of γ of the Lion and ε of the Lyre, of which the period exceeds a thousand, and even several thousand years. Others move more slowly still. Thus the double stars are so many *stellar dials* suspended in the heavens, marking without stop, in their majestic silence, the inexorable march of time, which glides away on high as here, and showing to the earth, from the depth of their unfathomable distance, the years and the centuries of other universes, the eternity of the veritable Empyrean! Eternal clocks of space! your motion does not stop; your finger, like that of Destiny, shows to beings and things the ever-turning wheel which rises to the summits of life and plunges into the abysses of death! And from our lower abode we may read in your perpetual motion the decree of our terrestrial fate, which bears along our poor history and sweeps away our generation like a whirlwind of dust flying on the roads of the sky, while you continue to revolve in silence in the mysterious depths of Infinitude!

In the systems of double stars we notice a great variety of magnitudes as well as of distances between the components: several pairs are formed of two absolutely equal suns, while in others the satellite is very small, and gives the idea of a simple planet still luminous. It is probable that in the latter case these are planetary systems which are before our eyes. Thus, the satellite of Sirius, discovered in 1844 from the analysis of the perturbations observed in this star, and in 1862 by the progress of optics, may be to Sirius what Jupiter is to our sun; there would be nothing impossible in its being enormous and dark, and only shining because it is illuminated by its dazzling sun.[1] But there are many systems composed of two equal suns. Most are white or yellow; but we know 130 in which the two suns have different colours, and among them eighty-five where the contrast is remarkable, the principal sun being orange and the second green or blue.

An idea may be formed of the annual motion observed and calculated in the rapid systems by the examination of fig. 267, which shows the apparent orbit of the double star ζ Herculis as we see it from the earth: this is among the most rapid and the most certainly determined. The

[1] [I have shown elsewhere (*Journal of the British Astronomical Association*, March 1891) that this is impossible in the case of Sirius and its satellite.—J. E. G.]

inner star (A) being taken as the centre of comparison, the position of the second (B) is fixed by taking the North as 0°, the East being at 90°, the South at 180°, and the West at 270°. We thus see that in 1838 the second star of this double system passed to the south of the principal star; following its course, you will see it pass to the east in 1851, to the north in 1862, to the west in 1865, to return to the south in 1872. Since

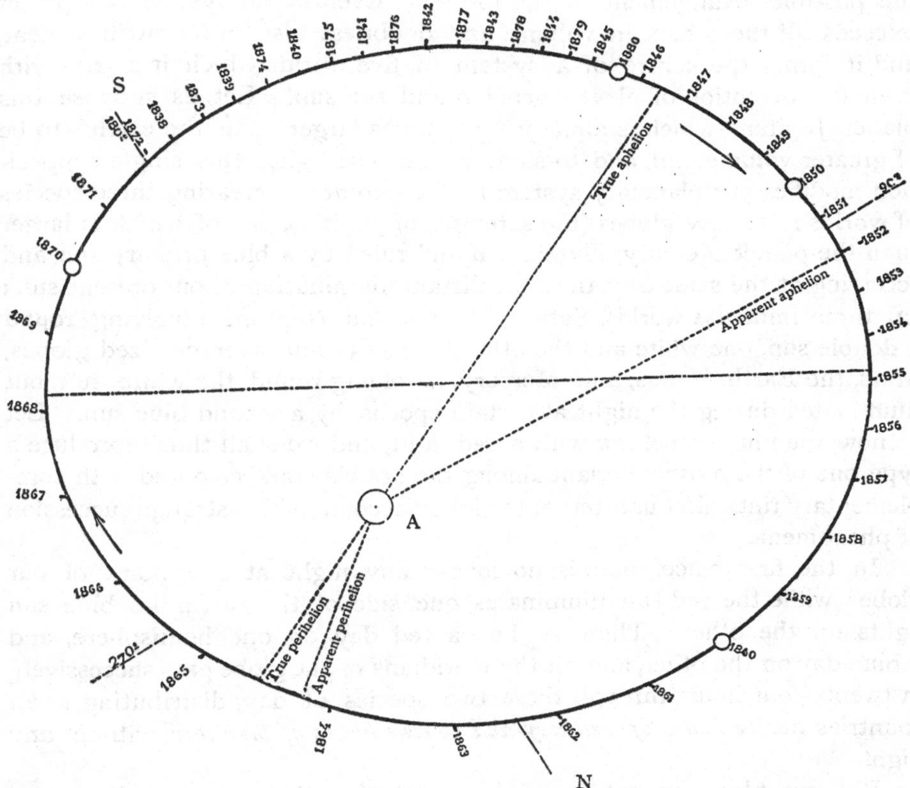

FIG. 267.—APPARENT ORBIT OF THE DOUBLE STAR ζ HERCULIS AS SEEN FROM THE EARTH

its discovery in 1782 by Herschel this star has already described more than three revolutions; its period is 34½ years.

As we see these motions in perspective, the orbit thus drawn does not represent the true form of the motion seen full face: it is necessary to calculate the inclination of the orbit, more or less inclined to our visual ray, in order to determine the absolute orbit. We thus find all kinds of ellipses, from the circle to those with great eccentricities.

What is the nature of the orbits described by the worlds belonging to these singular systems? Do these unknown planets revolve at the same time round both suns as a centre? and have they for the focus of their motions the centre of gravity of these twin suns, or has each of these suns its own planetary system? The latter arrangement should be the most probable and the most general.

Notwithstanding the essential difference which exists between these systems and ours, we may use the disposition of the latter to divine the possible arrangement of the former. Even in our system, one planet exceeds all the others in volume, and doubtless also in its intrinsic heat, and it forms the centre of a system of five worlds which it carries with it in its revolution of eleven years round the sun. Let us suppose this planet, Jupiter—which is already 1,240 times larger than the earth—to be of greater volume still, and to shine with a blue light; this single supposition modifies our planetary system to the extent of creating three species of worlds: (1) Five globes (the satellites of Jupiter), one of which is larger than the planet Mercury, illuminated and ruled by a blue primary sun, and receiving at the same time the more distant illumination of our present sun; (2) three immense worlds, Saturn, Uranus, and Neptune, revolving round a double sun, one white and the other blue; (3) four average-sized globes, Mars, the Earth, Venus, and Mercury, revolving round the white sun, but illuminated during the night, at certain epochs, by a second blue sun. Let us now illuminate the *sun* with a red light, and we shall thus reproduce a type, one of the most abundant among the double stars, coloured with complementary tints. Let us attempt to give an account of this strange succession of phenomena.

In the first place, there is no longer any night at any point of our globe: while the red sun illuminates one side of the earth, the blue sun lights up the other. There is thus a red day on one hemisphere, and a blue day on the other, and all the meridians of the globe pass successively in twenty-four hours through these two species of day, distributing to all countries *twelve hours of red day and twelve hours of blue day*, without any night.

But our blue sun, not remaining motionless in space, itself revolves slowly round the red sun. Presently it rises before the first has set, and appears above the eastern horizon when the celestial ruby is not yet extinguished. The blue day then succeeds, but, the sapphire sun setting in its turn before the rising of its scarlet rival, we have a night of some moments adorned with two magnetic auroras of a new kind—one reddish in the east, the other bluish in the west. The duration of this night increases day by day, and at the same time that of the double daylight, illuminated by two suns at once, the *blue hours* and *red hours* diminishing in the same proportion. Finally, at the epoch which corresponds to the conjunction of Jupiter, the blue sun approaches the red sun, and there is no longer a red day or a blue day, but double daylight, followed by complete night. The light of the double day is of course formed by the union of the colours of the two suns; it is *violet,* but might be quite white if the colours were complementary. Always carried along by its own motion, the secondary sun passes to the west of the first, and soon produces blue mornings, followed by a white or violet day, a red evening, and a night becoming shorter and shorter, until the blue sun returns to opposition, as we placed it at the beginning of this description.

In most of the systems of double stars the small star revolves round the larger, not in a circle, but in a very elongated ellipse. The stability of the systems requires that this small star should not approach too close to the large one; for in this case—supposing, as is natural, that the planets revolve in the same plane as the star itself—they might be attracted by the central sun at the moment of the perihelion passage, and abandon their primitive sun, to the great detriment of their inhabitants, who would doubtless be dead from the heat before the astronomers of these regions were able to ascertain the desertion. It is indispensable that these systems should be very much drawn in round each of their suns, and that the obedient planets should gravitate closely under the protecting wing of their own sun. But in any case the most singular alternations of heat, of light, and of seasons, are the cosmological consequence of these motions.

Thus, in every planetary system ruled by a double sun our double alternation of day and night is replaced by a quadruple alternation: (1) a double day illuminated by two suns at a time; (2) a simple day lit by a single sun; (3) another simple day illuminated by another sun; (4) some hours of complete night, when both suns are at the same time below the horizon.

The splendour of these natural illuminations can hardly be conceived by our terrestrial imagination. The tints which we admire in these stars from here can give but a distant idea of the real value of their colours. Already, in passing from our foggy latitudes to the limpid regions of the tropics, the colours of the stars are accentuated, and the sky becomes a veritable casket of precious stones; what would it be if we could transport ourselves beyond the limits of our atmosphere? Seen from the moon these colours would be splendid. Antares, *a* Herculis, Pollux, Aldebaran, Betelgeuse, Mars, shine like rubies; the polar star, Capella, Castor, Arcturus, Procyon, are veritable celestial topazes; while Sirius, Vega, and Altair are diamonds eclipsing all by their dazzling whiteness. How would it be if we could approach the stars so as to perceive their luminous discs, instead of merely seeing brilliant points destitute of all diameter?

Blue days, violet days, dazzling red days, livid green days! Could the imagination of poets, could the caprice of painters picture on the palette of fancy a world of light more astounding than this? Could the mad hand of the chimera, throwing on the receptive canvas the strange lights of its fancy, erect by chance a more astonishing edifice? Hegel has said that 'all which is real is rational,' and 'all which is rational is real.' This bold thought does not express the whole of the truth. There are many things which do not appear to us rational, but which, nevertheless, exist in reality in the numberless creations of Infinitude.

The most beautiful contrasts of colouring[1] are not presented by the systems in rapid motion, but by the systems in slow motion, and even in those which have remained motionless since their discovery. This curious fact does not prevent the planets which gravitate round these latter suns from being subject to the most singular alternations of illumination, of seasons, and of years. Our white and solitary sun, our solar system formed with a single focus round which revolve obedient worlds, following regular orbits, does not constitute the type and the model of universal creation. The multiple suns which we study here sometimes unite their light, sometimes oppose each other, sometimes alternate successively in the same sky; suns of dissimilar volumes and masses, acting often in contrary directions and distorting the singular orbits of the unknown worlds which gravitate in their power. No spectacle is more magnificent than the telescopic contemplation of these strange suns. When in the silent night, during the sleep of terrestrial nature, in those nocturnal hours when humanity around us is asleep in anticipated death, our gaze and our thoughts are elevated by the aid of the marvellous telescope towards these celestial lights which are lit up on high for other worlds, and radiate around them heat, activity, and life, the contrast is so great that we think we dream. Here night, above light; here lethargy, above motion; here shadows, above splendour; here heavy and dark matter, above the devouring flame and the sidereal life. How poor is our sun compared to his great brothers, his elders in space! How miserable is our world compared with those which sail on high on the rapid and multiplied wings of such attraction! What delightful hours pensive spirits and inquisitive minds might pass in pointing a telescope towards the heavens, if the best informed men, if the most highly educated women of the world were not universally ignorant of the most elementary truths of astronomy, and if they did not always revolve in a circle more or less monotonous, without suspecting the wonders which nature keeps in reserve for those who understand them.

And what shall we say of systems of triple and quadruple suns, of which the worlds never know night, where astronomy could not be born, since

[1] The most beautiful coloured double stars are: β Cygni, 3rd and 5th magnitudes, golden yellow and sapphire blue; γ Andromedæ, 3rd and 5th, orange and green, the latter divided into green and blue; γ Delphini, 4th and 5th, golden yellow and greenish blue; α Herculis, 3rd and 6th, orange yellow and sea-blue; α Canum Venaticorum, 3rd and 5th, gold and lilac; ε Boötis, 3rd and 6th, topaz and emerald; Antares, 1st and 7th, orange and clear green.

they never see the starry sky, and of which the inhabitants know not sleep?[1]

There are, possibly, human beings out yonder who contemplate daily these peculiarities! Who knows? The thing is probable enough. They doubtless agree but with ordinary observation, and, accustomed, like us, to the same life from their cradle, they do not appreciate the picturesque value of their abode. Thus are men constituted: the novel, the unexpected alone interests them; as to the natural, it seems that this might be an eternal, necessary, fortuitous state of blind nature which does not merit the trouble of observation. If human beings from yonder could come to us, although recognising the simplicity of our little universe, they would not fail to notice our indifference with surprise and astonishment.

If, like our moon, which gravitates round the globe, like those of Jupiter and Saturn, which unite their mirrors on the dark hemisphere of these worlds, the invisible planets which move out yonder are surrounded by satellites which continually accompany them, what must be the aspect of these moons illuminated by several suns? That moon which rises from the distant mountains is divided into parts differently coloured, one red, another green; one shows an azure crescent; another is at its full—it is green, and appears hung in the heavens like an immense fruit. Ruby moon, emerald moon, opal moon, what singular lustres! O nights of the earth which our solitary moon but modestly silvers, you are very beautiful when contemplated by a calm and pensive mind, but what are you compared to these wonderful nights?

And what are the eclipses of the sun on such worlds? Multiple suns, to what infinite sports must not your mutually eclipsed lights give birth? A blue sun and a yellow one approach each other; their combined light produces green on the surfaces illuminated by both, yellow or blue on those which receive but one light. Presently the yellow approaches the blue; it passes on to its disc, and the green colour spread over the world grows pale, fades till the moment it dies, blended into the gold, which pours into space its crystalline radiation. A total eclipse colours the world in yellow! An annular eclipse shows a blue ring encircling a disc of translucent gold. Gradually, imperceptibly, the green reappears and recovers its empire. We may add to this phenomenon that which is produced when some dark moon comes exactly in the middle of this golden eclipse to cover the yellow sun itself and plunge the world into darkness. We might add more—but no, this is the inexhaustible treasure of nature: to dive with our hands full is to take nothing.

[1] It is our habit, doubtless, but it is none the less odd, to see all human beings—on account of the earth's motion of rotation and the physiological organisation which results from it—at a certain hour each day undress and place themselves horizontally, closing their eyes in an annihilation of seven or eight hours, and losing a good third of their existence (twenty years in sixty) in anticipated death!

These descriptions suffice to give an idea of the nature of the subject and of the captivating interest which is attached to these questions. Science is only beginning to penetrate into the starry immensity. Even yesterday we were ignorant of the number of the real double stars now observed, the diversity of their motions and their proportion in the organisation of the heavens. We may estimate that about one-fifth of the suns of which the universe is composed are not single like that which illuminates us, but associated in binary, ternary, or multiple systems. Thus the double stars are veritable suns, gigantic and powerful, governing, in the regions illuminated by their splendour, systems different from that of which we form part. The sky is no longer a gloomy desert; its ancient solitudes have given place to regions peopled like those in which the earth gravitates; the darkness, the silence and death which reigned in these depths have given place to light, to motion, and to life; thousands and millions of suns pour out in great waves into space the energy, the heat, and the different undulations which emanate from their foci; the universe is transfigured to our thoughts; suns succeed to suns, worlds to worlds, universes to universes; tremendous proper motions carry all these systems through the endless regions of immensity; and everywhere, out to and beyond the farthest limits where the fatigued imagination may rest its wings, everywhere is developed in infinite variety the Divine creation in which our microscopical planet is but an insignificant province.

CHAPTER IX

THE PROPER MOTIONS OF THE STARS

Translation of all the Suns and Worlds through the Infinite Immensity. Secular Metamorphosis of the Heavens

THE ideas which we have formed up to the present of the stars and the sky must now undergo a complete transformation, a veritable transfiguration. *There are no longer any fixed stars.* Each of those distant suns lit up in Infinitude is carried along by immense motions which our imagination can hardly conceive. Notwithstanding the billions of miles which separate us from these suns, and which reduce them for our sight to little luminous points (although they may be as vast as our own sun and thousands or millions of times larger than the earth), the telescope and calculation grasp them, and prove that they are all in motion in all possible directions. The heavens are not immutable; the constellations represent to us no longer the symbol of absolute and indestructible order; the spectacle of the starry night no longer shows us repose and inertia. No; all the stars are fiery suns, foci of heat and of light, laboratories of unheard-of combustions, torches of travelling humanities, which incessantly dart around them waves of inexhaustible light, distribute the essence of life to the planets which surround them, and which move rapidly in space, carrying with them the systems of which they form the centres of gravity.

These tremendous movements are only visible from here by the minute displacements of the stars, which are measured by fractions of seconds of arc.

Can we imagine the smallness of this measure? Let us remember that a second is the sixtieth part of a minute, which is the sixtieth part of a degree, which is the 360th part of a great circle making the circuit of the sky. To form a comparison, the sun and moon are presented to us under the form of discs measuring on an average 31 minutes in diameter; these 31 minutes make 1,860 seconds. The displacement, then, of a star of which the proper motion would be one whole second a year, would be only the 1,860th part of the apparent diameter of the sun. In other words, it would take 1,860 years for a star to be displaced by this amount. As the proper motion of most of the stars is not even one second a year, we see that since the time of Christ and of Tiberius they have not accomplished even that journey. A certain number of stars are animated with more rapid motions, which rise to several seconds; but even for these exceptions we see that, relatively to our measures of daily estimation, these motions are still infinitely small to our eyes, although they may be infinitely great in reality. We might call them at the same time microscopic and telescopic.

What must be the velocity of these translations, however, in order that we may perceive them from here, distant as we are at many billions of miles! If we take, for example, Arcturus, of which the proper motion is almost three seconds a year, we find that its velocity through space is not less than 4,400,000 miles a day! And it requires 800 years to show us a displacement equal in length to the apparent diameter of the moon and sun! We are 200 billions of miles distant from this star: the course which it describes in a straight line during a year, at the rate of over 4 millions of miles a day, would be hidden by the thickness of a wire of one millimetre thick placed at 68 metres of distance from our eye [or $\frac{1}{25}$th of an inch seen at a distance of 223 feet].

The most remarkable star of the whole sky is, from this point of view, a small star of the 7th magnitude—that is to say, invisible to the naked eye—which has no particular name, and is denoted by a simple catalogue number. It bears the number 1,830 in the Catalogue of Groombridge, and it is by this denomination that it is known. It is a small star of the Great Bear, situated at $11^h\ 45^m$ of right ascension and 50° 21′ of polar distance; it shows annually the greatest displacement which has been observed. Its annual variation is 7″.

If we estimate this motion by the measure which we have used just now, we see that, in order to displace it in the sky by an amount equal to the apparent width of the sun, 255 years would be necessary. This motion is so rapid that it reaches 17 millions of miles a day. This is a velocity more than ten times greater than that of the earth in its course, our planet moving round the sun at the rate of 1,595,000 miles a day.

Here we see a star, a sun lost among the myriads of suns which fill the heavens, and which is carried along in space by a power so wonderful that it passes over not less than 2,480 millions of miles per annum, and this line of 2,480 millions of miles can only be perceived from here by the aid of the most minute and careful micrometrical measures! Look at the beautiful star Arcturus, which sails in the sky at the rate of 1,600 millions of miles a year, and for a thousand, two thousand, three thousand years and more since it has been observed and its place marked on astronomical charts it does not seem to have stirred! And still these are not the exact velocities of the celestial bodies. In order that these measures might be absolute, it would be necessary that the course followed by the star observed should be seen full face—that is, should be perpendicular to the visual ray which goes from here to the star. There is nothing to prove that this is exactly the absolute direction of the star's course, and it is extremely probable that its motion is more or less oblique. Whatever may be the absolute path of a star, we never see but the projection of its route on the apparent sphere of the sky.

Among stars of the 1st magnitude which are animated with a proper motion greater than the general average we find, after Arcturus, the two fine stars Procyon and Sirius. The proper motion of the first is nearly half that of Arcturus, and measures $1''\cdot27$. That of Sirius is $1''\cdot34$.

See Map of Proper Motions (Plate), in which the length of the arrow represents the motion in fifty thousand years.

. . . .

Amidst the variety of the proper motions of the stars we notice a general effect tending to remove them from a point situated in the constellation Hercules, and to direct them towards the opposite point, situated in the southern hemisphere. This general perspective motion, of which we have already spoken, proves that our solar system itself travels in space and proceeds towards the point indicated. According to the labours of the astronomers who, since William Herschel, have devoted themselves to this complicated analysis, the point of the sky towards which we are moving is situated in the following position :—

Right ascension $= 266°$; polar distance $= +31°$.

This point is a little to the north of the star μ of the constellation Hercules.

One of the most curious and astonishing results of these stellar motions is to modify slowly, inexorably, the aspect of the constellations, and to bring about what we may call the dislocation of the heavens.

See, for example, the Great Bear. Each of the stars which compose it is carried along by a proper motion. It follows that in the course of ages this figure will change in form. At present it somewhat recalls the outline of a vehicle, and it is this resemblance which has given it the popular name of the *chariot*. The four stars arranged in a quadrilateral are considered as representing the position of the wheels, and the three which precede them mark the place of the horses. Now, proper motion will change this arrangement; it will bring back the first horse behind, while it will carry the two others in front. Of the two hind wheels, the first will be drawn to one side and the second to the other. Knowing the annual value of the displacement of each of these seven stars, we can calculate their future respective positions. Here are the curious results to which these calculations lead :—

In fig. 268 the arrows indicate the direction towards which each of the stars moves. We see that of the seven stars, the first and the last, Alpha and Eta, point in one direction, while the five others point in the opposite direction. Moreover, the velocity is not the same for each of them.

In virtue of these proper motions, the relative distances of these stars change in time. But as the change is but a few seconds per century, many centuries are necessary for the difference to become perceptible to the naked eye. Our human generations, our dynasties, even our nations, do not exist sufficiently long for this measure. Here the question is of astronomical quantities, and in order to estimate them it is necessary to choose terms which correspond. Let us suppose 50,000 years; in this interval—which is not, however, enormous in the history of the stars, since the little earth where we are itself dates from several *millions* of years—all the constellations are modified.

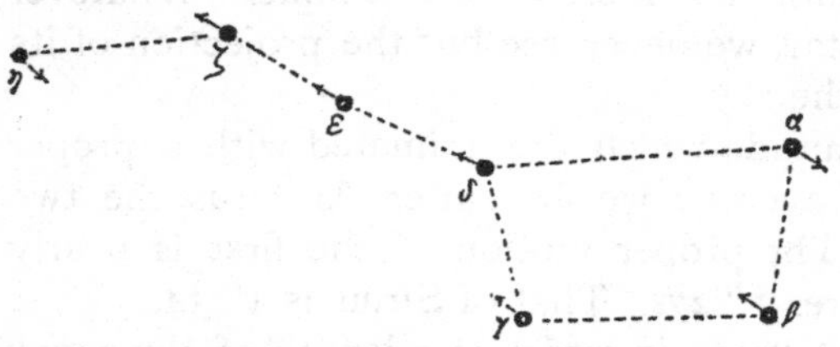

Fig. 268.—The Seven Stars of the Great Bear as they really are

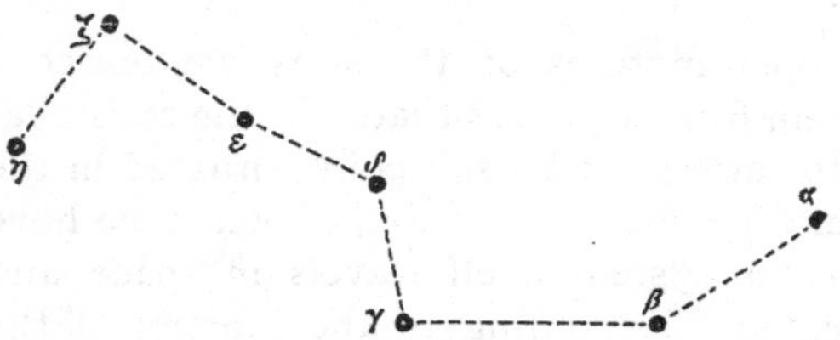

Fig. 269.—The Great Bear 50,000 Years hence

The following drawing (fig. 269) indicates the geometrical result of calculation for 50,000 years. We see that the figure will then have completely lost its present aspect. It is in vain that we look for traces of a chariot in this new figure. The seven famous stars will be distributed along a broken line, Alpha having descended to the right of Beta, and Eta, at the other end, having descended below Zeta.

Seeing what a profound alteration this constellation will undergo in the ages to come, we may also ask for how long it has had the form under which we now know it, and what aspect it showed in past ages. To find the position of each of these seven stars 50,000 years ago it is sufficient to carry them back by the same quantity by which they were carried forward on their direction in the preceding example. This calculation gives quite another figure (270), which in no way resembles the first or the second. Fifty thousand years ago these stars were aligned in such a way as to form a true cross, more exact and even finer than the Southern Cross which now shines near the South Pole, and which is itself also so rapidly becoming distorted that in 50,000 years its four branches will be completely dislocated. In this *cross of the North* the star Alpha formed the left side, Gamma the right side, Beta the top, and Delta, Epsilon, and Eta the upright. Zeta had not yet arrived in the assembly of the six others. However, on analysing the march of these stars, we become convinced that the five companions, Beta, Gamma, Delta, Epsilon, and Zeta are associated in their destiny by a common

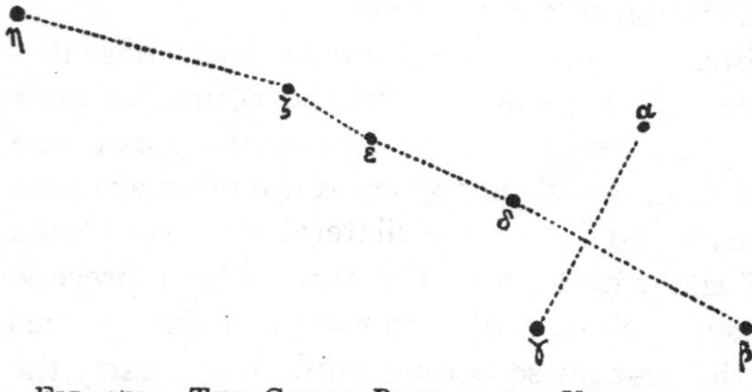

Fig. 270.—The Great Bear 50,000 Years ago

bond. They are like a group of friends ; they march with a common accord, and keep, as we see, the same position with regard to each other, while Alpha on the one side and Eta on the other are two intruders, which are now found to make part of the association, but which are entirely strangers.

If the Great Bear is the most characteristic and most universally known of the northern constellations, Orion is unquestionably the finest of the constellations of the south, and indeed of the whole sky. Curious to know what transformations the proper motions of the stars will cause in future ages in the aspect of this asterism, as well as in the respective situations of the three fine stars which surround it, Sirius, Aldebaran, and Procyon, I have treated it as I have the Great Bear, and calculated what changes of aspect time will produce in the respective positions of these stars.

Fig. 271 shows the present state of the constellation of Orion, with the position and respective distance of Sirius, Aldebaran, and Procyon. A little arrow attached to each star indicates the direction of its motion. Fig. 272 represents the positions of these stars 50,000 years hence.

Mythology represented Orion running after the Pleiades and the Bull. It is

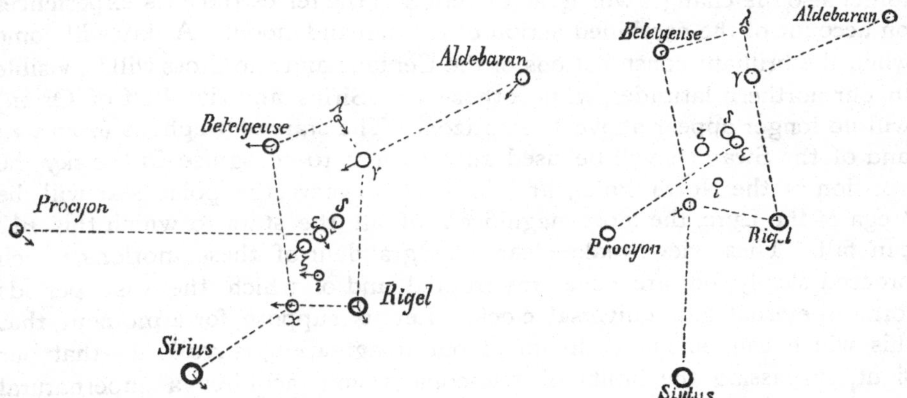

FIG. 271.—THE CONSTELLATION OF ORION AS IT IS NOW

FIG. 272.—THE CONSTELLATION OF ORION AS IT WILL BE 50,000 YEARS HENCE

Aldebaran, on the contrary, which falls towards Orion. The *Three Kings* will not long remain united.

But of the secular variations with which these stars are affected, the two most striking are those of Procyon and Sirius. Procyon, now so distant from Orion, will approach and come to form part of it, and the astronomers of the years 50,000 to 80,000 will consider it as belonging to that constellation ; it will form the south-east angle, and, joined by an ideal line to Betelgeuse and Rigel, it will represent, much better than the star κ, the left leg of the Giant. Carried along by a proper motion less than that of Procyon, Sirius will come to be placed at the foot of Orion, and will seem to further lengthen out this figure, already so gigantic. The Little Dog runs after the Great Dog, but it will never reach it, the latter itself fleeing, from age to age, in a direction oblique to the other. We see what the respective positions of these twelve stars will be in 50,000 years (fig. 272)—provided, however, there is no unforeseen combination.

The work which has been done for the secular variation of the constellations of the Great Bear and Orion might be applied to most of the other constellations. The proper motions for nearly all the stars visible to the naked eye have already been determined.

There are stellar systems formed of stars which, although very distant from each other, are nevertheless united by a common destiny. The five (β, γ, δ, ε, ζ) of the Great Bear show us an example which has been already described by Proctor. I have found many others.[1]

Thus all the stars are in motion. 'From numerous, incessant causes, which make the relative positions to vary, the brightness of different regions of the sky and the general appearance of the constellations may after thousands of years' (we may say with Humboldt) 'impress a new character on the grand and picturesque aspect of the starry vault. Besides these causes, we should add here the sudden apparition of new stars, and the enfeeblement, and even the extinction, of some ancient stars. Nor should we forget also the changes which the direction of the terrestrial axis experiences on account of the combined action of the sun and moon. A day will come when the brilliant constellations of the Centaur and the Cross will be visible in our northern latitudes, while other stars (Sirius and the Belt of Orion) will no longer appear above the horizon. The stars of Cepheus (γ and α) and of the Swan (δ) will be used successively to recognise in the sky the position of the North Pole; and in 12,000 years the polar star will be Vega of the Lyre, the most magnificent of all the stars to which this *rôle* can fall. These views make clearer the grandeur of these motions, which proceed slowly, but are never interrupted, and of which the vast periods form an eternal and universal clock. Let us suppose for a moment that this, which can be but a dream of our imagination, is realised—that our sight, surpassing the limits of telescopic vision, acquires a supernatural power; that our sensations of duration permit us to comprehend and to contract, so to say, the longest intervals of time: the apparent immobility which reigns in the vault of the heavens immediately disappears. Numberless stars are carried along like whirlwinds of dust in opposite directions; wandering nebulæ condense and dissolve; the Milky Way divides like an immense girdle torn into shreds; everywhere motion reigns in the celestial spaces, as it reigns on the earth in the life of animals and men.'

Like the dust of our roads, the whirlwinds of stars fly on the paths of the sky. It is an immense life, a perpetual swarming; the mind which could set aside time would cease to contemplate during the silent night an inert and motionless sky, and would see in its place myriads of fiery suns darting in all directions through immensity, and strewing in Infinitude multiplied forms of universal and irrepressible vitality. The knowledge of the proper motions of the stars absolutely transforms our ordinary ideas

[1] See the *Comptes Rendus de l'Académie des Sciences*, 1877.

[*Plate* III.

PROPER MOTIONS OF EACH STAR IN THE HEAVENS

NORTHERN HEMISPHERE

SOUTHERN HEMISPHERE

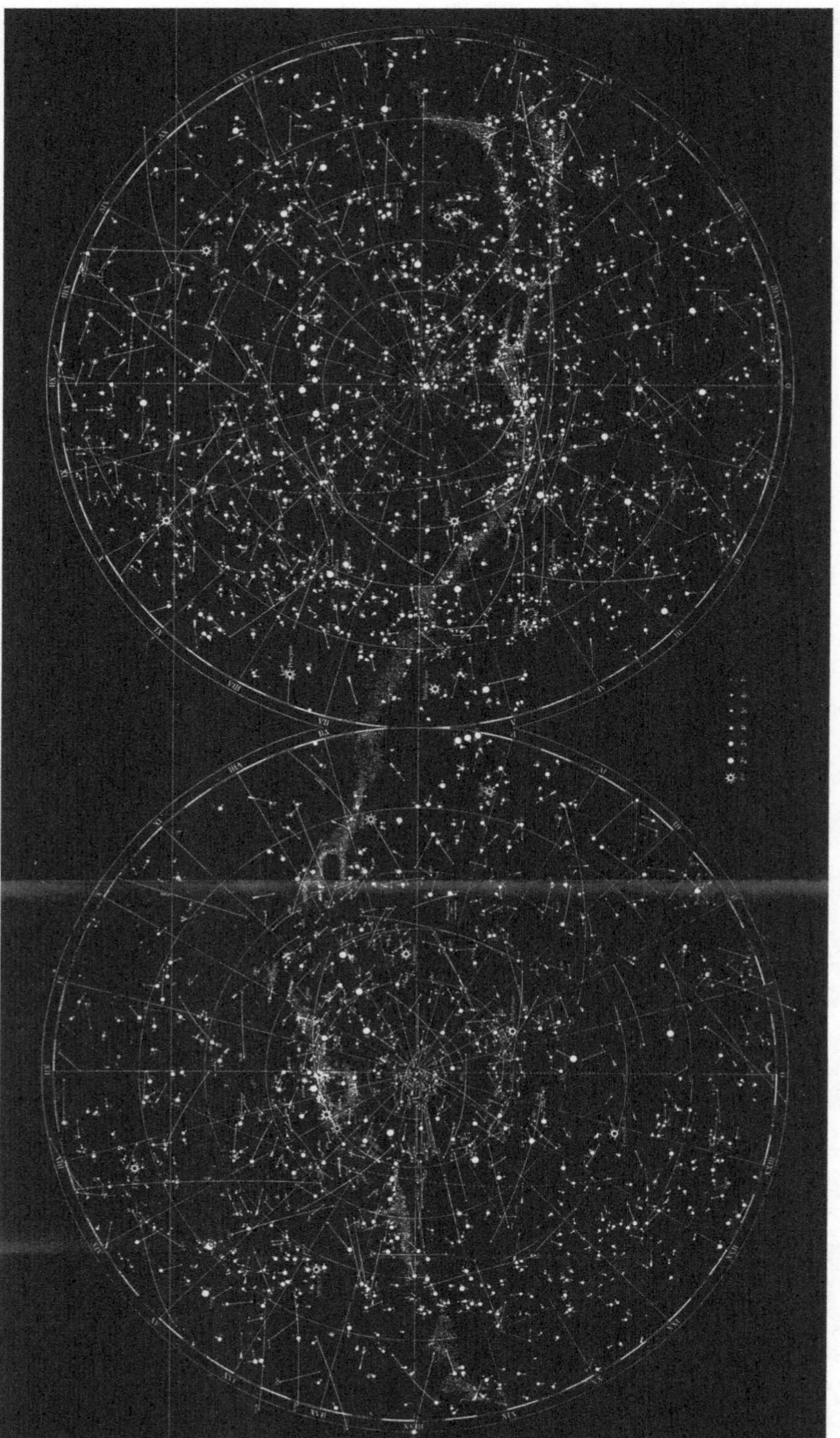

The material originally positioned here is too large for reproduction in this reissue. A PDF can be downloaded from the web address given on page iv of this book, by clicking on 'Resources Available'.

of the immutability of the heavens. The stars are carried along in all directions through the endless regions of immensity, and, like terrestrial nature, celestial nature and the constitution of the universe change from age to age, undergoing perpetual metamorphoses. (See Plate III.).

All these proper motions, *inferred from the positions of catalogued stars*, are necessarily perpendicular to our visual ray; but as there is no probability that the stars are displaced in this direction rather than in all the other directions possible, it is certain that most of the lines which we thus trace are only the projection of the oblique paths. Let us suppose that, unknown to us, all the stars were placed at the same distance from us, like brilliant points on the interior of a vault; we refer all the motions observed to lines traced in the same spherical plane along this vault; our paths thus traced are consequently shorter than the real path in all cases where the real path is not parallel to the celestial vault. Fig. 273 presents and explains all these projections according to the obliquity of the line followed by the star.

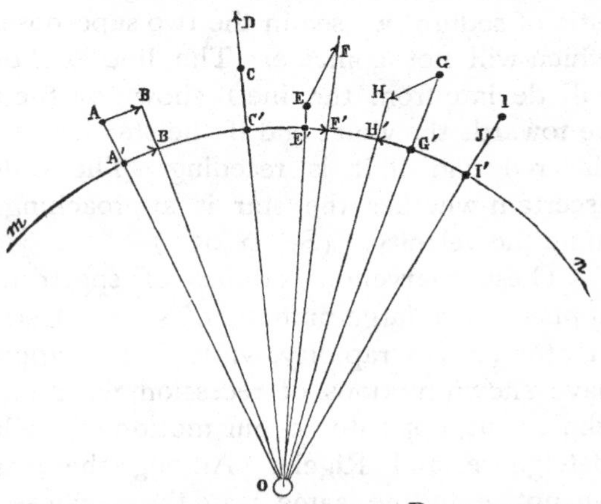

FIG. 273.—REAL AND APPARENT DISPLACEMENTS OF STARS

Can we know whether a star follows a course exactly parallel to the vault of the sky, or whether it is receding from or approaching the earth on an oblique line of which we observe only the projection? Given a star which appears even absolutely fixed, does there exist any method of discovering whether it is in motion in the direction of the visual ray and, in this case, whether it is receding from or approaching the earth?

He who should in former times have propounded such a question would have received only a smile for reply. However, human science has now made this new conquest in Infinitude.

Not only can we, notwithstanding their smallness and their imperceptibility, ascertain and measure the displacements of the stars in the sky, but we can even ascertain and measure their motion of recession or approach, even when it is directed along the visual ray and is not shown by any displacement in astronomical observations!

The method employed to arrive at these results has no relation to the process of comparisons by which the annual proper motion is measured;

it is based on the principles of optics and on the analysis of the rays of light. If we receive through a prism the luminous ray which comes from a star, we see that it produces a little spectrum, as we have explained (p. 321). We can produce a similar spectrum by receiving on another prism the luminous ray proceeding from an electric gleam traversing a tube filled with gases coincident with those recognised in the star studied.

This being granted, if the star is motionless, the two spectra are simply superposed, and nothing extraordinary is noticed in this superposition; but if the star is approaching or receding, the motion is reflected in the spectrum in a singular way. Let us suppose that it approaches. The lengths of the waves, which give rise to the diversity of colours, diminish, and the refrangibility of each colour increases. If, then, we observe with a spectroscope two luminous sources, the one fixed (the electric tube), the other moving (the star), both giving, for example, the line ('D') so characteristic of sodium, we see in the two superposed spectra the rays of this metal, which will not coincide. The line D shown by the spectrum of the star will deviate from the line D shown by the tube, and the displacement will be towards the violet end if the star is approaching the earth, and towards the red end if it is receding. The difference will serve not only to ascertain whether the star is approaching or receding, but even to determine the velocity. (See p. 651.)

These marvellous studies of spectrum analysis have already been applied to a large number of stars. Certain stars recede from us with a greater or less rapidity, while others approach. Among the stars which have shown motions of recession, we remark several of those which are, like Sirius, opposite to our motion of stellar translation, such as Procyon, Betelgeuse, and Rigel. Among the stars which are approaching us we notice in the same way those whose situation is near the celestial region towards which we are proceeding, as Arcturus, Vega, α Cygni. We might expect this; but this double fact does not in any way influence the opinion which we have set forth above relative to the real displacement of all the stars in Immensity. They have determined motions of recession and approach in all parts of the sky, as well as on the side of Hercules and the opposite side. The influence of our own translation on the general perspective is perceptible; but it does not prevent all the other suns of space from having their own personality, their distinct course, and their peculiar destiny.

According to the industrious researches of Mr. Huggins at his own observatory, and of Mr. Christie at the observatory of Greenwich, the motions of the stars studied are as follows:

Stars receding from us	Velocity per second[1]	Stars approaching us	Velocity per second[1]
α Coronæ	48 miles	α Ursæ Majoris	46 miles
Castor	28 ,,	α Andromedæ	45 ,,
Procyon	27 ,,	Vega	44 ,,
Capella	27 ,,	Arcturus	41 ,,
Regulus	23 ,,	γ Leonis	41 ,,
Sirius	22 ,,	Pollux	40 ,,
α Orionis	22 ,,	α Cygni	40 ,,
β Pegasi	20 ,,	η Ursæ	32 ,,
Aldebaran	19 ,,	α Herculis	31 ,,
β Orionis	19 ,,	δ Cygni	23 ,,
β, γ, δ, ε, ζ Ursæ Majoris	19 ,,	γ Cygni	20 ,,

The delicacy of these measures prevents us from attaining a rigorous precision in the translation of the slight displacement of the spectral lines into velocities in miles, and the figures of this little table can only be considered as provisional. We cannot help remarking, however, that the stars which are approaching us appear in general animated with more rapid velocities than those which are receding.

These velocities of course represent the proper motion of the star combined with that of the solar system in space. Their variety shows, on the other hand, that our translation through immensity forms but a part of the observed displacements. Thus, Vega of the Lyre approaches us with a probable velocity of forty-four miles per second, or we approach it with this velocity, or, to speak more correctly, our motion added to its motion reaches this velocity, as this sun is no more at rest than ours. On the other hand, Castor recedes from us with a velocity estimated at twenty-eight miles a second as the resultant of its motion and ours. It is rather curious to notice that the twins, Castor and Pollux, are not really associated as they appear to be ; one is receding from us, while the other is approaching. Each goes its own way, and they are not acquainted.

Thus the real motion of every star in space may now be interpreted by compounding the proper motion inferred from the observed positions with the motion in the line of the visual ray.[2]

[1] [From recent measures of photographed stellar spectra some of these velocities have been modified.—J. E. G.]

[2] Take, for example, Sirius. At the distance at which we are from this star its annual proper motion, which subtends an arc of 1″·33, indicates a displacement of 616 millions of miles measured perpendicularly to the visual ray. As it is receding in the same interval of time by a quantity which we have estimated at 666 millions of miles, this velocity is to the first in the ratio of 166 to 100 [108 to 100]. It follows that, although the annual recession may be indicated by the number we have quoted, the oblique march of the star raises it in reality to 738 millions of miles per annum. The meridian observations have discovered the displacement AB (see fig. 274) perpendicular to our visual ray ; the spectral comparisons have discovered the displacement AC in our visual ray. The true motion of Sirius is performed along the line AD.

The distance which separates us from Sirius increases annually by 666 millions of miles—more than 70,000 leagues (174,000 miles) a day.[1] And for four thousand years at least, since men's eyes have been fixed on this star, it has not diminished in brightness. Its fires still sparkle with incomparable splendour, and it still attracts our attention in the silent night like a radiant and unalterable sun. These thousands of years of observation represent, however, thousands of millions of miles, and the difference between the distance of Sirius four thousand years ago and its present distance may even rise to billions of miles—that is to say, reach the units of interstellar measurements ; and, notwithstanding such a difference, Sirius does not appear to have diminished in brightness, and it still reigns as sovereign in the midst of the eclipsed constellations !

FIG. 274.—MOVEMENT OF SIRIUS IN SPACE

We have seen that the motions with which the suns are animated are presented obliquely to our terrestrial observation. There are motions, however, which are presented quite full face, so that the star appears neither to recede nor approach ; γ Orionis, α Virginis, and α of the Eagle are examples. [Recent measures show that α Virginis and α Aquilæ are approaching.—J. E. G.]. There are others, on the contrary, of which the displacement is nearly nil on the celestial sphere, and which move exactly along our visual ray ; such is the star α Cygni, which comes in a straight line towards us with a velocity of 64 kilometres (40 miles) per second, 3,840 kilometres per minute, 230,400 per hour, and 1,382,400 leagues a day, or *more than twelve hundred millions of miles a year.* With this velocity the sun of the constellation Cygnus will arrive near us in 200,000 years, illuminating our sky with a brightness incomparably superior to that of Sirius, and coming, perhaps, to unite its light

[1] [Measures of recently photographed stellar spectra indicate a considerably less velocity—less than one-half.—J. E. G.]

with that of our sun himself; but there is nothing to prove that this motion will continue in a straight line, and besides, at that distant epoch, we shall not be at the same spot in space where we are now.

[These observations of motion in the line of sight have recently revealed the existence of very close double stars, of which the components are so close that the largest telescopes in existence fail to show them as anything but single stars. Among these interesting objects may be mentioned the telescopic double star Mizar, of which the brighter component is shown by the spectroscope to consist of two close components, which revolve round each other in a period of 104 days; β Aurigæ, which has a period of only four days; Spica, which has also a period of four days; and Algol, already referred to (p. 626). From the velocity of motion observed it is possible to calculate the mass of these stars in terms of the sun's mass, and it follows that the mass of Mizar is about forty times the mass of the sun, that of β Aurigæ about five times, that of Spica about two and a-half times, and that of Algol about two-thirds of the sun's mass.—J. E. G.]

Such are the stupendous motions which carry every sun, every system, every world, all life, and all destiny in all directions of the infinite immensity, through the boundless, bottomless abyss; in a void for ever open, ever yawning ever black, and for ever unfathomable; during an eternity without days, without years, without centuries or measures. Such is the aspect, grand, splendid, and sublime, of the universe which flies through space before the dazzled and stupefied gaze of the terrestrial astronomer born to-day to die to-morrow on a globule lost in the infinite night!

What do all these motions, all these distances, all these aspects teach us on the last and greatest problem which remains for us to solve—the structure of the universe?

[1] Have you ever happened to notice from a balcony of the Paris boulevards the crowd of busy beings who move in all directions? Where are they going? Why such a throng? Because each of them is trying to outrun the other and arrive the sooner. A hundred years ago there was the same crowd; a hundred years hence there will be the same ant-swarm. And where do they all run? To death! Thus do all the worlds rush rapidly through space! But can we suppose that all march to universal death?

CHAPTER X

STRUCTURE OF THE VISIBLE UNIVERSE

The Milky Way. The Nebulæ. Clusters of Stars. Secular Metamorphoses. Infinity and Eternity

THE continuous development of astronomical contemplations presented in this work places us at this moment at the summit of a universal panorama, and puts to us the greatest of questions which the study of nature can present to the human mind. Our sun is but a star; it carries us rapidly along—earth, moon, planets, satellites, comets—towards a point in space which we have determined. Each star is a sun, and bears along in the same way through the starry heavens the numberless worlds, the varied humanities which gravitate in their attraction and their light. Shall we some day strike against an extinct sun, hidden like a reef in our passage? Do we all go, nations of Infinitude, towards a common direction where all the powers, all the riches of nature, will one day be collected? Are we to be darkly destroyed without being aware of it? and do all these tremendous motions drive all humanities without an aim into the eternal abyss? Do the suns which surround us form a system with that which illuminates us, as the planets form one round our solar focus? and does our sun revolve round an attractive centre? Does this centre, the point of the revolutions of many suns, itself revolve round a preponderating centre? In a word, is the visible universe organised in one or several systems? No Divine revelation comes to instruct men on the mysteries which interest them most, their personal or collective destinies. We have now, as ever, but science and observation to answer us.

A problem so vast as this is still far from receiving even an approximate solution. From whatever point of view we consider it, we find ourselves face to face with the infinite in space and time. The present aspect of the universe immediately brings into question its past and its future state, and then the whole of united human learning supplies us in this great research with but a pale light scarcely illuminating the first steps of the dark and unknown road on which we are travelling. However, such a problem is worthy of engaging our attention, and positive science has already made sufficient discoveries in the knowledge of the laws of nature to permit us to attempt to penetrate these great mysteries. What is it that the general observation of the heavens, what is it that sidereal synthesis teaches us on our real situation in Infinitude?

In the calm and silent hours of beautiful evenings, what pensive gaze is not lost in the vague windings of the Milky Way, in the soft and celestial gleam of that cloudy arch, which seems supported on two opposite points of the horizon and elevated more or less in the sky according to the place of

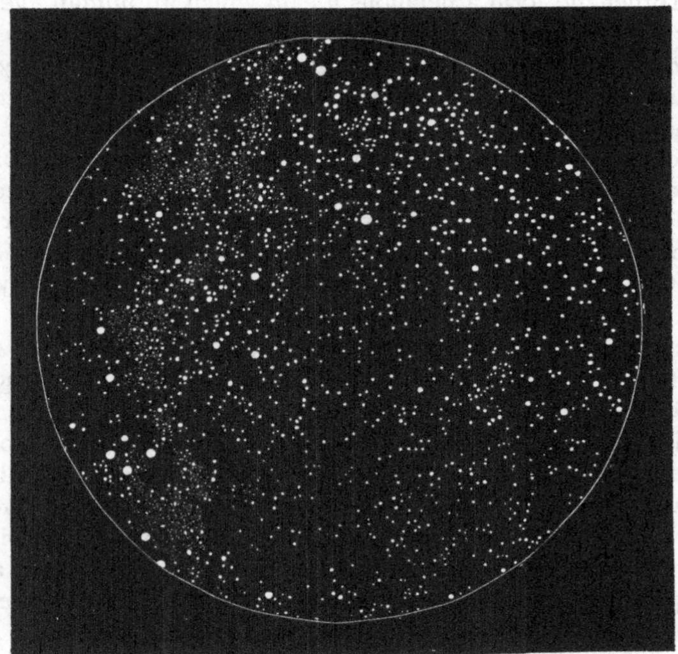

FIG. 275.—A TELESCOPIC FIELD IN THE MILKY WAY

the observer and the hour of the night? While one-half appears above the horizon, the other sinks below it, and if we removed the earth, or if it were rendered transparent, we should see the complete Milky Way, under the form of a great circle, making the whole circuit of the sky. The scientific study of this trail of light, and its comparison with the starry population of the heavens, begins for us the solution of the great problem.

Let us point a telescope towards any point of this vaporous arch: suddenly hundreds, thousands of stars show themselves in the telescopic field like needle-points on the celestial vault. Let us wait for some moments, that our eye may become accustomed to the darkness of the background, and the little sparks shine out by thousands. Let us leave the instrument pointed motionless towards the same region, and there slowly passes before our dazzled vision the distant army of stars. In a quarter of an hour we see them appear by thousands and thousands. William Herschel counted 331,000 in a width of 5° in the constellation Cygnus, so nebulous to the naked eye. If we could see the whole of the Milky Way pass before us, we should see 18 millions of stars.

This seed-plot of stars is formed of objects individually invisible to the naked eye, below the 6th magnitude, but so crowded that they appear to touch each other and form a nebulous gleam which all human eyes directed to the sky for thousands of years have contemplated and admired. Since it is developed like a girdle round the whole circuit of the sky, we ourselves must be in the Milky Way. The first fact which impresses our mind is that *our sun is a star of the Milky Way.*

Now, does this agglomeration of stars really form a sort of circular framework at a distance from us? There is no reason which leads us to imagine it thus, for its aspect to us would be the same whether it were a ring or a layer, a sheet or a plane in which thousands of stars were scattered. The way in which we should naturally represent the Milky Way is, then, *a plane in which the stars are accumulated out to immeasurable distances.* They appear to us to touch because they are projected one before the other. But we need not on that account suppose them to be equally spaced.

Thus the first stars of the Milky Way are near us. The sun is one; *a* Centauri is another; and so, in all directions, for billions and billions of miles, stars succeed each other, arranged chiefly in this remarkable plane. Between two stars of the Milky Way, which appear to touch each other, there may often exist billions of miles in the direction of our visual ray. Many, on the other hand, may be less distant from each other, and form double, triple, quadruple and multiple systems. We shall see directly that there are in reality systems formed of several thousands of stars.

Now, it is found that if we make a telescopic review of the sky for the stars invisible to the naked eye, we ascertain that these stars are more numerous the nearer we approach the plane of the Milky Way: the number of stars from the 10th to the 16th magnitude increases wonderfully and regularly from both poles of the Milky Way up to this zone itself. Thus, the same telescope which counts 122 stars in its field of 15′ diameter (half that of the sun) only includes 30 at 15° from the Milky Way, 10 at 45°, 16 at 60°, and only 4 at the poles of this stellar plane.

We may form an idea of the distribution of the stars in space by an attentive examination of the admirable and curious projection of the stars in the celestial atlas of Argelander made by Mr. Proctor on a planisphere of which fig. 276 is a reduction.[1] Each of these little points represents a

[1] *A Chart of the Northern Hemisphere.* Manchester: A. Brothers. 1871.

star, a sun analogous to ours; there are 324,198, the union of forty charts (one of which has been reproduced above, fig. 259) of this great atlas, which contains all the stars of our northern hemisphere down to the 10th magnitude inclusive, such as may be observed with a telescope of 7 centimetres

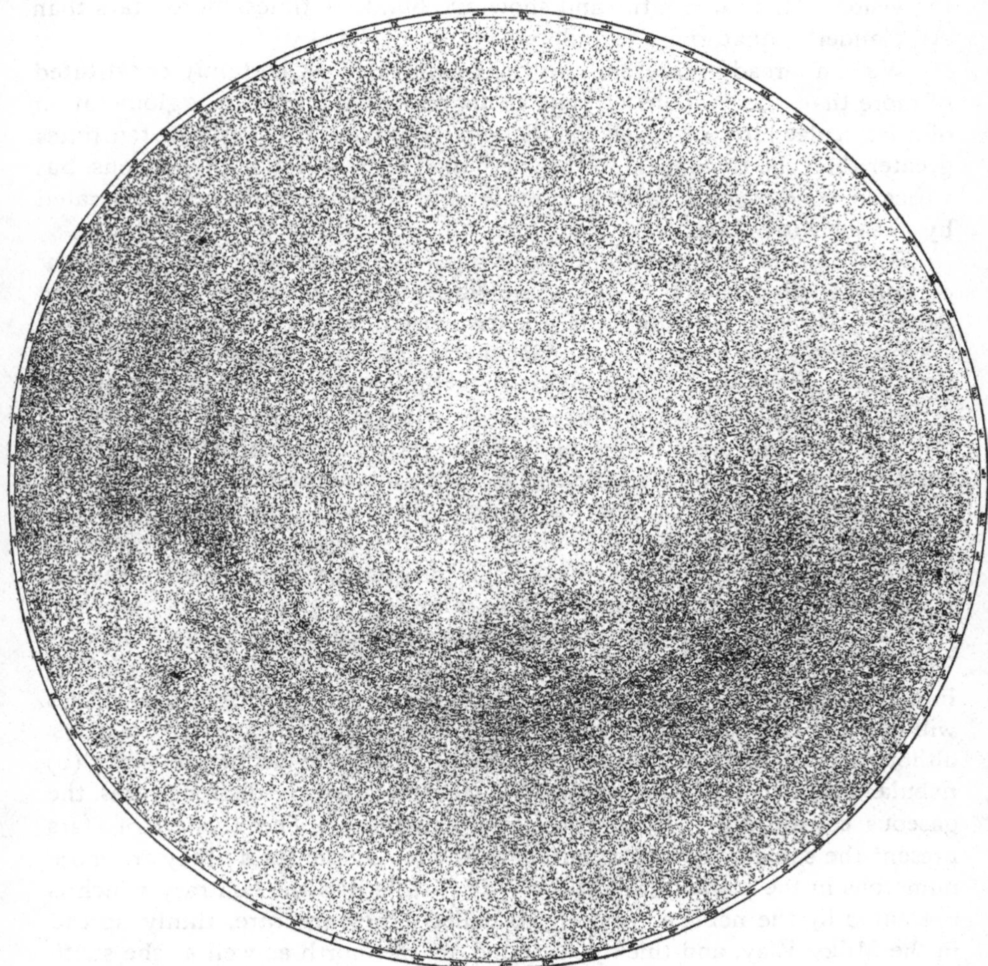

FIG. 276.—DISTRIBUTION OF 324,198 STARS, CATALOGUED ONE BY ONE FOR THE NORTHERN HEMISPHERE, SHOWING A PROGRESSIVE AGGLOMERATION TOWARDS THE MILKY WAY

($2\frac{3}{4}$ inches) in diameter. Each of these stars has its name or catalogue number. We see, on the one hand, a progressive agglomeration of stars towards the Milky Way, on the other hand curious irregularities, regions where the stars are singularly scarce, especially between the Pleiades and the Milky Way.

If our eye were equal to an object-glass of $2\frac{3}{4}$ inches, it is thus that we should see the heavens with the naked eye. But we have here only the first telescopic stars, about a hundred times more than we can count with the naked eye. Our present large telescopes again develop this astounding vision in the same ratio, and show one hundred times more stars than Argelander's equatorial—30 millions in half of the sky!

We can already represent the visible universe as certainly constituted of more than 100 millions of suns arranged in an immense agglomeration of a lenticular form, of which the diameter appears to be eight or ten times greater than the thickness.[1] This agglomeration is not homogeneous, but varies in condensation, and is composed of diverse associations separated by irregular intervals.

Are the suns in general uniformly separated from each other? are they of uniform dimensions, of the same light, the same mass, and the same power? No. An infinite variety is manifested among them. Many form associations, as is proved by the fact that a common proper motion carries them through space, although they may be billions of miles from each other; others are agglomerated in clusters, and perhaps separated by millions of miles only; some are millions of times larger than the earth: others may be, if not of an infinite smallness, at least not exceeding in volume the largest planets of our system, such as Jupiter and Saturn. There may exist suns absolutely isolated. Our sun, which appears so to us, is certainly subject to the attractive influence of his neighbours, and perhaps several march in consort with him towards the same end.

We know in the sky 1,034 clusters of stars and 4,042 irresolvable nebulæ. The former are composed of associated stars; the latter may be divided into two classes: (1) nebulæ which the ever-increasing progress of optics will one day resolve into stars, or which in any case are composed of stars, although their distances may be too great to enable us to prove it; (2) nebulæ properly so called, of which spectrum analysis demonstrates the gaseous constitution. Here is an instructive fact. The clusters of stars present the same general distribution as the telescopic stars: they are more numerous in the plane of the Milky Way; while it is the contrary which is presented by the nebulæ properly so called: they are rare, thinly spread in the Milky Way, and thickly scattered to the north as well as the south of this zone up to its poles. The constitution of the Milky Way, not nebulous but stellar, is a very significant fact. The nebulæ properly so called are distributed, in a sense, contrary to the stars, being more numerous towards the poles of the Milky Way and in regions poor in stars, as if they had absorbed the matter of which the stars are formed. William Herschel had already remarked this; when with his eye to the telescope he saw

[[1] This 'disc theory' of the visible universe is now very improbable, if not wholly untenable.—J. E. G.]

stars begin to become rare, he was in the habit of saying to his secretary, 'Prepare to write ; nebulæ are coming.'

Star clusters present all gradations in the number as well as the condensation of their components. Some are composed of but few stars ; others present associations of tens, hundreds, or even thousands. Among the star clusters visible to the naked eye, the best known —that which humanity has contemplated for so many centuries, and which formerly ruled the astronomical and climatological year of our ancestors—that of the Pleiades, may serve us for the first type, the first

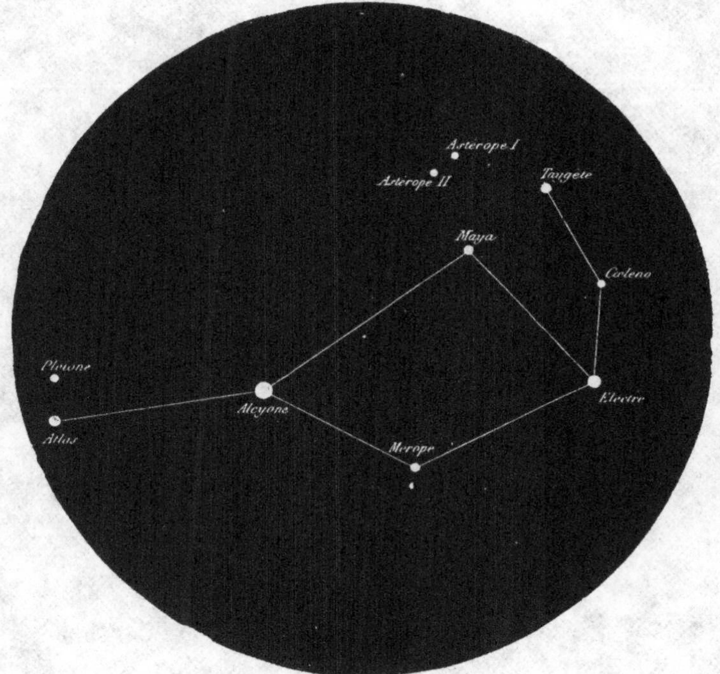

FIG. 277.—THE PLEIADES AS SEEN BY THE NAKED EYE

example, for penetrating into this new world of sidereal richness. Weak sight discerns but a nebulous and indistinct cluster. Ordinary sight distinguishes six stars—Alcyone, of the 3rd magnitude ; Electra and Atlas, of the 4th : Merope, Maia, and Taygeta, of the 5th. Good sight can distinguish a seventh, Pleione, of the 6th magnitude ; very good sight discerns Asterope, a star of the 7th magnitude ; excellent sight can double this star and distinguish Cœleno ; some extraordinary sights have gone as far as thirteen. The seventh (Pleione) appears to have diminished in brightness, for the Greek and Latin historians assert that it vanished at the time of the Trojan War ; but perhaps this legend is due only to the difficulty there always is in distinguishing it.

This cluster of stars, so modest when seen with the naked eye, becomes splendid in a telescope of even small power. We seem to see luminous

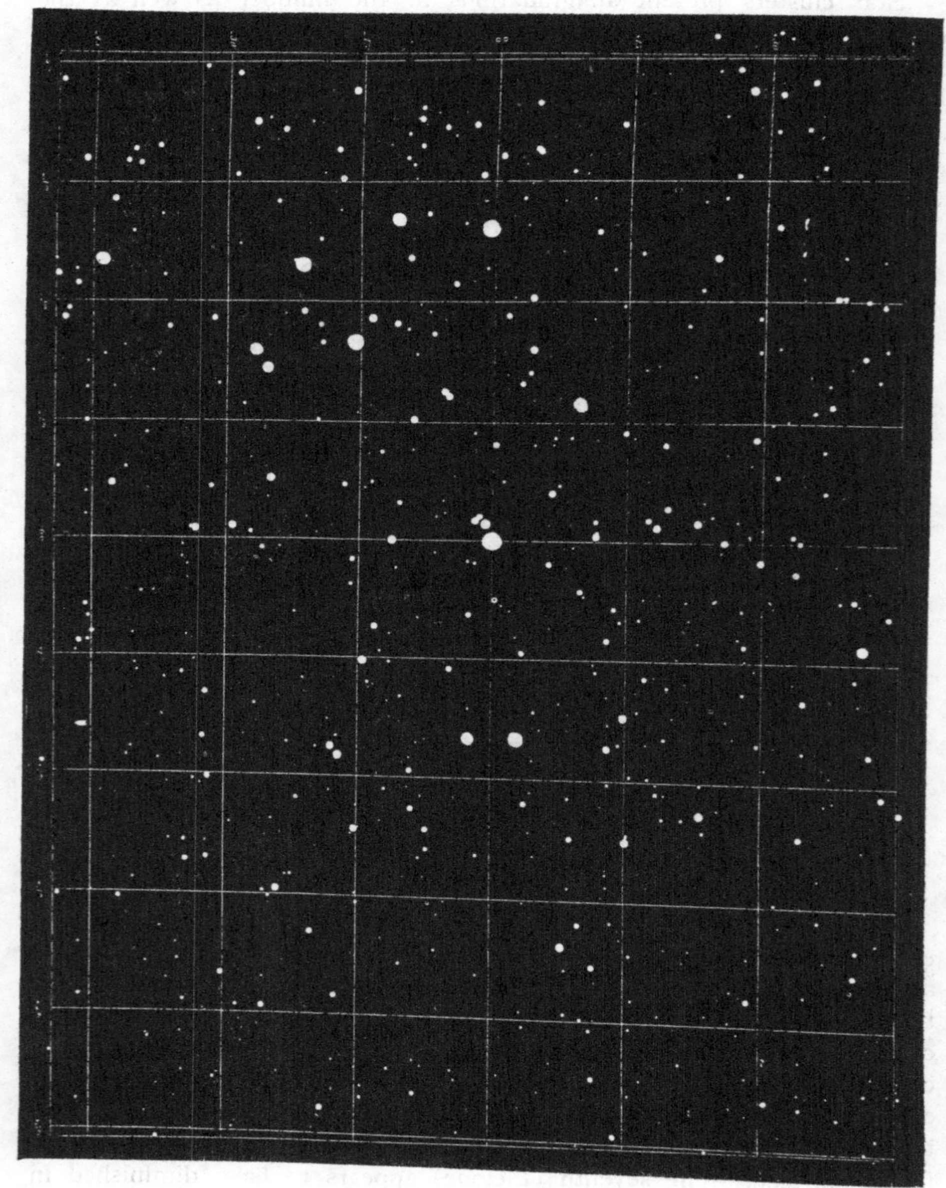

Fig. 278.—The Pleiades as seen in the Telescope

diamonds sparkling on the black background of the heavens ; the weaker the eyepiece, the less the magnifying power is raised, the larger and more luminous the field, the more vivid is the impression experienced by the observer. We imagine there suns of the nature of ours, doubtless surrounded by systems of inhabited planets, from which the nocturnal sky appears as black as seen from here, and which, nevertheless, gravitate in the midst of an agglomeration of nearly 600 suns[1] immensely distant from each other. In a powerful instrument we distinguish a nebula across certain regions of this brilliant cluster of stars. Already the exact position and the precise magnitude of each star is determined, so that after some centuries it can be decided whether notable variations are accomplished in this distant creation. We see this telescopic aspect on the chart which we reproduce in fig. 278, which was constructed at the Observatory of Paris.

The Hyades, which we also admire with the naked eye, near Aldebaran; the cluster of Cancer, which we distinguish in that constellation; the cluster in Gemini, the cluster in Perseus (fig. 279), the cluster in Canes Venatici, and the cluster in Hercules (fig. 281), show us richer and richer collections of suns visible in different regions of the sky. But this is still merely a prelude to that which telescopic vision reserves for us. Who can, for example, contemplate without emotion, even in the incomplete reproduction of a cold engraving, the stellar clusters in the Centaur (fig. 280), or Toucan, formed of *several thousands of stars*. The first is beyond comparison the richest and largest in the whole sky, and shows near its centre a dazzling condensation; the second, which is also visible to the naked eye in the vicinity of the smaller Magellanic Cloud, in a region of the southern sky almost entirely void of stars, extends farther, with stars less condensed; a double star is projected on the cluster, but it is probable that this is much nearer to us, and has no connection with the cluster.

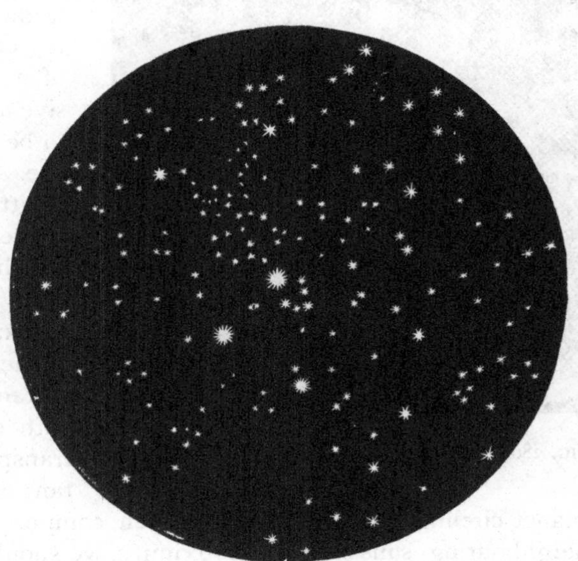

FIG. 279.—A CLUSTER OF STARS IN PERSEUS

[1] [A photograph recently taken at the Paris Observatory shows over 2,000 stars in the Pleiades group.—J. E. G.]

We may estimate that light takes from ten to fifteen years to come from there. In the Southern Cross we admire with unspeakable wonder a brilliant cluster of 110 stars of the 7th magnitude and fainter, of which the most luminous shine with all colours—ruby-red, emerald-green, sapphire-blue ; it is like a casket of glittering gems.

The contemplation of the heavens affords no spectacle so grand and so eloquent as that of a cluster of stars. Most of them lie at such a distance that the most powerful telescopes still show them to us like star dust. 'Their distance from us is such that they are beyond, not only all our means of measurement,' says Newcomb, 'but beyond all our powers of estimation. Minute as they appear, there is nothing that we know of to prevent our supposing each of them to be the centre of a group of planets as extensive as our own, and each planet to be as full of inhabitants as this one. We may thus think of them as little colonies on the outskirts of creation itself, and as we see all the suns which give them light condensed into one little speck, we might be led to think of the inhabitants of the various systems as holding intercourse with each other. Yet, were we transported to one of these distant clusters, and stationed on a planet circling one of the suns which compose it, instead of finding the neighbouring suns in close proximity, we should see a firmament of stars around us, such as we see from the earth. Probably it would be a brighter firmament, in which so many stars would glow with more than the splendour of Sirius as to make the night far brighter than ours ; but the inhabitants of the neighbouring worlds would as completely elude telescopic vision as the inhabitants of Mars do here. Consequently, to the inhabitants of every planet in the cluster, the question of the plurality of worlds might be as insolvable as it is to us.'

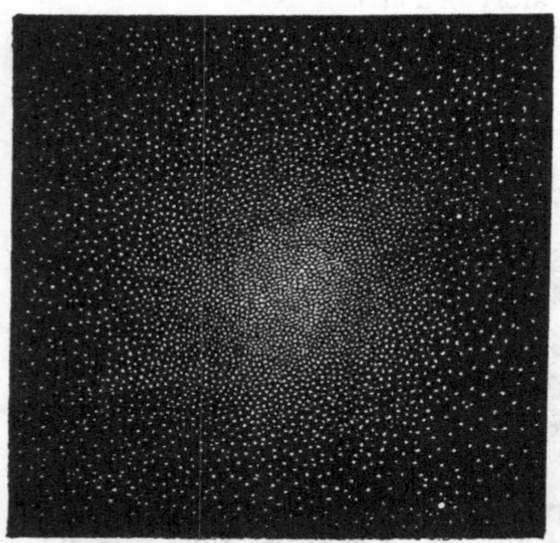

Fig. 280.—A Cluster of Stars in Centaurus

These are clusters of stars of regular form in which attraction appears to mark its secular stamp. Our mind, accustomed to order in the cosmos, anxious for harmony in the organisation of things, is satisfied with these agglomerations of suns, with these distant universes, which realise in their *ensemble* an aspect approaching the spherical form. More extraordinary, more marvellous still are the clusters of stars which appear organised in spirals, and among them appears, splendid and tremendous, the astonishing nebula (fig. 282) situated in the constellation Canes Venatici (quite near the

Fig. 281.—Star Cluster in Hercules (13 Messier)
(From a Photograph by MM. Henry)

Fig. 282.—Spiral Nebula in Canes Venatici

star η of the Great Bear, 3° to the south-west), of which the great telescope of Lord Rosse has revealed the singular structure (a structure confirmed by photography). It seems that the hand of ages has outlined this universe, that the numberless suns collected there are lengthened out into a line and directed towards the central focus, that a second focus is condensed near the confines of this universe, and that the whole is displaced in space, leaving a luminous train behind it. The imagination is confounded in the presence of such a grand spectacle. On the hypothesis of a complete resolvability into stars, the mind is lost in numbering the myriads of suns, the agglomerated individual lights of which produce these nebulous fringes of such different intensities. What must be the extent of this universe, of

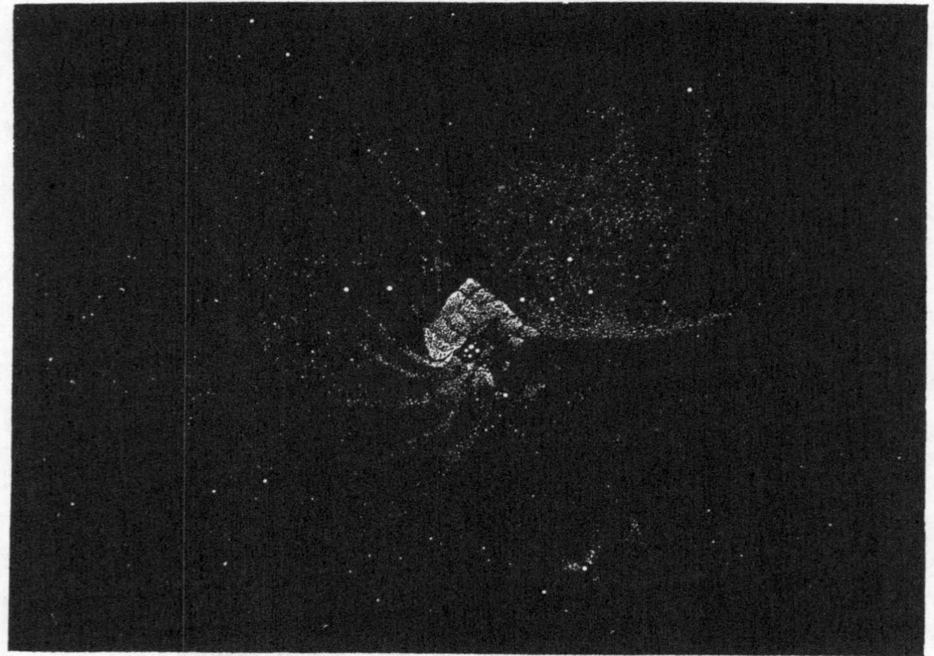

Fig. 283.—Nebula in Orion

which each sun is no more than a grain of luminous dust! Into what a deep abyss does our gaze plunge when we contemplate this distant creation! To what a profundity of time do we go back in regarding it! Is it fifteen thousand years, is it thirty thousand, is it a hundred thousand years of the past which we have now before our eyes? This nebula certainly no longer exists in the state in which its photograph comes to us to-day.

But here we penetrate into a still more mysterious world of nebulæ. Since the epoch when William Herschel expressed the opinion that these clusters are portions of the primitive cosmical matter which has served for the formation of stars now existing, and that in studying them we study at the same time the phases through which suns and planets have passed—especially since the ingenious processes of spectrum analysis permit us to study the chemical composition of these stellar clouds—the interest with which they inspire the astronomer and thinker has increased tenfold, or even one hundred-fold. On a very clear and transparent night of winter, at midnight in December, look below the Belt of Orion and you will distinguish the

FIG. 284.—NEBULA IN SOBIESKI'S SHIELD

mass of nebulous light (fig. 283) which glimmers in that constellation. Take a telescope, even of small power, and you remark the beautiful quadruple star (it is even sextuple), θ Orionis, surrounded by the most curious of nebulæ. Here is no cluster of suns: it is luminous gaseous matter, a little greenish. The spectroscope shows in its spectrum three bright lines sharply defined, and separated by dark intervals. A spectrum of this nature can only be produced by light which emanates from matter in the state of gas. What is this cosmical gas? Its spectrum recalls that of nitrogen; it is probable that *nitrogen* predominates in its constitution, or rather a substance still more elementary, which analysis has not yet discovered. [Recent researches show that this nebula contains *hydrogen*, but not nitrogen.—J. E. G.]. This immense nebula, the finest in the heavens, occupies a space much vaster than our whole planetary system!

Among the nebula of irregular form we also admire that of the Shield of Sobieski (fig. 284), a mysterious creation, on which a great number of suns seem to flash out; we might believe all these stars to be nearer to us than the nebula, and projected on it by perspective, if their odd grouping did not indicate a singular connection with the forms assumed by the nebula itself

In the Great Bear is a round and brilliant nebula (fig. 286), which presents at its centre two stars, each surrounded by a black circle; it resembles the head of an owl. Sometimes one of these two stars ceases to be visible, and the head appears blind of an eye. We see also in the constellation of the Lion

FIG. 285.—DOUBLE NEBULA IN AQUARIUS

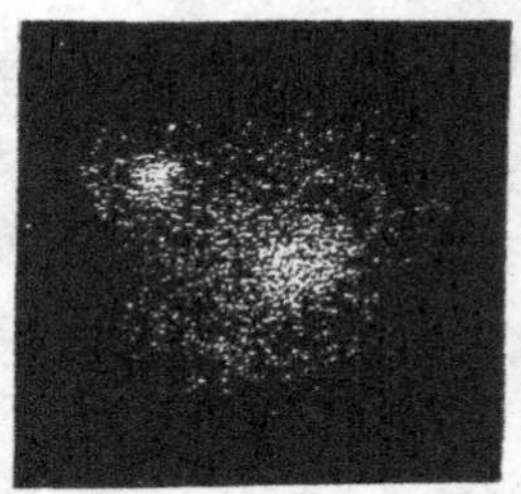

FIG. 286.—DOUBLE NEBULA IN THE GREAT BEAR

an elliptical nebula with a central nucleus surrounded by cloudy envelopes. We may mention also a nebula in the constellation of the Dragon, resembling a brilliant ring surrounded by a vague nebulosity.

The aspect and the chemical analysis of these nebulæ have brought again into favour the hypothesis of cosmical matter originally scattered through all space. A first condensation of this diffuse matter produces clouds of vapours or simple nebulæ. By a subsequent condensation one or more nuclei are formed in these nebulosities. These nuclei, attracting the surrounding matter, gradually increase and become stars, which afterwards by their mutual attraction approach each other, and group themselves into stellar clusters. We thus see nebulæ at all ages of their organisation. In order to develop in the gases lines as clear and sharp as those revealed by spectrum analysis, ordinary combustion accompanied by a feeble disengagement of heat would not suffice; on the contrary, a very elevated temperature is necessary, like that produced by the electric focus. We may conclude that the fluids which constitute the nebulæ are in a state of vivid incandescence, at a temperature at least as elevated as those which we can raise. The depths of space, which are usually presented to our mind as the seat of a glacial silence, similar to that of death, are then, on the contrary, in a state of tremendous activity which our imagination can hardly conceive. Thus suns are prepared which one day, when sufficiently condensed and cooled, will rule and illuminate a certain number of planets. The planetary nebulæ seem to be bodies already very far advanced in this way of formation. We know a compound body, of which the position is 19 hours 40 minutes of right ascension and 50° 6′ of northern declination; this is a star surrounded by a nebulous atmosphere, presenting at the same time two spectra—which seems to indicate an intermediate phase of sidereal formations.

Several nebulæ present forms which correspond to the transformations which we have studied in treating of the origin and the end of worlds. We have represented three, among others (p. 73), which show the phases of condensation, of rotation, and of detachment of rings, through which solar and planetary creations should pass according to the most probable cosmogonic theory. The spectrum of these nebulæ indicates in the first place the presence of nitrogen and hydrogen [hydrogen, but not nitrogen. —J. E. G.]

Mysterious objects, voices of the past, prophecies of the future, these soft and pale gleams open to the mind new perspectives in Infinitude; the first telescopic observers of the sky, who treasured the memory of the Empyrean, described them as openings through the celestial vault, permitting our gaze to penetrate to the light of Paradise. The types on which we have fixed our attention give still but an incomplete idea. We should add the lenticular and elliptical nebulæ; the perforated nebulæ; nebulous rays; the great cloud of Magellan, at 20° from the South Pole, which contains 291 nebulæ, 46 stellar clusters, and 582 stars, and covers 42 square degrees of the sky; the smaller cloud, which occupies 10 square degrees, contains 200 stars, 37 nebulæ, and 7 clusters; and, not far from that, the 'coal sacks,' regions entirely void of stars, yawning openings in the sidereal universe, as if a waterspout had devastated them; and, again, the faintest nebulæ lost in the depths of the sky—whose light would take, according to the Herschelian estimates, two millions of years to reach us!

Some nebulæ have varied perceptibly in form and brightness in less than a century since they have been observed with attention. One of the most curious examples of this fact is that which is presented by the nebula discovered in 1852 in the constellation Taurus by Hind. My lamented friend Chacornac, who had examined it at the Paris Observatory in 1854, was quite surprised not to find it again in 1858 and 1862. However, in 1865 and 1866 it was again observed with ease by D'Arrest; and now it has again disappeared so completely that it is quite invisible. A star near it is subject to the same phases. What can be the explanation of

FIG. 287.—SIR JOHN HERSCHEL'S DRAWING OF THE GREAT NEBULA ABOUT η ARGUS AS SEEN WITH HIS 24-FOOT REFLECTOR AT FELDHAUSEN, CAPE OF GOOD HOPE, IN THE YEARS 1834–1837. (Copied and reversed from the Plate in 'Cape Observations')

such a metamorphosis? This nebula is doubtless as vast as our entire solar system. Does it shine only by light reflected from a neighbouring

FIG. 288.—THE η ARGUS NEBULA. (From a Photograph taken by Mr. H. C. Russell with the Astro-photo Telescope of 13-inch aperture used for the International Survey of the Heavens)

sun, and is this sun variable like those which we have already studied? Does an immense opaque cloud revolve round this nebula and hide it from us periodically?

This is not the only case of the kind. Another nebula, situated in the Whale, was observed by both Herschel and Lord Rosse; then it became completely invisible in an instrument superior to that which had shown it five years previously. It was re-observed in 1863 and 1864; but it disappeared again in 1865. It was again seen in 1868 and 1877. Winnecke observed it without difficulty at Strasburg. Can it be a periodical variable nebula?

Another nebula, situated in the Dragon, observed for the first time by Tuttle in 1859, appeared very bright in 1862, less brilliant in 1863, and it became invisible in the finder in which it was admirably seen in 1862.

Another fact no less curious: a nebula in Scorpio, which bears the number 80 in the catalogue of Messier, was transformed into a star between May 9 and June 10, 1860; then, at the latter date, it had become again nebulous. Three different observers, Pogson, Luther, and Auwers, verified the change.

The observations and drawings of the singular nebula which surrounds the variable star η of Argo (figs. 287, 288); of the nebula in Orion (fig. 283), which sometimes appears agitated like the surface of the sea; of the nebula in the Shield of Sobieski, of which the form as drawn by Sir John Herschel in 1833 recalls the Greek letter Ω, while that of our fig. 284, due to Lassell in 1862, gives quite another picture, seem to indicate also notable variations in these distant creations. But these variations do not show the same degree of certainty as the preceding.

The nebulæ thus open to the fields of the imagination a space no less vast than that of the world of stars. We have met with *variable nebulæ*; we now come to *double nebulæ*. These cosmical gaseous masses are doubtless the original state, the primitive chaos, the genesis of double stars and of multiple systems, the constitution of which we have studied above. We there see, doubtless, new creations of worlds; but, as we have already remarked, light, that active messenger, brings us but tardy news of celestial phenomena, and now, perhaps, these nebulæ are condensed into suns and planets. Already some of these double nebulæ manifest traces of a slow orbital motion round each other, or of a relative displacement in space.

These are lights which glimmer on the frontiers of creation; they are the beginnings which show us the birth of other universes; they are the voices of the past which speak to us from the depths of the vanished ages. The heavens show us both cradles and graves: here humanities are born; there, arrived at their apogee, they measure the infinite in their vast contemplation; farther on they become consumed in the celestial fire or are lulled into the lethargy of the final ice ; this is the great history of the heavens—the veritable *universal history*.

Having reached these upper heights, we can now attempt to represent the constitution of the heavens as a whole.

In infinite space the stars are strewn in immense clusters, like archipelagoes of islands in the ocean of the heavens. To go from one star to another in the same archipelago light takes years ; to pass from one archipelago to another it takes thousands of years. Each of these stars is a sun similar to ours, surrounded, doubtless, at least for the most part, by worlds gravitating in its light ; each of these planets possesses, sooner or later, a natural history adapted to its constitution, and serves for many ages as the abode of a multitude of living beings of different species. Attempt to count the number of stars which people the universe, the number of living beings who are born and die in all these worlds, the pleasures and pains, the smiles and tears, the virtues and vices ! Imagination, stop thy flight !

Now, should we consider the whole of the visible universe, the solar system, single stars, double stars, multiple stars, clusters of stars, and nebulæ, as forming one vast sidereal system, composed of partial systems ? Seeing the planets revolving harmoniously round the sun, the philosophers Kant and Lambert, in last century, advanced the hypothesis that the stellar universe may be constructed on the same plan, and that each star may describe in space a closed orbit. This was a theory of which observation alone could estimate the value. William Herschel and William Struve undertook the work, and the result of their observations is opposed to this view—doubtless too simple. No star presents itself showing a preponderance sufficient to serve as a central sun ; and, on the other hand, if this central sun were dark (which would be difficult to admit), the motions of the stars round it would be interpreted for us by a certain regularity in the proper motions. Such is not the general case. If, on the other hand, we examine in detail the motions of some particular stars, we find that the hypothesis of these regular orbits is the most improbable of all.[1]

[1] Let us consider, for example, the star of which the proper motion is the most rapid (1830 Groombridge). Its parallax is certainly less than a tenth of a second—that is to say, the sun's distance from the earth is reduced, as seen from there, to less than a tenth of a second. But the proper motion of this star is seven seconds a year—that is to say, more than seventy times its parallax. It follows that this star is displaced in the sky every year by a space at least seventy times greater than the distance which separates us from the sun ; it would pass over this distance in five days, so that its velocity certainly exceeds 32,000 metres (198 miles) a second. But we have seen that the velocity which a body acquires in falling towards a centre of attraction can be calculated for each point of its course. For example, a body coming from infinity towards the earth, and attracted by the earth alone, would reach us with a velocity of only 11,300 metres (7 miles) in the last second. Conversely, shot from the earth with this velocity, it would never fall back. We have already gone into these considerations several times in the course of this work, and we have discussed and elucidated them. If we knew the masses of all the stars and their arrangement in space, we could

Moreover, the general result appears to be that the stellar universe does not possess in itself the conditions of stability which we have recognised in the solar system; all appears to rush through the boundless infinite. If there were no motion in the stars, they would all fall in time towards a common centre, would all unite in a single mass, and this would be the universal and definitive ruin of the entire universe. But the motions with which we see them animated prohibit such a catastrophe, since each star has in store a quantity of force sufficient to prevent it from submitting passively to the attraction of its neighbours. If, then, any star falls towards a centre of attraction, the velocity which it acquires by this fall throws it back in a new direction, and thus it continues to sail through Infinitude without our being able to foresee a certain collision.

We may add here that there may exist round our visible universe an immense space, absolutely void and desert, beyond which, at immeasurable distances, lie other universes. And so on.

Yes, the visible universe, with its hundred millions of suns, represents but an infinitesimal part of the total universe, of the infinite; it is a village in a province, and even less. On the other hand, the millions of years, or even the millions of centuries, by which we attempt to express the progressive development of the nebulæ, of suns, and of worlds, represent but a short moment in the eternal duration. We can, then, in attempting to imagine these grandeurs, only recognise the insufficiency of our field of observation, and be impressed with the conviction that the universe is incomparably vaster, more amazing, and more splendid than all that science can reveal to us and all that the imagination can dream of.

If all these suns were really fixed, motionless, sphinxes of eternity, immutable and unchangeable, each a king in his own imperishable domain, I do not know whether the aspect of the universe would not be as imposing and as grand. But it would be less living. *Mens agitat molem.* All these stars as vast as our sun, separated from each other by unfathomable distances,

calculate in the same way the maximum velocity which a body would acquire in falling from an infinite distance towards any point of the stellar system; and if we found that a star moves more quickly than this velocity, we should conclude that this star does not belong to the visible universe, and that it is simply a visitor coming from Infinitude and incapable of being stopped by the combined attraction of all the known stars.

Now, let us suppose that there are 100 millions of suns in our universe, and that on the average each of them is five times heavier than ours, and that our universe has a diameter equal to the distance passed over by light in 30,000 years. A body falling from infinity to the centre of this stellar system would be animated with a velocity of 40,000 metres (24·8 miles) a second, according to the calculations of Newcomb. Now, this is only the eighth of the probable velocity of the star 1830 Groombridge, and to produce eight times this velocity would require an attractive mass 64 times stronger than that which we have allowed for. From this simple consideration results the following dilemma: either the stars which compose our universe are more numerous and heavier than the telescope seems to indicate, or the star 1830 Groombridge does not belong to our universe; it is passing through it, and the united attraction of all these bodies cannot stop it. We cannot decide between these two vast hypotheses.

succeeding each other to infinity in the immensity of space, are in motion in the heavens. Nothing is fixed in the universe; there is not a single atom in absolute repose. The tremendous forces with which matter is animated universally rule its action. These motions of translation of the suns of space are imperceptible to our eyes because they are performed at too great a distance; but they are more rapid than any velocity observed on earth. To the eye which could neglect time as well as space, the sky would be a veritable swarming of different stars, falling in all directions of the eternal void. The star which is our sun comes from the constellation Columba, penetrating farther and farther every day, every year, every century, into the immensities for ever open in space.[1]

It is a strange and unexpected fact, but absolutely true, that each sun of space is carried along with a velocity so rapid that a cannon-ball represents rest in comparison; it is at neither a hundred, nor three hundred, nor five hundred yards per second that the earth, the sun, Sirius, Vega, Arcturus, and all the systems of Infinitude travel: it is at ten, twenty, thirty, a hundred thousand yards a second; all run, fly, fall, roll, rush through the void—and still, seen as a whole, all seems in repose. Let us take a stone, a block of granite, or, say, a block of solid iron; each of the molecules of this piece of iron is displaced, vibrates, moves with a velocity incomparably greater than a star, that sidereal molecule. If we represented, in a system as large as Paris, the sun and the stars whose distance is known, and should set in motion stars, planets, satellites, comets, each on the scale adopted, all would appear at rest, even to the

[1] Since they have been observed, these motions have been performed exactly in a straight line. If each star moves in an orbit, these orbits are of such extent that we cannot yet perceive any curvature in the short arc described since the beginning of observations. From rigorous observation no orbit of any kind is divined. The German astronomer Mädler has placed in the Pleiades the supposed centre round which are performed the orbits of the sun and of the stars which surround us. This would be the centre of gravity of the universe; but this theory does not rest on any serious foundation. The stars appear to move in all directions and with the most varied velocities. Many centuries must ye elapse before we shall be able to form any theory on this point.

The Milky Way appears to indicate the plane towards which the telescopic stars are accumulated. It is not quite the same with the bright stars. Take a compass, open it at a right angle, place one point on Fomalhaut, and draw an arc of a great circle; it cuts the Milky Way in the constellation Perseus, passes near Capella, crosses Hercules near the point to which the sun is moving, almost touches Vega, Aldebaran, β Centauri, cuts the Southern Cross, and passes between Sirius and Canopus. This zone contains the principal stars of the first four magnitudes. This might well be the plane of the sun's orbit, and if we describe an orbit, it might well be round Perseus.

Quite recently Mr. Maxwell Hall, an astronomer in Jamaica, took up the problem, and rejecting the Pleiades and Perseus as centres, concluded in favour of a point situated near the orbit of the double star of the 6th magnitude, 65 Piscium. The angular velocity of the sun would be 0″·066 a year, and its complete revolution would require no less than 20 millions of years for its performance; the total mass of the attractions which the sun obeys would be 78,000,000 times greater than its own, and composed of millions and millions of stars. All the stars which we know would revolve with the sun round the same centre, and thus constitute a single sidereal system—an hypothesis to be verified.

microscope! Where is the great? Where is the little? Where is motion? Where is rest? A marble is as large as the universe. A cubic centimetre of air is composed of a sextillion of molecules. If in thought we lay these in a line, a millimetre apart, there will be a thousand along a metre, a million in a kilometre, a thousand millions for a thousand kilometres, and our sextillion of molecules will occupy a length of 250 trillions of leagues (620 billions of miles), reaching from here to the stars, and those not the nearest! Now, these molecules of a cubic centimetre of air really exist, are agitated, vibrate, rotate, rush along, like suns in space; they also form a universe. Man is placed between two infinites; we live without reflecting in the midst of the sublime.

How do such contemplations enlarge and transfigure the vulgar idea which is generally entertained of the world! Should not the knowledge of these truths form the first basis of all instruction which aims at being serious? Is it not strange to see the immense majority of human beings living and dying without suspecting these grandeurs, without thinking of learning something of the magnificent reality which surrounds them?

Let us now re-examine rapidly the road we have passed over in this volume. We live on the earth, a floating, rolling, whirling globe, the sport of over ten incessant and varied motions; but we are so small on this globe and so distant from the rest of the universe that all appears to us motionless and immutable. Night, however, spreads its veil, the stars are lit up in the depths of the heavens, the evening star is resplendent in the west, and the moon pours out in the atmosphere her rosy light. Let us go, let us rush with the velocity of light. In a little more than a second we pass in view of the lunar world, which spreads before us its yawning craters and reveals its alpine and savage valleys. We do not stop. The sun reappears, and permits us to cast a last look at the illuminated earth, a little inclined globe slowly shrinking in the infinite night. Venus approaches, a new earth, equal to ours, peopled with beings in rapid and animated motion. We do not stop. We pass sufficiently near the sun to perceive his tremendous explosions; but we continue our flight. Here is Mars, with its mediterraneans with a thousand indentations, its gulfs, its shores, its great rivers, its nations, its strange towns, and its active, busy populations. Time presses: we cannot stop. An enormous Colossus, Jupiter approaches. A thousand worlds would not equal it. What rapidity in its days! what tumults on its surface! what storms, what volcanoes, what hurricanes in its immense atmosphere! what strange animals in its waters!—humanity has not yet appeared on the scene. Let us fly, for ever fly! This world as rapid as Jupiter, girdled with a strange ring, is the fantastic planet Saturn, round which revolve eight globes with varied phases; fantastic also appear to us the beings which inhabit it. Let us continue our celestial flight.

Uranus, Neptune, are the last-known worlds which we meet in our voyage. But let us fly, for ever fly! Wan, dishevelled, slow, fatigued, glides before us the wandering comet in the night of its aphelion; but we still distinguish the sun like an immense and brilliant star in the midst of the population of the sky. With the constant velocity of 186,000 miles a second, four hours have sufficed to carry us to the distance of Neptune; but we should take several days to fly through the cometary aphelia, and for several weeks, several months, we continue to traverse the solitudes which surround the solar family, only meeting with comets which travel from one system to another, shooting stars, meteorites, débris of ruined worlds erased from the Book of Life. We fly, still fly—for four years!—before reaching *the nearest sun*, grand furnace, double sun, gravitating in cadence and pouring out around it in space a more intense light and heat than those of our own sun. But let us not stop: let us continue our voyage for ten years, twenty years, a hundred years, a thousand years, with the same velocity of 186,000 miles a second! Yes, for 1,000 years, without stoppage or delay, we traverse, we examine on our journey these multiple systems, these new *suns* of all magnitudes, prolific and powerful foci, stars whose light is lit up and extinguished, these innumerable families of *planets*, varied, multiplied, distant worlds, peopled with unknown beings of all forms and nature, these many-coloured *satellites*, and all these unexpected celestial landscapes. Let us observe these sidereal nations; let us greet their labours, their works, their history; let us divine their morals, passions, and ideas; but let us not stop. A thousand more years are presented to us to continue our voyage in a straight line. Let us accept and employ them, let us pass through all these clusters of suns, these distant universes, these blazing nebulæ, this Milky Way torn into shreds, these tremendous systems which succeed each other through the ever-yawning immensity; let us not be surprised if the suns which approach, or the distant stars, rain down before us tears of fire, falling into the eternal abyss. We are present at the destruction of globes, the ruin of decaying earths, and the birth of new worlds; we behold the fall of systems towards the constellations which call them; but we do not stop! Another thousand years, another ten thousand, another hundred thousand years of this flight, without slackening, without dizziness, always in a straight line, always with the same velocity of 186,000 miles a second. Let us imagine that we thus sail during 1,000,000 years. Are we at the confines of the visible universe? See the black immensities we must cross! But yonder new stars are lit up in the depths of the heavens. We push on towards them: we reach them. Again a million of years: new revelations, new starry splendours, new universes, new worlds, new earths, new humanities! What! never an end, no vault, never a sky which stops us! for ever space! for ever the void! Where, then, are we? What road have we surveyed? We are *at the vestibule of the infinite!* We have not advanced *a single step*; we are always at the same point: the centre is everywhere, the circumference nowhere. Yes, see opened before us the infinite, of which the study is not yet begun! We have seen nothing; we recoil in terror; we fall back astounded, incapable of continuing a useless career. Well, we might fall, fall in a straight line into the yawning abyss, fall for ever, *during a whole eternity*; never, never should we reach the bottom, any more than we have attained the summit. What do I say? Never shall we even approach it! The nadir becomes the zenith. Neither above nor below, neither east nor west, neither right nor left. In whatever way we look at the universe, it is INFINITE IN ALL DIRECTIONS. In this Infinitude the associations of suns and of worlds which constitute our visible universe form but an island of a great archipelago, and in the eternity of duration, the life of our proud humanity, with all its religious and political history, the whole life of our entire planet is but the dream of a moment!

Let us stop before these contemplations. We are still, it is true, but at the threshold of the temple. The sidereal riches only begin to be unfolded before our eyes; the wealth of the heavens surrounds us, the constellated universes open to our steps, the panoramas of celestial nature charm and captivate our studious contemplation; but the conclusion of this work here comes to put its *veto*, and, like the curator of a splendid museum, who, deaf to the admiration of the visitor, pitilessly drives him out when closing the door at the appointed hour, so this last page of 'Popular Astronomy' takes advantage of its position to say to us: 'You shall go no farther!' But it is mistaken. The museum is not entirely visited; there are secret doors, hidden outlets, and annexes where are kept in reserve precisely the most interesting, and often the most-wished-for, curiosities. Let us go out, then, because we must leave; but we find ourselves again under the azure cupola. The constellations, the maps of the sky, the catalogues of curious stars, variable, double, and coloured, the description of instruments accessible to the student of the heavens, and useful tables to consult, are so many important chapters which have only been touched upon in our scheme. The reader whose scientific desires are satisfied by the possession of the *elements* of our beautiful science may stop here; he whose passion is more lively, and who still thirsts for the grand spectacles and the divine beauties of nature, he who wishes to enter into a more intimate connection with truth, can go farther and complete his astronomical instruction. It is sweet to live in the sphere of the mind; it is sweet to despise the rough noises of a vulgar world; it is sweet to soar in the ethereal heights, and to devote the best moments of life to the study of the true, the infinite, and the eternal!

CHAPTER XI

THE PROGRESS OF SCIENCE

Permanent Study of the Sky. Observation. Instruments. Foundation of a Monthly Review of Popular Astronomy. Foundation of the Astronomical Society of France and of the British Astronomical Association

MANY persons would like not only to learn astronomy, to be initiated into the knowledge of the universe by the reading of suitable works, but even to observe for themselves, by the aid of modest but sufficient instruments, the principal wonders of the heavens. During the course of publication of this work I received a great number of letters manifesting this desire.

In fact, it is difficult to understand that of all the normal schools, all the colleges, lyceums, seminaries and convents, none of these establishments possesses a small observatory where students could interest themselves in the things of the sky.[1] There are, however, professors there who should love the sciences in general and adorc astronomy in particular. It is also difficult to understand how so many fortunate men who live under our sky, and who have often too much leisure, so seldom think of giving themselves the pleasure of observing the celestial wonders, instead of using their wealth invariably in the same circle ; uselessly increasing incomes already superfluous, in order to run horses or entertain actresses. We must believc that no one suspects the captivating interest which is attached to the study of nature, nor the pleasure which the spirit experiences in putting itself in correspondence with the divine mysteries of creation. And yet, where is thc intelligent being, where is the being accessible to the emotions inspired by the contemplation of the beautiful, who can regard, even in a telescope of very feeble power, the silvery indentations of the lunar crescent quivering in the azure, without experiencing the most lively and

[1] Children are quite ignorant of them. The other day, at the school of the Legion of Honour at St. Denis, a young girl having said to her companions that there were coloured stars, none of them would believe it, and the mistress, being appealed to, declared that it was pure imagination. At the Polytechnic School the greater part of the course of astronomy is spent in geodesy. In the lyceums the pupils learn nothing of astronomy.—*Note to the Hundredth Edition*, 1889.

agreeable impression, without feeling himself transported towards this first stage in celestial voyages and detached from the vulgar things of earth? Where is the thoughtful mind that can see without admiration the brilliant Jupiter, accompanied by his four satellites, entering the field of a telescope inundated with its light, or the splendid Saturn moving along surrounded by his mysterious ring, or a double sun, scarlet and sapphire, revealing itself in the midst of the infinite night? Ah, if men knew, from the modest cultivator of the fields, from the laborious worker in the towns, up to the professor, the man of property, the man raised to the most eminent rank of fortune or of glory, or the woman of the world apparently the most frivolous—yes, if they knew what intimate and profound pleasure awaits the contemplator of the heavens, France, the whole of Europe, instead of being covered with bayonets, would be covered with telescopes, to the great advantage of peace and universal happiness!

But we have not yet reached this stage. However, I have received so many inquiries relative to the most simple means to be used for observing the principal curiosities of the sky, that I do not think I can be wrong in completing 'Popular Astronomy' with some practical hints, commencing with the most elementary instruments, and graduating them progressively in order to satisfy the appetite, 'which comes by eating,' as an old proverb says. The intellectual appetite is even more insatiable than the material, for the latter always ends by being appeased sooner or later, while the former develops as it feeds. The mind is never satisfied.

It is not necessary to possess complicated and costly instruments in order to begin this study, and we may even remark that a large number of discoveries in physical astronomy have been made by amateurs with the aid of very modest instruments. Moreover, the progress of manufacture has been so rapid that we can now obtain good instruments at a price very much less than is generally supposed. The following are those by the aid of which the study of the heavens can be easily commenced.

Before referring to optical instruments, properly so called, we may mention that with the aid of a good opera-glass we can observe certain truly remarkable celestial spectacles, especially the extremely rich agglomerations of stars in the white regions of the Milky Way, Berenice's Hair, the Hyades, the Pleiades, the cluster in Cancer (the Beehive), the nebula in Andromeda, the clusters in Perseus and Hercules, and the brilliant neighbouring stars, or the fine and very wide double stars.

Astronomical and Terrestrial Telescopes

No. 1

Diameter of object-glass	2¼ inches
Focal length	33 "
A terrestrial eyepiece magnifying . .	35 times
A celestial eyepiece magnifying . .	90 "
Mounting of copper and tripod of iron.	

Principal Uses

Observation of the moon—craters, seas, and mountains; satellites of Jupiter; great solar spots; ring of Saturn (quite small); phases of Venus; Pleiades, clusters in Hercules and Perseus; nebulæ in Orion and Andromeda; stars to the 8th magnitude. The separation of double stars may be attempted down to 2″·3 of distance; but in order to divide them clearly in ordinary atmospheric conditions and to have a fine view of them we should not descend below 5″ nor choose pairs of which the principal star is of the 1st or 2nd magnitude and its companion of the 7th or fainter.

No. 2

Diameter of object-glass	2·4 inches
Focal length	35 „
A terrestrial eyepiece magnifying . .	40 times
A celestial eyepiece magnifying . .	100 „

Principal Uses

Observation of lunar rings, seas, mountains, and craters; satellites, belts, and flattening of Jupiter; spots on the sun; ring of Saturn; phases of Venus; Pleiades, clusters Hercules, Perseus, Gemini, Canis Major and Serpens; nebulæ of Orion, Andromeda, Virgo, Taurus, and Leo; stars to 8½ magnitude. The separation of double stars may be attempted down to 2″·0 of distance; but we shall not succeed below 4″·6 if the atmosphere is not excellent, and if the principal star is too bright or the second too small.

No. 3

Diameter of object-glass	3 inches
Focal length	39 „
A terrestrial eyepiece magnifying . .	50 times
Two celestial eyepieces magnifying . .	80 and 150 times
Mounting of copper and stand of iron.	

Principal Uses

Observation of lunar rings, craters, peaks, and special characters of selenographical topography; satellites of Jupiter—flattening, belts, and clouds of that planet; sun spots; Saturn, ring and one satellite; phases of Venus and Mercury; Mars—polar spots; Uranus—small disc; Pleiades—clusters in Hercules, Perseus, Gemini, Canis Major, Serpens, Auriga, Ophiuchus; nebulæ of Orion, Andromeda, Virgo, Taurus, Leo, Lyra, Canes Venatici, and Berenice's Hair; stars to the 9th magnitude [10th or 11th.—J. E. G.] Double stars may be attempted down to 1″·7; but we shall not succeed below 4″·0 if the atmosphere is not excellent, and if the principal star is too bright and the second too small.

No. 4

Diameter of object-glass	3·75 inches
Focal length	51 „
Terrestrial eyepiece magnifying . .	60 times
Three celestial eyepieces magnifying .	80, 150, and 200 times
Mounting of copper and stand of cast iron.	

Principal Uses

Study of lunar topography—craters, peaks, details of landscapes and rills; variable aspects of Jupiter—clouds and spots; the sun—spots, penumbræ, faculæ; Saturn—division of the ring into two, two satellites; phases and indentations of the crescent of Venus; phases of Mercury; polar snows and principal spots on Mars; small planets; disc of Uranus; principal clusters of stars; principal nebulæ; stars to the 10th magnitude [11½.—J. E. G.] Double stars may be attempted down to 1″·3; but we shall not succeed below 3″ unless under exceptional conditions.

No. 5

We can make—and, indeed, they have been made—very interesting observations by the aid of small telescopes. But the proper instrument of study for the amateur astronomer who wishes seriously to begin the practice of astronomy is the telescope of 4¼ inches, which has become the principal equipment of every private observatory. This telescope, of which the focal length is 1·60 metre (63 inches), is mounted on a solid stand of cast iron, and is furnished with a finder to bring the desired star into the field. Three celestial eyepieces magnifying 100, 160, and 250 times, and a terrestrial eyepiece magnifying 80 times. This telescope enables the observer to travel in the moon in the midst of a spectacle always new: the craters are discerned, the peaks cast their fantastic shadows, and the smallest details are revealed to the astonished eye; Saturn is dazzling to the thoughtful mind, and we can well distinguish the two rings; Jupiter permits us to perceive the details of his atmosphere; Mars, the observations of his principal spots and his polar snows; Uranus shows a perceptible disc; all the important nebulæ of the heavens, all the really interesting clusters of stars, are visible, and the double, triple, multiple stars may be studied to those as close as 1″. The eye penetrates the sidereal sky to the 12th magnitude.

We see that in our time the practice even of the most beautiful of the sciences is accessible to all; we have now, so to say, but the perplexity of selection. To the five instruments just described we may add a sixth, as powerful as the third, although smaller and much more convenient for handling—the Foucault telescope of 3·94 inches aperture and 24 inches in length. There is but one essential point to recommend to those who may wish to purchase it; this is to learn to re-silver the mirror themselves, an operation inexpensive in itself (a few shillings), but rather delicate, and to be renewed every two or three years. Some readers may perhaps wonder at our here entering into such small details; but practice proves that they all have their importance and their relative value, and I have wished to neglect nothing to render astronomy *truly popular*. 'Honi soit qui mal y pense!'

Such are the first steps to be taken in the direct and practical study of the universe. It is now no longer a hidden science; the roads of the sky are open to everyone; everybody can study the splendid reality in the midst of which most men have lived up to the present like blind persons. Astronomy is the true integral science; it alone enables us to live in Immensity and makes us indulgent to human littleness; it alone makes us appreciate the insignificance of material life, the grandeur of the intelligence, and the intellectual beauty of the universe.

It has been remarked that from the time of the first edition of this 'Popular Astronomy' the taste for the study and even the practice of astronomy have made considerable development, to which the founding of the 'Revue mensuelle d'Astronomie' has rendered the greatest service.

. . . .

England especially has numerous private observatories and enthusiastic observers of the sky.[1] But in France also the taste for the science develops in all classes of society, even among the highest. The *bourgeois* begin to smile less at those who interest themselves, telescope in hand, in enjoying the celestial visions. Not very long since the mere idea of being occupied with astronomy—that is to say, with the study of the universe and the knowledge of truth—would have appeared extravagant, so unintellectual is our planet.

Now, when a curious celestial phenomenon presents itself, it is observed on all sides. Since the total eclipse of the moon on January 28, 1888, we have received no less than sixty-four accounts or drawings of the eclipse (*see* 'L'Astronomie,' March 1888). This association for the perpetual observation of the sky, this immense fraternity of minds, without distinction of countries, is it not better than all the political divisions which form the glory of ninety-nine hundredths of humanity?

Perhaps we may here repeat the first lines by which this general description of the universe was commenced : 'This work is written for those who wish to hear an account of the things which surround them, and who would like to acquire, without hard work, an elementary and exact idea of the present condition of the universe.' It is for the reader to decide whether this programme has been carried out ; the author has no other ambition but that of being useful in raising a corner of the veil which still hides from almost all eyes the true splendours of creation. We are at an epoch when the errors of ignorance, the phantoms of the night, and the dreams of human infancy should vanish. The dawning day brings back its pure light, the sun rises on awakened humanity ; let us all stand up before the heavens, and henceforth let us have but one and the same motto : THE PROGRESS OF SCIENCE.

[1] [The British Astronomical Association, especially intended for the cultivation of astronomy among amateur observers, was founded in 1890, and now numbers over 800 members. It publishes a journal, and holds monthly meetings during the session. A library is being formed. Ladies are admitted as members. An Astronomical Society has also been founded in France (1887). The first President was M. Camille Flammarion, the author of the present volume.—J. E. G.]

INDEX

PRINTED BY
SPOTTISWOODE AND CO., NEW-STREET SQUARE
LONDON

Printed in the United States
By Bookmasters

Printed in the United States
By Bookmasters